NEUTRON RADIOGRAPHY

Commission of the European Communities

NEUTRON RADIOGRAPHY

Proceedings of the Second World Conference
Paris, France, June 16-20, 1986

Edited by

JOHN P. BARTON

Neutron Radiography Consulting, La Jolla, California, U.S.A.

GÉRARD FARNY

CEA-CEN Saclay, France

JEAN-LOUIS PERSON

CEA-CEN Saclay, France

HEINZ RÖTTGER

C.E.C., Joint Research Centre,
Petten Establishment, The Netherlands

D. REIDEL PUBLISHING COMPANY

A MEMBER OF THE KLUWER ACADEMIC PUBLISHERS GROUP

DORDRECHT / BOSTON / LANCASTER / TOKYO

Library of Congress Cataloging in Publication Data

Neutron radiography.

Includes index.
1. Neutron radiography—Congresses. 2. Nuclear fuels—Inspection—Congresses. I. Barton, John P. (John Peter), 1934– . II. World Conference on Neutron Radiography (1986 : 2nd : Paris, France)
TA417.25.N4762   1987   621.48'37   87–4920
ISBN-13:978-94-010-8221-1   e-ISBN-13:978-94-009-3871-7
DOI:10.1007/978-94-009-3871-7

Lay-out: Reproduction service J. R. C. PETTEN

Publication arrangements by
Commission of the European Communities
Directorate-General Telecommunications, Information Industries and Innovation, Luxembourg.

EUR 11021
© 1987 ECSC, EEC, EAEC, Brussels and Luxembourg
Softcover reprint of the hardcover 1st edition 1987

LEGAL NOTICE
Neither the Commission of the European Communities nor any person acting on behalf of the Commission is responsible for the use which might be made of the following information.

Published by D. Reidel Publishing Company,
P.O. Box 17, 3300 AA Dordrecht, Holland.

Sold and distributed in the U.S.A. and Canada
by Kluwer Academic Publishers,
101 Philip Drive, Assinippi Park, Norwell, MA 02061, U.S.A.

In all other countries, sold and distributed
by Kluwer Academic Publishers Group,
P.O. Box 322, 3300 AH Dordrecht, Holland.

CONFERENCE ORGANIZATION

Organizers
Commissariat à l'Energie Atomique
 - Institut de Recherche Technologique et de Développement Industriel
   - Services des Piles de Saclay (CEA/IRDI/SPS/CEN Saclay)
Commission of the European Communities
 - Joint Research Centre
   - Petten Establishment (CEC/JRC Petten)
Comité Français des Essais Non Destructifs (COFREND)

Co-sponsors
International Atomic Energy Agency (IAEA)
American Society for Non Destructive Testing, Inc. (ASNT)
American Society for Testing and Materials (ASTM)
British Institute of Non-Destructive Testing
Deutsche Gesellschaft für Zerstörungsfreie Prüfung e.V. (DGfZP)
Japanese Society for Non Destructive Inspection (JSNDI)

Programme Committee

BARTON John (Co-Chairman)	U.S.A.
BAYON Guy	France
BERGENLID Ulf	Sweden
BERGER Harold	U.S.A
DOMANUS Jozef C.	Denmark
DUCROS Gérard	France
FARNY Gérard (Co-Chairman)	France
HARMS Archie A.	Canada
HAWKESWORTH Michael	Great Britain
KATSURAYAMA Kosuke	Japan
MATSUMOTO Gen-ichi	Japan
NEWACHEK Richard	U.S.A.
OKAMOTO Koichi	I.A.E.A. Vienna
PERSON Jean-Louis (Co-Secretary)	France
RANT Joseph	Yugoslavia
RÖTTGER Heinz (Co-Secretary)	Commission of the European Communities
RUAULT Pierre	France
von der HARDT Peter	Commission of the European Communities

Host Committee

Mrs.	M. CHMIEL	COFREND
	D. JEAU	COFREND
	C. PARIS	CEN Saclay
	F. PRUDON	CEN Saclay
	D. SARDA	Palais des Congrès
Messrs.	A. ALBERMAN	CEN Saclay
	R. BARBALAT	CEN Saclay
	G. BAYON	CEN Saclay
	J. DAVASSE	Palais des Congrès
	J.L. PERSON	CEN Saclay
	R. PONSOT	CEN Saclay
	H. RÖTTGER	JRC Petten

Local organization in cooperation with DARO-American Express Agency

Mrs.   S. NADJAR
        J. OLIVER
        D. POLANO

Simultaneous translation (by courtesy of the Commission of the European Communities)

Philippe GRATIER	CEC Brussels
Danielle MARTIN-ROZENBLAT	CEC Brussels
Carole MONTGOMERY	CEC Brussels
Hubert TOUBEAU	CEC Brussels

ANNOUNCEMENTS

Sound recordings of the oral presentations at this Second World Conference on Neutron Radiography are available.
For purchase please contact Mr. J.-L. Person (CEA-CEN Saclay).

Next Conference
The next Conference will probably be held in 1990 in Japan (see closing remarks).

ACKNOWLEDGEMENTS

Special thanks are due to Mr. Jean-Louis Person and Mr. Heinz Röttger, the joint secretaries of this conference, for the excellent organizational work in arranging this conference and the publication of the proceedings.

Special thanks are also due to the many hard working members of the host committee in Paris, and particularly to Mrs. Colette Paris, Françoise Prudon, Martine Chmiel and Mr. Roger Ponsot, whose kindness and effectiveness might have been appreciated by everybody.

Last, but not least, thanks are due to the authors and participants, many of whom travelled large distances to this information exchange.

J.P.B.

G.F.

NOTE : Papers sometimes discuss several subjects.
       For details see subject index at rear of this volume.

LIST OF CONTENTS

PART II : REACTOR SYSTEMS

X

PART III : SMALL SOURCE SYSTEMS

PART IV : GENERAL APPLICATIONS

PART V : NUCLEAR APPLICATIONS

PART VI : CORROSION APPLICATIONS

PART XIV : WORKSHOP REPORTS

H.C. ADERHOLD

O. AIZAWA

J.J. ANTAL

Ph. ATTWOOD

J.P. BARTON

A. BAYÜLKEN

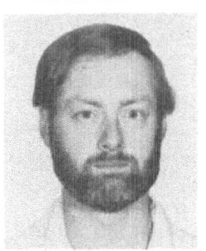

L.G.I. BENNETT

U. BERGENLID

H. BERGER

J.P. BOULOUMIE

J.S. BRENIZER

K. BRYHN-
INGEBRIGTSEN

T. BUCHBERGER

W.E. DANCE

A. DE VOLPI

J.C. DOMANUS

E. DÜHMKE

G. FARNY

C.-O. FISCHER

S. FUJINE

L. GREIM   M. GREIM   A.A. HARMS   D.H.C. HARRIS

E. HEIBERG   Y. IKEDA   A. ITAKA   D. KEDEM

H. KOBAYASHI   Mr. KUSMINARTO   B. LARSSON   P. MAIER

  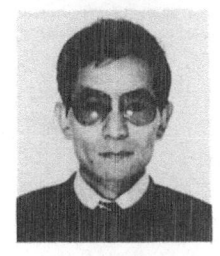 

J.F.W. MARKGRAF   G. MATSUMOTO   T. MATSUMOTO   R.L. MOSS

T. NAKANII   R.L. NEWACHECK   K. OKAMOTO   A. ONO

V.J. ORPHAN

K. van OTTERDIJK

N.N. PAPADOPOULOS

J.-L. PERSON

C. PIRART

J.J. RANT

H. RAUCH

V. REIS CRISPIM

A. RIDAL

H. RÖTTGER

K. SAUERWEIN

K.A. SCHMIDT

W. SCHULZ

W. SCHUMACHER

W.A.J. SEYMOUR

B. SHAPIRO

H. SILVA

H.G. SMITH

E. STEICHELE

G. STEYRER

V. STULENS

P. SWIFT

B. TATTEGRAIN

S. TAZAWA

P.E. UNDERHILL

J.-M. VULCAIN

N. WADA

J. WALKER

W. WHITTEMORE

M. WRÓBEL

K. YONEDA

ZHOU YONG MAO

MA ZHEN-ZE

OPENING OF THE CONFERENCE

ALLOCUTION DE M. FERRY
Directeur de la D.E.R.P.E.

Mesdames, Messieurs,

Bienvenue à Paris!

Au nom du Commissariat à l'Energie Atomique et des autres Institutions organisatrices, je souhaite la bienvenue à tous les Participants à cette Conférence Mondiale sur la Radiographie aux Neutrons qui va durer pendant 3 jours, y compris la séance Poster, et qui sera accompagnée de visites techniques, notamment au Centre d'Etudes Nucléaires de Saclay, d'une visite de la Centrale de Creys-Malville organisée autour du réacteur Superphénix et de certains voyages techniques et touristiques : je souhaite donc à tous un séjour à la fois utile et agréable dans notre pays.

Je vous présente les excuses de M. Michel RAPIN, Directeur de l'IRDI, qui a eu un empêchement tardif comme l'indiquait M. FARNY: il est actuellement en déplacement à l'étranger et il ne pouvait donc pas présider la séance d'ouverture de cette Conférence.

Je m'efforcerai de le remplacer de mon mieux, bien que je n'aie pas la même expérience que lui dans le domaine des essais de qualification de matériel et notamment des essais non-destructifs.

Je vous confirme l'importance que le Commissariat à l'Energie Atomique attache à vos travaux ; c'est pourquoi il a décidé de participer activement à l'organisation de cette Conférence.

En tant que responsable général de l'exploitation d'un certain nombre de Réacteurs Prototypes et Expérimentaux du CEA, je suis heureux de vous accueillir, et je crois qu'il faut, en quelques mots, insister sur l'importance de ce système de contrôle non-destructif, objet de cette Conférence.

Le patronage d'une organisation telle que le Commissariat à l'Energie Atomique peut être interprété de diverses façons : il est certain que nous sommes au service de l'Etat, du Public et de l'Industrie, mais, en exploitant nos propres reacteurs et en y faisant un certain nombre d'expériences, nous pouvons donner à certains observateurs extérieurs, l'impression de tourner parfois en circuit fermé.

Le CEA exploite des réacteurs : il maitrise, avec l'appui de ses filiales, le cycle des combustibles ; il est donc facile de se procurer de la matière fissile, d'en faire des combustibles nucleaires, de les placer dans un réacteur, puis de les examiner après irradiation afin de vérifier qu'ils ont les caractéristiques attendues. En fait, ce n'est pas le seul désir de faire fonctionner ces appareils pour notre propre satisfaction qui nous anime, car les réacteurs construits et exploités par le CEA ont des objectifs précis : certains, en tant que prototypes, doivent préparer le développement d'outils industriels, notamment électrogènes, dont aura besoin Electricité de France, éventuellement d'autres clients, pour la satisfaction des besoins du pays : d'autre part, la mise en oeuvre des techniques de qualification des matériaux et des matériels est ouverte à d'autres utilisateurs, et permet aussi de faire bénéficier l'ensemble de la communauté nationale, sous diverses conditions, notamment de financement, de confidentialité de certains résultats, de l'expérience qui peut être acquise sur nos réacteurs. Evidemment, cette expérience n'est pas limitée au Commissariat et à son groupe, ni à l'industrie française, et comme vous le savez, le Commissariat à l'Energie Atomique qui fait en même temps partie d'un certain nombre d'Organisations Internationales, a développé des accords bilatéraux avec de nombreux pays et, que ce soit sur le plan de l'échange de données en matière scientifique ou d'échanges de données technologiques, ou que ce soit dans des domaines qui sont parfois plus commerciaux, les échanges existent avec de nombreux pays dont la plupart sont d'ailleurs représentés ici.

Le nombre des participants et la variété de leurs professions, telle que je l'ai lue dans le Programme de la Conférence, montrent d'ailleurs qu'il ne s'agit pas d'une auto-consommation et d'une auto-satisfaction propres au CEA mais qu'il s'agit vraiment d'une Conférence Internationale avec des représentants de nombreux pays, notamment d'Amérique et d'Asie. D'autre part, seront évoqués dans cette Conférence, un certain nombre d'applications qui n'ont rien à voir avec l'Energie nucléaire, ni même avec certaines industries dites "avancées", dont je vais parler dans quelques minutes : on trouve également ici des médecins, des antiquaires, des gens d'un peu toutes les professions qui sont, ou qui seront, intéressés par l'application des neutrons et, éventuellement aussi, d'autres sources radioactives, pour faire du contrôle non-destructif et de l'expertise sur des matériaux aussi différents que des tableaux, ou des objets archéologiques, ou le corps humain, s'agissant des applications en médecine ou dans la science dentaire.

Nous vivons une civilisation évolutive : il n'y a pas tellement d'années, l'activité humaine avait encore en grande partie un caractère artisanal : la qualité d'un produit était basée essentiellement sur la compétence et sur la conscience professionnelle de l'artisan qui le produisait; mais l'activité industrielle a beaucoup évolué, et en quantité, et en qualité ; elle s'est étendue quasiment au monde entier, et là, amenée à produire des objets en grande série, elle a exigé une certaine division du travail. Tout cela exige une organisation, je le répète, évolutive, car les idées qu'on peut avoir aujourd'hui sur l'organisation du travail, en prenant en compte les facteurs humains, sont assez différentes des systèmes "Taylor" et autres qui étaient en vigueur encore au début du siècle.

De toute façon, la complexité des objets produits et la mise en oeuvre de chaînes de fabrication exigent des contrôles stricts à tous les niveaux, et ces contrôles ont tout intérêt à être faits d'une façon non-destructive.

Evidemment, on continue à pratiquer des controles destructifs par des methodes d'echantillonnage à divers stades de fabrication, mais je pense que les contrôles non-destructifs ont deux interêts principaux :

- le premier, c'est que, dans bien des cas, c'est notamment le cas dans l'industrie nucleaire, dans l'industrie spatiale, dans la navigation aérienne, dans la navigation sous-marine, le contrôle non-destructif est indispensable parce qu'on ne peut pas appliquer aux objets précieux, et dont on a besoin, le principe de la boite d'allumettes, dont on verifie l'une après l'autre que toutes les allumettes sont valables jusqu'à avoir craqué la dernière...,donc, ce contrôle non-destructif a des nécessités techniques évidentes.

- par ailleurs, même dans les cas où on pourrait satisfaire aux besoins de qualité par des contrôles destructifs, le contrôle non-destructif est très intéressant du point de vue économique, dans la mesure où il permet de faire l'économie de la destruction d'un certain nombre d'échantillons.

Alors, dans notre monde industriel où l'assurance de qualité devient indispensable, soit par la nature même des activités, soit par le souci de la rentabilité, tous les procédés physiques possibles doivent être étudiés, mis au point, et si possible industrialisés, pour aboutir à ce résultat.

Dans cette compétition entre divers procédés physiques, les neutrons ne sont pas arrivés les premiers : ils ont été précédés par les rayons "X" ou "γ"; d'autres méthodes, telles que les contrôles par ultrasons, les contrôles par courants de Foucault, et de nombreux autres types de contrôles sont utilisés ; la particularité des neutrons, comme elle a été signalée par M. BARTON au cours de la Conférence qui s'est tenue il y a quelques années à San Diego, c'est qu'il y a 50 ans, l'homme ne savait pas que le neutron existait, tandis que les rayons "X", par exemple, avaient été découverts dès la fin du XIXème siècle.

Or il existe toujours un certain temps entre la découverte physique et son application industrielle ; malgré l'accélération du progrès, malgré la possibilité de mettre en oeuvre des moyens puissants dans notre société qui est très riche et très performante par rapport à celle de nos grands-parents, il faut du temps, et cela est particulièrement vrai dans le cas des neutrons : en dépit des caractéristiques tout à fait particulières des neutrons (la nature même de leurs interactions avec la matière qui permet de faire avec les neutrons des choses qu'on ne peut pas faire avec d'autres types de rayons), il a fallu quelques dizaines d'années pour que cette activité devienne industrielle.

Les neutrons ont d'abord été mis en oeuvre dans les années suivant la deuxième guerre mondiale, en Grande-Bretagne, et à peu près en même temps aux Etats-Unis, et se sont développés dans le monde entier: je dirai quelques mots de la façon dont cela s'est passé en France, et du point ou nous en sommes arrivés.

La France a commencé par avoir, dans les années d'après-guerre, un programme basé sur le Commissariat à l'Energie Atomique, créé en 1945 par le Général de GAULLE ; celui-ci a construit un certain nombre de Réacteurs Prototypes et Expérimentaux : assez vite, en s'appuyant sur l'expérience déjà acquise dans les pays Anglo-Saxons, il a commencé à développer des dispositifs de contrôle aux neutrons ; je ne parle pas seulement de l'utilisation des neutrons à l'intérieur du coeur et de leur application immédiate sur le contrôle en piscine ou à la sortie de la piscine, des éléments combustibles irradiés, mais de la méthode qui consiste à sortir des faisceaux de neutrons, spécialement pour faire du contrôle non-destructif sur des matériaux qui n'étaient pas forcément radioactifs.

Le premier réacteur equipe d'une façon quasi-industrielle, fut TRITON, installé à Fontenay-aux-Roses (ce réacteur a été arrêté en 1979 et il est aujourd'hui démantelé) ; mais dans la foulée, d'autres réacteurs, notamment MELUSINE à Grenoble puis d'une façon plus récente, ISIS et ORPHEE à Saclay ont été équipés de faisceaux de neutrons sortis qui permettent de faire la Radiographie de nombreux matériels, notamment des matériels contenant des matériaux hydrogénés afin d'en assurer la fiabilité. Cette activité continue à se développer, peut-être pas aussi vite que nous le souhaiterions, en raison d'un certain nombre de difficultés et de servitudes, mais continue à se développer depuis quelques années.

Parallèlement, sur les réacteurs prototypes, surtout sur les réacteurs de la filière à neutrons rapides, se développaient des systèmes de Neutronographie dont la source n'était pas basée sur la réaction en chaîne d'un réacteur au fonctionnement continu, mais sur des sources appelées "pulsées", de neutrons qui se produisent d'une façon plus brève, et en particulier les premiers réacteurs de la filière à neutrons rapides RAPSODIE d'abord à Cadarache puis PHENIX à Marcoule, ont été équipés de sources de rayonnement distinctes du réacteur et permettant de faire l'examen du combustible irradié.

Ce point était particulièrement important dans le cas des combustibles des réacteurs à neutrons rapides, où vous savez que l'irradiation peut entraîner la formation puis l'évolution d'un trou central, ce qui a une importance considérable pour la physique : cela est vrai, non seulement pour la neutronique, mais pour la thermique du coeur. Donc, en même temps que le CEA équipait ses réacteurs expérimentaux, il associait à ses prototypes, des installations de Neutronographie dont la source était externe et n'était pas à l'intérieur du réacteur.

Aujourd'hui, nous sommes confrontés à de nouveaux problèmes, étant donné le développement des industries, notamment des industries aéronautiques et spatiales, qui font appel de plus en plus à des matériaux hydrogénés, soit comme combustible, soit pour fabriquer des pièces mécaniques, par exemple certains types de joint, de telle sorte qu'actuellement nous voyons défiler dans nos réacteurs un grand nombre d'objets, que nous passons au bout d'un long guide à neutrons sortant du réacteur.

Nous pensons que des applications nouvelles sont probables : on ne peut jamais en être sûr, mais on le sent venir dans les années qui viennent, non seulement d'ailleurs pour qualifier les matériels à l'état neuf, mais pour s'assurer de leur état après un certain temps de service ; et cela est vrai pour des matériels démontables comme les vannes, cela peut être vrai aussi pour des matériels beaucoup moins mobiles comme les cellules et les structures de l'industrie aéronautique : c'est pourquoi nous pensons qu'un développement important, qui fait l'objet de chapitres entiers de cette Conférence (une session complète), c'est l'utilisation de sources que l'on peut appeler "portatives", bien qu'au stade actuel, le caractère "portatif" soit encore relatif : on ne peut certainement pas les transporter dans une valise : il faut plutôt un camion.

Nous pensons qu'il y a là un champ d'applications importantes auxquelles, non seulement le CEA, mais un certain nombre d'industriels français et étrangers, peut-être dans le cadre "EUREKA" (nous avons déposé une proposition en ce sens), éventuellement dans d'autres cadres, pourront développer ces nouvelles méthodes.

Je pense que ces méthodes devraient pouvoir se développer assez rapidement et je souhaite qu'elles aient fait des progrès sensibles d'ici votre prochaine Conférence.

Je voudrais remercier les gens qui ont été à l'origine du développement de cette technique nouvelle et qui ont fortement contribué, non seulement à son développement mais à un échange international dont nous pouvons mesurer les effets, et je voudrais particulièrement citer M. Harold BERGER dont les travaux ont fait et font encore autorité en la matière ; je voudrais également citer M. J.P. BARTON qui est ici présent, et qui a été, je crois, un des principaux auteurs de la coopération internationale puisqu'il a travaillé tant en France qu'en Grande-Bretagne et aux Etats-Unis ; il a été le promoteur de la 1ère Conférence qui s'est tenue à San Diego en 1981, et M. BARTON a aussi beaucoup aidé M. FARNY à l'organisation de cette Conférence.

Je remercie pour leur participation à l'organisation et leur support, le COFREND, (le Comité Français des Essais Non Destructifs) qui sera représenté lors d'une prochaine session par M. EVRARD, et bien entendu, la Commission des Communautés Européennes, et plus spécialement son Centre Commun de Recherches dont l'établissement de Petten a également patronné cette Conférence ; j'en remercie M. VON DER HARDT qui n'est pas là ce matin mais qui est représenté par M. RoTTGER.

Messieurs, je crois qu'il faut laisser la parole aux spécialistes : je vous souhaite encore une fois une bonne Conférence, et je vous remercie de votre attention.

Allocution de M. G. FARNY

Après l'exposé de M. FERRY, j'aimerais ajouter quelques mots sur une première Conférence qui a eu lieu antérieurement à celle de San Diego qui a été citée, à savoir celle à Birmingham en Grande-Bretagne en 1973 : si mes souvenirs sont exacts, Michael HAWKESWORTH ici présent, avait été le principal organisateur de cette Conférence qui avait réuni alors une douzaine de Pays, et je garde personnellement un excellent souvenir de ces deux journées, d'autant que cette méthode constituait alors l'essentiel de mes occupations.

En outre, je crois pouvoir dire que cette Conférence est à l'origine de nombreux échanges entre les pays concernés : c'était l'époque des premières "Newsletters", publication informelle animée par John BARTON. Ces échanges ont donné naissance à des groupes de travail sur des sujets divers en importance, mais je dirais, ô combien utiles, comme:

- les Indicateurs de Qualité d'Images : on suivait un peu ce qui se faisait en radiographie,
- la mise en place de Normes : la Neutronographie a démarré, on peut dire, sans norme particulière : On s'est rendu compte ensuite qu'il fallait introduire des normes,
- l'emploi du nitrate de cellulose: un film bien entendu qu'on ne rencontre pas en rayons "X",
- la création d'un Atlas de défauts pour combustible nucléaire irradié, Atlas qui existe maintenant et qui rend service aux métallurgistes lorsque l'on interprète différents examens.

Appelé à d'autres fonctions en 1975, j'ai comme on dit, tout au moins en Français, tourné la page, sans savoir que 8 ans plus tard, je tournerai une autre page qui me permettrait de retrouver cette activité, mais cette fois-ci parmi d'autres.

C'est donc avec grand plaisir que je partage la Présidence de cette Conférence avec John BARTON, un ami depuis bientôt 20 ans.

Et je remercie les présentateurs pour l'originalité de leurs communications que la lecture des résumés fait d'ores et déjà ressortir.

Avant de donner la parole à John BARTON, je voudrais insister auprès des Conférenciers pour que le temps imparti à leur communication soit respecte, à savoir 15 mn de presentation et 4 mn de discussion : les journees semblent très chargées et bien que l'on ait enregistre 2 ou 3 desistements ce matin, il ne nous sera pas possible de compenser le moindre glissement.

Je vous en remercie.

J.P. Barton, Co-Chairman

Sincere appreciation is due to our hosts, including numerous individuals from Saclay and Petten for the valuable accomplishment of this Second World Conference on Neutron Radiography (WCNR-2).

Neutron radiography differs from most other forms of non-destructive testing in several ways. There are relatively few of us, and what we do is frequently unique or expensive. Usually there is only one group in anyone country working in a particular area. International communication has always been excellent, and hopefully will remain that way as the community gets bigger.

A few remarks may be appropriate looking in turn at the past, present and future.

In 1961, when working at Birmingham University Physics Department, I could find only one other person in Europe interested in using neutrons for radiography : a Professor Radwan of the Metallurgy Department at Warsaw University. We communicated immediately. Later I discovered the activity in the USA : Harold Watts, Dan Polanski, Harold Berger, Waren McGonnagle. In spite of the distance, we were soon to meet. A few years later I was welcomed in France, where, in spite of the language problem, we worked together. Thus, international communication has been central to the field since the earliest days of continuous development. Thanks are due to many individuals present today who helped in this early international collaboration. I would mention for example Professor Walker, UK; Mr. Farny, Mr. Cornuet and Mr. Boutaine, France and Mr. Berger, USA.

For the present, those new to Paris are urged to take particular note of the visits offered to Saclay. The neutron radiography facilities in Orphée, Osiris and Isis are impressive indeed.

For the future, please, will each person who has suggestions concerning a possible Third World Conference in this series bring your comments to myself or committee members Mr. Farny and Mr. Matsumoto.
Thank you.

PART I

NATIONAL REVIEWS

ACTIVITÉS DE NEUTROGRAPHIE
AUPRÉS DU RÉACTEUR VVR-S EN TCHÉCOSLOVAQUIE

Z. Hrdlička

L'Institut de Recherche Nucléaire

Řež, Tchécoslovaquie

RÉSUMÉ

L'histoire du développement, les conditions de méthode propre et d'application ainsique les perspectives
de travaux sont décrites. Deux domaines éssentielles
sont étudiés: l'analyse des éléments combustibles non
irradiés et l'analyse des paramètres choisis de la qualité des matériaux composites de construction au point
de vue du hydroïsolement.

INTRODUCTION

L'activité planifiée de recherche ainsique d'application de
neutrographie en Tchécoslovaquie s'est déroulée depuis plus que
15 ans (1). Actuellement plusieurs lieux de travail en sont enga-
gés, dont l'Institut de recherche nucléaire (IRN) joue un rôle le
plus important.

Dans la période 1971-1975 une base fondamentale expérimentale sur le réacteur VVR-S était réalisée, en 1976-1980 les travaux pour développement de l'amplificateur électronique de l'image
se déroulaient. En outre les deux périodes comprenaient les éssais
systématiques de presque toutes applications de la neutrographie
publiées ailleurs (sauf le combustible irradié) le diagnostic du car-
cinome osseux y compris. La période 1981-1985 était caracterisée

par une réduction du volume des sujets étudiés. Actuellement on s'occupe de deux domaines essentiels: de l'analyse des éléments combustibles non irradiés et de l'analyse des paramètres choisis de la qualité des matériaux composites de construction au point de vue du hydroisolement. En plus la méthode est utilisée comme un outil de l'analyse quantitative plutôt que celui de la défectoscopie qualitative. Une capacité limitée est toujours réservée pour les autres applications.

CONDITIONS DE MÉTHODE PROPRE ET D'APPLICATION

Les possibilités de recherche et d'application de la neutrographie dependent de la capacité ainsique du niveau téchnique de la base expérimentale. C'est pourquoi on en donnera d'abord une description.

La Base Expérimentale

L'installation pour la neutrographie au réacteur VVR-S est située au faisceau horizontal de diametre 100 mm. La partie de la source contient un collimateur convergent en graphite avec un diaphragme en bore de diam. 8 mm et un filtre en monocristal de bismuth de l'épaisseur 90 mm. Les caractéristiques radiographiques en sont: L/d = 360, Φ_{th} = 1,2·10^{11} n.m^{-2}.s^{-1}, R$_{Cd}$ = 16, φ_γ = = 10 nA.kg^{-1} (10 r/h) pour N$_t$ = 6000 kW. Une longueur importante ainsiqu'un petit diametre du canal nécessitaient un déplacement du foyer-source effectif vers la sortie du faisceau pour augmenter l'angle de sa divergence. Un collimateur-collecteur conique convergent en graphite permet de déplacer le foyer dans la position de demi-longueur du canal (fig. 1). On en profite une augmentation de flux neutronique du facteur 5, une réduction du rayonnement γ par absorption dans le graphite ainsique une thermalisation supplémentaire de neutrons (R$_{Cd}$ = 16). Le collimateur est un brevet tchéque.

La partie de l'objet et de la représentation radiographique se comporte
- d'une table à trois mouvements définis à distance,
- d'une console au mouvement vertical portant la cassette radiographique maintenue sur dépression,
- d'une console à deux mouvements portant l'amplificateur electronique de l'image fourni d'un convertisseur en Gd$_2$O$_2$S.

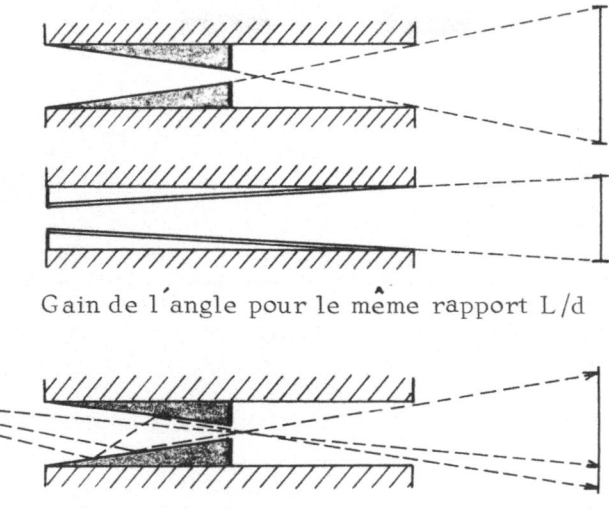

Gain de l'angle pour le même rapport L/d

Gain de flux neutronique

Fig. 1. Le collimateur convergent

L'installation entière est située dans un tunnel de briques en polyéthylène boré. Elle est logée sur un rail. Les parties du tunnel
sont mobiles permettant l'accès à l'installation. Divers dispositifs
supplémentaires sont à la disposition. Comme un exemple important
ou peut nommer un four cylindrique situé dans l'axe du faisceau
dont l'isolement thermique sur les deux bases est transparent pour
neutrons. Comme un dispositif expérimental externe on utilise le
microdensitomètre JOYCE LOEBL commandé par un ordinateur
HEWLETT PACKARD, où on fait les évaluations quantitatives.

Programme d'Application

Les deux domaines essentiels d'application sont basés sur
une compréhension spécifique de la neutrographie, comme une composante particulière d'une méthode générale − de l'analyse de
transmission neutronique. Celle-ci est définie comme la méthode
de l'analyse de tels paramètres et de telles propriétés de matière
et de matériaux, qui dependent de transmission neutronique à travers l'objet analysé. On apprécie alors l'image neutrographique
comme une matrice de données, codées dans la densité photographique pour l'évaluation de transmission neutronique.

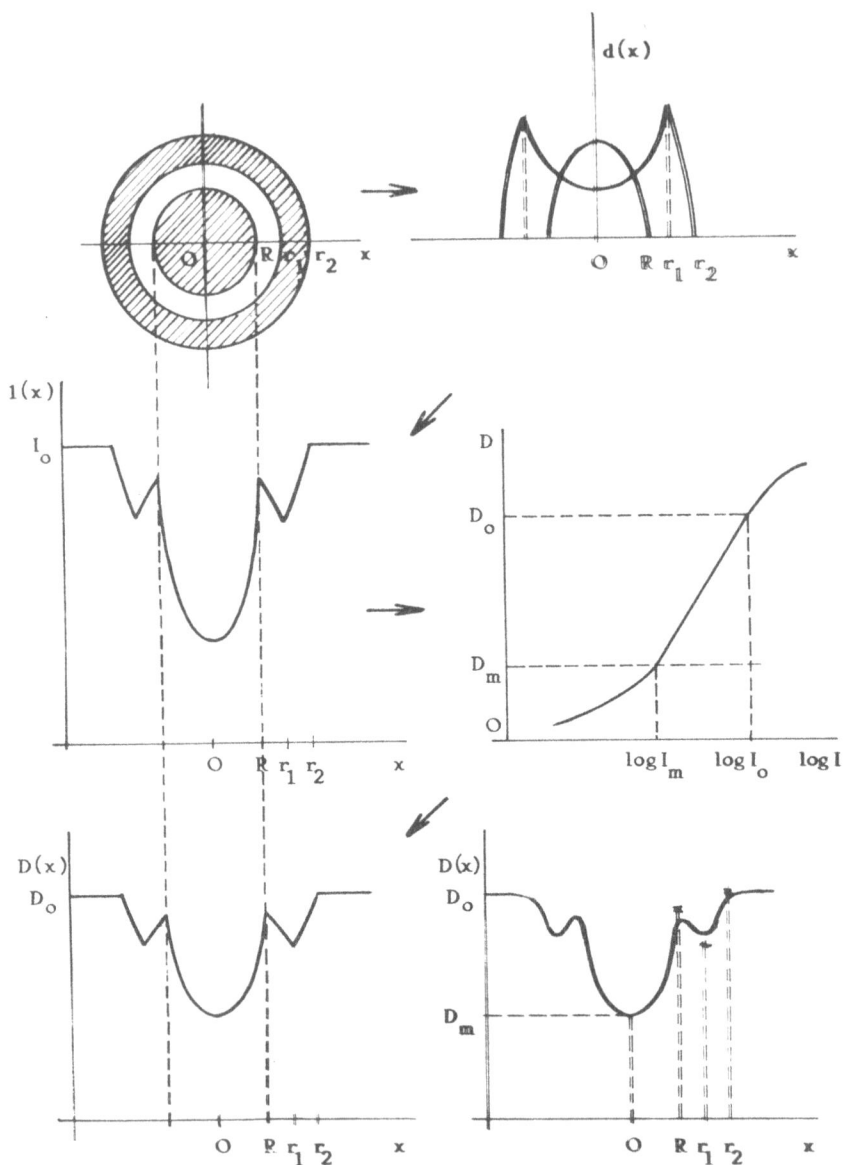

Fig. 2. Obtention de la courbe densitographique

La diagnose de qualité des éléments combustibles. Comme un indicateur important de qualité du combustible non irradié on apprécie la qualité neutronique. On la définit comme l'ensemble des propriétés de l'élément par lesquelles les propriétés nucléaires du coeur sont créées ainsiqu'influencées. On a montré ailleurs (2), qu'on peut considérer la mesure de respecter les valeurs de la section efficace Σ $_{UO_2}$, du diamètre de la pastille R et de la densité de bioxyde d'uranium ϱ $_{UO_2}$, comme la mesure de qualité neutronique. L'analyse de la fonction de densité photographique D d'une section de l'image neutrographique de l'élément permet d'évaluer les valeurs Σ $_{UO_2}$ et R dans l'intervalle de la longueur consideré. Les liaisons entre toutes les grandeurs qui interviennent dans le problème sont démontrées dans fig. 2. La netteté de l'image neutrographique n'étant pas idéale, on doit corriger la fonction densitographique mesurée. Grâce à l'éxistence d'une zone de même allure de la courbe idéale et mesurée toute l'allure de la courbe idéale en peut-être calculée. Comme niveau de référence dans la zone d'identité on considère

$$D_{ref}(x_{ref}) = D_o - 0,4(D_o - D_m) \quad , \tag{1}$$

Puis on obtient

$$R = \left\{ \left[\frac{0,46}{K.\Sigma_{UO_2}} (D_o - D_m) - \frac{\Sigma_{Zr}}{\Sigma_{UO_2}} \left(\sqrt{r_2^2 - \frac{d_{ref}^2}{4}} - \sqrt{r_1^2 - \frac{d_{ref}^2}{4}} \right) \right]^2 + \frac{d_{ref}^2}{4} \right\}^{1/2} \tag{2}$$

En connaissant la dimension R on peut déterminer l'épaisseur effective de la pastille $d_{ef}^{UO_2}$ et puis on obtient

$$\Sigma_{UO_2} = \frac{1}{d_{ef}^{UO_2}} \left(- \Sigma_{Zr} d_{ef}^{Zr} - \ln \frac{I}{I_o} \right) \quad , \tag{3}$$

la plupart des symboles étant expliquée dans fig. 2. K signifie la caractéristique du film.

Un programme d'évaluation densitographique ainsique du calcul de l'équation (2,3) pour le microdensitomètre commandé par l'ordinateur était élaboré (3). On a éprouvé, que la précision de l'évaluation de R fait $\pm 0,015$ mm pour les pastilles de diam. 7,5 mm en UO_2 enrichi à 3%.

L'analyse de cinétique de l'eau et des composés hydrogenés. Dans les procédés technologiques et dans la fonction des matériaux composites de construction l'eau joue deux rôles importants. D'une part comme un facteur constructif, produisant des matériaux solides par la réaction avec des constituants friables (y compris par celle-ci à température elevée), d'autre part comme un facteur destructif pénétrant dans la structure poreuse en la détruisant. Dans les deux cas, c'est la cinétique de l'eau, qui est l'objet de l'intérêt. En outre, dans le deuxième cas, cette cinétique est souvent influencée par les agents d'hydroïsolement, dont on observe la pénétration dans la structure ainsique l'efficacité de leur fonction (4).

Avant tout la neutrographie était utilisée pour l'étude de migration de l'eau pendant la prise du béton par le traitement thermique. Le champ de température influence le transport de l'eau ainsique sa distribution finale. Sur un modèle miniaturisé, la distribution de la température etait mésuré à l'aide de thermocouples l'objet étant neutrographié simultanément. Les neutrographs étaient évalués avec une sensibilité $\Delta \varrho H_2O = 0,01$ g.cm^3.

Une autre application était l'analyse du procédé de déhydroxylation de la kaolinite entre 450 $\div$ 550 °C la réaction $Al_4(OH)_8Si_4O_{10} \rightarrow 2(Al_2O_3 \cdot 2SiO_2) + 4H_2O$ s'effectuante. Le traitement thermique était réalisé dans le four mentionné à 2.1 en isothermes diverses. Quelques phases de la série des analyses sont demontrées dans fig. 3. Les courbes densitographiques étaient évaluées sur le dispositif déjà mentionné. Le procédé méthodique utilisé était appelé l'analyse thermique neutrographique (5).

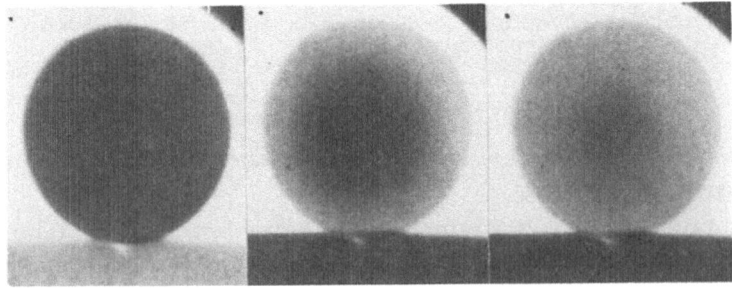

Fig. 3. Trois phases de déhydroxylation de la kaolinite dans un échantillon cylindrique

Pour l'analyse de qualité d'hydroïsolement (6) on examine habituellement sur les échantillons standards la cinétique de la pénétration

a) de l'eau dans le materiau non traité,

b) de l'agent hydroïsolant dans le matériau,

c) de l'eau dans le matériau traité.

Le procédé de mesure ad c) pouvant être effectué dans les phases diverses de la maturation du traitement donne l'information aussi sur le développement de la qualité finale. Quelque fois la maturation peut durer plusieurs mois.

Le mesurage ad b) est rationnel pour les agents dont le contenu des composés hydrogenés ne se sépare pas du dissolvant. Sinon, la pénétration effective de l'agent hydroïsolant est déterminée par le mesurage ad c) dont on fixe la limite de pénétration de l'eau. L'évaluation de l'image peut-être d'une part qualitative (visuelle), où on observe les propriétés de densité et geometriques de la limite de pénétration, d'autre part quantitative, où on evalue la relation de temps

a) du déplacement du front du liquide,

b) de l'augmentation de la concentration du liquide dans les points choisis de l'échantillon,

c) de l'allure des courbes de la concentration du liquide le long de l'échantillon.

Comme le mécanisme standard de pénétration on utilise l'ascension capilaire.

L'évaluation quantitative du contenu de l'eau a le procédé suivant. L'échantillon humide de l'épaisseur x transmet le flux I_o selon

$$I = I_o \, e^{-(\mu_M + \varrho_{H_2O} \cdot \bar{\mu}_{H_2O})x} \qquad (4)$$

où μ_M concerne l'échantillon sec, les autre symboles étant couramment connu.

Puis

$$\varrho_{H_2O} \, g \cdot cm^{-3} = -\frac{1}{\bar{\mu}_{H_2O}} \left(\mu_M + \frac{1}{x} \ln \frac{I}{I_o} \right) \qquad (5)$$

Ayant mesuré D/D_o on détermine la valeur I/I_o par l'évaluation de la courbe densitographique du film. La courbe se construit à partir de l'échelle de densité photographique obtenu par l'exposition simultanée de deux échelles d'abaissement neutronique en

Rh et Cd (l'epaisseur 0,1 ÷ 1 mm le pas 0,1 mm). Le mesurage, la construction de la courbe ainsique l'évaluation de l'équation (5) s'éffectue à l'aide de microdensitomètre commandé par l'ordinateur. La correction au neutrons dispersés est inutile pour les valeurs supérieures à 80 mm de distance objet-détecteur.

LES PERSPECTIVES

Dans le domaine de la diagnose de la qualité du combustible on prévoit élaborer une version méthodique pour analyser la distribution du poison consommable (Gd) dans les pastilles de combustible.

Dans le domaine des problèmes des matériaux composites de construction on prépare une tâche du plan de l'état de recherche, dont le but doit être une méthode standarde, une installation expérimentale spécialisée ainsique une proposition d'un standard pour la vérification de qualité des moyens réglants le régime de l'humidité dans les matériaux poreux de construction par la neutrographie.

REFERENCES

1. Z. Hrdlička, "Progrès de la neutrographie à l'IRN" (en tchéque). Nukleon, 3, 1983.

2. Z. Hrdlička, Possibilités de la neutrographie et de l'analyse de transmission neutronique pour determiner la qualité des éléments combustibles. Rapport IRN No. 6811-T, 1983 (en tchéque).

3. V. Krupa, Z. Hrdlička, M. Rychlik, La méthode neutrographique de l'évaluation de la qualité neutronique des éléments combustibles. Rapport IRN No. 7591-T, 1985 (en tchéque).

4. Z. Hrdlička, "Utilisation of the Neutronography in the Building Material Technology". Special Session on Nuclear and Related Methods for Nondestructive Testing in Civil Engineering. Leipzig 26 ÷ 28 September 1984.

5. Z. Hrdlička, J. Vachuška, "Neutron Radiographic Thermal Analysis". 8th International Conference on Thermal Analysis (ICTA '85). 19.÷ 23. 8. 1985, Bratislava.

6. Z. Hrdlička, F. Peterka, Le procédé et l'application de la neutrographie pour la solution des problèmes spéciaux de la technologie ainsique de la fonction du hydroïsolement. Rapport IRN No. 6812-T, 1984 (en tchéque).

NEUTRON RADIOGRAPHY FACILITY FOR AEOI NUCLEAR RESEARCH CENTER

K. KAMALI MOGHADAM, Z. TABATABAEIAN, N. MIRHABIBI

Nuclear Research Center, Atomic Energy Organization of Iran

P.O.Box 11365-8486, Tehran-Iran

ABSTRACT

Recently a Neutron Radiography System has been designed, constructed and installed at the research reactor of Tehran Nuclear Research Center (TNRC). The 6 inch through tube was chosen because of its low gamma intensity. A collimator was made of two parts, Iron and lead, which were covered thoroughly with cadmium. Cylindrical slabes of Bismuth and Graphite with calculated length were made to reduce gamma intensity and to thermalize fast neutrons respectively. The collimated thermal flux is uniform in direction and intensity, and at the reactor wall and thermal power of 1 MW, the beam has following chracteristics: Average thermal flux of $6.1*10^4$ n/cm^2-sec, cadmium ratio (thermal per total activity) of 98% and neutron-gamma ratio of $5*10^5$ n/cm^2/mR.

INTRODUCTION

The use of neutron in radiography started around 1942. Today by expansion of science and technology this method has found many applications in different fields. For example in nuclear reactors this method is used for determination of cracks, swelling and burn up of fuel rods. Neutron radiography can be very helpfull in taking radiographs of hydrogenized material, like measuring the quantity of grease inside the oilpipes of aircraft engines or the level of hydrogen in Zircunium as a fuel cladding. In general for neutron radiography (N.R) a neutron source is needed and the emitted neutrons are thermalized and then a narrow, parallel and uniform beam is produced, which is directed toward the target sample. While penetrating the sample, due to absorption and scattering the flux of neutron changes. By use of a proper emulsion, from these variations in neutron flux, the internal structure of the sample is observed. In this article the design and installation of a N.R. system, using 5 MWth research reactor of TNRC are discussed(1,2,3,4).

THE CHOICE OF BEAM TUBE

The schematic of beam tubes in 5 MWth research reactor of TNRC are shown in figure 1. Since the through tube is positioned on the lead shield of thermal column, it has less gamma background and therefore it is more suitable for this program. The specifications are given in figure 2.

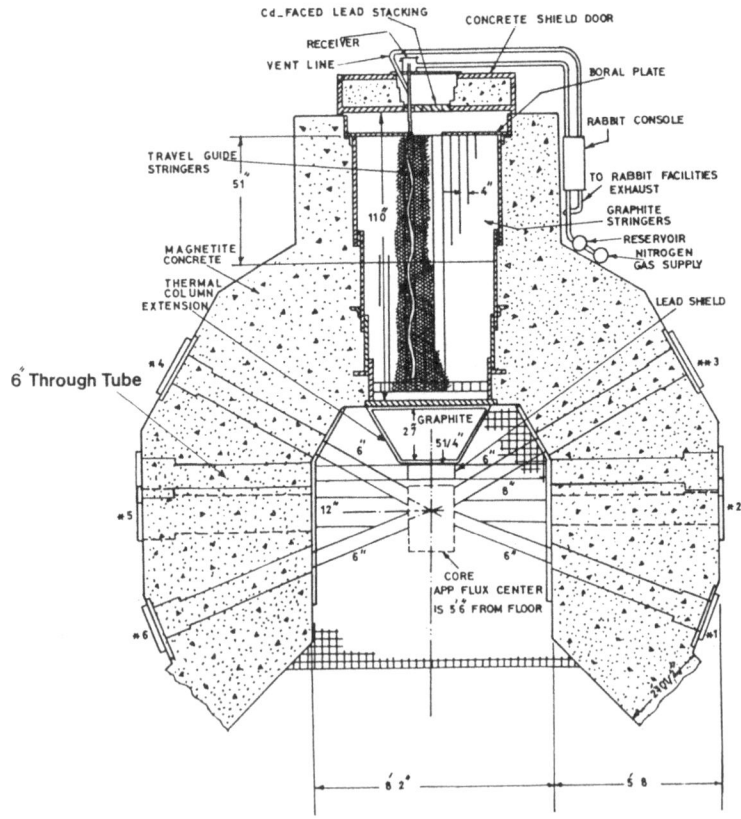

Fig. 1. Schematic of beam tubes in 5 MW research reactor of TNRC.

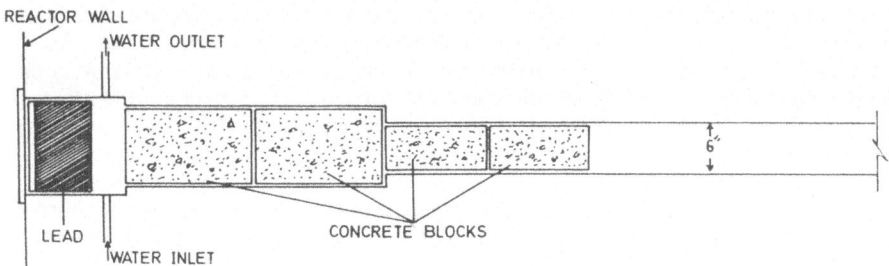

Fig. 2. Schema of through tube.

Facilites Which Were Inserted Inside The Through Tube

As it is shown in figure 3 these parts include Bismuth, Graphite, main collimator, Lead parts and beam sutter.

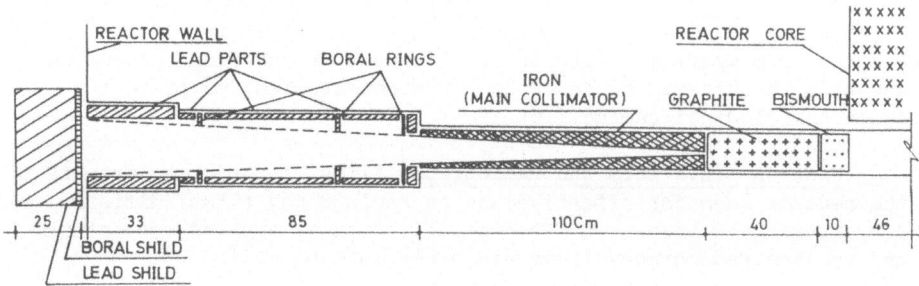

Fig. 3. Facilities which are inserted inside the through tube.

Bismuth Part. Because of high penetration of gamma ray through the radiography films, it was necessary to reduce gamma intensity at the inlet of collimator. This was done by using a piece of 10 cm by 15 cm Bismuth. Bismuth has a smaller neutron absorption cross section than lead and reduces the intensity of gamma beam effectively.

Graphite Part. Distibution of neutron flux in the through tube has been measured with wire scanning technique by using copper wire at 2 kw reactor power. The maximum thermal neutron flux was about $1.2*10^9$ (n/cm^2-sec) with cadmium ratio (thermal per total activity) about 20%. Because of high level of epithermal flux it was necessary to use a piece of Graphite with calculated dimensions of 40 cm by 15 cm, in order to thermalize the epithermal beam as much as possible.

Main Collimator. The main collimator is made of Iron (St-60)

27

with 110 cm length and 15 cm diameter with following impurities:
0.17% Si, 0.5% Mn, 0.05% P, and 0.05% S. Inside the collimator is
conically cut with 1.8 cm inlet and 10 cm outlet diameter. The cone
angle has been chosen 2.06 degree. To prevent the backscattering
of unwanted thermal neutron to the narrow beam, inside and outside
of the collimator has been covered with 0.5 mm thickness of cadmium
using the electro-deposition method. To prevent the possible scra-
tching of the outer cadmium layer, the collimator is covered by 95%
purity Alminium sheet containing 0.12% Cu, 0.1% Cr, 0.62% Fe impurities.

Lead Part. The second part of the collimator is made of cylen-
drical lead parts with 20 cm inner diameter. Also these parts has
been covered by 0.5 mm thick cadmium and inserted in front of the
main collimator, up to the end of the through tube outlet. In order
to remove the unwanted scattered neutrons from the 'main beam, three
different rings of Boral (0.4 cm thick) has been installed at distances 4, 27.5 and 79 cm from the main collimator(5).

Beam Shutter and Concrete Compartment. A lead beam shutter,
25 cm thick was made and 2 sheets of Boral were put in front of it.
A concrete compartment with 60 cm thickness at the front of the beam
and side thicknesses of 45 cm has been constructed, to protect
against biological damage(5,6).

Neutron Converters and Radiography Films. In order to prepare
the neutron beam for effective use on radiography films, different
converters have been used. The specification of applied converters
and related radiography films are given in table 1(4).

Converter Material & Composition	Useful Reactions	Cross Section (barns)	Half-life	Emission Type	Emission Max. Energy	Coresponding Films
Gadolinium Metal foil,100 μm	$^{155}Gd(n,e)Gd^{156}$	58000	Promt	e	.14Mev	X-ray(CRT7-3M)
	$^{157}Gd(n,e)Gd^{158}$	240000	Promt	e	.13 "	
Boron,B^4C,100 μm	$B^{10}(n,\alpha)Li^7$	3837	Promt	α	2.3 "	Kodak(CN-85)
Litium,(Lif-Zns) 250 μm	$Li^6(n,\alpha)H^3$	935	Promt	α	4.7 "	Fuji(MI-NC)

Table 1. Specification of applied converters & corresponding films.

RESULTES AND DISCUSSION

Characteristic of induced thermal beam, thermal to fast ratio, beam uniformity, L/D ratio and neutron to gamma intensity at reactor wall and 1.8 meter far from reactor wall will be discussed bellow.

<u>Beam Outlet Dimensions</u>. Diameter of the beam at:
1- Reactor wall is :   (without shadow)   18.6 cm
2- Reactor wall is :   (with shadow)   19.2 cm
3- 1.8 meter far from the wall is :   (without shadow)   31.7 cm
4- 1.8 meter far from the wall is :   (with shadow)   36.3 cm

<u>Neutron Beam Characteristic</u>. Thermal neutron flux distribution at 1 MW power has been measured in different points of beam outlet at reactor wall and 1.8 meter far from the wall. The resultes are shown in figure (4-a) and (4-b). Cadmium ratio has been estimated about 98% (thermal per total activity). Uniformity of the beam also has been tested by exposing Gadolinium converter and related film, and then measuring the density of the beam hole image on the film by photo optical densitometer. The result is shown in figure (4-c).

<u>L/D Ratio</u>. Collimator length per diameter of inlet aperture is indicated by L/D ratio. High value of L/D for a collimator re-persents good resolution and convergence of the narrow beam. Usually L/D ratio must be greater than 50. For Tehran research reactor neu-tron radiography collimator, L/D ratio is 114 at reactor wall(4).

<u>Thermal Neutron Per Gamma Intensity</u>. Gamma ray intensity of emerged beam at reactor wall was about 0.5 R/hr (at 1 MW reactor power). Considering the thermal neutron flux at this point, the value of neutron per gamma intensity can be calculated as $5*10^5$ $n/cm^2/mR$.

The observed results, also the radiographs taken from different objects, reveal the capability and high quality of this system, com-paired with similar neutron radiography facilities throughout the world. The main problem of our system is the low level power of reactor which increases the exposure time up to 2.5 hours. We are trying to overcome this problem by suitable converter and cooling the film casset.

Figure 5 shows the picture of four different cadmium plates each containing holes of different dimensions also a lighter and a chronometer. This figure shows the good resolution of our system.

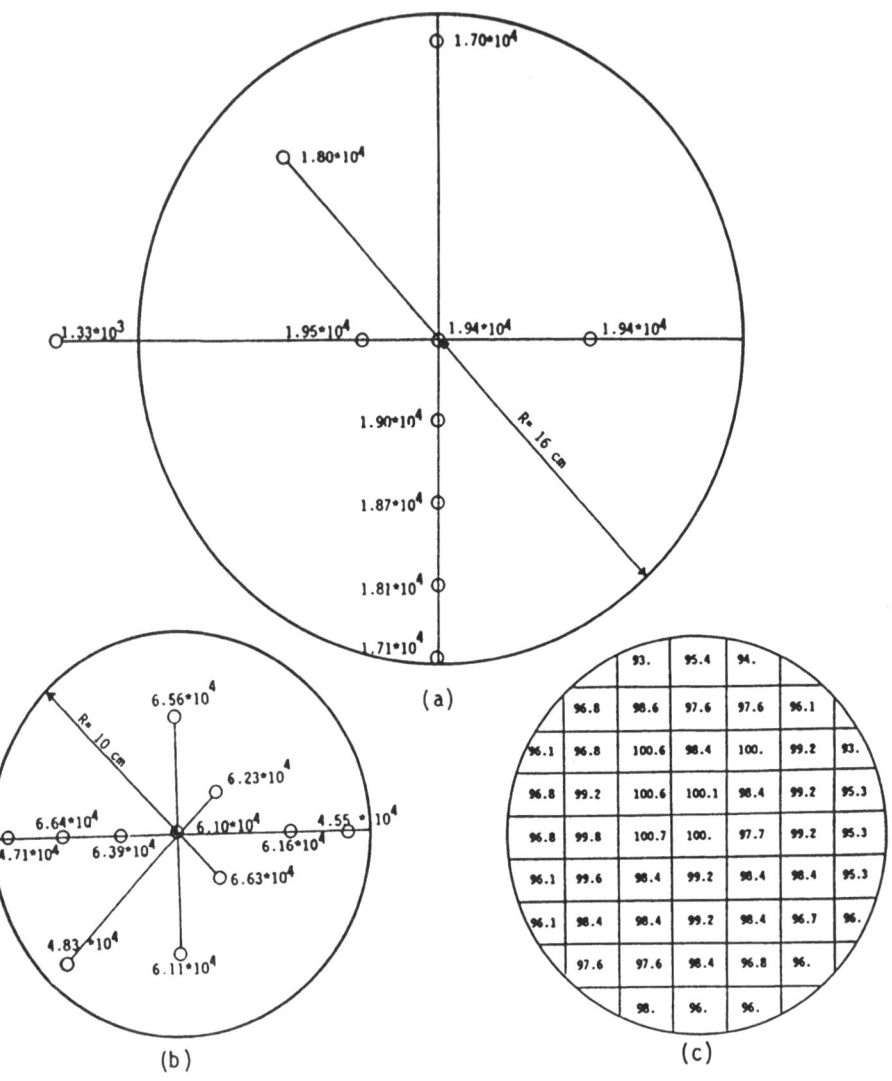

Fig. 4. a) Thermal Neutron Flux distribution at 1.8 meter from the wall. b) Thermal Neutron Flux distribution at the reactor wall. c) Beam uniformity consideration by using film-converter technique at the reactor wall.

Fig. 5. Radiograph which is taken from different objects.

ACKNOWLEDGEMENT

Thanks are due to our colleagues in reactor operation, Health Physics and the workshop sections of TNRC, also other government agencies that helped us in this respect.

REFERENCES

1. M. Hawkesworth, Ed. Proc. Conf. of Radiography With Neutrons, Univ. of Birmingham, pp.165-167, 1973.

2. H. Berger, Practical Applications of Neutron Radiography and Gaging, ASTM pub, 1975.

3. M. Hawkesworth, Neutron Radiography Equipment and Methods, Atomic Energy review, Vol.15, No.2, 1977.

4. P. Vonder Hardt, H. Rottger, Neutron Radiography Handbook, D. Reidel, pp.45-65, 1981.

5. H. Etherington BD,,Nuclear Engineering Handbook, McGraw-Hill, Ch.7, pp.61-112, 1958.

6. R.G. Jaeger, Engineering Compendium on Radiation Shielding, IAEA, Vienna, Vol.II & III, 1968.

PRESENT STATUS OF NEUTRON RADIOGRAPHY IN JAPAN

Kosuke Katsurayama

Research Reactor Institute, Kyoto University

Kumatori-cho, Sennan-gun, Osaka 590-04, Japan

ABSTRACT

In Japan, a series of domestic symposium on neutron radiography have been periodically held at Research Reactor Institute, Kyoto University (KURRI) every two years since November 1970. In April 1984, the Research Committee on Neutron Radiography has been organized by the Irradiation Development Association and the Science and Technology Agency of the Japanese Government, which is held every three months.

In this paper, the research activities and tentative targets of neutron radiography in Japan are described.

INTRODUCTION

Application of neutrons for the nondestructive testing (NDT) is of wide variety, for example, neutron radiography, neutron logging, neutron moisture meter and other applied measurements. This paper describes about neutron radiography in Japan.

Since November 1970, a series of domestic symposium on neutron radiography have been periodically held at Research Reactor Institute, Kyoto University (KURRI), every two years. The seventh symposium was held in August 1985. In this series of symposium, more than hundred researchers take part in every time for presentations of the latest papers and information exchange. In early 1970s, the research works performed in the Japan Atomic Research Institute were dominant, but gradually the other institutions such as universities and industrial laboratories have been joined in research and development of neutron radiography.

RESEARCH COMMITTEE ON NEUTRON RADIOGRAPHY

The Research Committee on Neutron Radiography has been organized by the Irradiation Development Association and the Science and Technology Agency of the Japanese Government in April 1984, which is held every three months.

This committee is chaired by Prof. Katsurayama and consists of the members from Kinki University, Kyoto University, Musashi Institute of Technology, Nagoya University, Rikkyo University, University of Tokyo, Japan Atomic Energy Research Institute, Institute of Physical and Chemical Research, National Space Development Agency of Japan, Radiation Center of Osaka Prefecture, Hitachi Ltd., Japan Air Line Co., Ltd., the Japan Steel Works Ltd., Mitsubishi Heavy Industries, Ltd., Nissan Motor Co., Ltd., Sumitomo Heavy Industries, Ltd., TESCO Co., Toshiba Co., Ltd., and staff of the Science and Technology Agency and the Irradiation Development Association as shown in Table 2.

The committee is very beneficial to exchange information, to standardize the neutron radiography techniques and others.

TENTATIVE TARGETS OF NEUTRON RADIOGRAPHY IN JAPAN

The neutron radiography facilities available in Japan are listed in Table 1.

So far, the tentative targets of neutron radiography in Japan are as follows:

(1) To obtain a high resolution image with low neutron flux:

There is a contradiction between the application of a low flux neutron source to neutron radiography and the obtaining of images of high resolution. However, we are trying to solve it by developing some new emulsion and by improving converters. Through them, the industrial applications are attempted.

(2) To develop a portable neutron radiography facility:

For wide applications, small and light portable neutron radiography facilities by using neutron sources, such as an accelerator and Cf-252 radioisotope, are under development in Japan.

(3) Real time operation:

34

Development of a real-time imaging system in combination with neutron television and computer tomography is under way. It will be applied not only for industry but also medicine.

(4) Standardization of neutron radiography:

When the neutron radiography technique is used for a way of inspection, the standardization is required by users. It is an international issue, and in Japan the Japanese Society for Non-Destructive Inspection must take charge of it.

For this purpose, the intercomparison of the indicators have been done by use of the neutron fields in the KUR, which is presented in Session XIII[1], and also in Musashi Institute of Technology Reactor. This kinds of test will be done in the YAYOI (Fast neutron source reactor, University of Tokyo) and in an accelerator.

CONCLUSION AND ACKNOWLEDGEMENTS

This paper just reviewed the frame work on neutron radiography in Japan, but so far the practical applications are limited in the inspection of space rockets and some nuclear fuels. To apply the neutron radiography in general, the establishment of the above standardization, and some other technical developments are inevitable.

The author expresses his thanks to all members of the organizations listed in Table 1 for preparation of its data.

REFERENCE

1. K. Katsurayama et al., "Intercomparison of Neutron Radiography Indicators Using KUR", presented in this conference.

Table 1. Neutron Radiography Facilities in Japan

Organization	Neutron sources	Thermal neutron flux at sample position (n/cm^2·sec)	Cd ratio of gold
Japan Atomic Energy Research Institute	JRR-2	1.1×10^6	4.8
	JRR-3 (Under planning)	1.0×10^7	170
		1.0×10^6	170
		2.0×10^5	---
	JRR-4	3.2×10^7	4.6
	NSSR	1.0×10^{10}	8.7
	Cf-252	$(1.9 \sim 3.4) \times 10^3$	1.7(by In)
Kyoto Univ.	KUR	1.2×10^6	400
Musashi Inst. Tech.	TRIGA-II	2.0×10^5	11
Rikkyo Univ.	TRIGA-II	6.9×10^5	5.6
Kinki Univ.	UTR-KINKI	1.2×10^4	4.0
		3.4×10^3	2.0
Nagoya Univ.	Sealed-Tube Neutron Gener.	8×10^3	---
	Van der Graff	5×10^4	2.5
Rad. Center of Osaka Pref.	Van der Graff	$1.2 \sim 9 \times 10^3$	2.2
	Linac	$1.5 \sim 7 \times 10^5$	3.5
Toyota Central R&D Lab.	Van der Graff	1.8×10^5	4.4
Tohoku Univ.	Cyclotron (Model 680)	4.5×10^5	3.0
		1.5×10^6	2.5
Sumitomo Heavy Industry Co.	Subcompact cyclotron	4.5×10^5	3.5
		1.1×10^6	3.5
The Japan Steel Works L.t.d.	Baby cyclotron	3×10^5	3.5
		4.4×10^5	3.5

Table 1. (continued)

L/D	The size of irradiation field (mm)	n/γ ratio (n/cm²·mR)	Remarks
70	80×80	1.5×10^5	10 MW
			20 MW
178	115×432	3.5×10^6	R_1 1st section
245	255×305	3.5×10^5	R_2 2nd section
200	20×60	----	R_3 R_3CNRF
67	60 φ (dia.)	6.4×10^5	3.5 MW
67	200×200	5×10^4	pulsed
12.5~50	200 φ	$(1.9 \sim 2.9) \times 10^4$	1mg=20GBq
100	160 φ	1.0×10^6	5 MW
250	250 φ	5.0×10^6	100 KW
108	100 φ	6.1×10^5	100 KW
10	200 φ	9.8×10^5	1 W
22	200 φ	2.7×10^5	
28	260×260	----	
17	125×165	----	
10~30	120×120	----	
10~30	120×120	----	15 MeV
30	200×250	7.6×10^5	3 MeV D.C. 300 μA
50	330×330	1.3×10^5	17 MeV
50	330×330	0.8×10^5	30 MeV
50	356×432	1.5×10^5	18 MeV 50μA
30	356×432	1.5×10^5	18 MeV 50μA
70	356×932	3×10^5	16 MeV 50μA
50	356×932	2.5×10^5	16 MeV 50μA

Table 2. Members of the Research Committee on Neutron Radiography

Name	Organization

Chairman

Kosuke Katsurayama — Research Reactor Institute, Kyoto University

Coordinator

Keiji Kanda — Research Reactor Institute, Kyoto University

Gen-ichi Matsumoto — Nagoya University

Junichiro Sekita — TESCO Co.

Nobuo Wada — Japan Atomic Energy Research Institute

Member

Otohiko Aizawa — Atomic Energy Research Laboratory, Musashi Institute of Technology

Yukio Fukushima — National Space Development Agency of Japan

Eiichi Hiraoka — Radiation Center of Osaka Prefecture

Isao Kohno — Institute of Physical and Chemical Research

Chikara Konagai — Toshiba Co., Ltd.

Hisao Kobayashi — Institute for Atomic Energy Rikkyo University

Takao Maniwa — Nissan Motor Co., Ltd.

Yasuhiko Miyasaka — Japan Atomic Energy Research Institute

Masaharu Nakazawa — University of Tokyo

Jiro Okamoto — Japan Air Line Co., Ltd.

Funimobu Takahashi — Hitachi, Ltd.

Shuichi Tazawa — Sumitono Heavy Industries, Ltd.

Takao Tsuruta — Kinki University

Teruo Yamada — Japan Steel Works, Ltd.

Tomio Yasui — Mitsubishi Heavy Industries, Ltd.

Observer

Takuya Naruse,
Kyoji Murakami — Science and Technology Agency
Yoichi Ito,
Masato Nakamura

Secretariat

Shigeo Tsujimura
Masatoshi Sekine — Irradiation Development Association
Tomoko Ogawa

LES DISPOSITIFS POUR LA RADIOGRAPHIE AUX NEUTRONS

ET LEURS APPLICATIONS EN TURQUIE

I. Alkan et A. Bayülken

T.A.E.K.-Ç.N.A.E.M.[1] et İ.T.Ü.-N.E.E[2]

Istanbul - TURQUIE

RESUME

Le premier système turc de radiographie utilisant
un faisceau sorti du premier réacteur de recherche et le
second système utilisant une source de neutrons isotopiques, ne sont plus en fonction. Le troisième système
fonctionnant avec un faisceau sorti du réacteur TRIGA,
va être utilisé pour des besoins industriels. Enfin le
dernier système qui est en cours de montage, est un système immergé dans la piscine du réacteur TR-2. Son but
est de servir comme moyen de radiographie pour les éléments combustibles des réacteurs, aussi bien que pour la
radiographie des objets ne craignant pas d'être immer-
gés dans l'eau.

LES DISPOSITIFS ARRETES

Quand les neutrons ont commencé à intéresser les chercheurs
turcs comme un moyen de recherche très efficace, à la suite de la
création d'un Centre de Recherche Nucléaire à Istanbul, on avait
construit dans ledit Centre un réacteur de recherche de type piscine qui avait 6 tubes à faisceau neutronique. En 1973, on avait
commencé à étudier l'installation d'un système de radiographie sur
l'un de ces tubes.

L'étude fût menée de façon à trouver la réponse aux problèmes
suivants:

- Obtenir un flux de neutrons nécessaire et suffisant pour une courte durée d'exposition,

- Obtenir une bonne collimation,

- Obtenir un grand rapport n/γ permettant l'utilisation de la méthode directe.

Pour ne pas apporter une grande réactivité au coeur du réacteur, le collimateur cylindrique B_4C, gainé d'aluminium, était séparé du coeur par un cylindre d'aluminium contenant du graphite; cette disposition avait pour résultat une légère augmentation du flux neutronique.

Ce dispositif qui fût en place en 1974, n'a pas pu servir longtemps, dû à l'arrêt du TR-1 en 1977, en vue d'augmentation de puissance. La seule étude significative qui y fût faite était la mesure du spectre des neutrons thermiques du coeur. (1).

Le rapport caractéristique L/D du système était égal à 95. (Fig.1), (2).

A la suite da l'arrêt du TR-1, le Centre de Recherche d'Ankara avait voulu expérimenter un dispositif utilisant une source isotopique. A ce propos, on avait fait l'acquisition d'une source de 5 Ci de (Pu-Be) qui pouvait produire un flux neutronique de l'ordre de $1,3.10^7$ n/cm^{2}sec. A la sortie du collimateur cylindrique formé de blocs de polyéthilène avec différents pourcentage de Li, B et Pb, et troué au centre d'un trou de 6 cm de diamètre, ce flux de neutrons diminuait à 70 n/cm^{2}sec. Le convertisseur était le scintillateur Nuclear Enterprise NE425. Avec ce système de test, où le film Agfa Curix RP1 était placé entre le convertisseur et le flux de neutrons incidents, on avait pris la radiographie d'une plâque de cadmium troué de plusieurs trous de différents diamètres. A la fin d'une durée d'exposition de 24 heures, on avait constaté que tous les trous étaient bien visibles. (3).

Ce dispositif, qui n'a jamais dépassé le stade de test, fût fonctionnel entre les années 1979-1981.

LES DISPOSITIFS ACTUELS

Après l'arrêt du TR-1, deux autres réacteurs sont construits à Istanbul. Le premier réacteur qui fût critique en 1979, est construit dans l'Institut d'Energie Nucléaire de l'Université Téchnique d'Istanbul. Ce réacteur, de type TRIGA, de 250 kw de puissance nominale, est un réacteur de type pulsé. Comme cette caractéristique est très recherchée pour l'étude des objets en mouvement ou pour les objets craignant une longue durée d'exposition aux neutrons,

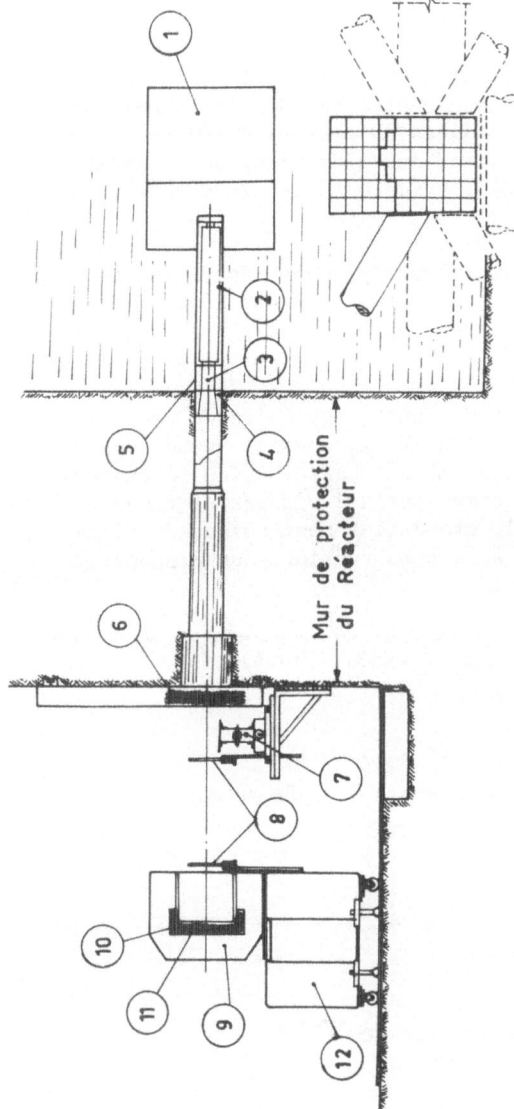

1 – Cœur du réacteur
2 – Collimateur en B$_4$C
3 – Bismuth (8 ∅ × 15)
4 – Cylindre de plomb
5 – Cylindre d'aluminium
6 – Obturateur de plomb

7 – Chariot mobile d'objet
8 – Cassette de film
9 – Béton lourd
10 – Plomb
11 – Paraffine
12 – Table portant le bloc de protection

Figure – 1. Le système de radiographie
aux neutrons du TR–1

on a installé sur le tube tengentiel de faisceau de ce réacteur un
système de radiographie.

Le système ,dont les tests continuent, consiste en un collimateur conique, construit avec des disques de fenêtres circulaires
de diamètres divergents. Ces disques sont formés par deux couches
différentes: la première couche est destinée à servir comme filtre
pour les neutrons thermiques incidents et la seconde est un filtre
contre les rayons gamma. (Fig.2).

Les résultats de ce dispositif dont le rapport caractéristique L/D est de 130, sont assez concluants. (Fig.3), (4). Les éprouvettes de cette figure sont irradiées 30 minutes à 100 kw. Elles
contiennent, respectivement, de l'eau, du colémanite, de la paraffine, de l'hydroxyde de litium, de l'acide borique, la dernière
étant vide.

Le but de ce dispositif est de servir comme instrument de
test, par exemple, au génie civil, et de chercher une réponse aux
problèmes de l'industrie où les autres moyens de test s'avèrent in-
compétents.

TR-2 est le second réacteur du Centre Nucléaire de Çekmece
et, il est construit dans la piscine du TR-1, dans le grand compar-
timent. Comme le TR-1 qui avait des tubes à faisceau, ne fonctionne
plus, faute de combustibles, et comme le TR-2 n'a pas de tubes à
faisceau , pour pouvoir utiliser le grand flux neutronique du réacteur, les responsables ont pensé à la construction d'un dispositif
immergé dans la piscine du réacteur.

Le système fût conçu d'après le système immergé qui se trouve
dans le réacteur GENRA II, à Geesthacht GKSS. (Fig.4), (5).

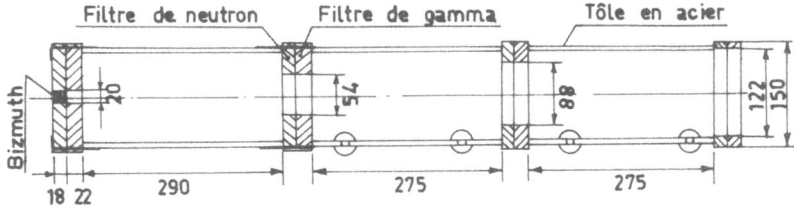

Figure-2. Le collimateur de TRIGA

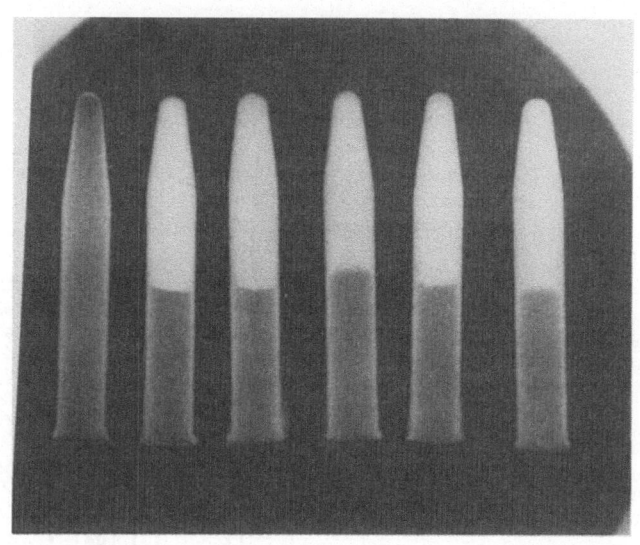

Figure-3. Les éprouvettes de test radiographiées dans TRIGA

Les caractéristiques de ce système peuvent être résumé comme il suit:

- Le collimateur divergent a une section rectangulaire. Il est construit avec des plâques de cadmium et d'indium, recouverts de plâques d'aluminium. Pour une ouverture de diaphragme de 1 mm, le rapport L/D est de 225. (Fig.5).

- Le diaphragme peut être changé de 0,5 mm à 4 mm. Ce changement peut être obtenu avec des plâques d'aluminium, ou, pour l'obtention des neutrons de différentes énergies, avec des filtres de cadmium ou de cadmium plus indium.

- Le flux des neutrons thermiques qui est de l'ordre de 6.10^{13} n/cm^{2}sec à l'entrée du collimateur, a une valeur de $7,4.10^7$ n/cm^{2}sec sur l'objet. (6). Les dimensions de ce flux sont de 40 cm de hauteur avec une largeur de 15 cm.

Comme cela, en 3 étapes, on peut prendre la radiographie des objets ayant 1 m de hauteur et placés dans un panier spécial, conçu à cet effet.

- L'eau de la piscine qui se trouve dans le nid de l'objet à radiographier est chassée à l'aide d'un compresseur et le niveau de l'eau est contrôlé automatiquement.

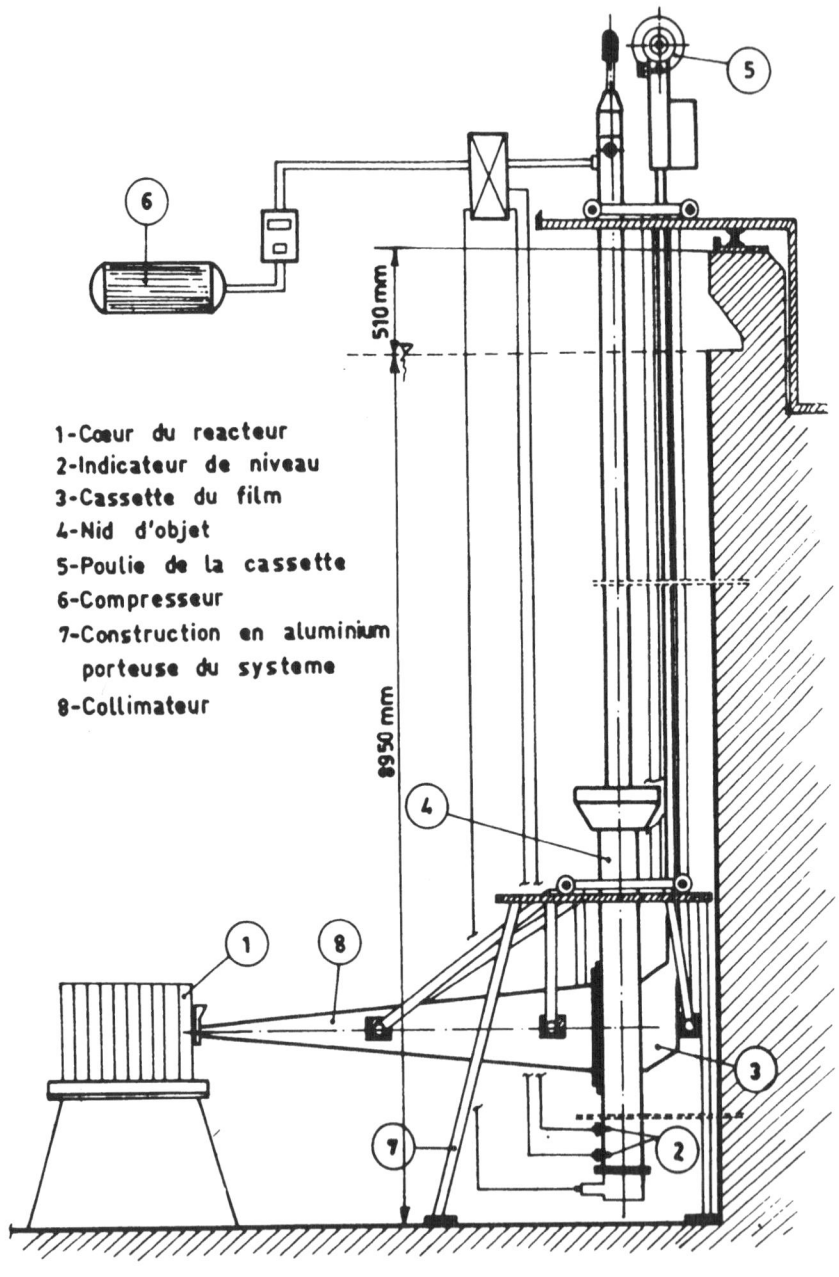

1-Cœur du reacteur
2-Indicateur de niveau
3-Cassette du film
4-Nid d'objet
5-Poulie de la cassette
6-Compresseur
7-Construction en aluminium
  porteuse du systeme
8-Collimateur

510 mm

8950 mm

Figure-4. Le systeme de radiographie du TR-2

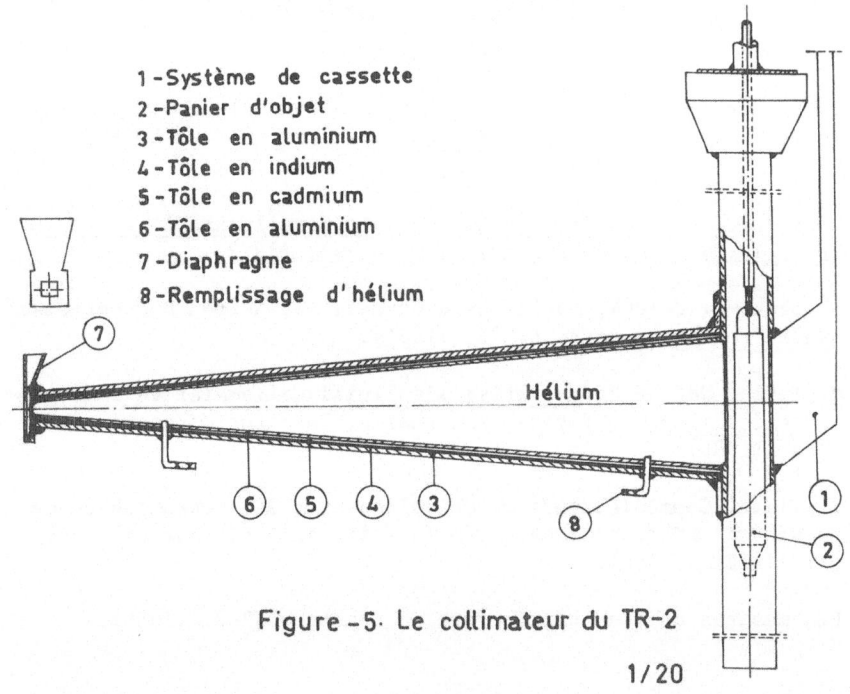

1-Système de cassette
2-Panier d'objet
3-Tôle en aluminium
4-Tôle en indium
5-Tôle en cadmium
6-Tôle en aluminium
7-Diaphragme
8-Remplissage d'hélium

Hélium

Figure -5. Le collimateur du TR-2

1/20

  - Avec ce système, on va utiliser la méthode directe avec un film en nitrate de cellulose CN85, ou bien la méthode indirecte avec un convertisseur en Dy.

  Le but principal de ce dispositif est de pouvoir prendre la radiographie des objets radioactifs comme les éléments combustibles de réacteur ou les objets irradiés dans le réacteur.

CONCLUSION

  Depuis 1974, les chercheurs turcs qui se sont intéressés à la radiographie aux neutrons ont contribué à l'accumulation d'un savoir important sur ce système de test non destructif.

  Les deux derniers systèmes, l'un en fonction, l'autre qui le sera en Octobre 1986, vont bientôt venir en aide, avec une précision assez suffisante, aux problèmes de l'industrie turque.

Pour un pays qui va bientôt entrer dans la technologie nucléaire, cette aide est sûrement de celles qu'on ne doit pas négliger.

REFERENCES

1. Z. B. Saraçoğlu, "Determination of TR-1 Thermal Spectrum Using Wedge Shaped Absorber", Atomkernenergie, Bd. 31, p. 256-258 (1978).

2. D. Öner, TR-1 Reaktörü'nün Kalp İçi ve Etrafındaki Işınlama Düzenleri, ÇNAEM raporu, No: 106, p. 44-47, (1972).

3. F. Özek, I. Çelenk, "5 Ci Pu-Be Kaynağı ile Nötron Radyografisi" ANAEM Progress Report, p. 25, (1981).

4. M. Tekin, Nötron Radyografisi ile İlgili Çalışmalar ve Demet Karakteristiklerinin Tayini, Ms. thesis, Institut d'Energie Nucléaire, (1986).

5. I. Alkan, Communications personnelles avec les responsables du Beesthacht GKSS, à la suite de la suite scientifique patronnée par l'AIEA. (1985).

6. Les mesures du flux neutronique du réacteur TR-2. (1986).

[1] Organisation Nationale Pour l'Energie Atomique. Centre de Recherche Nucléaire de Çekmece. P.K.1 Havaalanı ISTANBUL

[2] Institut d'Energie Nucléaire. Université Technique d'Istanbul. Yeni Levent. ISTANBUL-TURQUIE.

INDUSTRIAL NEUTRON RADIOGRAPHY

IN THE UNITED STATES

Paul E Underhill

Marketing Manager-Aerotest Operations, Incorporated

San Ramon, California, U.S.A.

ABSTRACT

This paper covers the neutron radiographic activities in the industrial community as experienced at Aerotest Operations over the last five years. A brief description of the Aerotest facilities, the neutron beam, the physical configuration and special capabilities are discussed. Several unique neutron radiographic applications are described, including a specific weld inspection of the vernier control engines for the space shuttle, the inspection of rubber diaphragms used to pressurize the fuel for operating engines under zero gravity conditions, the inspection of graphite phenolic blocks from which artificial hip joints are machined, and special fixture designs.

AEROTEST FACILITIES

The Aerotest Research and Radiography Reactor (ARRR), one of the two major neutron radiography production facilities in the United States, is shown in Figure 1. The source of neutrons is a 250 kw-thermal reactor located at the bottom of a 10 foot (3 meter) diameter by 23 foot (7 meter) deep, water filled tank. The beam of neutrons is brought to the water surface by a tapered aluminum, helium filled, rectangular cross sectioned duct. Figure 2 illustrates the 22 inch (55cm) by 30 inch (76cm) exposure area at the neutron radiographic location, which allows two neutron radiographic films to be exposed simultaneously.

Standard industrial 14 inch (35cm) by 17 inch (43cm) single sided fine grain film is used in a vacuum cassette. The neutron to electron conversion screen used is a 0.001 inch (0.0025cm) thickness of gadolinium metal that has been vapor deposited onto an aluminum plate.

The unique features of this facility are the long source-to-object distance, 250 inches (635cm); large exposure area; sliding horizontal double-ended shuttle tray and the availability of five switch selectable apertures. These apertures provide L/D ratios from under 100 to 500. The L/D ratio is the length from the source to the object being neutron radiographed, divided by the apparent source size. The apertures have been calibrated according to the ASTM document E 803-81 "Standard Method For Determining The L/D Ratio Of Neutron Radiographic Beams." The long source-to-object distance provides a near parallel beam of neutrons, while the five different apertures provide variable depths of fields to examine areas of interest at various distances from the film or to provide sharp definition for small components. The Aerotest facility also includes a separate, tightly collimated beam of neutrons for neutron gaging. This equipment has been instrumental in providing narrow beam thermal neutron attenuation coefficients for most of the materials that are commonly examined with neutron radiography.

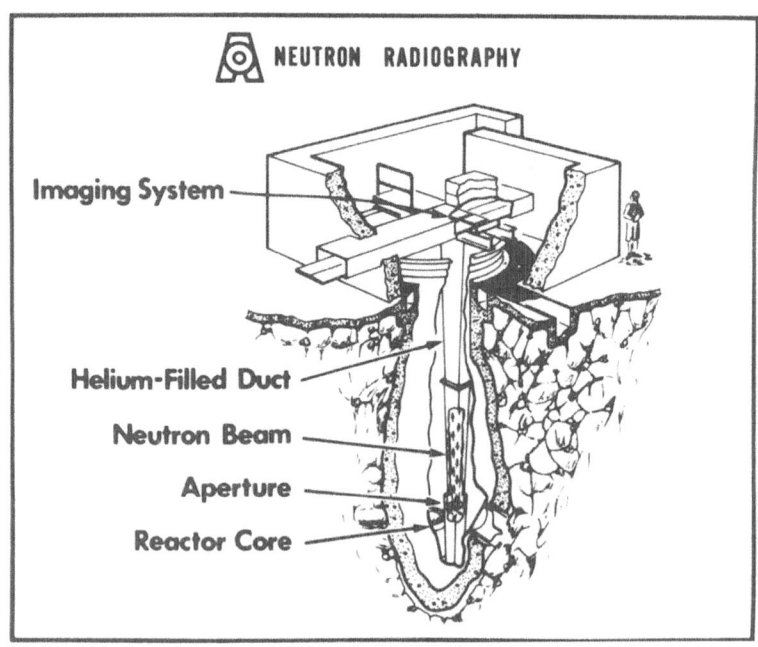

Figure 1 - Aerotest Reactor Facility

Figure 2 - Neutron Radiography Area

WELD INSPECTION OF THE VERNIER CONTROL ENGINE

Since 1981, the date of the last World Conference, production work has included examining explosive components, turbine blades, electronic components, and many mechanical components. While neutron radiographic inspection of explosives and turbine blades makes up a major portion of the production business, the more challenging and interesting tasks arise from the mechanical components area. For example, we have examined the titanium-niobium welds on the vernier control engines of the space shuttle. This application is unique in that the two metals have very different densities (4.5 and 8.57gm/cc respectively), but similar neutron attenuation characteristics. As illustrated in Figure 3, the quality of the weld is easily displayed in a neutron radiograph. Figure 4 shows the location of the vernier engines on the Space Shuttle.

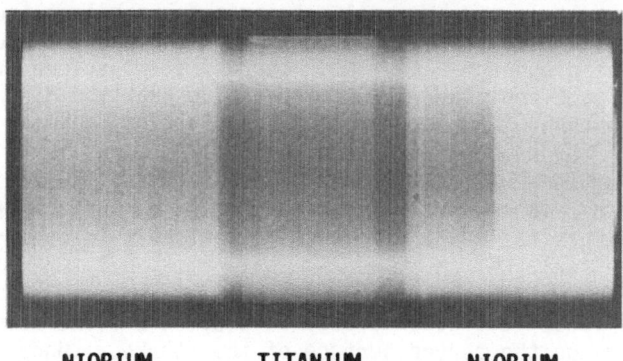

2.4 CM
DIAMETER

NIOBIUM   TITANIUM   NIOBIUM

Figure 3 - Neutron Radiograph of a Titanium-Niobium Weld

Figure 4 - Vernier Engines on the Space Shuttle

INSPECTION OF PROPELLANT TANK DIAPHRAGMS

Another space satellite application is the inspection of the rubber diaphragm seal that is used in the liquid propellant tanks of many satellites. Under zero gravity conditions, liquid fuel is delivered to the engines by pressurizing one side of a rubber diaphragm attached at the inside circumference of the titanium tank. The attachment provides a seal between the propellant and the pressurizing gas. Figure 5 shows the tank, Figure 6 shows a neutron radiograph of the diaphragm seal, and Figure 7 shows several tanks in place on a space vehicle. It is imperative that the diaphragm be in position with no flaws at the seam for a successful long life in outer space. By using neutron radiography, the quality of the rubber seal and its exact position can be critically examined. There have been times when the heat sink, used during the final welding process, has failed to remove sufficient heat, damaging the diaphragm at the seal.

A flexible vacuum cassette is used to provide a close film-to-object coupling on the curved surface of the tanks. Gadolinium metal is coated onto a 0.012 inch (0.030cm) thick piece of aluminum metal sheet and enclosed in a flexible vacuum cassette along with the film. A cassette bend radius of 4 to 6 inches (10 to 15 cm) can be achieved.

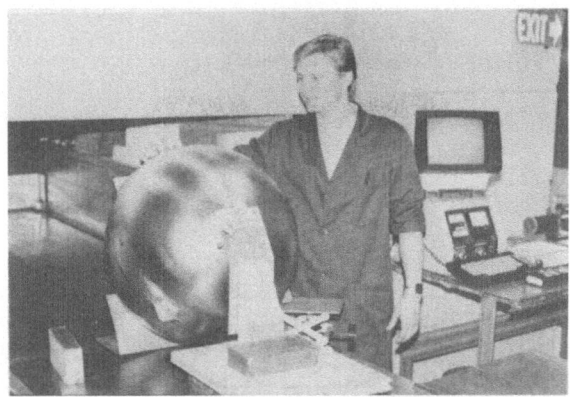

Figure 5 - Titanium Propellant Tank

Figure 6 - Neutron Radiograph of Diaphragm Seal

Figure 7 - Tanks in Place on a Space Vehicle

INSPECTION OF GRAPHITE PHENOLIC

One medical application of neutron radiography is examining carbon phenolic blocks for defects before machining, and subsequently examining the artificial hip joints that are machined from those blocks. Since carbon phenolic materials have a high scatter cross section, this program required experimentation to minimize the scattered neutrons. Cadmium lined "egg-crate" boxes were used in conjunction with separation of the cassette from the parts to provide the best neutron radiograph image. Figure 8 shows a photograph of the block and the hip joint, while figure 9 is a neutron radiograph of the finished hip joint. These prosthetic devices are currently being tested in laboratory animals.

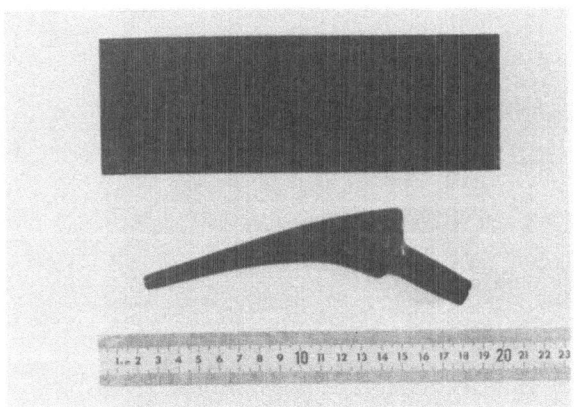

Figure 8 - Photograph of Phenolic Block and Hip Joint

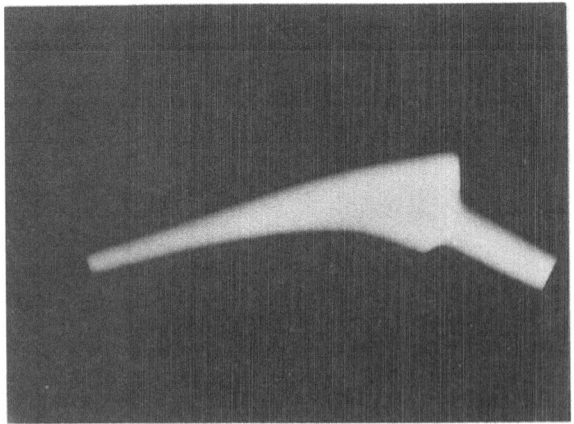

Figure 9 - Neutron Radiograph of the Hip Joint

SPECIAL FIXTURE DESIGNS

Several explosive devices, initiators, detonators, thrusters and the like, require precise alignment to assure that the position of the bridgewire to the adjacent explosive is accurately defined. The special fixtures are shown in Figure 10. Each shelf is tilted so that the axis of the part is exactly perpendicular to the beam. Each holder along the shelf is positioned rotationally to provide a precise radial orientation on a line from the source prior to being locked in place at the zero degree orientation. These specially designed fixtures provide a perpendicular orientation of all parts on the entire film as well as a method of easily changing from a zero to a ninety degree orientation. Figure 11 shows a neutron radiograph of some high voltage detonators, all being exactly perpendicular to the beam. The precise positioning contributes to high quality neutron radiographic results.

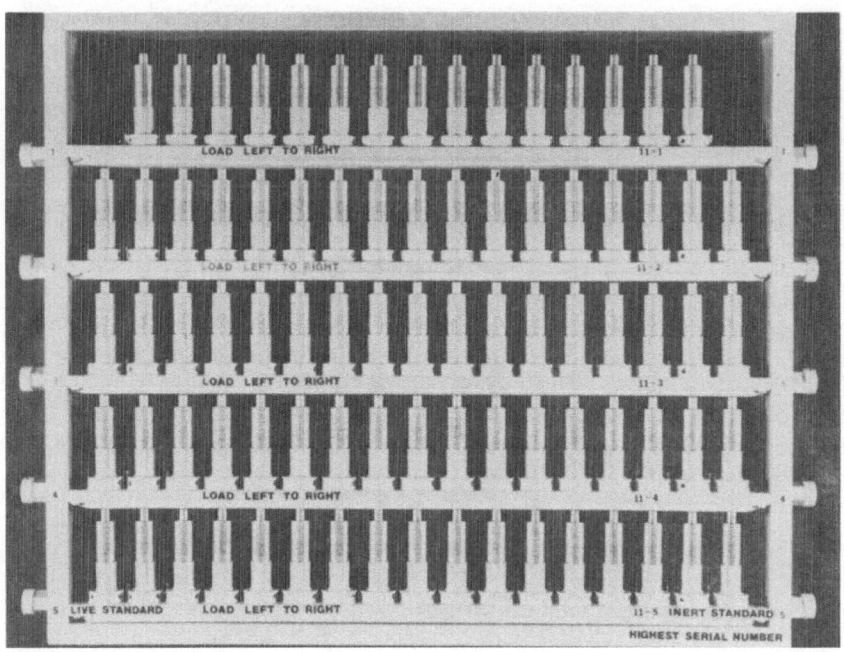

Figure 10 - Special Mounting Fixture for High Voltage Detonators (HVD)

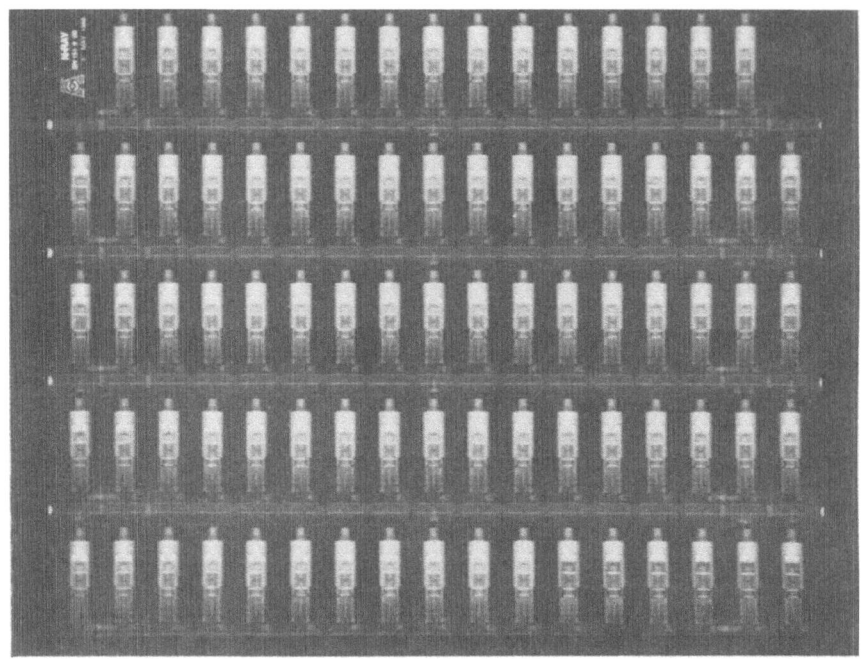

Figure 11 - Neutron Radiograph of a Loaded HVD Fixture

Neutron Radiography continues to be an important NDT tool. It has been shown to be effective in aiding the quality control and quality assurance engineering personnel in maintaining the reliability of products. The types of applications have grown slowly over the years; however a steady yearly increase in demand for neutron radiography services reflects confidence in the value of this nondestructive technique.

INTERNATIONAL
NEUTRON RADIOGRAPHY
NEWSLETTER

J. C. Domanus
Editor

Abstract

At the First World Conference on Neutron Radiography it was decided to continue the "Neutron Radiography Newsletter", published previously by J. P. Barton, as the "International Neutron Radiography Newsletter" (INRNL), with J. C. Domanus as editor.

The British Journal of Non-Destructive Testing (BJNDT) has agreed to publish the INRNL in its column "NDT Bookcase".

The Revue Pratique de Control Industriel has also agreed to publish the French version of the INRNL.

Up till now 12 issues of the INRNL were published in the BJNDT. They are reviewed below.

1. INTRODUCTION

For several years Dr. J. P. Barton has published the "Neutron Radiography Newsletter" (NRN) which has been printed courtesy of the American Society for Nondestructive Testing.

Unfortunately, this valuable publication, giving current news about the activities of various neutron radiography centers throughout the world has been discontinued.

There was a common feeling that such a publication is still necessary and this problem was discussed at the recent First World Conference on Neutron Radiography. As a conclusion of this discussion it was decided to resume the publication of the NRN in a new, enlarged form as an "International Neutron Radiography Newsletter" (INRNL), with J. C. Domanus (Risø National Laboratory, DK-4000 Roskilde, Denmark) as editor.

The Euratom Neutron Radiography Working Group (NRWG) (to which J. C. Domanus is a chairman) is particularly interested in the publication of the INRNL and will sponsor it.

The newsletter encloses short reports about the activities of various neutron radiography centers throughout the world, reports of the activities of the NRWG and, when appropriate, contributions about special items of interest to all active in the field of neutron radiography.

A French version of the INRNL is also published in Revue Pratique de Controle Industriel.

To facilitate and speed-up the task of collecting information about the activities of NR centers throughout the world, co-editors were chosen: Dr. J. P. Barton (N-Ray Engineering Co., 5709 Waverly Ave, La Jolla CA 92037, USA) for information originating from outside of Europe and Mr. G. Bayon (Services des Piles de Saclay, CEN-Saclay, F-91191 Gif-sur-Yvette, France) for information from the French speaking countries. All information worth publishing in the INRNL ought to be directed to one of the above mentioned persons.

While starting the continuation of publishing of the current news on neutron radiography special thanks must be expressed to Dr. J. P. Barton who has started the pioneering activity in that field . It seems most appropriate to start the new edition of the INRNL with a report on the previous activity of the NRN.

2. THE NEUTRON RADIOGRAPHY NEWSLETTER

Between 1964 and 1976 fifteen volumes of the Neutron Radiography Newsletter were prepared and distributed free of charge by direct mail to requestees from around the world.

The purpose, in those pioneering days, was primarily to provide a format whereby researchers in this new and exciting field could keep abreast of each other's progress.

In 1977, in response to the substantial requests for back copies, a combined edition of Volumes 1-15 was prepared.

In contains news submitted by one hundred centers around the world, a classified bibliography of seven hundred reports, and a hundred-page section of review articles.

3. THE INTERNATIONAL NEUTRON RADIOGRAPHY NEWSLETTER

The first issue of the INRNL appeared in Vol.26, No. 2 of the BJNDT in February, 1984. It has described the purpose and scope of the INRNL, as mentioned above. It was prepared by J. C. Domanus and J. P. Barton.

The second issue of the INRNL (prepared by J. C. Domanus) was describing the activities of the Euratom Neutron Radiography Working Group. It appeared in Vol. 26, No. 4 of the BJNDT in May 1984. Items like organisation and procedures of the NRWG, classifications of defects, reference neutron radiographs, image quality, accuracy of dimensional measurements, the NRWG Test Program, recommended practices, application of nitrocellulose film and publications and meetings of the NRWG were described.

The third issue of the INRNL (prepared by J. P. Barton) contained news from the USA. It was published in Vol. 26, No. 5 of the BJNDT in July 1984. Information was given on standard practices, terminology, film quality standards, system, specification and personnel qualification.

Although the fourth issue of the INRNL, prepared by J. P. Barton on aircraft inspection systems, was sent to the BJNDT on 20.06.84 it was not published yet.

The fifth issue of the INRNL (prepared by H. Tourwè and V. Stulens) describes the NR facilities at Mol, Belgium. It was published in Vol. 26, No 7 of the BJNDT in November 1984. The design and operation of two NR facilities at the BR1 and BR2 reactors are reviewed.

In the sixth issue of the INRNL H. P. Leeflang describes NR facilities at Petten, Netherlands. It was published in Vol. 27, No 2 of the BJNDT in March, 1985. It describes the pool and dry NR facilities of the High Flux Reactor as well as the optical evaluation equipment.

The seventh issue of the INRNL (prepared by G. Ducros), describes the NR facilities at Grenoble, France. It was published in Vol. 27, No 3 of the BJNDT in May, 1985. The NR facilities at Grenoble are of dry type (reactor Melusine) and pool type (reactor Siloe).

In the eihth issue of the INRNL D. H. C. Harris and W. A. J. Seymour describe the NR facility at the Dido reactor at Harwell, England. It was published in Vol. 27, No 4, of the BJNDT in July, 1985. Here the Dido reactor itself and the thermal and cold neutron NR facilities are described.

The ninth issue of the INRNL describes the NR facility at the DR1 reactor at Risø, Denmark. It was published in Vol. 27, No 6 of the BJNDT in November, 1985. Here J. C. Domanus describes the double beam NR facility, the radiographic procedure used there and the application of NR.

The tenth issue of the INRNL contains the description of NR activity at NDE laboratory (Department of Nuclear Engineering) at Technion – Israel Institute of Technology in Haifa, Israel. It was contributed by Y. Segal and published in Vol. 28, No 1 of the BJNDT in January, 1986.

In the eleventh issue of the INRNL J. Rant describes the NR and autoradiography activities at the J. Stefan Institute in Ljubljana, Yogoslavia.

The contribution of G. Bayon about the NR facility at the Orphee reactor in Saclay, France, prepared as the twelfth issue of the INRNL, was sent for publication in the BJNDT. It will appear in Vol. 28, No 4, July, 1986. After the description of the facility itself (lay-out, exposure room, adjoining rooms) operation of the facility is explained and its applications summarized.

NEUTRON RADIOGRAPHY WORKING GROUP (NRWG)

Summary of activities and publications

J.F.W. Markgraf

Joint Research Centre of the Commission of the European
Communities, Petten Establishment
Postbus 2
NL 1755 ZG Petten, The Netherlands

ABSTRACT

This contribution briefly summarizes the activities and
publications of the Neutron Radiography Working Group
(NRWG).

This group was constituted in 1979 under the auspices of
the Commission of the European Communities. The members
are experts in neutron radiography associated with nuclear
research centres within the European Communities.

The main tasks of NRWG are the coordination of common
interest activities in the field of neutron radiography
and the promulgation of information and knowledge on
neutron radiography, e.g. by sponsoring international
conferences and research projects, publication of books,
reports and an international newsletter on neutron radiography.

---

1. INTRODUCTION

The NRWG was constituted in 1979 by experts in neutron
radiography from nuclear research centres within the European
Communities (under the auspices of the Commission of the European
Communities, represented by the Joint Research Centre, Petten
Establishment, Petten, The Netherlands).

This contribution summarizes in brief the activities and
publications of the NRWG /1/ as displayed on a poster at the
conference.

2. ORGANIZATIONAL

- NWRG membership is open to experts in neutron radiography from the member states of the European Communities.
- NWRG observers are invited experts in neutron radiography from non-EC countries to participate in NRWG activities.
- Present NWRG members are from : Belgium, Denmark, France, Federal Republic of Germany, Italy, The Netherlands, United Kingdom and the Joint Research Centre of the Commission of the European Communities.
- NRWG observers are from : Canada, Sweden, Switzerland, United States of America.
- Chairman of NWRG : Mr. J.C. Domanus, Risø National Laboratory, Roskilde, Denmark.
- NWRG secretariate at : Joint Research Centre, HFR Division, Mr. H. Röttger, Petten Establishment, Postbox 2, NL 1755 ZG Petten, The Netherlands.
- Meetings
  o NWRG annual meeting : Two or three days meeting at various participating laboratory sites.
  o NWRG sub-group meetings : Coordinative meetings at various participating laboratory sites; frequency depending on activity of sub-group concerned.

3. MAIN TASKS OF NRWG

- Coordination of common activities in the field of neutron radiography within the European Communities.
- Promotion of information exchange and neutron radiography know-how extension, e.g. by sponsoring international conferences and research projects, editing of topical reports and books on neutron radiography and publishing of an international newsletter.

4. MAIN TOPICS OF NRWG ACTIVITIES AND ITS SUB-GROUPS

- Dimensional measurements of neutron radiography images
  o Development of measuring techniques and image enhancement methods
  o Development of reference objects, e.g. calibration fuel rod
  o Evaluation and qualification of reference objects
- Characterization and qualification of neutron radiography devices and standards used in neutron radiography

60

     o Development and investigation of beam purity and beam quality
       indicators
     o Development and qualification of image characterization methods
     o Testing and evaluation of new standard and qualification
       techniques, e.g. L/D detector
- Nitrocellulose film
     o Compilation of presently existing experience with nitrocellulose
       film (topical report)
     o Comparative study on the utilization of nitrocellulose and
       silver halide film
- International neutron radiography newsletter
     o Periodical information about activities in NWRG and its member
       laboratories
- Editing of topical reports or books
  presently in preparation :
     o Topical report on nitrocellulose film in neutron radiography
     o Book on Practical Neutron Radiography
- Contacts with other organizations interested in neutron
  radiography, e.g. ASTM.

5.   INTERNATIONAL CONFERENCES WITH NRWG SPONSORSHIP

o First World Conference on Neutron Radiography, San Diego,
  California, USA, December 1981
o Second World Conference on Neutron Radiography, Paris, France,
  June 1986

6.   INTERNATIONAL NEUTRON RADIOGRAPHY NEWSLETTER

issued since 1984 in
o British Journal for Non-Destructive Testing and
o Revue Pratique de Contrôle Industriel

7.   NRWG PUBLICATIONS AND EDITIONS

o 1979 : Neutron Radiography Findings in Light Water Reactor Fuel
           (Topical report) /2/
o 1981 : Neutron Radiography Handbook /3/
o 1982 : Neutron Radiography, Proceedings of the First World
           Conference on Neutron Radiography in San Diego /4/
o 1984 : Reference Neutron Radiographs of Nuclear Reactor Fuel
           (Book)
           Neutronogrammes de Référence pour le Combustible
           Nucléaire /5/

Under preparation :
o 1986 : Nitrocellulose Film in Neutron Radiography (Topical
           report)
o 1986 : Neutron Radiography, Proceedings of the Second World
           Conference on Neutron Radiography in Paris
o 1987 : Practical Neutron Radiography (Book)

Publishers of NRWG Books and Proceedings
D. Reidel Publishing Company, Postbox 17, NL-3300 AA Dordrecht,
The Netherlands

Publishers of NRWG Topical Reports
Commission of the European Communities, Joint Research Centre,
Petten Establishment

8.   CONTENTS IN BRIEF OF NRWG PUBLICATIONS

8.1. Neutron Radiography Handbook /3/

Radiographic testing with neutrons can yield important
information not obtainable by more traditional methods.

In contrast to X-rays, thermal neutrons are particularly
attenuated by certain light materials such as hydrogen, lithium
and boron, whereas they penetrate easily through iron, lead and
uranium.

By inspecting nuclear materials, different isotopes of the
same element can often be differentiated by neutrons and sharp
radiographs can be obtained even if the object is highly
radioactive.

Contents of the Neutron Radiography Handbook :
- Introduction
- Principles and practice of neutron radiography
- Recommended practice for the neutron radiography of nuclear fuel
- NRWG indicators for testing of beam purity, sensitivity and
  accuracy of dimensions of neutron radiographs
- Atlas (compact version) of defects revealed by neutron radiography
  in light water reactor fuel
- Neutron radiography installations in the European Communities.

8.2. Proceedings of the First World Conference on Neutron
     Radiography in San Diego, December 1981 /4/

The 140 papers from 20 countries that were presented at the
Conference show a striking diversity in the applications of neutron
radiography.

A large proportion of the papers concern applications for
inspection of nuclear fuel, whereas high precision dimensional
measurements and track-etch imaging are receiving growing attention.

Other topics dealt with include :
- Reactor facilities with general applications
- Applications in general industry
- Applications in nuclear industry (fuel)
- Applications in nuclear industry (general)

- Non-industrial applications
- Real-time applications
- Small source systems
- Special techniques
- Neutron radiography standards

8.3. Reference Neutron Radiographs of nuclear reactor fuel /5/

The collection of reference neutron radiographs of nuclear
reactor fuel has been prepared by the NRWG on the basis of radiographs from nuclear research centres within the European
Communities.

It contains 158 radiographs of fuel rods from light water and
fast reactors (on film and in enlarged print).

The radiographs are arranged in accordance with a
classification system listing the components of fuel and cladding
and illustrating defects which can be revealed by neutron radiography.

Introductory chapters (in English and French) describe the
classification system in relation to a typical fuel rod and explain
how to identify defects and to use the collection.

All terms used in the classification together with additional
useful terms are listed with their equivalents in six languages :
French, Danish, Dutch, English, German and Italian.

The main technical data and addresses of neutron radiography
installations in the European Communities are listed in a final
chapter.

8.4. Practical Neutron Radiography (book)
     (Publication is anticipated for 1987)

As a follow-up to the Neutron Radiography Handbook the NRWG
is now preparing a completely new book on Practical Neutron
Radiography.

It will provide basic information on the principles of
neutron radiography and emphasizes practical aspects of neutron
radiography and its applications.

Titles of the main chapters in Practical Neutron Radiography
are :
- Principles of Neutron Radiography
  (Mathematical and physical foundations of neutron radiography,
  neutron sources, neutron beams, imaging techniques, imaging
  recorders)

- Neutron Facilities
  (Choice of sources, filters, collimators)
- Exposure Facilities
- Detectors and Recorders
  (Methods, description of detectors and recorders)
- Cassettes
- Processing
- Viewing
- Track-Etch Technique
- Radiographic Techniques
  (Geometry of exposure, exposure factors, dimensional measurements, radiation safety)
- Application of Neutron Radiography
  (Nuclear applications, non-nuclear applications, dynamic neutron radiography, other applications)
- Neutron Radiography Indicators - Description and Use
  (ASTM BPI and NRWG BPI, NRWG BPI-F, ASTM BI, AFNOR IQI, Calibration fuel rod, evaluation of neutron radiography indicators)
- Neutron Radiography Installations in the EC
- Terminology
- Useful Data

8.5. Nitrocellulose film in neutron radiography (topical report) (Publication anticipated for 1986)

The utilization of nitrocellulose film in neutron radiography is advantageous because of its direct imaging capability of radioactive objects and of its potential for sharper images.

In view of the increasing utilization of nitrocellulose film in neutron radiography, the NRWG is at present preparing a compilation of detailed information on this film type.

The following subjects will be treated :
- Film
- Converters
- Comparison between Nitrocellulose and Silver Halide Film
- Technique of Exposure and Processing
- Image Quality Indicators
  o IQI's used in industrial and neutron radiography
  o Calibration fuel pin
- Assessment, Interpretation, Use of Radiographs
- Dimensional Measurements on Film
  o Methods and equipment
  o Test objects
  o Results obtained
- Nuclear Applications
  o Radiography of irradiated items
  o Radiography of fuel

64

- Non-nuclear Applications
  o Radiography of non-nuclear items

9.   AN EXAMPLE OF A NRWG ACTIVITY :

    The NWRG Test Programme for investigation of image quality
methods and dimensional measurements techniques /6/

9.1. NWRG Test Programme

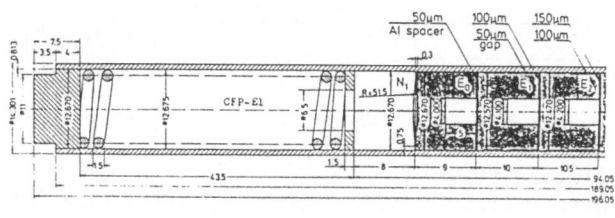

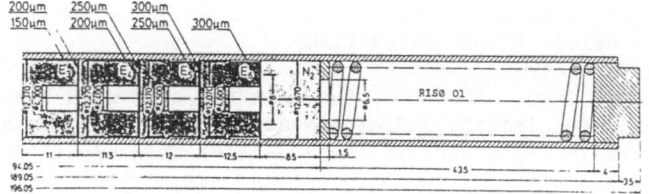

NRWG Calibration Fuel Pin, CFP-E1

9.2. Objectives
    - Investigation of direct and indirect neutron radiographic
      imaging methods of nuclear fuel
    - Evaluation of different methods for dimensional determination
      from neutron radiographic images of nuclear fuel (pins)

9.3. Methods
    - Development of a NRWG Test Programme directed towards image
      quality checking and dimensional measurements of nuclear fuel
    - Fabrication of a NRWG Calibration Fuel Pin and indicators for
      beam purity and sensitivity

65

9.4. Assessment and measurements

- 30 individual checks and measurements are performed on each image set (see table)

	Processed at																	
	NR center								RISØ									
Converter	Gd			Dy			B	BN1	Gd			Dy			B		BN1	
Film	SR	D4	M	SR	D4	M	CNB	CN	SR	D4	M	SR	D4	M	CNB		CN	
Code No	1	3	5	7	9	11	13	22	2	4	6	8	10	12				
Etched at															20°C	50°C	20°C	50°C
Code No															16	19	25	28
Copy on							S0015								S0015			
Code No							14	23							17	20	26	29
Viewed through						Polarizing filters												
Code No							15	24							18	21	27	30

- Selection of materials to be investigated by direct and indirect imaging methods
  o Nitrocellulose film
  o High contrast film
  o Polarizing filters
  o Metal conversion screens
  o Silver halide films (X-ray film)
  o Converter screens
- Exposure of NRWG Calibration Fuel Pin together with beam purity and sensitivity indicators at 11 participating NRWG neutron radiography facilities (2 sets of images/participant)
- Processing and evaluation of one image set at the participating NRWG laboratories
- A second image set is processed and evaluated (under JRC contract) at the Risø National Laboratory including the following additional activities :
  o Assessment of image quality
  o Determination of dimensions from images
    - Axial dimensions, e.g. fuel stack length, pellet length, length of central void, pellet to pellet gaps, dishing distances
    - Radial dimensions, e.g. pellet diameters, pellet to pellet gaps, cladding tube wall thickness, diameter of central void
  by
    - Travelling microdensitometer and
    - Profile projector
  o Comparison of measurements and true dimensions from the participating laboratories

9.5. Preliminary results (from assessment of image quality only)
- Image quality based on visual assessment
  o Images taken on nitrocellulose film are generally not
    better than images derived from a Dysprosium foil with
    single coated silver halide film
  o Utilization of polarizing filters does not improve image
    quality of nitrocellulose film
  o Etching of nitrocellulose film at higher temperature
    (e.g. 50°C) for shorter periods (e.g. 45 minutes) leads
    to higher image quality
- Image quality based upon comparison of dimensional
  measurements
  o Dimensions determined from nitrocellulose film are more
    accurate than obtained from silver halide film
  o Most measurements yielded larger values than the true
    dimensions
  o Single coated silver halide film always gave the best
    results among all silver halide films compared
  o Nitrocellulose film with coated converter on both sides
    provides better measuring accuracies.

10. REFERENCES

/1/ J.C. Domanus
     Euratom Neutron Radiography Working Group
     This conference, session V

/2/ J.C. Domanus
     Neutron radiographic findings in light water reactor fuel,
     Risø National Laboratory, Metallurgy Department, June 1979

/3/ P. von der Hardt, H. Röttger, editors
     Neutron Radiography Handbook, D. Reidel, Publishing Company,
     1981, EUR 7622e, ISBN 90-277-1378-2

/4/ J.P. Barton, P. von der Hardt, editors
     Neutron Radiography, Proceedings of the First World
     Conference, 1982, EUR 8296 EN, ISBN 90-277-1528-9

/5/ R. Barbalat, J.C. Domanus, J.F.W. Markgraf, F. Michel,
     D.J. Taylor
     Reference Neutron Radiographs of nuclear reactor fuel,
     D. Reidel, Publishing Company, 1983, EUR 8912 EN EP,
     ISBN 90-277-1717-6·

/6/ J.C. Domanus
     Assessment of Radiographic Image Quality by Visual
     Examination of Neutron Radiographs of the Calibration Fuel
     Rod
     This conference, session XIII

Status of Neutron Radiography in Malaysia

Abdul Ghaffar Ramli, Azali Muhamad,
Rosly Jaafar and Sheriffah Noor Khamseah
Unit Tenaga Nuklear,
Kompleks PUSPATI,
43000 Kajang, Malaysia.

ABSTRACT

The development of neutron radiography (NR) in Malaysia began with the availability of the country's first research reactor. The Reaktor TRIGA PUSPATI (RTP) was commissioned in June 1982 and plans to construct the neutron radiography facility commenced soon after that. In late 1983, a NR test facility (NuR 1) was constructed out of small modular concrete blocks. Tests carried out in this facility enabled the design of the permanent facility (NuR 2) to be finalized. The construction of NuR 2 took place in November 1984 and it was completed in February 1985. NuR 2 is now actively used in developmental work in NR to ensure that an efficient local NR service can be given in the near future. A number of shortcomings in the design of NuR 2 have been identified, and efforts to rectify these are under way.

INTRODUCTION

The development of neutron radiography (NR) in Malaysia has been a challenge against the lack of materials, manpower and expertise. The 1 MW Reaktor TRIGA PUSPATI (RTP) is the country's first and only research reactor. Since its commissioning at the Tun Ismail Atomic Research Centre (PUSPATI) in June 1982, efforts have been made to develop a number of applications with potential commercial interest. To date, RTP is sufficiently equipped to conduct analyses through neutron-activation analysis (NAA) and delayed-neutron analysis (DNA), isotope production and neutron radiography.

Efforts to install a neutron radiography facility on RTP began in 1983. A neutron radiography test facility (NuR 1) was constructed in order to verify calculations and design before a permanent facility (NuR 2) was made. Both NuR 1 and NuR 2 were used to develop imaging techniques suitable to their designs, and these have been reported elsewhere (1). These two facilities and the work associated with them are described in the ensuing sections.

THE TEST FACILITY (NuR 1)

The installation of the test facility took place in December 1983. NuR 1 was built from small modular concrete blocks, thus making it easy to modify the shape or thickness of the shield required. The radial (non-piercing) beam-port #1 of RTP was chosen for the test NR work. A simple collimator design was adopted in order to overcome the materials limitation. The collimator consisted of an aluminium cylinder 143 cm long (Fig. 2a) at the end of which was 20 cm paraffin (for neutron moderation) and 10 cm lead (for gamma attenuation). A 1.7 cm hole was drilled in the lead block to reduce neutron attenuation without significantly increasing the gamma intensity. Imaging was done using CN85 track detector (with BN1 converter foil) due to its simplicity and the inavailability of light-tight aluminium cassettes and suitable converter foils for photographic imaging to be realized.

NuR 1 produced the first successful neutron radiograph in April 1984. Typical of images produced from track detectors, the radiograph showed fine detail but lacked contrast. A number of attempts were made to increase the contrast through the use of polarizers and computerized image-analysis (2). The latter showed a good potential but the lack of sufficient hardware limited its capability.

NuR 1 was dismantled in December 1984 after all the necessary verifications for the construction of NuR 2 had been obtained.

THE PERMANENT FACILITY (NuR 2)

The construction of NuR 2 started in May 1984 with the fabrication of the shielding walls. NuR 2 was installed at beam-port #3 in the reactor hall in December 1984 and the work was completed by February 1985. NuR 2 (Fig. 1) is more refined than NuR 1. Common to both designs are neutron and gamma shutters outside the reactor wall. In the case of NuR 2, the 1 tonne shutter was designed as a beam trap. This allows effective shut-off of the neutron beam to enable on-line loading of objects to be radiographed. Photographic imaging was developed with the availability of NuR 2. It is now possible to produce neutron radiographs by direct photographic technique, although the contrast and quality need be improved. This will be done through the use of a better collimator and will be discussed in the next section.

COLLIMATOR DEVELOPMENT

Both NuR 1 and NuR 2 were extensively used for collimator development. Figure 2 shows the sequence of the collimator development throughout this work. The first collimator (Fig. 2a) could only achieve L/d of about 18. Exposure time using CN85 detector and BN1 converter was typically 6 hours at 500 kW (3000 kW-h). In an attempt to improve the geometric unsharpness, a conical collimator was made (Fig. 2b). The inlet aperture to the collimator was made to be 3.5 cm in order to provide a sufficiently high neutron intensity. The wall of the collimator was filled with Borax powder (boric acid) to absorb out-scattered neutrons. Even though this collimator achieved L/d of approximately 60, it suffered from a much lower intensity than the first. This finding caused the decision to construct NuR 2 on beam-port #3 instead of #1. In theory, the port should yield a six-fold increase in neutron flux due to the absence of the graphite reflector at that position. However, it was later found that this did not show a true advantage since the neutron spectrum was harder (more fast neutrons) in beam-port #3, resulting in the need for a thicker moderator. The intensity at the radiographic position was thus only marginally higher.

The solution to this came in the form of the direct photographic technique. This method would also provide a higher contrast. The conical collimator, however, had to be modified in order to reduce the gamma intensity sufficiently. A neutron to gamma dose (in absorbed-dose units) ratio of approximately 1 is needed to produce a neutron radiograph with little gamma contribution (3). A 20 cm lead plug inserted into the conical collimator was found to be sufficient for this purpose (Fig. 2c). The gamma dose decreased by two orders of magnitude (fig. 3). Unfortunately, the plug also reduced the neutron dose by one order of magnitude. This collimator, although capable of producing neutron radiographs, did not provide the exposure-time reduction needed even with the use of a fast film such as Agfa Structurix D7.

From the experiences gained with the three collimator designs, it was then possible to design a final collimator which should provide a solution to all the shortcomings encountered. This design is shown in Fig. 2e. This collimator will be longer than any previous ones. This enables the moderator block to be placed closer to the reactor core, thus increasing the source intensity. This would have an effect of increasing the flux by approximately 15. The gamma attenuation material shall be made of bismuth, while the moderating block shall be polyethylene. The choice of these materials would result in a lower neutron attenuation by 1 order of magnitude. Hence, the neutron intensity at the sample position would be 2 orders of magnitude higher than at present. This should shorten the exposure time to a few minutes or allow the use of slow, fine-grain film such as Agfa Structurix D2. This collimator is now being fabricated and it is expected to be installed during the December maintenance period at the end of this year (1986).

In the meantime, to enable NR to be performed adequately, a temporary pin-hole collimator has been made and is presently in use. This collimator, shown in Fig. 2d, has the same length and uses the same materials as the first three designs. The pin-hole and outlet aperture are being defined by cadmium plates. The use of cadmium results in a slight increase in gamma intensity due to the Cd(n,γ) reaction. It was found that with an exposure of 1500 kW-h using Agfa Structurix D7 with gadolinium oxysulphide screen (sufficient to produce D=2.5), the gamma image was very faint. The radiograph (Fig. 4) given in this paper is an example of the image obtained with this collimator.

APPLICATIONS

Throughout the development period, there was very little actual application of NuR 1 and NuR 2. The earliest application of NuR 1 was due to an unexpected event when a plastic sample tube lodged itself stuck at the end of RTP's pneumatic transfer system. Both X- and neutron-radiographs were taken. The X-radiograph showed the presence of a bent metal piece, while the neutron-radiograph clearly showed the location of the plastic tube in relation to the metal piece. The stuck tube was subsequently removed.

Inspired by the work of Couchat and Moutonnet (4) and Willatt and Struss (5), some effort had been put in the development of the application of NR for in situ study of root growth in plants. This was done to assist the Mutation-Breeding Group at the PUSPATI. This technique has been demonstrated to be feasible but the technique badly needs the new collimator to shorten the exposure while at the same time provide a higher resolution. The last point is very important because of the Mutation-Breeding Group's emphasis on padi, a monocotyledon, which has a fibrous-root system.

Efforts have been made to popularize NR by conducting visits to and discussions with local industries, research institutions and other organizations. Potential customers are the Malaysian Airline System (e.g examination of emergency-door thrusters), a local bullet-manufacturing company, the Defence Science and Technology

Research Centre, the National Museum and a few other organizations.

The National Museum was our first and still is our only external customer. Metallic archeological artifacts have been sent to us in the Museum's attempt to elucidate any internal structures in them. The 1000-year old bronze statues from the ruins of a Hindu Temple in Kedah (a northern state of Peninsular Malaysia) did not reveal any exciting features in them. The tin-ingots from the recently-discovered excavation site of the ancient Malay Empire of Gangga Negara near Bruas, Perak (another northern state) gave a more interesting radiograph. Common to all the ingots, a hole was found originating from its flat side. The ingot, whose neutron radiograph is shown in Fig. 4, had the hole blocked with some fibrous material. The radiograph shows clearly the extent of the hole, as well as the tight fit of the fibrous material in it. As to why the ingots were made this way and what the fibrous material was doing in it is still unclear and is being analysed by the Museum staff.

SUMMARY

There is a Malay proverb which states "berjagung-jagung dahulu sebelum padi masak". Translated, it means 'live on maize while waiting for padi to ripen". This proverb summarizes the way in which the neutron-radiography project had been performed at PUSPATI. While waiting for the availabilty of bismuth, lead was used. Similarly, paraffin was used while solid polyethylene was still unavailable. Temporary collimators were made while the permanent one is being fabricated. The adoption of this strategy managed to keep the NR developmental work active, and provided enormous experience to those involved. The NR project has made significant advances in slightly more than 2 years since it was initiated. A few applications of the technique have been attempted and it is hoped that more would be conducted in the future.

Acknowledgement

We would like to thank all those who have given their cooperation, in one way or another, to enable us to proceed with this developmental work. In particular, our gratitude goes to Mr. T. Wall, AAEC Australia, for the help and loan of aluminium cassettes and converter foils to enable us to initiate photographic imaging. Likewise, we are indebted to Prof. S. Harasawa of Rikkyo University, Japan, for the gift of an aluminium cassette and a gadolinium oxysulphide screen. To Mr. Adi Hj. Taha, Curator of Excavation, Muzium Negara, Malaysia, we thank him for his unfailing and keen interest in our work and his trust on us in the provision of priceless archeological items to be subjected to uncertain fate in our experiments.

Figure 1: A view of the permanent NR facility (NuR2). The concrete shield seen here is 6.7m across and 1.2m high.

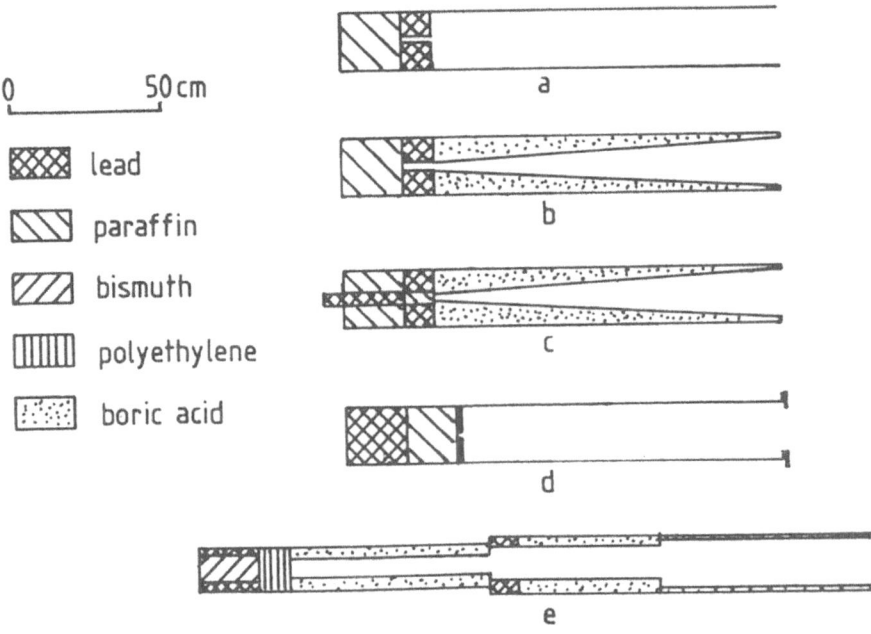

Figure 2: Sequence of collimator development in this work. The designs started simple and used easily available materials. The final collimator (e) will be installed soon.

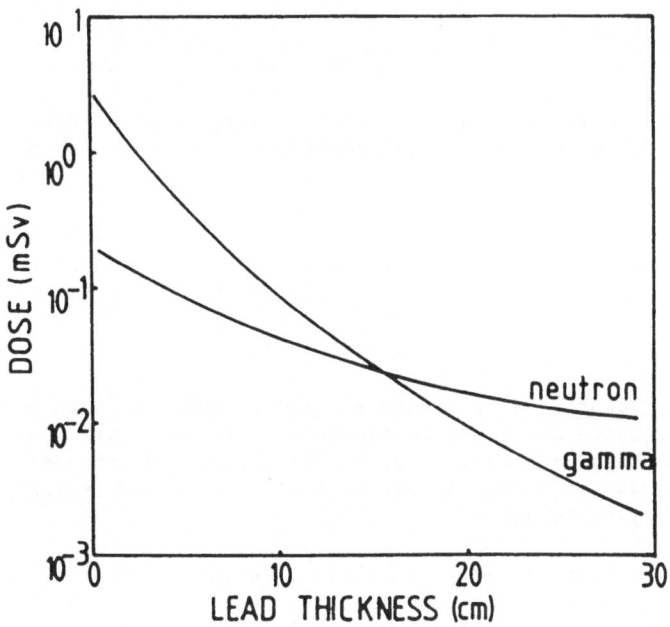

Figure 3: The insertion of a lead plug into the collimator causes a drop in both neutron and gamma dose. Beyond 15cm, the neutron dose is higher than the gamma dose.

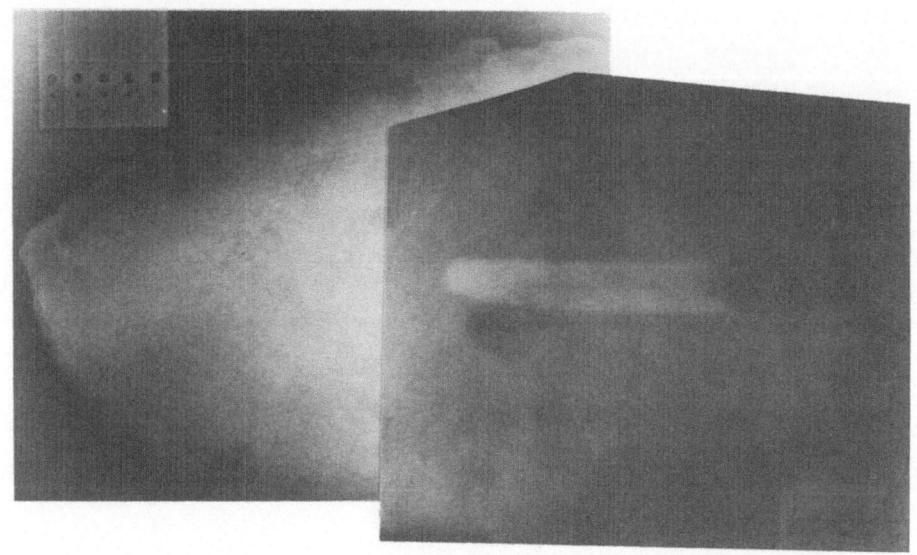

Figure 4: A neutron radiograph of an archeological specimen (tin ingot) shows clearly the presence and extent of the fibrous (wood/rattan) material in it.

REFERENCES

1. A. B. Ramli, A. Muhamad, W. R. Yusof, A. S. A. Razak,
   J. K. Ibrahim and R. Jaafar (1984), Percubaan Radiografi
   Neutron di PUSPATI, J. Sains Nuklear Malaysia 2,2 pp 26-41.

2. A. B. Ramli (1985), Image Analysis for Increasing the
   Contrast of Neutron Radiographs, In Activities of the
   Research Division 1984, PUSPATI Internal Report PPA/AR/13.

3. T. Wall (1985), Private Communication.

4. P. Couchat and P. Moutonnet (1974), Une Application
   Originale de la Neutronographie: L'etude en place du
   Developpement Racinaire, In Proc. Symp. on Isotopes
   and Radiation Techniques in Soil Physics and Irrigation
   Studies, Istanbul.

5. S. T. Willatt and R. B. Struss, Neutron Radiography, a
   Technique for Studying Young Roots Growing in Soil,
   IAEA Publication SM-235/1.

PART II

REACTOR SYSTEMS

A NEUTRON RADIOGRAPHIC FACILITY ON A RESEARCH REACTOR AND SOME RESULTS OBTAINED IN NDT

Ma Zhenze, etc.

Southwest Institute of Nuclear Physics and

Chemistry, P.O. Box 515-22, Chengdu, China

ABSTRACT

A neutron radiographic facility on a research reactor and its characteristics are briefly described. A collimator of the facility is placed in a horizontal thermal column channel of 150mm in diameter. The converters of Gd or Dy metal foils, or painted Gd_2O_2S paper plate combined respectively with x-film are used as image detector and indicator. Some performances of the system, such as resolution and sensitivity, have experimentally been tested. Some practical applications of the facility and better results obtained in NDT for the quality inspecting and controlling of several industrial products are illustrated.

INTRODUCTION

In China the study of neutron radiographic technology began in the late 1970s. Research reactors are used as the neutron source for all neutron radiographic facilities. The research reactor mentioned in this paper is a pool type reactor (PTR). It has been operated in our institute since June 1979. The neutron radiography is only one aspect of nuclear technical applications at the reactor, and has been developed for NDT of some industrial products, such as several assemblies including explosive, nuclear fuel elements and its components, etc. In recent years because of improving divice and adjusting parameters several times, and making inspect for testing specimens or product parts of different materials, some better results were obtained. This technology has already become an important means for testing and controlling quality of some products that can not be examined by x or γ-ray radiography.

THE FACILITY AND ITS PROPERTY

1. Source and Beam Energy

PTR is a thermal neutron research reactor with a power of 3MW. It utilizes light water as coolant and neutron moderator, beryllium and graphite as neutron reflector. The highest thermal neutron flux in the core is 6.5×10^{13} n/cm^2.s. The neutron beam for the radiography is derived from a horizontal thermal column channel with 150mm-diameter by 2200mm-long. A 1100mm-long collimator is inserted into latter half of the channel, as shown in Fig.1. Several different long graphite plugs can be also inserted into front part of the channel to moderate neutrons, and thereby to change the beam energy (experssed in cadmium rate'Rcd').As the plug lengths are accumulated from 0 to 1000mm the values of Rcd (for gold foil) are also changed from 4 to 150, but the highest thermal neutron flux at radiographed objects is contrarily decreased from 4.2×10^7 to 2.5×10^5 n/cm^2.s.

2. Collimator and Collimation Ratio

The collimator of the facility is a similar conical-like structure.[1] On the beam passageway a few annuli are placed, the hole diameters of which are enlarged step by step along the beam direction. A minimum hole size (D) for introducing beam into the collimator is 50mm in diameter. Both the annuli and the inner wall of the collimator are made of gadolinium foils, boron--carbide plastic and lead plates. this kind of structure can largely decrease the effect to radiography from scattered neutrons and γ-ray emitted by inner surface of the collimator.

In order to adjust the collimation ratio (L/D) the L values can be changed by means of moving a handcart with a support of the cassette and the objects to be radiographed. In this way the ratio L/D can continuously chosen in the range of 50-110.

3. The Ratio of Neutrons to γ-Ray

For rasing ratio of neutrons to γ-ray (n/γ), in

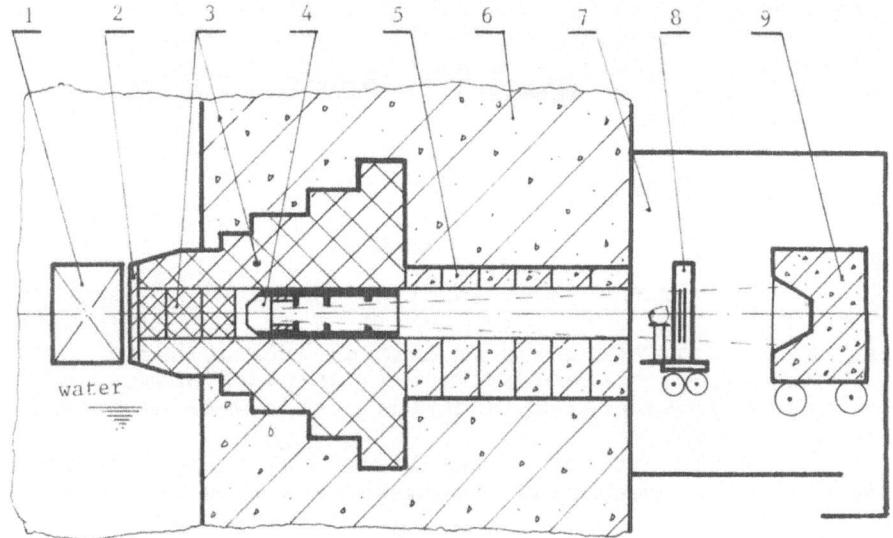

1. core 2. lead plate 3. graphite plugs and blocks
4. collimator 5. shutter 6. biological shield
7. shield room 8. converter and film in cassette
9. beam stop

Fig.1 Neutron Radiographic Facility on PTR Reactor

addition to a lead plate of 55mm-thick between the core
and thermal column-head, a 44mm-thick lead disc was
covered on the collimator-head for stopping γ-ray.[2]
Thus, as the length of plugs is 210mm, a maximum n/γ
value will be 2.1×10^7n/cm^2.mR. If the length exceeds
210mm the n/γ value will decrease with increasing of
the plug-length.

4. Image Detector and Indicator
    In the sysem a combination of converter and x-film
was used as image detector as well as indicator. It was
experimentally shown that the converter of the metal Gd
foil about 0.05mm-thick (only installing 'back' screen)
in direct exposure imaging can attain better contrast
and moderate speed. In indirect imaging method the Dy
and In foils of about 0.1mm-thick was suitable.[3] Thus
far the services performed on the system for several
users were only statical tests of products, and so we
have paied attention to improving inspection resolution
and sensitivity rather than advancing the rapidity. So

the 0.02mm-0.07mm-thick metal Gd foils were generally used as converter for non-active objects, and a few minute-grain x-films were used as far as possible, such as China-made 'Shanghai-x', Belgium-made D_4 and D_2 type, and Kodak-SR type made in U.S.A. These kinds of films can be used to reduce the influence of γ-ray on radiographic details, and to raise the testing level.

BEAM PURITY AND IMAGE QUALITY

1. Examination According to ASTM E545-75 Standard

According to the ASTM E545-75 standard promulgaed by U.S.A.[4] Several data of the system were measured, as follows:

(1) Four contents in the beam to film exposure were determined by measuring the image densities of a beam purity indicator (BPI), and they are respectively:

thermal neutron content C=91%,
scattered neutron content S=3%,
epithermal neutron content E=5.6%,
low-energy gamma content γ=0.3%.

(2) Image quality level obtained by observing the radiograph of a type 'A' image quality indicator (IQI) can reach: sensitivity level R=8, and so the image quality level (IQL) of the system is N91-3-8 in terms of the designation NC-S-R of the E545-75 standard.

2. Two Methods for Determination of Resolution

(1) Observe minimal hole size method.[5]

On a cadmium strip with 0.5mm-thick, a line of holes in different diameter of 1.0-0.02mm were drilled through in proper order, and then it was radiographed at the system. The resolution determined by visible minimal hole image was 0.02mm.

(2) Klasens' method[6]

A thin opaque knife edge made of cadmium with 0.5mm -thick was radiographed, and then the image was measured and treated complying with Klasens' method, the system unsharpness obtained in the method was 0.04mm.

3. Testing Sensitivies of Two Typical Materals.

Stainless steel and explosive are taken as typical metal and nonmetal materials respectively.As the thickness of the steel was 25mm the sensitivity tested by means of a steel-pin penetrometer was 1.6%, and as the explosive disc thickness was 6mm, the sensitivity inspected by observing image of several holes with different deepths on the disc was about 4.2%.

SOME APPLICATIONS AND RESULTS

This facility has been utilized in recent years for performing NDT of some industrial products to detect internal structures or defects, and some better results have been obtained. Several examples are only illustrated, as following:

1. Tecting Nuclear Fuel Elements

It has important significance for raising fuel burnup that different amount of Gd_2O_3-powder as burnable poison is doped into nuclear fuel elements with cosine distribution. The content and distribution of Gd_2O_3 in the elements must be determined for controlling fuel quality. Several specimens of UO_2 fuels with the same thickness (3mm) and different content of Gd_2O_3 were taken to be together radiographed. The results measuring the image densities on the radiograph showed: the content of Gd_2O_3 change of 0.3% could be inspected while the total amount of Gd_2O_3 did not exceed 5% of UO_2 in weight.

In another case, the UO_2-elements doped with coase-grains of Gd_2O_3 were radiographed, and the Gd_2O_3 grain size and their distribution can be clearly observed in the image, as shown in Fig.2.

The Fig.3 is a neutron radiograph of two components (a part) of UO_2-fuels with magnesium powders as disperser. It can be seen in the images that the different gaps between the both fuel parts, and the Mg-grains and their distribution can be also seen clearly.

2. Inspecting Remnant Boron in Brazed Joints

In the brazed joints of the drills, the amount and distribution of remnant boron in metal flux must be inspected for examining quality of the drills, but for this purpose, x or γ-radiography can not be used. However the remnants are easily inspected by neutron radiography. Fig.4 is a neutron radiograph for a drill, and clearly showed the boron and its distribution in the drill.

3. Detecting Parts Including Explosive

Neutron radiography has particular advantage to detect the explosives and its discontinuous defects in metal conduit, on account of metal outer-covers as compared with explosives almost look upon as transperent for neutron penetration. Fig.5 is a radiograph of explosive network (a section sample) with some artificial defects in it, and minmum spece of 0.02mm can be observed. Fig.6 is a radiograph of two bullets.The grain-size and filling level of the explosives can be clearly shown in the images.

4. Checking Adhesives

Fig.7 is a neutron radiograph of a part of the hollow ball. The ball was glued by means of a adhesive between inner and outer shells made of two kinds of metal materials. In the radiograph the dishomogeneity and even heap up of the adhesives are distinctly revealed.

DISCUSSION AND CONCLUSION

The neutron radiographic facility on research reactor has higher beam intensity and larger adjustable range of the beam energy, L/D and n/γ. So it has better suitability to NDT of different materials and thick objecls. Under the conditions, using the thiner Gd metal foil converter and the minute-grain x-films in static tests is profitable to improve testing precision, and making neutron radiography of radiative objects by the indirect imaging method with Dy-converter is available, and so is even the real-time neutron radiography, if a emit-light-screen of high sensitivity combined with electron multiplier and TV pickup system are used. A shortcoming of the system, however, is not easy to move at will to worksite.

At present the existing problems in the facility are to improve the quality of Gd and Dy foils, such as flatness, smoothness and homogeneity of impurity in it. The cassettes should be further ameliorated for decreasing neutrons scattered and absorbed by the front-cover of the cassettes, and for raising all contact between the converter and the x-film closely.

REFERENCES

[1] R.S.Matfield, Neutron Radiography Services in Research Reactor Division at Harwell,AERE-R6372(1971)
[2] W.J.Richarda, et al. Neutron Radiographic Facility at the 3-MW Livermore Pool-Type Reactor,UCRL-51906, 4-14 (1975).
[3] J.C.Domanus, R.S.Matfield, et al. Neutron Radiography Handbook, EUR7622, 19-40 (1981).
[4] Standard Method for Determining Image Quality in Thermal Neutron Radiographic Testing, ASTM E545-75.
[5] H.Berger, et al. Neutron Radiography-I Trans, V.1, 525-528 (1971).
[6] The same with the [3], 31-33.

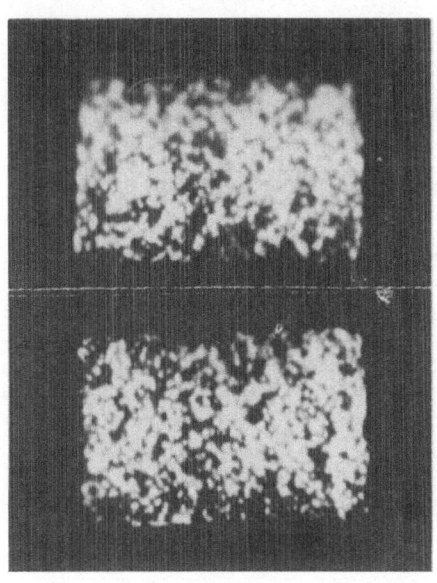

Fig.2 Neutron Radiograph
Showing Gd_2O_3-Grains
in UO_2-Fuel Elements

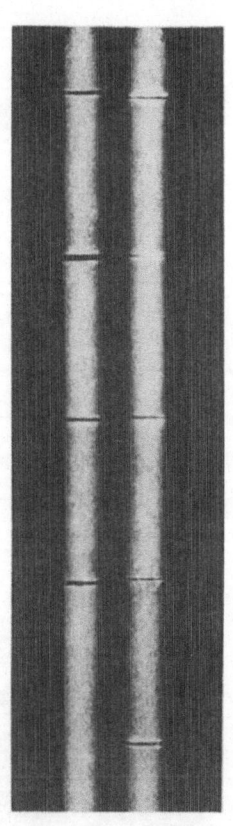

Fig.3 Neutron Radiograph
of UO_2-Mg Fuel
Components for
Inspecting Gaps
Between Both
Elements

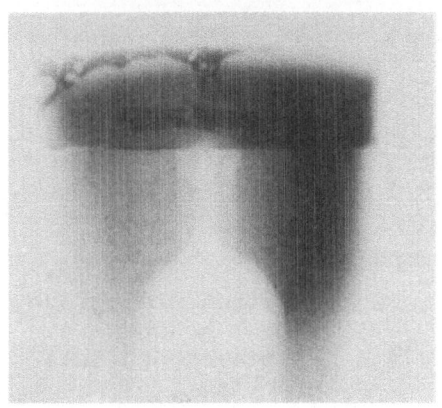

Fig.4 Neutron Radiograph
Showing Remnant
Boron in a Drill

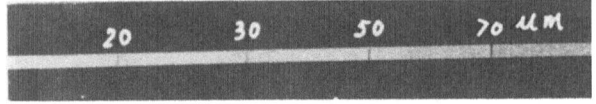

Fig.5 Neutron Radiograph Showing
Explosive Interruptive
Defects in a Explosive
Network

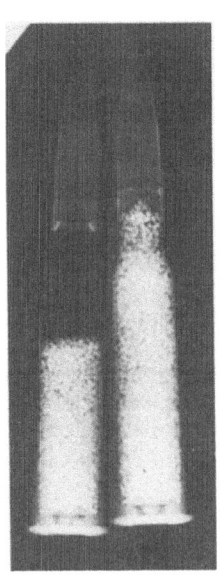

Fig.6 Neutron Radiograph
Showing Filling
Level of Explosive
in Two Bullets

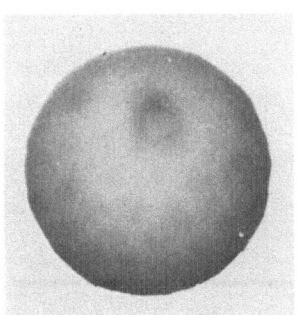

Fig.7 Neutron Radiograph
Showing Adhesive
Between Both Ball
Shells

NEUTRON RADIOGRAPHY STUDY IN TSINGHUA UNIVERSITY

Mo Da-Wei   Liu Yi-Si   Guo Zhi-Ping
An Fu-Liu   Zhang Chao-Zong   Miao Qi-Tian
Institute of Nuclear Energy Technology
Tsinghua University
          P.O.Box 1021, Beijing, China

Abstract

The work of Neutron Radiography described
in this paper was performed at the research reactor of Tsinghua University,which is a pool-type
light water reactor with a power of 2 MW. This
neutron radiography system reached the level N80-
10-7 of the ASTM E545-75 standard. A MTF programme
was developed for making neutron radiography
convertors. Neutron radiographs of corn roots
growing in soil were obtained with resolution of
about 0.3 mm. The research of water permeation in
concrete using neutron radiography technique was
also tested. A preliminary real-time neutron
radiography system usable to observe two-phase
flow was installed.

Introduction

Neutron radiography, younger than X-radiography, has
followed a development pattern very similar to that of
X-radiography[1]: Discovery of neutron radiation (1932);
Proof of the principle of radiography (late 1930's);
Availability of reliable radiation sources -- thermal
neutron reactor (1955); Widespread industrial applica-
cion (1968).

In China the research of neutron radiography are

carring on by several groups including the one of Tsinghua University since 1980. The purposes of our work are
the training of graduate students and the study of possible applications of neutron radiography in practical
fields. As a portable neutron source, being adequate for
neutron radiography, is not available now, we are oblig-
ed to do our preliminary experimental works at the
research reactor of Tsinghua University.

Neutron Radiography Facility

The experimental arrangement is shown schematically
in Fig.1

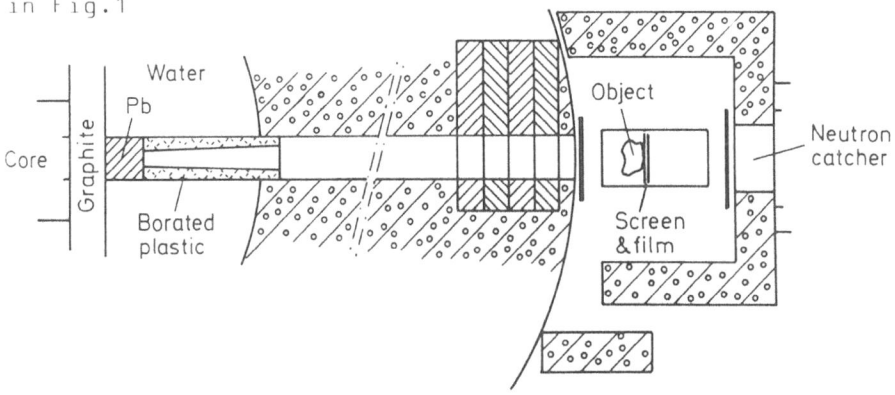

Fig. 1

Neutron source is a pool-type light water research
reactor with power of 2 megawatts. The thermal neutron
beam was derived from the NO.4 horizontal hole, 100mm--
diameter. The neutron flux at the object location was $5 \times 10^5 n/cm^2.s/kw$ without using additional beam collimator,
that provided a collimator ratio L/D of 30. While the
beam collimators 1 and 2 were used the neutron flux was
$2 \times 10^3 n/cm^2.s/kw$ with L/D of 60, and $0.7 \times 10^3 n/cm^2.s/kw$
with L/D of 100 respectively. A reactor power level of
50-200kw was often selected. The gamma ray contamination
was 45 mR/H at the film cassette location for the
collimator 2 at 200kw.

The collimator 2 was composed of a lead cylinder in
the front, two lead cylinders with cone-shaped hole and
three borated plastics washers, which were fixed in a
3.5 meter long aluminium pipe with boron carbide lining on the inner surface to decrease scattered neutrons
(see Fig.2)

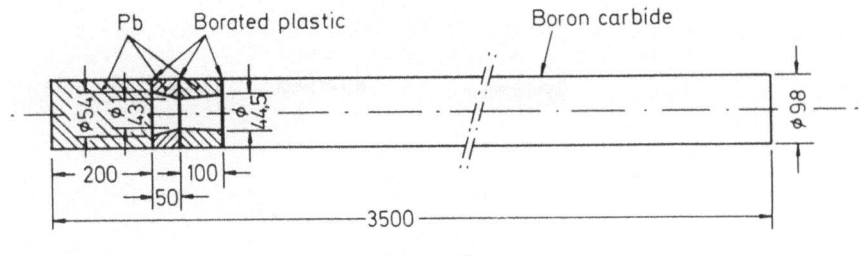

Fig. 2

Both light-emitting screens (⁶LiF-ZnS, Gd₂O₂S) and particle-emitting sreens (In, Dy) were used. Although metal screens were characterized by better spatial resolution, LiF-ZnS (NE426) was preferential, considering low intensity portable neutron sources will be used in the future. Most of our radiographs were taken by NE 426 screens. As single-emulsion X-ray films were not available, double-emulsion films were used for neutron radiography, that degraded the spacial resolution of neutron radiography.

Quality of Neutron Radiography

ASTM E545-75 IQI[2] were used to measure the image quality. The level of total system reached N80-10-7. where

80 is the percentage of contribution of thermal neutrons to the film density.

10 is the percentage of contribution of scattered neutrons to the film density, and

7 is the level of detail resolution.

The advantages of IQI method are being small in size, being able to use repeatedly and providing result directly and visually. But IQI may only give a half-quantitative result with 16 levels, so that it can not measure the detail resolution ability precisely.

For comparison, the MTF method was used, namely, imaging a knife-edge, differentiating the knife-dege density distribution function to obtain the line-spread function (LSF), and Fourier transforming the LSF. We developed a MFT computor programme and measured the MTF using In, LiF-ZnS and Gd₂O₂S converters at different exposure conditions.

Figure 3 shows a group of the MTF curves of an In
screen at different total exposure. The knife edge was
made of a cadmium slice 0.2mm thick, the collimator ratio
L/D was 30.

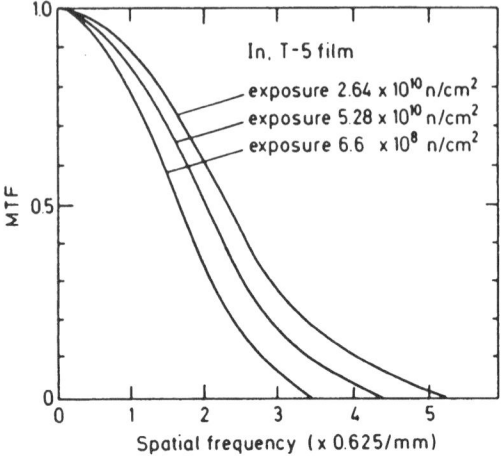

Fig. 3

Type Tianjin V double emulsion X-radiography film was
used in the measurement. Optimal exposure 2.64×10^{10} n/cm^2
was given from Fig.3. The curve 1 was about equivalent
to level 8 of ASTM E545-75 standard and curves 2 and 3
were equiv alent to level 7. We also used the MTF
programme in the manufacture procedure of sintillator-
intensifying screens to help with selecting optimal
processing.

Neutron Radiography for studying young corn roots growing in soil *

The agronomists have long been seeking a method,
which allows observing the process of plant roots
growing continually without disrupting root system, but
no method is satisfactory. Neutron radiography has provided a new method which successfully resolved this problem.

The higher concentration of water in the roots
attenuate more neutrons by scattering and absorption than
the soil, thus a neutron image may be produced, but the
water in soil removes neutrons from the collimated beam

*   This work was completed by cooperation with Ms. Yang
Xiu, associate professor of the Agricultural Academy of
China.

"with the image of roots", therefore the image becomes blurred by the water in soil. So, to control the water content in soil during the exposure is the key point of the radiography quality.

We have taken the neutron radiographs of the same young corn roots serially. Fig.4 shows one set of them.

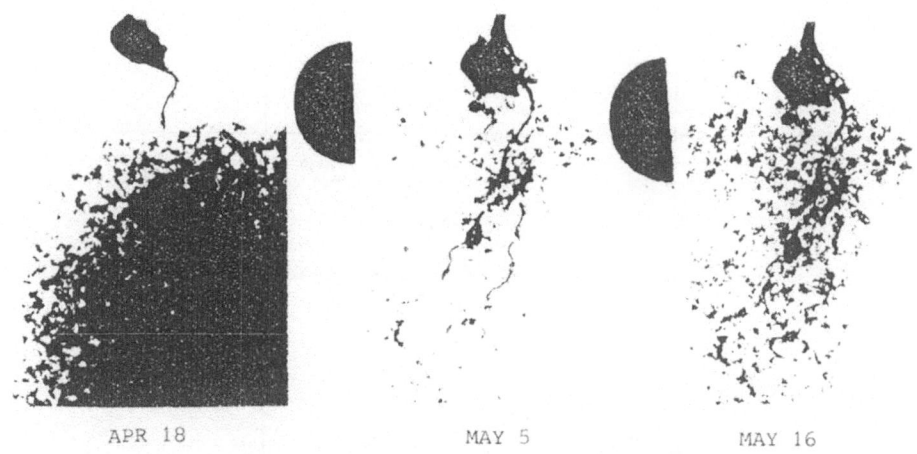

APR 18   MAY 5   MAY 16

Fig. 4   This illustration is printed
              from the best available original

In the experiments 200x200x20 mm aluminium boxes filled with sandy soil were used. the bottom of these boxes were shaped sifter-like, so that it were be able to connect them to a vacuum air pump to contral the water content in soil. The collimator 2 (L/D-100) and a converter of NE426 were used. At 3% water in soil 0.25 mm-diameter roots were observed clearly. The black background inhomogeneity of photographs was caused by the inhomogeneity of the water content in soil after draw-drying. The 3% water content in soil during a short period of exposure and about 1 rad total dose (of n and) were unlikely to influence the plant growing notablely.

There were three methods used to compare the spatial resolution.

1) Indicator method

A cadmium strip was placed in the soil. The 0.3 -mm hole was visible clearly through 2 -cm soil thickness.

2) MTF method

The calculation showed that the 0.28mm-diameter roots were able to be resolved at 8% water content in soil.

3) Monte-Carlo method

A programme of Monte-Carlo method was developed to estimate the influence of the water content in soil on the image quality. A calculated result is shown in Fig.5

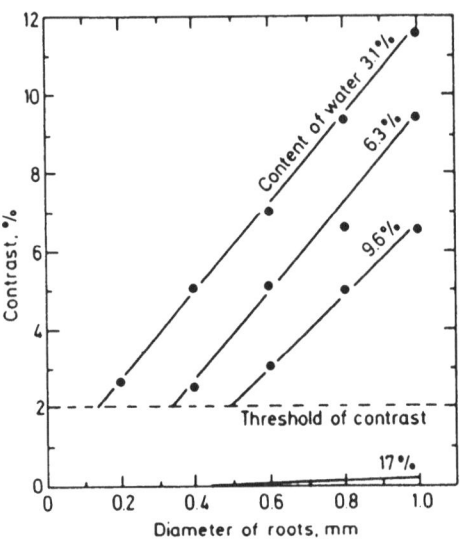

Fig. 5

The ordinate represents the contrast of image and the abscissa represents the water content in soil. The line parallel to abscissa represents the contrast of 2% which is the mimiaum for human's eye. From Fig.5 it is seen that 0.2mm-diameter roots may be resolved as the water content is about 3.1% and the roots of any size may not be resolved as the water content comes up to 17%.

Neutron Radiography for Measurement
of the water-permeability of concrete*

The water-permeability of concrete has significance for dam, offshore platfom, and underwater basement of

---

* This work was completed by cooperation with the Civil Engineering Department of Tsinghua Univesity.

bridge etc. The traditional method is as follows: making samples, mounting them into a special water-tight device to prevent from leakage of water onto the flanks, pressusing with water from one end for a certain period, cutting and determining the position of water-permeation. The disadvantage of the method is that the permeating process for the same sample can not be observed continually and that the same experimental curve is oblained from different samples. It is obvious that the non-identity of samples leads to increase of amount of them and makes analysis complex. In addition, we are unable to know if there are some defects in the samples and if they have any influence on the experiment data.

Because the wet part of concrete sample will attenuate thermal neutrons greater than the dry part. We may use the neutron radiography technique to observe the water-permeating process in concrete from beginning to end. This method overcomes above-mentioned dis-advantage of the traditional method. Fig.6 is a photograph of water-permeating in a concrete sample.

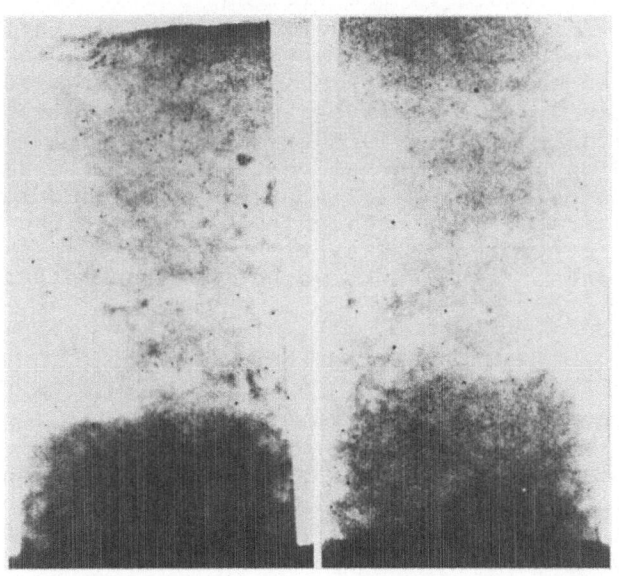

Fig. 6

A Real-Time Neutron Radiography Viewing System

A remote viewing system was made up of a NE426 screen

and a silicon electron multiplier target vidicon TV
camara (see Fig.7).

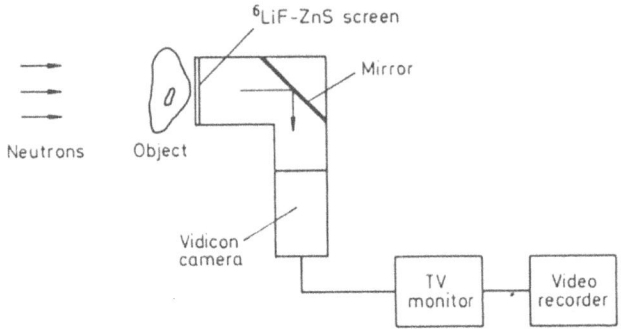

Fig. 7

Between the screen and the vidicon a mirror was set
at an angle of 45° to the incident neutron beam to avoid
the radiation damage to the vidicon. The incident neutron
flux was 10^5-10^6n/cm^{2}sec. The motive neutron radiography
images of two-phase flow of water with bubble in a cast
iron tap and boiling water in a thin aluminium box were
investigated by using this system. It is unsuitable to
reduce the neutron flux further for statisticl reason,
but providing such a thermal neutron beam intensity
either by a californium-252 source or by a portable
neutron generator is practicable. Generally, the spatial
resolution of vidicon is lower than film's, so the final
image sharpness is somewhat worse than the latter's.

Future direction

We intend to use a sealed neutron tube as a portable
thrmal neutron source and to connect such a real-time
radiography system to a computer image systam, and
finally, to install a portable neutron radiography system,
usable for general practical applications. The work is
still in progress.

Reference

(1) J.P.Barton Neutron Radiography -- An Overview 1975
(2) ASTM Designation E545-75
    Standard Method for Determining Image Quality in Ther-
    mal Neutron Radiographic Testing

THE MINIATURE NEUTRON SOURCE REACTOR (MNSR)

A PRIVATE NUCLEAR TOOL

Zhou Yongmao, Wu Fuxing and Gu deming

China Institute of Atomic Energy and

China Nuclear Energy Industry Corporation

In view of the problems faced by large and medium
research reactors, we have developed and constructed
the prototype of Miniature Neutron Source Reactor in
1984. During 2 years operation, the MNSR has demonstrated
its high reliability, unique safety, facilitation of
operation, tiny radiation effect, and cost attraction.

Based on the experience from prototype, we have
been designing the new type of commercial MNSR to
improve the performance and have been investigating
to extend its application on neutron radiography.

GENERAL FEATURES

Using the technology and experience from High Flux Engineering
Test Reactor(1), in 1980 we began to work on the Miniature Neutron
Source Reactor -- MNSR to satisfy the user's demand on nuclear
analysis and nuclear application. The prototype of MNSR(2) is
located in the Institute of Atomic Energy at the southwest suburb
of Beijing. The construction was completed and the reactor was
started up in 1984. In the first stage, it is designed for neutron
activation analysis, short-lived isotopes production and education.

In early experiments for HFETR(3) (4), the exciting status
showed that a core with highly enriched uranium and light water,
surrounded by sufficiently thick beryllium, exhibited a low
critical mass. The MNSR has adopted 90% enrichment uranium -
aluminium alloy as fuel, light water as moderator and coolant,

and beryllium as reflectors. The volume of right cylinder core is only 11 litres. The total loading of U235 mass is less than 1 kg.

The neutron flux with high energy from a compact core experienced sufficient slowdown in the peripheral Be annulus to form available thermal neutron flux of 1×10^{12} n/cm^2.s, during 27 kw nominal power. The irradiation tubes are located in the position where the value of neutron flux is available. There is only one control rod in the centre of core to start up and shut down the reactor, to regulate the neutron flux level, and to compensate the consumption of reactivity. The reactor vessel is suspended from two flanged beams placed across the pool with inside diameter of 2.7m and depth 6.5m. (fig. 1). The summary of specifications and performance of MNSR is shown in table 1. Up to now the MNSR has operated nearly two years. The total energy generated has been estimated about 19,000 kW/hr. More than 12,000 different samples have been analyzed, 68 elements in the periodic table have been quantitatively detected, including F with half-lifetime 11.03 s. and $_2$Co with half-lifetime 5.26yr., detecting limit is from less 10^2ppm to over 10^{-2}ppm of detectable elements. ^{64}Cu, ^{72}Ga, ^{82}Br and ^{197}Hg have been produced. Besides, about 100 students, engineers and researchers have received training on the MNSR. The successful operation of MNSR has demonstrated its high running reliability and stability, its unique safety, its facilitation of operation and control, its extremely low radiation effect on persons as well as its low operating cost.

Inherent Safety

Undermoderated core and reasonable inlet size of coolant to core formed the large negative temperature coefficient of reactivity. The 3.5 mk excess reactivity and -0.1 mk/$^\circ$C (at 15 -- 40°C) temperature coefficient make the reactor inherently safe. The transient behaviour during adding 3.6 mk positive reactivity described in fig. 2.

The possibility of nuclear accident or nuclear contamination would be eliminated in the MNSR. The core is immersed in water with dual containers of vessel and pool. The danger of coolant loss would be excluded in MNSR. Fission heat in core is transfered by natural convection to the atmosphere without pump and motor, the incidents from interrupting power or component failures would not exist in MNSR. So the MNSR can be located in densely populated or economically developed areas.

Do Not Worry About Radiation Effect on People

Because of the surprisingly low overall surface contamination level of less 2×10^{-9} g-^{235}U/cm^2 of fuel cladding in fabrication, and the sufficiently thick water protection of 4.67m, radiation dose rate in the working area of the reactor room is 1/50 - 1/100 of permissible value, outside the reactor building, it is tha same as natural background. Radiation effect on staff personnel is negligible in the MNSR.

Attraction of Cost

There is no component of considerable power consumption in the MNSR, the normal consumption of U235 is only a few grams per year, and no special reactor engineering operators are needed in the MNSR. These factors make the construction as well as the operation of the MNSR inexpensive. The estimated cost of the MNSR is only about 1/10 of the cost of a typical medium research reactor.

The End Words

Using the prototype experience, we have been developing a new type pf MNSR to improve the performance and to extend the application on neutron radiography according to user's requirement.

Compensated some disadvantages and troubles of large and medium research reactors, the MNSR would be expected to form a convenient private nuclear tool for utilization of user in place and in time.

REFERENCES

1. Xu Chuanxiao, Zhou Yongmao, Qian Jinhui, etc.
   High Flux Engineering Test Reactor
   Nuclear Power Engineering, Vol. 2, No. 3, 1981

2. Zhou Yongmao
   The MNSR Reactor
   Paper for Consultants' Meeting on Technology and Use of Low
   Power Research Reactors, organized by IAEA Beijing, China
   April 30th - May 3rd, 1985

3. Phy. Lab. of 2nd department of southwest Designing Institute
   of Reactor Engineering Research
   0-power Physical Experiments of HFETR Reactor Core
   Collected Papers of First work Exchange Conference on Reactor
   Physics. Unpublished, June 1979, Atomic Energy Print Bureau,
   Beijing, china

4. Xu Jiangqing, Xu Hanming, etc.
   Analyzing Calculation of Critical Experiments of Be-H_2O Reactor
   Collected Papers of First Work Exchange Conference on Reactor
   Physics, unpublished, June 1979, Atomic Energy Print Bureau,
   Beijing, China

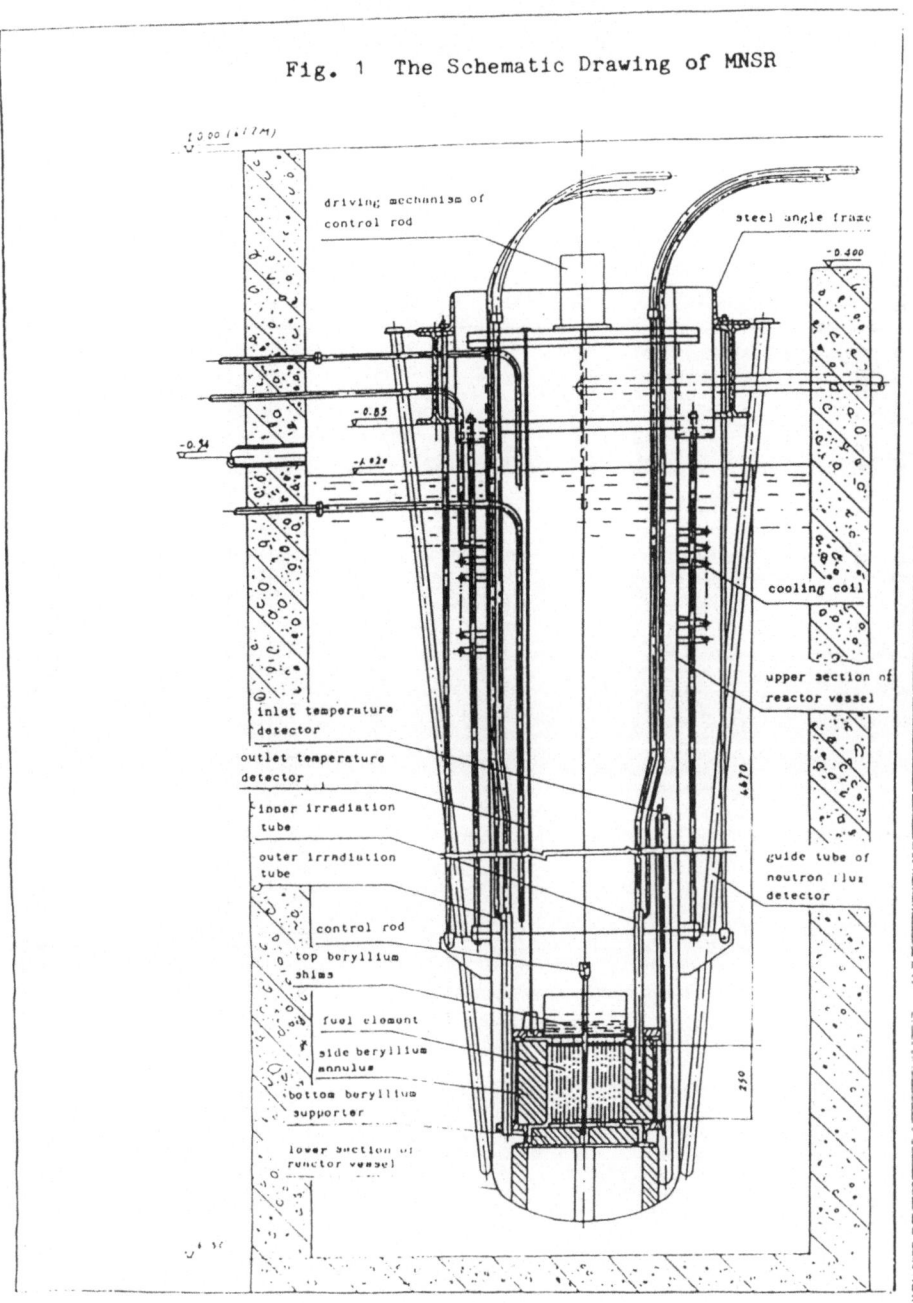

Fig. 1 The Schematic Drawing of MNSR

99

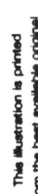

Fig. 2 The Transient Behaviou of MNSR adding in +3.6 mk

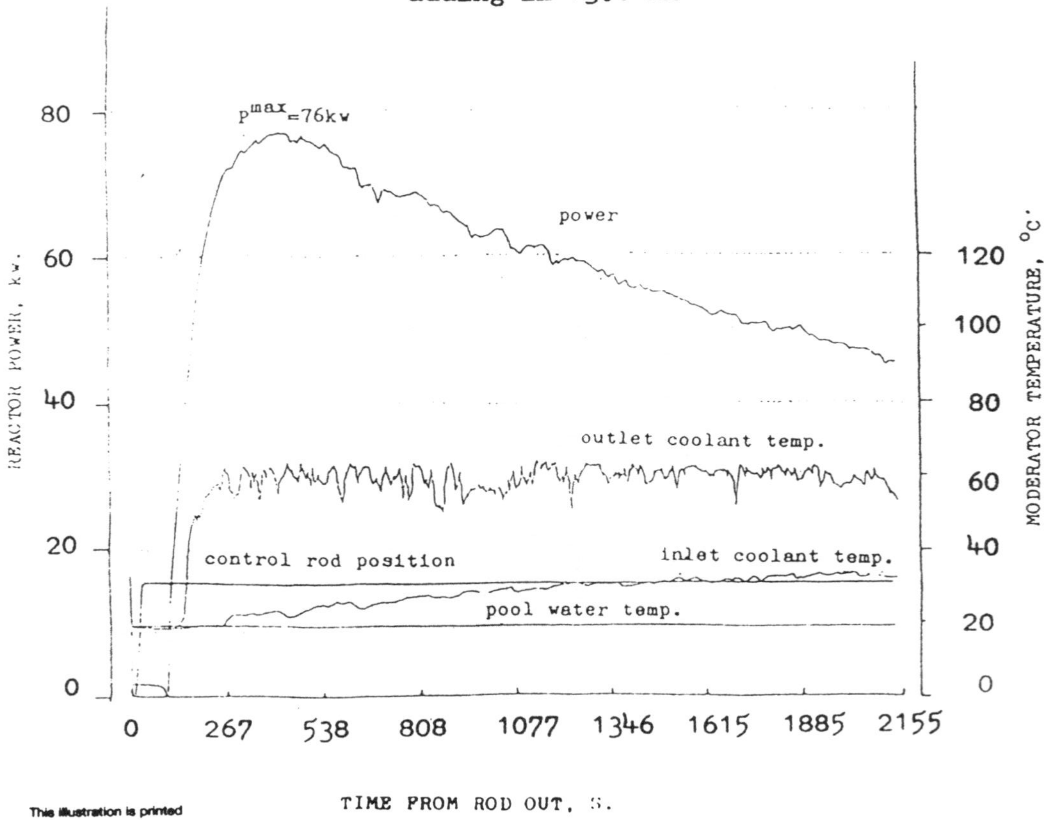

This illustration is printed from the best available original

PERFORMANCE OF MNSR   Date of information 1986

Feature	Detail Specification (in 27kw, routine oper. 20°C start-up temp.)
Availability for Use	5 inner sites in Be annulus, 7c.c capsule vol. 5 outer sites adjacent to outside of annulus, 25c.c capsule vol. Neutron flux (n/cm².s), daily operation time (hr), flux fluctuation in operation: in inner sites: 1×10^{12}, 2.5. 5×10^{11}, 5-6. $< 2\%$ of the setting value in outer sites: 5×10^{11}, 2.5. 2.5×10^{11}, 5-6. Irradiation temperature (°C), with irradiating flux (n/cm².s) in inner sites: <45 (1×10^{12}), <40 (5×10^{11}) in outer sites: <36 (5×10^{11}), <32 (2.5×10^{11})
Inherent Safety	Total excess reactivity: 3.5 mk in cold clean system Core temperature coefficient: -0.1 mk/°C (15-40°C) Results of 3.6 mk reactivity transient: peak power: 76kw time to peak: 360s. outlet cool. temp.: 62°C max. wall temp. of fuel element: 92°C
Low Radiation Level	Overall surface contamination level of fuel element: $<2 \times 10^{-9}$ g·U-235/cm² Depth of water above core top: 4.67 m. Radiation dose rate (mRem/hr): in reactor room working area: <0.05. ≤ 0.2 at max. power of 3.6mk transient in area adjacent to reactor room: 0.01 in enviroment outside reactor building: natural background
Economical Operation Costs	Estimated as less 1/10 of SPR's annual operation expense per year including the depreciation of fuel, Be, and the permanent equipments and the consumption of water and power.

**"LES INSTALLATIONS ET PROGRAMMES DE RADIOGRAPHIE AUX NEUTRONS
AUPRES DES REACTEURS DES CENTRES CEA"**

Contribution du Centre de Saclay à la 2ème Conférence Mondiale sur
la Radiographie aux neutrons - Paris, 16-20 Juin 1986

G. BAYON - J.L. PERSON

Les Services des Piles du CEN/Saclay disposent des
installations associées aux 3 réacteurs OSIRIS, ISIS et ORPHEE :

OSIRIS (70 MW) . un banc de neutronographie immergée dans la
piscine du réacteur

ISIS (700 kW) . un banc identique au précédent
. un banc implanté sur faisceau sorti radial
. un dispositif récemment installé sur faisceau
sorti tangentiel

ORPHEE (14 MW) . une installation de neutronographie industrielle
implantée à l'extrémité d'un guide à neutrons

I - INSTALLATION IMMERGEE A OSIRIS

I.1 - Description

Le dispositif est placé au fond de la piscine (face N.O.)
de telle sorte que le nez du collimateur, dirigé vers le centre du
coeur, soit situé dans le plan médian de ce dernier. Ce dispositif
comporte une crémaillère permettant une translation horizontale
de 50 cm.

I.2 - Collimation

Le collimateur, long de 2,50 m, a été réalisé en AG3. Il
est divergent : les surfaces internes sont recouvertes de Cadmium.
Une précollimation convergente (manoeuvrée pneumatiquement) peut
être commandée à distance. Ce collimateur est pressurisé à
l'hélium.

Les caractéristiques physiques sont rappelées ci-dessous :
. surface d'entrée du collimateur (Cd, Boral, Indium): 16 x 16 mm^2
. surface d'entrée du pré-collimateur : 73 x 31 mm^2
. surface utile rectangulaire   h : 650 mm
                                      l : 100 mm
. rapport L/D   : 156
. flux thermique au niveau de l'objet : $6,5.10^7$ n.cm^{-2}.s^{-1}
. R(Cd) $\simeq$ 4 .

I.3 - Méthode utilisée
       Actuellement, seule la méthode de transfert est pratiquée
en raison du flux gamma du réacteur : la cassette contenant le
convertisseur de Dysprosium est mise en place manuellement à
l'aide de cordes. Elle est ensuite plaquée par 2 vérins pneumati-
ques, contre la face de la chaussette dans laquelle est descendu
l'objet à examiner.
       A la partie supérieure de la chaussette, un dispositif
spécial (tête froide) permet de réaliser un joint de glace par
circulation d'azote liquide : l'eau de la chaussette peut alors
être purgée par une surpression d'air. Le collimateur est transla-
té vers l'avant le plus près possible du coeur. Le temps d'exposition est de l'ordre de 12 à 15 mn (dans le cas d'un développement
automatique du film argentique).
       Après récupération, le convertisseur de Dysprosium est
appliqué pendant 90 mn contre un film KODAK INDUSTREX "type M"
(bi-couche), puis pendant une vingtaine d'heures contre un "type
R" (mono-couche).
       Le premier film est développé instantanément, c'est-à-dire
environ 2 heures après l'irradiation, ce qui permet de prendre
très rapidement connaissance des résultats de l'expérience.
       La méthode directe à l'aide d'un film de nitrate de cellu-
lose est en cours d'adaptation : indépendamment d'une plus grande
facilité d'emploi, cette dernière permettra d'effectuer dans de
bonnes conditions des copies sur papier positif, très souvent
réclamées par les utilisateurs.

I.4 - Utilisation
       Les contrôles sont effectués sur de très nombreux disposi-
tifs. En général, un cliché est réalisé avant l'irradiation, ce
qui permet de visualiser les évolutions suivantes :
- fracture, espaces entre les billettes combustibles,..
- entrée d'eau (fissure),
- consommation de poison,
- déformations,
- positionnement : combustible ou système d'instrumentation.

II - INSTALLATIONS A ISIS

I.1 - Installation immergée

Le dispositif est en tout point comparable à celui d'OSIRIS qu'il peut suppléer dans de nombreux cas. La surface utile est de : 150 x 600 mm^2.

. Rapport L/D : 137
. Flux thermique : $1,3.10^7$ n.cm^{-2}s^{-1}.

D'accès plus facile, et en appoint des contrôles de dispositifs liés à ISIS, il peut être utilisé pour l'optimisation du procédé.

II.2 - Neutronographie par défilement

Cette installation, réalisée depuis bientôt 15 ans, a été décrite de nombreuses fois /1/.

Le dispositif est automatisé pour effectuer, derrière un faisceau sorti radial, des prises de vue séquentielles par défilement d'un cadre contenant des crayons combustibles irradiés, en synchronisation avec le déroulement du film sensible.

La méthode directe est appliquée, à l'aide d'un film de nitrate de cellulose contenu dans une caméra spéciale et de convertisseurs au Bore naturel ou au Lithium enrichi. La surface utile est de 100 x 150 mm^2 à l'extrémité d'une collimation de 4 m de long.

A la puissance nominale de 700 kW et pour les neutrons thermiques, nous avons :

- flux incident au nez du collimateur : $2,0.10^{12}$ n.cm^{-2}s^{-1}
- flux émergeant en sortie : $1,2.10^7$ n.cm^{-2}s^{-1}
- Rapport $\frac{L}{D} > 100$
- Rapport Cadmium : 2,5 à 3,0.

Les crayons combustibles (5 crayons au maximum jusqu'à 4 m de long) sont placés dans un cadre et transportés à l'intérieur d'une hotte protectrice de 15 à 20 cm de plomb. Cette dernière est disposée verticalement sur la partie supérieure d'un puits d'une profondeur de 12 m.

Chaque cliché demande une exposition de 6 à 7 mn : l'examen total du cadre de 4 m est réalisé en 4 heures (35 clichés environ).

Le cadre étant équipé d'un repère longitudinal, des mesures dimensionnelles très précises sont possibles en complément de tous les contrôles visuels habituels.

II.3 - Faisceau sorti tangentiel

La réalisation de ce dernier est à peine achevée et les essais de qualification doivent être effectués au cours du dernier trimestre 1986.

. surface utilisable   : 240 x 180 mm^2

. flux thermique probable : 10^7 à 10^8 n.$cm^{-2}s^{-1}$.

Cette installation sera, en outre, équipée de façon à compléter la neutronographie industrielle d'ORPHEE (durcissement du spectre).

III - INSTALLATION DE NEUTRONOGRAPHIE INDUSTRIELLE A ORPHEE

Réalisée depuis 5 ans maintenant en remplacement du premier dispositif implanté à TRITON (Fontenay-aux-Roses), cette installation a fait l'objet de nombreuses publications et communications antérieures /2/.

Ses caractéristiques essentielles sont dues à son exploitation à l'extrémité d'un guide à neutrons prélevant des neutrons de très basse énergie au sein d'une source froide située dans le réflecteur d'eau lourde du réacteur.

. section du guide: 25 x 150 mm^2

. flux thermique de la cassette 10^9 n.cm^{-2} s^{-1}.

Le contrôle s'effectue par défilement des pièces devant le faisceau, à des vitesses règlables.

La méthode utilisée est exclusivement la méthode directe avec un convertisseur de Gadolinium plaqué contre le film argentique monocouche dans une cassette à vide.

Les films sont développés automatiquement, immédiatement après l'exposition.

Au cours de l'année 1986, 2500 clichés ont ainsi été réalisés ; 80 % de ces contrôles sont effectués sur les éléments pyrotechniques des lanceurs d'engins.

REFERENCES

/1/ "Neutron Radiography Handbook" D. Reidel Publishing Company

R. BARBALAT et G. FARNY Kerntechnik 18, Jahrgang (1976) - N°2

R. BARBALAT - Séminaire sur l'Exploitation et l'Utilisation de Réacteurs de Recherche, Julich (Septembre 1981)

/2/ "Neutron Radiography Handbook" D. Reidel Publishing Company

1ère Conférence Mondiale sur la Radiographie aux Neutrons, San Diego (Décembre 1981)

10ème Conférence Mondiale sur les Essais non Destructifs, Moscou (Août 1982)

"LES INSTALLATIONS ET PROGRAMMES DE RADIOGRAPHIE
AUX NEUTRONS AUPRES DES REACTEURS DES CENTRES CEA"

Contribution du Centre de Grenoble à la 2e conférence Mondiale
sur la radiographie aux neutrons - Paris 16-20 juin 1986

G. DUCROS

Le Service des Piles du CEN/G dispose de 2 installations de
neutronographie :

- Un banc de neutronographie immergé dans la piscine du réacteur
  SILOE, réacteur fonctionnant à 35 MW, destiné plus spécifiquement au contrôle, après chaque phase d'irradiation, des dispositifs expérimentaux contenant le plus souvent des échantillons
  de combustibles nucléaires.

- Un banc de neutronographie sur faisceau sorti implanté sur le
  réacteur MELUSINE, réacteur de type piscine fonctionnant à 8 MW ;
  c'est une installation à vocation industrielle destinée au contrôle de pièces non radioactives.

I - INSTALLATION IMMERGEE DE SILOE

I.1. Description

Le banc de neutronographie de SILOE est implanté au fond du
bassin principal de la piscine, le nez du collimateur étant
situé dans une boîte à eau en périphérie du coeur, dans le
plan médian de celui-ci.

Collimation

Le collimateur, long de 2 m est réalisé en (B_4C + In) pour le
corps, (Cd + In + Gd + Au + Dy) pour le nez. Il est divergent
à section rectangulaire, translatable d'avant en arrière sur
30 cm ; la distance variable du collimateur au coeur permet
une adaptation du spectre et du flux de neutrons, ainsi que

le stockage de l'appareillage dans une position où il s'active peu :

- Ouverture au nez du collimateur : $\emptyset$ 6 mm
- Dimension du faisceau de sortie : largeur 120 mm
  hauteur 400 mm
- Rapport L/D : 380
- Flux thermique au niveau de l'objet : $8,5 \times 10^7$ n/cm^2.s
- Flux épithermique au niveau de l'objet (> 0,5 eV) :
  $7,9 \times 10^6$ n/cm^2.s

Méthode utilisée

La présence d'un flux γ important au niveau de l'objet (2×10^4 rad/h) conduit à l'utilisation en routine de la méthode de transfert. Le choix du convertisseur est adaptable au spectre neutronique à privilégier :

- clichés "Dysprosium" pour le spectre thermique
- clichés "Indium" ou "Or" pour le spectre épithermique.

La méthode directe avec utilisation de films à traces en nitrocellulose est également employée. Présentant d'indéniables avantages (plus grande souplesse d'emploi, meilleure reproductibilité des clichés) sa généralisation est en cours d'étude, notamment pour résoudre les problèmes posés par la tomographie immergée.

Manutention des objets

Les dispositifs sont suspendus à une potence spéciale, l'extrémité inférieure à contrôler étant introduite dans une cloche, vidée de son eau, et appliquée contre le collimateur d'une part, contre la cassette contenant le convertisseur d'autre part, par un vérin pneumatique.
L'étanchéité de la cloche est assurée soit par un joint diabolo (pour dispositifs à section cylindrique de diamètre compris entre 30 et 70 mm), soit par un joint de glace (pour dispositifs de·section quelconque inscriptible dans 80x170mm^2).

I.2. Utilisation

Combustible nucléaire

La principale utilisation concerne le contrôle des dispositifs contenant des échantillons de combustible nucléaire avant et après leur irradiation dans le réacteur SILOE. Les objectifs poursuivis par ces contrôles sont essentiellement :

- la visualisation du combustible (pastilles saines, fracturées, fondues...) et les mesures dimensionnelles qui lui sont associées (longueur, diamètre).

- la visualisation des gaines, notamment lors des expériences de simulation de séquences accidentelles.

- Les vérifications de positionnement du crayon combustible dans le dispositif, ainsi que des divers systèmes d'instrumentation (Thermocouples, capteurs de mesures...).

Barres absorbantes

Un important programme de contrôle de barres absorbantes en carbure de bore est en cours depuis plusieurs années. Ces examens s'inscrivent dans un double programme de surveillance des barres de contrôles du réacteur PHENIX et d'amélioration des performances des futures barres de contrôles de SUPER PHENIX.

Développement

Une étude concernant la tomographie à partir de clichés neutronographiques (clichés "nitrate de cellulose") est en cours de qualification expérimentale. Les aspects théoriques étant maintenant bien cernés (voir communication FR2 de Ph. RIZO), l'application de cette nouvelle technique va débuter sur un dispositif irradié dans SILOE contenant un faisceau de 7 aiguilles.

II - FAISCEAU SORTI DE MELUSINE

II.1. Description

Le banc de neutronographie de MELUSINE est implanté dans une casemate où débouche un faisceau de neutrons alimenté par la cuve à eau lourde adjacente au coeur du réacteur (puissance 8 MW). Il est destiné à l'examen d'objets non radioactifs, pour toutes les applications nucléaires et industrielles.

Collimation

Le collimateur convergent puis divergent est équipé de diaphragmes mobiles permettant de passer d'un rapport de collimation de 125 à 400 selon les besoins ; le premier cas correspond à un rythme élevé de clichés (quelques minutes par cliché) très bien adapté à la mesure industrielle d'objets de faible épaisseur ; le deuxième cas correspond à la réalisation d'un examen de qualité exceptionnelle à une rythme d'opération plus lent (environ 30 minutes par cliché).
Le canal de collimation est en outre équipé d'un filtre γ permanent (monocristal de Bismuth) et d'un filtre béryllium mobile offrant la possibilité de travailler en neutrons froids :

- Dimension du faisceau de sortie : $\emptyset$ 400 mm
- Flux thermique au niveau de l'objet :

    2.10^6 n/cm^2.s pour le diaphragme de 16.2 mm

    2.10^7 n/cm^2.s pour le diaphragme de 50 mm.

Méthode utilisée

Le flux γ étant très réduit (quelques rad/h), la méthode directe est systématiquement utilisée : convertisseur gadolinium + film Kodak type M, MX ou R monocouche, ou possibilité de films à traces en nitrocellulose.

Manutention des objets

Les objets à contrôler sont placés à l'intérieur d'une casemate d'exposition en béton (longueur 3 m, largeur 2 m, hauteur 2,2 m). L'ouverture/fermeture du faisceau est assurée par un "beam catcher" d'une tonne monté sur rails et commandé par vérin pneumatique.

Ce système permet un accès complètement libre à la casemate entre deux examens, facilitant ainsi grandement la mise en place des objets quels que soient leur forme et leur encombrement.

II.2. Utilisation

La vocation industrielle de cette installation permet à la neutronographie sur faisceau sorti d'être un outil indispensable et complémentaire de la radiographie X ou γ pour le contrôle non destructif d'objets non radioactifs.

Les principales applications actuelles à MELUSINE sont les suivantes :

- Contrôle industriel de composants pyrotechniques destinés au secteur aérospatial : cordeaux découpants, cordeaux de transmission, relais pyrotechniques...

  La majeur partie de ces objets est destinée au programme européen ARIANE.

- Contrôles divers pour l'industrie ou la recherche, notamment contrôle d'homogénéité d'un absorbant (Uranium, Bore, Lithium...) dilué dans une matière non absorbante sous forme de plaque (plaque d'aluminium...).

- Une installation complémentaire de neutronométrie (mesure directe de la transmission du faisceau de neutrons par le biais d'une unité de comptage et traitement informatique des résultats) équipe le faisceau et permet dans le cas évoqué ci-dessus d'accéder à la détermination quantitative de la teneur d'un absorbant dilué dans une plaque.

"LES INSTALLATIONS ET PROGRAMMES DE RADIOGRAPHIE
AUX NEUTRONS AUPRES DES REACTEURS DES CENTRES CEA"

Contribution du Centre de Cadarache (IPSN/DERS/SES/SRME) à la 2e Conférence Mondiale
sur la radiographie aux neutrons - Paris 16-20 juin 1986

B. TATTEGRAIN

DESCRIPTION DE L'INSTALLATION DE RADIOGRAPHIE AUX NEUTRONS AUPRES DU REACTEUR SCARABEE AU CEN.CADARACHE

B. TATTEGRAIN
IPSN/DERS/SES/SRME
CENTRE D'ETUDE NUCLEAIRES DE CADARACHE
13108 SAINT PAUL LEZ DURANCE CEDEX - FRANCE

RESUME

Dans le cadre des examens non destructifs la radiographie aux neutrons est très performante en tant que technique d'examens sur des combustibles irradiés.

Une installation de neutronographie a donc été développée auprès du réacteur SCARABEE dans le but d'examiner des combustibles irradiés fortement dégradés provenant d'essais de sûreté, réalisés sur les installations avoisinantes : SCARABEE, CABRI et PHEBUS.

INTRODUCTION

Dans le cadre du Département d'Etudes et de Recherches en Sécurité (DERS), trois programmes d'essais de sûreté des réacteurs sont menés au Service d'Essais de Sûreté (SES), au sein de l'Institut de Protection et Sûreté Nucléaire (IPSN).

Ces programmes permettent d'évaluer les risques liés aux accidents et d'apprécier le degré de la sûreté des réacteurs refroidis soit au sodium, soit à l'eau pressurisée. Les accidents envisagés se résument toujours à un déséquilibre entre le niveau de puissance du réacteur et le débit du fluide caloporteur.

Après chaque essai, des examens non destructifs sont réalisés sur les combustibles irradiés fortement dégradés.

Le but de ces examens est de donner l'état final du combustible et de la section d'essai dans un délai relativement bref, après l'essai lui-même.

Les techniques mises en œuvre pour atteindre cet objectif sont actuellement :
- la radiographie par rayons X
- la radiographie aux neutrons
- la gammamétrie.

INSTALLATION DE RADIOGRAPHIE AUX NEUTRONS AUPRES DU REACTEUR SCARABEE

la radiographie aux neutrons utilise comme source de neutrons le cœur du réacteur Scarabée.

L'installation est entrée en service en juillet 1984. Le principe général est donné en annexe 1.

La source neutronique

La source est définie par un diaphragme de 10 mm de diamètre.

Le faisceau neutronique est acheminé vers l'extérieur du bloc réacteur par un collimateur, empli d'hélium, d'une longueur de 2160 mm (annexe 2). Il est obturé par une porte mobile commandée automatiquement à distance.

Le faisceau transite dans l'air puis pénétre dans le puits d'examens par le canal neutronique obturé en non fonctionnement par une porte blindée mobile commandée automatiquement à distance.

Les principales caractéristiques de l'installation sont les suivantes :

- collimateur (voir annexe 3)
  . Longueur — L = 2160 mm
  . Diaphragme de diamètre — D = 10 mm
  . Ecrans absorbants sur 250 mm — $\begin{cases} \text{Dy} & e = 0.2 \text{ mm} \\ \text{Au} & e = 0.2 \text{ mm} \\ \text{Gd} & e = 0.5 \text{ mm} \end{cases}$
  . Ecran absorbant sur toute la longueur — Cd e = 0.7 mm
  . Rapport L/D à la sortie collimateur — L/D = 216
  . Distance Source - Image — = 4700 mm
  . Rapport L/D dans le plan image — L/D = 470

- Dimensions du faisceau neutronique — = 170 x 215 mm^2
    (à l'entrée du puits)

- Distance objet - image $\leqslant$ 50 mm

- Diamètre maximum de l'objet examiné $\emptyset$ = 120 mm

- Longueur maximum observable   = 2 300 mm

- $\emptyset$ thermique à la sortie collimateur   = 9.10^6 n/cm^2/s/MW

- $\emptyset$ thermique dans le plan de l'image   = $1,5 \; 10^6$ n/cm^2/s/MW

Prise de vue

La prise de vue est réalisée par une caméra à nitrate de cellulose, commandée automatiquement à distance par un microprocesseur. Les mouvements du film sont synchronisés avec les mouvements du dispositif à examiner.

Le nitrate de cellulose présente l'avantage d'être insensible aux gammas. Il permet donc de travailler en méthode directe en s'affranchissant des gammas du combustible irradié, et des gammas du faisceau issu du réacteur Scarabée.

Le principe de fonctionnement de la caméra est basé sur la détection des neutrons après transmission au travers du dispositif à examiner. Le nitrate de cellulose détecte les particules α créées par réaction (n,α) dans les convertisseurs au bore 10 et au lithium 6 entre lesquels il est pris en sandwich par un contact très étroit.

L'image obtenue est donc fonction des sections efficaces des constituants traversés par les neutrons dans le dispositif examiné.

La section efficace de l'UO2 étant importante, l'image obtenue sera donc très représentative de l'état final de la répartition du combustible après les essais en pile, auxquels le combustible aura été soumis.

Résultats

Un grand nombre d'examens par radiographie aux neutrons a été réalisé, à ce jour sur des combustibles hautement irradiés ayant été soumis à des accidents en géométrie monoaiguille. Nous en donnons une illustration en annexe 3.

Des examens ont été aussi réalisés sur des grappes d'aiguilles en géométrie 19 ou 37 aiguilles, un exemple en est donné en annexe 4.

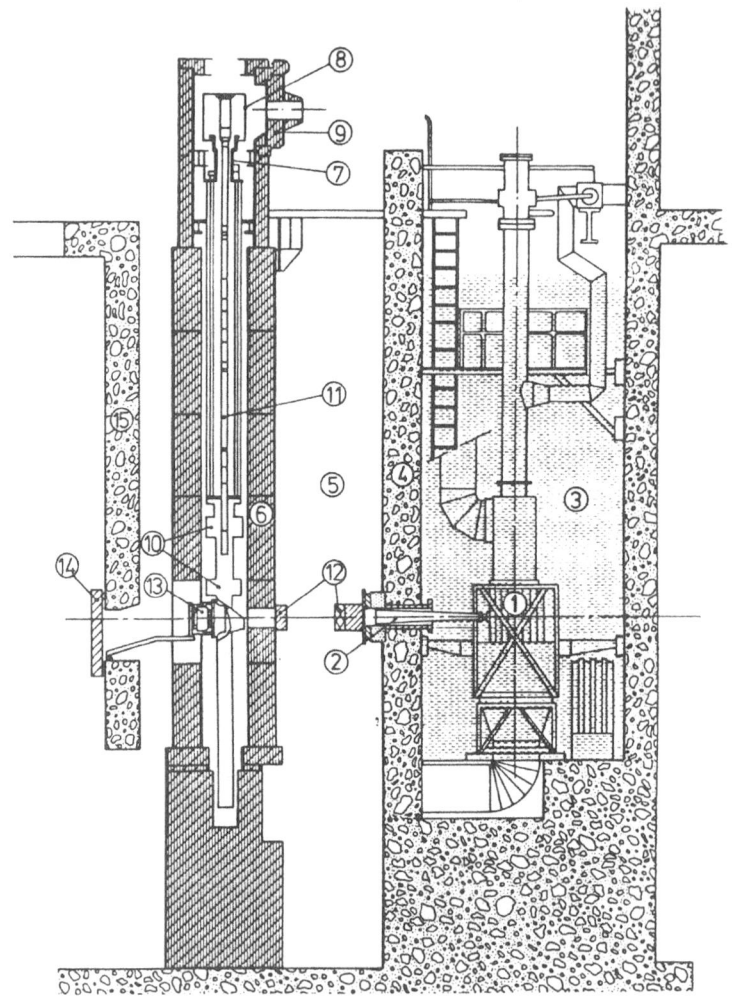

1 Cœur réacteur SCARABEE	8 Boite à gants
2 Collimateur empli d'Hélium	9 Protection en plomb de la boite à gant
3 Piscine réacteur	10 Mécanismes de centrage
4 Protection du réacteur SCARABEE	11 Conteneur étanche mobile pour aiguille
5 Air	combustible irradiées
6 Protection béton	12 Portes mobiles
7 Mécanismes de translation et	13 Caméra neutronographique
rotation	14 Protection en plomb mobile
	15 Mur en béton

Annexe 1

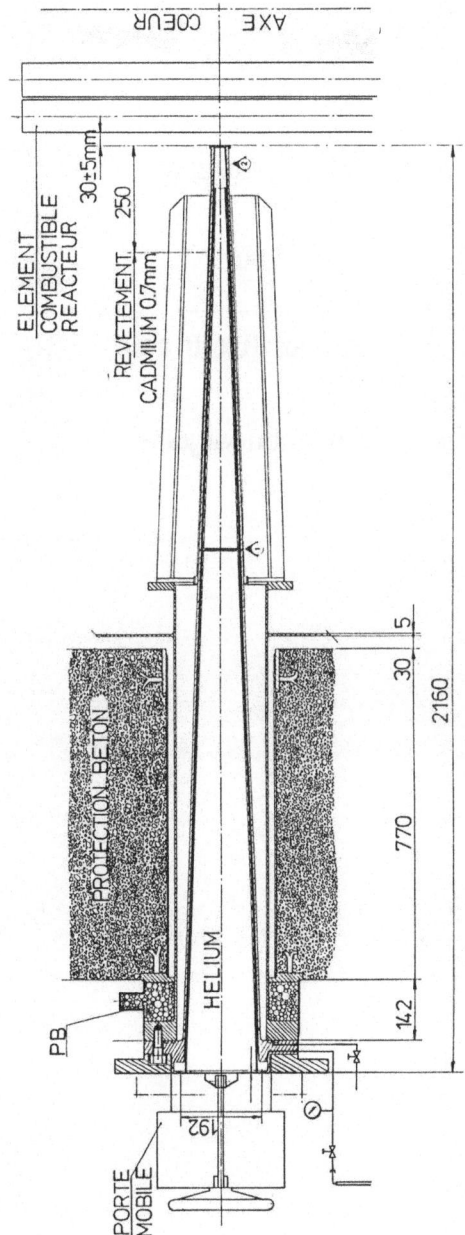

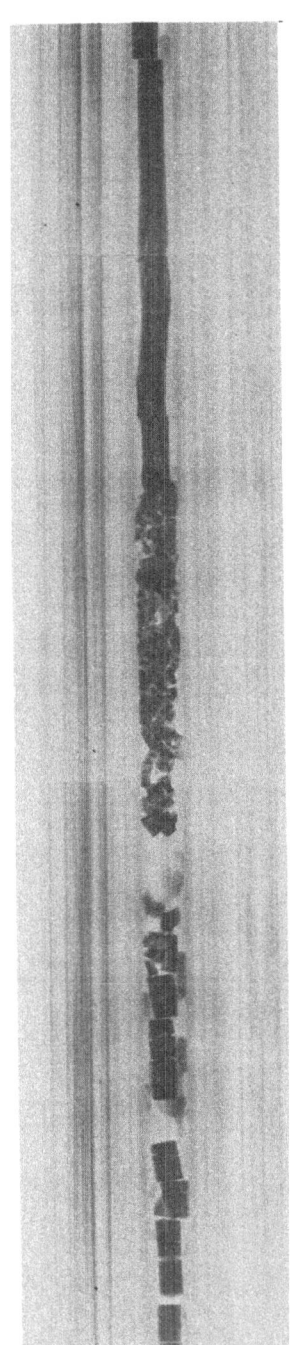

Neutronographie

sur

combustible irradie

en monoaiguille

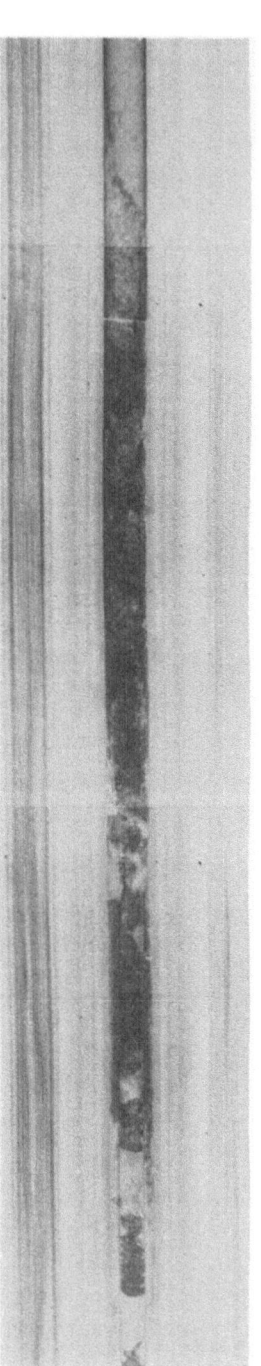

Annexe 3

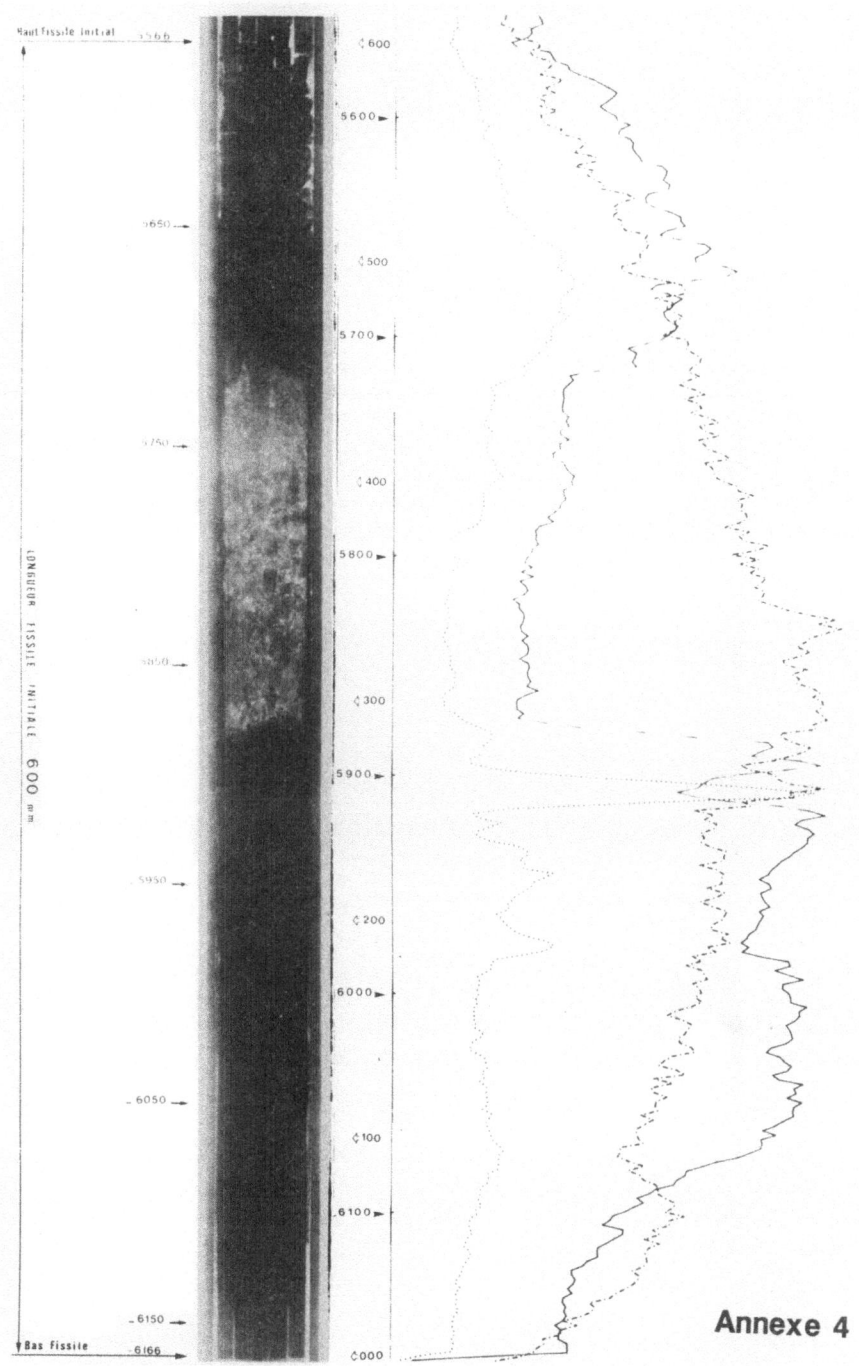

NEUTRONOGRAPHIE SUR GRAPPE 19 AIGUILLES

Annexe 4

"LES INSTALLATIONS ET PROGRAMMES DE RADIOGRAPHIE
AUX NEUTRONS AUPRES DES REACTEURS DES CENTRES CEA"

Contribution du Centre de Cadarache (DECPu/SLHA/LDAC) à la Conférence Mondiale
sur la radiographie aux neutrons - Paris 16-20 juin 1986

M. PELLETIER - F. GIBIAT

INSTALLATIONS DE NEUTRONOGRAPHIE DU LABORATOIRE DE DECOUPAGE ET D'EXAMENS APRES IRRADIATION D'ASSEMBLAGES COMBUSTIBLES (LDAC) ET DE L'A CELLULE DES ELEMENTS IRRADIES (CEI) A PHENIX

M. PELLETIER et F. GIBIAT
DECPu/SLHA/LDAC
CENTRE D'ETUDES NUCLEAIRES DE CADARACHE
13.115 SAINT PAUL LEZ DURANCE FRANCE

INTRODUCTION

Ces installations disposent de réacteurs pulsés de petites dimensions produisant une salve de neutrons, elles sont utilisées principalement pour le contrôle non destructif des combustibles irradiés des Réacteurs à Neutrons Rapides (PHENIX).

Ce contrôle important associé à d'autres examens non destructifs permet d'avoir une bonne idée statistique du fonctionnement de l'élément combustible.

INSTALLATIONS DE NEUTRONOGRAPHIE DU LDAC ET DE LA CEI

Principe du Réacteur

La production d'une salve de neutrons est obtenue en laissant évoluer librement une solution fissile après l'avoir placée dans un état surcritique par apport d'un excès de réactivité. La réaction en chaîne s'étouffe d'elle-même par réduction de la réactivité due en particulier à l'élévation de température de la solution.

Description Sommaire du Réacteur

Les installations du LDAC et de la CEI sont de conception générale similaires.

Les principaux éléments du réacteur sont :

- la cuve cylindrique contenant la solution fissile de nitrate d'uranyle enrichie à 93 %,

- la barre de sécurité en cadmium,
- le réflecteur mobile, fournissant l'excès de réactivité, actionné par un verin,
- le dispositif de refroidissement ou de rechauffage de la solution,
- le circuit de recombinaison des gaz de radiolyse.

Le réacteur, entièrement télécommandé est placé à l'intérieur d'une protection en béton assurant une protection biologique efficace.

Installation pour la Neutronographie

Les neutrons de la salve émise sont en partie canalisés dans un collimateur intercalé entre la cuve du réacteur et un puits de neutronographie qui peut recevoir les éléments combustibles à neutronographier.

Le collimateur est de forme pyramidale en polyéthylène et recouvert de cadmium.

Selon l'installation les dimensions de la fenêtre d'exposition sont les suivantes :

$$100 \times 500 \text{ mm pour le LDAC}$$
$$125 \times 1000 \text{ mm pour la CEI}$$

Cette dernière installation peut couvrir la totalité de la colonne combustible d'aiguilles PHENIX en un seul cliché.

Pour le réacteur du LDAC, le rapport de collimation $\frac{L}{D}$ est d'environ 14 tandis que la fluence de neutrons au niveau de la fenêtre d'exposition est de $2\,10^9\,n/cm^2$ par salve avec un rapport Cd conduisant à une composante épithermique importante convenant parfaitement aux matériaux fissiles à fort enrichissement en Plutonium.

RESULTATS DE NEUTRONOGRAPHIE

Les principaux examens neutronographiques effectués au LDAC et à la CEI reposent sur des assemblages et aiguilles combustibles issus de PHENIX. L'installation de la CEI permet d'obtenir des clichés sur un assemblage entier, selon différentes orientations. L'épaisseur de l'objet et en conséquence l'importante absorption par le combustible mixte UPu 0_2 ne permet pas de visualiser l'intégralité des lits d'aiguilles combustibles, néanmoins les clichés montrent sans ambiguité, les déformations du faisceau d'aiguilles et les intéractions éventuelles de celles-ci avec le tube enveloppe.

Cependant l'essentiel des examens porte sur les aiguilles combustibles des assemblages standard et expérimentaux.

l'examen neutronographique joue le rôle de contrôle statistique du comportement du combustible en situation normale ou dans des conditions particulières de fonctionnement. Ainsi environ 250 à 300 aiguilles appartenant à une vingtaine d'assemblages fissiles sont neutronographiées par an dans les installations du LDAC et à la CEI.

Les clichés fournissent des renseignements qualitatifs sur :

- les aspects du combustibles UPu 0_2 et des couvertures axiales,
- la présence ou non d'interpastilles, d'un trou axial et de fractures dans le combustible,
- la présence éventuelle de produits de fission dans la cavité centrale

ainsi que quelques valeurs quantitatives :

- allongement de la colonne combustible,
- largeur des interpastilles. leur nombre et leur répartition axiale,
- valeur maximale du diamètre du trou central.

Ces observations sont habituellement corrélées avec les autres contrôles non destructifs couramment effectués sur les aiguilles combustibles qu'il s'agisse de métrologies, de contrôles par courants de Foucault ou de spectrométrie gamma.

En condition incidentelle de fonctionnement (surpuissance de l'élément combustible) l'examen neutronographique a permis de visualiser la fusion plus ou moins importante du combustible en fonction de la puissance linéaire atteinte. On a ainsi pu mettre en évidence la répartition des cavités et bulles dans le combustible consécutives de la fusion partielle de celui-ci.

EN CONCLUSION :

Sans toutefois atteindre la qualité des neutronographies réalisées sur les installations des gros réacteurs, les clichés obtenus montrent que des petits réacteurs comme ceux du LDAC et de la CEI peuvent apporter des renseignements utiles à l'examen des combustibles irradiés.

LES INSTALLATIONS ET PROGRAMMES DE RADIOGRAPHIE
AUX NEUTRONS AUPRES DES REACTEURS DES CENTRES CEA"

Contribution du Centre de Valduc (Service de Recherche en Sûreté Criticité) à la Conférence Mondiale sur la radiographie aux neutrons
Paris 16-20 juin 1986

G. COLOMB - R. RATEL

M I R E N E

MINI REACTEUR POUR NEUTRONOGRAPHIE

Le mini réacteur pour neutronographie MIRENE est composé (voir fig 1)

- d'une cuve cylindrique en acier inoxydable de 30 cm de $\emptyset$ et de 36 cm de hauteur contenant 1 kg d'uranium enrichi, sous forme de solution fissile, entourée d'un réflecteur de neutrons,fixe.

- un second réflecteur de neutrons, mobile, actionné par un vérin pneumatique et appliqué contre le fond de la cuve, déclenche la réaction en chaîne et la salve de neutrons. Son éloignement arrête la réaction.

- un dispositif de conditionnement de la solution permet de ramener celle-ci aux conditions initiales de température, dans la plage des limites admises pour le fonctionnement.

- un système de recombinaison des gaz de radiolyse dégagés pendant le fonctionnement, permet de conserver les caractéristiques de la solution fissile.

- les circuits de la solution et de gaz sont étanches, le bloc réacteur et les circuits sont placés dans un caisson métallique ventilé. L'ensemble est entouré par une protection en béton de 1,50m d'épaisseur.

- la conduite du réacteur s'effectue à partir du pupitre de commande situé à quelques mètres.

- le réacteur peut produire une salve de neutrons toutes les quinze minutes.

Caractéristiques techniques

Une solution de 20 litres de nitrate d'uranium enrichi à 93% en U^{235} à la concentration de 59g/l est nécessaire.

Réactivité apportée par l'approche du réflecteur mobile $\frac{\Delta k}{k}$ = 0,5646%(564,6 pcm).

Période de montée en puissance 0,872 s

Pic de puissance 161,1 kw.

Energie dégagée dans la salve 2,09 MJ (6,84. 10^{16} fissions) durée de la salve 3 minutes.

Applications

- Un grand nombre d'applications de la neutronographie découle de sa capacité de visualiser des constituants à base d'éléments légers (hydrogène, bore, adhésifs, matières plastiques, caoutchouc) et ce à travers des épaisseurs même importantes de métal.

- MIRENE peut répondre aux besoins de contrôle rencontrés dans un vaste champ d'applications couvrant les études de laboratoire et le domaine industriel, que ce soit la pyrotechnie, la mécanique, l'électronique, l'aéronautique et l'industrie aérospatiale

Exemples

Domaine industriel

Un exemple concerne le contrôle de pièces métalliques collées (voir fig. 2). La densité des métaux assemblés est relativement élevée (uranium - acier) et pourtant on peut parfaitement mettre en évidence une mauvaise répartition de la colle dont l'épaisseur à certains endroits est de l'ordre de 1/100 mm.

Domaine aérospatial

Le recours à un gaz traceur neutrophage (He.3) permet de contrôler l'étanchéité de volumes emboîtés (type "poupée russe") comme le montre la fig.3, capsules en quartz conditionnées dans un tube soudé en titane(dispositif expérimental embarqué sur SPACELAB par le CNES)

Agronomie (voir fig. 4)

La neutronographie est un bon exemple de technique nucléaire mise au service de la recherche agronomique.

Elle a été appliquée à l'observation du développement de racines de maïs, de tournesol et de riz tropical sans porter atteinte à l'intégrité du système racine sol.

La visualisation des racines à travers le sol est basée sur le fait qu'elles contiennent une quantité importante d'eau, dans une proportion voisine de 80%,en tout état de cause supérieure au taux d'humidité du sol. Elles présentent par la même un pouvoir absorbant aux neutrons plus élevé que ce dernier et produisent une image fortement contrastée par rapport à leur environnement.

MINI REACTEUR POUR NEUTRONOGRAPHIE
MIRENE

1. CUVE CONTENANT LA SOLUTION FISSILE
2. REFLECTEUR FIXE
3. REFLECTEUR MOBILE
4. VERIN DESCENTE & MONTEE
5. CONDITIONNEMENT DU CŒUR
6. CIRCUIT DE RECOMBINAISON
7. COLLIMATEUR AXIAL
8. COLLIMATEUR TANGENTIEL
9. RESERVOIR DE STOCKAGE
10. CHASSIS
11. CAISSON
12. PROTECTION BIOLOGIQUE
13. PORTE DE VISITE
14. GENERATEUR D'EAU FROIDE
15. ECHANGEUR
16. RECHAUFFEUR
17. PUPITRE DE CONTROLE COMMANDE
18. B.A.G. DES TRANSFERTS
19. CASEMATE BOITE A GANTS
20. PORTILLONS

FIG. 1

FIG. 2 *Contrôle de la qualité d'un collage entre deux pièces métalliques*

127

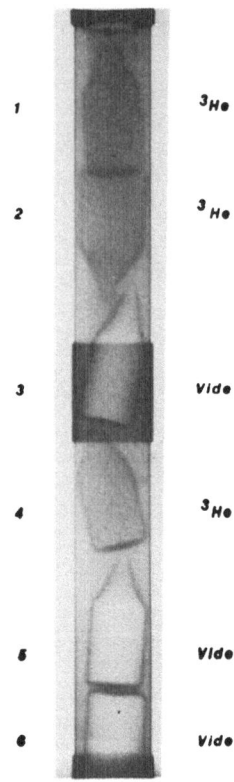

1   ^3He

2   ^3He

3   Vide

4   ^3He

5   Vide

6   Vide

FIG.3 Capsule Spacelab

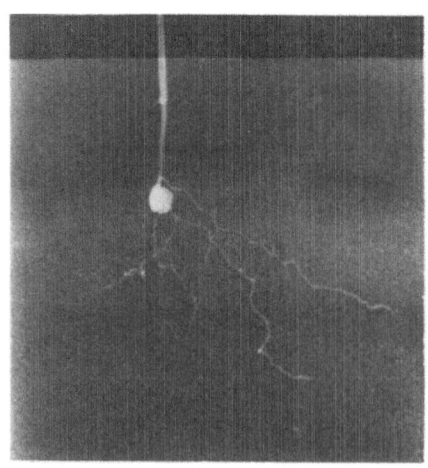

FIG. 4

NEUTRON RADIOGRAPHY AT RIKKYO TRIGA-II REACTOR

Hisao KOBAYASHI, Susumu HARASAWA, Kenji TOMURA, Yasukiyo TAKAMI,
Fumio SHIRAISHI, Manabu HATTORI, Teruaki NAGAHARA,
Tatsuo MATSUURA, Iwao OGAWA
Institute for Atomic Energy, Rikkyo University,
2-5-1 Nagasaka, Yokosuka, Kanagawa, 240-01 JAPAN.

Yusuke YAMAMOTO
Department of Radiology, Komazawa Junior College,
1-23-1, Komazawa, Setagaya-ku, Tokyo, 154 JAPAN.

ABSTRACT

A vertical diverging collimator is installed in the
reactor tank of Rikkyo 100 kW TRIGA-II. The configuration of the collimator and the beam characteristics are
summarized. The ASTM E545-81 sensitivity and beam purity
indicators have shown that the collimator is qualified
to be the category I. Inherent unsharpness for the
combination of the 25 μm Gd screen and the KODAK-SR film
is determined experimentally and found to be 48.0±0.5
μm. An application is given for the inspection of mold
cavities in carburetors.

INTRODUCTION

Neutron Radiography(NR) Facility Installation Project was
initiated in 1984 to extend the utilization of TRIGA II reactor
(100 kW) at Institute for Atomic Energy, Rikkyo University. And
the related basic research and data acquisition for NR technique
development were started. In 1985, a vertical diverging collimator
with neutron imaging area of 100 mm in diameter at the top end was
installed in the reactor tank to accumulate preliminary data for
the Project. A detailed description of the studies is presented
in this paper, covering experimental results of unsharpness in NR

129

imaging as well as measured parameters of the NR facility. In addition, NR photographs of a carburetor are also shown as one of examples to utilize of the NR facility in our laboratory. They successfully demonstrate mold cavities in the aluminum die-cast product which can not be identified by other means such as X-ray radiography.

CHARACTERISTICS OF NR FACILITY

Schematic diagram of the vertical type NR facility, which is consisted of an aluminum pipe assembly and of a set of diverging collimators, is illustrated in Fig. 1. The pipe assembly has dimensions of 110 mm in outer diameter, 100 mm in inner diameter and 5265 mm in length. In order to facilitate the installation and removal, the pipe assembly is divided into three sections. The diverging collimator is made with a wooden neutron collimator and a gamma ray shield. The wooden neutron collimator has a shape of coaxial cylindrical ring (50 mm in inner diameter, 87 mm in length) and is placed at the bottom of the pipe assembly. The inside cylindrical face and the top flat face of the ring are covered with Cd plates of 1 mm thick. The configuration results in effective collimator aperture, D, of 48.0±0.5 mm. The gamma ray shield consists of eight lead blocks with penetrating holes of 50 to 57 mm in diameters and are set on the neutron collimator for shielding unfavorable gamma-rays. A film cassette was set perpendicularly to the neutron beam, 105 mm away from the top end of the pipe assembly. Thus, film distance to the effective center of the neutron collimator (⊕ marking in Fig.1), L, is 5321±4 mm, and L/D is estimated to be

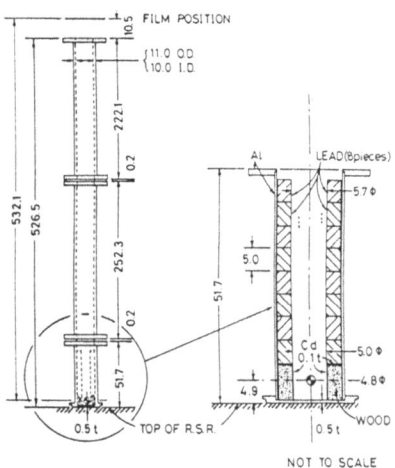

Fig.1 A sectional view of the aluminum pipe assembly and a detail of the diverging collimators (not to scale).

$$L/D = 110.9 \pm 1.2. \tag{1}$$

Thermal neutron flux was measured by gold foil activation methods applying a beta-gamma coincidence technique and also by a gamma ray peak analysis using a calibrated Ge(Li) detector. The

130

thermal neutron flux on the film position at 100 kW was 6.9×10^5 n/cm²·s. Cadmium ratio at the point was 4.3. Gamma dose rate measured by an ionization chamber at the same position was 4.1 R/h. So that, neutron to gamma ratio was 6.1×10^5 n/mR/cm². These results are summarized in Table 1.

NR images were photographed by a combination of a Gd screen and a film. The 25 μm thick Gd screen in chemical form of Gd_2O_3 was evaporated on a 2 mm aluminum plate and was coated by 85 Å thick Al_2O_3 (Research Chemicals, NUCOR). The film was a single coated KODAK industrex R SR-5 film. Optical film densities of developed film images were measured mainly by a manually operated densitometer with a circular slit of 0.5 mm in diameter (SAKURA PDA-15). A computerized microdensitometer was also used to scan and to analyze images. The microdensitometer has a capability of changing the slit dimensions from 1 μm x 1 μm to 1 mm x 1 mm (SAKURA PDM-5) and can scan two-dimensionally on the film surface.

Relationship of optical film density against irradiation time at 100 kW is illustrated in Fig. 2. Optical film density in Fig. 2 is the mean value of 10 to 20 measuring points inside 90 % radius of the irradiation area. Standard deviation of each points gives ±4.1 % (i.e. 95% confidence interval: ±8.2 %). Short dashed and long dashed curves in Fig. 2 are the 95 % confidence interval of regression line and that of data points in the range of optical film density less than 3.5, respectively. A linear relation is shown for the optical film density in the region of 1 to 3.5 and also for neutron fluence region of 1.2 to 5.0×10^9 n/cm².

Table 1. Characteristics of 100 mm φ vertical type collimator of Rikkyo university TRIGA-II reactor.

REACTOR POWER	100 kW
SOURCE	TOP OF THE REFLECTOR
SIZE	100 mm φ x 5265 mm
L / D	110.9 ± 1.2
n_{th} FLUX	6.9×10^5 n/cm2-s
GAMMA DOSE RATE	4.1 R/h
n-GAMMA RATIO	6.1×10^5 n/cm2/mR
Cd RATIO	4.3

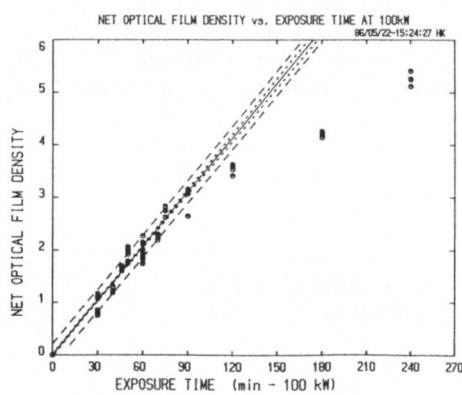

Fig.2 Optical film density as a function of neutron exposure time.

Optical film density image in the irradiation area against neutron flux distribution is illustrated in Figs. 3 (a) and (b). Figure 3(a) shows a contour map of optical film density, and in Fig. 3(b) distribution of optical film densities along A-A' and B-B' as defined in Fig. 3(a) are illustrated in a solid line (and solid circles) and a dashed line (and open circles), respectively. Relative neutron fluence measured by the gold foil activation method along A-A' are also plotted in the figure with x marks, and along B-B' with + marks.

Reliability of an indicator in determination of NR image quality is to be presented in another session of this Proceedings (1). The quality indicators used in this work were the beam purity indicator and the sensitive indicator, both in accordance with ASTM-E545 specification of 1981 (2). Image qualities determined by these indicators are described in Table 2. Given results were at neutron fluence of 2.1×10^9 n/cm^2 (50 minutes irradiation at 100 kW). The numerical figures show that the Rikkyo NR Facility fulfills the criteria of Category I.

DETERMINATION OF INHERENT UNSHARPNESS

NR image quality is mainly determined by the configuration of an irradiation facility, but it also depends on the photographic imaging tech-

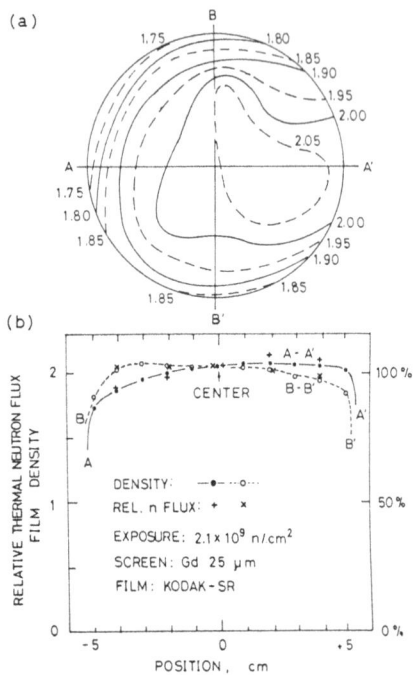

Fig.3 (a) Contour map of film density on the 100 mmφ exposure field at the top of the collimator, and (b) film density distributions along the line A-A' (solid circles and straight curve) and the line B-B' (open circles and dashed curve).

Table 2. Image quality for 100 mm φ vertical type collimator of Rikkyo university TRIGA-II reactor by using BPI and SI indicator (ASTM-E545 1981 ed.)

	RESULTS	CATEGORY I	DECISION
neutron component (%)	70.2 ± 1.5	≥ 65	+
scattered neutron (%)	0.9 ± 0.2	≤ 5	+
gamma component (%)	2.8 ± 0.9	≤ 3	+
pair production (%)	0.5 ± 0.5	≤ 3	+
number of gaps	7 ± 0	≥ 6	+
number of holes	9 ± 1	≥ 6	+

SCREEN: Gd 25 μm,
FILM: KODAK SR-5,
EXPOSURE: 2.1×10^9 n$_{th}$/cm^2 (50 min x 100 kW).

nique. The image unsharpness is consisted with (1) geometrical unsharpness, U_g, which is the main component, when a separation gap between the screen and an object, l, is sufficiently larger than a critical length, l_c, (to be described later) and (2) inherent unsharpness, U_I, of imaging equipment, if l is much smaller than l_c. As another influencing factor, there is scattering effect in thick samples which has been analyzed by Whittemore et al. (3) However, the scattering effect can be neglected, if a thickness of an object is sufficiently small.

Inherent unsharpness U_I is to be discussed. In the analysis below, one dimensional model is assumed and the position on the screen is expressed by the coordinate x. When a narrow collimated neutron beam irradiates x_0, resultant image is referred to as the line spread function (LSF), and is expressed as $f(x, x_0)$. If optical film density, $I(x)$, is proportional to neutron fluence per exposure, $\Phi(x)$, then $I(x)$ is given by a following equation;

$$I(x) = C \int_{-\infty}^{+\infty} dx_0 \int_{x_0-w}^{x_0+w} dx\, \Phi(x_0) \cdot f(x, x_0), \qquad (2)$$

where C is a constant to convert neutron fluence into film density, and w is a slit width of a microdensitometer.

When a neutron beam irradiates on the screen with a step-like spacious distribution at x=0 as shown in Fig.4(a), let us define measured film density $I(x)$ as the edge spreading function (ESF). For a rectangular shape of LSF as shown in Fig.4(b), ESF takes a form of a step with a slope around the edge (x=0) of $\Phi(x)$ as shown in Fig.4(c). The slope width of the step is exactly equal to U_I and is also equal to LSF width. As more generalized treatment, if a slope width of ESF is defined as the difference of x coordinates of two points, which correspond to two crossed point of following three lines; the tangential line of $I(x)$ at the midpoint of $I(+\infty)$ and $I(-\infty)$, and the two horizontal lines $y=I(+\infty)$ and

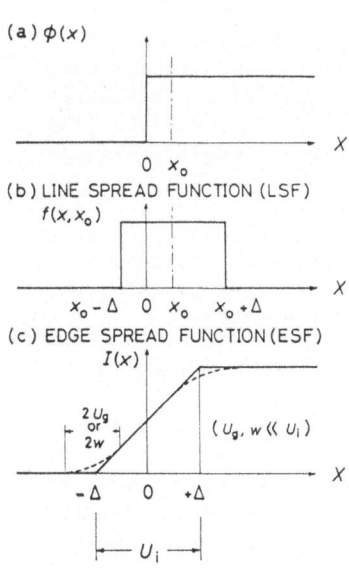

Fig.4 A schematic illustration of (a) neutron flux distribution $\Phi(x)$, (b) line spread function $f(x,x_0)$, and (c) edge spread function $I(x)$.

$y=I(-\infty)$, the slope width can be approximated to be U_i. And then, it can be shown that U_i is approximately equal to FWHM of LSF. For example, if LSF is in an equilateral triangle shape, U_i is exactly equal to FWHM; and if $f(x,x_0)$ is Gaussian, U_i equals to 1.06 FWHM. Therefore, defining U_i as above and regarding it as FWHM of LSF, can virtually be justified.

Generally, w and U_v are not equal to zero and have a finite numerical value, where U_v is defined elsewhere(3-5) by

$$U_v = 1 \cdot D/(L-1) \simeq 1 \cdot D/L, \qquad 1 \ll 1_c. \qquad (3)$$

However, if both w and U_v are sufficiently small compared with U_i, those contribution to ESF are not significant. As shown in dashed line of Fig.4(c), the effect is merely smoothing of the slope edges within the region of $\pm$w and/or $\pm U_v/2$, and give little influence in determination of U_i. Now we can introduce a critical length, 1_c, which corresponds to the 1 value for $U_i=U_v$, and then

$$1_c=U_i \cdot L/(D+U_i) \simeq U_i \cdot (L/D), \qquad 1 \ll 1_c. \qquad (4)$$

In order to realize the step-like neutron beam experimentally, a Cd plate of 0.5 mm thick was set in contact with a vacuum cassette and I(x) was experimentally measured by the microdensitometer. Typical recorder output of the scan image, viz., ESF for 2w=5 μm is shown in Fig.5. Measurements were also made for 2w=3 μm and 4 μm. Averaging them,

$$U_i = 48.0 \pm 0.5 \text{ μm}, \qquad (5)$$

was obtained. The value of U_v was estimated as 4.3 μm from the experimental results. Therefore, both w and U_v are much smaller than U_i and practically give no effect on the estimation of U_i.

Internal conversion electrons (CE) of 79.5 keV from $^{157}Gd(n, \gamma)^{158}Gd$ nu-

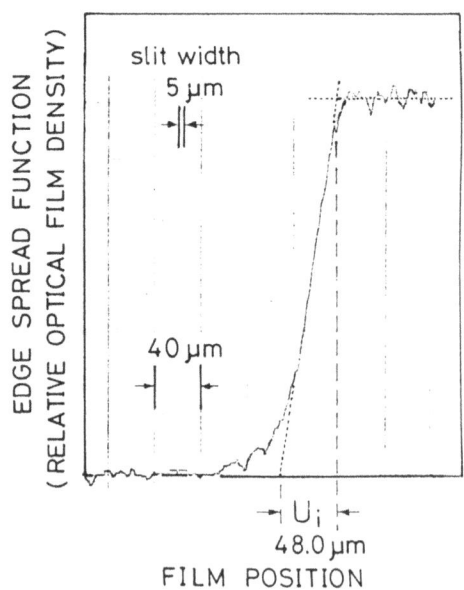

Fig.5 Observed edge spread function for the combination of the 25 μm Gd screen and the KODAK SR-5 film.

clear reaction(6) contribute mostly to make latent images in film emulsion. The electron range in Gd_2O_3 is 13 µm(7) and smaller than the Gd screen thickness of 25 µm. The Al_2O_3 layer is only 85 Å which can be neglected. Therefore, maximum width of experimentally determined LSF can be approximated by doubling CE range in film emulsion(~45 µm). If the LSF is similar to rectangular, the maximum width can be approximated as U_i. And the value is in good agreement with measured U_i value of 48 µm.

APPLICATIONS

Carburetors made by aluminum die-casting, often show mold cavities, resulting low yielding rate. The yield rate of less than 90 % is frequent in the manufacturing process. Cavities in the aluminum alloy were practically impossible to be recognized by X-ray radiography and any other methods. However, after filling water into the mold cavities of carburetors, NR pictures were successfully taken. An example of the NR images is illustrated in Fig. 6. The carburetor cavities sensitized by water was photographed in four different angles. We can clearly find three dimensionally distributed cavities in the images. Other applications,

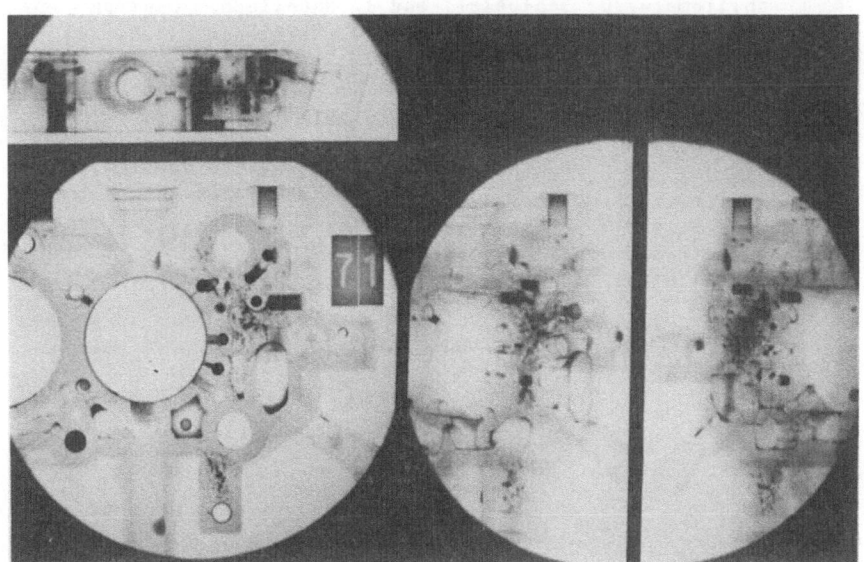

Fig.6 Typical neutron radiography images to inspect molded cavities in a carburetor. The carburetor after being soaked in water is photographed in four different angles. We can clearly find three dimensionally distributed cavities in the images.

such as tooth image analysis which has complex structure with different atomic composition, are to be presented at the session XII in this Proceedings(8).

CONCLUSIONS

The NR test facility used in this experiment was inexpensive, easy to install or remove. The characteristics, as shown in Table 2, were sufficient to be Category I. In combination of the 25 μm thick Gd screen and the Kodak SR-5 film, the inherent unsharpness was experimentally shown as 48.0 ± 0.5 μm. Some applications were made, including successful inspection the test of cavities in carburetors. We have used the wooden collimator covered with the Cd plates. The efficacy of the collimator will be verified experimentally and the results would appear elsewhere.

REFERENCES

1. K. Katsurayama et al. (in this Proceedings).
2. ASTM-E545-81 "Standard Method for Image Quality in Thermal Neutron Radiographic Testing",(The 1981 Annual Book of ASTM Standards, Part II, 1981)p.527.
3. W. L. Whittemore, G. Schlueter, and J. Shoptaugh, "Neutron Radiography", J. P. Barton and P. v. d. Hardt eds., (D.Reidel Publ.Co., Dordrecht, 1983)p.885.
4. R. Matfield, ibid. p.817.
5. J. Olsen and P. Gade-Nielsen, ibid. p.891.
6. J. H. Hamilton, A. V. Ramayya, B. v. Nooijen, R. G. Albridge, E. F. Zganjar, S. C. Pancholi, J. M. Hollander, V. S. Shirley, and C. M. Lederer, Nucl. Data Sec. A 1, 521 (1966).
7. C. M. Lederer and V. S. Shirley ed. "Table of Isotopes", 7th ed. (John Wiley & Sons Inc. New York,1978)
8. H. Kobayashi et al. (in this Proceedings, JP14, 1986).

DEVELOPMENT OF A NEUTRON RADIOGRAPHY SYSTEM

AT THE MUSASHI INSTITUTE OF TECHNOLOGY RESEARCH REACTOR

Tetsuo Matsumoto, Otohiko Aizawa and Shigeru Watanabe

Atomic Energy Research Laboratory, Musashi Institute of Technology

Ozenji 971, Asao-ku, Kawasaki-shi, 215 Japan

ABSTRACT

A neutron radiography(NRG) facility has been designed and installed at the one of the horizontal beam port attached to the Musashi Institute of Technology Research Reactor(Musashi reactor).   An X-ray radiography(XRG) apparatus using a portable generator was also equipped for the comparison with the NRG.   Good characteristics of the neutron field were obtained by using a single-crystal bismuth and a divergent collimation system.   The result of resolution measurements for the Image Quality Indicators(IQI) satisfied the grade 1 values of ASTM E545-81 category using a vacuum cassette with a metallic gadolinium converter. Various combinations of the converters and X-ray films were evaluated in order to obtain a high-quality photograph. A real-time NRG has also been installed for investigation of the moving objects and for study of a neutron computed tomography(NCT).   A high-quality radiographic image was obtained by an image orthicon camera and a digital image processing system.

INTRODUCTION

The Musashi reactor, located near Tokyo, is a 100 kW TRIGA-II reactor.   The four horizontal beam ports and one thermal column have been effectively used for many purposes: Brain tumors treatments and preclinical melanoma studies of boron neutron capture therapy have been performed with the remodeled thermal column(1).

Total neutron cross sections of single and polycrystal materials in the thermal region have been measured with a time-of-flight (TOF) method and also been measured in the keV region with silicon and iron filtered methods using the beam ports(2),(3).

For study of applications of NRG in biology and industry, we have designed and installed a NRG facility with a real-time imaging system as well as a conventional film system at the one of the beam ports. The present report describes:
1) Outline of the NRG facility and characteristics of the irradiation field.
2) Improvement of image quality with a real-time imaging system.
3) Quantitative evaluation for the NRG and neutron TV systems.
4) Test of a NCT using a real-time imaging system.

REACTOR AND NRG FACILITY

Figure 1 outlines four beam facilities available for neutron experiments denominated Chopper-TOF Facility, Irradiation Room, Filtered Beam Facility and Neutron Radiography Facility.

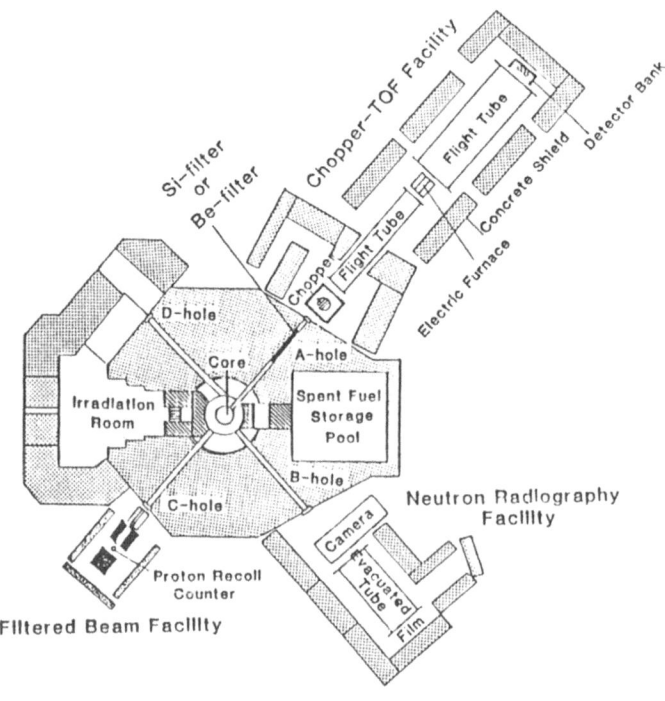

Figure 1. Beam facilities of the Musashi reactor.

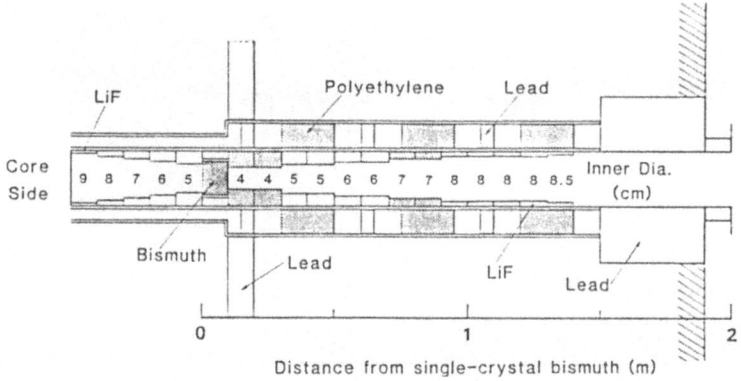

Core
Side

Inner Dia.
(cm)

Distance from single-crystal bismuth (m)

Figure 2. Divergent collimation system including
a single-crystal bismuth filter.

Figure 2 shows the final collimation system improved at the B hole
including a single-crystal bismuth filter.   As shown in Fig.3,
single-crystals of bismuth and silicon have small cross sections
in the cold and thermal neutron ranges.   These materials can
produce an adequate thermal neutron beam source and reduce the
gamma rays and fast neutrons background.   The divergent
collimation system consists of the cylindrical polyethylene,lead
and LiF tile collimators with different diameters.

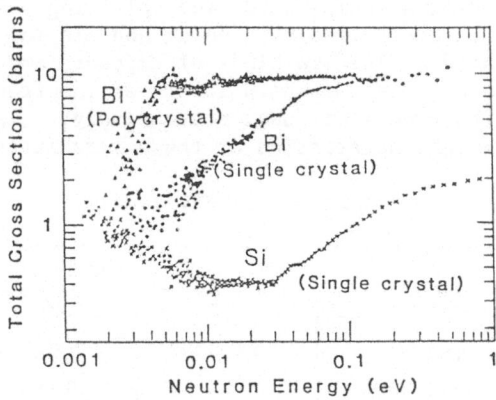

Figure 3. Neutron cross sections of single-crystal
bismuth and silicon.

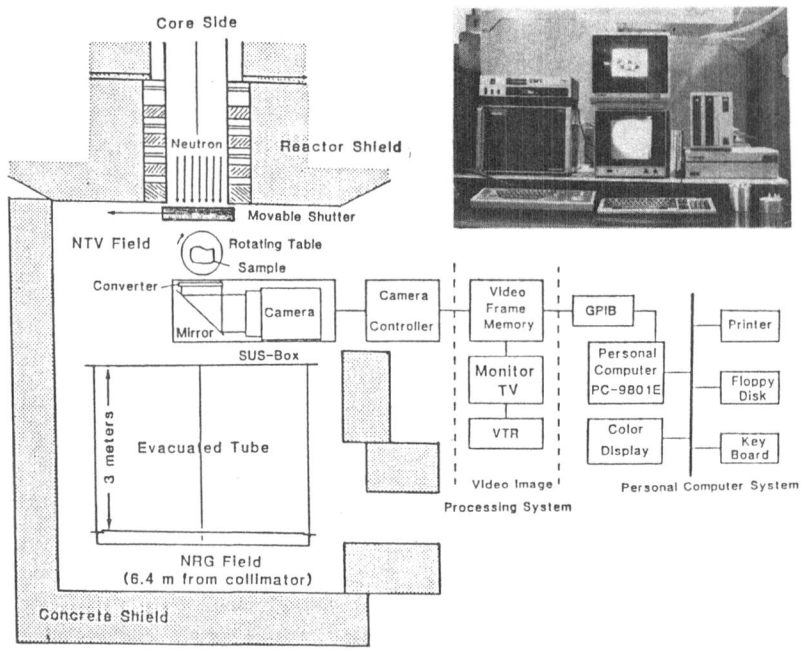

Figure 4. NRG facility including a real-time
imaging system.

Figure 4 illustrates the arrangement of NRG system outside the
reactor face including a beam shutter and shielding walls. The
exposure of films is controlled by opening and closing a shutter
electrically operated. The two kinds of exposure are available
at the NTV and the NRG fields, which locate at the front-side and
the back-side of the evacuated tube, respectively. Table 1
summarizes the beam characteristics of these irradiation fields.

Table 1. Characteristics of the NRG facility.

	NTV Field	NRG Field
Geometric Resolution (L/D)	65	160
Nutron Flux (n/cm²/s)	1.7×10^6	2.0×10^5
Gamma Dose Rate (R/h)	1.1	0.25
N/γ Ratio (n/cm²/mR)	3.3×10^6	2.9×10^6
Cadmium Ratio of Gold Foil	8	8
Film Size Available	15 cm	38 cm

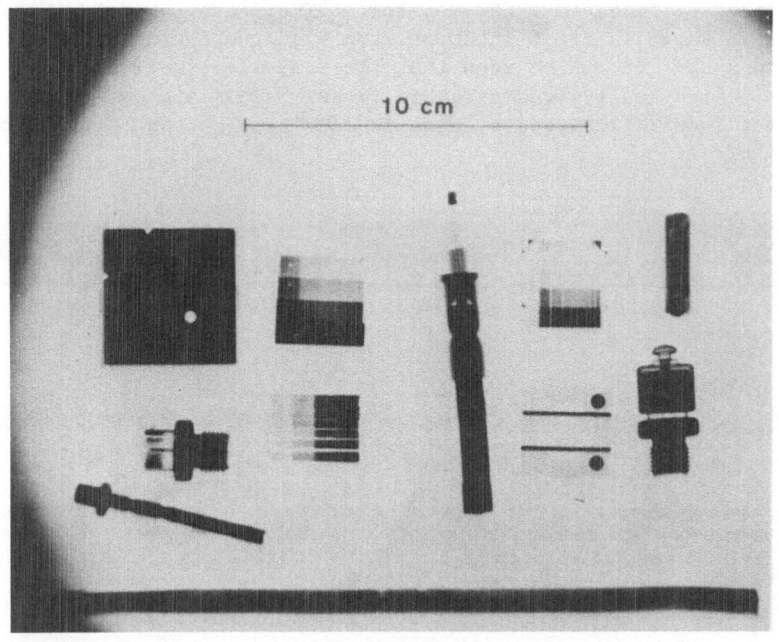

Figure 5. IQI photograph of the film method at
the NRG field.

Good characteristics of the neutron field were obtained using a
bismuth filter and a divergent collimation system.   It is
possible to obtain a large size photograph of 38 cm in diameter
at the NRG field, which is shown in Fig.5, and also possible to
investigate the moving objects at the NTV field.

REAL-TIME IMAGING SYSTEM

A real-time NRG has been installed for investigation of the
moving objects and for study of a NCT.   As illustrated in Fig.4,
a real-time imaging system consists of a neutron TV system, a
video image processing system and a personal computer system.
The neutron TV system has a high-quality TV camera with a camera
controller.   The TV camera directly takes a radiographic image
from a NE-426 neutron scintillation screen with an image orthicon
made for X-ray use.   In order to obtain a higher-quality radiographic image, a digital image processing with a video frame

memory(Hamamatsu C1901) was adopted in the video image processing
system.   This system offers image integration, image subtraction,
image reversal, slice operation and enhance operation, and feeds
the processed image to a CRT monitor.   The image taken from this
system is shown in Fig.6 in comparison with the photograph of the
NRG and XRG.   It can be seen that the real-time imaging system
offers a high-quality radiographic image.   This system can be
used for industrial purpose, because many samples can be treated
in a short time.

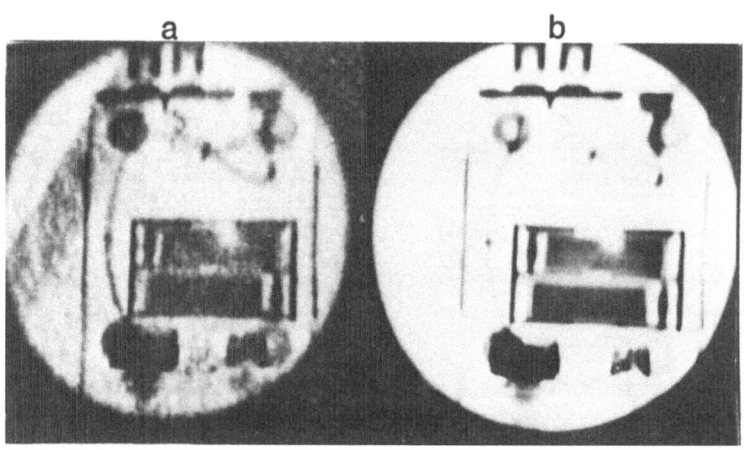

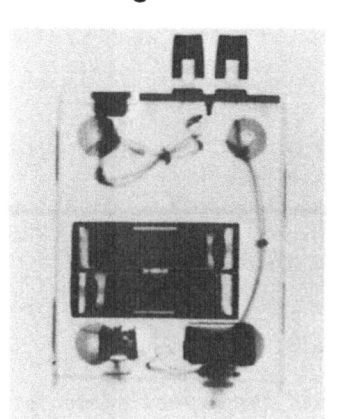
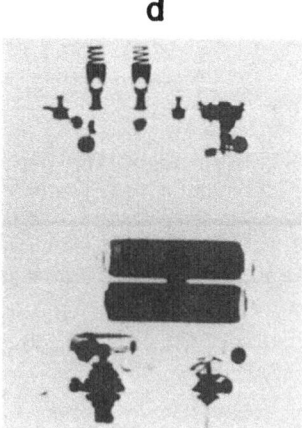

Figure 6. a) Direct image from a real-time imaging system.
          b) Integrated image of 400 frames.
          c) Photograph of the NRG film method.
          d) Photograph of the XRG film method.

QUANTITATIVE EVALUATION OF NRG AND NTV SYSTEMS

Quantitative evaluation of films, converters and neutron beam

In order to obtain a high-quality photograph by a conventional film method, various combinations of films and converters were quantatively evaluated using IQI and ASTM designation. Table 2 summarizes the results. It is shown that a combination of the metallic gadolinium converter and the Kodak SR film is an appropriate choice for a high-quality photograph.
The evaluation of the neutron beam was also performed with the combination using a vacuum cassette. The result satisfies the grade 1 values of ASTM E545-81 category.

Evaluation of the neutron TV

The sensitivities of several converter materials used in the neutron TV system were measured by the digitalization of the image intensity from an image orthicon camera. A resolution was also evaluated using resolution indicators such as a cadmium line-pairs. The NE-426 converter showed 2-times larger sensitivity and better resolution than the Sakura KH and Fuji G8 converters.

Table 2. Evaluated results of various combinations.

Converter	Gd			KH			G8			ASTM grade 1
Film	HS	FG	SR	HS	FG	SR	HS	FG	SR	
NC (%)	45.1	65.6	78.7	62.2	77.8	84.7	58.1	73.9	83.0	>65
S (%)	0.6	0.1	0.6	1.3	2.2	0.0	0.1	0.1	0.0	<5
γ (%)	17.3	0.1	0.0	0.0	0.0	1.7	0.0	-0.1	0.0	<3
P (%)	4.6	3.2	0.6	2.5	0.1	3.7	1.9	1.6	1.5	<3
Hole	4	7	9	7	7	7	5	7	6	>6
Gap	7	7	7	7	7	7	7	7	7	>6

Gd : Chemical Research, KH : Sakura, G8 : Fuji   HS, FG : Fuji softex , SR : Kodak

NC : Thermal neutron content, S : Scattered neutron content, γ : Gamma content, P : Pair production content

NEUTRON COMPUTED TOMOGRAPHY

By using the neutron TV system, a projection data can be obtained in a single measurement. A processed projection image can be observed on a CRT monitor using the video imaging system. The image can be stored in a personal computer system. A step motor is used to obtain the rotated projection data. The reconstructed image can be produced by using the Fourier-convolution technique with a host computer connected to the personal computer. This work is still in progress.

SUMMARY

A high resolution NRG facility with a single-crystal bismuth and a complete collimation system has been developed at the one of the horizontal beam port attached to the conventional TRIGA-II reactor. The result of resolution measurements with IQI satisfied the grade 1 values of ASTM E545-81 category, and also various combinations of the films and converters were evaluated. A high-quality real-time NRG with a digital image processing system and a performance personal computer has also been installed for investigation of the moving objects and for study of a NCT. These systems will be very useful for industry and biology applications.

REFERENCES

1. T.Matsumoto, O.Aizawa:"Depth-dose evaluation and optimization of the irradiation facility for boron neutron capture therapy of brain tumors," Phy.Med.Biol., Vol.44(1985)897-907

2. O.Aizawa, T.Matsumoto, H.Kadotani:"Total neutron cross sections of Magnesium, Aluminium, Zirconium, Niobium and Molybdenum in energy range from 0.001 to 0.3 eV," J.Nucl.Sci.Technol., Vol.20(1983)713-721

3. O.Aizawa, T.Matsumoto, H.Kadotani:"Measurements of total cross sections at 24 keV by means of iron filter method," J.Nucl.Sci.Technol., Vol.20(1983)354-356

4. ASTM designation E545-81(1981)

5. S.Fujine, K.Yoneda, K.Kanda:"Digital processing to improve image quality in real-time neutron radiography," N.I.M.,Vol.228(1985)541-548

THE FACILITY FOR DRY THERMAL AND SUBTHERMAL NEUTRON RADIOGRAPHY

AT THE PETTEN HIGH FLUX REACTOR (HFR)

E.J. Bleeker, H.P. Leeflang

Netherlands Energy Research Foundation ECN

1755 ZG   Petten, The Netherlands

ABSTRACT

This paper describes the facility for neutron radiography at beam
tube HB8 of the HFR at Petten. The earlier existing facility for
thermal neutron radiography of long fuel rods with a direct beam
from the reactor in an internal exposure station has been modified
and extended. It has now options for external radiography for
general purposes with subthermal neutrons applying a beam filtered
by cooled Be and Bi and with thermal neutrons from a beam filtered
with Si and Bi. A He filled flight tube allows for radiography in
an external station.

INTRODUCTION

For the purpose of inspection of spent fuel rods from nuclear
power reactors, at ECN a dry thermal neutron radiography facility
has been successfully in operation during a number of years. This
facility exploits a polychromatic beam extracted from the 45 MW
HFR at Petten.
To increase the output of the facility and to cover the interest
for non-destructive testing, (especially for hydrocarbons in heavy
metal encasings, hydrogen accumulation in steel and material
science purposes) the installation is in the process of being
extended with a facility for subthermal neutron radiography with
the incorporation of beryllium, bismuth and silicon filters and a
flight tube towards an external exposure station.

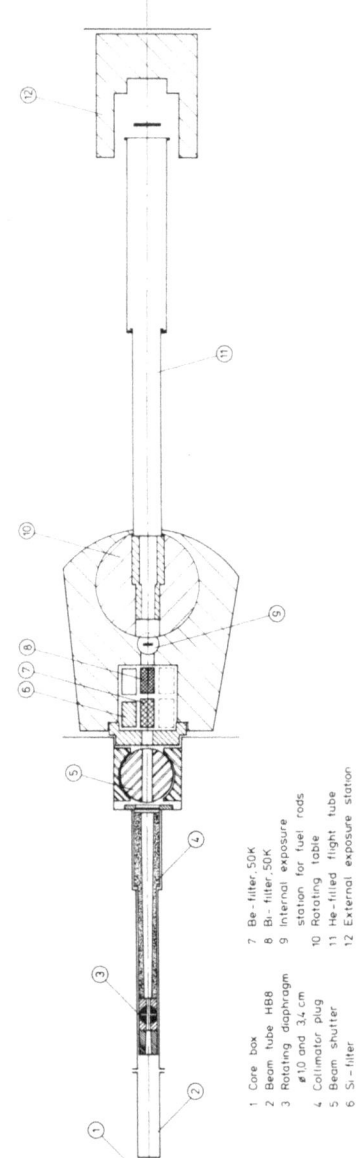

1 Core box
2 Beam tube HB8
3 Rotating diaphragm
  ⌀10 and 3,4 cm
4 Collimator plug
5 Beam shutter
6 Si-filter

7 Be-filter,50K
8 Bi-filter,50K
9 Internal exposure
  station for fuel rods
10 Rotating table
11 He-filled flight tube
12 External exposure station

Fig 1 Neutron Radiography Facility at HB 8 of the HFR, Petten

THE FACILITY FOR THERMAL NEUTRON RADIOGRAPHY OF FUEL RODS

On this facility has already been reported previously (1).
Therefore it now suffices to focuss the attention particularly on
those components of the thermal facility which have been modified
for the sake of dual purpose of the neutron beam.

The In-pile Collimator

A new aluminium in-pile collimator plug has been constructed
according to a design, which is essentially comparable with the
one of the previous plug. This new plug, which has been inserted
in beam tube HB8, carries on the reactor side a solid aluminium
pre-collimator. Behind this a disc is located which has been
provided with two apertures, which serve as diaphragm openings and
which have diameters of 1.0 and 3.4 cm for internal and external
radiography, respectively. Each of the apertures can be rotated
from outside in line with the beam axis with the aid of a manually
operated teleflex wire. The neutron diaphragms are defined by
boral plates which are preceded by pieces of tungsten or lead. The
diaphragm disc rotates on a graphite plate and possesses also a
closed position.
The conical neutron flight path, which lies behind the diaphragm
disc is lined with boral sheet. A water-flooded compartment
surrounds the flight path, which itself is flushed by helium gas.
A window flange, which closes the beam tube, forms with its
penetrations for water, gas and diaphragm wires part of the third
containment of the reactor.

The Internal Neutron Radiography Station

Via an open channel of the beam shutter the neutron beam is
allowed to enter the internal object area for thermal neutron
radiography. This area is situated in the middle block of three
vertically stacked shielding blocks outside the reactor.
The fuel pins, of which four can be examined simultaneously, are
lowered with a lifting device from a transport container into a
vertical object channel in front of the position, where the camera
is located.
Radiographs are recorded on track-etch film with an automatic
camera. From outside this camera can be brought into position
behind the fuel rods via a rotating table. This table also stops
the beam from the reactor.
At the internal object plane the cross section of the beam is
10×16 cm^2 and the thermal neutron current density amounts
1.4×10^7 $cm^{-2}s^{-1}$. The L/D ratio here is 400.
In order to exploit the possibility to discriminate against fast
and epithermal neutrons a silicon single crystal filter of 25 cm

length and 12 cm diameter can be brought into the beam (see next section).
For a more comprehensive description of the other components of this facility one is referred to (1).

THE FACILITY FOR EXTERNAL NEUTRON RADIOGRAPHY

For external radiography it is intended to design a facility, which possessed the highest possible flexibility. To this end the filters, which are needed to tailor the neutron beam, have been positioned in a space in the middle shielding bloc in front of the internal object area. In principle this area can house four filters, which very easily can be positioned into the beam separately or in pairs.

The Filters

For the time being three filters have been selected:
- A polycrystalline beryllium filter with a length of 30 cm and a diameter of 12 cm. Polycrystalline Be allows passage of subthermal neutrons with an energy below 5 meV (a wavelength above 0.4 nm) (2).
- A monocrystalline bismuth filter of a total length of 25 cm and 12 cm diameter. The Bi-filter serves to limiting the gamma-dose rate in the neutron beam and to let pass simultaneously the subthermal neutrons coming from the Be-filter through a wide energy window (2).
- A monocrystalline silicon filter (mentioned above) with a length of 25 cm and 12 cm diameter. This filter is needed for shaping a pure thermal beam (2,3,4).
The transmission of the desired neutrons through the Be- and Bi-filters can be increased considerably on cooling, because the losses by thermal diffuse scattering are very much reduced.
Two cryogenerators of the closed-cycle expander type with a capacity of 13 W at 50 K are accomodated with their respective filters in separate cryostats. In this way the intended flexibility is maintained, such that a particular neutron beam can be selected within a minute by means of an electro-pneumatic system.
The transmission of such combined filters of Be and Bi is about 30% for the subthermal neutrons, which together form only a fraction of less than 2% of the thermal beam.

The External Flight Tube and Exposure Station

The rotating table in the shielding contains a cylindrical plug, which is removed in a simple way, when external radiographs are to

be made. From the filter region the neutrons continue their flight
via the opening in the shielding through a tube, which connects
the internal with the external exposure station. This aluminium
flight tube is lined with boroflex and is equipped with thin Al
and Mg/Zr end‑ windows. In order to limit neutron absorption and
scattering it is filled with helium gas. The tube construction
consists of two telescoping parts. This enables to position the
external exposure station at 11 m or 9 m from the reactor core. In
the last position transport on the reactor floor is not blocked by
the flight tube.

The exposure station contains a beam stop and is shielded on all
sides, preventing scattered radiation to enter the hall. The
shielding allows easy access to the object area.

The expected subthermal neutron current density with 3.4 cm
diaphragm and gamma-dose rate at the object area is $(1.5-3) \times 10^5$
$cm^{-2}s^{-1}$ and 0.15 mSvh^{-1}, respectively. The diameter of the beam at
11 m distance is 25 cm and the L/D ratio is 270.

For the thermal beam after having passed a combination of silicon
and bismuth filter the values will be 1.7×10^6 $cm^{-2}s^{-1}$ and
0.11 mSvh^{-1}, respectively.

SAFETY OF THE INSTALLATION

In order to avoid that improper operation could lead to damage or
to an intolerable situation, when e.g. a not-filtered beam would
enter the reactor hall, several precautions have been taken. To
prevent such accidents the status of several components of the
facility are monitored by sense and end switches. Any action,
which might result in an undesired situation is suppressed
automatically.

In case of an emergency situation in the reactor hall it is
possible to deblock the pass way by pushing the telescoping flight
tube back manually.

REFERENCES

1. H.P. Leeflang, "Neutron Radiography Installation for Long Fuel
   Rods at the HFR Petten", Neutron Radiography, Proceedings of
   the First World Conference, San Diego, California, USA, 1981.
2. A.K. Freund, "Cross-sections of Materials as Neutron
   Monochromators and Filters", Nuclear Instruments and Methods,
   213, 495 (1983).
3. R.M. Brugger, "A Single Crystal Silicon Thermal Neutron
   Filter", Nuclear Instruments and Methods, 135, 289 (1976).
4. A.K. Freund, H. Friedrich, W. Nistler, R. Scherm, "Neutron
   Transmission Properties of Perfect Silicon Crystals", Nuclear
   Instruments and Methods, A234, 116 (1985).

A CONICALLY DIVERGENT GUIDE TUBE

AS A COLLIMATOR FOR NEUTRON RADIOGRAPHY

E. Steichele and E. Gutsmiedl

Fakultät für Physik, E 21, Technische Universität München

D-8046 Garching

ABSTRACT

Neutron guides are useful collimators for radiography, because they produce pure neutron beams of high intensity. For wavelengths beyond 5 Å, however, the divergency of neutrons from a nickel coated guide tube is too large for high resolution radiography. An essential improvement can be obtained with a conically divergent guide. Due to garland- and zig-zag-reflections in a curved guide differences in the neutron divergency of different wavelength can be equalized.

INTRODUCTION

Intensity and resolution in neutron radiography are basically determined by the collimation of the neutron beam. High resolution needs strong collimation and thus leads to low intensity. For high intensity work the collimation has to be relaxed and then the spatial resolution, especially with thick objects, is poor. An optimum choice of the collimation seems therefore appropriate.

The classical collimator for neutron radiography is a parallel or conically divergent tube with absorbing walls. The collimation is determined by the L/D ratio. L is the length of the collimator, D is its diameter near the neutron source. The far end cross section of the collimator determines the area, which can be radiographed by one setting. Typical L/D ratios are in the range of 50 to 500 (1). It is a main drawback of such a collimator that it cannot separate thermal neutrons from the background of γ-rays

and filters of bismuth e.g. have to be used. These, however, reduce the neutron intensity by a factor 5 - 10. Such a loss of intensity is certainly a disadvantage, especially for time resolved radiography.

A more elegant way to filter out the γ-rays and fast neutron
background is the use of a curved neutron guide tube (2). A neutron radiography station at the end of a guide tube is installed
at the ORPHEE reactor in Saclay (3). The guide tube there has been
originally designed for neutron scattering experiments and does
not offer the optimum conditions for neutron radiography work. In
the present paper we want to show how a neutron guide could be optimized for radiography applications.

THE NEUTRON GUIDE AS A COLLIMATOR

Guide tubes of 30 to 80 m length are nowadays standard components in neutron research work. Neutrons are transmitted by total
reflection at the inner side of usually rectangular tubes (typical
cross section 15 x 3 cm^2). The index of refraction is less than 1
for most materials and total reflection occurs, when the angle
between the neutron's direction and the mirror surface is less
than the critical angle γ_c, given by $\gamma_c = \lambda \sqrt{Na}/\pi$. N is the particle
density of the mirror, a the coherent scattering length and λ the
wavelength of the neutrons. The critical glancing angle increases
with wavelength and is 1×10^{-3} radian/Å for a glass mirror and 1.7
$\times 10^{-3}$ radian/Å for a nickel mirror of natural isotope composition.
The beam divergency at any point of the exit of a straight guide
is $2\gamma_c$ in both horizontal and vertical direction independent of the
dimensions of the guide. With a moderator of T = 300 K and an isotropic flux ϕ at the entrance of a straight guide the current density j at the exit is $j = 1.75 \times 10^{-6} \times \phi$ (n/cm^2 s). With a typical
flux $\phi = 10^{14}$n/cm^2 s the usually needed 10^8n/cm^2 for high resolution neutron radiography (4) are obtained within 1 second and a
picture of an object with cross section area F needs about F/f
seconds exposure time, where f is the beam area of the guide. For
a straight guide looking to a thermal moderator the mean wavelength is about 2 Å and hence the beam divergency about 7×10^{-3}
radian. This estimate is, yet, slightly too optimistic, as it
holds for a straight guide. In order to clean the neutron beam
from γ-background the guide tube has to be curved. The guide is
then a low-pass filter in the sense, that it cuts away the γ-rays
and short wavelength (high energy) neutrons. If the curvature is
matched to the maximum of the moderator spectrum, then the reduction of the total intensity due to the curvature is not more than
a factor 2.

The neutrons of different wavelengths have different diver-
gencies and these can be transformed into L/D ratios as shown in

λ (Å)		1	3	5	10
γ_c (10^{-3} radian)		1.7	5.1	8.5	17
L/D		300	100	60	30

Table 1: Critical glancing angles of a nickel mirror for neutrons of different wavelength and equivalent L/D ratios.

Table 1.

The guide tube collimation for neutrons between 1 and 3 Å corresponds to L/D ratios as normally used for neutron radiography. For long wavelength neutrons between 5 Å and 10 Å the collimation is, however, poor. This is the more a drawback as the decrease of the scattering cross-section at the Bragg cut-off of many materials favours radiography of thick objects like motors etc. with long wavelength neutrons (4). Therefore the standard neutron guide is not an ideal collimator for long wavelength radiography. In principle the situation could be improved by using a material with a smaller critical glancing angle. A more efficient way is, however, a beam transformation as shown in the following section.

THE CONICALLY DIVERGENT GUIDE

The mirror reflection of neutrons is governed by Liouville's theorem, which states, that phase space density remains constant under the action of conservative forces. Applied to the optical problem of transmission through a conically divergent guide it means, that the product of neutron solid angle and guide tube area is a constant along the guide. For a straight divergent system it is evident from fig.1 how the angle of the neutrons relative to

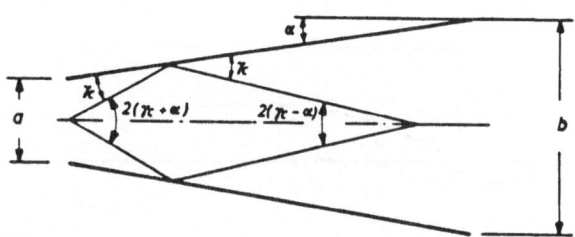

Figure 1: Reduction of divergency by reflection at a divergent mirror system

the system axis decreases by 2α after each reflection, where α is the inclination angle of the mirror plane to the center line. The angle α is normally much smaller than the critical glancing angle and then the neutrons are homogeneously distributed across the exit area after many reflections. The angle of acceptance at the beginning of the guide is now $2(\gamma_c + \alpha)$, which is marginally larger than with a parallel system. If the mirror inclinations for the horizontal and vertical planes were different, then the beam divergency at the exit would be also different in the two directions. Though the number of neutrons per cm^2 is less now at the guide tube end, there is no increase of exposure time as long as the area of the object is still larger than the now increased exit area of the guide. One thus creates an ideal situation for neutron radiography: One starts with a small guide area near the reactor, what means only a small disturbance of the core's reactivity, and ends with a highly collimated, large area beam at the radiography station outside the reactor.

The straight conical guide does, however, not separate the γ-ray background from the thermal neutrons. The guide has to be curved as shown in fig.2. A curved guide has a "characteristic" wavelength, for which the critical angle is equal to the glancing angle for neutrons on path AB in fig.2. For a conical guide of width a at the entrance and width b at the exit the following relations hold for the "length of direct sight" L_1, the radius R of curvature and the critical angle γ*for the characteristic wavelength λ:

$$\gamma^* = \sqrt{(a + b)/R} \qquad\qquad L_1 = \sqrt{4(a + b)\,R}$$

The length L_1 for a guide with parallel walls is $\sqrt{8aR}$ and it is quite obvious, that a conical guide has to be made longer than a parallel one to suppress the background radiation.

For a straight conical guide one expects the same reduction of divergency by a factor b/a for all wavelengths. This is different for a curved guide. Besides the "zig-zag"-reflections as in a straight guide one has also "garland"-reflections, where neutrons

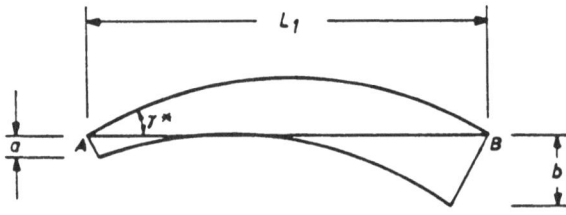

Figure 2: Conically divergent, curved guide tube, L_1 "length of direct sight".

are reflected only at the outer wall. For those neutrons the angle of incidence remains constant and there is no divergency transformation as discussed above. Roughly speaking, the neutrons with wavelengths longer than λ^* are mostly zig-zag-reflected (with divergency transformation), whereas the short wavelength neutrons are mostly garland-reflected (with no divergency transformation). One therefore can expect, that by transmission through a curved, conically divergent guide the differences in the divergency of neutrons of different wavelengths are reduced.

These ideas have been checked by Monte-Carlo simulations. Fig.3 shows the results obtained for the distribution of the neutrons at the exit as a function of the z-coordinate across the guide and the angle relative to the guide axis at the end. The guide tube has a length of 80 m, is designed for a characteristic

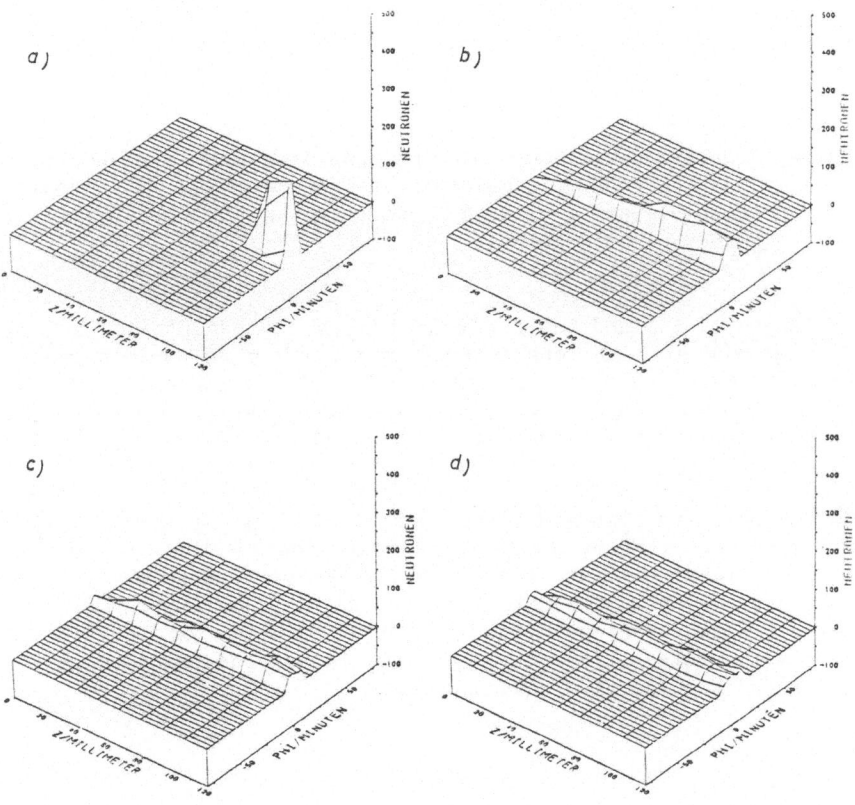

Figure 3: Angular (PHI) and spatial (Z) distribution at the exit of a conically guide for neutrons of different wavelength a) 1 Å, b) 3 Å, c) 5 Å, d) 7 Å

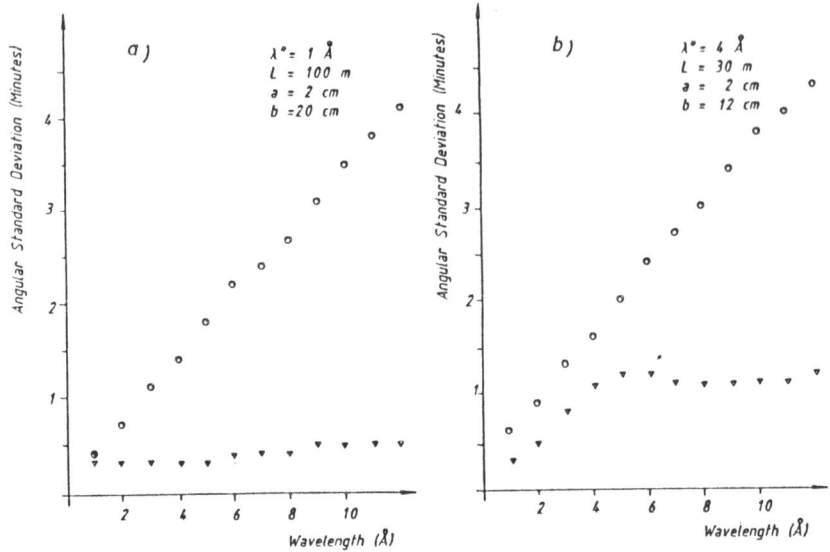

Fig. 4: Standard deviation of the angular distribution of neutrons
as a function of wavelength at the end of a parallel guide
(circles) and a divergent guide (triangles). a) Thermal
guide, b) Cold guide.

wavelength of 1 Å and has a width of 2 cm at the entrance and of
12 cm at the exit. As characteristic for a curved neutron guide,
the short wavelength neutrons at 1 Å are mainly concentrated near
the outer wall (fig. 3a), whereas the longer wavelength neutrons
are more or less homogeneously distributed across the guide (fig.
3b, 3c, 3d).

Fig. 4 shows the standard deviations of the angular distributions
as a function of neutron wavelength after passage through a parallel guide (open circles) and a conically divergent guide (triangles). Fig. 4a shows the data for a guide of 100 m length and 1 Å
characteristic wavelength and fig. 4b for a guide of 30 m length
and 4 Å characteristic wavelength, the latter one being representative for a guide at a cold source.

CONCLUSIONS

The Monte-Carlo simulations clearly demonstrate, that by passage
through a conically divergent guide tube highly collimated neutron
beams can be obtained, which have no contributions of γ-radiation.
Such beams are ideal for neutron radiography work and are superior
to beams from conventional neutron guides. A guide for thermal

156

neutrons could be as long as 100 meters, a guide at a cold source would have a length of 30 - 50 meters depending on the "opening ratio" b/a. A neutron guide is certainly more complicated and more expensive than a conventional collimator. But it also makes available a large experimental area outside the reactor with easy access. This may be an important point of view in times, when neutron radiography becomes more and more attractive and safety regulations at reactors become more and more rigorous.

REFERENCES

1. P. von der Hardt, H. Röttger, Neutron Radiography Handbook, D. Reidel Publishing Company, Dordrecht, Holland 1981.

2. H. Maier-Leibnitz, "Neutron Conducting Tubes", p. 93 in Neutron Capture Gamma-Ray Spectroscopy, International Atomic Energy Agency, Vienna, Austria 1969.

3. M. Zaccheo, J. Berthon, G. Uzureau and A. Laporte, "Use of a Neutron Guide in Industrial Neutron Radiography", p. 99 in Neutron Radiography, J. P. Barton and P. von der Hardt, Editors, D. Reidel Publishing Company, Dordrecht, Holland 1983.

4. M. R. Hawhesworth and J. Walker, "Basic Principles of Thermal Neutron Radiography", p. 5 in Neutron Radiography, J. P. Barton and P. von der Hardt, Editors, D. Reidel Publishing Company, Dordrecht, Holland 1983.

PART III

SMALL SOURCE SYSTEMS

RECENT CUSTOM DESIGNED SYSTEMS COMPARED

John P. Barton
N-Ray Engineering Company
5709 Waverley Avenue
La Jolla, CA   92037

ABSTRACT

Comparisons between nine recent neutron radiography system designs will be provided. N-Ray Engineering Company has been under contract to assist in the planning and acquisition process for each of these systems during the period 1983-1986. They include (a) a stationary, high sensitivity, high throughput system (reactor); (b) two stationary, high resolution systems (isotope); and (c) four movable, low resolution systems (accelerator and isotope). The nine systems feature in total thirteen neutron beams of which nine are electronic imaging, and four film imaging. Robotic manipulators for source, imager, or object, and advanced digital image analysis systems are also common to most designs. The source selection process included studies of alternative reactor types, sub-critical multipliers, cyclotrons, van de Graaffs deuteron accelerators and isotopic sources. Much progress has depended on the conviction that (1) if the initial aviation industry demonstrations can be successfully implemented the eventual industry wide implications could be significant; (2) high contrast, low resolution images (film or electronic) can be obtained with exposures one thousand times less than exposures necessary for high resolution imaging; (3) high resolutiion, reactor quality images can be obtained with sources one thousand times less intense than a reactor, provided the required daily throughput is not high.

INTRODUCTION

An institution requiring an in-house x-ray system can normally chose from equipment manufactured in series and

161

available "off the shelf". Acquisition of an in-house n-ray system, on the other hand, requires a different process. Typically a specialist is required to analyse the needs, select from the currently available technologies, and write specifications so that companies can compete for the custom design and manufacture of a turnkey system on a fixed price basis. Prior to writing the specifications it is necessary to perform a conceptual design and, in this, the consideration of neutron source is essential. The variety of sources is large, and the situation is constantly changing. Types to be considered include various isotopic sources, small accelerators, cyclotrons, sub-critical multipliers, and reactors. This company has participated recently in the conceptual design, specification, and procurement review of neutron radiography systems to meet the needs of several centers. The objective of the paper is to briefly survey some system designs, compare, and draw some deductions of general interest. The review will best meet its purpose if presented in general terms; details of potentially proprietary material will be omitted.

Type 1: Stationary, High Contrast, High Throughput (S, HC)

The requirement is inspection of detached aircraft panels to identify low level aluminum corrosion. The sensitivity requirement is down to 0.002 inches of corrosion product, which is achievable by the use of sophisticated electronic imaging and digital image processing; including field flattening, contrast stretching, frame averaging, and x-ray/n-ray comparison. The throughput requirement is a minimum of 500 square feet of aircraft panel per day. The parts, which include steeply curved items a few feet in dimension, and large items such as wings, 36 feet long, must have special positioners and parts handling equipment.

The neutron source selected, after careful consideration of all types, is a TRIGA reactor. Power will be 250 kW, upgradeable to 1 MW. Three complete beam systems will operate simultaneously, two shifts per day, for this application. Each beam system will consist of beam collimator (L/D=60/1) shutters, component positioners (programmable), neutron image intensifiers, and image interpretation system. The systems will be capable of continuous scan in time real-time (30 frames per second).

Type 2: Stationary, High Resolution, High Throughput (S,HR)

The requirement is inspection of aircraft parts including reusable pyrotechnics for detection of small cracks. Resolution must be equal to that given by a 100:1 collimator ratio, using

Kodak single R fine grain film held in vacuum cassettes against thin, flat gadolinium. Throughput requires an exposure time of 10 minutes or less for each standard size (14x17 inch) film.

This capability will be provided on the TRIGA reactor, described above, by extraction of a fourth beam. Because the reactor must support four beams simultaneously, it cannot be leaded to one side, such as at the 250 kW Aerotest TRIGA reactor. The mean flux at the input end of the collimator is calculated to be $3x10^{11}$ n/cm^2-s. The collimated flux corresponding to this at 100:1 collimation is $3x10^6$ n/cm^2-s.

Type 3: Stationary, High Resolution, Low Throughput (S,HR)

The requirement is for an in-house neutron radiography system that can provide "reactor quality" radiographs (i.e. approximately equal in resolution to the best US commercial service neutron radiograph). Throughput required varies up to 16 per day but is usually between 4 and 9 per day. The system should be low cost, reliable, and simple to operate.

The source selected is an isotopic source, californium-252, of strength in the range 100 - 150 mg. This isotope is produced as a by product of research programs in the USA, and is normally stored unused until the californium-252 has decayed to curium. It is available on loan to US government centers for the cost of encapsulation and shipment. The key to utilization of this source is the fact that the throughput site of up to 16 films per day can be arranged, using the divergent collimator, in a single exposure array (i.e. 2x2, 3x3 or 4x4). The high source reliability make long exposures acceptable (24 hours or more). Thus using a collimator ratio of 70:1 and single R film with gadolinium, a source of 100 mg will require an exposure time of under 24 hours. Collimator ratios of 100:1 can be used by doubling either the source strength or exposure time.

Type 4: Stationary, High Contrast, Low Throughput (S,HC)

The requirement is for a rapid, electronic imaging check to determine the correct orientation of objects using the source described under Type 3 above.

This is simply provided by a second beam position with collimator ratio reduced to 24:1. The neutron flux from a 100 mg californium source is $2x10^5$ n/cm^2-s at the object so that near real-time electronic imaging with exposure times in the range of 25 seconds is possible.

Type 5: Maneuverable, High Contrast, Low Throughput (M,HC)

The requirement is inspection of intact aircraft for detection of corrosion in wings and tail using electronic imaging. The inspection capability must be available in all weather, and will therefore be housed in a special building.

The source selected is 50 mg of californium-252. The source does not have to be switchable on and off, as for a transportable system. Advantages over the on/off KAMAN-711 accelerator include (1) a useful thermal neutron intensity that is approximately five times greater than the accelerator (2) a lighter source head for the robotic positioner due to the compact moderation possible (3) a lesser shield requirement for a given thermal neutron flux due to the thermalisation rates and the avoidance of 14 MeV neutrons.

Type 6: Maneuverable High Contrast, Low Throughput-Film
(M,HC)

This requirement is like Type 5 above except that film imaging must be used because the access is not available for the larger electronic imaging devices. (Certain areas of aircraft fuselage).

The source selected is an additional 50 mg of californium-252. The two systems (5 and 6) will be operable simultaneously provided precautions are taken to preload film and limit interference. The film-screen combination will be high speed (low resolution) requiring fluences of 10^8 n/cm^2-s and exposure times of 20 minutes or less.

Type 7: Maneuverable, High Contrast, Minimum Cost (M,HC)

The requirement is inspection for aluminum corrosion in cylindrical or conical shaped metal objects. The object itself is to be rotatable such that all parts can be scanned. The neutron source is to be maneuverable on the outside of the object with the electronic imager on the inside. For reasons concerning licensing and uncertainty about loan availability the use of californium-252 is excluded. The selected source is a deuteron-tritium source (KAMAN-711) with an optional small uranium booster. At a collimator ratio of 24:1 this system requires exposure times of about 30 seconds for near real-time imaging.

Type 8: Transportable, High Contrast, Minimum Cost (T,HC)

The requirement is for inspection of intact aircraft for corrosion with a system that can be switched off and transported to other locations as needed. It is to operate outdoors using a defined perimeter to ensure radiation doses do not reach significant levels. Imaging is both by electronic and film methods (1).

The source selected is the deuteron-tritium type, which, incidentally, uses a uranium booster.

Type 9: Transportable, High Contrast, Increased Capability

The interest is for a transportable neutron radiography system for aircraft inspection, comparable to the requirements for Type 8 above, but with significantly increased neutron yield to provide improved capability. A project was initiated in 1983 to pursue a so-called "advanced accelerator for neutron radiography". A detailed study undertaken by another company under US government funding considered in detail first a Cockcroft-Walton type accelerator, then a Van de Graaff Type 1.

At the time of writing this situation remains open. No single source has been identified as the best choice. Ideally the accelerator should provide a thermal neutron flux five or ten times greater than the KAMAN-711, yet be light weight at the target head and reliable. Accelerators to be considered in addition to the electrostatic and Van de Graaff types are the small linear accelerator (MINAC), a light weight cyclotron, or a compact radio frequency quadrupole (RFQ) source.

COMPARISONS

Custom designed neutron radiography systems that have reached the conceptual design stage in the USA include the nine types discussed above, grouped into three system categories: stationary, maneuverable, and transportable. Source types considered include $D-T_2$ Cf-252, D-Be, cyclotron, sub-critical resolution (2×10^9 n/cm^2-s at 100:1) to low resolution (10^5 n/cm^2-s at 30:1).

For stationary systems where exposures of 24 hours or more are acceptable the isotopic source has received particular interest because of low capital and operating cost. For maneuverable systems both isotopic and accelerator sources have received acceptance in different circumstances. The isotopic source cannot be switched off, but does offer significantly higher neutron yields than proven light-weight alternatives. For transportable (or mobile) systems the D-T accelerator for which the KAMAN-711 is dominant in the USA, is the only type planned for use at this time.

SUMMARY

Of the nine types of system described, the planned Type 1 system will have three similar Type beams, and there will be two systems containing Type 3 and Type 4 beams. Thus, in total, plans for thirteen neutron radiographic beam-image combinations have been surveyed. Of these, nine are planned primarily for electronic imaging and four for film imaging.

If high resolution, high constant and high throughput are required a small nuclear reactor appears the most favored source candidate.

However, for corrosion inspection where high resolution is not required, neutron image fluences of 10^6 n/cm^2-s become acceptable instead of 10^9 n/cm^2-s. This means that available accelerator or isotopic sources can be useful to a limited extent.

Moreover, for high resolution neutron radiography where the required throughput is low, exposure times of 10^5 seconds (24 hours) may be used instead of 100 seconds typical for a 250 kW reactor beam. Thus an isotopic source such as 200 mg californium could produce typically 16 radiographs per day of the same quality as that provided by a reactor.

Areas in which particular interest exists include alternatives to californium-252 for centers that cannot benefit from the US loan program and alternatives to the KAMAN-711 accelerator for situations where higher neutron yield is required from an on/off source.

REFERENCES

1. Science Applications International Corporation, San Diego (unpublished work, 1986).

COMPARISON OF PROPOSED SYSTEMS

Type Number	Application	Source	Coll. ratio	Neu. Flux	Imaging Method	Fluence used	Exposure time
1 Stationary (S, HC)	Honeycomb Panels Removed from Aircraft	Reactor 250 kW	60:1	10^7	Electronic	10^6	10 fps
2 Stationary (S, HR)	Pyrotechnics etc.	Reactor 250 kW	100:1	3.10^6	SR film gadolinium	2×10^9	10 min
3 Stationary (S, HR)	In-house items	Calif 250 kW	70:1	2×10^4	SR film	2×10^9	24 hrs
4 Stationary (S, HC)	Orientation of Parts	Calif 100 mg	24:1	2×10^5	Electronic	5×10^6	25 sec
5 Maneuverable (M, HC)	Intact aircraft scan for corrosion	Calif 50 mg	24:1	10^5	Electronic	5×10^6	50 sec
6 Maneuverable (M, HC)	Intact aircraft film radiography	Calif 50 mg	24:1	10^5	Fast Film scintillator	10^6	20 min
7 Maneuverable (M, HC)	Space program parts rotated thru N & X-ray	D-T-U KAMAN-711	24:1	3×10^4	Electronic	10^6	30 sec
8 Transportable (T, HC)	Intact aircraft scan for corrosion	D-T-U KAMAN-711	24:1	3×10^4	Electronic	10^6	30 sec
9 Transportable (T, HC)	Aircraft	D-Be Advanced	-	3×10^5	Electronic	10^6	3 sec

SOURCE PHOTONEUTRONIQUE POUR LA RADIOGRAPHIE AUX NEUTRONS AUPRES D'UN MICROTRON.

Č. Š i m á n ě

Ecole Polytechnique

P r a g u e, ČSSR

R é s u m é .

On décrit un dispositif pour la radiographie aux neutrons utilisant des neutrons de la photofission de l'uranium naturel et installé auprès d'un microtron circulaire de 23 MeV. Un cône double est utilisé pour la collimation avec un rapport L/D = 30. Les problèmes posés par les photons de grande énergie capables do provoquer les réactions photonucléaires dans les détecteurs de neutrons sont traités. Deux neutronogrammes présentés ont été obtenus avec du dysprosium comme détecteur par la méthode indirecte pendant une exposition de 2 heures.

Différents types de générateurs à la base d'accélérateurs des particules sont utilisés actuellement pour la radiographie aux neutrons. Nous avons essayé de faire la radiographie avec les neutrons de la fission de 1 uranium naturel, provoquée par les photons de freinage des électrons accélérés dans un microtron circulaire à 23 MeV du type développé à l'Institut Uni de la Recherche Nucléaire a Dubna, URSS, et à l'Ecole Polytechnique de Prague, ČSSR /1,2/. Les expériences ont été exécutées auprès le microtron à Dubna.

Comme la cible pour la génération simultanée des photons et des neutrons de fission, on s'est servi d'un cylindre en uranium naturel de 25 mm de diamètre et 25 mm long, refroidi par l'eau avec un rendement estimé à environ 5.10^{16} neutrons/μA du courant cible des électrons. Le conduit des électrons avec la cible au bout entrait au centre d'un bloque moderateur par un

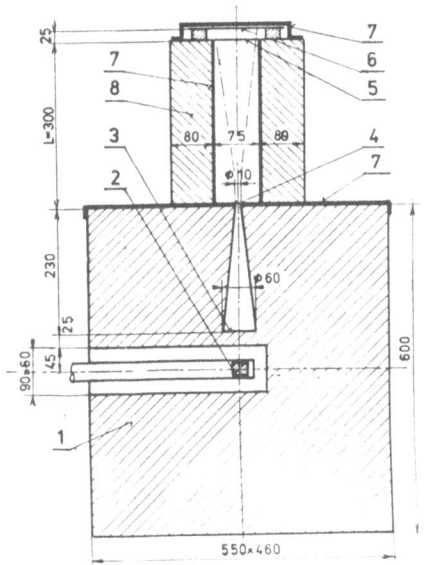

Fig. 1 - Dispositif neutronographique: 1-bloque modérateur
en polyéthylène, 2-cible en uranium, 3-surface radiateur, 4-diaphragme, 5-support pour les objects
à radiographier, 6-plan de détection avec une feuille en dysprosium, 7-tole de blindage en cadmium,
8-écran en polyéthylène, contenant B_2O_3.

canal latéral.

Le bloque modérateur /fig.1/ était assemblé des briques de
polyéthylène, dont les dimensions fixaient ses côtés. Pour des
raisons amenées plus tard, on a préféré l'orientation verticale
de l'axe du canal neutronographique, perpendiculaire par rapport
au plan d'accélération du microtron et aux plans de déflection
des dipoles magnétiques du conduit des électrons. On a choisi
le collimateur des neutrons sous forme d'un cône double, proté-
geant mieux la partie centrale du modérateur contre les fuites
des neutrons. L'inconvénient de cette solution consiste dans le
fait, que toute l'inhomogénéité de la distribution du flux de
neutrons sur la surface radiateur 3 se projète par le diaphragme
4 - un trou de diamètre D = 10 mm dans la tôle de cadmium de blindage - sur le plan 6 de détection des neutrons. Il est donc pré-
férable de choisir la distance entre la surface 3 et le diaphragme 4 si grande seulement, que son accroissement progresif n'a-

mène plus à l'une augmentation significative du flux de neutrons
autour du centre du bloque modérateur, éventuellement d'utiliser une forme plus étudiée de la surface radiateur. Dans nos
essais, portant un caractère préliminaire, nous n'avons pas procédé à ce genre d'études. L'épaisseur de la couche du modérateur
entre le canal cible et la surface radiateur 3 - du diamètre de
60 mm - a été prise égale à 25 mm. Le rapport de collimation
L/D = 30 était choisi pour les premières expériences. La densité du flux des neutrons sur le plan 6 est donnée par la même expression que pour un cône simple, $L = \emptyset D^2/16L^2$.

Le côté supérieur du bloque modérateur ainsi que la surface intérieure de la partie supérieure du collimateur ont été
couverts par une tôle de cadmium d'épaisseur de 0,5 mm. Un écran
supplémentaire 8 en briques de polyéthylène, contenant B_2O_3, pro-
tégait la partie supérieure du collimateur en plus.

Un problème spécial posent les photons capables de provoquer
des réactions photonucléaires dans le détecteur, dont le résultat
sont les activités parasites par rapport à celles dues aux neutrons. Il s'agit des photons créés dans les parois internes des
parties diverses du microtron et du conduit des électrons au cours
de freinage des électrons perdus du faisceau, ces photons pouvant
être diffusés par l'effet Compton vers le détecteur.

L'angle de diffusion Θ , qui est aussi approximativement é-
gal à l'angle entre la direction d'incidence de l'électron éme-
tteur du photon primaire et la direction du photon diffusé, les
énergies E_1 du photon primaire et E du photon diffusé, sont
liées par la relation

$$\cos \Theta = 1 - \frac{1 - E/E_1}{E} \qquad \qquad 1/$$

représentée sur la fig. 2. En substituant pour E_1 la limite supérieure du spectre du rayonnement de freinage, égale à l'énergie
des électrons, et pour E l'énergie E du seuil de la réaction
photonucléaire donnée, on obtient à l'aide de la formule 1/ l'angle Θ_0 , au delà duquel les photons diffusés ne sont plus capables de provoquer cette réaction. Si on place alors le détecteur
sur le plan 6 hors de tous les cônes ayant le demiangle au sommet
égal à Θ_0 et partant des points de création potentielle du rayonnement de freinage, on n'a pas besoin du tout d'une protection
contre les gamma de la première diffusion Compton. Les gamma de la
diffusion multiple, même s'ils possèdent l'énergie au dessus du
seuil, étant moins nombreux sont aussi moins graves. Ce raison-
nement nous a mené à l'orientation verticale du collimateur des
neutrons.

Les matériaux utilisés comme détecteurs de neutrons, présentent une sensibilité relative au rayonnement gamma par rap-

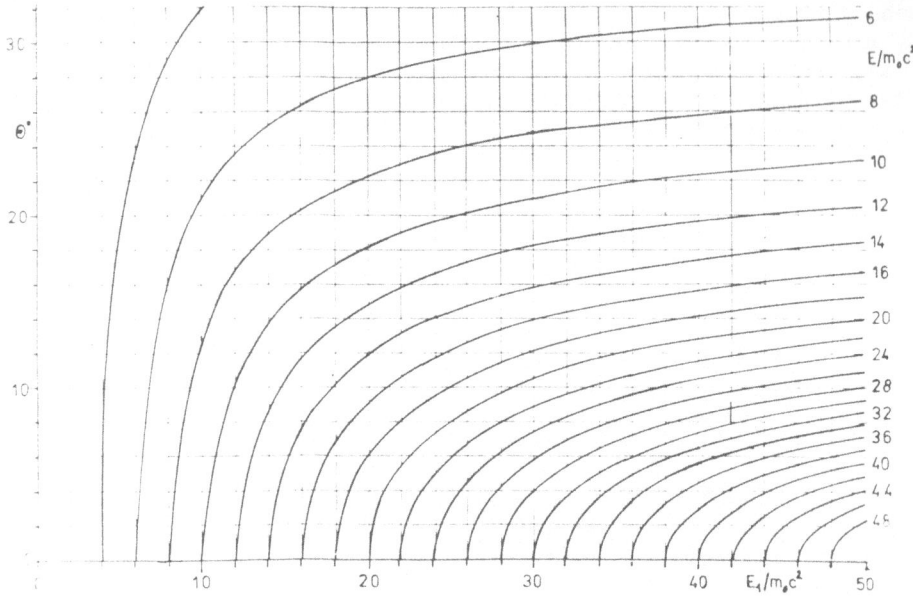

Fig. 2.

port aux neutrons, variable dans grandes limites. Désignons par
T_n la demipériode de désintégration du produit principal de la
réaction /n,γ/ et par N_n et N_γ les nombres intégrals des désin-
tégrations des produits des réactions /n,γ/ et des réactions
photonucléaires pendant l'interval du temps T_n suivant immédiatement après l'irradiation du détecteur, aussi d'une durée de T_n.
On peut alors définir la sensibilité relative comme le rapport
des fluences des neutrons et des photons résultant en $N_n = N_\gamma$.
Cette définition ne respecte pas les différences dans le caractère spectral des rayonnements des produit radioactifs formés
au cours d'irradiation, qui influencent l'efficacité de leur détection, néanmoins elle peut etre utile pour le choix initial
des détecteurs.

Les différences dans les valeurs de la sensibilité relative
des détecteurs individuels sont dues avant tout à des différences
dans les valeurs des sections efficaces pour les neutrons - les
sections efficaces des réactions photonucléaires variant dans les
limites beaucoup plus étroites - à des différences dans les demi-
périodes de désintégration des radioisotopes formés et à des compositions isotopiques des élements utilisés pour la détection.
On a évalué la sensibilité relative pour les cas du dysprosium
et de l'indium en considérant les réactions suivantes:

Réactions /n,γ/: ^{164}Dy/n,γ/^{165}Dy/139m/

^{115}In/n,γ/^{116}In/54m/

172

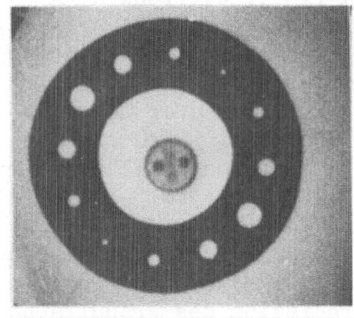 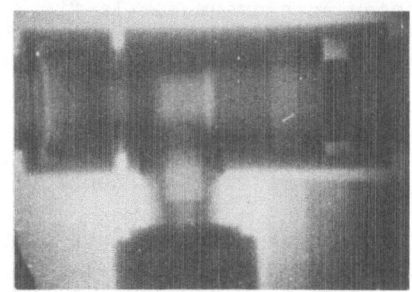

Fig. 3   Fig. 4

Réactions $/\gamma,n/$: $^{156}Dy/\gamma,n/^{155}Dy/10,2h/$

$\qquad\qquad\qquad\quad ^{158}Dy/\gamma,n/^{157}Dy/8,1h/$

$\qquad\qquad\quad ^{113}In/\gamma,n/^{112}In/14m/$ resp. $^{112}In\ /20,7m/$

La sensibilité relative est de l'ordre de 10^{-6} pour le cas dy dysprosium et 10^{-4} dans le cas de l'indium. Les sections efficaces intégrales ont été prises pour le spectre de freinage, ce qui n'est pas tout à fait correct, mais ce qui permet de se faire une idée sur le degré de la nécessité de protéger le détecteur contre les gamma. En choisissant l'emplacement convenable du détecteur hors de flux de photons de la première diffusion Compton et en ajoutant éventuellement un écran supplémentaire, on peut en principe arriver à une suppréssion presque complète des réactions photonucléaires dans les détecteurs.

Naturellement, il faut se limiter à la méthode indirecte de la radiographie aux neutrons, le flux intense de gamma de basse énergie autour du microtron empêchant l'utilisation du matériel photographique au cours d'irradiation.

Nous avons travaillé avec une feuille de dysprosium de 0,1 mm d'épaisseur. Petits échantillons de cette feuille servaient pour mesures directes par activation du flux de neutrons dans le plan 6 du collimateur. Le rapport des activités avec le diaphragme 4 bouché par cadmium et ouvert se montrait égal à 0,015, ce qui prouve, qu'il n'y avait presque pas de neutrons parasitiques. Le rapport de cadmium pour le dysprosium, déterminé par irradiations des échantillons nus et couverts par cadmium, était égal à 0,025.

Les figures 3 et 4 représentent des neutronogrammmes pris

avec un temps d'exposition de 2 heures. L'énergie des électrons
était 20,25 MeV, le courant cible 10 à 11 μA. Les trous dans
l'anneau de cadmium sur la fig.3 sont de diamètre de 1 à 4 mm,
au centre se trouve un transistor. La distance entre les objects
et la feuille en dysprosium était 25 mm environ. Sur la fig. 4
on peut voir la partie d'une armature en laiton avec des éléments
en matière organique. Les feuilles de polvéthylène en bas ont
des épaissers progressives de 0,3 à 1,2mm. Le matériel photographique utilisé était un Kodak D7.

Nous nous ne sommes pas occupés au cours de nos essais, qui
ne portaient pas un caractère de travail de série, du problème
de protection du personnel contre la radiation des produits de
fission dans 1 uranium irradié, qui peut être résolu de plusieures
manières.

Conclusion.

On est arrivé avec un dispositif simple à faire la radiographie aux neutrons auprès d'un microtron avec un rapport de
collimation L/D = 30. On peut augmenter ce rapport ainsi que les
dimensions du champ en exploitant des réserves existantes. En
travaillant aux paramètres nominaux de 23 MeV avec le courant
cible de 20 μA, on gagne un facteur à peu près trois dans le flux
de neutrons. On peut aussi réduire des pertes des neutrons en
diminuant la section du canal de rentrée de la cible, dictée dans
notre cas par les dimensions transversales d'une cible éxistante.
Par une étude plus approfondie de la forme de la surface radiateur
3, de sa position relative à la source de neutrons et de sa distance du diaphragme, on pourra aussi augmenter le flux. Ainsi
un facteur considérable de l'accroissement du flux peut être ex-
pecté pour un dispositif futur.

De toute façon, le coût modéré d'un microtron circulaire du
type utilisé, avec la possibilité de son exploitation aussi pour
la radiographie aux gamma, fait une installation neutronographique du type décrit dans cet article très attractive.

A la fin, 1 auteur aimerait remercier M.Flerov G.N., membre
de l'Academie des Sciences de l'Union Sovietique et directeur
du laboratoire des réactions nucléaires de l'IURN à Dubna de son
interêt et le support de ce travail, ainsi que MM.Belov et Tete-
rev du personnel du microtron de leur assistance.

Références:

1. Č.Šimáně et al.: Jaderná energie 27/1981/421, 28/1982/14.
2. A.G.Belov et al.:Mikrotron MT-22, Communications of the Joint
                    Inst.for Nuclear Research 89-82-301,
                    Dubna 1982.

LES TUBES NEUTRONIQUES SCELLES
ET LA NEUTRONOGRAPHIE MOBILE

Michel DUBOUCHET, Pierre BACH, Serge CLUZEAU.

SODERN : 1 AVENUE DESCARTES

BP 23 LIMEIL-BREVANNES 94451 CEDEX

Après avoir rappelé quelques ordres de grandeur
concernant le flux nécessaire pour faire de la
neutronographie et comparer ce flux avec celui émis
par un tube neutronique scellé, on décrit le tube
neutronique SODERN TN 46, et on analyse sommairement
son fonctionnement. On donne ensuite quelques
indications sur le projet de développement d'un
dispositif intégré et automatique de neutronographie
(DIANE), projet conçu par la Société SODERN, en
collaboration avec d'autres Sociétés européennes.

Le haut flux délivré par les réacteurs nucléaires permet d'obtenir des clichés neutronographiques d'excellente qualité, mais les contraintes spécifiques limitent considérablement le domaine d'application de la méthode.

L'avenir industriel de la neutronographie suppose l'existence d'un équipement transportable, de mise en oeuvre rapide, sans problèmes importants de radio protection et de maintenance, et dont les coûts d'investissement initiaux et de maintien en condition restent mineurs eu égard au service rendu.

Pour évaluer cet objectif, il convient de se poser deux questions :

- technique : quelle valeur de flux neutronique peut-on atteindre avec une source collimatée portable et y a t'il compatibilité entre cette valeur et les performances nécessaires pour qu'un système d'imagerie par neutrons présente un intérêt industriel?

- commerciale : quelles configurations faut-il donner au système pour qu'il offre un débouché commercial suffisant?

Ce sont des éléments de réponse à ces deux questions que le présent article se propose de présenter.

ELEMENTS DE DIMENSIONNEMENT

Le flux de neutrons nécessaire pour obtenir une image neutronographique dépend de la sensibilité du détecteur utilisé (figure 1).
En associant film et écran de gadolinium, la dose nécessaire pour obtenir une densité optique de 1 est de l'ordre de 10^8 neutrons par cm^2, cette combinaison est bien adaptée aux réacteurs nucléaires.
En utilisant un scintillateur on peut se limiter, toujours pour la même densité optique, à quelques 10^5 neutrons par cm^2; enfin si l'on fait appel à une détection électronique avec tube intensificateur, un flux intégré de l'ordre de 10^4 neutrons par cm^2 est suffisant.

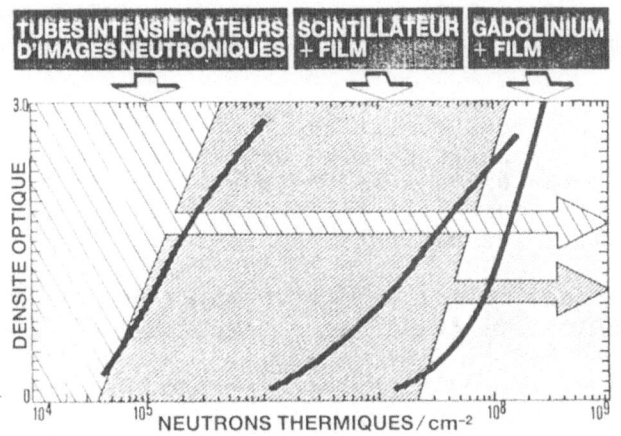

Fig 1 : Sensibilité des détecteurs

Le flux de neutrons thermiques est émis par la tête d'analyse comprenant une source de neutrons rapides, un thermaliseur et un collimateur. Le schéma d'un tel dispositif est donné par la Fig (2).

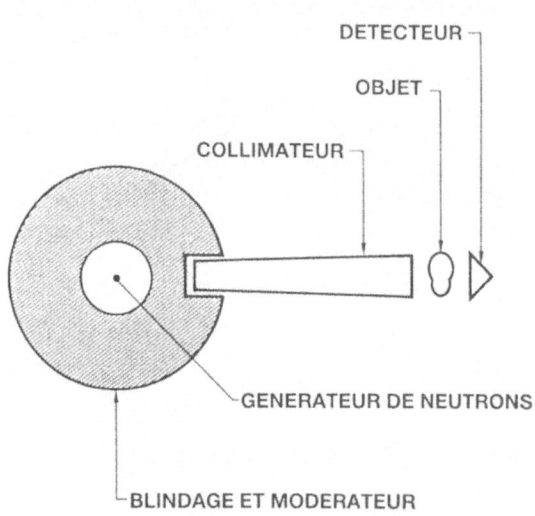

Figure 2 : Principe de la neutronographie

Un rapide calcul d'ordre de grandeur montre qu'il faut faire appel à un flux initial d'environ 10^{12} neutrons rapides émis par seconde dans 4 T pour obtenir un temps de pose industriellement admissible. En effet :

- la figure 3 montre que le flux de neutrons thermalisés, à environ 50 mm de la source rapide, varie de 10^9 à 4.10^9 n/cm^2-s selon la position relative cible-collimateur et l'utilisation ou non d'un multiplicateur de neutrons en uranium.

- le coefficient de transmission du collimateur est proportionnel au carré du rapport de collimation, comme le montre la figure 4.

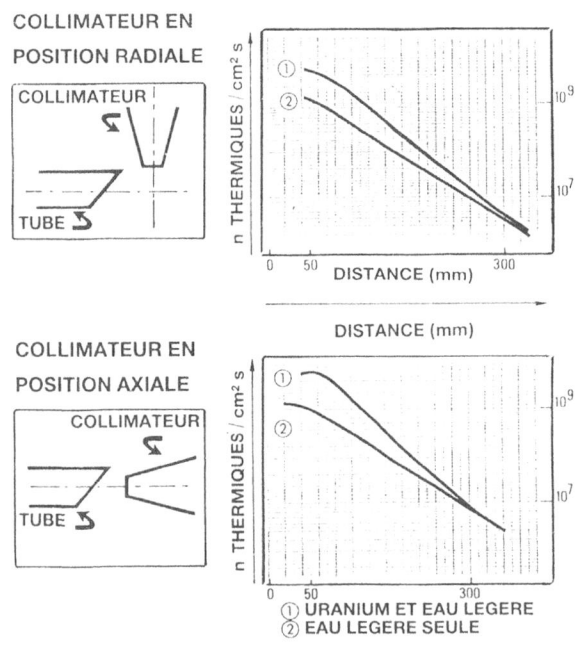

COLLIMATEUR EN POSITION RADIALE

COLLIMATEUR EN POSITION AXIALE

① URANIUM ET EAU LEGERE
② EAU LEGERE SEULE

Figure 3 : Thermalisation des neutrons

Il en résulte qu'à partir d'un générateur de neutrons rapides de 10^{12} n/s, l'objet reçoit un flux de l'ordre de 6.10^5 n/cm^2-s.

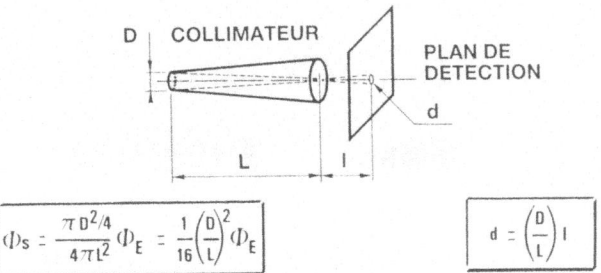

$$\Phi_s = \frac{\pi D^2/4}{4\pi l^2}\,\Phi_E = \frac{1}{16}\left(\frac{D}{L}\right)^2 \Phi_E$$

$$d = \left(\frac{D}{L}\right) l$$

AVEC 10^{12} NEUTRONS DE 14 MeV PAR SECONDE DANS 4π sr

• IL EST POSSIBLE DE CONCEVOIR UN MODERATEUR

   • DELIVRANT: $\Phi_E = 4.10^9$ NEUTRONS THERMIQUES$/$cm^2 s.

   • AVEC D/L = 1/20

   • SUR L'OBJET : $\Phi_s = 6.10^5$ n. cm^{-2}. s^{-1}

Figure 4 : Collimation du flux de neutrons

En se reportant au schéma de la figure 2, on peut en conclure qu'un dispositif de neutronographie dont les performances sont du même ordre de grandeur que les données présentées ci-dessus, permet d'obtenir des images avec des temps de pose allant de quelques secondes à quelques minutes selon la nature du détecteur.

LE TUBE TN 46
DESCRIPTION, FONCTIONNEMENT SOMMAIRE

Description :
Le tube TN 46 est constitué de 3 parties principales (figure 5) :

- la source d'ions,
- le corps du tube,
- le boîtier de cible.

La source d'ions est montée sur un manchon isolant qui constitue l'interface avec le connecteur THT.
Le corps du tube, métallique, constitue l'enveloppe extérieure à laquelle sont raccordés :

- d'une part, le manchon isolant de la source d'ions, qui délimite l'espace d'accélération des faisceaux issus de la source d'ions,
- d'autre part, le boîtier de cible qui ferme le volume du tube.

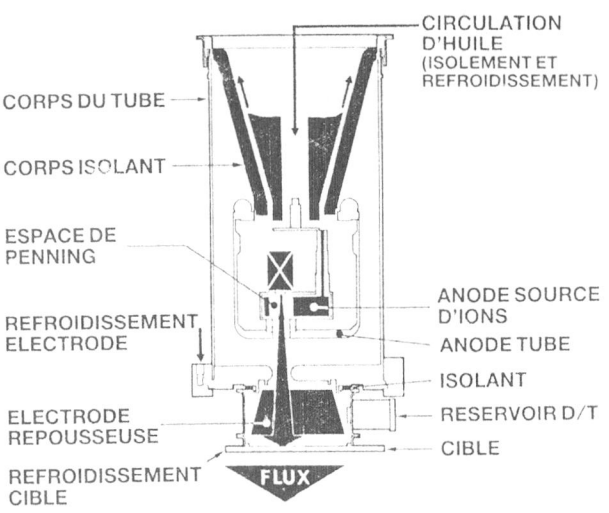

CIRCULATION D'HUILE (ISOLEMENT ET REFROIDISSEMENT)

CORPS DU TUBE

CORPS ISOLANT

ESPACE DE PENNING

REFROIDISSEMENT ELECTRODE

ANODE SOURCE D'IONS

ANODE TUBE

ISOLANT

ELECTRODE REPOUSSEUSE

RESERVOIR D/T

CIBLE

REFROIDISSEMENT CIBLE

FLUX

Figure 5 : Tube TN 46

Les principales caractéristiques du tube sont présentées dans le tableau 1.

SPECIFICATIONS		NOMIN.	MAXI.
EMISSION NEUTRONIQUE			
TAUX D'EMISSION	(n/s)	1.10^{12}	
ENERGIE DES NEUTRONS	(Mev)	14	
ALIMENTATIONS ELECTRIQUES			
THT (TENSION D'ACCELERATION)	(kV)	300	350
COURANT CIBLE	(mA)	15	25
TENSION SOURCE D'IONS	(kV)	5	10
COURANT SOURCE D'IONS	(mA)	60	100
REFROIDISSEMENT			
REFROIDISSEMENT CIBLE (par eau désionisée)			
DEBIT	(l/mn)	50	
PRESSION	(Atm)	4	
REFROIDISSEMENT SOURCE (par huile)			
DEBIT	(l/mn)	10	
PRESSION	(Atm)	2	

Tableau 1 : Caractéristiques du tube TN 46

Fonctionnement

Après la mise en pression du tube, des ions sont créés dans la source d'ions, extraits de celle-ci accélérés par l'électrode d'accélération et, après un transit de quelques centimètres, sont projetés, avec une grande énergie, sur la cible pour produire les neutrons.

La mise en pression du tube en mélange deutérium - tritium (D_2-T_2) est assurée par chauffage d'un hydrure de titane jusqu'à une température correspondant à la pression de fonctionnement recherchée; celle-ci est ensuite régulée.

La création des ions est assurée par une configuration d'électrodes de type PENNING. la quantité d'ions produite est contrôlée par action sur la pression de fonctionnement.

L'application d'une THT de 250 à 350 kV entre la source d'ions et la partie accélération cible, entraîne une extraction des ions par l'orifice d'extraction; l'ensemble des deux électrodes (extraction et accélération) a été conçu de façon à constituer une optique ionique, permettant au faisceau d'ions de pénétrer dans l'espace cible sans interception; la forme, l'état de surface, la géométrie et la nature des matériaux ont été étudiés pour limiter les risques de claquages dans le tube.

Les réactions nucléaires entre les ions projectiles et les atomes "cible" ont lieu sur une épaisseur superficielle de l'ordre de 1 à 2 microns. Les atomes "cible" sont en fait les projectiles implantés par les fonctionnements antérieurs. Le rendement de l'émission neutronique est fonction de la densité d'atome cible, on utilise donc un matériau hydrurable jusqu'à des taux atomiques élevés (de l'ordre de 1,5 à 2 pour le titane, le scandium ou le zirconium). Lorsque ce taux est atteint sur l'épaisseur complète du matériau hydrurable, le mélange D.T est désorbé et participe à nouveau à la création d'ions. Compte tenu de ce processus de "recyclage" du gaz à ioniser et de la nécessité de disposer d'une émission neutronique stable, les deux isotopes de l'hydrogène, deutérium et tritium, nécessaires à la réaction nucléaire, sont introduits en quantité équivalente dans le tube.

Un système de Neutronographie Mobile : DIANE

DISPOSITIF INTEGRE ET AUTOMATIQUE DE NEUTRONOGRAPHIE

Pour réaliser un système de neutronographie mobile, la difficulté technique majeure à surmonter est le développement de la source de neutrons. Après avoir vérifié la faisabilité technique et technologique d'un tube neutronique scellé capable d'émettre 10^{12} n/s, la Société SODERN en collaboration avec les Sociétés DORNIER et SENER entreprend le développement d'un système de neutronographie mobile.

Le premier objectif est de réaliser un équipement prototype polyvalent, très versatile et entièrement mobile de telle sorte qu'il puisse être évalué dans toutes les situations envisageables. Un équipement répondant à cette conception est présenté en figure 6. Ses performances et caractéristiques prévisionnelles font l'objet du tableau 2.

Cet équipement prototype est destiné à ouvrir la voie à des dispositifs plus spécialisés qui viendront compléter les moyens actuels existants auprès des réacteurs, pour mieux répondre aux besoins industriels et soutenir le développement de nouvelles technologies.

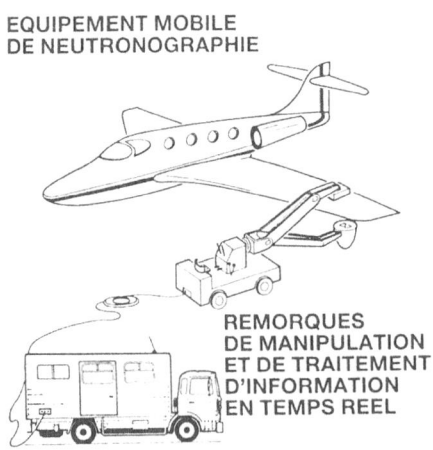

EQUIPEMENT MOBILE
DE NEUTRONOGRAPHIE

REMORQUES
DE MANIPULATION
ET DE TRAITEMENT
D'INFORMATION
EN TEMPS REEL

GENERATEUR DE NEUTRONS A TUBE SCELLE		SODERN TN 46
TAUX D'EMISSION (4 'Tsr)		10^{12}ns^{-1} – 14 MeV
ENERGIE DES NEUTRONS		
DIMENSION (ϕ x L)	mm	200 x 800
DIAMETRE FAISCEAU	mm	3 x 13
DIAMETRE CIBLE	mm	50
DUREE DE VIE (A Q n / 2)	h	500
ALIMENTATIONS ELECTRIQUES		
THT TENSION	kV	300
COURANT	mA	25
SOURCE D'IONS	kV	5
MASSE	kg	1500
REFROIDISSEMENT		
CIBLE		EAU DESIONISEE
SOURCE		HUILE
SYSTEME DE NEUTRONOGRAPHIE		
MASSE DU MODERATEUR (VALEUR APPROXIMATIVE)	kg	400
RAPPORT DE COLLIMATION	(D / L)	1/20
FLUX THERMIQUE A LA SORTIE DU COLLIMATEUR	n.cm^{-2}S^{-1}	6.10^5
DIMENSION DU FILM	mm	360 x 430
TEMPS D'EXPOSITION	mn	1 – 5

Figure 6 : Equipement Mobile de Neutronographie DIANE : vue d'ensemble.

Tableau 2 : Equipement Mobile de Neutronographie DIANE : caractéristiques générales.

REACTOR-QUALITY NEUTRON RADIOGRAPHS WITH A NEW MOBILE MACHINE

- A SUPERCONDUCTING MAGNET CYCLOTRON

M.R. Hawkesworth,

Department of Physics,
University of Birmingham,
Birmingham B15 2TT, UK.

ABSTRACT

The neutron radiographic performance of a new mobile cyclotron from Oxford Instruments Limited, which uses superconducting magnet technology, is assessed. It is concluded that thermal neutron beams comparable with those from some reactors used for neutron radiography are achievable ($10^5 n.cm^{-2}s^{-1}$ at L/D = 250), and radiographs of reactor quality will therefore be available from a mobile source for the first time. There are, moreover, no major technical disadvantages to militate against the cyclotron's clear neutronic advantage over other mobile sources. The machine, moderator system and immediate shielding are readily transportable as a whole, the technologies involved are all proven and reliable, and the use of the Be(p,n) reaction offers a durable target and beams with low photon contamination.

INTRODUCTION

Neutron radiography using intense low energy (epi-thermal, thermal and cold) beams from nuclear reactors has been an indispensible technique of non-destructive examination for certain classes of subject, notably nuclear fuel, small explosive devices and investment-cast turbine blades, for a number of years. It is undesirable to transport nuclear fuel and explosives, however, so there has always been interest in the comparative performance of mobile non-reactor sources of

183

neutrons - particle accelerators and radioisotopes - for radiography. A further attraction of such sources, if they can offer sufficiently intense beams, is that they could open two important additional areas to neutron radiography: the examination of aircraft for corrosion and other water-related damage, and the study of lubricant and fuel flow in operating engines and similar devices.

Cyclotron particle accelerators have long been known to provide some of the most intense non-reactor neutron fluxes. They are also extremely reliable and have proved to be cost-effective machines for research and isotope production throughout the World for over three decades. Hitherto, however, they have not been considered for neutron radiography in the field because their size and weight made the cost of mobile systems of adequate performance prohibitive. Recently this situation suddenly changed with the announcement by Oxford Instruments Limited, the World's leading manufacturer of superconducting magnets, of a compact and powerful 17 MeV cyclotron incorporating the latest superconducting magnet and negative-ion source technology[1].

The first Oxford Instruments cyclotron is optimised for radionuclide production and only capable of accelerating protons. It has a negative-ion source to facilitate the extraction of an external beam; at its extreme radius the beam passes through a thin (25 μm) carbon foil to strip away all the electrons leaving a pure proton beam which curves cleanly out of the machine. Its total weight will be about 1250kg so, with an external proton beam expected to be over 200 μA at 17 MeV, this cyclotron must be considered as the basis for a transportable neutron radiographic facility. What nuclear reaction should be used? What will be the source intensity and energy spectrum? What moderator material should be selected? What will be the neutron flux at the base of the beam tubes in this moderator? What basic target design is sensible, and are there any other factors militating against the use of a compact cyclotron in a mobile system?

CHOICE OF NUCLEAR REACTION

For proton beams of modest energy, two nuclear reactions are well ahead of all others in terms of yield, Be(p,n) and Li(p,n). Reliable yield data around 17 MeV are not as readily available as might be expected, however, and the curves presented in figure 1 are based only on the careful experiments of Atta and Scott[2] using the vanadium bath technique upto a maximum proton energy of 10.2 MeV, and calculations by Lone et al[3] based on published yields as a function of angle. It should be noted that all the measurements involved were made

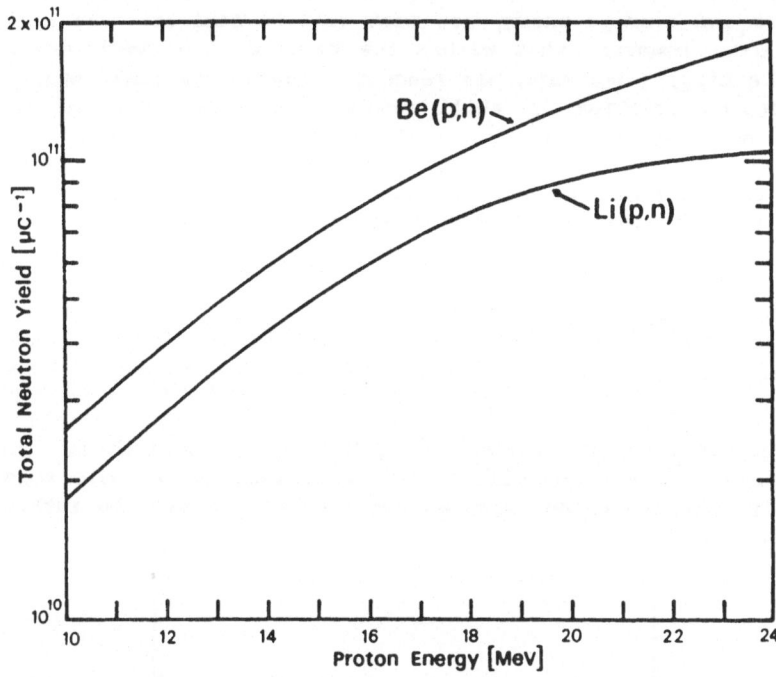

Figure 1. Total neutron yields from thick beryllium, and lithium metal targets bombarded by protons.

using cyclotrons so there can be no question that the beams were composed only of protons and did not contain charged molecules, for example. The yield from Be(p,n) is usefully higher than Li(p,n) at 17 MeV and beryllium is expected to provide the better target combining the high melting point of 1280°C with excellent thermal conductivity and mechanical strength. Moreover, the use of lithium would add the complexity of a liquid metal target so, evidently, the initial choice is Be(p,n) with a thick target of beryllium metal, yielding 9.2 x 10^{10}n.μC^{-1}.

SOURCE NEUTRON ENERGY SPECTRUM

Like the neutron yield, the energy spectrum for 17 MeV protons is only well known in the forward direction (0°), since most measurements have been made with neutron therapy in mind. Moreover, the spectrometers available for the early measurements were incapable of covering the important region for n-r — low energy neutrons. Even today, the latest data[3] are limited to neutron energies >300 keV. The 0° spectra at 18 MeV measured by Lone et al[3] for Be and Li are reproduced in figure 2. Both show the dominance of low energy neutrons and suggest the 4π

185

spectrum and average energy for each will be similar. It will be noticed, however, that within the range of the spectrometer used the Li(p,n) spectrum has reached a low energy peak, whilst the Be(p,n) spectrum is still rising. The abundance of low energy neutrons and lack of features in the spectra are the result of the wide variety of three-body break-up reactions which are possible with a projectile energy of 17 MeV. In addition to the main reaction for protons on beryllium:-

$$^9\text{Be} + p \rightarrow {}^9\text{B} + n \quad -1.9 \text{ MeV}$$

these include [3]:-

$$^9\text{Be} + p \rightarrow {}^5\text{Li} + {}^4\text{He} + n \quad -3.5 \text{ MeV}$$
$$^9\text{Be} + p \rightarrow {}^8\text{Be} + p + n \quad -1.7 \text{ MeV}$$
$$^9\text{Be} + p \rightarrow {}^5\text{He} + {}^4\text{He} + \acute{p} \quad -2.5 \text{ MeV}$$
$$\rightarrow {}^4\text{He} + n \quad +0.9 \text{ MeV}$$

The situation is similar for 17 MeV protons on lithium. In such cases the spectra will soften with increasing angle more strongly than otherwise expected and the outputs will be rather more isotropic.

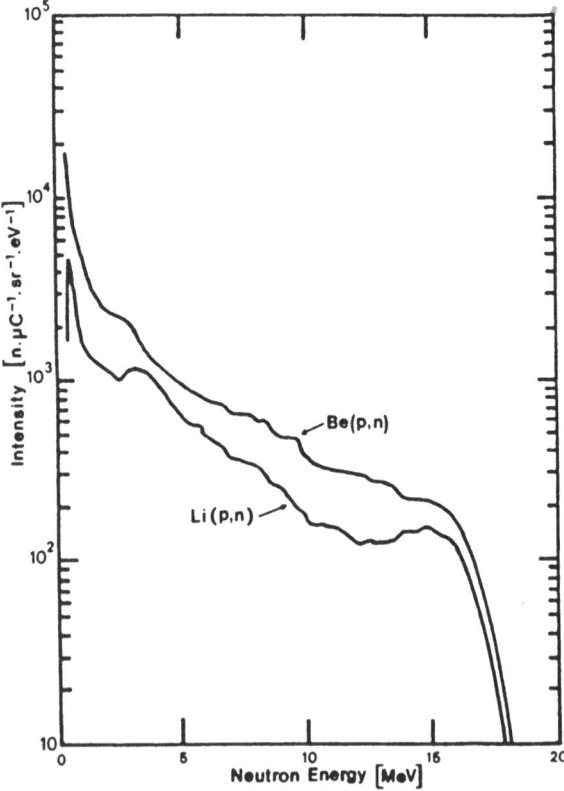

Figure 2. Neutron energy spectra at 0° for the Be(p,n) and Li(p,n) reactions (ref. 3).

Livingstone and Blewett[4] have published the 4π spectrum for Be(p,n) at 16 MeV drawing attention to its similarity to a uranium fission spectrum shifted upwards by about 0.6 MeV. Their spectrometer was limited to energies >1 MeV however so, accepting the abundance of lower energy neutrons observed by Lone et al[3], it seems reasonable to assume that the average neutron energy from Be(p,n) at 17 MeV will be close to that of a fission spectrum and that the thermalisation factors (the ratios of source intensity to the peak thermal fluxes they produce) for various moderators will be similar also. Thus, the peak thermal neutron flux produced in a water moderator is expected to be numerically less than 100 times smaller than the source intensity[5] in a well designed moderator-target assembly. Note, the absence of accurate spectra at low energies and all angles makes sensible computation of thermalisation factors impossible.

MODERATOR SELECTION AND DESIGN

It is now well known that the highest thermal neutron flux from a fast point source is produced in the moderator with the highest hydrogen concentration, assuming no strong neutron absorbers are present. Thus, if for flexibility a liquid moderator is necessary, there is no need to look further than demineralised water, and once the target and probe-tube designs are fixed the choice should be high density polyethylene to give a theoretical improvement in peak flux of 20 to 40%.

The size of the moderator should be kept to the minimum of, say, five thermal neutron mean free paths in radius (~15 cm) to reduce contamination from 2.2 MeV gammas from neutron capture by hydrogen. Many fast neutrons will escape from such a small moderator of course, but they would have had little chance of contributing to the thermal beam had the moderator been larger.

In most, though not all, applications envisaged for mobile n-r equipment it will be desirable to provide a neutron shield round the moderator for personnel, equipment and the neutron imaging station. This shield should also be primarily of hydrogenous material, and should contain a strong neutron absorber emitting as little capture gamma energy as possible. A lithium compound such as lithium fluoride is ideal, but most workers use a boron-containing material such as boric acid at, say, 5% by weight[6]. There is little point in making the shield larger than a metre in overall diameter for a cyclotron-based facility, since operators will not be expected to approach the equipment closely during operation and the imaging station can be shadow-shielded if necessary.

There is as yet no consensus on the best design for probe

tubes, or collimators, beyond that they be conical and as few as possible in number to displace as little moderator as possible. To achieve the latter objective consistent with sytem flexibility the moderator may be rotated round the proton flight tube[1], or the whole cylotron-moderator assembly tilted forward or back. Generally a truncated cone of bismuth or lead will be needed near the base of each collimator tube to reduce the effect of capture gammas, and they will be lined with strong thermal and epi-thermal neutron absorbers upto 7.5 cm from their bases to improve collimation. The precise design of these features is a matter of opinion and trial-and-error on each facility.

TARGET DESIGN

At the full beam current of 200 μA, the target will need to dissipate 3.4 kW. For the same proton beam a conical target surface gives a reduced thermal loading per unit area compared with a disc target and also lends itself to cooling in such a way as to make water ingress into the vacuum system unlikely in the event of target failure. A reasonable design aim is <500 W cm^{-2}[7], and a possible design is outlined in figure 3. Such a target will be more expensive than a simple disc, of course, and whether it can be justified or not probably depends on the efficiency with which it can be periodically detached and annealed. The target assembly should not become particularly radioactive since none of the main nuclear reaction and neutron-capture products have an appreciable lifetime, so successful annealing for the removal of hydrogen gas deposited during operation and radiation damage should be straightforward and permit an indefinite life. If no annealing is carried out beryllium will eventually spall from the surface producing a potential health hazard and ultimately the target will crack open.

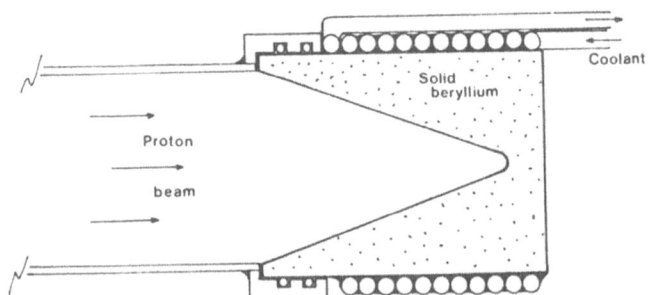

Figure 3.   Sketch of a high power beryllium metal target cooled in such a way as to protect the particle accelerator in event of target failure. Typically the beryllium will be around 50mm in diameter.

THE COMPLETE MOBILE SYSTEM

An artist's impression of a mobile cyclotron-based n-r system has been presented by Wilson and Finlan[1]. It is not necessary, of course, to couple the moderator closely to the cyclotron and, even without focussing magnets, the proton flight tube could be several metres long. Nor is it necessary to have a single straight flight tube though this simplifies the design of the basic equipment. Other details are provided by Wilson and Finlan[1], but it is worth restating here that neither the need for helium (once per month), the provision of electrical supplies (35 kW), nor problems caused by stray magnetic fields should inhibit the use of the new cyclotron in truly mobile n-r equipment.

Finally, the performance expected: as previously indicated the neutron yield at 17 MeV and the full beam of 200 μA is expected to be 1.84×10^{13}n.s^{-1}. It is probably unrealistic to expect a thermalisation factor of less than 100 in a practical system however, even with a polyethylene moderator, but the target and probe tubes need displace little moderator and produce little parasitic neutron absorption. Thus, a thermalisation factor approaching 100, and certainly no worse than 200, can be confidently expected. The figure of 180, giving a source thermal flux for the probe tubes of around 10^{11}n.cm^{-2}s^{-1}, has been used for the comparison with other n-r equipment in table 1. It is concluded that mobile equipment based on the new cyclotron should provide radiographs of reactor quality (L/D = 250) with convenient exposure times, albeit times somewhat longer than those needed with a typical reactor beam.

REFERENCES

1. M.N. Wilson and M.F. Finlan. The superconducting cyclotron as a transportable neutron source. These proceedings.
2. M.A. Atta and M.C. Scott. The neutron yield from Li, Be, Co and Cu under proton bombardment at energies from 5 MeV to 10 MeV. J. Nucl. Energy 27 (1973) 875-884.
3. M.A. Lone et al. Thick target neutron yields and spectral distributions from ^{7}Li(^d_p,n) and ^{9}Be(^d_p,n) reactions. Nucl. Instrum. Meth. 143 (1977) 331-344.
4. M.S. Livingstone and J.P. Blewett. Particle Accelerators, McGraw Hill Book Co., New York (1962).
5. M.R. Hawkesworth. Neutron radiography: Equipment and methods. Atomic Energy Review 15 2 (1977) 169-220.
6. D.H. Stoddard. ^{252}Cf shielding with water-extended polyester. Du Pont de Nemours report DT-1339 (1974).
7. L. Baker. MRC Neutron Therapy Unit, Hammersmith Hospital, London. Private communication.

Table 1   Comparative performance of neutron radiography systems

SOURCE	SOURCE THERMAL FLUX	L/D*	TYPICAL BEAM		TYPICAL EXPOSURE TIME
			U_g/cm^+	ø	
Radioisotope ^{252}Cf, 10^{10} ns^{-1}	10^8	50	0.2 mm	3×10^3	Hours
Sealed-Tube + U booster T(d,n), 7×10^{10} ns^{-1}	3×10^8	25	0.4 mm	3×10^4	Mins → hour
Van de Graaff Be(d,n), 8×10^{11} ns^{-1} (400μA d$^+$ of 3 MeV)	3×10^9	50	0.2 mm	8×10^4	Tens of minutes
Cyclotron Be(p,n), 2×10^{13} ns^{-1} (200μA p$^+$ of 17 MeV)	$>10^{11}$	100 / 250	100 μm / 40 μm	10^6 / 10^5	Secs → mins / Tens of minutes
Radiography reactor	10^{13}	250	40 μm	10^7	0.1s → secs
	n.cm^{-2}s^{-1}			n.cm^{-2}s^{-1}	

* L/D = ratio of probe-tube length (to imaging station) to its base diameter

+ U_g = geometric unsharpness of image.

NEUTRON RADIOGRAPHY WITH VAN DE GRAAFF NEUTRON SOURCE

BY
JIN SI-KWON
DAEJEON MACHINE DEPOT
Daejeon, Korea
AND
W. L. WHITTEMORE
GA TECHNOLOGIES
San Diego, California

- ABSTRACT -

An NR system (NRM-100*) was installed in 1983 to produce routine high quality neutron radiographs with a Van de Graaff neutron source. Two divergent beam ports emerge on opposite sides of the moderator. Two shielded walk-in cells are provided for radiography. Values of L/D can be chosen in the range of 20 <L/D <100.

Special features of the moderator are of interest. The moderator is specially contoured and constructed to permit at least two modes of operation: in one, a high quality NR beam is produced with a very low content of gamma rays; in the other, the NR beam contains a greatly increased fraction of very highly collimated gamma rays. For some NR objects, the compound beam compared to an essentially pure neutron beam offers particular advantages for radiographic imaging.

*U.S. Patent has been granted.

INTRODUCTION

The neutron radiographic system for the Daejeon Machine Depot consists of a Van de Graaff (VdG) neutron source, a neutron moderator to produce the required neutron radiographic beams, and the necessary shielding to protect the personnel and to assure adequately low background for neutron radiography (NR). The purpose of the installation is to provide a proper NR facility to examine a wide range of objects that consist of a variety of metals and plastics. Figure 1 shows a sketch of the manner in which the major components are arranged.

The neutron radiographic system functions in the following manner. A high energy beam of deuterons from the VdG impinges on a robust beryllium target. A current of 300 microamperes of 3 MEV deuterons produces about 4.6×10^{11} fast neutrons per second in the beryllium target. The beryllium target is located internal to the NR moderator near its center (and can be moved for different modes of NR operation). Consequently, a large fraction of these fast neutrons are thermalized with a peak thermal neutron intensity of about 2.6×10^9 n/cm^2-sec occurring near the center of the moderator. Two carefully designed NR beam ports extract the two divergent beams of neutrons with the desired purity with respect to gamma ray content and the desired beam resolution. Complete flexibility has been provided in the degree of focussing of neutrons and gamma ray content. A large amount of lead shielding has been incorporated strategically within the moderator assembly to control the amount of gamma radiation that enters the NR beam. The entire NR facility is enclosed within a room having 4-foot concrete shielding walls.

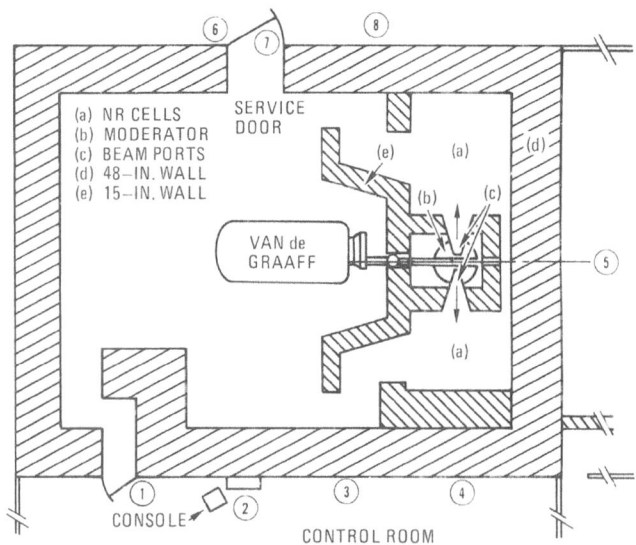

Fig. 1. Major Components of NR Facility and Areas Where Radiological Survey was Conducted.

Facility shielding is an important consideration for high quality NR. It is not sufficient simply to surround the accelerator and NR moderator with a room having 4-foot thick walls to protect personnel. Exposure areas for performing NR require protection from the bremsstrahlung radiation from the VdG and from leakage and albedo fast neutrons. To this end, concrete shielding walls were installed separating both the VdG and the moderator assembly from the NR exposure bays. Finally, the exposure bays were protected from albedo gamma rays from above with a concrete shield on top. These bays were lined on all internal walls with a two-inch layer of industrial borax to control albedo thermal neutrons emanating from the concrete shield walls. The effectiveness of the above shielding was to reduce background gamma and neutron radiation in the NR beams to a tiny fraction of the beam intensities.

COMPONENTS OF NR SYSTEM

The reader will find below a brief description of the important components of the Daejeon Machine Depot neutron radiographic system. Others have published related material for Van de Graaff accelerator applications to NR (1, 2, 3).

Accelerator

The accelerator is a 3.0 MEV VdG (High Voltage Engineering Corp., Model KN 3000) used to produce neutrons from a (d, Be) reaction. The accelerator operating at 3.0 MEV with 300 microamperes of deuteron beam current produces fast neutrons with energies ranging from about 1.5 to 5 MEV. The beryllium target (0.030 in. thick) is very robust and will operate without attention for hundreds of hours if care is taken to keep it clean and free from pump oil vapor. Experiments performed with a similar KN 4000 VdG accelerator have given data on the neutron output as a function of deuteron energy E for accelerator voltages up to 4.0 MEV. These data fit the following function:

$$\text{Yield} = KE^{2.3 \pm 0.2}$$

which closely represents the data published in the literature (4).

Neutron Moderator

The heart of the NR system is the NRM-100.* The design of the moderator, its internal shielding, the large size diameter [48 inch (122 cm)] of the moderator, and the surrounding concrete

---

*U.S. Patent has been granted.

shielding blocks all work to reduce unwanted gamma rays from the non-beam locations to less than on a few percent of the NR beam. The gamma rays in the beam are focussed with the same care as for the neutron beam. Therefore, the gamma component of the NR beam will not provide a general decrease in image contrast as would be the case for a generalized, unfocussed gamma beam.

The design and construction has provided two very flexible beam ports with capability to accept a variety of beam apertures. The basic aperture is 2 inches in diameter, contains neutron absorbers of elemental boron power and foils of gadolinium, indium, dysprosium and cadmium, and is configured with the system to produce a basic L/D of 20. In addition, a lead and boron shield is included on the outside portion of the aperture assembly. Within the aperture region between the aperture and the beryllium source is a series of polyethylene rings each containing a central hole thus providing a reentrant neutron port. Finally, a 50-Kg shield containing lead and boron provides an internal, divergent, beam port. All of the above fits within the 48 inch (122 cm) diameter moderator tank. A flared horn extends the divergent beam outward to make a basic length of 40 inches. Its outer surface is covered with gadolinium paint. The concrete block and sand shield around the moderator fits tightly around the flared horn.

The moderator, the two beam ports, and the concrete block shield around the moderator are mounted on a movable assembly. Enough motion in this assembly is provided so that the assembly can be moved with respect to the fixed target from one position in which the target is in the central NR beam to a variety of other positions in many of which the beryllium target and its associated gamma ray source are hidden from direct view in the NR beam. The former position provides maximum gamma ray content in the beam with the gamma rays extremely well focussed (L/D ≈1000). The latter positions provide minimum gamma rays in the beam and have about the same focus as the neutron beam.

NR PERFORMANCE

With both NR ports filled with high density polyethylene, the peak thermal flux Φ_{th} and the cadmium ratio were determined for VdG operation at 3 MEV and 300 μa current.

$$\Phi_{th} = 2.6 \times 10^9 \, n/cm^2 - sec$$
$$Cadmium \ ratio = 3.6$$

With the system fully assembled, all shielding in place, and both beam ports in service, NR operation was observed for the two distinctly different modes of operation: namely (1) with the minimum gamma ray content in beam and (2) with maximum gamma ray content. These data are presented in the Table 1 and demonstrate that the system is capable of producing excellent neutron radiographs.

As noted above, shielding was a very important consideration in the design of the facility. Not only was it necessary to shield within the NR exposure cells for the desired NR results, but it was also necessary to provide shielding sufficient to maintain the level of radiation within the control room area (1-4 in Fig 1) at essentially background. Also, the radiation level in an adjacent fluoroscopy area (5 in Fig 1) due to full power operation of the NR facility should be very low. In addition, the radiation levels in a rear, closely controlled service

Table 1. Performance Data for Minimum and Maximum Gamma Rays in NR Beam.

Parameter	L/D	Minimum Gamma in Beam	Maximum Gamma in Beam
Φ_{th}	30	2.4×10^5 nv	2.9×10^5 nv
	20	5.4×10^5 nv	6.3×10^5 nv
Cad. Ratio	30	1.9	1.9
$D_\gamma/D_{\gamma+n}$ (1)	30	18%	35
Gamma Dose in Beam	30	4.2 R/hr	18.2 R/hr
% Variation Edge to Center	30	± 6	± 5
BPI (2)	30	68%	48%
IQI (3)	30	10	9-10

(1) $\underline{D}$ is the density measured on Kodak SR film when it is exposed for gamma rays only or gamma rays plus neutrons.

(2) BPI is the ASTM Beam Purity Indicator, ASTM E545-75

(3) IQI is the ASTM Image Quality Indicator, ASTM E545-75

Table 2. Summary of Radiation Measurements for NR
Facility Operated at Full Power
(See Fig. 1 for Locations).

| Location | Description | Dose Rate | |
		Neutrons (mrem/hr)	Gamma (mr/hr)
1, 2 3, 4	Control Room	0.0 0.0	0.025 0.0
3	"	10.0	no moderator or moderator shielding
5	Fluoroscopy Room	0.0	0.25
6	Service Door Area	1.6	15
7	"	0.0	0.7
8	"	0.0	0.7

area were required also to be low. The results in Table 2
demonstrate that all shielding goals were fully met for full
power operation. It may be interesting to note that full power
operation of the Van de Graaff and its neutron target within the
room shielded with 4-foot thick concrete walls but without the
moderator or moderator shield gave a 10-millirem/hr neutron dose
rate at location 3 in the control room.

CONCLUSION

With adequate shielding against (1) the bremstrahlung radiation from the Van de Graaff, (2) the unmoderated fast neutrons
from the beryllium target, and (3) the general gamma ray and
neutron background, a high quality neutron radiography facility
can be constructed. The resulting neutron radiographs can rival
the quality of those produced in a reactor based NR facility.
The main limitation for a Van de Graaff based facility is the

very low NR beam intensity compared to that from a reactor. An advantage of the VdG NR system can be the incorporation (when desirable) of very highly focussed (L/D≈1000) gamma rays within the neutron beam to provide simultaneously the benefits of neutron and gamma radiography.

REFERENCES

1. J. Stokes, et.al, "A New Accelerator-Based Neutron Radiography System" p 717f in NEUTRON RADIOGRAPHY, Proceedings of FIRST WORLD Conference, D. Reidel Publishing Co., Holland, 1983.

2. J.P. Cassidy, "Use of a Low-Energy Van de Graaff Accelerator in Neutron Radiography of Encased Explosives," p. 117f in NEUTRON RADIOGRAPHY AND GAGING, H. Berger, Editor; ASTM Special Technical Publication 586, Philadelphia, Pa, 1976.

3. F.R. Swanson, F.J. Kuehne, "Neutron Radiography with a Van de Graaff Accelerator for Aerospace Applications," p. 158f Ibid.

4. H. Berger, NEUTRON RADIOGRAPHY, Elsevier Publishing Co, New York 1965, P. 14-16.

THE SUPERCONDUCTING CYCLOTRON AS A TRANSPORTABLE NEUTRON SOURCE

* M N Wilson and + M F Finlan

* Oxford Instruments Ltd   Osney Mead   Oxford   OX2 ODX   England

+ Amersham International PLC   Amersham Place   Little Chalfont HP7 9NA   England

ABSTRACT

Oxford Instruments, working in close collaboration with Amersham International, are developing a compact lightweight superconducting cyclotron. Although primarily intended for use as a local generator of short lived isotopes for PET, the same machine could also be made into a neutron generator by directing the extracted proton beam onto a beryllium target. With a proton energy of 17 MeV and a beam current of 200 μA, the expected total neutron yield is 1.8×10^{13} neutrons/sec. Using a moderator of high density polythene, this total output may be transformed into thermal neutron flux of about 1.8×10^{11} n cm^{-2} s^{-1}. With a weight of just 2000 kg, the source will be fully transportable using quite simple mechanical handling equipment.

INTRODUCTION

The use of superconducting magnets is now well established in many areas of research, development and medicine. Particular mention should be made of superconducting particle accelerators for basic research, superconducting nuclear magnet resonance (NMR) spectrometers for molecular structure analysis and superconducting magnetic resonance imaging (MRI) systems for medical diagnosis. Oxford Instruments are leading suppliers of equipment for the latter two applications and maintain an ongoing interest in new uses of superconductivity. In line with this interest, Oxford has recently begun, in collaboration with Amersham International PLC, the development of a small superconducting cyclotron.

Cyclotrons have been widely used over many years for research, isotope production and other applications in nuclear medicine. With a few notable exceptions, they have used conventional magnets consisting of a massive iron yoke, energised by a set of water cooled copper coils. Modern superconducting magnets are able to work at much higher current densities and thereby produce higher fields without the need for an iron yoke. A superconducting cyclotron magnet can thus be much lighter than a conventional one, typically a factor $\approx$ 20. Furthermore, by working at higher fields, the orbit radius may be reduced, with consequent reductions in overall size and savings in rf power.

With these advantages in mind, Oxford and Amersham have studied several superconducting cyclotron designs spanning the range 10-40 MeV and have found significant advantages in every case. Following on a survey of market prospects, we have decided to concentrate, in the first instance, on protons at an energy of 17 MeV. The main application forseen for this cyclotron is in the production of short lived isotopes (particularly Carbon 11, Nitrogen 13, Oxygen 15 and Fluorine 18) for positron emission tomography PET. In this regard, it is envisaged that the compact size, ease of installation and reduced demand for services of the superconducting cyclotron will be attractive to hospitals requiring a dedicated local source of isotopes for routine clinical use.

During the course of our study, it has become clear that a fully transportable cyclotron could also be used in areas other than isotope production. Neutron radiography is one such area and we present here an adaptation of the 17 MeV PET cyclotron for use as a thermal neutron source. Already, this cyclotron produces fluxes which are significantly higher than existing transportable sources. Furthermore, there are clear possibilities for using the same general style of cyclotron to accelerate either protons or deuterons to higher energies and thereby produce even higher levels of neutron flux.

THE COMPACT SUPERCONDUCTING CYCLOTRON

Fig 1 presents an artists impresssion of the 17 MeV PET cyclotron and Fig 2 shows a cross section of the same machine; Table 1 lists the key parameters. Referring to Figs 1 and 2, the vertically directed magnetic field, needed to induce the necessary spiral orbits, is produced by means of four solenoid coils which are split equally above and below the orbit plane.

Design of the magnet follows along fairly standard Oxford lines - indeed it is very similar to an existing product for NMR spectroscopy and imaging. The magnet coils will be immersed in a liquid helium tank, forming part of a fairly conventional vacuum-insulated cryostat, with double radiation shields to intercept any thermal radiation coming from room temperature. These shields will be cooled by a small closed-cycle cooler of the kind commonly used with cryopumps.

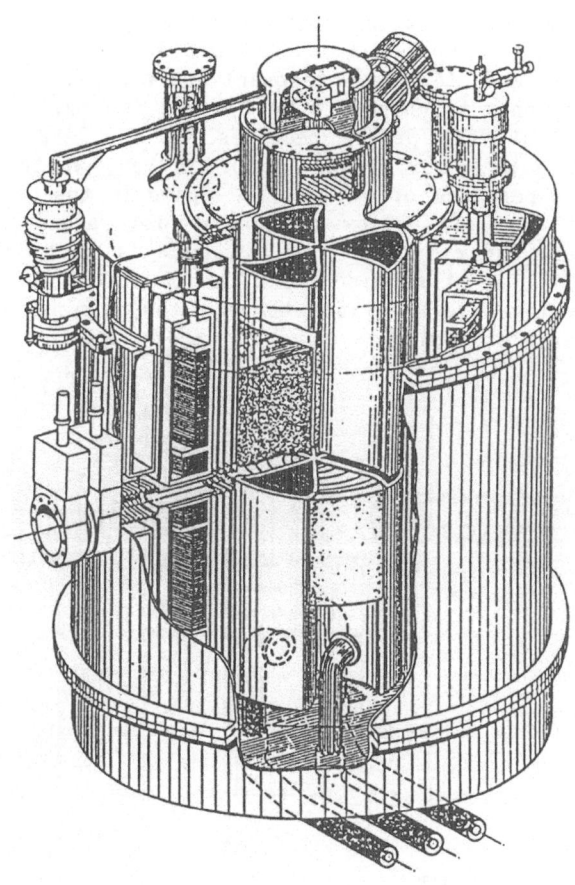

Fig 1: Artists
Impression of the
Superconducting
Cyclotron

Within the magnet cryostat bore will be situated the main cyclotron vacuum tank. For ease of servicing etc, this vacuum space will be kept quite separate from the cryogenic vacuum and all components within it will be at normal room temperature. Vacuum pumping will be provided principally by a cryopump, located immediately on top of the main vacuum tank and using a second closed-cycle cooler. In addition, there will be a small diffusion pump for differential pumping of the ion source.

Inside the main vacuum tank are the iron pole pieces needed to produce an azimuthal flutter of field level and thereby provide a vertical focussing force. As shown in Fig 1 these pole pieces are in the form of 3 pairs, located symmetrically above and below the orbit plane and having a spiral sector cross section.

Between the iron poles are located the rf accelerating cavities. Like the pole pieces, these are in 3 pairs, located above and below the orbit plane. Each cavity takes the form of a pair of nested sector shaped shells, joined electrically at their ends to form a resonant quarter wave cavity. The resonant frequency is chosen to be 3 times the particle orbit frequency, so that each proton receives six accelerating kicks per turn. For a peak cavity voltage of 30 keV, we thus expect an energy gain per turn of ≈ 150 keV. Space in the centre of each cavity will be left hollow to permit good vacuum access between beam space and cryopump. Good vacuum is particularly important in this type of machine which, as described below, will accelerate negative hydrogen ions because they can be easily extracted by charge stripping.

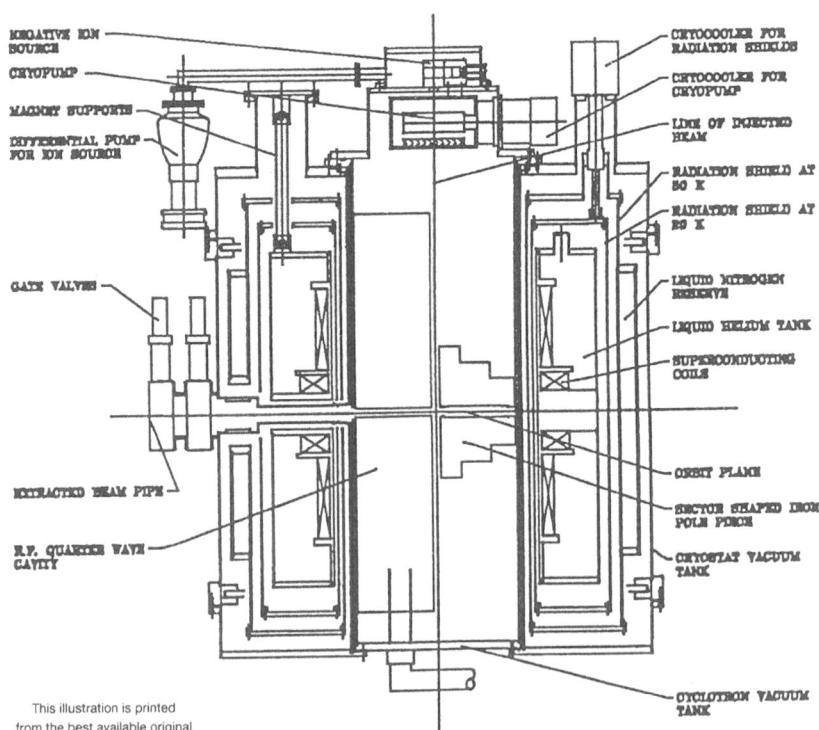

This illustration is printed
from the best available original

Fig 2: Cross section of the cyclotron.

Negative hydrogen ions will be produced in an Ehler's source, or a cold cathode source operating in the axial magnetic field of the main magnet. As shown in Fig 2 the source will be situated outside and above the vacuum tank to permit differential pumping of neutral hydrogen gas emerging from the source. After extraction from the source, the ion beam will be deflected 90° to pass through the cryopump and thence down the centreline of the vacuum tank, taking advantage of the strong focussing action of the axial magnetic field. On reaching the orbit plane, the ion beam will be electrostatically deflected through 90° to begin its first cyclotron orbit.

After being accelerated to their maximum energy, the ions will be made to pass through a thin carbon foil which will strip away two electrons leaving protons, which the magnetic field will bend in the opposite direction, thereby extracting the beam. Extraction by charge stripping is preferable to other methods because it is very efficient and therefore permits high currents to be extracted without heating or activation of components. Good vacuum is particularly important to avoid unwanted stripping of the negative ions by residual gas molecules before they reach the extraction foil. In this regard, the open pumping geometry of the superconducting machine's main vacuum tank and the differential pumping of the external ion source should be particularly advantageous.

Table 1: Key Parameters of the Superconducting PET Cyclotron

Energy (protons)	17 MeV
External beam current	50 μA for isotopes
	200 μA for neutrons
Ion source	H^-
Extraction	Stripper foil
Beam extraction radius	170 mm
Mean field	3.5 Tesla
Field flutter amplitude	1.5 Tesla
Iron spiral pole pieces	3 pairs @ 60° each
Orbit frequency	53 MHz
Rf accelerating cavities	3 @ 60° each
Rf frequency	159 MHz
Rf voltage	30 kV
Energy gain per turn	150 keV

NEUTRON PRODUCTION

Hawkesworth (1) presents data on neutron production rates and concludes that, over the energy range of interest, beryllium is the preferred choice of target material. At the PET cyclotron energy of 17 MeV, neutron yield is expected to be 9 x 10^{10} neutrons per μ coulomb. For PET isotope production, it is envisaged that an extracted beam current of 50 μA will be quite sufficient. However experience at Amersham has shown that negative ion machines are capable of much higher currents and we therefore take 200 μA as a reasonable expected value, ie a neutron yield of 1.8 x 10^{13} n s^{-1}.

Fig 3 sketches the general arrangement for neutron production, described in more detail by Hawkesworth. The extracted beam passes from the cyclotron along a short beam pipe to the water cooled beryllium target. This target is surrounded by a spherical block of high density polythene $\approx$ 1 m diameter, which acts as a moderator, reflector and fast neutron shield. For a total neutron production rate of 1.8 x 10^{13} n s^{-1}, Hawkesworth predicts a useful thermal neutron flux of 1.8 x 10^{11} n cm^{-2} s^{-1}. Table 2 summarizes the neutron output parameters.

Table 2: Neutron Output

Extracted proton beam energy	17 MeV
Extracted proton beam current	200 μA
Total neutron yield	1.8 x 10^{13} n s^{-1}
Thermal neutron flux	1.8 x 10^{11} n cm^{-2} s^{-1}

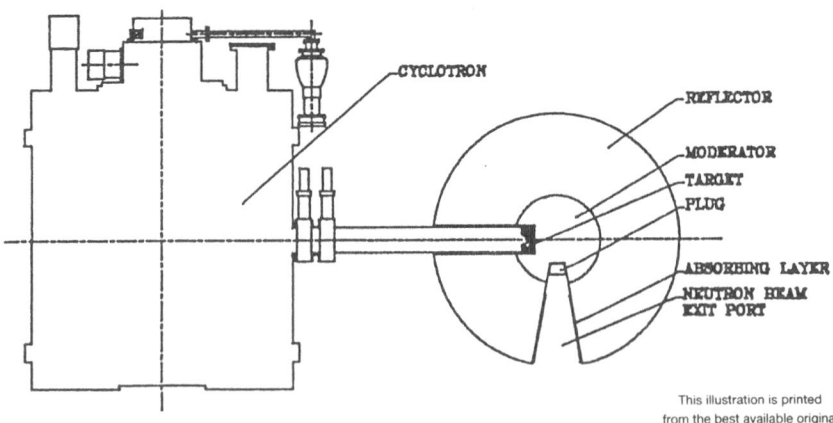

Fig 3: General arrangement of target and moderator.

TRANSPORTATION AND OPERATION

The cyclotron is expected to weigh about 1250 kg and the polythene moderator will be ≈ 750 kg, ie a total system weight of 2 tonnes. It thus becomes possible to envisage a very mobile and versatile installation for pointing the neutron beam in any direction. as sketched for example in Fig 4.

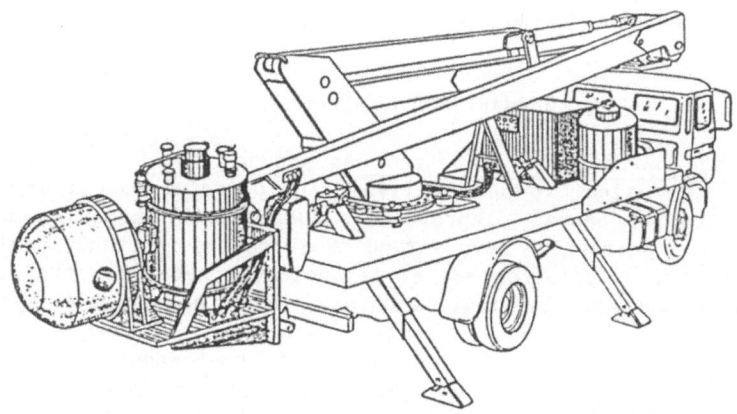

Fig 4: A possible mobile arrangement for the neutron source.

Table 3 lists the main installation parameters for the cyclotron moderator package. The superconducting magnet will usually be kept permanently cold, topping up with liquid helium every month. Liquid nitrogen is used in the outer radiation shield as a heat transfer medium and also to provide buffer storage of 'cold' against the unlikely eventuality of cooler failure. No topping up with liquid nitrogen will normally be required.

Table 3: Installation and Utility Requirements

Weight of cyclotron	1250 kg
Overall diameter of cyclotron	1.15 m
Overall height of cyclotron	1.76 m
Diameter of moderator/reflector sphere	1.0 m
Weight of moderator/reflector sphere	750 kg
Rf power consumption	25 kW
Miscellaneous power	7 kW
Cooling water	50 litre/min
Radiation shield endurance	
without cryocooler	48 hours
Liquid helium consumption	0.2 litre/hour
Liquid helium capacity	180 litres
Liquid helium endurance	37 days
'Topping-up' intervals for liquid helium	1 month

For the cyclotron itself, two components require periodic attention. In the ion source, cathodes must be replaced after 300 hours of continuous operation at full current. The external location of our ion source will make this a rather simple and speedy operation. Extraction stripper foils will need to be changed after 200 hours at full current but, by using a carousel of 5 foils, it is planned to increase the service intervals of this component to 1000 hours.

SUMMARY AND CONCLUSIONS

The conventional cyclotron has proved itself, over many years of operation, to be a reliable and powerful source of dc accelerated ion beams. Improvements such as AVF focussing and negative ion extraction have served to improve performance quite substantially. Cyclotron magnet technology however has changed greatly since the earliest machines, with the result that cyclotrons are so heavy as to be essentially immoveable. By changing to the new magnet technology of superconductivity, it has proved possible to reduce weight by a factor more than 20, thereby transforming the cyclotron from a fixed installation to a portable source.

The presently preferred Oxford design, formulated with a view to providing a local isotope source for PET, is for a 17 MeV proton machine with negative ion extraction. Adapting this machine for use as a neutron source, by means of a water cooled beryllium target, should give a total neutron yield of $\approx 1.8 \times 10^{13}$ n s^{-1} - considerably in excess of other portable neutron sources. After moderation, for example in a block of high density polythene, the useable flux of thermal neutrons will be $\approx 1.8 \times 10^{11}$ n s^{-1}. With a weight of about 2 tonnes the source will be quite portable by conventional means.

In the longer term, it is also possible to envisage even more powerful sources via the same technique, for example by using deuterons at higher energy.

Reference

(1) Hawkesworth M R. Reactor Quality Radiographs with a New Mobile Machine - A Superconducting Magnet Cyclotron, Session III A at this conference.

Cyclotron-based Real-Time Neutron Radiography System

Eiichi HIRAOKA, Ryoichi TANIGUCHI and Masatoshi FUJISHIRO
Radiation Center of Osaka Prefecture
Shinke-cho, Sakai, 593 Japan

Shuichi TAZAWA and Takehiko NAKANII
Sumitomo Heavy Industries, Ltd.
Sobiraki-cho, Niihama, 792 Japan

Atsuo ONO and Koichi SONODA
College of Liberal Arts, Kobe University
Nada, Kobe, 657 Japan

Yujiro SUZUKI and Norio MIURA
Kasei Optonix, Ltd.
Narita, Odahara, 250 Japan

Kenji YONEDA, Shigenori FUJINE, Kosuke KATSURAYAMA
Research Reactor Institute, Kyoto University
Kumatori, Osaka, 950-04 Japan

Kazutaka KOBAYASHI
Tokyo Electronics Industries CO., Ltd.
Hino, Tokyo, 191 Japan

Abstract

A real-time neutron radiography system was installed by using
a sub-compact cyclotron as a neutron generator. We utilized the
neutron radiography facilities of Sumitomo Heavy Industries, Ltd.
in which nominal neutron flux is 1.1×10^6 n/cm^2/sec at L / D = 30
with field size of 14" x 17".

Preliminary tests were done for various kinds of colimators,
scintillators, cameras and lenses, and the system was chosen to
give the highest contrast image. Especially for scintillation
screen, screening tests were done systematically on the two

compounds, LiF/ZnS(Ag) and Gd_2O_3/ZnS(Ag), by radiographic film method by changing particle size, composition ratio and thickness. And finally a scintillation screen of LiF/ZnS(Ag) was adopted at a size of 14" x 17". It has a MTF value of 90% to Gd converter at 1.0 lp/mm.

As a TV camera, SIT tube was chosen. The video signal is directly displayed on CRT or processed with digital image processing devices. The resultant image resolution is more than 1.5 lp/mm accompanying a zoom lens.

INTRODUCTION

The neutron radiography is a similar technique to the X-ray radiography and has important advantage that higher contrast images can be obtained on the material containing hydrogen and other certain light elements.

In this conference we introduced a cyclotron-based neutron radiography facility which uses a sub-compact cyclotron as neutron source (1). This system can produce intense neutron fluxes comparative with the nuclear reactor.

Assisted by recent advance of digital image processing and electronics, the real time X-ray radiography has been established for the medical field and applied for some industrial field. If the real time neutron imaging is realized combining the cyclotron-based facility, the dynamic behavior of the fluid and plastic for actual hydraulic components, plastic injector etc, would be analyzed relatively more convinient than the other neutron source.

In this article the procedure to constitute a real-time neutron radiography system is described together with the characteristics of the components and the assembled system.

OUTLINE OF SYSTEM

The arrangement of the real time neutron radiography system is shown in Fig. 1. The specification and the characteristics of the cyclotron is shown in Table 1. The neutron is obtained through the reaction of Be (p, n) bombarding 18 MeV proton beams on a beryllium target, accelerated by a sub-compact cyclotron. The neutron emitted from the target is thermalized by a moderator and collimated travelling through a collimator. In this experiment the characteristics of neutron obtained at the collimator exit is shown in Table 2 by activation of gold foils. The field size was kept as 14" x 17" size.

After passing the objects, the neutron arrives on the fluorecent
converter plane and the photo-image is viewed by a TV camera.
A mirror is placed in front of the camera to alter direction of
the light path.
Also, this mirror is used for centering of the TV camera by
tilting and found a very useful tool when one magnifies the screen
images by a remote control of a zoom lens attached on the camera.
The video-signal is displayed on CRT directly or after treated
with digital image processing devices.

CONVERTER SCREEN

The fluorecent converter screens, had been examined, are
composed of two kind of compounds.
One is the mixture of granular particles of Gd_2O_3 and ZnS(Ag).
The other is LiF and ZnS(Ag) in which the neutron is converted to
electrons by Gd or alpha particles by Li, where ZnS acts as a
fluorecent material. A number of converter screens of prototypes
were fabricated changing the mixture ratio, granular size and
coating thickness.
One of the test results on the sensitivity is shown in Fig. 2a
where X-ray film is used as sensing device of the fluorescence.
The results of two screens of Gd and NE 426, which are commercial
available, are also shown. The resolution of the converters were
compared using a test chart of cadmium of 100 micron thickness in
which patterns are engraved corresponding spatial frequencies from
0.5 to 5 line pairs/mm.

As the result of the screening test, LiF-4 was found highest
sensitive and the resolution was more than 4.0 lines pair/mm and a
large size screen of 14" x 17" was produced using the same
material component and fabrication procedure.

The result of square   MTF measurement of LiF-4 by film
method is shown in Fig. 2b, for both reflection (N) and
transmission (R) case. Gd film converter of 25μ thickness is also
shown in it for comparison.
Also, analysis of ASTM-81 IQI is shown in Fig. 2c and 2b for
sensitive level and thermal neutron content, respectively.

TV CAMERA

Using a RCA silicon intensifier target (SIT) tube of 4808 H,
a high sensitive noct-camera was fabricated.
This camera is equipped with automatic level adjustment and image
roll-over of which latter circuit was found useful for correction
of the mirror images from reverse to normal ones.
The resolution was obtained as 550 TV lines in the center and 400

TV lines at the corner at an object field size of 240 x 180 mm^2. This values correspond 1.0 and 0.6 line pair/mm for the spatial frequency, respectively. Making comparison with several types of multi-channel plate (MCP) camera, the resolution of the SIT camera was far excellent and 250 TV line of the resolution was obtained even at low light level of 1 x 10^{-3} lux.

SOFTWARE OF IMAGE PROCESSING

The real-time system has performances that the video-signal obtained from the TV camera can be stored in a video recorder or can be reconstructed by a set of digital image processors. The finally obtained CRT images contains noises and fadings arisen from several origins.

Specially, the CRT signal images of one frame of 1/30 sec were very poor comparing the image obtained from the film even using the same converter. These reason can be attributed to the deficiency of the neutron flux and statistical distribution of the image points causes the noises (2). The image reconstruction process, convenient for the real-time neutron image, are classified as shown in Table 3.

The softwares, herein developed, have characteristics as follows;
(1) Median filtering
    This filter acts for elliminating noises by replacing the value of a given pixel to the median value of surrounding L x L pixeles. This technique has advantage over the averaging filtering by the point that conserving the edge of shapes.
(2) Contrast stretching (3)
    This technique utilizes an algorithm which enhances low contrast details in images by adjusting local means and variances of the picture elements through the total image, as taking L x L region for computing sections.
    It was designed to identify slight change of contrast due to defect of material for non-destructive testing purpose.
(3) Matrix filtering is well known in image reconstruction and one of typical technique is Laplacian operator, which is often used for edge enhancement.

EXPERIMENTAL RESULTS

A typical example of one frame TV image is shown in Fig. 3a in which turbine blades(upper) and the standard test chart(lower) are represented. There are shown many noises on the image and an example of 256 intergration processing is shown in Fig. 3b in which improvement of image quality is apparent.

Final resolution of 1.0 line pair/mm was obtained at the object field size of 250 x 200 mm^2. Comparing the resolution of the converter and the TV camera, it is concluded that the final resolution is mainly limitted by the performance of the camera. For example, the resultant resolution (R) in lp/mm is approximated in the equation assuming random stochastic process as follows;

$$1/R = 1/4 \text{ (LiF converter)} + 1/5.8 \text{ (lens)} + 1/1.8 \text{ (TV camera)}$$
$$= 1/1$$

An example of the matrix filtering image processing is shown in Fig. 4. The original picture of the upper left is markingly improved in   proper choice of matrix element.

In this system we tested increment of the resolution by magnifying converter image by the use of a zoom lens (4).

CONCLUSION

The real-time imaging method by use of electronic technique has many advantages over the film method as follows;
(1)   application to dynamic objects
(2)   real-time inspection observing at different angle direction by moving specimen.
(3)   no use of film and development process

If this system is connected to a digital computer system equipped with such mass storage memory as optical memory, much more improvement on the image processing and storing image data would be expected. The disadvantage of this system lies on the low resolution, but the development project of high quality TV system will soon solve this problem.

Finally, neutron radiography would applied for various fields by using cyclotron based neutron source, as it was found that the image quality and the neutron intensity were rather comparative to the nuclear reactor source by this study.

This work was supported in part by a Grant-in-Aid from the Ministry of Science and Technology Agency.

REFERENCES

(1)   S. Tazawa, M. Yano, T. Nakanii et al.
      Proc. 2nd World Conference on Neutron Radiography (JP. 15)

(2)   R. Taniguchi, E. Hiraoka et al.
      Proc. 2nd World Conference on Neutron Radiography (JP. 17)

(3) R. A. Wallis
    Proc. 11th Asilomar Conf. on Circuits, System and Computer,
    240 (1977)

(4) A. Ono, K. Sonoda et al.
    Proc. 2nd World Conference on Neutron Radiography (JP. 18)

Table 1. Characteristics of Cyclotron

Energy and particle	--- 18 MeV, proton
Beam current	--- 50 μA
Pole Dia.	--- 880 mmϕ
Ext. Rad.	--- 370 mm
Magnet Wt.	--- 17 ton
RF Freq. (p)	--- 24 MHz
Dee Angle	--- 180°
RF Power	--- 25 kW
Dee Voltage	--- 40 kV

Table 2. Characteristics of neutron

Neutron Sources	Neutron Flux at Object (n/cm^2.sec)	Cd Ratio	L/D	Irradiation Field Size (mm)	n/γ Ratio (n/cm^2.mR)	Remarks
Subcompact Cyclotron	4.5 x 10^5	3.5	50	356x432 (14"x17")	1.5 x 10^5	18MeV50μA
Model 370	1.1 x 10^6	3.5	30	"	"	"

Table 3. Classification of image processing

Image Reconstruction Method	Image Enchancement Method
. Integration (addition of frames)	. Median filter
. Correction of edge (matrix filter)	. Contrast stretching

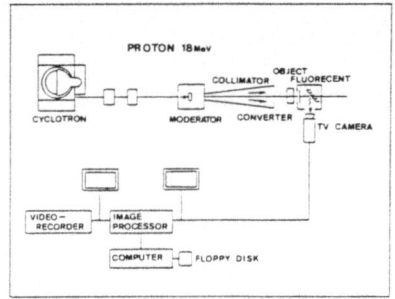

FIG. 1.

REAL-TIME IMAGING SYSTEM OF NEUTRON RADIOGRAPHY
BY THE USE OF CYCLOTRON-BASED NEUTRON SOURCE

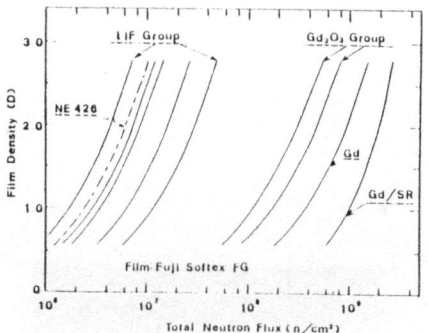

FIG. 2A.

COMPARISON OF NEUTRON SENSITIVITY BY FILM METHOD

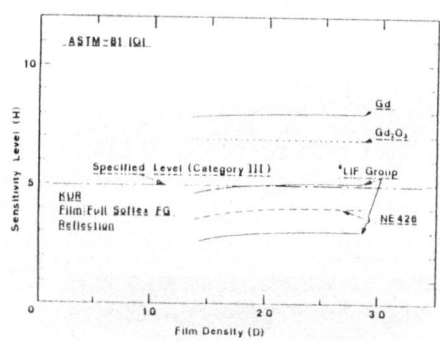

FIG. 2C.

ANALYSIS OF SENSITIVITY LEVEL(H) OF ASTM-81 IQI
FOR VARIOUS CONVERTERS

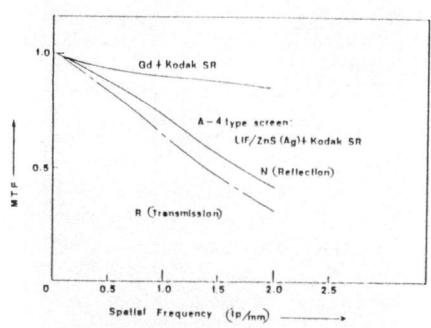

FIG. 2B.

SQUARE MODULATION TRANSFER FUNCTION(MTF)
ANALYSIS FOR CONVERTER

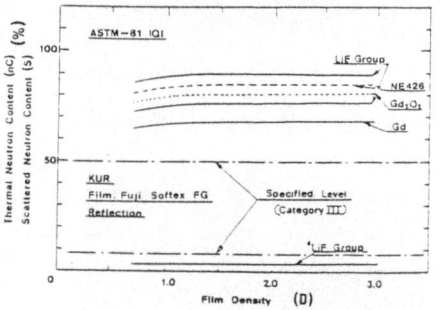

FIG. 2D.

THERMAL NEUTRON CONTENT ANALYSIS OF ASTM-81 IQI

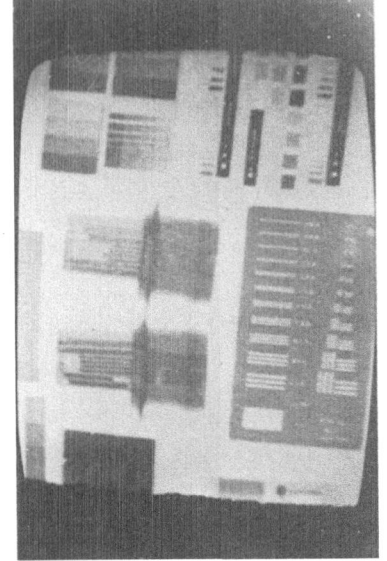

FIG. 3B.
INTEGRATION OF 256 FRAMES

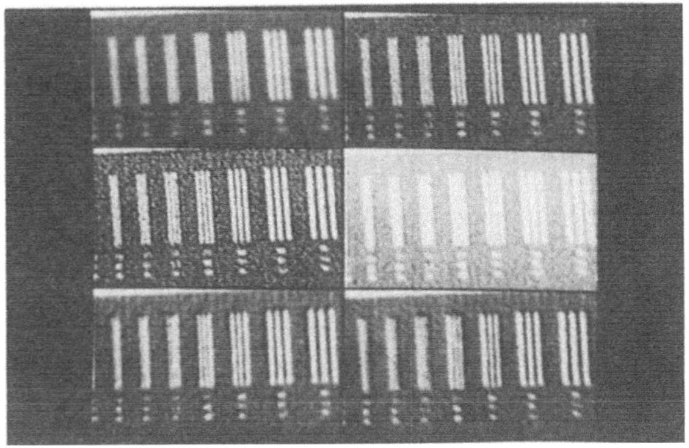

FIG. 3A.
ONE TV FRAME IMAGE OF REAL-TIME NEUTRON RADIOGRAPHY
AT FIELD SIZE OF 250 x 200 MM^2

FIG. 4.
EDGE CORRECTION USING MATRIX FILTERING

ORIGINAL

$$\begin{bmatrix} -1 & -1 & -1 \\ -1 & 9 & -1 \\ -1 & -1 & -1 \end{bmatrix} \quad \begin{bmatrix} -1 & -1 \\ -1 & 5 & -1 \\ -1 & -1 \end{bmatrix}$$

$$\begin{bmatrix} -1 & -1 & -1 \\ -1 & 4 & 4 & -1 \\ -1 & 4 & 4 & -1 \\ -1 & -1 & -1 \end{bmatrix}$$

$$\begin{bmatrix} 2 & 0 & 0 \\ 0 & -1 & 0 \\ 0 & 0 & 0 \end{bmatrix} \quad \begin{bmatrix} 0 & 0 & 0 \\ 0 & 0 & 0 \\ 0 & 0 & 2 \end{bmatrix}$$

NEUTRON RADIOGRAPHY USING ULTRA-COMPACT CYCLOTRON

Yukio FUKUSHIMA[*1], Tomihisa NAKAMURA[*1],
Eiichi HIRAOKA[*2], Junichiro SEKITA[*3], Hiroshi YOKOCHI[*4]
Teruo YAMADA[*5], and Shinichi YAMAKI[*5]
[*1] National Space Development Agency of Japan (NASDA)
[*2] Radiation Center of Osaka Prefecture
[*3] TESCO Co.,   [*4] Nissan Motors, Co. Ltd.,
[*5] Japan Steel Works, Ltd. (JSW)
[*1], [*3], [*4] Tokyo, [*2] Osaka, [*5] Muroran, Japan

ABSTRACT

Neutron radiography test was performed with ultra-compact cyclotron named "BABY CYCLOTRON" which produces 16MeV proton. $^9Be(p,n)^9B$ reaction was applied for neutron soruce. The collimation system has two collimators arranged horizontally, each having cross section 14in. x 17in., L/D ratio : 52, and thermal neutron flux of 3 x $10^5 n/cm^2 \cdot s$ was applied at the object. Radiographic image qualities measured with ASTM indicator have shown very useful results and JSW carried out inspection of exprosive devices for NASDA.

INTRODUCTION

Because of its high sensitivity to detect hydrogeneous materials, radiography using thermal neutron is given attention by them who desire to see organic materials and explosive charges enclosed in metallic structures.

Recent years, the needs of non-destructive testing of explosive devices have arisen from astronautical industries in Japan. However, in Japan, neutron radiography of industrial level had not fully developed partly because neutronic facilities fit for use of radiography were not popular.

Meanwhile an ultra-compact type of cyclotron named "BABY CY-CLOTRON" has been developed by Japan Steel Works, Ltd., and feasibility of accelerator-based neutron source using the cyclotron was pointed out.

Since 1980, authors have carried out the study and testing of practical neutron radiography of the explosive devices using BABY CYCLOTRON and established the apparatus and procedure of neutron radiographic inspection.

CYCLOTRON

The accelerator which is used to generate energetic proton beam to produce neutron in $^9Be(p,n)^9B$ reaction, is an ultra-compact cyclotron model BC168 produced by Japan Steel Works, Ltd.

Because the model is designed originally as a small machine suitable to be installed in a hospital to produce short-lived radionuclides for clinical use, the machine has the features as follows.
   - Energy of extraced beams are fixed; 16MeV proton and 8MeV deuteron.
   - The dimensions of the cyclotron proper is small enough to be installed in a shielding vault of $20m^2$.
   - Magnet yoke surrounding the major part of the machine itself reduces radiation leakage.

The specification of the machine is shown in Table 1.

The appearance of the machine is shown in Photograph 1.

A beam transport system is placed between the cyclotron and the beryllium target to adjust the beam quality e.g. profile and direction. The transport is 5m long and has one Q-singlet, one set of Q-triplet and a pair of steering magnets.

MODERATOR AND DUAL COLLIMATOR SYSTEM

The properties required to moderator material are; large average logalithmic energy loss ξ, large scattering cross section Σs, low absorption cross section Σ_a and, therefore, large slowing down ratio $\xi \Sigma_S / \Sigma_a$. Among the materials commonly known as moderators, the authors decided to use polyethylene being light, solid, machinable and inexpensive.

The authors evaluated the spacial distribution of thermal

216

Photograph 1   BABY CYCLOTRON BC168

Table 1   SPECIFICATION OF BC168

BEAM ENERGY	PROTON	16MeV
	DEUTERON	8MeV
BEAM CURRENT	PROTON	50μA
	DEUTERON	50μA
AVF 4 SECTOR MAGNET	EXTRACTION RADIUS	375mm
	AVERAGE MAGNETIC FIELD	1.5T
RF ACCELERATING SYSTEM	DEE	45° x 2DEE
	FREQUENCY	47MHz
	ACCELERATING VOLTAGE	30kV
ION SOURCE	TYPE	HOT CATHODE PENNING TYPE
	ARC VOLTAGE	500V max.
	ARC CURRENT	3A max.
DEFLECTOR	TYPE	ELECTROSTATIC
	VOLTAGE	50kV max.
VACUUM SYSTEM	PRESSURE	10^{-6}Torr
	MAIN PUMP	700 1/sec OIL DIFFUSION PUMP
TOTAL POWER CONSUMPTION		120kW

neutron flux in a spherical moderator with two group diffusion equation and the expression of flux distribution derived by N.D. Tyufyakov et al. The diameter of polyethylene sphere which is necessary and sufficient to obtain thermal neutron flux distribution by no means inferior to that of infinite geometry and to reduce flux near the surface of moderator to 1/10,000 is about 800mm. So the basic geometry of the moderator is cubic having a side of 800mm. To reduce the leakage of thermal neutron, the surface of moderator is covered with cadmium sheets of 0.5mm thickness.

At the center of moderator, a beryllium target assembly is placed. Being colled by water flow, the target is durable under the heavy bombardment condition of 50µA x 16MeV for over 500 hours.

The collimator is divergent type made of polyethylene with lining of 0.5mm cadmium sheets. The maximum photographing area was suited to film of 14in. x 17in. and angle of divergence is less than 10deg. To facilitate experimental exposure in various collimation ratio L/D, sleeves of collimator inlet having different sizes and collimator which length is changeable on 3 steps are prepared. L/D is changeable from 15 to 74.

Preliminary testing proves that it takes 60min. to obtain sufficient film density 2.5 with collimator having L/D = 52. To raise the productivity of radiographic testing with a limited neutron source, the authors decided to use dual collimator system.

In order to put a dual collimator system to practical use, it is necessary to equalize the qualities of pictures taken at ends of each coolimator in same exposure time. For this reason, two collimators arranged symmetrically are prefered. However, because of the arrangement of existing building and cyclotron system in the facility, the authors use two collimators arranged in 0deg. and 90deg. directions with respect to the beam line. So the composition of filters placed near the target and the positions of extraction of thermal neutron flux are adjusted to obtain same image qualities for two directions. The arrangement of dual collimator system is shown in Photograph 2 and Figure 1.

The results of the dual photographing are shown in Photograph 3 and Table 2. The thermal neutron flux at object in dual collimator system is about 10% lower than flux in single collimator system.

NEUTRON RADIOGRAPHIC INSPECTION OF EXPLOSIVE DEVICES

Before starting the inspection of practical explosive divices, the authors made specificatins of photographing to standardize the

Photograph 2   DUAL COLLIMATOR SYSTEM

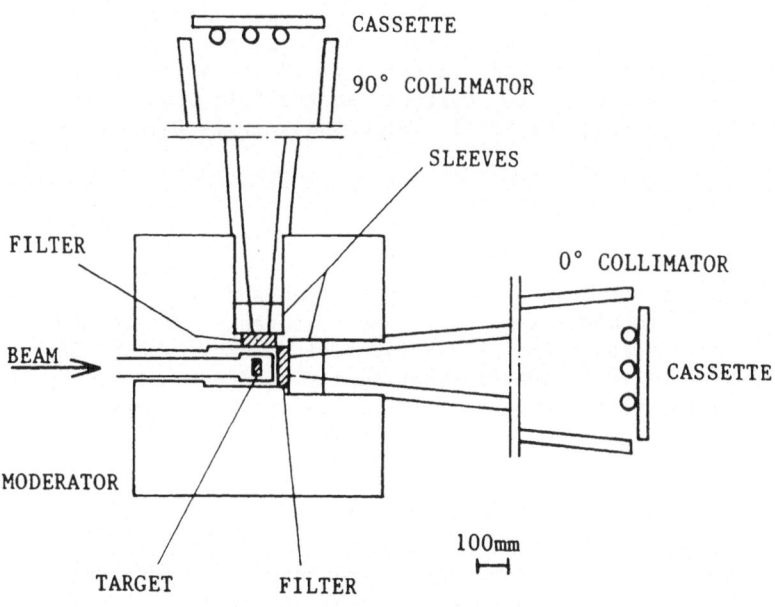

Figure 1   MODERATOR AND COLLIMATORS

qualities of inspection.

To evaluate the qualities of image, the authors used BPI and IQI prescribed in ASTM E 545 and explosive indicators which are stainless steel capsules containing real explosives. The evaluation standards were established by means of these indicators. The standards for the base density, the fogging density and L/D ratio were also created. The details of these standards of photography are described by K. Yamawaki et al.

Using the cyclotron and the dual exposure system described above and conforming to the standards, neutron radiographic inspection of explosive devices has been carried out.

Film cassettes with gadolinium screen are set upright at the ends of the collimators. Aluminum frames to support the explosive devices are fastened on the film cassettes. One cassette can carry, for example, 26 small detonating fuses at a time. The indicators are also mounted on cassettes. An example of setting of objects and cassette is shown in Photograph 4.

To start and stop the exposure. The ion source of the cyclotron is turned on and off. To obtain the base density of 2.5 with normal, L/D = 52 collimators, proton bombardment of 50uA x 60min. into the target is required. Neutron yield of target is stable and proportional to proton beam current, so the film density is controlled by adjusting the total charge (µA x min.) injected into the target.

CONCLUSION

The neutron radiography testing system and standards established by the authors have been found to be an effective measure for the quality control of explosive devices developed in the project of National Space Development Agency of Japan and have proved that NRT using ultra-compact cyclotron is of utility value in industrial fields.

REFERENCES

1. TYUFYAKOV, N.D., SHTAN, A.S., Principles of Neutron Radiography, Atomizdat Publishers, Noscow, 1975, P76

2. YAMAWAKI, K. et al., Neutron Radiography Facility for Explosive Devices of H-1 Launch Vehicle and its Application, these proceedings.

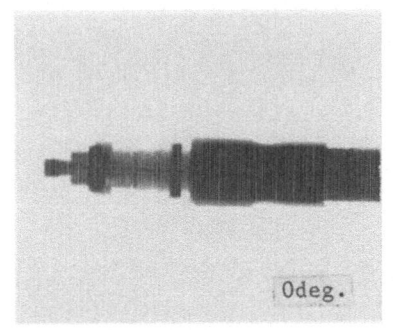 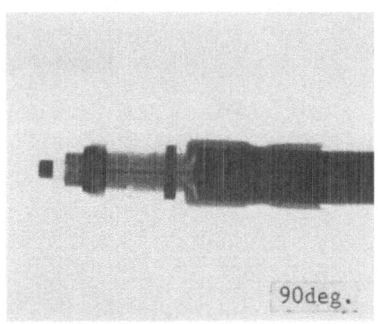

Photograph 3 IMAGES OF DUAL NRT

Table 2 RESULTS OF DUAL NRT

	AVG DENSITY	ASTM C & S LEVEL %					n-FLUX n/cm²sec	Cd RATIO	n/γ
		NC	S	γ	P	H			
0°	2.45	53.5	3.1	0.8	1.2	7	2.75×10^5	3.2	2.8×10^5
90°	2.44	55.7	1.9	1.2	1.9	7	2.72×10^5	2.9	2.9×10^5

16MeV proton 50µA   Exp. 60min.

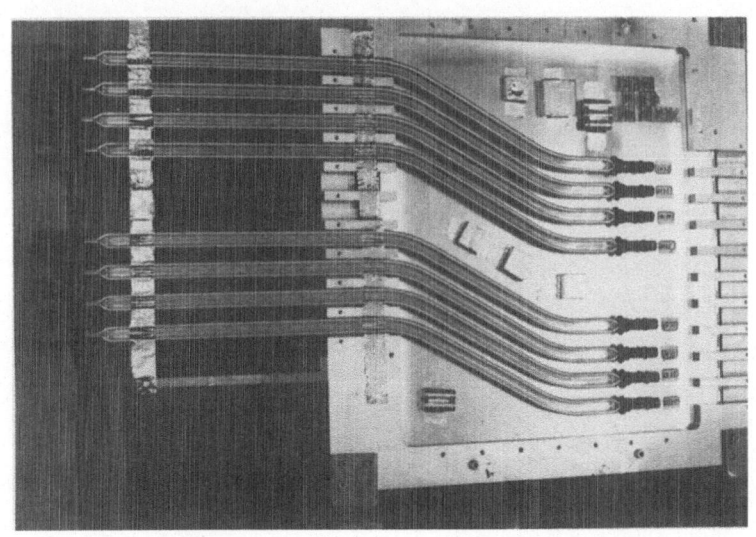

Photograph 4 SETTING OF EXPLOSIVE DEVICES

LOW-FLUX NEUTRON IMAGING FOR ^{252}Cf-BASED
THERMAL NEUTRON RADIOGRAPHY

N. Wada

Department of Radioisotopes
Japan Atomic Energy Research Institute
Oarai-machi, Ibaraki-ken, Japan

ABSTRACT

A transportable exposure device capable of using 1 mg of ^{252}Cf was designed and constructed to evaluate the practical effectiveness of the ^{252}Cf-based thermal neutron radiography with a small source. More refined neutron beams are extracted through three diverging collimators with large outlet aperture of 200 mm in diameter. A thermal neutron flux of $3.0 \times 10^3 \text{cm}^{-2}\text{s}^{-1}$ was obtained at a collimation ratio of 25. The neutron/gamma ratio was 2.9 $\times 10^4 \text{cm}^{-2}\text{mR}^{-1}$. A high sensitivity scintillation converter, ZnS-^{6}LiF was selected because of the limited thermal neutron flux from the exposure device. Various combinations of the converter and X-ray films were evaluated in order to provide resonable results. Radiographic images were analyzed by Klasens' method with the aid of a computerized microdensitometer. It was shown that a combination of the converter and Kodak-SR film is appropriate choice for the high quality radiographs. The combination was effectively applied to the measurement of uranium enrichment in uranium dioxide pellets. On the other hand, a simple imaging device which consists of an image intensifier and Polaroid camera was developed to decrease exposure times in such low-flux neutron imaging.

INTRODUCTION

Three types of neutron sources, namely nuclear reactors, accelerators and radioisotopes can be used for neutron radiography.

Radioisotope neutron source, ^{252}Cf appears appropriate choice for both fixed and transportable radiography exposure devices. It has the advantages of reliability and lower cost. But those exposure devices do not readly provide high quality radiographs because of the somewhat limited thermal neutron flux. To effectively use such lower flux, development of appropriate exposure devices and imaging techniques is important. Therefore, a transportable exposure device using a ^{252}Cf was constructed and evaluated for low-flux neutron radiography. Optimization of a film imaging technique using a high sensitivity scintillation converter was also carried out with emphasis on low-flux neutron radiography. By the use of the exposure device, feasibility tests for inspecting nuclear fuel pellets were done under the optimum conditions of film imaging technique. On the contrary of the film imaging technique with a long exposure time, a simple imaging device which consists of an image intensifier and Polaroid camera was developed to reduce the exposure time.

TRANSPORTABLE EXPOSURE DEVICE

Figure 1 shows an exposure device for ^{252}Cf-based thermal neutron radiography. It was designed for portability and for a 1 mg source of ^{252}Cf. It weighs 1150 kg. A set of wheels allows hand maneuvering of the device on a floor surface. The device is also equipped with lifting ears for crane maneuvering. Leak radiation dose rates are 115 mrem/h at the surface and 10 mrem/h at 1 meter from the surface.

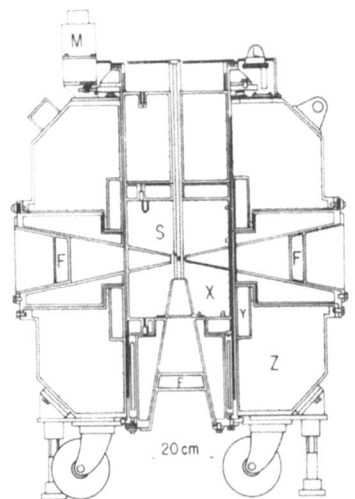

More refined neutron beams allowing a larger radiographic area may be extracted through three diverging collimators lined with cadmium, indium and lead plates. The collimators have respective inlet aperture of 10,20 or 40 mm in diameter. The maximum L/D ratio is 50 at the outlet of one of the horizontal collimators. because the collimators have the same length of 500 mm. Performance characteristics of neutron beams are summarized in Table 1 . Thermal neutron content measured with the ASTM beam purity indicator shows that neutron beams are well thermalized. However, thermal neutron flux and neutron/gamma ratio are comparatively low.

Fig.1 A transportable exposure device for ^{252}Cf-based thermal neutron radiography.
S:neutron source(^{252}Cf:1 mg),
X:high-density polyethylene moderator, Y:lead shield,
Z:boron-loaded paraffin shield,
F:bismuth filter, M:motor,L:lamp

224

Table 1. Performance characteristic of neutron beams

	Beam Ports		
	X; D=20 mm	Y; D=10 mm	Z; D=40 mm
L/D Ratio	25	50	12.5
Radiographic Size	200 mm	200 mm	200 mm
Thermal Neutron Flux	$3.0 \times 10^3 \mathrm{cm}^{-2}\mathrm{s}^{-1}$	1.9×10^3	3.4×10^3
Thermal Neutron Content	73 % (ASTM BPI)		
Gamma Ray Dose Rate	370 mR/h	355 mR/h	642 mR/h
n/γ Ratio	$2.9 \times 10^4 \mathrm{cm}^{-2}\mathrm{mR}^{-1}$	1.9×10^4	1.9×10^4

(^{252}Cf source was placed at a vertical position of
30 mm upper from the horizontal collimator axis)

Both can be improved by selecting ^{252}Cf source position or radiographic position, as shown in Figure 2 and Figure 3. It was also effective to insert bismuth filter in the collimator to avoid gamma contamination in the beam. Nearly a factor of 2 increase in neutron/gamma ratio could be achived with the use of a bismuth filter of 4 cm thick. But the bismuth filter was not used in the following experiments because a significant reduction in thermal neutron flux.

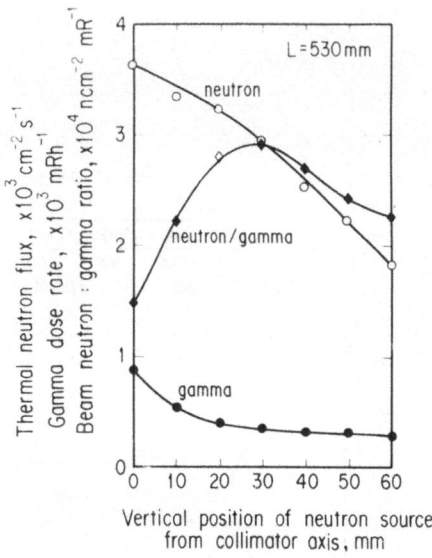

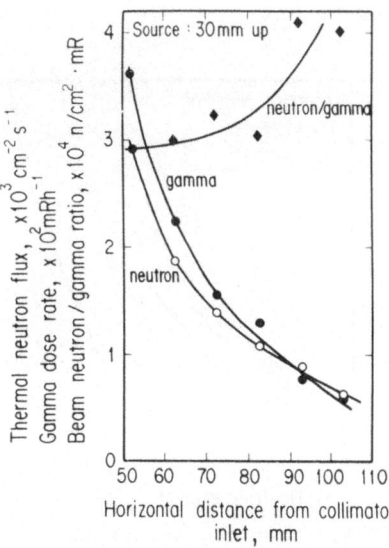

Fig.2 Effects of vertical position of ^{252}Cf source in the exposure device on neutron/gamma ratio.

Fig.3 Effects of radiographic position on neutron/gamma ratio. ^{252}Cf source was placed at a vertical position of 30 mm upper from the horizontal collimator axis.

THE OPTIMUM CONDITIONS IN FILM IMAGING TECHNIQUE

The direct exposure method using neutron scintillation converter in contact with X-ray film is the most attractive because of the somewhat limited beam intensity available from the exposure device. The most fastest scitillation converter, ^{6}LiF-ZnS(NE-426) which is relative insensitive to gamma ray contamination was used singly behind the film relative to the incoming neutron beam. Figure 4 shows density-exposure characteristic curves when X-ray films of different kinds were exposed in conjunction with the converter. Since the data used in this figure are not corrected for gamma contribution, the relative speed at a neutron/gamma ratio of 2.9×10^4 $cm^{-2}mR^{-1}$ is given for each combination.

In practice, it is necessary to make a trade-off between speed and image quality because the choice of film and converter combination determines the image quality. The factors relating to the image quality are contrast and unsharpness of radiographs. Therefore, the ratio of contrast to unsharpness is defined here as figure of merit of the image quality. Unsharpness was measured by Klasens' method with the aid of a computerized scanning microdensitometer. Contrast was given as density range across the image of a test object placed in contact with film. A 0.6mm thick cadmium knife edge was used as the test object. An example of the microdensitometer trace of the test object is shown in Figure 5.

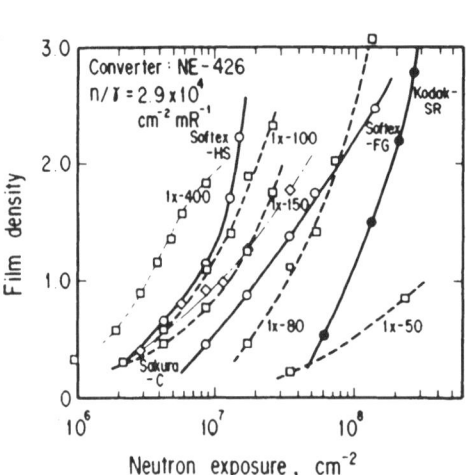

Fig.4 Film density versus neutron exposure characteristic curves for a number of films in conjunction with a NE-426 scintillation converter. Exposures were performed at a neutron/gamma ratio of 2.9×10^4 $cm^{-2}mR^{-1}$.

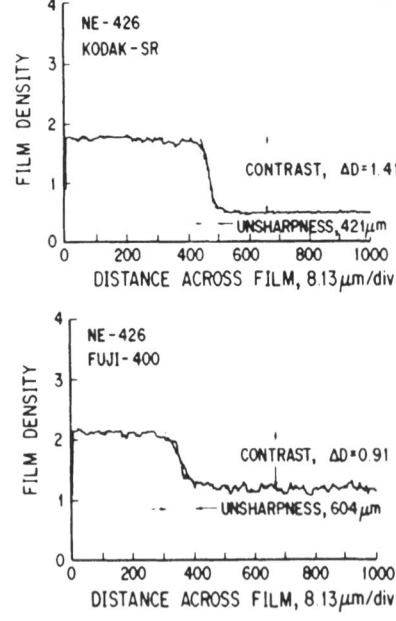

Fig.5 Microdensitometer traces across a cadmium knife edge. (upper) fine-grain film:Kodak-SR (lower) large-grain film:Fuji-400

Figure 5 shows that there are great differences in contrast and unsharpness depending the type of film used. This results from grain
size and thickness of the film. A series of the measurements was
carried out in the same manner for a number of films shown in Fig.4.
The figure of merit of each film is plotted on Figure 6 in relation
to neutron exposure required for a film density of 1.5. From this
figure, it is clearly shown that fine-grain film, Kodak-SR will
give the highest quality radiographs. But the neutron exposure required to obtain a radiograph is fairly large. Whilst large grain
film, Fuji-HS will give acceptable results with a significant reduction in neutron exposure. This may be usual for normal inspection purpose.

APPLICATIONS

Capability Test Using Caliblation Fuel Pin

To evaluate the practical effectiveness for inspecting nuclear
fuel pins, a caliblation fuel pin, CFP-E1 was radiographed under
the optimum conditions of film imaging technique described above.
A reproduction of the radiograph is shown in Figure 7. Perhaps,the
image is not as sharp as that from reactor neutron source. But
internal details are recognizable in some detail. A further study
to improve the image quality is scheduled with the aid of digital
image processing. A system to be used for image processing is
shown in Figure 8.

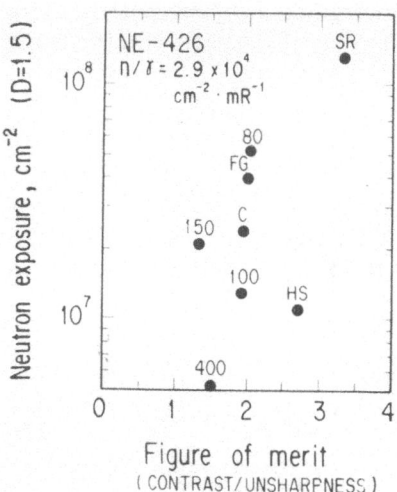

Figure of merit
(CONTRAST/UNSHARPNESS)

Fig.6 Film figure of merit
plotted in relation to neutron
exposure required for a film
density of 1.5.

Measurement of Uranium
Enrichment in Uranium
Dioxide Pellets

In relation to the capability
test for inspecting nuclear fuel
pins, an attempt was made to determine uranium enrichment in
uranium dioxide pellets. Uranium
dioxide pellets with different
uranium enrichment were used as
samples without cladding. The
size of the pellets was 14.3 mm
in diameter and 20 mm in length.
The pellets had a density of
10.34 g/cm^3.

Figure 9 shows microdensito-
traces of a radiograph of the
pellets. One can clearly distinguish each pellet with different
uranium enrichment. The relation
between uranium enrichment and

difference in film density is shown in Figure 10. A straight line was given from the method of least square. The lower limit of detection of uranium enrichment is dependent upon the noise in the density trace. From figure 10, the lower limit of detection seems to be smaller than 1.5 % by weight, since the film density can be measured with an accuracy of ± 0.01. On that account, it is concluded that ^{252}Cf-based thermal neutron radiography is a useful method for the quantitative measurement of uranium enrichment in uranium dioxide pellets.

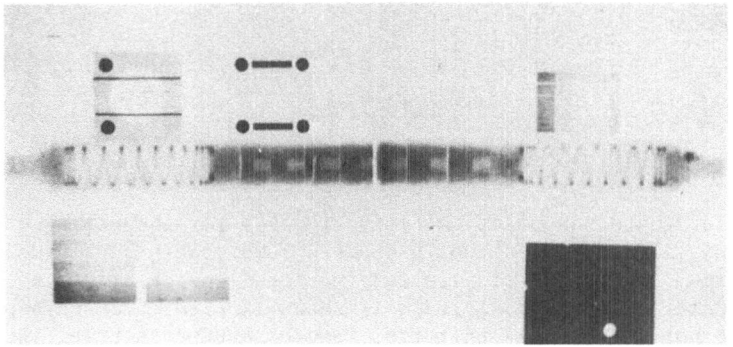

Fig.7 A reproduction of the radiograph of a calibration fuel pin (CFP-El). The exposure were performed at a L/D ratio of 45. NE-426 converter and Kodak-SR film combination were used.

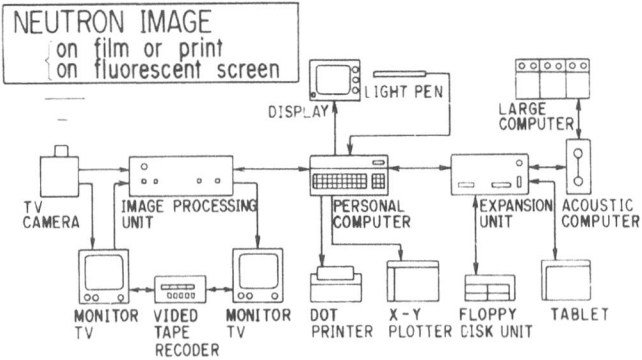

Fig. 8 Neutron image processing system. The image processing unit is a digital video frame memory which has a size of 512 x 512 x 16 bits.

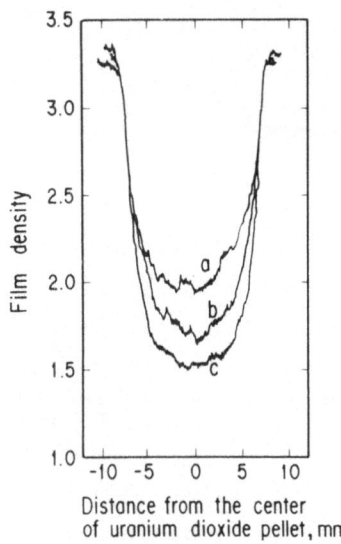

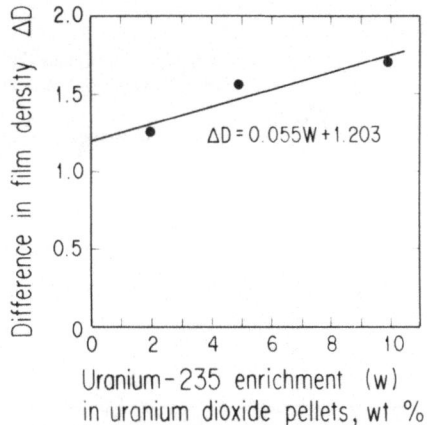

$$\Delta D = 0.055W + 1.203$$

Uranium-235 enrichment (w) in uranium dioxide pellets, wt %

Fig.10 Relation between film density difference and uranium enrichment.

Distance from the center of uranium dioxide pellet, mm

Fig.9 Microdensitometer traces of the radiograph of uranium dioxide pellets. a ^{235}U-1.999 %; b ^{235}U-4.969 %; c ^{235}U-9.994 %.

Simple Imaging Device Using Intensifier and Polaroid Camera

In case of film imaging technique using NE-426 scintillation converter combined with fine grain film, exposures of 30 mg-h were necessary to take the images of good quality.

A simple imaging device consisting of an image intensifier and Polaroid camera was designed and tested for its radiographic performance. Figure 11 shows the device. A microchannel plate image intensifier with a maximum gain of 2.5 x 10^4 was used between NE-426 scintillation converter and Polaroid camera. The converter was placed directly on the photocathode input face of the intensifire when small objects less than 25 mm in diameter were radiographed. But the converter was placed indirectly through a fiber optic magnifier on the photocathode face when radiographs of larger objects were taken. The radiographic size at this time was 50 mm in diameter.

Reproductions of thermal neutron radiographs taken using the simple imaging device and the exposure device shown in figure 1 are shown in Figure 12. Neutron exposures required at maximum gain was only 0.03 mg-h when Polaroid ASA 3000 film was used. This value is one-thousandth of that for the film imaging technique.

A further study on the use of a high sensitivity television camera in place of Polaroid camera is carried out to improve the apparent image quality, with the aid of digital image processing.

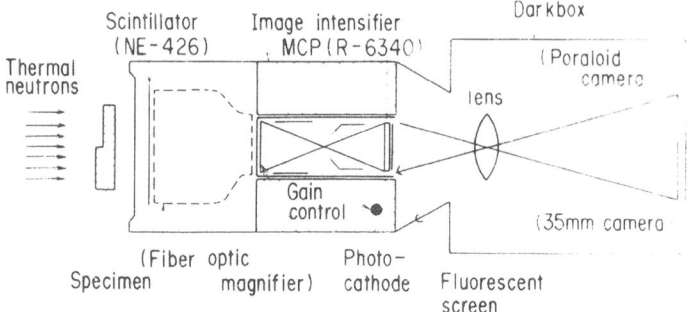

Fig.11 A simple neutron imaging device consisting of
an image intensifier and Polaroid camera.

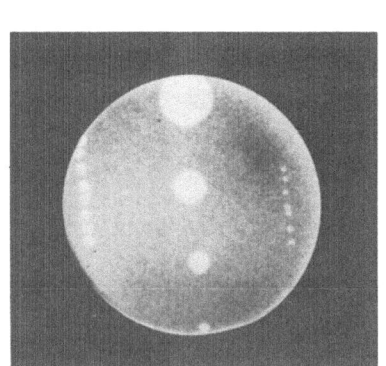

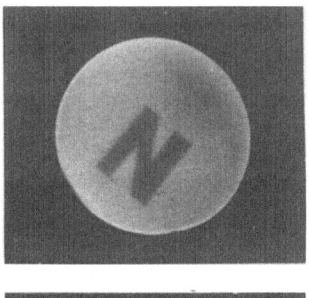

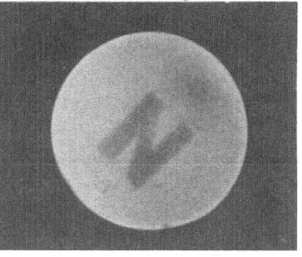

Fig.12 Reproductions of ^{252}Cf-based thermal neutron radiographs
taken by the imaging device shown in figure 11.
(left)   (right)
  0.5 mm thick cadmium plate   N-shaped cadmium plate of
  with holes   0.5 mm thick
    left - 1 mm   with the magnifier
    center- 5,3,2 and 1 mm   L/D=12.5,
    right - 0.5 mm   [upper] gain=min.,
  without the magnifier   exposure=0.6 mg-h
  L/D=25, gain=min.,   [lower[gain=max.
  neutron exposure=0.6 mg-h   exposure=0.03 mg-h

Cyclotron-Based Neutron Radiography Facility

Shuichi TAZAWA, Munehiko YANO and Takehiko NAKANII
Sumitomo Heavy Industries, Ltd., Toyo Works

Eiichi HIRAOKA, Ryoichi TANIGUCHI and Yukio Tsuji
Radiation Center of Osaka Prefecture

Junichiro FURUTA
Osaka Nuclear Science Association

Abstract

Neutron radiography is not widely used yet due to lack of high intense thermal neutron source convenient and economical for practical use. A new neutron radiography facility, installing a sub-compact cyclotron accelerating 18 MeV proton and equipped with both vertical and horizontal ports, is presented.

Typical neutron intensity at the collimator end of $L/D = 30$ is 1.1×10^6 $n/cm^2/sec$ with a field size of 14" X 17" and used mainly for real-time experiments. Prior to the construction, preliminary experiment was performed by using proton beam from high energy cyclotron to investigate thermal neutron intensity and neutron radiography image quality at several energy points.

By optimization of neutron intensity and cost, a new facility using a 30 MeV proton cyclotron with neutron flux of 5×10^6 $n/cm^2/sec$ at field size of 14" X 17" is proposed.

Introduction

Neutron radiography has been developed, mainly using high intense thermal neutron flux obtained by the nuclear reactor. Now, its applicability is widely acknowledged and it is strongly required to use more economical and convenient neutron source than it. One of the possibility to fulfill the requirement would be an accelerator-based system which had been proposed long time years ago and still seems remaining at rather laboratory level, although stastic type of accelerator as 3 MeV Van de Graff(1) using ^{9}Be(d,n)^{10}B reaction or D-T generator are widely used.

Now, several types of sub-compact cyclotron, which have higher acceleration energy than the static type, are available for the production of medical radio-isotopes.

We have investigated the applicability of cyclotron to neutron radiography.

Preliminary experiment with cyclotron

To investigate the characteristics of neutrons generated by cyclotron, preliminary experiment was conducted by the use of 680 cyclotron at Cyclotron Radio-isotope Center of Tohoku University, which was built by Sumitomo and CGR-MeV and can accelerate protons from 3 to 40 MeV variably. Two energy points of 17 and 30 MeV were selected for neutron production reaction of ^{9}Be(p, n)^{9}B.

The experimental condition is summarized in Table 1. The yield of thermal neutron at the L/D = 50 collimator end is shown in Fig. 1, together with the result of total neutron yield for ^{9}Be(p, n) and ^{9}Be(d, n) reaction[2]. In this experiment we used a vacuum cassette of Research Chemicals in which a 25 micron Gd converter was contained. Also, the kinds of film were Eastman Kodak SR, Fuji Film FG and 100S.

To investigate the neutron quality, ASTM E545-75 Beam Purity Indicator (BPI), Sensitivity Indicator (SI) and System Image Quality Indicator (SYQI) were used. The analysed results are shown in Table 2, comparing the result obtained at Kyoto University Reactor (KUR).

From this experiment the following results were conducted;
(1) Inspection of explosive dummies is comparable to the result at KUR.
(2) The analysis of BPI images suggests the necessity of improving thermal content of neutrons.
(3) To gain higher neutron yield, increasing proton energy is more efficient than beam current.

(4) Provision of vertical irradiation port would be convenient to
handle objective material.

Outline of neutron radiography facility

On the basis of the results obtained from the previous
experiments, a cyclotron-based neutron radiography facility has
been installed since 1983 at Toyo Works of Sumitomo Heavy
Industries, Ltd. in Ehime Prefecture. in Japan. The
characteristics of the cyclotron are shown in Table 2 and the
building layout is illustrated in Fig. 2, where model 370
cyclotron was employed, the extracted proton beams from the
cyclotron is conducted through a beam duct and hits a Be target.
The induced neutrons have mean energy of 3.4 MeV and are
thermalized in a polyethelene moderator. Two collimators of
vertical and horizontal are installed at the moderator and object
samples and a film cassette are placed at the end of the
collimator. Typical radiograph condition is shown in Table 4.

Setting of samples and a vacuum cassette for vertical
radiography is shown in Fig. 3.

Most of the neutron emitted from the target move forward
direction and the moderator design to obtain optimum neutron flux
was expected different for the vertical direction. A computer
simulation was carried out before its design.

From the analysis of the BPI, the neutron quality is more
improved than the previous experiment, specially for the vertical
extraction.

Typical neutron radiography facilities for various sources
are compared in Table 5. In regard to the intensity and quality
of neutron, the cyclotron based facility of Toyo Works could
comare favorably with the reactor based one.

Example of neutron radiography

One of the interesting radiography example is a bronze mirror
of 4 B. C. which was excaved at Kaneko-yama mound in Niihama city,
Ehime prefecture. Both neutron radiographs X-ray and are shown in
Fig. 4a and 4b, respectively. This mirror has four belles around
the fringe and such decoration has never been found anywhere else
Japan. The neutron radiograph indicates stone clearly locating
inside the bell. One estimate that the stone was placed before
casting the bronze and then, the sand, coated the stone, would be
taken away from the slit of the bell to keep the space for
ringing.

In Fig. 5, an example of stastical analysis of B_4C granules, contained few percents in a sheet of graphite, is shown. The granule image was taken in a film by ordinary Gd converter system and the film was enlarged by a TV camera. The electrically converted image was processed by an image processor for stastical analysis.

Conclusion and proposal

A cyclotron-based neutron radiography facility of Toyo Works, where a sub-compact cyclotron of 18 MeV proton is installed, is introduced. According to the experimental result of Tohoku University, it is concluded that the neutron flux is obtainable as 6×10^6 $n/cm^2/sec$ at the collimation ratio of 30, if one installs a cyclotron of larger size of model 480 in Sumitomo-CGR MeV series (Table 3).

If this system is realized, the exposure time can be reduced within few minutes and the cost performance of the facility would be expected to increase much more than model 370 and would be applicable for practical use.

(1) F. R. Swanson and F. J. Kuehne, Practical Application of Neutron Radiography and Gaging (American Society of Testing Materials, 1976)

(2) M. A. Lone et al., Nucl. Inst. Methods 143 (1977) 59.

Table 1. Experimental conditions at Tohoku University

		17	30
Energy (MeV)		17	30
Current (μA)	50 (Normalized)		
L/D, L (m)	30, 0.9	50, 1.8	50, 1.8
Neutrons ($\times 10^6$ $n/cm^2/sec$)	1.1	0.45	1.5
n/γ ($\times 10^5$ $n/cm^2/mR$)	0.77	1.3	0.76
Cd ratio	2.4	3.0	2.5
Film	Exposure Neutrons ($\times 10^9$ n/cm^2)		
Fuji FG	1.2	–	–
Fuji 100S	0.32	0.45	–
Kodak SR	–	1.6	2.2

Table 2. Analysis of beam purity indicator (ASTM E545-75)

Facility	KUR 5 MW Reactor	Tohoku University Cyclotron 17 MeV	30 MeV
L/D	100	50	50
Film	FG	SR	SR
Converter	Gd	Gd	Gd
Thermal (C) %	73.4	57.6	46.3
Scatter (S) %	15.9	20.1	22.1
Gamma (Γ) %	2.8	3.3	3.0
R	11	9	8

Table 3. Characteristics of cyclotron of Sumitomo - CGR MeV

Particle	Model 370 Energy (MeV)	Beam Current (μA)	Model 480 Energy (MeV)	Beam Current (μA)
Proton	18.0	0 - 50	3 - 30	100
External radius (mm)	370		480	
Magnet weight (ton)	17		30	
RF frequency (MHz)	24		15 - 30.5	
Dee angle (°)	100		90	
PF power (KW)	25		37	
Dee voltage (KV)	40		40	

Table 4. Parameters of cyclotron-based neutron radiograph facility at Toyo Works of Sumitomo Heavy Industries, Ltd.

Collimator	Horizontal	Vertical
Direction	Horizontal	Vertical
L/D	32 - 68	
Field (in.)	14 x 17	
Moderator		
Outer diameter (mm)	700 x 700 x 700	
Thick. of exit (mm)	30	-
Thick. of Bi (mm)	30	20
Neutron flux (n/cm^2/sec)	1.1×10^6	-
Cd ratio	3	5
Film	Kodak SR	
Exposure time[1] (min)	21	28
BPI characteristics	ASTM 545-75 (545-81)	
Thermal (%)	61.4 (53.9)	73.6 (60.0)
Scatter	9.9 (1.0)	8.6 (4.0)
E (P)	2.9 (1.5)	5.7 (3.0)
Gamma	1.2 (6.2)	3.5 (1.5)
R (H)	10 (7)	10 (7)

[1] at film density 2.0.

Table 5 Comparison of typical neutron radiography facility

Organization and Neutron Source		Neutron Flux at Object (n/cm^2.sec)	Cd Ratio	L/D	Irradiation Field Size (mm)	n/Γ Ratio (n/cm^2.mR)	Remarks
Rad. Center of Osaka Pref.	Vand de Graff.	1.2 - 9x10^3	2.2	10-30	120 x 120	---	1MeV (d,n)
	Linac.	1.5 - 7x10^5	3.5	10-30	"	---	15MeV (Γ,n)
Tohoku Univ.	Cyclotron	4.8 x 10^5	3.5	50	330 x 330	1.3 x 10^5	17MeV
	(Model 680)	1.5 x 10^6	2.5	50	"	0.8 x 10^5	30MeV
Sumitomo Heavy Industries, Ltd.	Subcompact Cyclotron	4.5 x 10^5	3.5	50	356 x 432 (14" x 17")	1.5 x 10^5	18MeV, 50μA
		1.1 x 10^5	3.5	30	"	"	" "
		(6 x 10^6)	-	30	"	"	30MeV, 100μA
Kyoto Univ. Reactor	KUR	1.2 x 10^6	400	100	160ϕ	1.0 x 10^6	5MW E-2

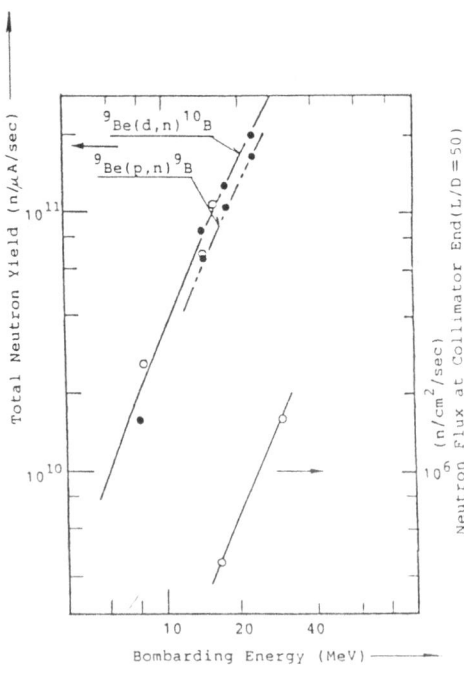

FIG. 1 TOTAL NEUTRON YIELD VERSUS PROTON BOMBARDING ENERGY AND NEUTRON FLUX AT COLLIMATOR END

236

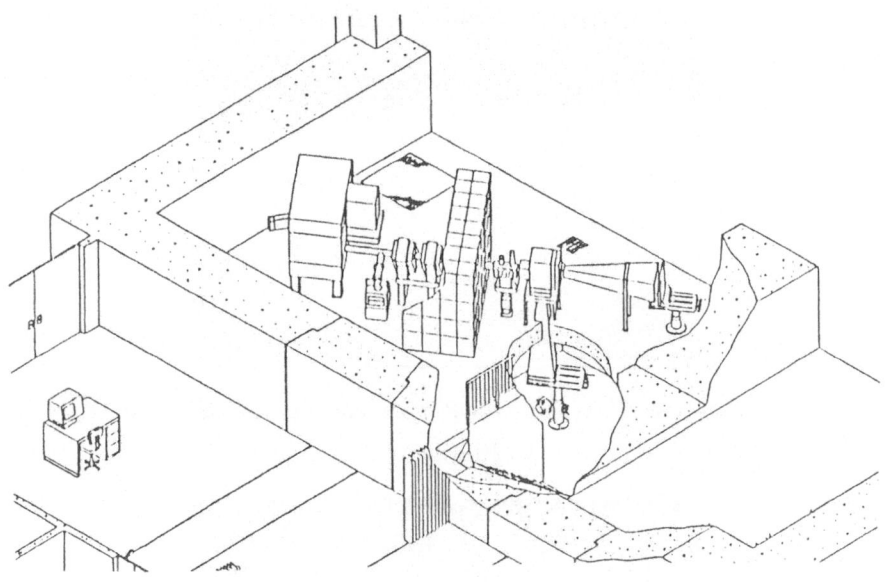

FIG. 2   LAYOUT OF CYCLOTRON-BASED NEUTRON RADIOGRAPHY FACILITY OF
TOYO WORKS, SUMITOMO HEAVY INDUSTRIES LTD.

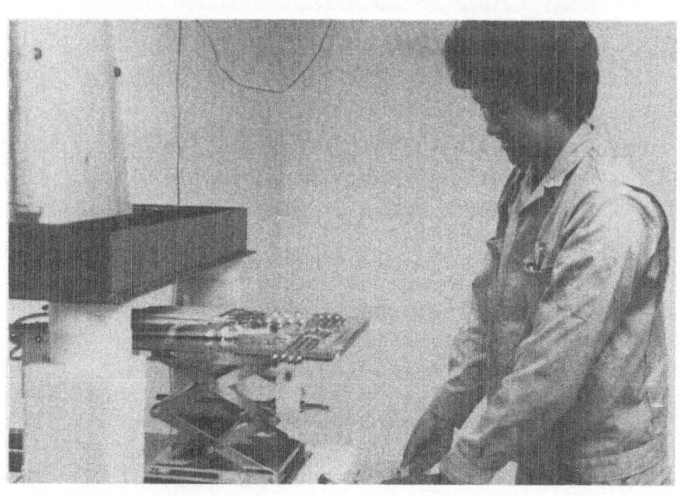

FIG. 3   SETTING OF SAMPLES AND VACUUM CASSETTE FOR VERTICAL
RADIOGRAPHY

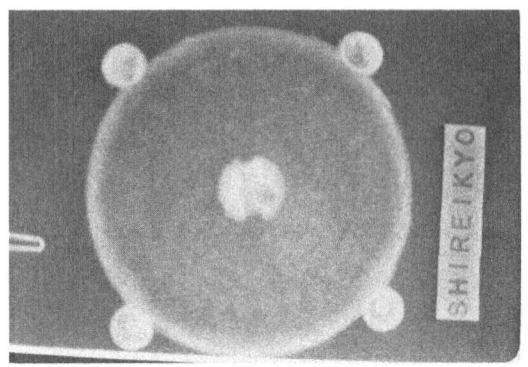

FIG. 4A. NEUTRON RADIOGRAPHY OF ANCIENT BRONZE MIRROR
(SHIREIKYO)

FIG. 4B. X-RAY RADIOGRAPHY OF SHIREIKYO

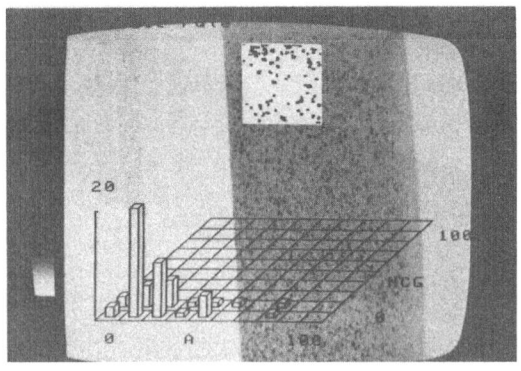

FIG. 5. STATISTICAL ANALYSIS OF GRANULES OF B_4C IN GRAPHITE
FOR AREA(A) AND GRAVITY CENTER DISTANCE(NCG)

KARIN AND KATRIN : A NEW TECHNOLOGY FOR HIGH POWER 14 MEV D-T FUSION NEUTRON
GENERATOR TUBES FOR RADIOTHERAPY, FAST NEUTRON ACTIVATION ANALYSIS, SAFEGUARDS,
NEUTRON RADIOGRAPHY AND LABORATORY APPLICATIONS

Karl A. Schmidt (Kernforschungszentrum Karlsruhe - KfK)

Neutron Radiography; P.O.Box 41905; Im Zeitvogel 5, D 75 Karlsruhe 41
Federal Republic of Germany

INIS DESCRIPTORS: ACCELERATORS, ACTIVATION ANALYSIS, D-T NEUTRON GENERATORS,
FAST NEUTRONS, FUSION, GAGING, NEUTRON RADIOGRAPHY, NEUTRON SOURCES, NEUTRON
THERAPY, NONDESTRUCTIVE TESTING, RADIOGRAPHY, RADIOTHERAPY, SAFEGUARDS,
TOMOGRAPHY, TRITIUM TARGET

ABSTRACT

The Karlsruhe Ring Ion Source Neutron Generator (KARIN) and the
Karlsruhe Target Ring Neutron Generator Proposal (KATRIN) are compact,
sealed high power, high voltage gas discharge accelerator tubes designed for
the generation of 14 MeV neutrons from the fusion reactions of deuterium and
tritium ions. The ions originate in a peripheral ring ion source low
pressure discharge volume and are guided and accelerated to the center to
impinge at a coaxial solid state ScDT metal hydride target sheath on a
central watercooled electrode, to release fusion neutrons which may be
utilized for various irradiation purposes at the target center or externally
near the axis of the system.

Medical and technical applications have been covered partially by the
commercialisation of a KARIN tube prototype and its operational system with
a source strength warranty up to $5*10^{12}$ n/s and 300 h minimum tube life.

Technical neutron radiography has been performed at the Heidelberg
(DKFZ) neutron therapy unit by application of a Pb-converter and
PE-moderator and a adjustable Cd-aperture placed into the radiator head exit
channel. At a distance of 240 cm from the 12 cm diameter aperture a thermal
neutron fluence of $5*10^8$ n/cm² can be realised in one hour irradiation time.
Direct Neutron radiographs with the Gd foil method and track etch foils have
been made for several applications for nondestructive testing.

The proposed KATRIN tube having a ten fold increased target area, would
give a neutron source strength of 10^{14} n/s, which for neutron radiographic
applications is transformed by a central Pb-converter and PE-moderator to an
axially emitted thermal neutron emission flux density of $1.6*10^{11}$ n/(s*cm²)
over a emission area of 12 cm max diameter to be used for high intensity,
large distance neutron radiography and tomography with a movable radiator
head.

THE TUBE "KARIN"

KARIN (s.Fig.1-4) is a 6 ltr volume, 100 kW gas discharge tube with 30 kW target ion beam power with negative polarity 200 kV target voltage, designed and operated to produce a neutron output of $6.5*10^{17}$ n/s from a cylindrical 4.5 cm diameter, 8 cm long central mantle target, onto which the ions impinge radially.

(1) V-M 250 kV feedthrough with watercooling and pneumatic probe "rabbit" guide tubes into the target electrode

(2) Radial DC ion accelerator

(3) Peripheral ring ion source

(4) C-M 20 kV feedthrough with watercooling tubes for ion source ring electrode

(5) DT-P.I.G.-pressure control cell/Ni-leak fill valve

(6) Thinned window for axial neutron beam exit

Fig.1: KARIN Accelerator Vacuum Tube

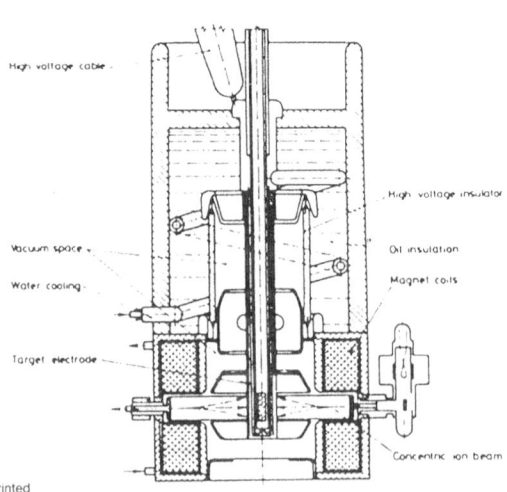

This illustration is printed from the best available original

Fig.2: KARIN Activation Tube Assemblage

240

Applicational Variations

For fast neutron activation purposes the fast neutron flux inside the central hollow target cylinder is utilised. The neutron flux density distribution in this region is nearly isotropic and very homogeneous, because its close proximity to all the neutron emitting target surface elements. From the center saddle point the flux density radially increases, in axial directions it falls. This internal irradiation position with a fast flux density of more than $5*10^{10}$ n/(s*cm²) is accessible by means of a max 25 mm diameter, 50 mm long probe carrier moved pneumatically by an insulated conveyor ("rabbit") system (s.Fig.2). |1,2,4,5,8|

Fast neutron therapy is performed by a modified version with the tube's target being a truncated cone with its apex like a penceltip in the axial direction, facing toward the center of the external irradiation field, thus opening the way for the useful neutron beam to emerge axially, by avoiding shielding of the fast neutron beam by the target struture itself. Here internal access into an internal irradiation position normally is not provided for, because the maximum diameter accessible only is 12 mm (s.Fig.3/4).

This version of the tube also is appropriate for use in external fast neutron, medical, in vivo activations and fast and thermal neutron medical and technical neutron radiography. |3-7|

By means of a Pb-PE radiation transformer at the end of the tube a thermal neutron emission flux in excess of $1.6*10^9$ n/(s*cm²) over an area of max 12 cm diameter may be utilised for short distance (L=2.5 m), large area (D=2.5 m), medium intensity ($1*10^9$ n/cm² per hour) industrial neutron radiography (s.Fig.5)

For laboratory applications it is possible at one tube installation to utilize both options for activation at the internal irradiation position at the center of the target electrode, as well, as external irradiation on the axis outside the target end of the tube. This can be done alternatively or even simultaneously, e.g. together with a probe conveyor system, for Fast Neutron Activation Analysis and a external fixed beam moderator and collimator installation for Thermal Neutron Radiography.

"KARIN" TUBE INSTALLATIONS

Several 14 MeV KARIN neutron generator tube installations have been put in operation since 1976, at several scientific institutions.

Fast Neutron Activation Analysis

At the GKSS KARIN-installation "KORONA" a small, but ultra fast (140 msec) pneumatic probe transport system with a quadratic duct (16x16 mm) is in use (since 1980), which allows repeated activation and rapid transport for counting of fast decaying samples.|13-17|.

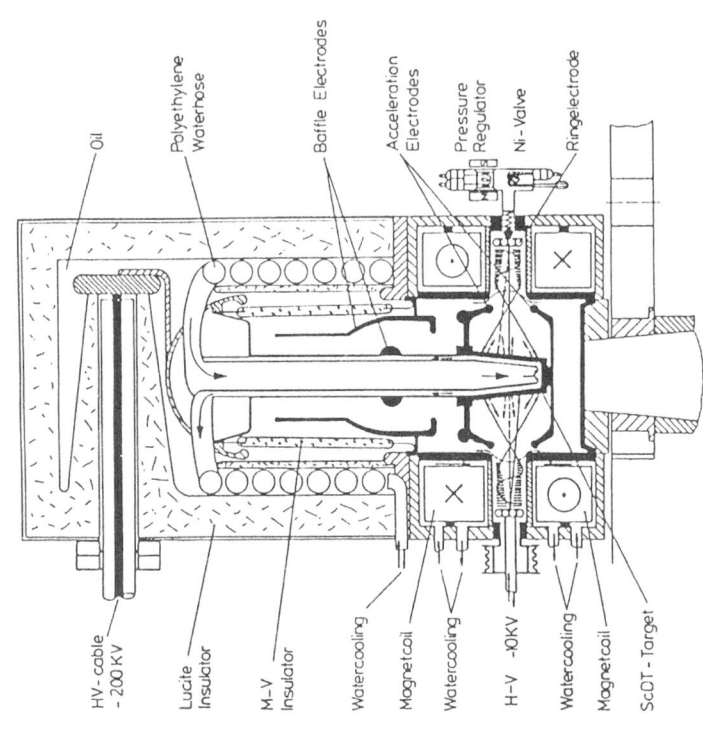

Oil

HV-cable − 200 KV

Lucite Insulator

M−V Insulator

Polyethylene Waterhose

Baffle Electrodes

Acceleration Electrodes

Pressure Regulator

Ni-Valve

Ringelectrode

Watercooling

Magnetcoil

Watercooling

H−V −10KV

Watercooling

Magnetcoil

ScDT-Target

Fig. 3/4: KARIN Therapy Tube Assemblage

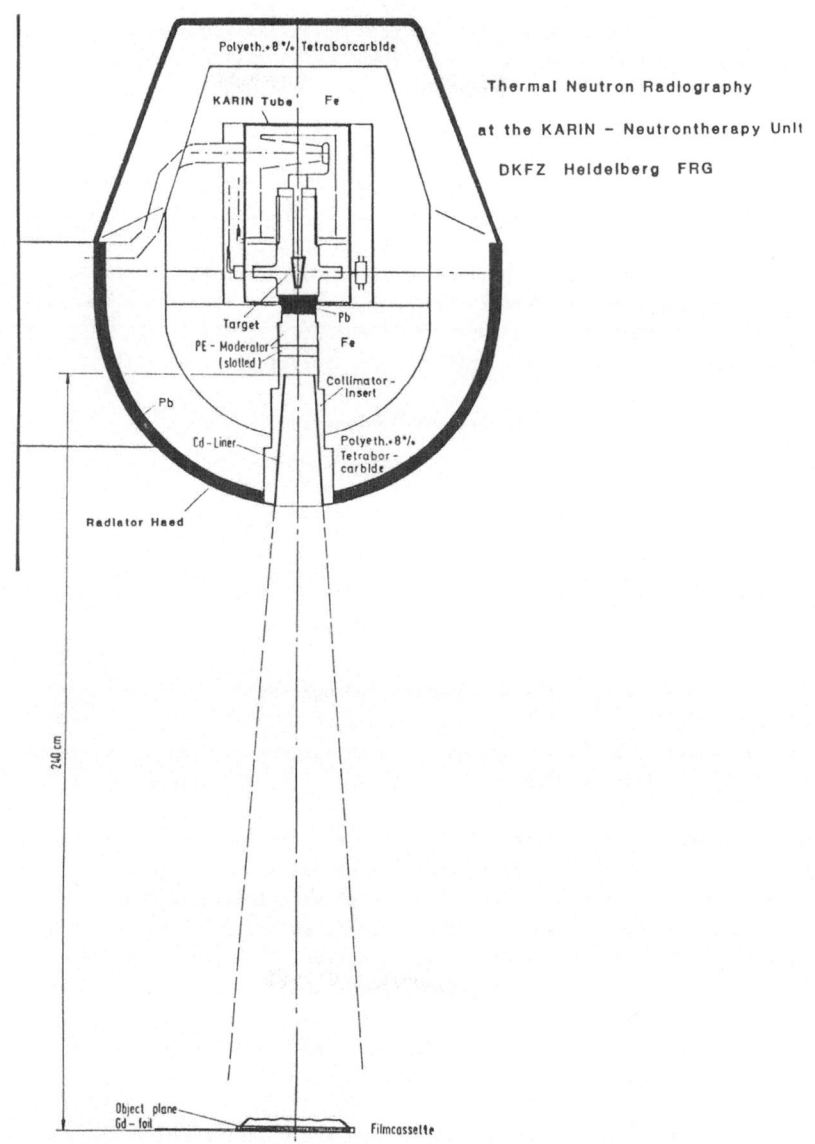

Thermal Neutron Radiography

at the KARIN – Neutrontherapy Unit

DKFZ Heidelberg FRG

Fig.5: Radiator Head Shield with Thermal Neutron Radiography
Accessories: Pb-n-Converter, PE-Moderator and Cd-liner
inserted into collimator receptacle channel exit
Al-X-ray film cassette with 25 μm Gd behind film:
(L/D)=20

243

Fast Neutron Radiotherapy

Clinical radiotherapy with fast neutrons at a KERMA-dose rate of 20 rad/min is performed with KARIN medical therapy tubes at a distance of 1 m from the target center, 200 mm outside of a 700 mm long collimator duct. A radiator head shield - the combined steel, borated PE and lead shell weights about 8.5 tons - is provided with automatically exchangeable steel collimator inserts - which define the actual irradiation field size - and is mounted on a counterweighted gantry isocentrically with respect to the couch, which is used for positioning the patient. The gantry thus is rotatable +/-110 degrees from vertical for moving arc therapy around the patient at the radiotherapeutic installations at the German Cancer Research Center (DKFZ) in Heidelberg (1977) and the radiation clinical installation at the university of Muenster (1984) and is rotatable over the whole circle at the Zuerich university radiological clinic (1980). |3-7|.

Laboratory Application

The neutron source at the "Lotus"-Fusion-Fission Hybrid Test Facility at the ETH Lausanne (1984) is provided for by an unshielded version of the medical tube KARIN.|18|

THE "KATRIN"-TUBE PROPOSAL

Thermal Neutron Radiography

KATRIN would be a 500 kW tube with 300 kW target ion beam power, to be operated with positive voltage from a 320 kV; 1.6 A; 12-pulse single-sided rectifier. It is designed to produce a constant neutron output of $1.*10^{14}$ n/sec from a cylindrical target with 250 mm diameter, 125 mm hight. The tube is of annular construction with an open central free region, which permits insertion of a lead converter combined with a coaxial polyethylene flux trap moderator to form an optimized neutron spectrum transformer, suited to multiply and moderate the fast source neutron emission down to a thermal flux density of about $1.6*10^{11}$ n/(s*cm²). A 120 mm diameter max aperture, yielding about $1.6*10^{13}$ n/s thermal neutron emission, can be utilized for long distance (L=6 m), large area (D=6 m), high intensity (fluence=$1*10^{10}$ n/cm² per hour), short time (6 min for $1*10^{9}$ n/cm²) industrial neutron radiography.

Activation, Radiotherapy, Laboratory Applications

For other purposes, the target and tube dimensions may be scaled down, for use with reduced beam power and source strength, accordingly, but with the higher beam power efficiency of this tube version. This may be done e.g. to rapidly activate to get short-lived nuclei or for fast neutron activation analysis purpose with a pass-through conveyor system, or with a truncated cone target for fast neutron radiation therapy.

Safeguards

A upscaled version with a 800 mm target ring diameter free central space
has been proposed for an active safeguard assay by delayed fission neutron
response (s.Fig.6).

Operat.data:250 KV; 2.5A.
Source strength:10^{14} n/s
Tubelife expectation:800 h

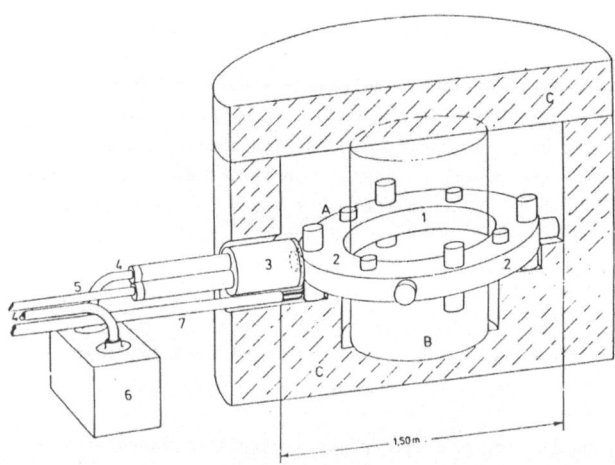

A.Neutr.Generator
  1.Target ring
  2.Ring ion source and
    radial accelerator
  3.High voltage feed-
    through oil cover
  4.High voltage cable
  5.Watercooling and
  6.Powersupply for the
    ion source
  7.Watercooling for
    target and tube body
B.Waste barrel (4001)
C.Graphite reflector for
  delayed fiss.neutron
  measurement.

Fig.6:Target Ring Neutron Generator KATRIN
for waste barrel delayed fission neutron response

PRINCIPLES OF OPERATION

Deuterons and tritons and ionized molecules(78%) are generated from a low
pressure 1:1 tritium deuterium gas mixture within a peripheral ring ion source
region by means of two trapped electron layers. These electrons drift in
radially extended magnetic and essentially axial electric crossed fields in
front of the discharge anodes, approximately with optimum velocity to ionize.
About 90% of the ions thus generated are accelerated and guided radially inward
by the applied acceleration potential, where they impinge on the mantle of a
watercooled central cylindrical target electrode and induce the
neutron-producing reactions. The remaining 10% of the ions are impingeing on the
peripheral ring electrode of the ion source; the negative potential (max 15 kV)
of this ring is used to control the ion current capability of the source by
adjusting the number of trapped electrons in the discharge layers.

Depending on the gas pressure, a minor fraction of the ion flow in the
acceleration gap suffers from charge exchange encounters with neutrals. If a

245

fast postive ion picks up an electron in-flight and becomes neutralized, it cannot be accelerated further and leaves a slow ion from the former neutral behind, which is now accelerated to the target by the remaining potential. Since the acceleration voltage is distributed between particles, in consequence practically no neutrons are released if these particles hit the target.

As the deuterons, tritons, molecular ions and fast neutrals are implanted into the saturated metal hydride target sheath, under stationary conditions, at the same rate neutral gas is leaving the sheath surface. This molecular gas freely flows back to the ion source and is ionized again within the ion source electron layers.

The ions which impinge on the target sheath release the emission of two secondary electrons, on the average, which in the case of the KARIN tube are accelerated back from the target and are guided onto the ion source ring electrode, which thus gets twice the power loading of the target electrode. This power loss is taken to be less expensive, than the costs would be for providing efficient secondary electron suppression means at the negative high voltage target electrode assembly at this tube.

At the tube KATRIN where the target electrode is operated at ground potential, efficient suppression of the secondary electrons is provided for, in order to increase the beam power efficiency accordingly.

Another source of electrons flowing back are photoelectrons released at the negative acceleration electrodes by the light which is emitted from the ion source gas discharge. These contribute at a rate of 30%, refered to the ion beam current, resulting in a power load of the ion source ring electrode and the anodes at the positive end of the acceleration gap which amounts to 30% of the target power.

During operation the equilibrium between gas flow and ion current is controlled by a fast acting acceleration voltage control and a ion current control by the ion source ring electrode voltage which keeps the target and tube power constant as well as the discharge gas pressure from a temperature controlled reversible titanium hydride gas reservoir. The pressure control time, depending on thermal and gas flow time constants being slower than voltage and current control, is used to reduce the ion current losses from charge exchange encounters by reducing the pressure as an secondary adjustment, which results in increased ion source ring electrode voltage for a given current.

For a given target power the steady state condition corresponds to a specific hydrogen deficiency from the saturated ScDT composition. By means of the fast acting current and power control immediate switch-on is acchieved into the same power level, for which this specific target hydrogen concentration had reached its equilibrium value during previous operation and had been frozen-in by fast power switch-off interruption.

246

Operational Experience

The KARIN tubes, constructed as compact, closed glass-ceramic-metal sealed UHV systems, each containing about 500 Ci of tritium, have been operated in about 35 units and have proved the tube's safe operability. Its initial life expectancy of several hundred hours also could be verified to be sound. The source strength of at least $5*10^{12}$ n/s could be utilized routinely for 300 h (warranty). Since the fast acting, tube-protecting power interruption circuitry could be commercialized, it would be possible routinely to apply 200 kV high voltage to achieve a source strength of $6.5*10^{12}$ n/s, and by combined action of reduced current, reduced specific sputtering rate and reduced charge exchange ion current losses the total neutron production per tube may be increased by a factor of two (figure of merit) as compared to the prototype warranty figures.

Tube Operational Times

The usefull tube operating times for both tube versions are limited by ion sputtering through of the target solid state metal hydride sheath, which is made from ScDT filled with 500 Ci (30 kW target power), respectively 5 kCi (300 kW target power) tritium and the same amount of deuterium gas. The sheath being many ion penetration depths thick initially, is approximately matched to the rather even ion current density distribution in the axial direction. Under the cited operational conditions, the life expectation of the target sheath is about 400 h. Operating life time depends on the operational regime and can be increased by reducing the current.

If the quoted output intensity e.g. from shielding-, area-, induced activity or mobility- considerations for a special equipment could be reduced generally, the power and cooling efforts should be reduced accordingly for economic reasons. Consequently, the useful operational tubelife will increase inverse with proportionality to the reduced current, as to allow at least the same total neutron output from the tube. A limitation with respect to the maximum shelf life stems from the radioactive decay of the tritium. More than two years should not be considered realistic for the tube, before it has to be replaced by a new one.

Whereas the evacuated and backed UHV tubes prior to the filling process may be stored unlimited, due to the decay of the tritium the filling in of the tube gas in practice is done in the factory shortly prior to the delivery.

"KARIN" NEUTRON RADIOGRAPHS of TEST SAMPLES
Preliminary Results (s.Figs.7-10)

Provisional neutronradiographic trials have been performed at the (DKFZ) Heidelberg neutron therapy unit, by replacing the long steel collimator duct insert by a 12 cm diameter 21 cm long polyethylene moderator insert, which is formed to have a 6 cm deep 1:1 slotted (1cm) thermal neutron emission end area.

The 6 cm thick lead shutter, which is located next to the target tube output window, and which in therapy operation is provided for radiation protection during patient set up, remains closed now and is used for multiplying and slowing down the neutrons by Pb(n,2n) nuclear reactions. Thus the average neutron energy is reduced to an evaporation neutron spectrum energy distribution, so as to increase the effect of the PE-moderator and to shield the beam exit better from the fast primary neutrons and from the gamma sources being induced by the neutrons in the target structure.

The thermalised neutrons leaving the moderator flow down the stepped hollow receptacle. The thermal neutron beam diameter is defined by the entrance aperture of a conical 12 cm inner diameter Cd-sheath (2 mm thick) liner located half way down the collimator receptacle before the beam leaves the 20 cm diameter opening in the radiator head shield (s.Fig.5).

At a distance of 240 cm from the Cd aperture the objects are placed in front of a X-ray film cassette made from Al in which the one-sided silver halogenid X-ray film sheath is held in close contact to the 25 μm Gd converter foil at its rear side.

At the film plane, 300 cm distant from the target, the primary fast neutron flux density is about 10^6 n/(s*cm²), being attenuated by a factor of .16 from the additional Pb and PE radiation transformer inserts.

Thermal neutrons, due to the compromized geometry are emitted with a (thermal flux density)/(source strength) conversion factor of 1/4000 only, giving an center emission flux density of $1.6*10^9$ n/(s*cm²).

At 240 cm Distance from the Cd aperture with an effective emission area of 100 cm² one gets a thermal flux density of $2.3*10^5$ n/(s*cm² at the film plane, an exposition fluence of $8*10^8$ n/cm² in one hour irradiation time.

The fast flux could be evaluated from the fast neutron dosimetric system to be equivalent to .36 rad/min or $9*10^5$ n/(s*cm²) in the film plane.

The gamma-ray content could be measured to about 1 rad/h, so the neutron to gamma ratio in the beam is about $9*10^5$ n/(cm²*mrad).

Since the change at the DKFZ therapy unit should be marginal only and easy to restore, so as not to affect its principal use, the moderator geometry at the therapy unit is merely provisional for demonstrating first possibilities for a strong, fixed or movable unit. Optimisation of the source moderator geometry would improve the thermal neutron output at least by a factor of two and would also result in a considerable reduction of the remainder of the primary fast neutron flux at the film plane.

FUTURE ASPECTS

Developmental improvements by the application of higher tube voltages are suitable to increase further the source strength as well as the total neutron output per tubelife as compared to the prototype KARIN considerably.

The proposed KATRIN tube version being optimized for high efficiency, as compared to a KARIN facility, would give at about 5-fold tube power a 15-fold fast neutron source strength and, due to the optimized radiation transformer geometry, a 100-fold increased thermal neutron emission flux density with considerably reduced fast neutron content, to be utilized for long distance, large area, high intensity neutron radiography.

Further work will also concern combining the sources with computerized picture-framing detectors to achieve high speed fast and thermal neutron radiographic high resolution mobile scanning systems.|19|

Figs.7:a,b,c,d: Turbine blade core spots doped with 1mg/cm² Gd-Oxide :
a,b: Even distribution / c,d uneven distribution / c:piece of Gd-sheath
25 μm inserted / pieces of CA 80-15B track etch film fixed to the blades

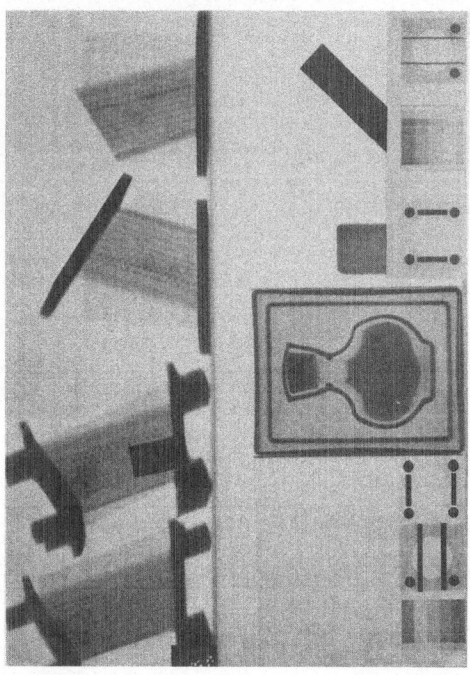

Figs.7:e,f,g,i,j,k: Sensitivity Indicators SI, Beam Purity Indicators BPI,BPI-F
e-f:BPI,SI,BPI-F(from RISO) / i-k:BPI-F,BPI,SI(homemade, on etch Film CA80-15)
f':Gd-sheath 25 μm / g':Rubber 6mm / h:* L'air du Temps * Paris 1986 *

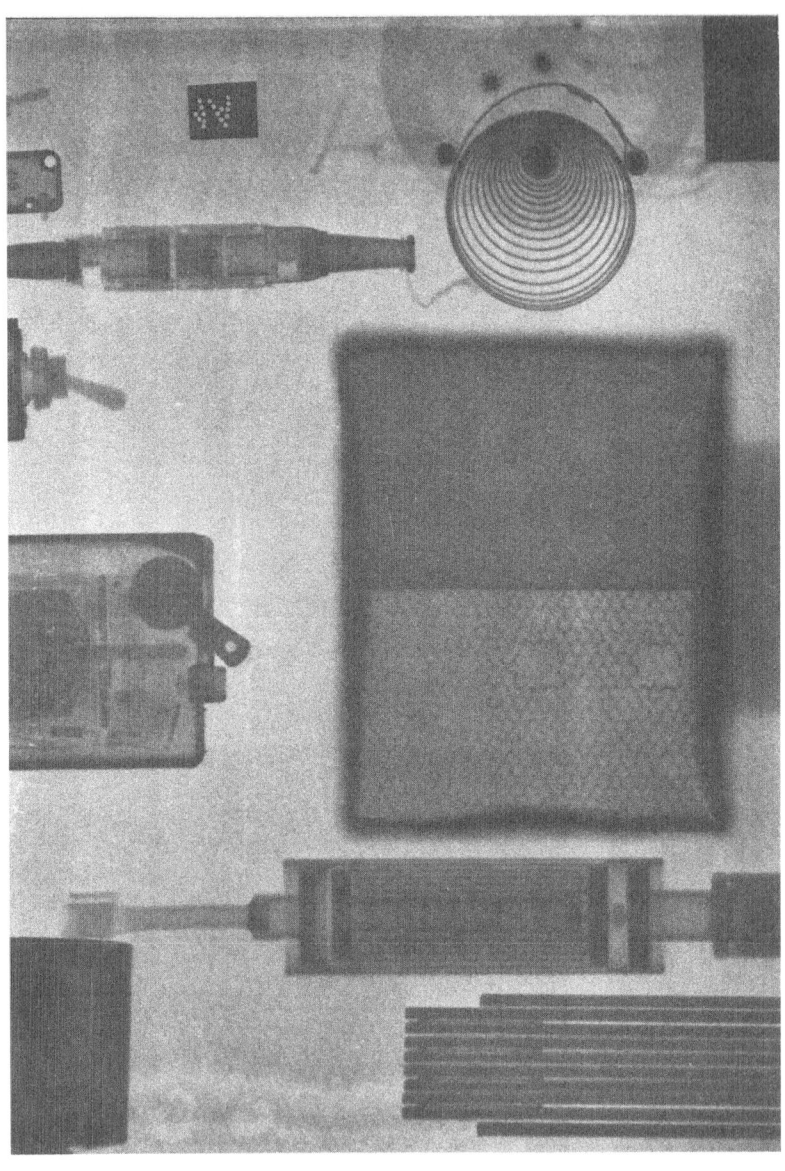

Figs.8 CFK-AL-Honeycomb Adhesive Test: At Spots Adhesion prevented by Teflon :
Predominantly the adhesive, the Al-honeycomb structure very faint only / Large
Fission Chamber FC1000 interior structure

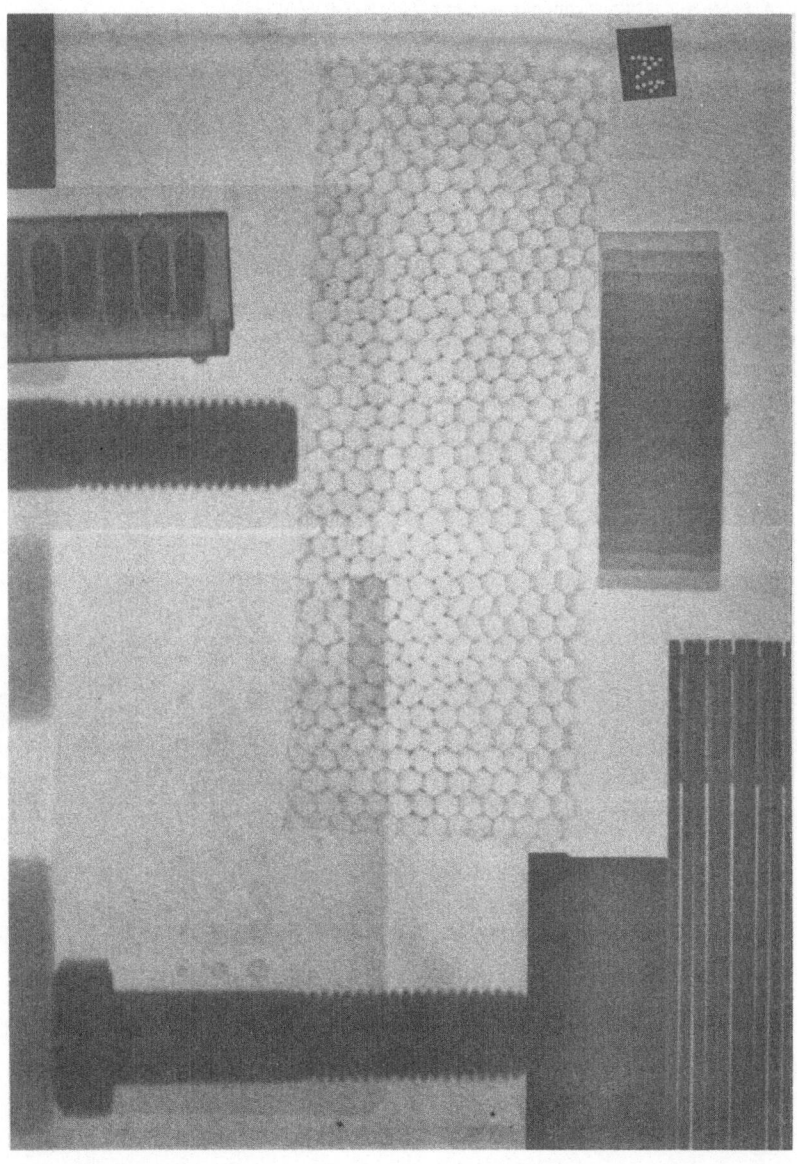

Figs.9: Al-Al(OH)3 corrosion model: grooves depth = .1-1mm, steps = .1 mm;
Diameter:1,2,4,6mm; Fill: powder sediment /Fe-Bolts / Hydrogen-metal: Ti sheath
10x40x1mm a)filled with H (TiH2), b) empty (Ti) / Al-Honeycomb, 10mm thick one
side open / Step wedge 8*1.25mm Perspex plateletts, at centre wirewrapp fix

REFERENCES

|1| K.A.Schmidt : "Neutronengenerator"(Germ.Pat.2112215)
                   (U.S.Pat.3786258)

|2| K.A.Schmidt :Proc.5th Int.Conf.Rad.Research(July1974)Seattle,Wa.
    Abst.E-31-2,299(The KARIN Neutron Generator Tube)

|3| K.E.Scheer : Strahlentherapie 148,440(1974)

|4| K.A.Schmidt : Proc.2nd.Symp.on Neutron Dosimetry in Biology and
    Medicine,Munich-Neuherberg 1974 (Abstr.Eur.5273d-e-f)

|5| K.A.Schmidt, H.Dohrmann : Atomkernenergie (ATKE) 27,159(1976)

|6| K.E.Scheer, K.A.Schmidt, K.H.Hoever : Proc.Conf.on the Inter-
    actions of Neutrons with Nuclei,Lowell,CONF-760715-P1,1162(1976)

|7| K.A.Schmidt,G.Reinhold : Int.J.Rad.Onk.Biol.Phys.3 ,373(1977)

|8| K.A.Schmidt, E.Freiberg : Proc.5th.Symp.Activation Analysis
    Oxford (July 1978)

|9| H.H.Barschall in: Neutron Sources for Basic Physics and
    Applications; S.Cierjacks(Ed.); Pergamon (1983)

|10| H.H.Barschall : AMERICAN SCIENTIST 64,668(1976)

|11| P.Koehler, W.Seifritz, J.Stepanek : Atomkernenergie-Kerntechnik
     (ATKE)41,80(1982)

|12| M.Kuechle, S.Taczanowski, Karlsruhe, KfK: Priv.Comm.

|13| H.U.Fanger, R.Pepelnik, W.Michaelis: J.Radioanal.Chem.61,3(1981)

|14| H.U.Fanger, W.Michaelis, R.Pepelnik: Phys.Bl.38,156(1982)

|15| R.Pepelnik et al:J.Radioanal.Chem.72,393(1982)

|16| B.M.Bahal, H.U.Fanger: Nucl.Instr.and Meth.211,469(1983)

|17| W.Michaelis: GKSS 83/E/77

|18| S.Sahin: Atomkerneneιgie.Kerntechnik(ATKE) 41,95(1982)

|19| L.Steinbock (Cadarache), Karlsruhe, KfK: Priv.Comm.(1986)

PART IV

GENERAL APPLICATIONS

THE APPLICATION OF NEUTRON RADIOGRAPHY TO THE

MEASUREMENT OF THE WATER-PERMEABILITY OF CONCRET *

MO DAWEI,   ZHANG CHAOZONG,   GUO ZHIPING,
LIU YISI,   AN FULIN,   MIO QITIAN
Institute of Nuclear and Technology,
Tsinghua University,
P.O.Box 1021, Beijing, China.
WANG ZHIMIN   LIAN HUIZHAN
Building Material Teaching Group
Civil Engineering Department of Tsinghua University

Abstract

The water-permeability of concret is significant for dam, offshore plateform and underwater basement of bridge etc. The traditional
measuring method of permeability is the fixed
pressure of water method in which the water-
permeating process in a concret block connot be
measured continuously. Owing to the obvious difference of hydrogen content in the permeated
regions of samples and the regions which have not
been permeated. A combination of the neutron
radiography and traditional method has been used
to study continuously the whole process of water
permeating. The combined method overcomes some
shortages of the traditional methods and helps to
gain more informations.

Introduction

The water-permeability of concret is one of the most
important characteristics of concret. Concret contains a

*   Projects support by the Science Fund of the Chinese
Academy of Science

large number of pore which joint one another. The porosity is about 10%-15% by volume. Under some conditions water, ions and other harmful substance may permeate into a concret block with water corroding the reinfocing steels, damaging of building when the water is freezed and shorting livetime of a building.

Hence, studying of the concret permeability, specially water-permeability, is important for dam, ports and hydrarlic engineering facilities. Studying of the water permeability also is significant for the durability of normal buildings.

History

How to measure water-permeability is a long standing problem for studying water-ermeability of concret. Several measuring methods have been adopted in which the fixed pressure permeating is an usual method.[1] This method has been accepted as a standard method in China. Its principle is shown in Fig.1 and Fig.2. The pump controlled by pressuring system supplies the pressured water to a water pot and maintsins pressure from several atm. To several tens atm. According to a predetermined program. The pressured water flowed through a distributor and controlled valves entered one end of several test sleeves. The concret samples were sealed with the test sleeves that the pressured water only can appear in the other end of the sleeves by permeating concret samples.

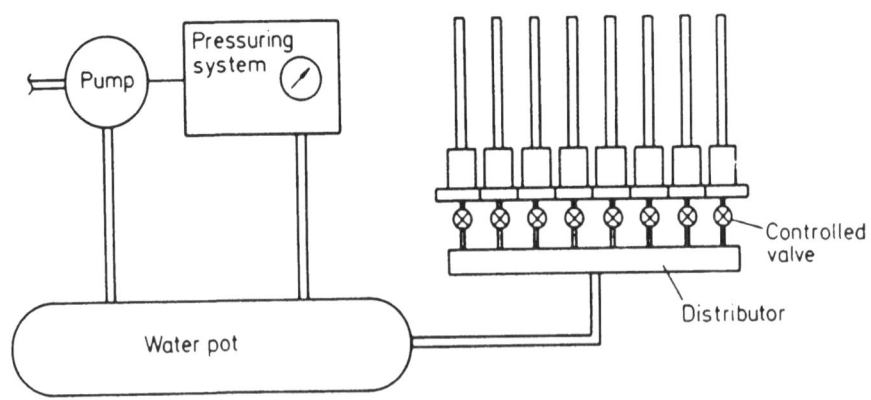

Fig. 1 Permeating water experimental facility

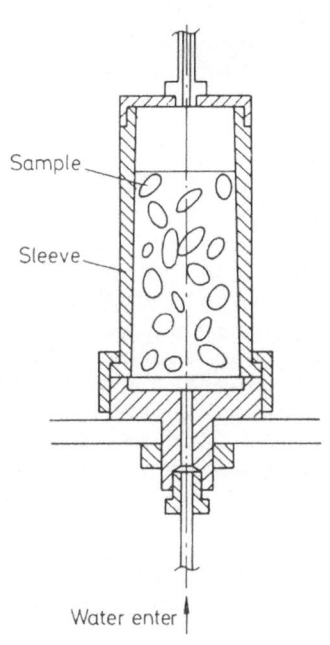

Sample

Sleeve

Water enter

Fig. 2 Details of sleeve

The goals of this kind of measurement are as follows:

A. The pressure of permeating water was increased for 1atm. Every 8hours, until the permeating water was discovered at other ends of three samples out of all tested six samples. At that time the pressure S was record. S-1 is defined as grade of watertighiness of concret. The pressure was increased up to 12atm. If water was not discovered at another end of any three samples, the concret was characterized as S$\geq$12.

B. Under the special pressure and time the samples were examined to see whether thy can satisfy this watertight condition.

C. If the permeating water forms a steady current, the flow rate was measured and permeating coefficient was caculated.

But this method has some difficulties.

A. According to Chinese national standard the sample length of permeating experement is 150mm. If the water-permeating deepth is shorter than this, the water-permeating will not be known.

B. There is no way to gather informations about the water-permeating process and permeating law.

C. If one try to measure the water-permeating deepth cutting away the sample is nomal procedure. In this case the experement can not go farther. If a group of data about water-permeating deepth was demanded, a large number of samples was needed, that arose dispersion of data.

D. To a cquire a group of data a huge amount of work and a long period of time are needed. So it is difficult and expensive to test samples in this way.

Feasibility

Considering all this, we look into the posibility of application of neutron radiography to study water permeability of concret. At 1972 H.Reijon etc.[3] Had studied carbonated layer of concret, using neutron radiography. Basing on his work, we estimated the feasibility as following:

We can see from Table 1. That thermo-neutron mass atteniuation coefficientof concret is dominated by the hydrogen contained. The contained hydrogen mainly exists in hydrated cement and cause background problem in the neutron radiography. To estimete, a concret sample was tested. The sample was baked to constant weight at less than 95°C the 'dry' weight is 408g. Then it was sunk in water. The saturatedweight was 429g. So soaked water is about 0.05g per gram concret. The diameter of sample mass about 4cm. So in the neutron radiography the neutrons have to pass through an additional $0.46g/cm^2$ water or a $0.0575g/cm^2$ of hydrogen. The amount of water approach to present permeating water into concret. Neglecting the contribution of scattered neutrons the density of neutron radiography of the sample without soaked water (D_{cement}) and that of the sample with water (D_{water}) can be calculated the density of film would change due to the water

$$D_{water}/D_{cement} \doteq e^{-\mu_H x_H - \mu_{cement} x_{cement}}/e^{-\mu_{cement} x_{cement}}$$
$$= e^{-\mu_H x_H} = e^{-28.9 \times 0.575} \doteq 19\%$$

The change certainly could be observed, although the scattered neutrons could make it a little worser, but the total absorb length is only $\Sigma\mu x = 6.3$, not big enough to sink the directly penetrating neutron. So it is possible to use neutron radiography to get the information of water permeating.

Table 1

Main elements composition of cement mortar and its total neutron mass attenuation coefficient cm^2/g

element	weight % of cement paste	thermo-neutron total elementofcement	mass attenuation cement mortar
H	1.4	28.9	0.404
O	55	0.158	0.870
Si	33	0.049	0.0163
Al	0.53	0.036	18.9×10^{-5}
Ca	0.89	0.050	4.4×10^{-5}
Fe	0.37	0.146	51.9×10^{-5}
total	99.2		0.5017

Equipment

A.Concret samples were truncated cones. Their upside and downside diameter of samples were 44mm and 51mm correspondingly. The hight was 103mm samples were made from 525 portland cement, UNF II superplasticizers, coarse aggregate 5-15mm and mid-coarse sand. Water/sand ratio was 0.36-0.65. Each cubic meter of concret contained 300-500kg cement by hand mixture and mechanical vibration. The samples had been cured for three monthes under the standard condition and baked at 95^{0}C unitll to constant weight.

B. Pressuring system was shown in Fig 2. The max pressure was $25kg/cm^2$. Eight samples could be pressured at the same time. Each time two samples were examed by neutron radiography.

C. The neutron radiography was got at horizontal hole of research reactor of Tsinghua University. Collimator ratio was about 60, neutron flux was $2 \times 10^3 n/cm^2$.s.kw. The radiography size was about 18x18cm. the converter screen was NE 426, Film was double coated Tian-V, and III (4).

D. Experement was performed at 15 atm. The water was saturated with 1atm air. Usually pressuring time was 6 hours, exposure time was 30 sec. According to the Chinese national standard, if over eight hours the water has not penetrated through the samples the grade of water tightness of cement can be marked as S>12, However the permeability of different S>12 samples is still different. The traditional method cannot observe this difference, neutron radiography can study it.

Result

We had tested permeability of seven group samples. Now we select three group data show on Fig 3. We compared the Experemental data with Darcy formular(4)

$$\frac{1}{A} \frac{dQ}{df} = K \frac{\Delta h}{L} \tag{1}$$

from it, we can obtained

$$L = \sqrt{K(\Delta h/q)} \ t \tag{2}$$

in which

    Q: Quantity of Permeating Water

    A: Section area of Sample

    K: Permeating Coefficient

    L: Permeating Deepth.

    Δh: Experemental Pressure of Permeating water.

    q: Permeatable Porosity of Concret.

In Fig (3) Experemental data was drawn as the points caculated result was drawn as the curves.

We considerated that the neutron radiography is a good way to evaluate concret permeability. This work is only preliminary work, we plan to study the permeability more precisely and posibility of using the small neutron source to take neutron radiography.

Conclusion

A. Technology of neutron radiography supplies a new way to evaluate the permeability of concret.

B. Using this method, a series of data is obtained with minimum dispersion.

C. Two-Dimension dynamic process of water permeating could be obtained.

D. This method prossess a large dynamic range and can measure the grade of water-fightness but this manner also has disadvatage. Due to containing water of concret, the samples cannot be large. We had tested the samples, If their diameters are more than 100mm, the picture will become blurred; Because of contribution of scattering neutron which is scattered by water which is difficultly moved out from concret.

Reference

1. Jacob, Bear, Dynamics of fluids in porous media American Elsevier Publishing Company Inc. 1972

2. 中华人民共和国交通部《港口工程混凝土试验方法》1 9 8 1 北京

3. H. reijonen, etc. "Cement and Concret Research" Vol.2 P.607.1972

4. Mo Dawei etc. "Neutron Radiography Study in Tsinghua University" Second world conference on neutron radiography. 1986.

5. A.M. Neille Properties of Concret Pitman Publishing Limited (London Great Britain, 1981)

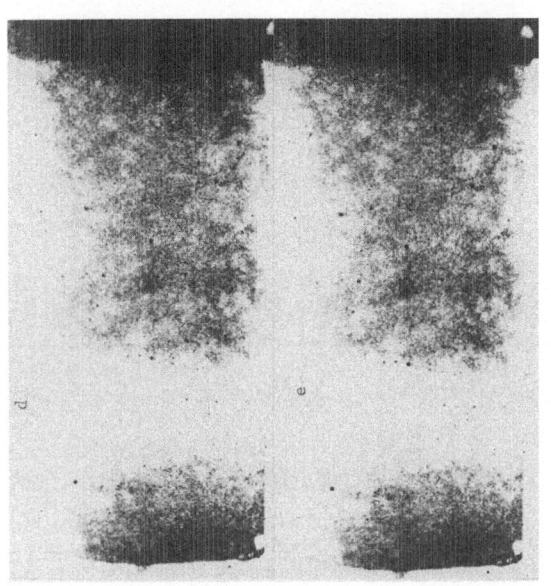

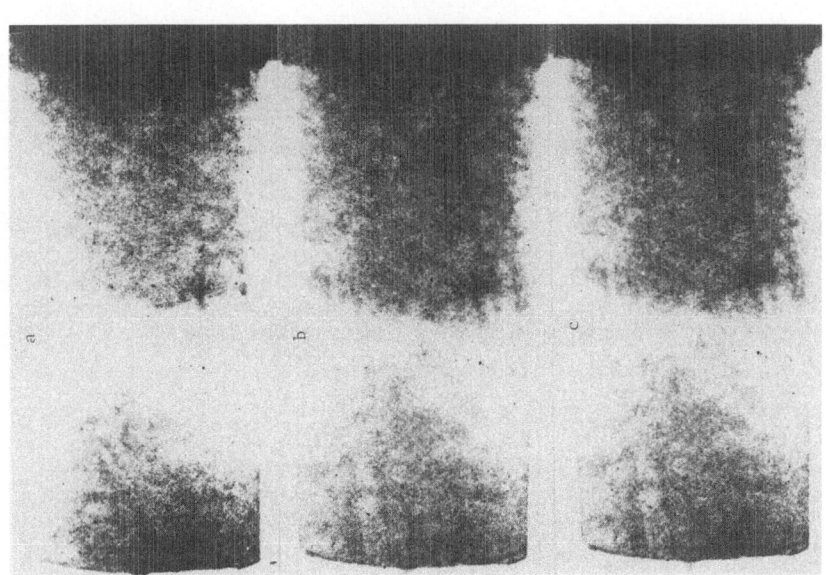

Fig. 3   A group of permeating neutron
radiography

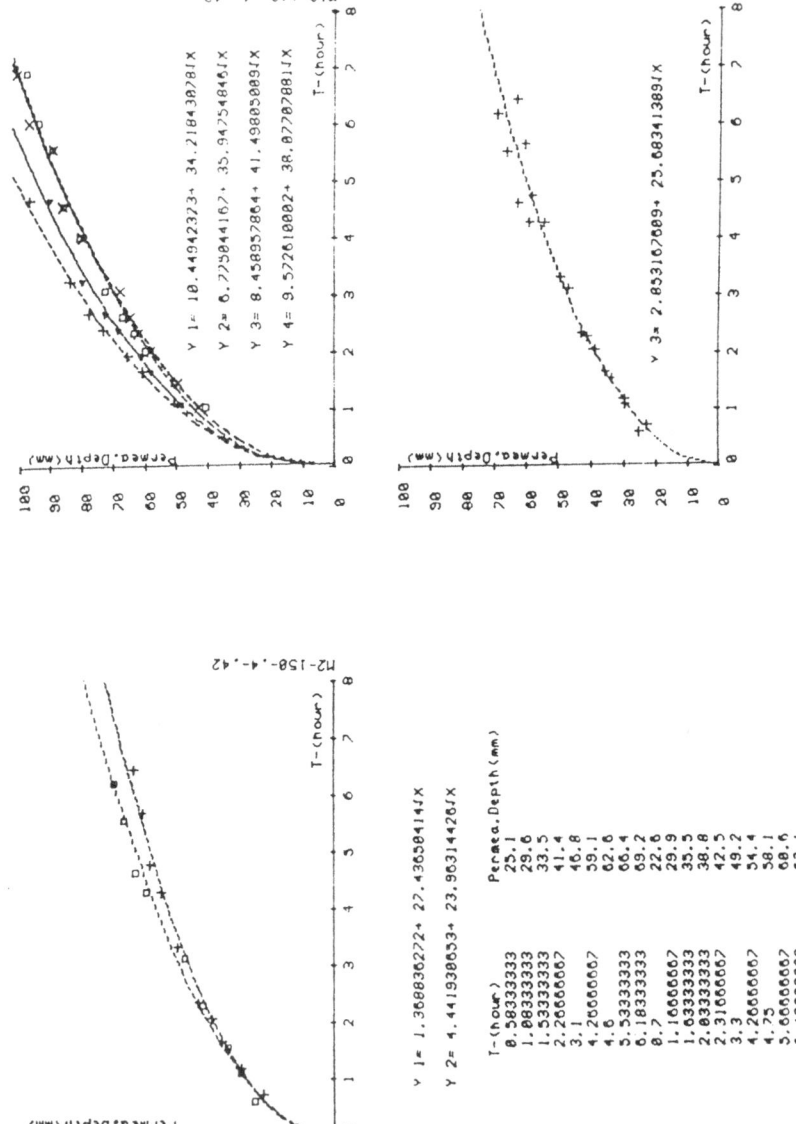

Fig. 4   The result of the experiment

262

ASPECTS QUALITATIFS ET QUANTITATIFS DU CONTROLE NEUTRONOGRAPHIQUE

APPLIQUE AUX CHAINES PYROTECHNIQUES DU LANCEUR ARIANE

Jean-Pierre BOULOUMIE

Centre National d'Etudes Spatiales

Toulouse   France

Le contrôle neutronographique est l'un des moyens d'accés privilégié aux caractéristiques intrinsèques des composants pyrotechniques. Cette méthode de contrôle non-destructif a été associée dés l'origine au développement des systèmes pyrotechniques du lanceur. Après 14 lancements réussis, et devant un besoin croissant de fonctions pyrotechniques sur les nouvelles versions du lanceur, il est fait un premier bilan de son utilisation.

1. LA NEUTRONOGRAPHIE DANS LE DOMAINE DE LA PYROTECHNIE SPATIALE

1.1. L'intérêt du contrôle neutronographique

La fiabilité intrinsèque des systèmes pyrotechniques spatiaux est étroitement conditionnée par la possibilité du contrôle de leur configuration interne et de leurs performances. La conception rigoureusement hermétique du conditionnement de ces composants, leur mode de fonctionnement irréversible et destructif et les faibles quantités développées, imposent une approche maximum de leurs caractéristiques physiques dans toutes les phases de leur utilisation. D'où l'importance des méthodes de contrôle non-destructif, et en particulier de type radiographique. La radiographie X apportait une information sur la structure métallique du composant, la neutronographie est venue la compléter en autorisant un accés direct au contrôle des compositions pyrotechniques, des matériaux plastiques et des dispositifs d'assemblage et d'étanchéité. La complémentarité de la neutronographie apparaît immédiatement dans une comparaison des coefficients d'absorption de quelques éléments pour les neutrons et pour les X (Cf. tableau 1).

COEFFICIENTS D'ABSORPTION

ELEMENT	NEUTRONS (1,08A)	X(0,098A)
Pentrite(PETN)	2,545	0,255
Hexogène(RDX)	2,894	0,266
Azoture de plomb	0,783	12,15
Bore/Nitrate K	8,588	0,344
Poudre GBSE	2,15	0,225
Eau	5,53	0,150

Tableau 1

1.2. La pyrotechnie du lanceur ARIANE

L'utilisation croissante de la pyrotechnie dans les systèmes spatiaux est liée à la spécificité des performances de ces composants: ils possèdent en effet une énergie potentielle de niveau trés élevé et un temps de réponse trés court pour un volume, une masse et une énergie d'activation trés faibles. Sur le lanceur ARIANE ils interviennent dans toutes les phases de la séquence de vol jusqu'à l'injection sur orbite de la charge utile:

- Fonctions d'allumage des moteurs (propulseurs d'appoint du 1° étage, impulseurs dé séparation, moteur à ergols cryogéniques, moteur d'apogée),
- Fonctions de séparation (ouverture des crochets de table de lancement, séparation des propulseurs d'appoint, séparation des étages, séparation des impulseurs d'accélération des étages, séparation de la coiffe, séparation des charges utiles),
- Fonctions de sauvegarde (destruction des 3 étages).

La localisation de ces différentes fonctions est matérialisée sur le schéma synoptique suivant (Cf. schéma 1).

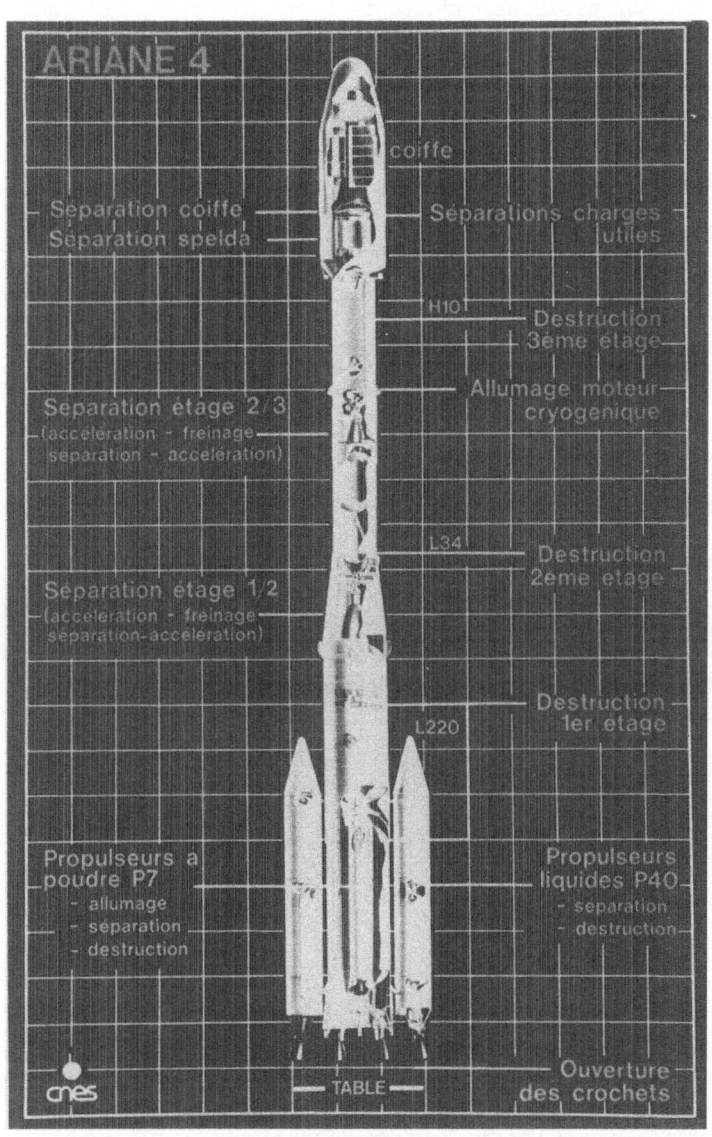

ARIANE 4

Séparation coiffe
Séparation spelda

coiffe

Séparations charges utiles

H10

Destruction 3ème étage

Séparation étage 2/3
(accélération - freinage
séparation - accélération)

Allumage moteur cryogénique

L34

Destruction 2ème étage

Séparation étage 1/2
(accélération - freinage
séparation-accélération)

L220

Destruction 1er étage

Propulseurs à poudre P7
- allumage
- séparation
- destruction

Propulseurs liquides P40
- séparation
- destruction

cnes

TABLE

Ouverture des crochets

Planche 1

FONCTIONS PYROTECHNIQUES

Avec les nouvelles versions d'ARIANE en cours de développement, le nombre des composants utilisés, et en conséquence soumis au contrôle neutronographique, est en constante augmentation. Le tableau suivant montre l'évolution des besoins sur quelques composants représentatifs (Cf. tableau 2).

COMPOSANT	ARIANE 2	ARIANE 4
Initiateur électrique	24	54
Initiateur à onde de choc	36	64
Centrale d'initiation	8	24
Elément de transmission à cordeau détonant	153/306 relais	240/480 relais
Raccord de distribution	12	32
Relais retard	12	20
Réglette de destruction	8	20
Réglette de séparation	24	42
Pyrotechnie/Coût lanceur	2,5%	4%

Tableau 2

1.3. La réalisation du contrôle neutronographique

Deux installations de neutronographie industrielle sont disponibles en France au Commissariat à l'Energie Atomique (C.E.A.):

- ORPHEE: Située sur le site du Centre d'Etudes Nucléaires de SACLAY,
Créée spécifiquement pour le contrôle industriel à l'extrêmité d'un guide à neutrons de 70 mètres sorti du coeur de ce réacteur expérimental.
Flux de neutrons thermiques : 10^9 n/cm^2.s
- MELUSINE: Située sur le site du Centre d'Etudes Nucléaires de GRENOBLE,
Constituée à partir de l'équipement du réacteur TRITON (FONTENAY-AUX-ROSES) utilisé entre 1970 et 1980 pour l'industrialisation du

contrôle neutronographique, et en service sur un
faisceau "sortie" de ce réacteur expériemental.

On dispose pour ces contrôles de normes AFNOR relatives
au contrôle de l'installation et des films, et de procé-
dures normalisées de contrôle définissant en particulier
des indicateurs de qualité d'image (I.Q.I.) spécifiques
pour chaque type de composant.

2. EXPLOITATION QUALITATIVE ET QUANTITATIVE

Le contrôle neutronographique intervient dans le
développement des systèmes pyrotechniques selon trois types
de processus:

- Au niveau de la conception de produits nouveaux en
  tant que moyen d'analyse fine avant et aprés le
  fonctionnement,
- Au niveau du contrôle systèmatique en réception de
  lots de composants ou de sous-ensembles (Cf.
  Spécifications Générales ARIANE SG-1-31 et
  AQ-1-33-120),
- Au niveau des expertises dans le cas d'analyses de
  défaillances ou de dérives de fabrication.

2.1. Synthèse d'une exploitation qualitative

L'utilisation systèmatique du contrôle neutronographique a conduit à définir pour la pyrotechnie un
certain nombre de critères d'acceptation des composants
ou des sous-systèmes:

- Homogénéité du chargement pyrotechnique : densité,
  cohésion, absence de corps étrangers, absence
  d'humidité ou de traces de migrations (collages),
- Continuité de l'âme explosive dans les cordeaux
  détonant,
- Jeux pyrotechniques (entre chargements ou aux
  interfaces de composants) conformes aux tolérances
  du dossier de définition,
- Homogénéité des systèmes d'étanchéité (joints,
  résine polymérisable, traversées étanches),
- Homogénéité des technologies d'assemblage (brasures,
  collages, sertissages),etc...

2.2. Exemples d'un utilisation quantitative

Deux exemples récents illustrent ce mode d'exploitation
utilisé lors des analyses de défaillances. Dans les deux
cas la localisation de la dérive des performances a pu être
établie au niveau du seul contrôle non-destructif.

ALLUMEUR PYROTECHNIQUE (Cf. planche 2)

Définition: Ce composant est constitué par un petit bloc
de propergol (40 g), de deux initiateurs électriques et de

deux charges relais en redondance. L'ensemble est condi-
tionné dans une enveloppe massive en acier inoxydable et
scellée par soudage par bombardement électronique.

Défaillance: Trois ans aprés la qualification du composant,
lors d'un contrôle de recette de lot, on a constaté une
valeur de résistance d'isolement totalement hors tolérances.

Expertise: L'analyse qualitative des clichés neutronographique n'apportait aucune information spécifique.
Les performances de l'allumeur, lors des essais fonctionnels,
étaient nominales.
Une analyse microdensitométrique comparative des clichés
neutronographiques au niveau de la charge d'initiation, sur
des allumeurs du lot incriminé et d'un lot plus ancien,
devait révéler une variation anormale de la densité
optique du chargement (40 mg) qui passait de 0,32 à 0,46.
A l'ouverture de l'allumeur, on constatait une bakélisation
du chargement pyrotechnique de l'initiateur dont l'origine
a été associée à une dérive de la procédure de soudage B.E.

EXPLORATION MICRODENSITOMETRIQUE SUR UN ALLUMEUR

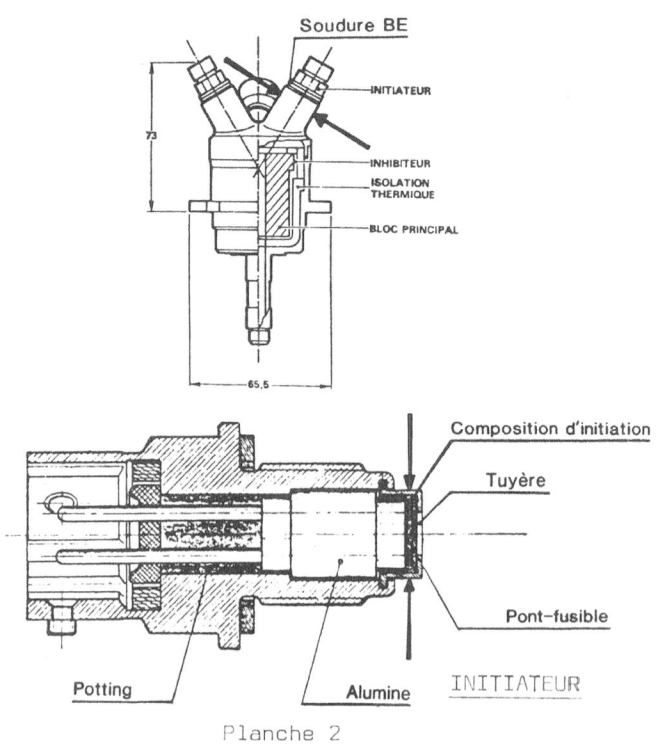

Planche 2

RELAIS DE TRANSMISSION DE CHAINE DETONIQUE (Cf. planche 3)

Définition: Ce composant est constitué d'un étui métallique
et d'une charge explosive à taux de compression élevé (50mg).
Il est contrôlé unitairement puis intégré aux éléments de
chaîne de transmission par sertissage et collage aux extré-
mités du cordeau détonant. Dans cette configuration il est
à nouveau soumis à un contrôle neutronographique unitaire.

Défaillance: Cinq ans après la qualification, lors des essais
de recette des lots 1985, on a constaté une diminution de
l'ordre de 30 à 40% des performances liées à la transmission
de l'onde de détonation.

ANALYSE MICRODENSITOMETRIQUE D'UN RELAIS DE TRANSMISSION

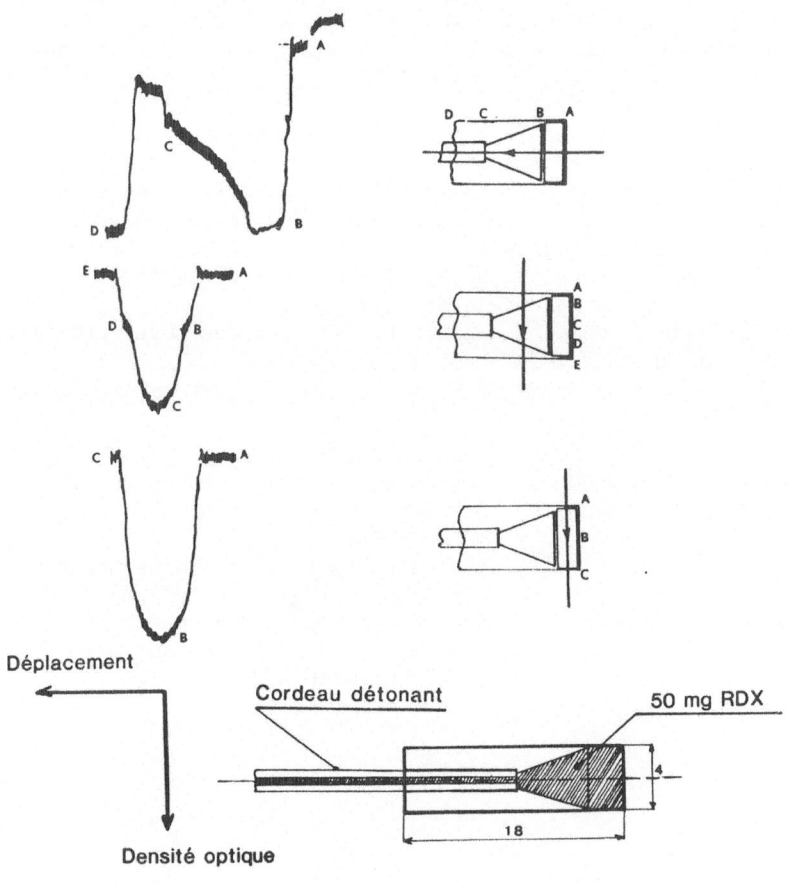

Planche 3

Expertise: L'analyse qualitative des clichés neutronographiques n'apportait aucune piste de recherche.
Une étude comparative des spectres de densité optique réalisés selon 3 axes du chargement et sur des composants des différents lots 1985 confirmait l'homogénéité des densités de chargement.
L'origine de la défaillance a été finalement identifiée à la suite d'une campagne d'essais comme une diminution du taux de compactage de l'explosif. La disponibilité de composants ou de clichés neutronographiques de lots antérieurs aurait probablement permis par analyse comparative de détecter beaucoup plus rapidement cette dérive de fabrication.

Dans les années à venir, avec en particulier l'apparition de vols habités, la compétition sera extrêmement serrée au niveau de la promotion des matériels spatiaux, compte-tenu des exigences de plus en plus impératives de fiabilité, de qualité et de sécurité attachées à la pyrotechnie.
La neutronographie apparaît à l'heure actuelle comme le seul moyen d'accés à caractère non-destructif au chargement actif des composants pyrotechniques. Il est donc fondamental de normaliser dés le stade de la conception l'utilisation de cette méthode de contrôle, sans exclure celle de la radiogrpahie X qui reste son complément naturel:
- dans un domaine qualitatif au titre du contrôla qualité à caractère systèmatique,
- dans un domaine quantitatif en tant que moyen privilégié d'expertise.

Bibliographie:

- A. LAPORTE   Etude CNES 74/723/CEA/CENFAR/TRITON 1974-1980,
- A. LAPORTE/J.P. BOULOUMIE   Jahrestagung ICT KARLSRUHE 06/1975,
- A. LAPORTE/J.P. BOULOUMIE   9th World Conference on NDT
                               MELBOURNE 11/1979,
                               Explosives and Pyrotechnics-Space
                               Applications TOULOUSE 10/1979,
- J.P. BOULOUMIE   3rd International Conference on
                               Reliability and Maintainability
                               TOULOUSE 10/1982.

DEVELOPMENTS IN THE USE OF THERMAL NEUTRON RADIOGRAPHY FOR STUDYING
MASS TRANSFER IN A PARTIALLY FROZEN SOIL
by

M A Clark, Dr R J Kettle and G D'Souza
respectively of
Department of Construction and Environmental Health
Bristol Polytechnic
(formerly of Department of Civil Engineering, Aston University)
Department of Civil Engineering, Aston University
Remote Sensing Unit, Bristol University

ABSTRACT

A technique for studying water movements in a partially frozen
soil, using thermal neutron radiography, has been devised by the
author. This paper is intended as an update of the work, including
recent experimental improvements and developments in analysis of
results. The work has been carried out at the Universities Research
Reactor, Risley, and has utilized a beam from a 200mm x 100mm
horizontal access hole of base flux 1.2×10^{12} n cm^{-2}s^{-1}, which
gives a flux of 7.5×10^{7} n cm^{-2}s^{-1} at the radiographic position.
The neutron beam is used to obtain radiographs of an artificial,
Snowcal, soil matrix subject to partial freezing. The effects of
sub-zero temperatures on soils, which are frost susceptible and
have available water, are complex and result in the formation of
bands of ice, ice lenses, in the soil. The radiographs show the
formation of ice lenses and their analysis has provided
quantitative data on relative water contents, within the soil matrix,
throughout the test. Thermocouple/psychrometer instrumentation in
the matrix has provided information on water potentials and by
combining this with the water content data, hydraulic conductivity
determinations for the soil matrix have been made.

INTRODUCTION

Large areas of the Earths land surface are subject to freezing
conditions. At the Earths middle latitudes most of the land is
subject to seasonal freezing and at the high latitudes permafrost
is predominant.

271

Problems associated with frost action are of particular importance in Canada, Alaska and the USSR where permafrost covers a total area of approximately 22 million square kilometers and where differential frost heave and thaw weakening cause damage to buildings, roads and pipelines.

To evaluate the mechanism of frost heave in partially frozen soil, hydraulic conductivities must be determined. To achieve this a quantitative assessment of water movements in the soil throughout the freezing process is essential.

The technique reported here employs neutron radiography as a means of determining relative water contents in a soil matrix subject to partial freezing. The radiographs are produced by the use of a thermal neutron beam and enhanced using 'Remote Sensing' facilities. Information from instrumentation incorporated in the matrix, in the form of thermocouple/psychrometers electrical resistance probes and thermocouples, is then combined with the water contents to determine hydraulic conductivities.

The various stages in the development of the technique have been reported on in previous papers (Clark & Kettle 1985 a, b). This paper gives the results of a final test using the latest in a series of experimental set ups.

APPARATUS

The snowcal (powdered chalk) matrix is housed in a rig constructed from materials with low neutron adsorption coefficients. Three hollow rectangular sections of PTFE, stiffened with aluminium plates make up the main cell. The matrix is located in the middle section sandwiched between a freezing head (top section) and a water bath with a water temperature regulation unit (bottom section).

Juxtaposed to the main cell is a reference cell constructed as for the main cell and containing the matrix at three known water contents. Attached to the reference cell is a watch on which the hands and hours are marked with sections of plastic tube.

A light-tight container (LTC) containing an X-ray film plate and gadolinium intensifying screen is located behind the cells at times of radiographic exposure.

Radiographs are obtained at regular intervals throughout the test. For each radiograph the time of exposure is shown by dots on the image formed by the plastic tube sections on the watch. The variations in grey shades in the matrix are compared with those of the reference cell to determine water contents in the matrix.

272

Temperature control of the matrix is facilitated by circulating ethylene glycol/water mixture through the water temperature regulation unit and freezing head using cryostats. Air temperature around the cells is maintained by circulating ethylene glycol/water mixture through a series of copper tubes surrounding the rig. The whole set up is then installed in a box made from plywood and paraffin wax sections for insulation. Access for the neutron beam to the cells is made possible by a PTFE window installed in the front section of the box.

The box is positioned in a cavity in the concrete shielding and accessed via two channels formed in the concrete. One channel carries the instrumentation leads and cooling tubes whilst the other, a narrow slit, is used to position the LTC in the box. Suspended from the ceiling of the slit are rubber strips, set so that the LTC disturbs the strips which then fall back into place to thermally seal the passage.

The experimental set up is depicted in figures 1, 2 and 3. Figure 1 shows a plan view through the shielding showing the cavity and access channels. Figure 2 gives a section through the insulated box revealing the test rig and figure 3 is an exploded view of the main cell.

INSTRUMENTATION

The water supply to the matrix is from a marriote vessel. Within the vessel a polystyrene float connected to a linear motion potentiometer, rests on the water. In this way any water uptake by the matrix is represented by a movement of the float which is registered by the potentiometer and recorded on a chart recorder.

Wescor PST 55-30 psychrometers and a Wescor PR55 psychrometric microvoltmeter are used to measure water potential in the soil matrix. The psychrometers incorporate a small chromel constantan thermocouple concentrically mounted at the centre of a hollow cylindrical stainless steel bulb. This design is chosen as it minimises the effects of temperature gradients which adversely affect the suction readings. The single junction thermocouple permits measurement of these thermal gradients to enable correction of the suction readings. A calibration model for this type of thermocouple psychrometer has been developed by Brown and Bartos (1982) and has been used in this test.

Water potentials are measured with the psychrometers using the peltier effect. Water from the matrix is conducted through the bulb to the inner surface where it evaporates until the humidity. approaches 100%. An electric current is passed through the thermocouple for fifteen seconds cooling the sensing junction slightly below ambient temperature so that water from the atmosphere condenses on the thermocouple.

273

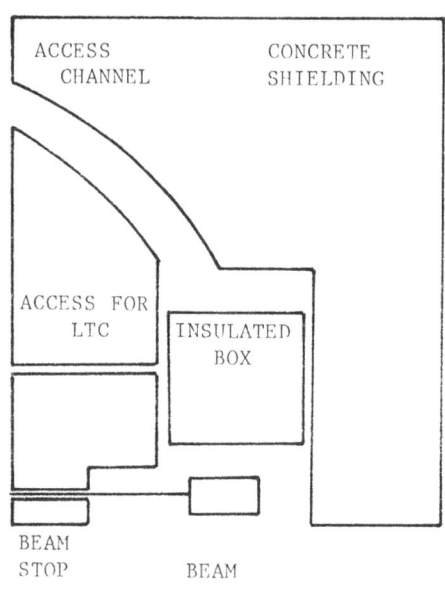

FIGURE 1: PLAN VIEW OF
THE CAVITY IN THE CONCRETE
SHIELDING

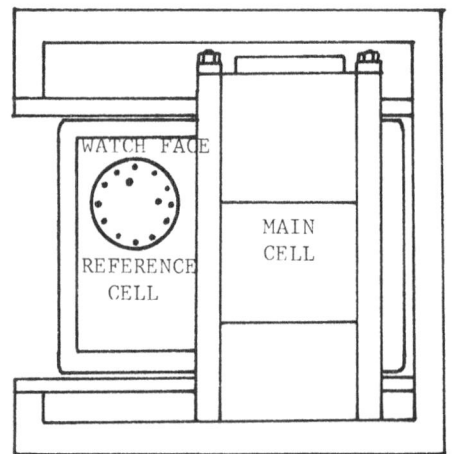

FIGURE 2: SECTION THROUGH INSULATED
BOX REVEALING TEST RIG.

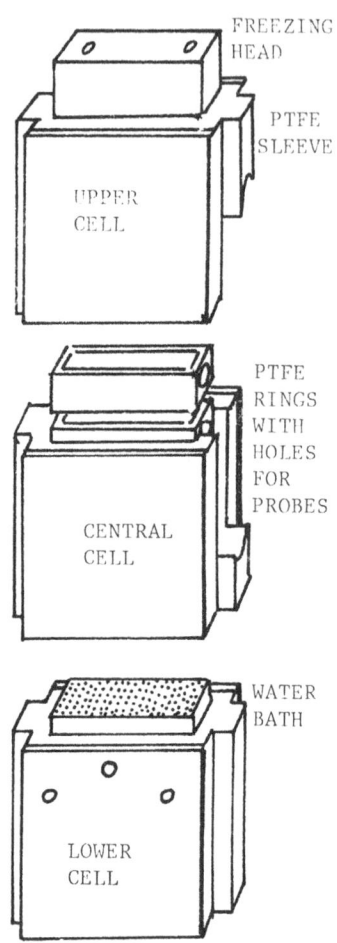

FIGURE 3: EXPLODED VIEW
OF MAIN CELL

After cooling the condensed water evaporates which again cools the junction. This time however, the cooling is a function of the rate of evaporation which is a function of the vapour pressure of the atmosphere which in turn is a function of the water potential of the matrix.

Copper-constantan thermocouples are located alongside the psychrometers to record the matrix temperature throughout the test.

The electrical resistance of the soil between the psychrometer probes is determined throughout the test to identify ice in the matrix. This is achieved through connections to the stainless steel shields of the psychrometers. The shields are insulated from the rest of the probe by a plastic casing and from each other by the PTFE rings which house them.

METHOD

The work has been carried out at the Universities of Manchester and Liverpool Research Reactor using a thermal neutron beam from a horizontal access hole, which gives a maximum neutron flux of $7.5 \quad 10^7 n \ cm^{-2}s^{-1}$ at the radiographic position. An exposure time of 18 seconds is used for each radiograph at a reactor power of 100kw.

Loading the Matrix

With the three cells of the main apparatus in position and all instrumentation in place the matrix is loaded, via the upper cell, into the six 12.5mm deep PTFE rings of the central cell. The matrix is compacted in ten equal layers. Each layer contains a specified quantity of matrix (dependent on the water content) and is compacted to a thickness of 7.5mm. In this way the compaction is standardised and no compaction plane is incident with a ring intersection.

Test Procedure

With the apparatus positioned in the cavity formed in the concrete shielding, and all connections made to instrumentation and cooling equipment, cooled ethylene glycol/water solution is circulated through the copper tubing and left on overnight to lower and stabilise the matrix temperature. An initial radiograph is then taken before freezing is initiated, by circulating cooled ethylene glycol/water through the freezing head. Subsequent radiographs are obtained at approximately half hour intervals throughout the test.

The same process is used to obtain each radiograph.

With the LTC positioned behind the cell the beam stop is withdrawn allowing the thermal neutron beam to fall on the cell, and replaced after the exposure time of eighteen seconds has elapsed. When the LTC has radioactively 'cooled down' it is removed and the photographic plate developed.

RADIOGRAPHIC MATERIALS

FILM   -   'KODAK' INDUSTREX CX, Medium Speed, fine-grain, high-contrast, direct exposure film.

DEVELOPER   -   'KODAK' LX24 X-RAY DEVELOPER

STOP BATH   -   'KODAK' LX INDICATOR STOP BATH

FIXER   -   'KODAK' FX-40 X-RAY LIQUID FIXER

HARDENER   -   'KODAK' HX-40 X-RAY LIQUID HARDENER

RADIOGRAPHIC ENHANCEMENT AND ANALYSIS

Contact prints are made of the radiographs(negatives)and 35mm slides (transparancies) made of these prints. The transparencies are used in a scanning microdensitrometer. The size of the soil matrix on the original negative is 72.5mm x 84mm and on the 35 mm negatives this is reduced to 15mm x 17.5mm. The 35mm negatives are scanned with a spot size of 100 µm (0.1mm) and hence the soil matrix area is covered by 150 x 175 pixels. Thus each pixel represents an area of 0.5mm x 0.5mm on the original negative.

The negatives are scanned at 3-D optical density, levels of 0 - 255 are assigned to the degree of brightness. The scanner sets white on the negative to zero and black to 255. Therefore the reference cell of 100% water content has higher digital count values than those of the 0% water content reference cell.

The scanned data are displayed on an 1^2S Image Analyser.

To account for any differences in exposure, printing and scanning, the frames are individually normalised using the 0% and 100% water reference cells.

For any one frame, the digital counts less than those within the 0% water reference cell are set to zero (black) and those equal to or greater than those within the 100% water reference cell are set to 255 (white) to produce an enhanced black and white image. Any values within these extremes are then linearly stretched to that the new digital count becomes:-

$$X_n = ((X - X_0) / (X_{100} - X_0)) \times 255$$

where:
X_n = New normalised digital count

X = Old digital count

X_{100} = Digital counts of 100% water content reference cell

X_0 = Digital count of 0% water content reference cell.

The soil matrix areas with water contents of 0% - 100% are thus enhanced to give a range of 256 grey levels. In order that the water content variations can be more easily seen the frames are density sliced in colour to give false colour images. The distribution of neutrons across the width of the beam is uneven, the intensity decreases towards the edges. This is accounted for in the analysis by dividing each image by the first image as the water content distribution in the matrix for the first image is uniform.

RESULTS

To illustrate the technique a radiograph, taken towards the end of a three hour freezing experiment is shown. The radiograph shows the outline of the constituant parts of the central cell, reference cell and the matrix with fully developed ice lenses. Also shown is the enhanced and normalised black and white image of the radiograph.

The graph shows the temperature, suction and electrical resistance profiles in the matrix for the instant recorded in the radiograph.

CONCLUSIONS

Variations in water contents in the matrix are discernable in the radiograph and with the enhancement distinct contrasts in grey shades can be detected facilitating accurate and repeatable determinations of water contents.

The development of ice lenses in the soil is reflected in the profiles shown in the graph. The increase in electrical resistance and suction in the top of the soil and the sub-zero temperatures coincide with the increased water content shown in the radiograph indicating the formation of ice lenses

The information on water distribution and ice lens formation can now be used in conjunction with the suction, electrical resistance and temperature for the soil through the test to establish hydraulic conductivities.

RADIOGRAPH WITH CORRESPONDING NORMALISED AND ENHANCED IMAGE

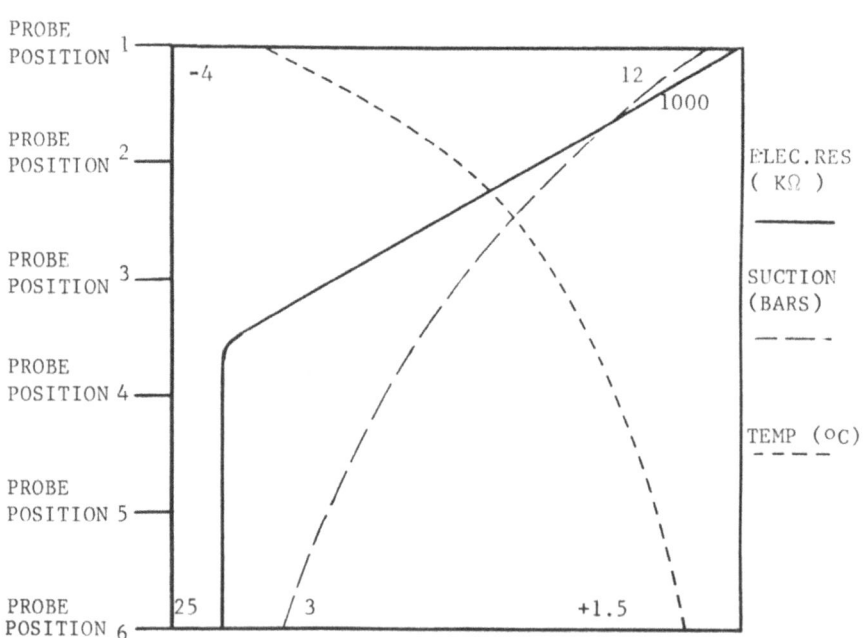

PROFILES OF SUCTION, ELECTRICAL RESISTANCE AND
TEMPERATURE IN THE MATRIX FOR THE INSTANT
RECORDED IN THE RADIOGRAPH

ACKNOWLEDGEMENTS

The authors wish to thank the University of Aston for funding this research through a Faculty grant and the staff of the Universities of Manchester and Liverpool Research Reactor for their continued support.

REFERENCES

Clarke M A and Kettle R J 1985 a - Thermal Neutron Radiography for Studying Mass Transfer in Partially Frozen Soil - Proceedings of 4th International Symposium on Ground Freezing - Sapporo, Japan, 1985 Volume II pp 168-173.

Clarke M A and Kettle R J 1985 b . Modifications to Equipment and Improvements in Facilities used in the study of Mass Transport in Partially Frozen Soil by Thermal Neutron Radiography. Proceedings of 2nd National Symposium on Ground Freezing, Nottingham, England 1985.

Brown R W and Bartos D L 1982. A calibration Model for Screen-Caged Peltier Thermocouple Phychrometers. US Department Agriculture, Forest Service, Research Paper INT-293 July 1982.

COLD NEUTRON RADIOGRAPHY OF MECHANICAL CONNECTORS

P.A. Attwood and P. Swift

Shell Research Limited

Thornton Research Centre, P.O. Box 1, Chester, England.

ABSTRACT

The cold neutron radiography facility at Harwell has been used to investigate possible causes of serious corrosion and mechanical damage to a number of mechanical connectors installed on Shell Expro North Sea platforms. To date, neutron radiography has revealed that the elastomeric O-ring seals which are designed to prevent seawater ingress into the connector, can be displaced from their groove during assembly, resulting in severe O-ring distortion and crushing, and thereby causing ineffective sealing. Split O-rings and incomplete O-ring compression are features which have also been observed during the course of the neutron radiographic studies. Furthermore, the technique can be used to detect internal joint corrosion located within the relatively thick metal walls of the joints.

1. INTRODUCTION

Corrosion induced failure and the loss of mechanical integrity has been recorded on several 75 cm diameter conductor pipe joints (Talon connectors) installed on North Sea production platforms. Failure of these mechanical connectors could lead to severe wave-load transfer and consequent rupturing of the production tubulars contained within the conductor pipe.

The Talon type connector is used offshore to couple together 14 metre lengths of steel conductor pipe which surround the production and water injection tubulars. The Talon connector was developed so that the conductor pipes (which run vertically from the platform into the seabed) could be more rapidly installed. Some 7000 mechanical connectors are now in use worldwide.

2. CONNECTOR DESIGN AND REMEDIAL TREATMENT

The Talon connector comprises a tapered pin located in a tapered box. A series of annular shoulders (teeth) and recesses (corridors) are machined along the tapered surfaces of both pin and box. O-rings are fitted at each end of the pin for the purpose of sealing the intermeshing surfaces of the pin and box from seawater penetration when the connector is in service. Half-sections of the Talon pin and box are shown in Figure 1.

The connector joint is made by stabbing the pin and box together under hydraulic pressure exerted through a fluid injection port on the outside of the box which leads to the annular space along the tapered surfaces between the pin and box. When the joint is made, (see Figure 2) the series of annular teeth engage generating considerable compressive force.

When in service any metal loss at the joint teeth, caused for example by seawater corrosion, results in a significant reduction in the strength of the joint and its ability to withstand severe wave loads. To combat the possibility of such corrosion-induced damage and thereby extend the service life of remaining Talon connectors to the desired 25 year period, a number of remedial measures were considered.

A novel remedial approach involved injecting each joint in service with a thixotropic gel containing a corrosion inhibitor which would act as both a reservoir for the inhibitor and a physical barrier to seawater ingress into the joint. Gel injection into each connector was achieved using the hydraulic injection port on the box. Since a crucial aspect of this treatment was to ensure that the entire intermeshing surface of the joint was uniformly coated with inhibited gel, it was necessary to identify a method of monitoring the gel flow and distribution during injection.

3. NEUTRON RADIOGRAPHY

It was recognised that cold neutron radiography could be used for the inspection of assembled Talon connectors because of the relative transparency of the 4 cm thick steel pipe and the opacity of the hydrogenous material, (e.g. O-rings, gel, corrosion product) present between the intermeshing surfaces of the pin and box, to slow (i.e. < 10 meV) neutrons. Therefore, a selection of Talon connectors were examined using the cold neutron beam radiographic facility of the DIDO reactor at Harwell.

Each Talon connector in turn was mounted on a rotatable platform which could be raised or lowered to provide alignment with the 20 cm diameter neutron beam. Real time imaging, using a video camera technique, was carried out on each Talon connector inspected in order to locate features of interest. Radiographs were then recorded in order to obtain images of better quality (i.e. higher contrast and resolution) to allow easier interpretation of these features.

4. EXAMINATION OF ASSEMBLED TALON CONNECTORS

4.1 Examination of an unused Talon connector

A neutron radiograph taken of the outside interface region of an unused, corrosion-free, assembled Talon connector is shown in Plate 1.   The position of the O-ring seal, located in its groove, and the succession of teeth and corridors are clearly identified.

The intermeshing surfaces between the teeth and corridors is seen to be of a uniform grey level, indicating the absence of corrosion products. Examination of the outside interface O-ring seal around the entire circumference of the joint did not reveal any defects.

Plate 2 provides a view of the inside interface region of the same Talon connector. In this case the radiograph shows that the O-ring seal has been displaced from its groove during assembly and crushed between the pin and box sections causing the O-ring seal to be squeezed to a width of about 2 cm. The radiograph also indicates that the majority of the elastomeric seal has been trapped in the first corridor of the joint. Furthermore, unlike the outside interface which forms a tight metal/metal seal, the gap formed between the pin and box sections at the inside interface is about 0.2 cm.

4.2 Examination of an unused pre-corroded Talon connector

A radiograph taken at the outside interface region of a pre-corroded connector is shown in Plate 3. The position of the O-ring seal and the succession of teeth and corridors are again clearly identified. It was apparent that, although the outside interface O-ring seal was located in its groove, it had been split, leaving a gap of about 3 cm in length. This demonstrates that assembly of the Talon connector joint (even under ideal conditions) can give rise to defective O-ring seals. Examination of the inside interface O-ring seal showed this seal to be intact and located in its groove around the entire circumference of the joint. The presence of corrosion between the intermeshing surfaces of the joint is signified by the mottled appearance of the radiograph.

4.3 Examination of an ex-North Sea Talon connector

Radiographs of the outside and inside interface regions of an ex-North Sea Talon connector were recorded. The mottled appearance of these radiographs indicated that seawater ingress into the joint had occurred during service and that corrosion of the intermeshing region of the joint had ensued.

Unlike the unused connector examined previously, both O-ring seals appeared to be located in their respective grooves. A split in the outside interface O-ring seal was detected and can be seen in Plate 4, which also shows the grey level intensity immediately above and below the O-ring seal to be very similar. Changes to the relative intensities of these areas will later be used to demonstrate successful gel injection, since areas coated with gel will appear darker as a result of increased neutron absorption by the gel.

5. GEL INJECTION

Injection of gel into the unused Talon connector described in 4.1 was monitored directly using the video camera in line with the cold neutron beam. The region around the hydraulic injection port was viewed continuously as the injection pressure was slowly increased. At a certain pressure, gel was seen to enter the longitudinal recess and flow out along the corridors and across

the intermeshing surfaces of the assembled connector. After
maintaining the injection pressure for a period of 60 seconds,
gel could be seen to flow around the empty inside interface
O-ring groove.

Detailed examination of radiographs taken at locations
around the connector after injection indicated that effective gel
distribution had been achieved over almost the entire
intermeshing interface region.

These experiments were successful in demonstrating that
effective gel injection and distribution around a corrosion free
Talon connector could be achieved using a single injection point.
Significantly, it was observed that the gel predominantly filled
regions of the connector which provide an easy access for
seawater ingress. The high viscosity of the gel coupled with
its water repellent properties would therefore provide an
extremely effective barrier to seawater ingress into connectors
that possess defective O-ring seals.

A significantly higher injection pressure was required to
achieve gel injection into the ex-North Sea connector described
in 4.3 compared with that required for joints with displaced
O-ring seals. This reflected the more efficient sealing obtained
when both O-rings are located in their respective grooves.

Comparison between the radiographs taken of the region
around the split O-ring seal of the ex-North Sea connector,
before and after gel injection (Plates 4 and 5 respectively),
showed that the gel had successfully plugged this defect and had
spread into the region between the O-ring and outside interface.
Further movement of gel had been prevented by the metal/metal
seal formed by the first tooth of the Talon connector. It is
noticeable that the presence of a uniform layer of gel over the
intermeshing surfaces of the joint has given rise to a more
uniform grey apearance of the radiograph seen in Plate 5 in
contrast to the mottled effect attributed to corrosion as seen in
Plate 4. Examination of radiographs taken at other locations,
remote from the injection port, clearly indicated that gel
distribution around the entire intermeshing surface of the joint
had been achieved.

6. CONCLUSIONS

Cold neutron radiography has provided a unique and invaluable opportunity to inspect the condition of assembled, large diameter mechanical connectors. The technique has revealed that O-ring seals can become displaced from their grooves during assembly, resulting in severe O-ring distortion and crushing, thereby causing ineffective sealing. Split O-rings have also been detected.

The use of a video system made possible the real time monitoring of gel injection and, coupled with conventional radiographic imaging techniques, demonstrated that successful gel injection and distribution can be achieved over the intermeshing surfaces of assembled Talon connectors.

During the course of the work it has also been shown that neutron radiography can be used to locate the presence of corrosion products only a few microns thick contained within relatively thick (e.g. 4 cm) metal components.

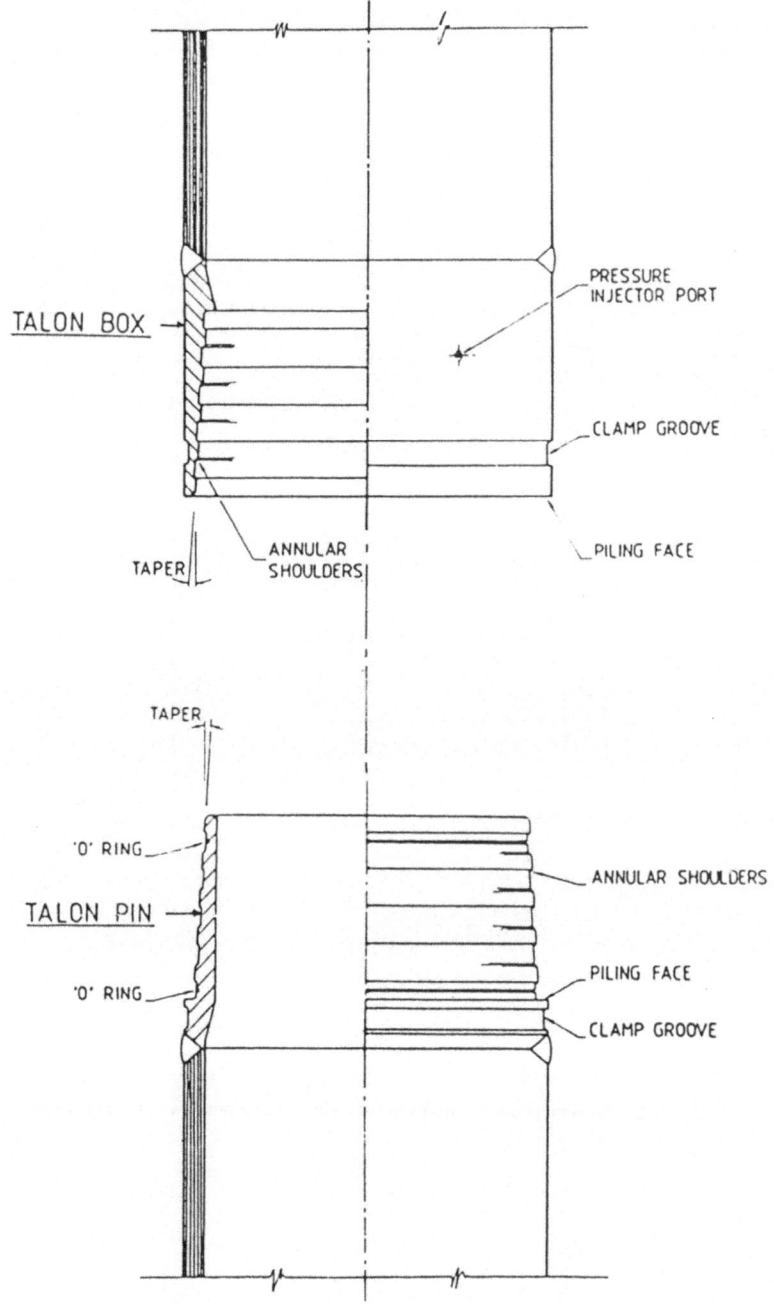

TALON BOX

PRESSURE
INJECTOR PORT

CLAMP GROOVE

PILING FACE

TAPER

ANNULAR
SHOULDERS

TAPER

'O' RING

TALON PIN

'O' RING

ANNULAR SHOULDERS

PILING FACE

CLAMP GROOVE

FIG. 1 – Talon pin and box – half section

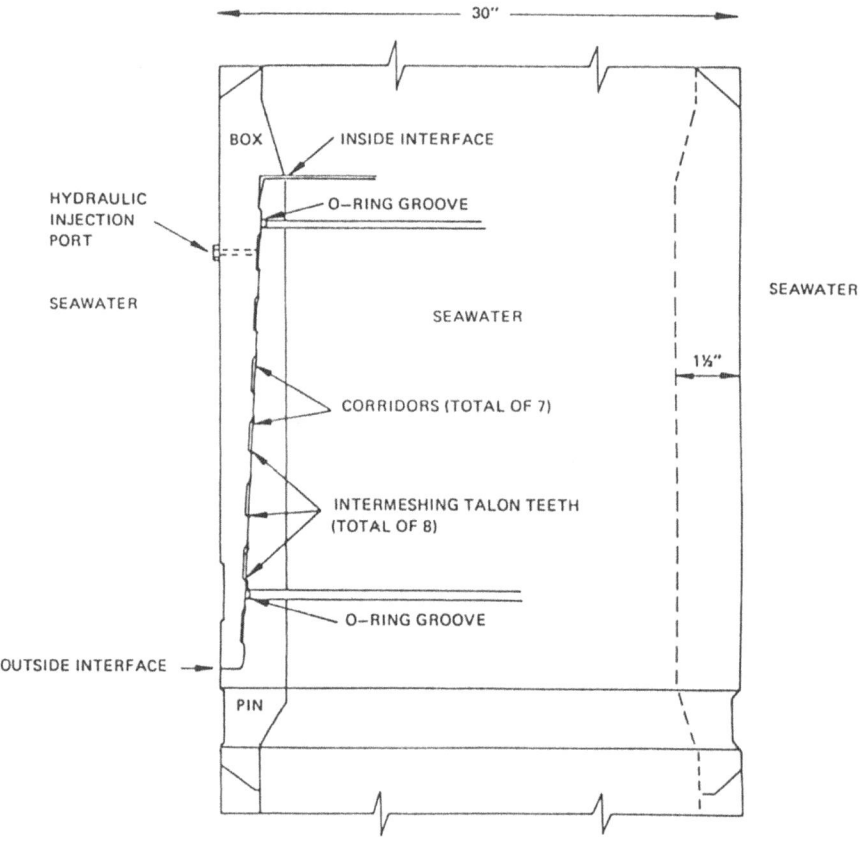

FIG. 2 – Sectional view of interfacing pin and box sections of a Talon connector

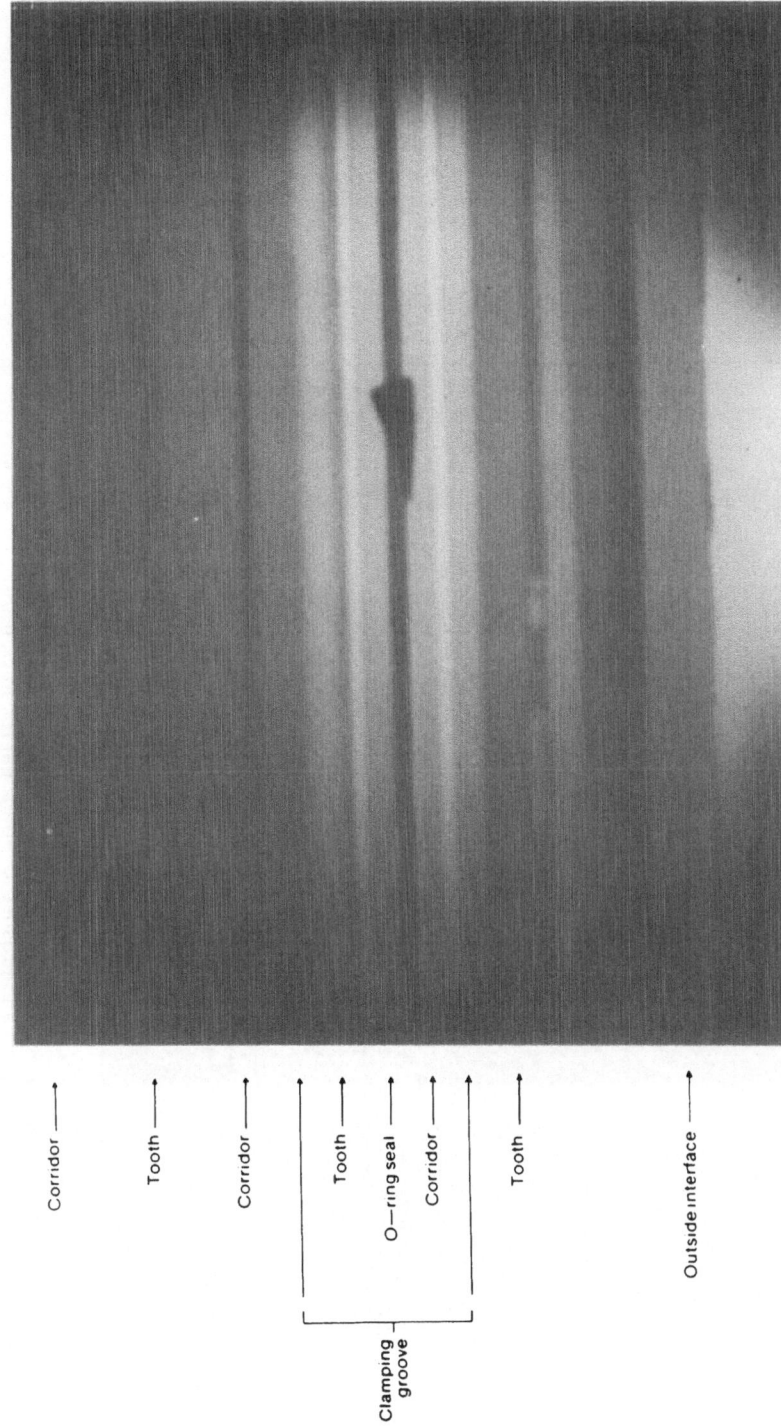

Corridor →

Tooth →

Corridor →

Tooth →

O—ring seal →

Corridor →

Tooth →

Outside interface →

Clamping groove

PLATE 1 — Neutron radiograph of the outside interface region of an unused Talon connector (triangular cadmium marker indicates position of O—ring groove) [positive print].

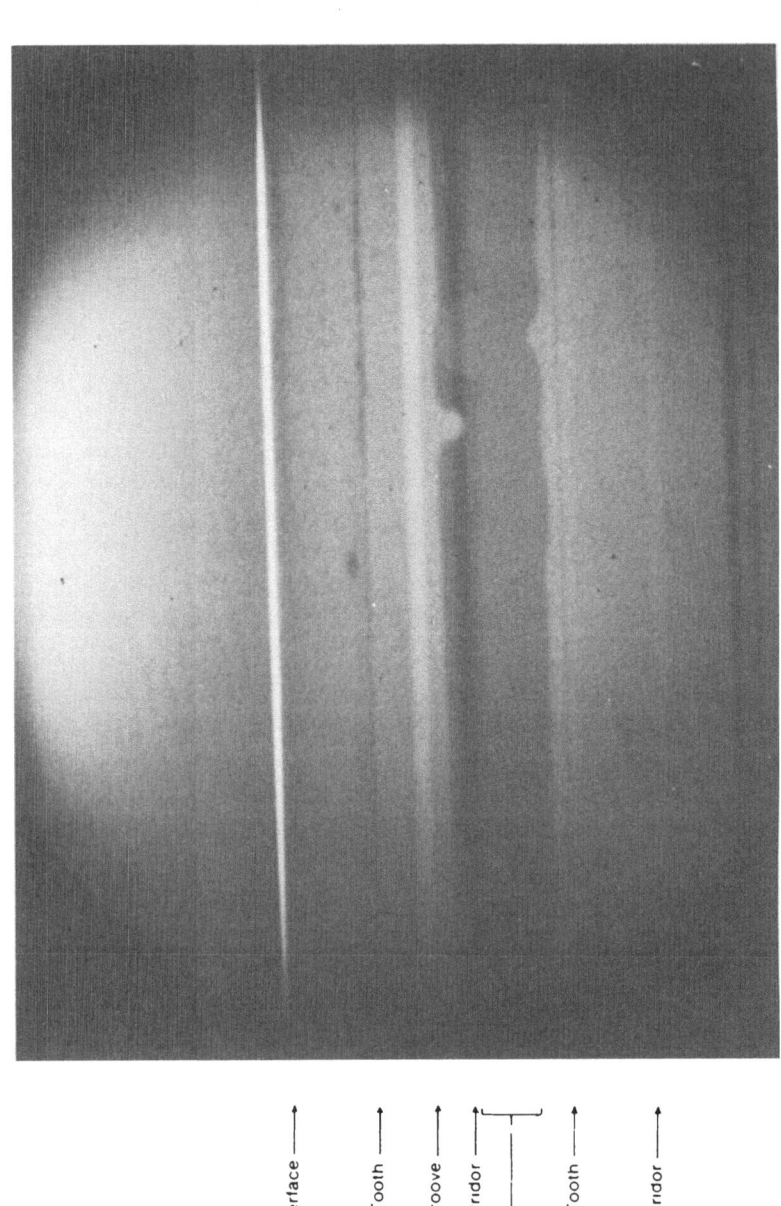

Outside interface →

Tooth →

O—ring groove →

Corridor →

Crushed O—ring seal

Tooth →

Corridor →

PLATE 2 — Neutron radiograph of the inside interface region of an unused Talon connector, showing displaced and crushed O—ring seal [positive print].

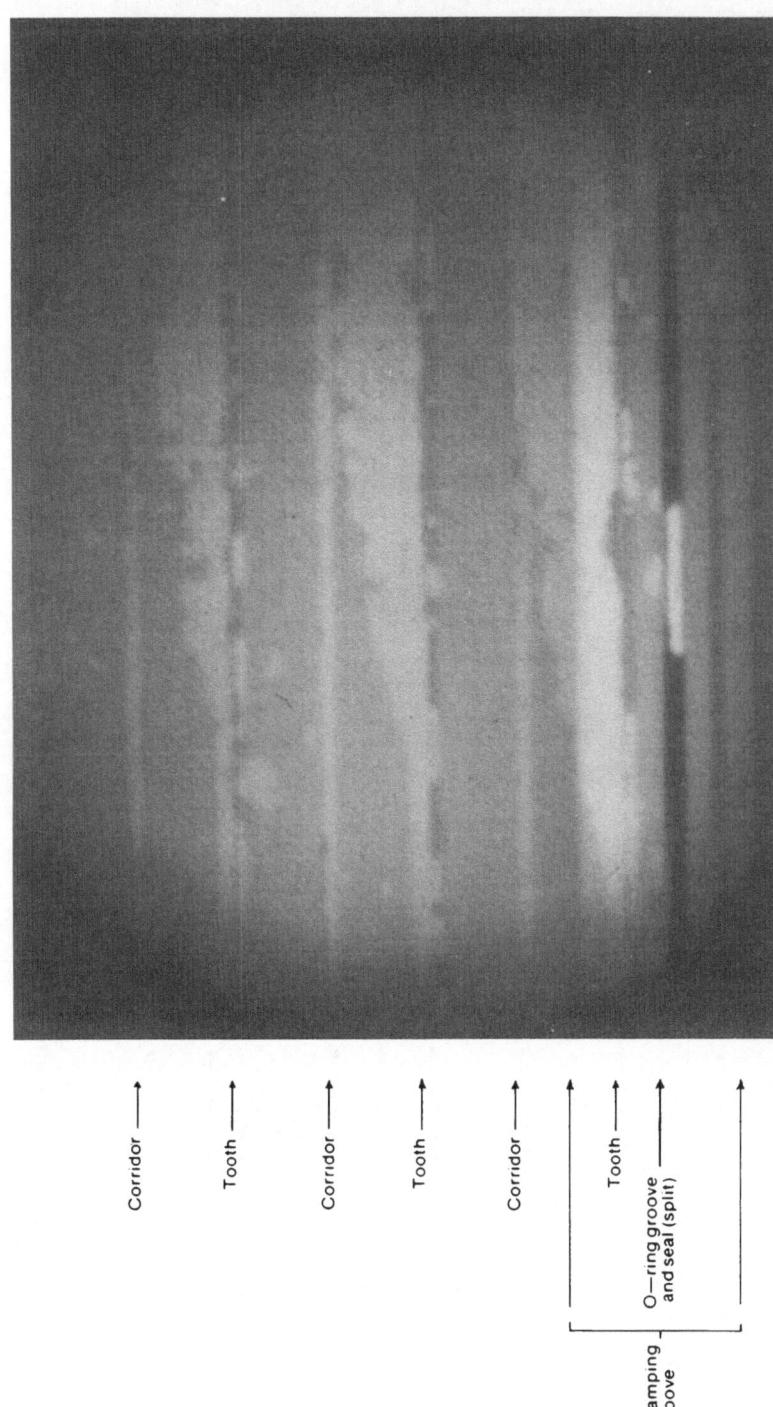

Corridor →

Tooth →

Corridor →

Tooth →

Corridor →

Tooth →

O—ring groove
and seal (split) →

Clamping
groove

PLATE 3 — Neutron radiograph of the outside interface region of a pre-corroded connector [positive print]. Corrosion indicated by mottled background.

291

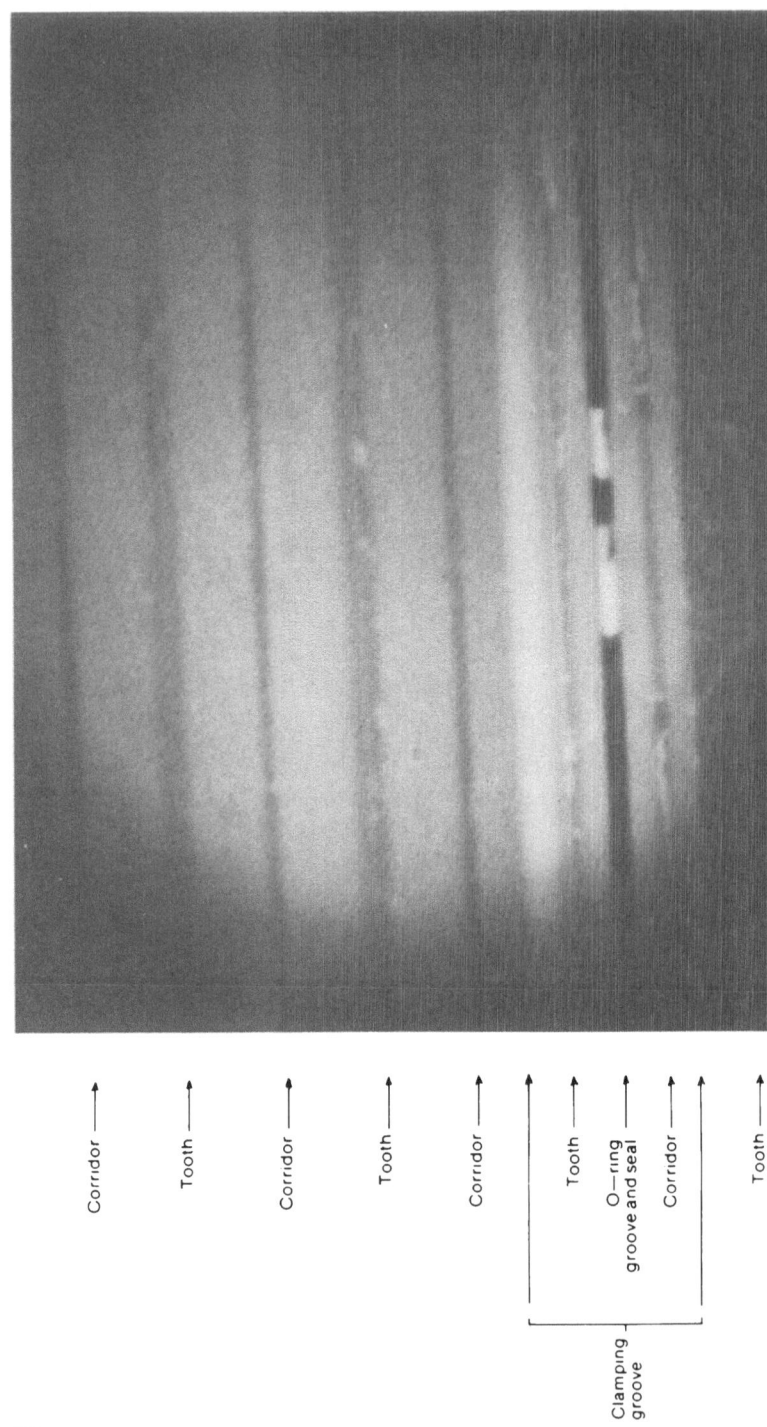

Corridor →

Tooth →

Corridor →

Tooth →

Corridor →

Tooth →

O—ring
groove and seal →

Corridor →

Tooth →

Clamping
groove

PLATE 4 — Neutron radiograph of the outside interface of an ex-North Sea Talon connector [Positive print].

Tooth →

Corridor →

Tooth →

Corridor →

Tooth →

O—ring seal
plugged with gel →

Corridor →

Tooth →

Clamping
groove

PLATE 5 — Neutron radiograph of the outside interface region of an ex-North Sea connector, after gel injection (compare with Plate 4) [positive print].

293

PRESSURE DETERMINATION OF NITROGEN GAS IN STAINLESS-STEEL CONTAINERS BY THE METHOD OF NEUTRON RADIOGRAPHY

Y. Nir-El, E. Yellin, B. Breitman and A. Gayer

Soreq Nuclear Research Center, Yavne, Israel 70600

ABSTRACT

Neutron radiography was utilized to develop a nondestructive testing technique for the determination of compressed nitrogen gas pressure in sealed stainless-steel containers. This technique provides a means for measuring pressures of small volumes of gas, where weighing or using a gauge is unfeasible. Containers were radiographed for 22 min in the neutron radiography facility of the Soreq Nuclear Research Center reactor. A gadolinium metal converter was used to image the containers on Kodak SR film. Optical densities were measured and gas pressures were evaluated by a calibration curve. A precision of ±5% was obtained for a single measurement in the region of 3000 psi.

INTRODUCTION

Industry has always had a need for a nondestructive technique to determine quantitatively the pressure of a compressed gas in a sealed and opaque container. Gravimetry (weighing) is unfeasible in cases where the containers and their integral assembly are not absolutely identical in mass. In addition, gravimetry does not yield accurate results when the mass of the gas is significantly smaller than that of the container. Connecting a pressure gauge is seldom feasible and is liable to change the pressure and composition of the gas. It is therefore worthwhile to utilize other techniques to measure such gas pressures. This paper describes the application of neutron radiography for this purpose.

This technique was used (1) to measure the time-resolved gas-gas interface location inside a gas transfer system composed of 3 pressure vessels connected by thick-walled tubing. The gas employed, ^{3}He, is neutron radiographed very efficiently because its total microscopic cross section of 5327 b (2) is large compared to that of other gases. In another application, this gas was employed as a penetrant to enhance contrast in neutron radiography, to detect a leakage path in a squib (3).

The present paper deals with the determination of compressed nitrogen gas pressure in stainless-steel containers. The neutron total microscopic cross section of this gas is similar to that of the structural material used to contain it. Therefore, moderate and even high pressures are needed to compensate for the higher density of the solid container and thus, obtain a reasonable contrast in the radiogram. This requirement is fulfilled in the present work as described by the experimental details.

EXPERIMENTAL

Nitrogen gas containers were made of 1 mm thick stainless steel in the shape of oblate ellipsoids, with inner diameters of 4.8 cm. Dry nitrogen gas of technical grade purity was compressed into the containers to pressures of the order of 3000 psi.

The containers were exposed to the thermal neutron beam of the neutron radiography facility (4) at the Soreq 5-MW swimming pool type nuclear research reactor. The thermal neutron beam passes through the circular beam aperture of diameter $D = 2$ cm. A thermal neutron flux of 1×10^6 n cm^{-2} sec^{-1} accompanied by an epithermal flux 17 times weaker, was measured at the position where exposures were made. The distance from the aperture to the front face of the container was $L = 480$ cm, thus giving an L/D ratio of 240. The radiographic image was obtained by means of the "direct exposure" technique; the converter was a 25 μm thick layer of gadolinium metal deposited on a 3 mm aluminum base and the photographic film used was Kodak Industrex R single coated. Both items were held in a vacuum cassette.

A digital readout densitometer (Macbeth, model TD-502 with an aperture diameter of 0.2 cm) was used to measure diffuse optical density.

In order to determine the flux distribution, the neutron beam at the cassette position was surveyed by exposing the loaded cassette without any object in front of it. This determination is important for the density normalization procedure.   A matrix of 5x5 cm squares was traced on the processed 43x43 cm film and densities were measured at the center of each square.   The characteristic curve of the system, which shows the optical density vs. the exposure, was determined by varying the exposure time and keeping the flux constant without changing the aperture opening.   Exposure time for unknown containers and 3 standards was 22 min.

Two standards were containers equipped with a Bourdon tube type pressure gauge and a needle valve to permit reducing the nitrogen pressure and obtaining several calibration points. The gauge precision provided by the manufacturer is ±2%.   A third empty container holding air, was used to obtain the zero pressure calibration point.

RESULTS AND DISCUSSION

Neutron Beam Distribution

Results were displayed in 2 dimensions, namely in horizontal and vertical sections over the film plane.   Figure 1 shows 2 horizontal and 2 vertical sections.   The general trend is identical in all sections ‑ the flux decreases from right to left, looking at the film from the reactor core, and from the top to bottom of film.   The extent of the decrease is section-dependent and was found to be 10 to 20% and 7 to 17% for the horizontal and vertical sections, respectively.   The decrease along the diagonal is 25 to 30%.   These findings show that in a 5 cm square, which is the diameter of the container, the density varies by   ∿1.5%.

The neutron beam distribution shows that the flux over the film plane is not flat and therefore, the measured local optical density should be normalized against the beam intensity at the same location.   For this purpose, the density in the proximity of the container was measured.

Characteristic Curve

Figure 2 shows the average optical density vs. the exposure time.   Averages were calculated from density measurements at the squares traced on the film.   Linearity is apparent up to a density value of 2.0. By linear regression in this region, the intercept 0.11 ±0.01 was calculated.   Hence, the net density, obtained by subtracting this value, is proportional to the neutron exposure E: $D = aIt$, where I is the neutron intensity and t is the exposure time.   The main contributions to the non‑zero intercept come from the film base and film fog.

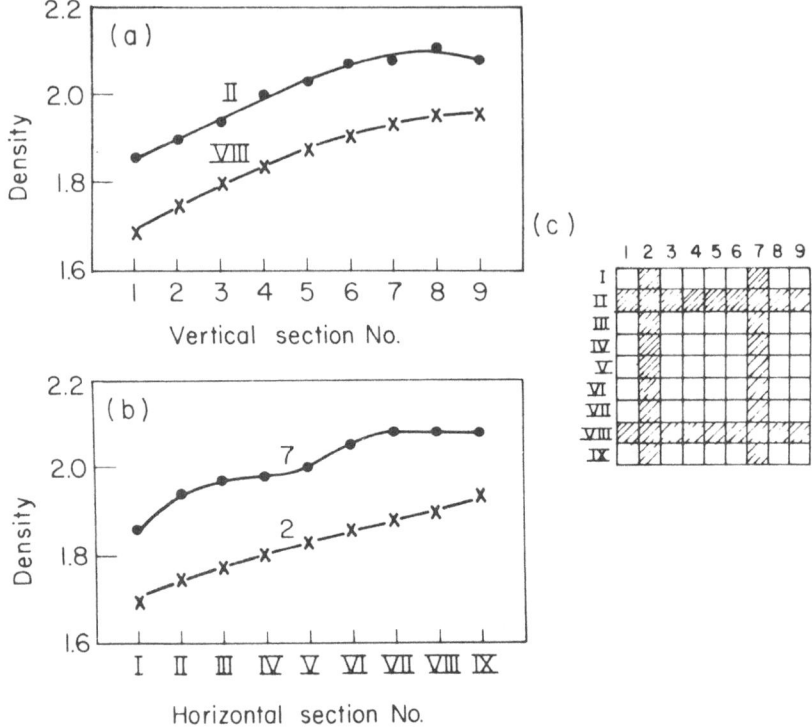

Fig. 1. Film optical densities: (a) horizontal sections;
(b) vertical sections; (c) scheme of numeric
designation of sections.

Analysis of Results

The ratio between the net optical densities at the center
and in the proximity of the container, is given by

$$r = \frac{D}{D_o} = \frac{I}{I_o} = \exp[-(\mu_1 \rho_1 x_1 + \mu_2 \rho_2 x_2)]$$

where: μ is the mass attenuation coefficient, ρ is the density
of the attenuating medium, x is its thickness, 1 and 2 denote the
walls and nitrogen gas respectively. The walls' attenuation is

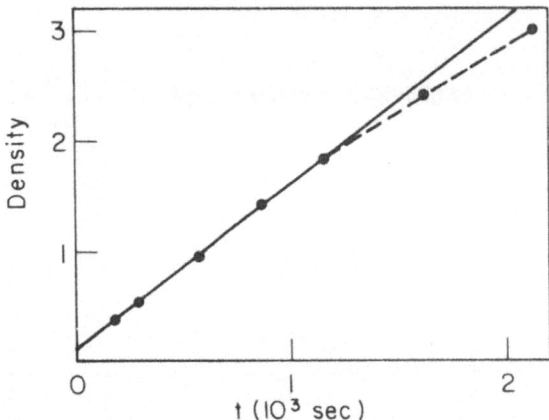

Fig. 2. Characteristic curve of the neutron radiography system. Straight line parameters: slope = $(1.506\pm0.026)\times10^{-3}$ sec^{-1}; intercept = 0.11 ± 0.01; correlation coefficient = 0.9996.

independent of gas pressure, i.e. it is constant and the transmission due to the walls may be denoted by the constant r_o, where

$$r_o = \exp\left(-\mu_1\rho_1 x_1\right)$$

The gas density ρ_2 can be expressed by using the general gas equation, $PV = nRT$, and then $\rho_2 = PM/RT$ where M is the molecular weight of nitrogen, P is the pressure, R is the gas constant and T is the absolute temperature. The Van der Walls' equation of state for real gases is a better approximation, but in the present region of nitrogen pressures ($\sim$3000 psi), the difference is only 3%. The ratio r is therefore given by

$$r = r_o \exp(-\lambda P)$$

where $\lambda = \mu_2 x_2 M/RT$ is constant for the present nitrogen containers assuming constant temperature.

The constants r_o and λ were calculated, using the thermal neutron total microscopic cross sections 12.5 b and 13.5 b, for nitrogen and iron, respectively (2). At T = 293K, r_o = 0.796; λ = 1.93×10^{-4} psi^{-1}. Experimental data may be analyzed by a least-squares fit with the exponential relation between P and r.

Calibration Curves

The calibration procedure was repeated 24 times during a 2 week period with the set of 3 standards, but without reducing the gas pressures. The constants r_o and λ were derived from these measurements and are shown in Fig. 3. In both distributions, the variation of values is quite random and no systematic trends are noticed. The average values are:

$$r_o = 0.830 \pm 0.004; \qquad \lambda = (1.32 \pm 0.05) \times 10^{-4} \text{ psi}^{-1}$$

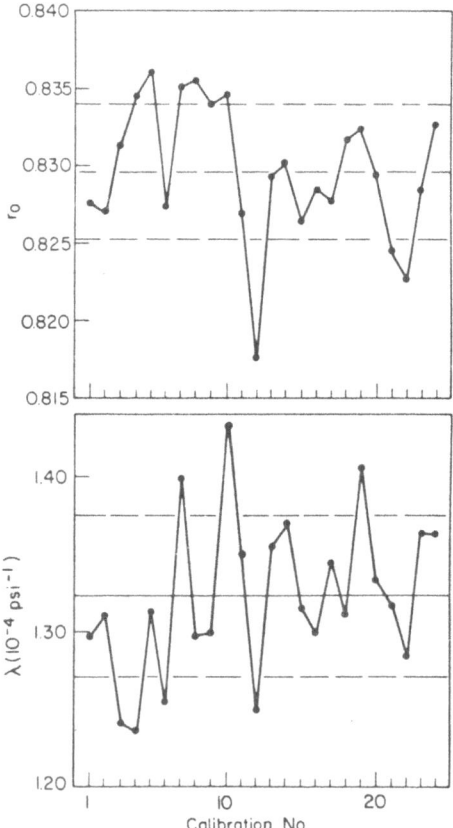

Fig. 3. Distribution of constants r_o and λ (see text) in the course of calibrations.

The ±0.5% dispersion in the r_o values is quite small. However, the λ values present a ±4% dispersion which may be explained by several effects. The main contribution is the pressure reading of the gauges. Fluctuations of ±50 psi in both gauges result in ±2% changes in the slope λ. Temperature deviations are less significant since a ±3°C deviation induces a ±1% change in λ.

A calibration curve, shown in Fig. 4, was obtained by reducing gas pressures in the standards. By linear regression (correlation coefficient 0.997), the following values were obtained:

$$r_o = 0.815 \pm 0.007; \qquad \lambda = (1.37 \pm 0.05) \times 10^{-4} \ psi^{-1}$$

The 2 ±4% agreement between the measured and calculated r_o values is satisfactory but measured λ values are 30% lower than the calculated value. A possible explanation for this discrepancy may lie in the value of the total microscopic cross section used for nitrogen.

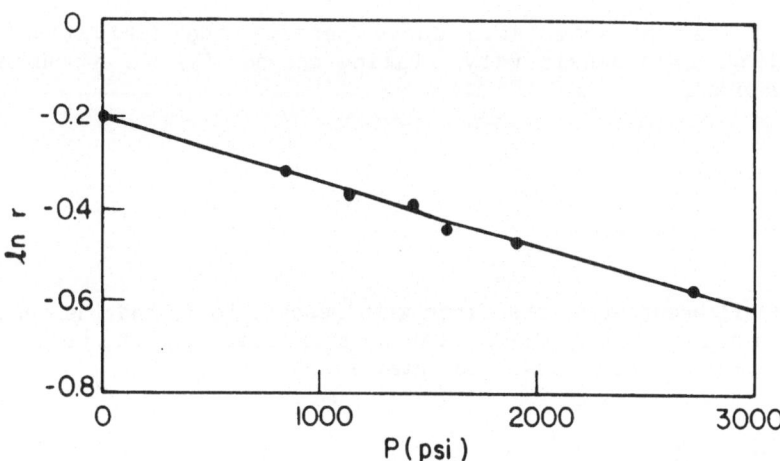

Fig. 4. Calibration curve obtained by pressure reduction in standards.

Experimental Error

The following factors contribute to the overall experimental error in the determination of pressure in a container. Optical densities are measured with a precision of ±0.01 (absolute), and using the intercept 0.11±0.01, the ratio r is obtained with a precision of ±1% at a gas pressure of about 3000 psi. The precision of the calibration curve is ±3% over each of the two parameters λ and r_o. The total accumulated error is ±5%. At a lower pressure, around 1500 psi, the error ±6%, is somewhat higher.

Two sources of errors were found to be unimportant:
- thickness tolerance of container walls: assuming a real ±0.04 mm tolerance, the corresponding pressure variation is ±35 psi.
- location where optical density was measured: assuming a ±2 mm deviation, the corresponding pressure uncertainty is ±20 psi.

Compared to the total error of ±5%, the contribution of both may be neglected.

Radiographic Sensitivity

The characteristic curve permits the derivation of radiographic sensitivity. Taking the density and attenuation equations,

$$D = aIt + b$$

$$I = I_o \exp(- \mu\rho x)$$

and differentiating the first with respect to I, and the second with respect to x, the following expression for the relative radiographic sensitivity is obtained:

$$\frac{\Delta x}{x} = \frac{\Delta D}{\mu\rho Dx}$$

The minimum measurable density change is $\Delta D = 0.01$. At a density reference level D = 2.00, and for nitrogen at 3000 psi and 293K in the present container, the value calculated for the relative sensitivity is 0.8%, or 7.9 atm-cm, which has been denoted as the lower detection limit, LDL (1).

REFERENCES

1. S. C. Johnston and L. W. Dahlke, Rev. Sci. Instrum. 42, 242 (1978)

2. S. F. Mughabghab and D. I. Garber, Neutron Cross Sections, BNL 325, 3rd edition (1973).

3. B. G. Holland, in Neutron Radiography, Proc. 1st World Conf., San Diego, Dec. 7-10, 1981. J.P. Barton and P. von der Hardt eds., D. Reidel Publishing Co., Dodrecht, Holland.

4. D. Kedem, in Irradiation Facilities for Research Reactors, Proc. Conf., IAEA, Vienna, 1973, p. 165.

AN EVALUATION OF NEUTRON RADIOGRAPHY FOR

NON-DESTRUCTIVE TESTING OF DEFECTS

Y. Ikeda, K. Ohkubo, H. Suzuki, Y. Tomatsu
and G. Matsumoto

Dept. of Nuclear Engineering,
Faculty of Engineering, Nagoya University

Chikusa-ku, Nagoya, Japan

ABSTRACT

Neutron radiography (NR) for non-destructive testing
(NDT) of defects in structural materials was studied.
Sensitivities and resolutions for various combinations of
converters and emulsion films were measured and evaluated. Some defect indicators were prepared and imaged with
NR to obtain practically usable data for NDT of steels
and plastics.   Very fine fatigued cracks in steels were
detected by using a simple image-enhance technique.   In
order to obtain clearer images from blurred ones, a
computed resoration technique was developed.

INTRODUCTION

Neutron radiography (NR) is expected to be useful for non-
destructive testing (NDT) of defects and cracks in structural
materials.   Because of limiting conditions for transportable
neutron sources, such NDT applications were rather restricted
hitherto.   However, recent developing of small but strong
accelerator neutron sources seem to change the previous situation
and to open wider NDT applications for NR.

Nowadays, many converters and emulsion films as well as real time radiography systems are available for NR. Therefore, it is essencially important to clear the characteristics, sensitivities and resolutions, of such NR devices for effective NDT workings. Further, it is practically important to know the limitations of devices when they are used in NDT of structural materials. In this work, we measured the characteristics of usable sets for converters and films, and studied limitations or degradations of NR by imaging some defect indicators prepared in the laboratory. The utilized NR facilities in this work are research reactors, such as KUR*[1] RTR, MITRR*[1] and UTR, and accelerator sources, such as a JSW baby-cyclotron and a Kaman sealed tube generator in this university*[2] .

CHARACTERISTICS OF NR CONVERTERS AND FILMS

In this work, several NR facilities were employed, whose neutron intensities were widely different, i.e. they range from the order of $10^6/cm^2$.s for KUR and MITRR to that of $10^4/cm^2$.s for UTR and the sealed tube neutron generator in Nagoya university. Then in order to select the most appropriate imaging devices for each NR facility, various combinations of converters and films were needed to be tested. The sensitivities of those NR devices are essencial factors in NDT, but they usually conflict with the resolutions of the devices. Therefore, both sensitivities and resolutions of them have to be measured and evaluated with combination.

The tested converters were a Gd evaporation film (thickness 12.5μm) used in a vacuum cassette, a Gd metal foil (thickness 25 μm) and some scintillation converters such as a NE426, a BN/ZnS(Ag) and a $Gd_2O_2S(Tb)$, the last two being prepared in this laboratory . Conventional spring type cassettes were used with exception for the Gd evaporation filmThe tested films were mainly the ones from Fuji film Co. (Japan) and Kodak Co.. These were soft X-ray type (Fuji HS and FG), industrial X-ray type (IX-50 and IX-100), medical use type (RX and RXO) and an optical microfilm (Fuji minicopy). The tested Kodak films were SR-5, M, and AA types.

Figure 1 shows a set of the typical results of the sensitivity measuements, which are the characteristic curves of various films combined with the Gd metal foil converter. In these cases, the neutron exposures on the films were changed with using a Cd step-wedge instead of changing exposing times of such films. It can be seen from the figure that the HS, IX-100 and AA show fairly high sensitivities to neutron exposure, but also somewhat high gamma-ray fogging. On the other hand, the FG, M and SR show lower sensitivities to neutron exposure with lower gamma fogging.

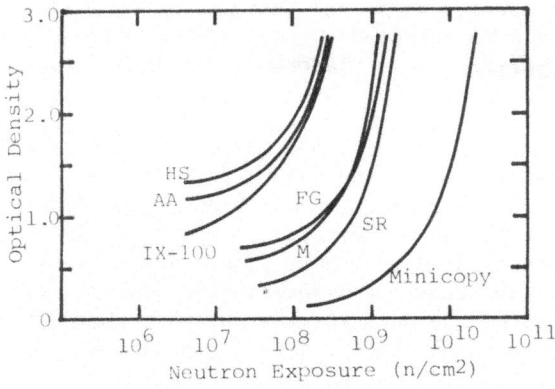

Fig.1

Characteristic curves
of various films
obtained by a Gd metal
foil converter
(Cd step wedge used)

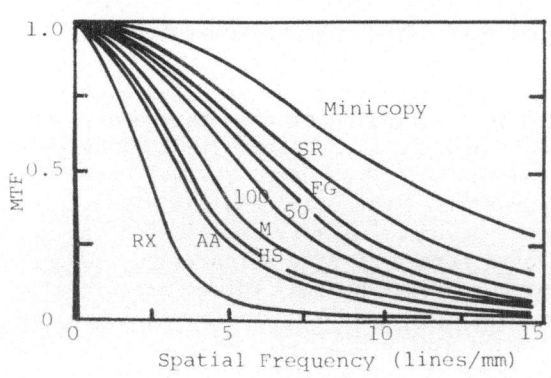

Fig.2

MTF curves of various
films obtained by a
Gd metal foil conver-
ter

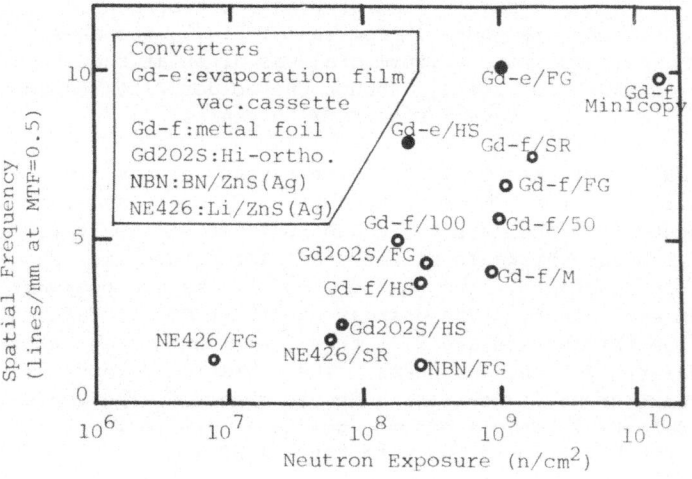

Fig.3.

Exposure vs
resolution

Figure 2 shows the results of MTF measurements for the combinaions of the Gd foil and various films. The MTF curves were obtained from the Fourie transforms of the line spread functions (LSF) for the NR images of a Gd sharp-edge. It is clear that the SR and FG have high MTF values in the higher spatial frequency region. The Minicopy film shows low gamma fogging and a high resolution, but it is too insensitive for NR. Then the film may not be used in usual NDT with NR. Including these two results, relationships between sensitivities and resolutions of various sets of converters and films are summarized and plotted in Fig.3. It is cleared that the Gd converters, especially the Gd evaporation film one with the vacuum cassette, have fairly high resolutions, but lower sensitivities than the scintillation converters. Because the covering ranges of the data for the neutron exposure and the resolution are both wide, this figure can be useful for selecting a appropriate set in each case of NDT with NR.

CONTRAST AND RESOLUTION TEST USING SOME DEFECT-INDICATORS

In order to investigate the variations of contrasts and resolutions of NR images with increasing of the thicknesses of NDT objects, such as steels and plastics, some kinds of defect-indcators were prepared and tested. For example, Fig.4 shows two illustrations of iron step-wedges, the STI-1 and the STI-2. The STI-1 is made of simple steps and is usually imaged with the W-1 indicator, which is made of a series of fine nylon wires. Whereas the STI-2 is made of the steps having three artificially induced defects on each, that is, a drilled vacancy hole (1.0mm dia.), a Cd rod filler(1.0mm dia) and a nylon wire one (0.8mm dia). Figure 5 shows the NR images of these indicators, (a) being the STI-1 with W-1, and (b) being the STI-2. As the thicknesses of steel steps increase, the contrasts between these defects and iron steps are poorer. From (a) of Fig.5, the observable limit of the finest wire (0.24mm dia) was given at the step of the 26mm thickness of steel, though the bolder wire (0.8mm dia) was detected at the step of till 32mm thickness.

The optical densities (ODs) of the steps and defects for the STI-2 were measured with a microphotometer. Figure 6 shows the results obtained from the data by SR and FG films. The variations of ODs with the thickness of the steel for both films show very similar inclinations, though the absolute ODs are somewhat different from each other. The decreasing of the ODs is almost linear when the thicknesses are less than 20mm, but it deviates from the linearity over the thickness. The deviation mainly came out from the neutron multi-scattering in the object. The detection of fine defects in steels becomes gradually difficult at the thickness larger than 20mm in NDT with thermal NR.

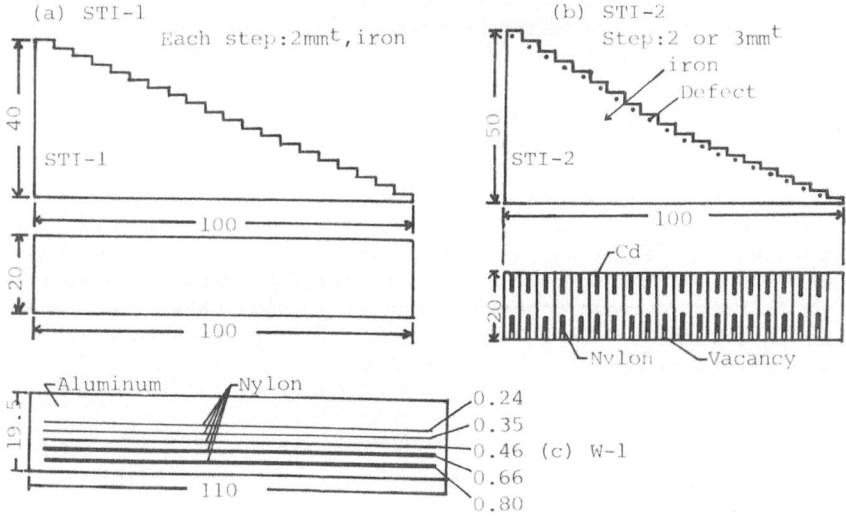

Fig.4 Illustrations of defect-indicators

(a) NR image of STI-1

(b) NR image of STI-2

Fig.5 NR of two iron defect-indicators

In order to check the degradation of resolutions for imaging defects when the thickness of steel objects becomes large, the image widths of the Cd rod fillers in the STI-2 were measured from the chart of the microphotometer. The measured widths are plotted in Fig.7, in which two obtained data are given, one being by KUR, and another being by JSW Baby-cyclotron. For the former case, the dimensional change is very small, but for the latter case, it becomes somewhat large at the steel thicknesses over 30mm. The cause of the latter image blur probably came out from the geometrical unsharpness of imaging due to somewhat low L/D of the collimator used in Baby-cyclotron. It may be said that the increasing of the thickness of steel objects does not change the resolutions of NDT so badly, but the large thickness yields the large distance between defects and films on NDT, then the image blurring will be caused by the restriction of neutron collimation of the system used.

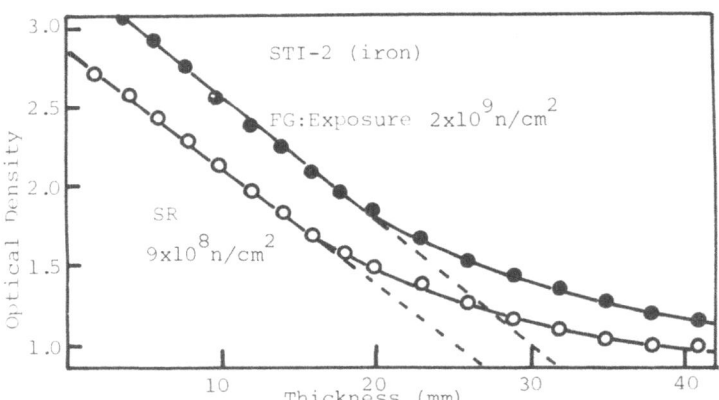

Fig.6 Variation of OD of the STI-2 with thickness

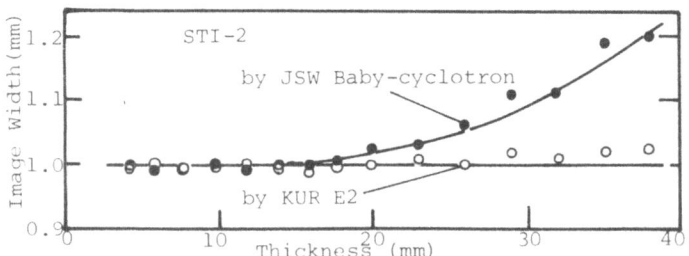

Fig.7 Variation of image width of Cd (1.0mm dia.)

IMAGE RESTORATION FROM GEOMETRICALLY DEGRADED NR

When a transportable NR equipment is used, the L/D of the system must be restricted to a low value in order to attain the neutron flux as high as possible. In this case, the image degradation caused by a poor neutron collimation will be sever. For such geometrically degraded NR images, computational image restoration techniques have been newly developed and applied to the NR of some small objects. The used restoration algorisms contain some image-restoration-filters, such as an inverse filter or a constraint-least-mean-square one, which were obtained from the point spread function (PSF) of the NR image itself. Figure 8 shows a degraded NR image and its restoration for a series of holes (0.25mm-2.0mm dia) opened in a Cd plate. The restorated image has fairly sharpness and better resolution.

(a) Original degraded NR image

(b) Restorated image with Constraint-least-mean-square filter

Fig.8 Computational image restoration of degraded original NR one

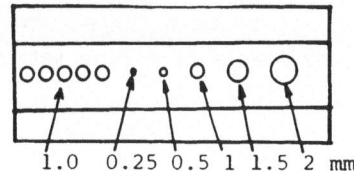

1.0 0.25 0.5 1 1.5 2 mm

NR IMAGES OF VERY FINE FATIGUED CRACKS IN STEELS

Figure 9 and 10 shows two NR images of very fine cracks generated in steel specimens during fatigue testing. The former specimens has the thickness of 40 mm, and the dimension of the crack is less than 50 μm in width and 8 mm in depth. The used NR film was Kodak SR-5. The required neutron exposure was very large such as 4.3×10^9 n/cm^2. Because of fogging by the gamma radiation (about 4R/h), any other X-ray films were unable to image the crack successfully. The latter fine cracks shown in Fig.10 are produced in a much thinner steel, being 3 mm in thickness, but the dimensions of the cracks are very small. The width of the smaller one is about 15 μm.   In this case, both SR and FG films were usable, but the expanded picture by the former is slightly better than the one by the latter.   In these cases, the neutron fluences were about 1.0×10^9 n/cm^2.   and a simple image-enhancement technique using a Gd solution was employed.

REFERENCES
1) The KUR, RTR and MITRR are reported in the present meeting as the report JP2, JP5 and JP6, respectively.
2) Matsumoto,G. et al., Nucl.Technol., 68, 94(1985)
3) Matsumoto,G. et al., Materials Evaluation, 42,1379(1984)

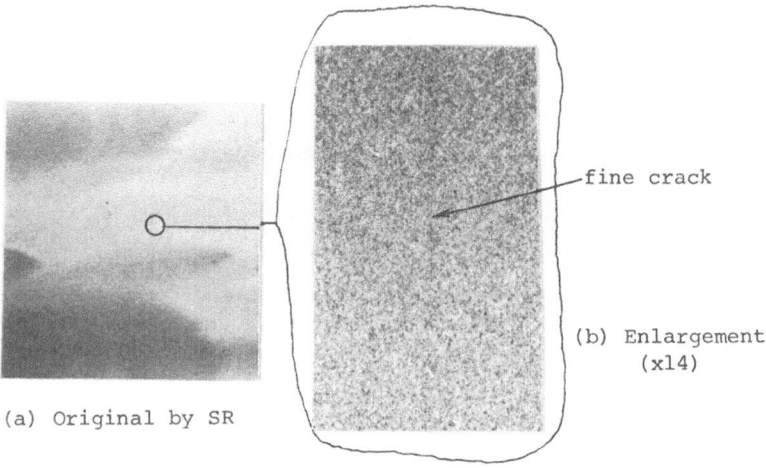

fine crack

(a) Original by SR

(b) Enlargement (x14)

Fig.9 NR images of a fine fatigued crack in a thick steel specimen

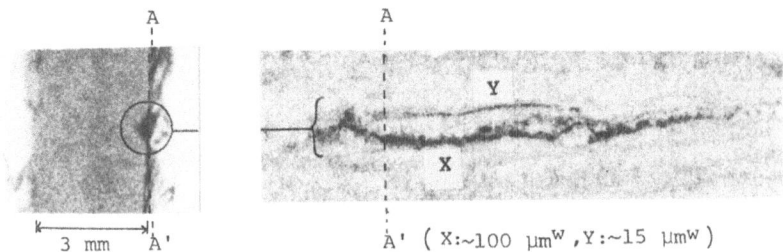

A

A

Y

X

3 mm   A'

A' (X:~100 μm^W ,Y:~15 μm^W)

Fig.10 NR images of very fine fatigued cracks in a steel plate

NEUTRON RADIOGRAPHY FACILITY FOR EXPLOSIVE DEVICES OF

H   LAUNCH VEHICLES AND ITS APPLICATION

*Kouichi Yamawaki, *Yukio Fukushima,
*Tomihisha Nakamura, **Junichiro Sekita
*National Space Development Agency of Japan,
**TESCO Corporation
Tokyo, Japan

ABSTRACT

In order to perform the inspection explosive devices
of H-I launch vehicles by neutron radiography, we employed
Baby Cyclotron as a neutron generator for NRT. As a
result of performance test utilizing moderator and a col-
limetor, approximate 3×10^5 n/cm^2 sec thermal neutron
flux were obtained on the film surface. The film size of
radiograph should be 14" × 17" and a fine and clear image
quality for the inspection of explosive devices was obtained. Furthermore, this apparatus is designed to emit
neutron beam in dual directions, and it has increased
radiographing efficiency.

---

1. PREFACE

The National Space Development Agency of Japan has the programs
of N-11, H-I and H-II launch vehicles as the launch vehicles meeting
the demand of increasing the scale of artificial satellites.

Among these programs, as a part of the development of H-I
launch vehicles, the explosive devices used for the H-I launch vehicles were developed. As for these explosive devices, since the
reliability of their functioning cannot be confirmed by the repeated
functioning, the establishment of the techniques depending on non-
destructive inspection at the time of manufacture is very important.

It has been well known that among nondestructive inspection
techniques, the inspection by neutron radiography is very effective
for explosive devices. However in Japan, the experience of applying neutron radiography to explosive devices has been none. Therefore in the National Space Development Agency of Japan, in order to
develop the neutron radiography apparatus conforming to the state
of affairs in Japan, the investigation and research have been carried out for putting the apparatus in practical use since 1980,
and the development of the apparatus using a super small cyclotron
as a neutron source was planned. The trial and evaluation were
repeated, and the good results have been obtained.

2. SELECTION OF NEUTRON SOURCE

Generally, as a source used for neutron radiography, it is
desirable to utilize a nuclear reactor. However in Japan, legal
and economical difficulties are forecast in carrying out neutron
radiography using a nuclear reactor as the nondestructive inspection
of explosive devices.

Therefore, the authors examined the use of an accelerator as
the neutron source substituting for a nuclear reactor, and it was
decided to use a very small baby cyclotron as the neutron source.
The baby cyclotron used by the authors was that of 16 MeV proton
beam energy and 50 μA beam current, beryllium was used as the target, and $^9Be(p,n)^{10}B$ reaction was utilized.

3. DESIGN CONCEPT FOR MODERATOR AND COLLIMATOR

When a moderator and a collimator were designed, the following
points were taken in consideration to examine the material, size
and structure.

(a) The averagè logarithmic energy loss ξ due to one elastic
    scattering shall be large.
(b) Neutron scattering cross section Σs shall be large.
(c) Neutron absorption cross section $\Sigma \alpha$ shall be small.

Accordingly, as $\xi\Sigma s/\Sigma \alpha$ is larger, a material is to be better
as the moderator. The authors referred to the values of $\xi\Sigma s/\Sigma \alpha$ of
moderators calculated by N. D. Tyufyakov et al., and designed a
moderator and a collimator using polyethylene which is superior in
the aspect of workability among various moderator materials.

As to the size of a moderator, the authors used the equation
for the spatial distribution of thermal neutrons $\phi_t(r)$ in a moderator
derived by N. D. Tyufyakov et al., and determined the thickness of
polyethylene required for attenuating thermal neutrons to about

314

1/10,000 in the polyethylene when the moderator was assumed to be a sphere, thus decided the moderator as a square having a side of 80 cm based on that thickness, and arranged a beryllium target at the center. Besides, the circumference of the moderator was enclosed with the cadmium sheets of 0.5 mm thickness for absorbing the thermal neutrons leaking outside. Photo 1.

The moderator was so contructed that beam ports were provided so as to be able to take thermal neutrons out in two directions, that is, in 0° direction in which proton beam straightly advances, and in 90° direction perpendicular to the beam. Moreover, these beam ports were so constructed that a moderator or an absorber such as polyethylene, paraffin, graphite or bismuth was able to be arranged.

The material of a collimator was polyethylene, similarly to the moderator. The collimator was made into that of divergent type having the angle of divergence smaller than 10°, and the size of the photographing area was decided so that a film of 14 in × 17 in was able to be used.

Since the collimation ratio L/D exerts large influence on the quality of images, by adopting the construction that the sleeves having the different diameter of collimator inlet were used, and the length of the collimator was changeable in three stages, L/D was made so as to be arbitrarily changeable from 15 to 74.

Photo 1. Arrangement of moderator and target

4. IMPROVEMENT OF APPARATUS FOR THE EFFICIENCY OF PHOTOGRAPHING

In the neutron radiography apparatus for photographing explosive devices manufactured on the basis of the design idea of the authors, the thermal neutron flux at the photographing plane was about 3×10^5 n/cm²·sec at L/D = 52. Accordingly, for obtaining the film density of 2.5, the exposure for about 60 min was required. Therefore, in order to heighten the efficiency of photographing, the authors attempted the dual photographing by attaching collimators simultaneously to 0° direction and 90° direction of moderators. The intensity of thermal neutrons is different in 0° direction and 90° direction owing to the structure of moderators, therefore, by performing the dual photographing under the condition that the neutron flux, cadmium ratio and n/γ ratio in both directions agreed by changing the kinds and thickness of filters, the heightening of the efficiency of photographing explosive devices became feasible. The results of the dual photographing are shown in Photograph 2 and Table 1.

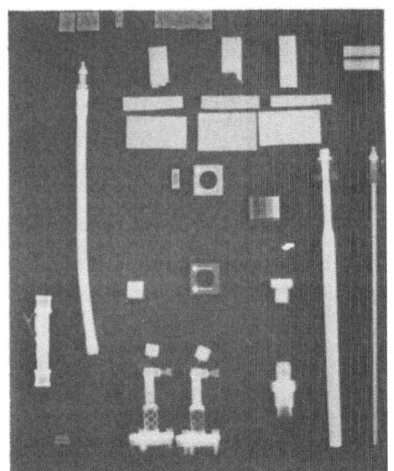 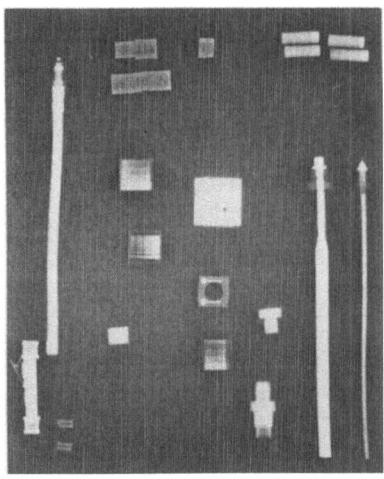

0° direction   90° direction

Photo 2. The results of the dual photographing

Table 1. Results of dual NRT

	Filter (mm)	$\overline{D}$	ASTM C&S Level					n-fl* x10⁵	Cd ratio	n/γ
			Nc*	S*	T*	P*	H			
0°	30P+20G+(10P)+10Bi +(10Bi)	2.45	53.5	3.1	0.8	1.2	7	2.75	3.2	2.8
90°	10P+10G+5Bi+(10Bi)	2.44	55.7	1.9	1.2	1.9	7	2.72	2.9	2.9

16Mev, 50μA, Exp.60min. * n/cm²·sec.
P: Poliethylene, G: Graphite, Bi: Bismuth

5. CONDITION OF PHOTOGRAPHING FOR EXPLOSIVE DEVICE INSPECTION

As the specification of the explosive device inspection using the neutron radiography apparatus put in practical use by the authors, the specification of general implementation and the test specification for each explosive device were drawn up.

As the evaluation of the image quality of films, the evaluation according to BPI and IQI prescribed in ASTM E 545 was adopted, and the evaluation standard as shown in Table 2 was established. Besides, in order to evaluate explosive devices, indicators were made. The indicators are for comparatively evaluating the charging pressure and the clearance of the charged part with the bulk head in the explosive-charged part of devices, and as shown in Figure 1, two types of 11 mm outside diameter × 35 mm length and 3.4 mm outside diameter × 15 mm length were made. The case of the indicators was made of stainless steel, and PETN explosive was charged inside. The charging pressure indicators are those charged with PETN explosive at four levels of charging pressure, 2,000 kgf/cm^2, 1,000 kgf/cm^2, 500 kgf/cm^2 and 100 kgf/cm^2, and the state of charging pressure in the charged part of explosive devices is to evaluated comparatively by film density. Besides, a blank indicator having the same dimensions as these indicators but without charging explosive inside was photographed by putting it side by side, and it was decided to evaluate by determining the density difference in both transmission images.

The clearance indicators are those in which an aluminum foil of 0.02 mm, 0.08 mm, 0.1 mm, 0.2 mm or 0.3 mm thickness is inserted between PETN explosive layers which are charged at the charging pressure of 1,000 kgf/cm^2 and sealed, and are used as the standard for evaluating clearance.

The evaluation standard for explosive device indicators is shown in Table 3.

Table 2. Evaluation level
of BPI & IQI

ASTM E545 BPI Level	Thermal neutron Content	Nc	More than 45%
	Scattered neutron Content	S	Less than 5%
	Gamma Content	γ	Less than 3%
	Pair production Content	P	Less than 3%
ASTM E545 IQI Level	Value of H		More than 6
	Value of G		7

Table 3. Evaluation level of
explosive indicator

Type I	Gap	G	Aluminum gap of 0.02mm in a clearance indicator shall be detectable.
	Contrast ΔD		The density difference in the part of 2,000 kgf/cm^2 of a charging pressure indicator and the density at the center of a blank indicator DBL are measured,and D2,000-DBL shall be more than 0.85.
Type II	Gap	G	Aluminium gap of 0.02mm in a clearance indicator shall be dtectable.
	Contrast ΔD		The density difference in the part of 2,000kgf/cm^2 of a charging pressure indicator and the density at the center of a blank indicator DBL are measured,and D2,000-DBL shall be more than 0.35.

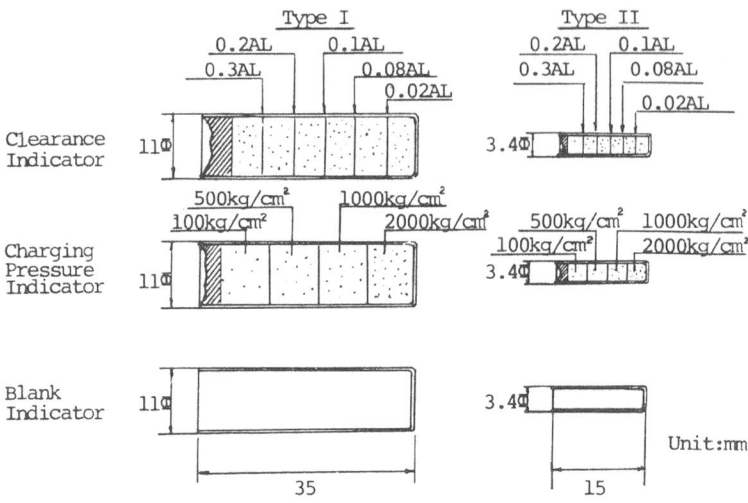

Fig. 1. Explosive indicator

The type 1 of the explosive device indicators is used for the inspection of ID, LSC, TBI, CA and so on, and the type II is used for the inspection of CDF, SCDF, SMDC, LDD, SDD and so on, and those were directly put on the face of a film cassette so as to be photographed.

Since the form of explosive devices is not uniform, it is difficult to bring the whole in close contact with a film surface, therefore, the standard of allowance was prepared for geometric unsharpness Ug, and when the standard of allowance was not satisfied, it was decided to change L/D so as to satisfy the standard value. The standard of the allowance for geometric unsharpness Ug regarding the thickness of explosive devices is shown in Table 4. In order to sufficiently satisfy the standard value in Table 4, it was decided to photograph by taking L/D more than 50 in the explosive devices having the thickness not more than 70 mm, and by taking L/D more than 70 for the explosive devices, thickness of which exceeds 70 mm.

As for film density, it was decided to photograph under the condition that the film base density becomes 2.5. Besides, the fog density of films was decided to be less than 0.3. As to the uniformity of the density of transmission photographs, the density at respective corners and at the center of a film was measured, and the average value was determined, and

Table 4. Geometric unsharpness

Material Thickness mm	Ug Max. mm
Under 35	0.8
35 through 50	1.0
Greater than 50	1.4

318

it was prescribed that the dispersion of the density at respective places should be within ±5% of the average value.

6. RESULTS OF INSPECTION OF EXPLOSIVE DEVICES

During the period from Octobrer, 1983 to February, 1984, neutron radiography was carried out on the explosive devices manufactured for trial immediately after the manufacture and after the environment test. The number of the explosive devices put to the test and the number of photographing are shown in Table 5. Moreover, the test results are shown in Table 6. As clearly seen in Table 6, there were several films which did not reach the prescribed value in the difference of density of explosive device indicators, but the images of very good quality were obtained in the others.

As to the inspected parts of explosive devices, some difference existed according to respective explosive devices, but the inspection of the clearance between charged explosive and bulk head, the cracks and voids in charged explosive and charging pressure, and the inspection of the presence of adhesive, O-rings and others were carried out, and the results as shown in Table 7 were obtained.

By this method, the inspection of the cracks and voids in charged explosive, which were not able to be inspected so far, became feasible, and the method became sufficiently usable for the quality control of explosive devices.

Table 5. The number of the
          explosive devices
          and photographing

Explosive devices	After manufacture	After environment test	Number of photograph
	Explosive devices/ photographing	Explosive devices/ photographing	
CDF	340/48	45/ 8	56
L8C	150/36	45/18	54
SMDC	50/16	30/ 8	24
ID	250/12	45/ 2	14
LDD	50/14	30/ 8	22
SDD	30/ 6	18/ 4	10
TBI	50/ 4	30/ 4	8
E/A	24/18	15/10	28
Total	944/154	258/62	216

Table 6. Results of photo-
          graphing test

		Average value	Maximum value	Minimum value	Standard value
	Average density	2.50	2.85	2.31	2.0~3.0
Image quality 73-METH	Thermal neutron Content C %	59.65	72.64	50.00	≧ 50
	Scattered neutron Content S %	6.78	12.99	0.80	≦ 15
	Low energy γ-ray Content Y %	1.54	2.49	0.78	≦ 5
	Sensitivity R		14	10	≧ 10
Image quality 81-METH	Thermal neutron Content nc %	52.59	61.09	43.70	≧ 45
	Scattered neutron Content S %	2.41	5.73	0.38	≦ 5
	Low energy γ-ray Content Y %	0.92	2.63	0	≦ 3
	Pair production Content P %	1.44	5.05	0	≦ 3
	Sensitivity M		8	6	≧ 6
	Sensitivity G		7	7	＝ 7
Explo. ID	Contrast of type I	1.01	1.38	0.73	≧ 0.85
	Contrast of type II	0.40	0.54	0.21	≧ 0.35

Table 7. Test results of explosive devices

	CDF	LSC	ID	TBi	SMDC	LDD	SDD
Clearance between charged explosive and bulk head	Ratio of defective products 0.9%	Those surpassing the prescribed value are many	No defect	Not discernible	Not discernible	Not discernible	Not discernible
Cracks in charged explosive	1.8%	Mainly haircracks 56%	Haircracks 6%	2%	No defect	No defect	No defect
Voids in charged explosive	No defect	No defect	2.4%	2%	4%	4%	3%
Adhesive, O-rings and others	Voids exist in adhesive	Imperfect adhesive	Normal	Normal	Normal	Normal	Normal

7. APPLICATION TO EXPLOSIVE DEVICES HAVING SPECIAL FORM

The form and size of the explosive devices for launch vehicles are diverse. In the case of the explosive devices having extreme curvature and the specimens having curved surfaces, the conventional cassettes are poor in the close contact with the test parts, therefore, a flexible cassette made of rubber, of which the inside can be vacuumized, was developed, and photographing was carried out be bringing it in close contact with a test part.

As the result, a very sharp image was obtained, and it was found that the developed cassette was sufficiently usable for photographing explosive devices.

8. CONCLUSION

For the inspection of the explosive devices for H-I launch vehicles, the development of a neutron radiography apparatus and its practical use were carried out, and as the means of increasing the reliability of explosive devices, it was able to be put to use sufficiently.

We acknowledged Muroran Plant, The Japan Steel Works, Ltd.; Aeronautical & Space Division, Nissan Motor Co., Ltd.; and Radiation Center of Osaka Prefecture.

REFERENCES

1. H. Berger, Neutron Radiography, Elsevier Publishing Company, 1965, p.14

2. N.D. Tyufyakov, A. S. Shtan, Principles of Neutron Radiography, Atomizdat Publishers, MOSCOW, 1975, p.67

THE CORNELL NEUTRON RADIOGRAPHY FACILITY AND ITS APPLICATIONS TO

THE STUDY OF THE INTERNAL STRUCTURE AND MICROCRACKING OF CONCRETE

Howard C. Aderhold[1]; Kenneth C. Hover[2]; and Walid S. Najjar[3]

(1) Reactor Supervisor, Ward Laboratory of Nuclear Science and Engineering; (2) Associate Professor of Structural Engineering; (3) Graduate Research Assistant and Ph.D. Candidate, Department of Structural Engineering, Cornell University, Ithaca, NY, 14853 USA.

ABSTRACT

The horizontal thermal column of the Cornell University TRIGA Mark II reactor has been adapted for neutron radiography and is described. It provides an L/D ratio of 140 and a beam area of 38 cm x 38 cm at the exposure point, where the neutron flux is 5.6 x 10^5 n/cm^2-sec at 480 kW. Both a gadolinium and a gadolinium oxy-sulfide screen have been used to study concrete, and results are compared. Internal discontinuities e.g. microcracks (small fractures) and air voids have been successfully identified due to their partial impregnation with a gadolinium nitrate contrast agent. Gadolinium oxide has also been used as a contrast agent, but it was successful only in identifying air voids. A neutron radiograph of a nitrate impregnated specimen is compared with x-radiographs of the same specimen with and without impregnation. Neutron radiography seems much more effective in identifying microcracks than either x-radiograph. Practical applications are suggested.

THE CORNELL FACILITY

The neutron source is a TRIGA Mark II nuclear reactor, shown in Figure 1a. Figure 1b shows the experimental setup at the exit of the thermal column; Figure 1c shows a top cross-sectional view of the horizontal thermal column; and Figure 1d shows an enlarged view of the collimator. The beam port for radiography is located on the axis

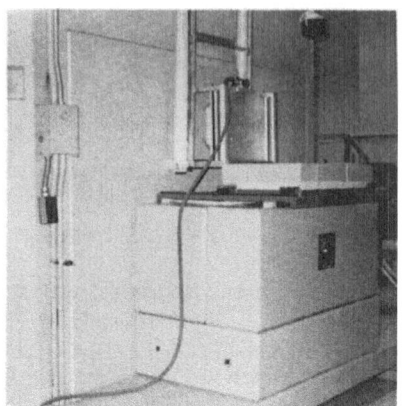

(1a)   (1b)

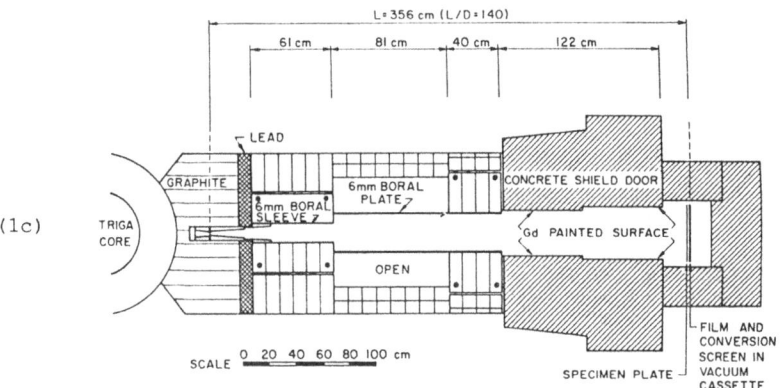

(1c)

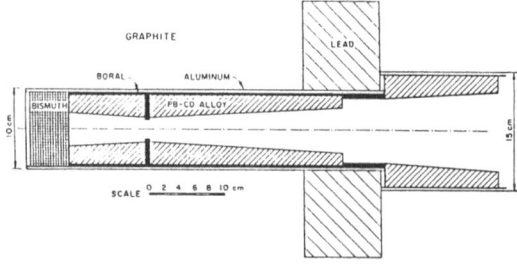

Figure 1: (1a) Cut-away view of the reactor facility.
(1b) Experimental setup at the exit of the thermal column.
(1c) Top cross-sectional view of the horizontal thermal column.
(1d) Enlarged view of the collimator.

of the thermal column; it has a two step circular cross-section near
the core, and a three step square cross-section near the exit. The
circular sections are cut through graphite and lead and are fitted
with an aluminum housing lined with Boral sleeves. A tapered
collimator of lead-cadmium alloy is enclosed within the inner sleeve,
and it partially extends through the inside end of the outer sleeve.
The inner section also includes a defining aperture which is formed
by a 2.5 cm diameter hole in a 6 mm Boral plate, and a 5.1 cm thick
bismuth absorber disk at the core end. The three step square
cross-section is formed as follows: a distance of 81 cm through an
open area (hohlraum), 40 cm through graphite at the exit end, and 122
cm through a concrete shield (rolling) door. An aluminum housing is
snugly fitted in this tapered hole, and is lined with a Boral plate
on the inner section, while the two outermost sections are painted
with a gadolinium bearing paint. A sliding aluminum frame is used as
a fixture for the vacuum cassette and the object to be radiographed,
as shown in Figure 1b. The object is mounted on a thin aluminum
(specimen) plate which fits into the frame. The system provides an
L/D of 140 and a beam area of 38 cm x 38 cm at the exposure point. At
480 kW, the neutron flux there is 5.6×10^5 n/cm^2-sec and the gamma
flux is 3.75 R/hr. The gold cadmium ratio is approximately 6. A Kodak
Type SR film is used in combination with both a gadolinium screen and
a gadolinium oxy-sulfide screen. See reference (1) for detailed
description of the reactor facility and its non-radiography
applications.

APPLICATIONS

Microcracking in concrete has been studied using the neutron
facility. Microcracks are small discontinuities which eventually
coalesce to cause macroscopic failures in concrete members. The study
of microcracking had been of particular interest in civil engineering.
In this application an aqueous solution of gadolinium nitrate is
applied as a contrast agent to a polished concrete specimen, thus
impregnating the microcracks with the solution. The specimens are thin
slices (0.38 cm) of concrete, cut from a cylinder 10 cm in diameter.
An external load was applied to the cylinder prior to slicing. Further
details of the preparation of the specimen and related information are
beyond the scope of this paper; see reference (2).

Figure 2a shows a neutron radiograph of a concrete test specimen.
Microcracks appear as white (line-like) images within a darker
concrete matrix, due to their partial impregnation with the contrast
agent. The white circular spots indicate the air voids which were also
impregnated with the contrast agent. Figure 2b shows a neutron
radiograph of the same specimen as in Figure 2a, using the gadolinium
oxy-sulfide screen; (note that all neutron radiographs in this paper,
other than Figure 2b, were done using the gadolinium screen). It is
evident that the radiographic sensitivity of the gadolinium

(2a) N-RAY,
using Gd
screen.

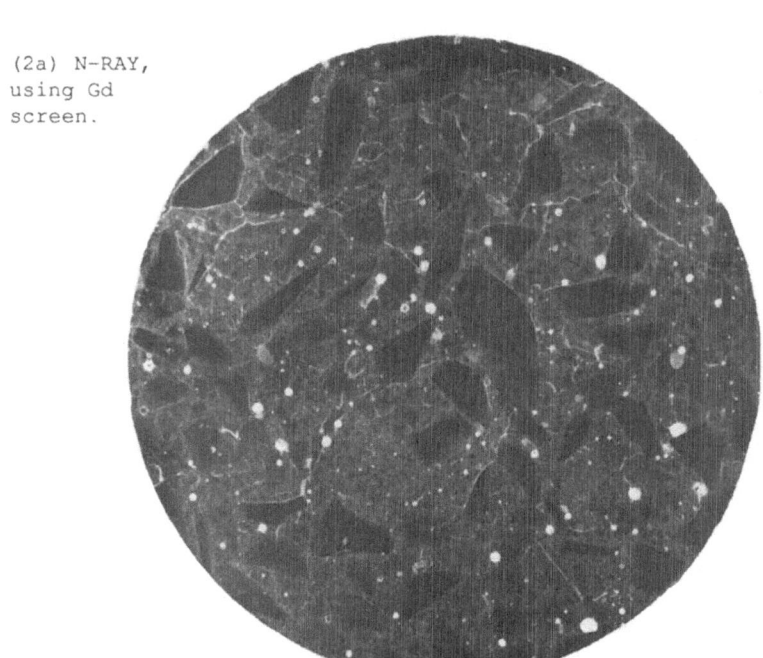

(2b) N-RAY,
using Gd_2O_2S
screen.

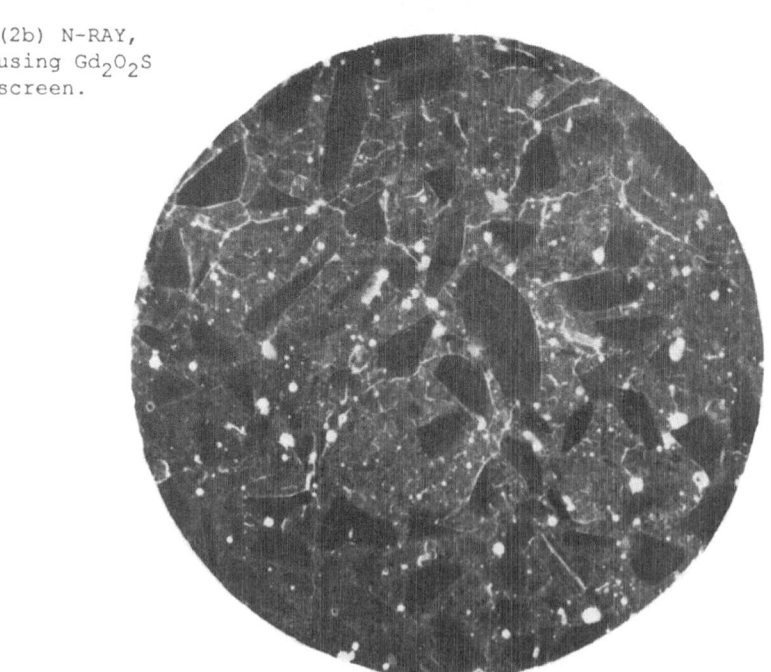

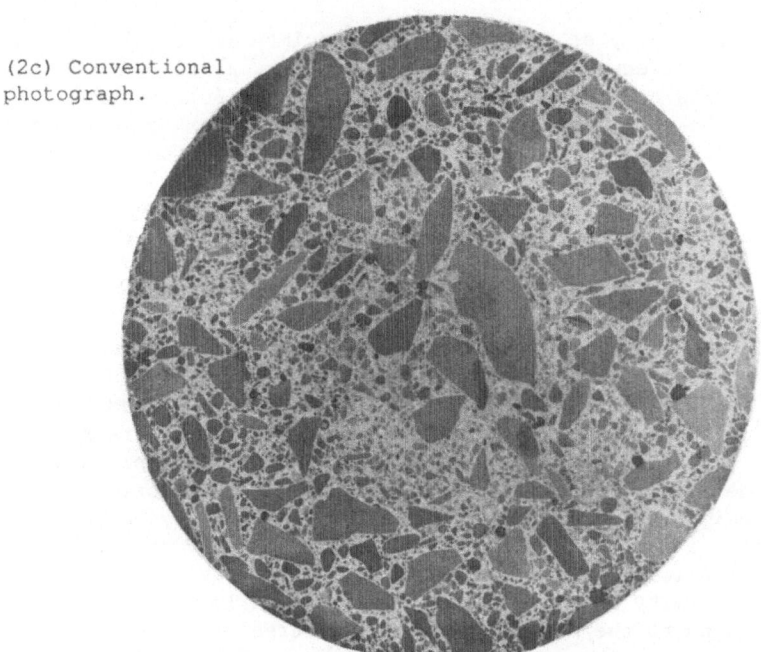

Figure 2: (2a) Neutron radiograph of a 10 cm diameter concrete
specimen that was loaded to 100% of its ultimate load (39 MPa), prior
to slicing. A gadolinium screen was used at 480 kW and 1 hr exposure.
(2b) Neutron radiograph of the same specimen as in (2a), using the
gadolinium oxy-sulfide screen at 480 kW and 17.5 min exposure.
(2c) Conventional photograph of the same specimen as in (2a) and (2b).

oxy-sulfide screen in detecting microcracks is comparable to the
gadolinium screen. Figure 2c shows a conventional photograph of the
same specimen as in Figures 2a and 2b, where few cracks (if any) can
be detected by the unaided eye. Examination of this specimen using an
optical microscope indicated that many microcracks which were visible
on the neutron radiograph were not visible under 40X magnification.

Figure 3a shows a neutron radiograph of a specimen that was
subjected to a higher compressive strain than the specimen in Figures
2, as is evident by the high extent of microcracking. Figure 3b shows
an x-radiograph of the same specimen as in Figure 3a. Figure 3c shows
an x-radiograph of the same specimen as in Figures 3a and 3b, prior to
impregnation with the contrast agent and prior to neutron exposure.
Microcracks should appear as black images in this application due to
the absence of the contrast agent (empty cracks). {Note that x-rays
were generated with a DINEX 150F unit operated at 40 kV, 5 mA, 91 cm
ffd, and 16 min exposure on a Kodak Type SR film}.

While x-radiography of an impregnated specimen (Figure 3b) seems to be at least as effective in identifying microcracks as x-radiography for a non-impregnated specimen (Figure 3c), neutron radiography of the same impregnated specimen (Figure 3a) is much more effective than either x-radiograph. A practical application of neutron radiography is the evaluation of concrete structures (e. g. reactor containment structures). This can be done by drilling core samples from suspect components in a structure and testing the samples. The extent of microcracking can be measured via digital image analysis or other means (2), and related to the load history of the structure.

A gadolinium oxide powder was also used as a contrast agent (Figure 4), but it was successful only in identifying air voids. Studies of the spatial distribution of air voids are suggested using the oxide agent. Having no interference from the images of cracks, the air voids can be easily isolated and studied. In another application using the oxide agent, concrete mixes were doped with the oxide powder at different weight percentages. It was shown that 1.0% by weight of gadolinium oxide in concrete can reduce the film density by about 45%. Two practical applications can be investigated with this concept: (1) Testing the uniformity of a concrete mix, particularly when an admixture is added to the mix; note that admixtures are generally added in relatively small amounts and therefore uniform mixing is important. (2) Testing the adequacy of proper grouting in post- and pre- tensioned tendons in prestressed concrete especially in high performance structures. Both admixtures and grouts may be doped with either the nitrate or the oxide agent.

ACKNOWLEDGEMENT

The authors would like to thank Prof. David Clark of Nuclear Science and Engineering for his review and comments on this paper.

REFERENCES

1. H. C. Aderhold, B. E. Blank, D. D. Clark, P. I. Craven, and L. J. Young, "New Facilities for Basic and Applied Research at the Cornell University TRIGA Reactor", Eigth European Conference of the TRIGA Reactor Users, Espoo, Finland, August 21-23, 1984, pp.5-1 thru 5-10.

2. Walid S. Najjar, Howard C. Aderhold, and Kenneth C. Hover, "The Application of Neutron Radiography to the Study of Microcracking in Concrete", to be published in Winter 1986 issue of Cement, Concrete, and Aggregates, Journal of the American Society for Testing and Materials.

(3a) N-RAY
[with contrast
agent].

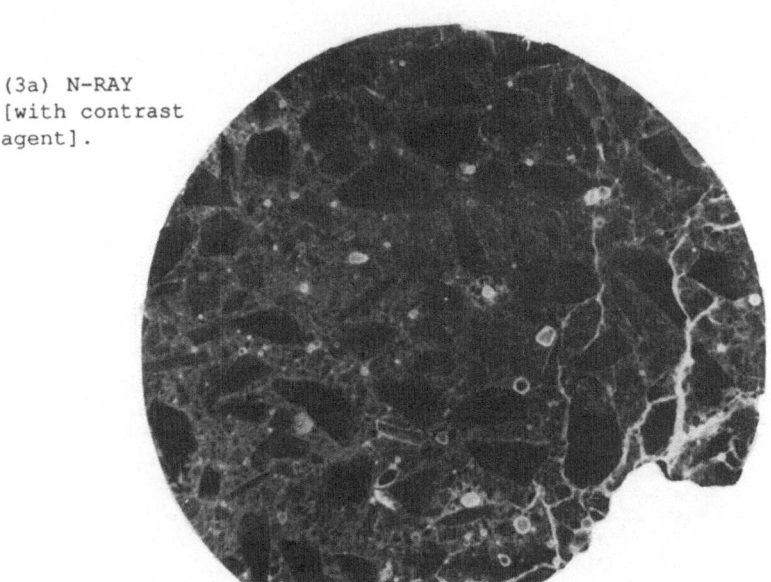

(3b) X-RAY
[with contrast
agent].

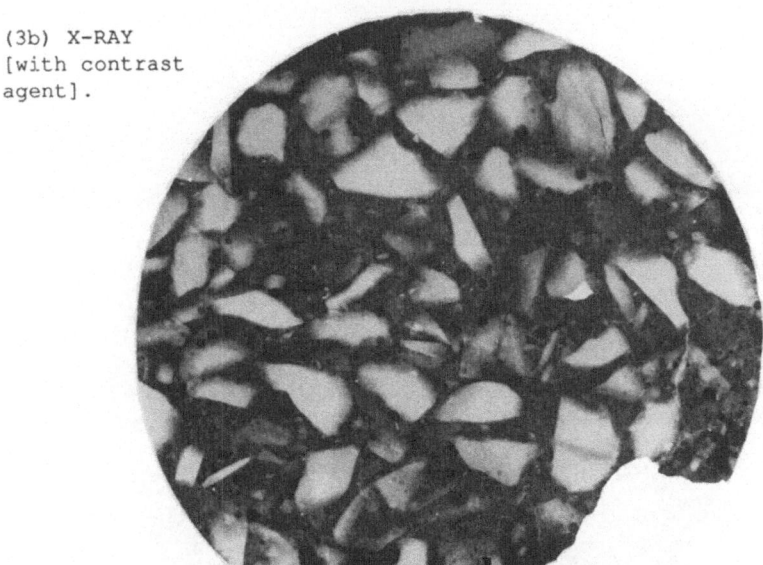

Figure 3: (3a) Neutron radiograph of a 10 cm concrete specimen that
was loaded to a strain (3000 micro-cm/cm) just beyond its ultimate
stress (39 MPa), prior to slicing.
(3b) X-radiograph of the same specimen as in (3a). (3c) X-radiograph
of the same specimen as in (3a) and (3b), prior to impregnation with
the contrast agent and prior to neutron exposure.

(3c) X-RAY
[without contrast
agent].

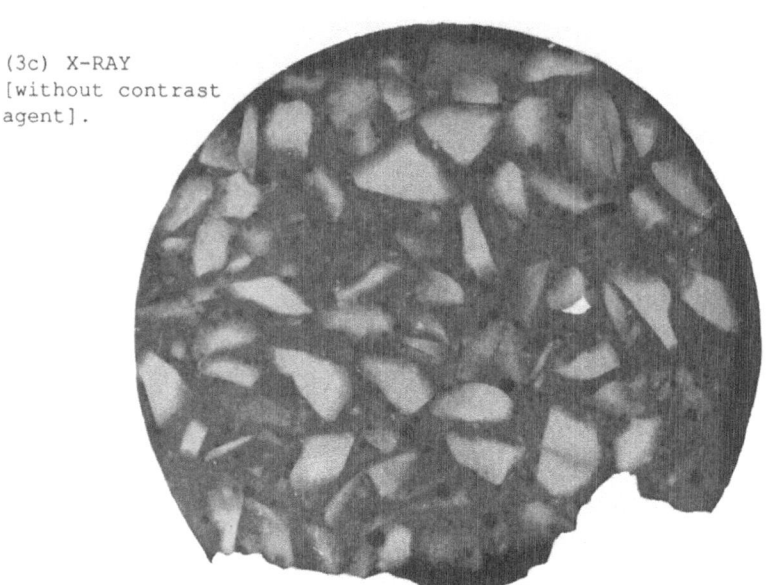

(4) N-RAY

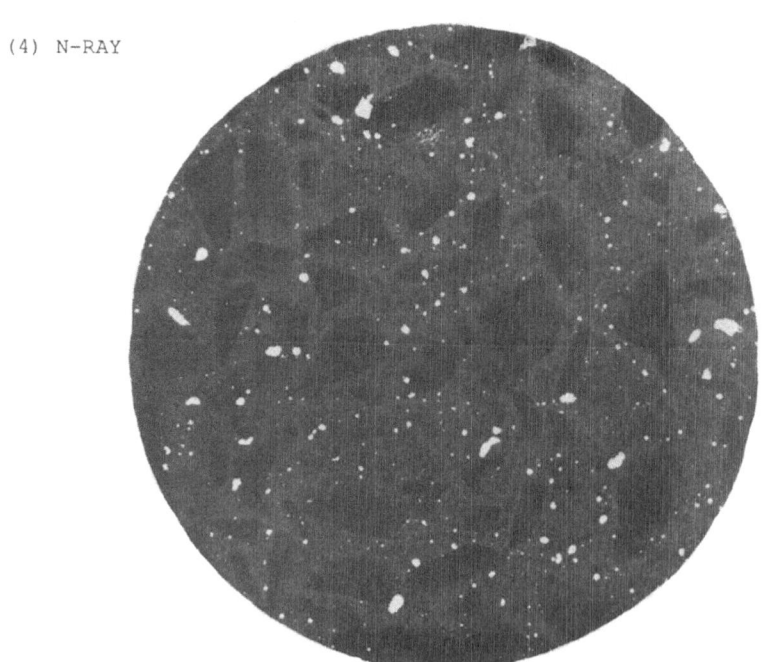

Figure 4: Neutron radiograph of a 10 cm concrete specimen that was impregnated with gadolinium oxide powder.

SENSITIVITY OF CORE DETECTION IN TURBINE BLADES

R. R. Tsukimura, P. E. Underhill, Aerotest Operations, Inc.,

San Ramon, California, USA; M. J. Wells, Howmet Turbine

Components Corporation, Dover, New Jersey, USA.

ABSTRACT

Neutron radiography has proven to be a valuable tool for
the detection of residual core material in cast turbine
blades. A series of experiments was designed to determine the
detection sensitivity for various combinations of core and
alloy thickness. The core material samples were of three
types; untreated, doped with gadolinium oxide and tagged with
gadolinium nitrate solution.

The untreated core is not readily imaged in thicknesses
expected to simulate residual core material. The detection
sensitivity for the doped and tagged core samples is fifty
micrograms, equivalent to a sphere of 0.033 cm diameter.
These results can be used to estimate the sensitivity for
residual core detection in other alloys with approximately
similar compositions.

INTRODUCTION

One of the more salient uses of thermal neutron radiography is
the detection of residual core material in turbine blades
fabricated by the investment casting technique. Whereas the
presence of residual core material in the internal passages of gas-
cooled turbine blades was not as critical in the past because of
simpler passages, the extremely high temperature environment in
current engines pushes the turbine blade material mechanical

properties to their limit. Without cooling, the turbine blades would be subjected to overheating and severe loss of durability, which would lead to shortened engine life.

Modern-day turbine engines have blades with convoluted or serpentine coolant passages which provide for greater heat transfer and, therefore, better engine efficiency. In these state-of-the-art engines, residual core material in the coolant passages is not tolerable. Use of borescopes and X-radiography have been ineffective in detecting residual core material, particularly in the newer cores which are mostly silicon dioxide based. The nickel-based superalloys mask the silica from the X-radiographer.

The value of neutron radiography is exemplified by the ability of the technique to highlight the residual core material within the turbine blades. However there is one requirement before the neutron radiographer can pinpoint the offending residue. The core residue must contain a neutron opaque material.

In the early sixties, Watt (1) improved neutron radiographic contrast in turbine blade internal passages by filling the passages with cadmium nitrate solution. Since this solution-filling method was not readily adaptable to production techniques, Mallon (2) experimented with the "doping" of the core bodies with one to ten weight percent of gadolinium oxide in the early seventies. Mallon concluded that this addition of gadolinium oxide is more than sufficient to permit detection of residual core material as small as 1000 micrograms or 0.090 cm in diameter.

Doping is defined as the process whereby neutron absorbing material, such as gadolinium oxide, is added to the core material before the core body is formed. This method, first used as a production process prior to 1976 (3), is the method of choice in terms of ease of use and detection sensitivity, particularly at the 3% doping level.

An alternative method, a proprietary technique for tagging residual core material after the core-leaching (autoclaving) process, has been developed at Aerotest Operations. The tagging technique, adapted for neutron application, had its roots in early 1971 when gadolinium tagging solutions were used to detect cracks in the internal surfaces of hollow metal bodies where borescopes could not probe. After a variety of uses, especially for containers and tubes, the method was applied, in 1974, to the inspection of turbine blades.

The current experimental program was conducted to determine the sensitivity for residual core detection during neutron radiography for both the doped core and tagging processes.

METHODOLOGY

The fabrication of turbine blades with core-free internal cooling passages blends the ancient process of lost wax casting with the modern, non-destructive testing technique of neutron radiography. The serpentine coolant passages and cast end tips of the new generation of turbine blades have led to a greater possibility of incomplete leaching of the core material.

The experimental program was designed to evaluate the dectectability of residual core material as a function of turbine blade wall thickness. In order to simplify the fabrication procedures, the cores were fabricated as wedges containing 0, 1, and 3% gadolinium oxide. The turbine blade wall thickness was simulated by injection casting three thicknesses of metal coupons. Thermal neutron radiographs were made of the core wedges placed upon the metal coupons. Detection sensitivity was established by taking densitometer readings through known thicknesses of the core-alloy combination.

Sample Fabrication

The rectangular cast coupons and core wedges for this study were fabricated by the Howmet Turbine Components Corporation, Dover Casting Division. The nickel-based superalloys are typical of the alloys used to manufacture turbine blades.

The coupons were fabricated by the injection method. Wax was injected into a hardened stainless steel tool, the mold was formed around the wax body using a ceramic slurry, and the coupons were cast in the molds using controlled solidification. The coupons are 2.5 cm by 14 cm long and were cast in thicknesses of 0.191 cm, 0.318 cm and 0.635 cm. The 1.27 cm samples used for the test were assembled by stacking two 0.635 cm coupons.

The core samples are 1.3 cm by 11.3 cm wedges of a high silica content ceramic material fabricated as a right triangle. The thicker section is 0.635 cm high and tapers to 0.033 cm, subtending an angle of 3 degrees. Since high silica content ceramics do not image readily in neutron radiographs, the techniques of doping and tagging are used to accentuate any residual core material.

The doped cores were fabricated by mixing 1 and 3 weight percent of gadolinium oxide with the ceramic core material before the core wedges were formed. The tagging process uses a 0.3 molar solution of gadolinium nitrate dissolved in water or methyl alcohol or various mixtures of the two solvents. The castings are immersed in the gadolinium nitrate solution, the solution is drained from

331

the turbine blades and the blades are rinsed lightly with ethyl alcohol. After the castings are dried, they are ready for neutron radiography.

Experimental Procedures

Neutron radiographs were made of coupons fabricated from the two similar nickel-based superalloys to screen the experimental samples for homogeneity. A similar neutron radiograph was made of the fabricated core samples.

From the screened samples, combinations of various alloy thicknesses and doped and tagged cores were neutron radiographed with each type of alloy. The alloy thicknesses were chosen to approximate the thicknesses found in actual turbine blade castings, except for the thick platform or shroud area.

RESULTS

The neutron radiographs of the coupons are shown in Figures 1 and 2 for Alloys No. 1 and 2, respectively. In Figure 3, the neutron radiograph of the cores shows the untreated wedges in the lower right corner. Above the untreated wedges are the tagged wedges; the 1% doped cores are in the center row and the 3% doped cores are on the right. The tagged and doped coupons are readily imaged along their full length. The edge effect seen in the tagged samples indicates that the solution does not fully penetrate the ceramic wedge. However, for typical residual core samples in the 0.035 cm diameter size, tagging would saturate the entire sample.

The mating of the four groups of core samples and the two types of alloys are shown in Figure 4, a photograph of the test plate, and Figure 5, the neutron radiograph of the test plate. Two important factors are readily apparent. The untagged core is barely discernible, especially in the thinner section of the core wedge. The entire length of the doped and tagged cores are visible, even for the tagged and 3% doped cores on the 1.27 cm thick coupons.

The coupons in the left half of Figure 4 are made from Alloy #1 and the right half from Alloy #2. The coupon thicknesses per set of four samples are 0.191 cm, 0.318 cm and 0.635 cm as shown·in the figure. To the right of the group of 12 coupons in the left half is the 1.27 cm thick Alloy #1 sample. Two 1.27 cm thick samples of Alloy #2, are located to the right of the group of Alloy #2 coupons.

The core sample locations per group of four coupons are the untreated core in the lower left, the tagged core in the upper left, the 1% doped core in the lower right and the 3% doped core in the upper right. For the 1.27 cm coupon of Alloy #1, a tagged core was used for sensitivity determination. For the two 1.27 cm coupons of Alloy #2, the coupon on the bottom has a tagged core and the one on the top, a 3% doped core. Each core sample is arranged with the thicker section of the wedge at the bottom and the 0.033 cm thick section on the top.

The optical density for a group of four cores on the 0.191 cm coupons is shown in Figure 6. The graph shows that untreated core has little change in optical density, within statistical limitations of the densitometer, from the average optical density of the coupon on which it rests. The most visible is the wedge containing 3% gadolinium oxide. Practically speaking, however, the residual core material is expected to be in the 0.033 cm range where the change in densities is similar for all doped and tagged cores.

These results show that the detection sensitivity for 1% and 3% doped cores and 0.3 molar tagged cores is a 0.033 cm diameter residual core object which translates to a 50 microgram sphere.

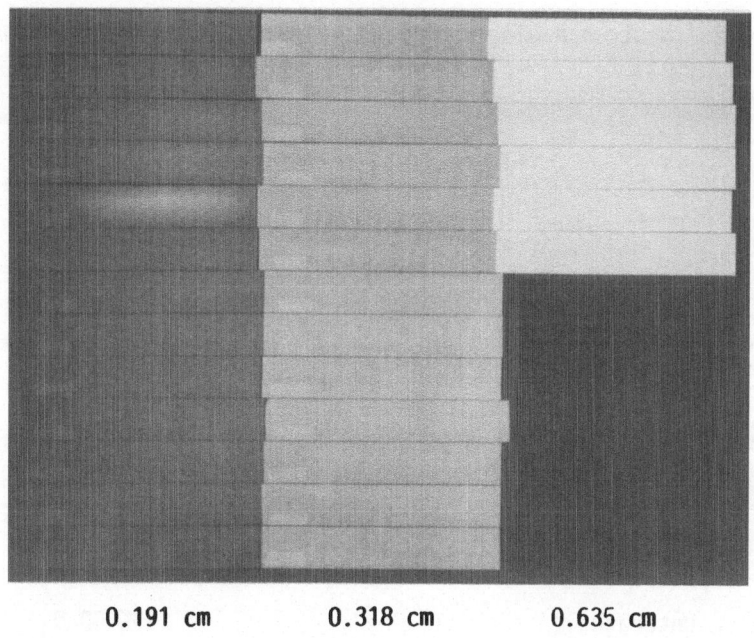

    0.191 cm   0.318 cm   0.635 cm

Figure 1 - Neutron Radiograph of Nickel Based Superalloy Number 1

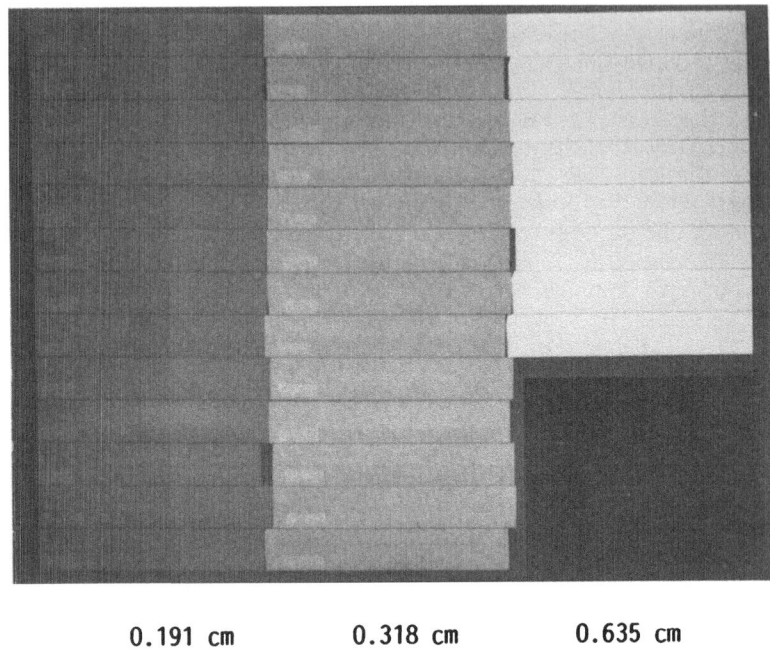

| 0.191 cm | 0.318 cm | 0.635 cm |

Figure 2 - Neutron Radiograph of Nickel Based Superalloy Number 2

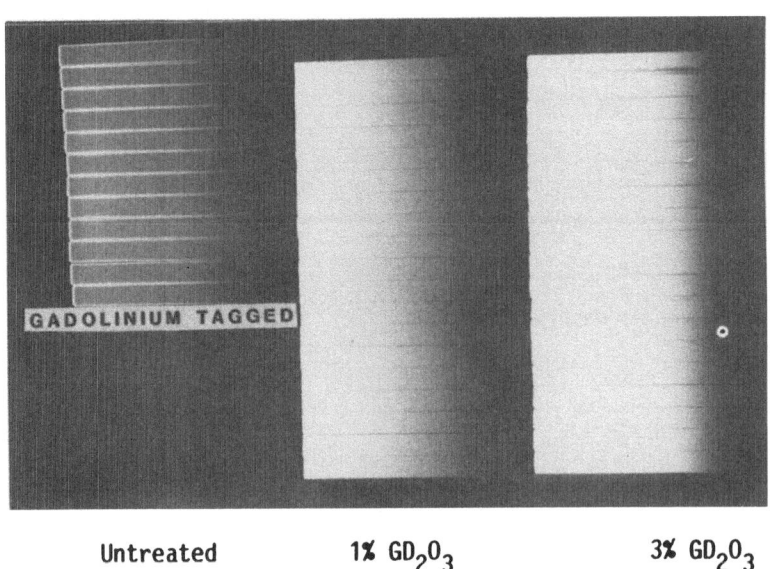

| Untreated | 1% GD_2O_3 | 3% GD_2O_3 |

Figure 3 - Neutron Radiograph of Wedge-shaped Core Samples

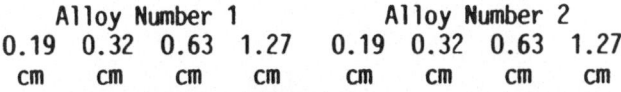

	Alloy Number 1			Alloy Number 2			
0.19 cm	0.32 cm	0.63 cm	1.27 cm	0.19 cm	0.32 cm	0.63 cm	1.27 cm

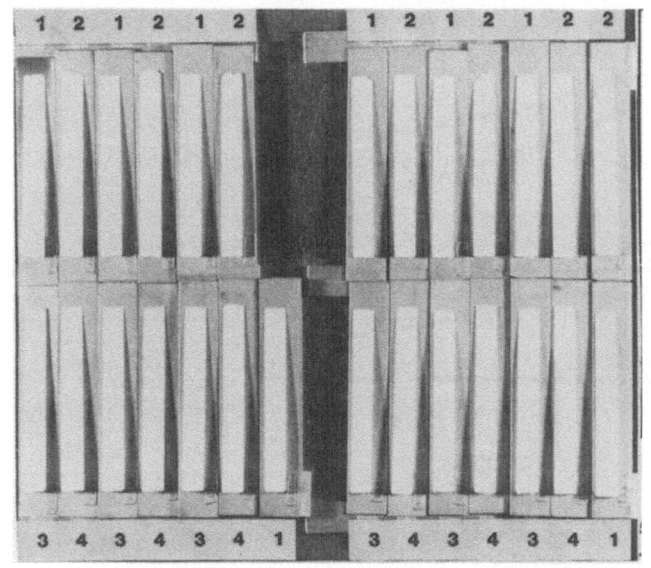

1=Tagged

2=3% GD_2O_3

3=Untagged

4=1% GD_2O_3

Figure 4 - Photograph of Test Plate

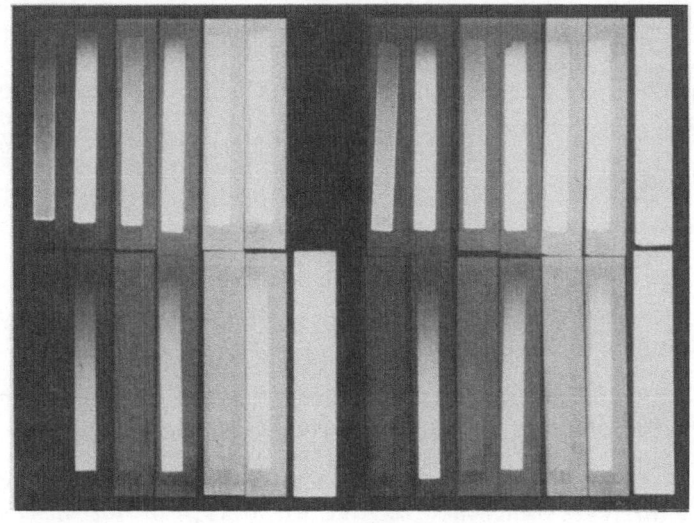

Figure 5 - Neutron Radiograph of Test Plate Above

335

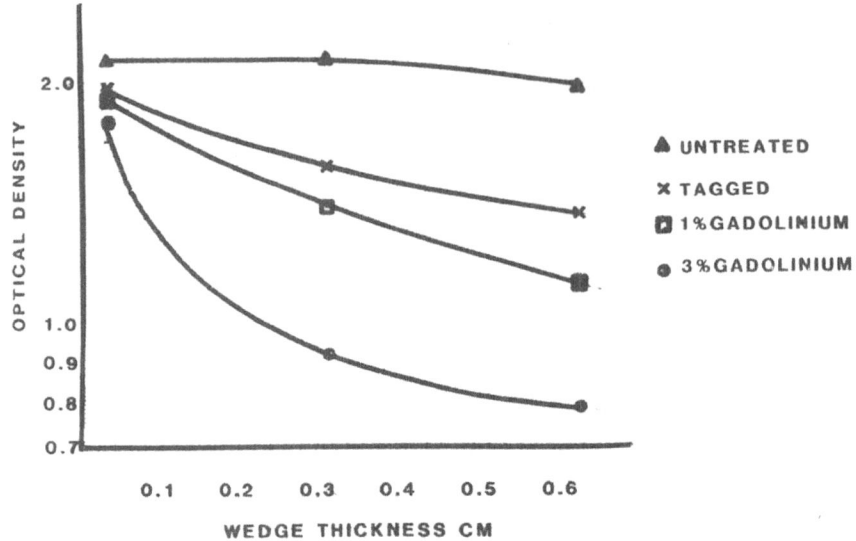

Figure 6 - Optical Density of Various Wedges on 0.191 cm
Thick, Nickel-based Superalloy Number 1.

The empirical determination of residual core detection
sensitivity has shown that core pieces as miniscule as 0.033 cm can
be imaged with reactor-based neutron radiography using gadolinium
enhancement techniques. Further experiments are planned with
ceramic chips down to 0.005 cm diameter and also with greater
coupon thicknesses.

REFERENCES

1. H. V. Watts, Report ARF 1164-27, Armour Research Foundation,
   Chicago, Illinois (1962).

2. D. J. Mallon, "Neutron Radiography," AGARD Advanced Technology
   for Production of Aerospace Engines (Sept., 1970).

3. N. B. Edenborough, "Neutron Radiography to Detect Residual
   Core in Investment Cast Turbine Airfoils," p. 152 ·in Practical
   Applications of Neutron Radiography and Gaging, H. Berger,
   editor, ASTM STP 586, American Society for Testing and
   Materials (1976).

PART V

NUCLEAR APPLICATIONS

EURATOM NEUTRON RADIOGRAPHY
WORKING GROUP

J. C. Domanus
Chairman, Euratom Neutron Radiography
Working Group

Abstract

In 1979 a Neutron Radiography Working Group (NRWG) was constituted
within Euratom with the participation of all centers within the European
Community at which neutron facilities were available. The main purpose of
NRWG was to standardize methods and procedures used in neutron radiography
of nuclear reactor fuel as well as establish standards for radiographic
image quality of neutron radiographs. The NRWG meets once a year in each of
the neutron radiography centers to review the progress made and draw plans
for the future.

Besides, ad-hoc sub-groups on different topics within the field of
neutron radiography are constituted. This paper reviews the activities and
achievements of the NRWG and its sub-groups.

1.INTRODUCTION

The significance of neutron radiography (NR) for nuclear industry applications in Europe was reviewed in a previous paper |1| presented at the First World Conference on Neutron Radiography (San Diego, 1981). There it was explained that Industry usually addresses requests for non-destructive testing with neutrons to the government-funded research centers. The major part of the work is carried out with research and test reactors.

Neutron radiography, like other NDT methods in the nuclear field, needs standardization in its many aspects. Unfortunately, very few standards relating to NDT exist in different nuclear fields |2|.

As many centers in the European Community perform NR of nuclear fuel it was felt that some standardization is needed in that field. Such standardization work was initiated at Risø National Laboratory, Denmark in the 70s, and in 1979 a formal Neutron Radiography Working Group (NRWG) was constituted by Euratom. In this working group all research centers of the European Community, where neutron radiography is performed, are participating (13 NR facilities in 8 centers).

The preliminary work done by NRWG was described in |3|. below a description is given of the main activities and achievements of the NRWG.

2. REFERENCE NEUTRON RADIOGRAPHS

The NRWG began its work by accepting a classification of defects occuring in light water reactor fuel, which was included in an atlas of reference neutron radiographs produced at Risø in 1979 |4|. This classification |5| and examples of defects revealed by NR were extended to light water and fast reactor fuel and a special sub-group of the NRWG has produced a collection of "Reference neutron radiographs of nuclear reactor fuel", published in 1984 |6|, and described in |7|.

This collection provides some typical examples of nuclear fuel pins and their components. Thereafter, a classification of different findings on neutron radiographs of nuclear fuel is given. Altogether 159 neutron radiographs are reproduced. (See poster "Reference Neutron Radiographs of Nuclear Reactor Fuel").

3. NEUTRON RADIOGRAHY HANDBOOK

The members of the NRWG have also prepared a "Neutron radiography handbook" |8| published just before the First World Conference on Neutron Radiography to which the NRWG has greatly contributed.

Now a special sub-group of the NRWG is preparing a new, revised and enlarged edition of this handbook. It is called "Practical Neutron Radiography".

4. QUALITY STANDARDS

Unlike the situation in other fields of industrial radiography, where standard methods and procedures are used to control the quality of the radiographic image, no such standards exist for neutron radiography of nuclear fuel. To fill that gap it was felt that standardization work ought

340

to be started in that field, too. This was done by the NRWG which has initiated the search for adequate quality standards for neutron radiography of nuclear fuel |9|.

At the time when NRWG standardization activity was started there was only one standard published in the field of neutron radiography |10|. Knowing that the ASTM standard |10| was under revision the NRWG has decided to follow the new design of the BPI and SI, which were thereafter recommended in the new issue of the ASTM standard |11|. The ASTM standard was prepared for general neutron radiography and it was felt that a specific standard is needed for neutron radiography of nuclear fuel. Therefore, the NRWG has designed a beam purity indicator almost identical as the ASTM BPI. As this BPI serves also to determine the gamma content in the neutron beam, and in neutron radiography of nuclear fuel this information is irrelevant, as the irradiated nuclear fuel itself emits a very strong gamma radiation, it was decided to design a special beam purity indicator for nuclear fuel: BPI-F.

The new design at the ASTM sensitivity indicator (SI) was also adapted by the NRWG.

All those indicators (described in |9|) were produced at Risø and distributed among all the participants of the NRWG. Even the possesion of those three indicators seemed to be insufficient for the image quality control of nuclear radiographs of nuclear fuel. Therefore, the NRWG has followed the statement of the ASTM E 545 that: "It is recognized that the only truly valid sensitivity indicator is a material or component, equivalent to the part being neutron radiographed, with a known standard discontinuity (reference standard comparison part)". Such a "reference standard comparison part" was designed by the NRWG for nuclear fuel. It was produced at Risø as a calibration fuel pin (CFP-El) and distributed among all the participants of the NRWG. It includes such "standard discontinuities" as pellet-to-pellet and pellet-to-cladding gaps (calibrated gaps from 50 to 300 μm). Further details about the CFP-El can be found both in |9| and |12|.

All the three indicators together with the calibration fuel pin are now being tested under a NRWG test program, described below.

It is worth mentioning that up till now 25 sets of those items were produced at Risø and that they are used also in neutron radiography centers outside the European Community (Israel, Japan, South Korea, Turkey, Iraq).

5. DIMENSIONAL MEASUREMENTS

In order to judge the behaviour of nuclear fuel pins after irradiation in a reactor, it is essential to quantitatively assess the dimensional changes occurring in the fuel itself and the cladding, and to compare this with preirradiation measurements. Neutron radiographs contain adequate information about such phenomena as swelling or cracking of the cladding or cracking or voidage formation of the fuel which can occur during irradiation. To extract this information from neutron radiographs, one must have an accurate method of measuring dimensions on the films on which neutron radiographs are taken.

A measurement of dimensions from neutron radiographs of nuclear fuel

341

pins consists actually of measuring distances between locally occurring maxima or minima in optical film densities on the radiographs. It is consequently a length measurement. In practice, the following dimensions are measured from neutron radiographs:

1) Outer diameter of the cladding tube.
2) Diameter of the fuel pellet.
3) Fuel-to-clad gap.
4) Cladding tube wall thickness.
5) Pellet length.
6) Pellet-to-pellet gap.
7) Dishing in the pellet.
8) Fuel column length.

Other dimensions can also be of interest, e.g. the central void diameter and length, the crack width in the pellets and many other.

Although it is comparatively easy to see even minute changes in dimensions on neutron radiographs, it is very difficult to measure them accurately. This problem was investigated at Risø |13| using different measuring techniques and is now under investigation by the NRWG.

The folowing devices and methods have been reported for dimensional measurements: travelling microscope and light table micrometer, optical projector, travelling microdensitometer, photographic enlargement, sharpening and image enhancement, electronic image processing. In practice two instruments are suitable for that purpose: the optical projector and the travelling microdensitometer. They are both described in |14|.

The underlying physical processes and mathematical methods used for dimensional measurements were treated in numerous publications e.g. |15|. A special book on that subject was recently published. |16|

6. NRWG TEST PROGRAM

The NRWG has developed a test program for checking the image quality and accuracy of dimensions measured from neutron radiographs of nuclear fuel pins |17, 18|. For that program indicators described above (BPI, BPI-F, SI) and calibration fuel pin (CFP-El) are used. They are neutron radiographed together at each of the NR facilities participating in the NRWG. Silver halide X-ray films are exposed with Gd and Dy converters by the direct and transfer method. Nitrocellulose film coated on both sides with a converter and without coating but between two converter screens, are also used. The radiographs are thereafter processed at the centers themselves and another set of identical radiographs centrally at Risø.

Altogether 30 visual evaluations, film density and dimensional measurements are being done for each set of neutron radiographs taken at each NR facility. (See fig. 1).

342

NRWG TEST PROGRAM

	Processed at																	
	NR center								RISØ									
Converter	Gd			Dy			B	BN1	Gd			Dy			B		BN1	
Film	SR	D4	M	SR	D4	M	CNB	CN	SR	D4	M	SR	D4	M	CNB		CN	
Code No	1	3	5	7	9	11	13	22	2	4	6	8	10	12				
Etched at															20°C	50°C	20°C	50°C
Code No															16	19	25	28
Copy on							S0015								S0015			
Code No							14	23							17	20	26	29
Viewed through							Polarizing filters								Polarizing filters			
Code No							15	24							18	21	27	30

Fig. 1. NRWG Test Program

From the radiographs of the BPI and BPI-F neutron beam constituents will be calculated. The sensitivity levels are found from the SI and from the CFP-El the image quality and accuracy of dimensional measurements are determined. The results of those measurements will be compared and conclusions drawn about the suitability of the test items for the purpose of controlling the quality of neutron radiographs of nuclear reactor fuel.

7. RADIOGRAPHIC IMAGE QUALITY

As stated in the ASTM standard |11|: "The judgement of the quality of a neutron radiograph is based upon the evaluation of images obtained from indicators that are exposed along with the test object". Such indicators (BPI, BPI-F, SI, CFP-El, are presently tested under the NRWG Test Program.

Although the Test Program is not finished yet, some preliminary conclusions were drawn already from the results available up till now. They can be found in |19, 20, 21|.

The problem of the usefulness of the SI for the assessment of radiographic image quality was further treated in |23| , presented at this conference.

Another approach to the assessment of radiographic image quality by visual examination of neutron radiographs of the calibration fuel pin was described in |24|, presented at this conference.

8. NUCLEAR BEAM COMPONENTS

The ASTM standard |11| describes the purpose of using of the beam purity indicator in the following way (§3.1): "The BPI is designed to yield information concerning neutron beam and image system parameters that contribute to film exposure and thereby affect overall image quality. In addition the beam purity indicator can be used to verify the day-to-day consistency of neutron radiographic quality". Furthermore formulas are given in the standard to calculate neutron beam constituents from density measurements of the BPI. Similar formulas are given for the NRWG BPI-F in |18|.

According to those formulas nuclear beam components are calculated for all 30 film/converter combinations (see fig. 1) for each NR facility participating in the NRWG Test Program. Those results will be furthermore used to assess usefulness of the beam purity indicators.

The conclusions about the usefulness of the beam purity indicators will be first formulated after the termination of the NRWG Test Program (as is the case for the radiographic image quality).

Also here some preliminary conclusions were drawn from the Test Program results available till now. They can be found in |19, 20, 21| and |22|.

This subject is also treated in |23| presented at this conference.

9. NITROCELLULOSE FILM

The use of nitrocellulose film for neutron radiography of irradiated nuclear fuel is steadily increasing. The advantages of the track-etch over the transfer method can explain this.

In Europe several NR facilities, where post-irradiation examination of nuclear fuel is done by neutron radiography, perform this control exclusively using the nitrocellulose film. Here the Service des Piles of the French center in Saclay as well as the Joint Research Centre at Petten, Holland, can serve as an example.

There are many factors that govern the correct exposure of nitrocellulose film. To obtain the best radiographic image quality one must choose optimum conditions for exposing the nitrocellulose film to neutrons (neutron fluence) as well as optimum conditions for etching (etching solution and its concentration, etching time and temperature). All those factors are under examination at several NR centers. The most extensive research in that field is done at Saclay, France, where experiments were done with NaOH and KOH as etching agents and different converter screens used with the Kodak-Pathé CN85 nitrocellulose film. The results were described in |25| and |26|. According to R. Barbalat |27| one can adopt as a rule of thumb that an underirradiation followed by an overetching will give improved sensitivity and contrast, whereas an overirradiation and underetching prduces an image of lowered contrast but richer in detail.

A similar investigation as at Saclay is being also made at Petten, Holland, now |28|.

At Risø a comparison was made of radiographic image quality and sensitivity of neutron radiographs taken on silver halide and nitrocellulose film |29, 30|. A more extensive comparison of that kind will soon be possible when all results from the NRWG test program will be available.

A sub-group on nitrocellulose film of the NRWG is currently preparing a special report on the use of nitrocellulose film in neutron radiography. The results from this report will thereafter be included in "Practical Neutron Radiography" handbook (see above).

10. INTERNATIONAL NEUTRON RADIOGRAPHY NEWSLETTER

To keep all concerned informed about the activities of different centers and organizations in the field of NR an "International Neutron Radiography Newsletter" (INRNL) is published (editor, J. C. Domanus) in English in the British Journal of Non-Destructive Testing and in French in the Revue Pratique de Control Industriel. It is the continuation of the previous Neutron Radiography Newsletter, edited by J. P. Barton and published by the ASNT (J. P. Barton is co-editor for America of the present INRNL).

REFERENCES

|1| J. C. Domanus & P. von der Hardt. Nuclear industry application of neutron radiography in Europe, 35-46 in Proceedings of the 1st WCNR.

|2| J. C. Domanus. Editorial. Nuclear Europe. Vol.4, No.11, November 1984, p.11.

|3| J. C. Domanus. Standardization activities of the Euratom Neutron Radiography Working Group. Risø-M-2356. June 1982.

|4| J. C. Domanus. Neutron radiographic findings in light water reactor fuel. Metallurgy Department, Risø National Laboratory. June 1979.

|5| J. C. Domanus. First attempt to classify defects revealed by neutron radiography in nuclear fuel for light water reactors. Risø-M-2171. April 1979.

|6| J. C. Domanus (editor). Reference neutron radiographs of nuclear reactor fuel. D. Reidel Publishing Company. 1984.

|7| J. C. Domanus. Neutron radiographic findings in reactor fuel. Proceedings of the 6th ASM International Conference on NDE in the Nuclear Industry, Zürich, Switzerland. 28.11-2.12.1983, pp 453-460.

|8| P. von der Hardt & H. Röttger. (eds.) Neutron radiography handbook. D. Reidel Publishing Company. 1981.

|9| J. C. Domanus. Search for adequate quality standards for neutron radiography of nuclear fuel: 1017-1024 in Proceedings of the 1st WCNR.

|10| ASTM E 545-75. Standard method for determining image quality in thermal neutron radiographic testing.

|11| ASTM E 545-81.

|12| J. C. Domanus. Calibration fuel pin CFP-E1. Risø Report B-499. Metallurgy Department, Risø National Laboratory. February 1981.

|13| J. C. Domanus. Accuracy of dimension measurements from neutron radiographs of nuclear fuel pins. Risø-M-1860. March 1976.

|14| J. C. Domanus. Euratom work on standard defects and dimensional measurements in neutron radiography of nuclear fuel elements. Risø-M-2318. October 1981.

|15| A. A. Harms. Physical processes and mathematical methods in neutron radiography. Atomic Energy Review. Vol. 15, No. 2. June 1977, 143-168.

|16| A. A. Harms & D. R. Wyman. Mathematics and physics of neutron radiography. D. Reidel Publishing Company.

|17| J. C. Domanus. Euratom test program for image quality and accuracy of dimensions. 1025-1033 in Proceedings of the 1st WCNR

|18| J. C. Domanus. Revised test program for testing of the CFP-E1; ASTM (revised) BPI and SI and BPI-F. Risø Report B-512. Metallurgy Department. Risø National Laboratory. August 1981.

|19| A. Laporte & G. Bayon. Determination of image quality in indu-
        strial neutron radiography. CEN–FAR. Service des Piles de Saclay
        - Section d'exploatation TRITON. 6.203.2.

|20| J. C. Domanus. Control of radiographic image quality in neutron
        radiography of nuclear fuel. Proceedings of the 6th ASM Interna-
        tional Conference on NDE in the Nuclear Industry. Zürich,
        Switzerland. 28.11.-2.12.1983, 447-451.

|21| J. C. Domanus, P. Gade-Nielsen & J. Olsen. How good are the
        standards for the image quality control in neutron radiography of
        nuclear fuel? Proceedings of the 7th International Conference on
        NDE in the Nuclear Industry. Grenoble, France. 29.1-1.2.1985.
        325-328.

|22| J. C. Domanus. Activities and achievements of the Euratom Neutron
        Radiography Working Group. Materials Evaluation, Vol. 44, No 1,
        January 1986, 114-119.

|23| J. C. Domanus. Can neutron beam components and radiographic image
        quality be determined by the use of beam purity and sensitivity
        indicators? Proceedings of the 2nd WCNDT.

|24| J. C. Domanus. Assessment of radiographic image quality by visual
        examination of neutron radiographs of the calibration fuel pin.
        Proceedings of the 2nd WCNR.

|25| R. Barbalat. Utilisation de l'installation de neutronographie sur
        materiaux radioactifs en service sur le reacteur ISIS ou
        CEN/Saclay. Mise en oeuvre et exploatation de la nitrocellulose
        en neutronographie. CEA. CONF. 5736. Paper prepared for the NRWG
        meeting at Grenoble. 2-3.6.1981.

|26| R. Barbalat. Use of cellulose nitrate for neutron radiographic
        testing of burned fuel elements. 747-753 in Proceedings of the
        1st WCNR.

|27| R. Barbalat in EUR report: "Neutron radiography on nitrocellulose
        film" (to be published in 1986).

|28| H. P. Leeflang. Private communication.

|29| J. C. Domanus. Comparison of image quality of nuclear fuel neu-
        tron radiographs taken on silver halide and nitrocellulose film.
        Risø-M-2170. April 1979.

|30| J. C. Domanus. How good is nitrocellulose film for neutron radio-
        graphy? 729-736 in Proceedings of the 1st WCNR.

Use of Epithermal Neutron Radiography to determine
the extent of melting in mixed oxide LMFBR fuel pins
irradiated in the High Flux Reactor, Petten

R.L. Moss and M. Beers

Commission of the European Communities,
Joint Research Centre, Petten Establishment,
The Netherlands

ABSTRACT

A series of mixed oxide LMFBR fuel pins are being irradiated in the High
Flux Reactor (HFR) at Petten, NL. One of the aims of the experiments is
to determine the extent of melting of the fuel under various operational
conditions, including transients.

By means of a cadmium filter inserted in the collimator of the reactor's
neutron radiography installation, it is possible to create a source of
predominately epithermal neutrons. This enables sufficient penetrations
of neutrons through the fuel and produces on film distinct images of
fuel melt patterns.

The paper highlights some of the more revealing results within the
present irradiation series.

1. INTRODUCTION

Neutron Radiography offers one of the better techniques in non-destructive testing of nuclear fuel materials. The method is extensively exploited at the High Flux Reactor (HFR) Petten to detect, amongst other things, fuel cracking, fuel relocation, central void formation, gap closure and related phenomena. In the context of LMFBR fuel, which is high enriched, the standard practice of using thermal neutrons is inappropriate because, like high density materials such as uranium oxide or carbide which is hardly transparent to X-rays, high enriched fuel is not transparent to thermal neutrons. Consequently, epithermal or fast neutrons are required. At the HFR, the former is utilised.

In sections 2 and 3, the HFR and the neutron radiography installation are discussed, with particular attention being given to the means of obtaining predominately epithermal sources. In section 4, a summary is given of the LMFBR experiments being carried out in the HFR, whilst in section 5, the epithermal neutron radiographs of one of the power-to-melt experiments is presented. A few concluding remarks are given in section 6.

2. THE HIGH FLUX REACTOR

The HFR is a 45 MW materials testing reactor, cooled and moderated by light water. A general lay-out of the reactor is shown in Fig. 1. The core is a 8 x 9 rectangular array, consisting of 33 fuel assemblies of the MTR type, 6 control rods, 16 Beryllium reflector elements along 3 sides of the core and 17 free positions into which experimental facilities may be placed for irradiation. Depending on the position in the core, a variety of nuclear conditions can be attained. In addition to the in-core positions, the reactor is equipped with a poolside facility (PSF) which has been designed to allow for power transient experiments to be carried out.

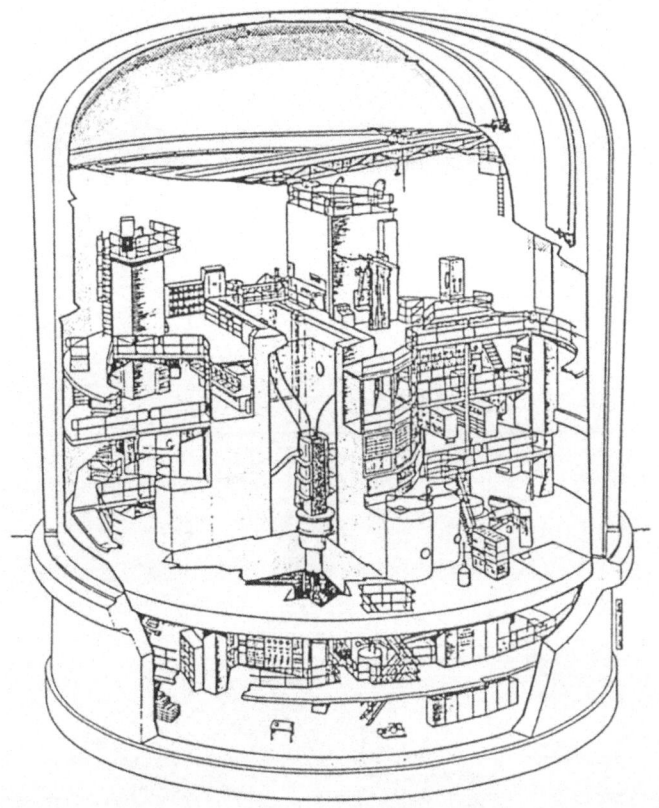

Fig.1 : HFR Petten, Isometric drawing of the reactor building

There are also available 12 horizontal beam tubes, which are primarily used for nuclear physics and solid state physics experiments.

For more details, the general characteristics of the reactor may be read in reference (1).

3. NEUTRON RADIOGRAPHY INSTALLATION

The neutron radiography installation is positioned underwater in the reactor pool and occupies the PSF position nr. 1. The

Fig. 2 Neutron Radiography Installation

installation, see Fig. 2, is described in more detail in reference (2). The installation consists essentially of:

- the collimator, which consists of a divergent rectangular aluminium tube covered on the inside with 1/4" thickness boral for shielding against outside neutrons. The L/D ratio is 220, resulting in a thermal neutron fluence rate of $1.5 \times 10^{11} \text{ m}^{-2}.\text{s}^{-1}$ at the position of the object,

- the diaphragm system, which has a fixed inlet aperture of 8 mm diameter,

- the object holder, which accepts objects of 15 x 20 cm, and 155 cm in length, and by means of an upper side flange may be drained by flowing pressurised air to contain the object in a dry state, thus preventing unsharpness on the detector,

- the cassette system, which contains the convertor foil or track-etch film, which in turn are placed between a front plate and cadmium-lined back plate to counteract back-scatter,

- the measuring and control equipment, which by means of a hydraulic system, provides control for the displacement of the cassette, evacuation of the object holder and displacement of the whole installation with respect to the reactor core.

For standard neutron radiography, using predominately thermal neutrons, the installation uses Kodak nitro-cellulose film, type CN85, with two separated (n,α) convertor screens (type BN1). The irradiation time is approximately 7-8 minutes, with approximately 30 minutes etching times at 46°C.

In the case of LMFBR fuel materials, the required predominately epithermal neutron source is achieved by placing a cadmium filter directly behind the collimator and in front of the object holder. The cadmium filter absorbs thermal neutrons below 0.4eV, thus creating a harder spectrum but of lower intensity. In this respect, best results are obtained using gold foils which, due to the much lower flux density, require exposure times of between 30-60 minutes. The activation of the gold foil is subsequently transmitted to a Kodak SR 54 film by a ten days exposure.

4.   LMFBR FUEL EXPERIMENTS IN THE HFR

Over the last years numerous LMFBR fuel pin experiments have been executed and at present, up to 6 experiments are in progress. The aim of the experiments is to investigate LMFBR fuel pin behaviour under operational transients and to study fundamental fuel material

Table I : Summary of Present LMFBR Fuel Pin Tests in the HFR Petten

Project no. name	Pins	Subject	Features	End of expt.
D183 KAKADU	120	In-situ transients	PSF, 2 full size pins, fresh & pre-irradiated	1987
D192 OPOST	60	Overpower steady-state	In-core, 3 pins(TRIOX), Cd screen	1987
D170 POCY	10	power cycling, pellet/clad inter-action	PSF, one pin, intermit. diameter measurement (axially)	1987
D184 POTOM	120	Power to melt	In-core, 3 pins (TRIOX) Cd screen	1986
E211 NILOC	6N	Overpower steady-state	In-core, 3 pins (TRIOX) Cd screen	1987
D215 RELIEF	50	Fuel/cladding axial elongation	PSF, 2 pins, instrum-ented probe	1989
E226 POM	S	Very high burn-up	In-core, mixed oxide discs (vertical), Cd screen	1987
E228 BUMMEL	S	Fission gas bubble mobility	In-core, UO2 discs (TRIOX), Cd screen	1987

nO indicates : n oxide pins
mN indicates : m nitride pins
 S indicates : special pellets

behaviour. An overview of the on-going experiments is shown in Table I.

One of these experiments, named POTOM (power-to-melt), is carried out to achieve partial fuel melting under start-up conditions. The results assist fuel behaviour experts to study and eventually to be able to predict such a phenomenum. By means of epithermal neutron radiography, an indication of the extent of melting, both radially and axially, can be observed soon after the completion of the experiment.

5.   SOME RESULTS

For the particular experiment under investigation, 3 LMFBR fuel pins are irradiated in a special TRIOX carrier. A typical fuel pin

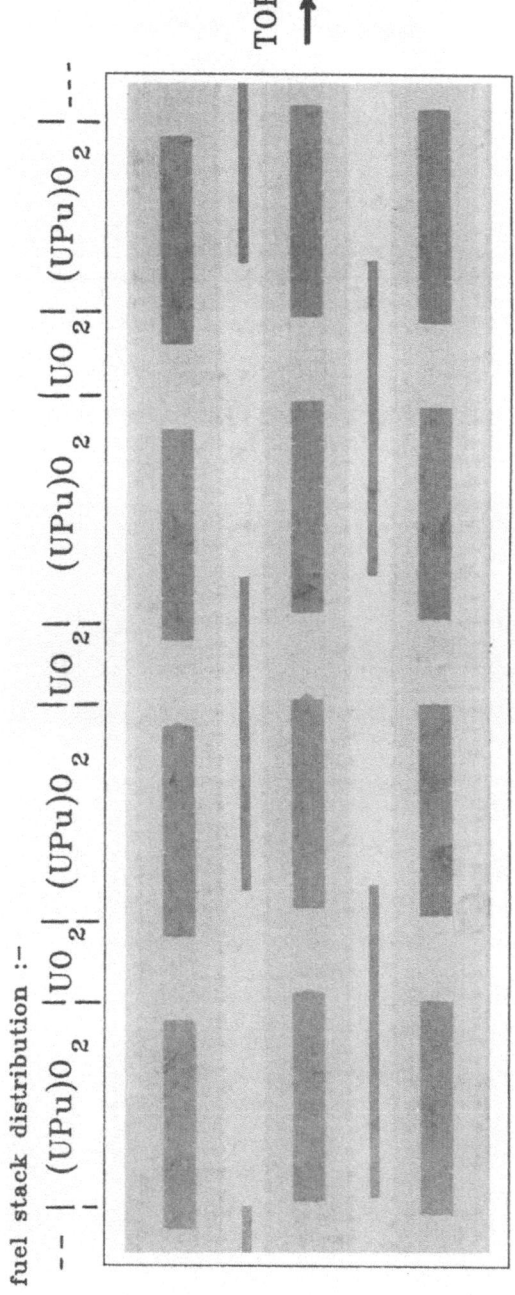

Fig 4 : Thermal Neutron radiograph of 3 LMFBR Fuel Pins after 1 hour irradiation at 700 W/cm maximum fissile power

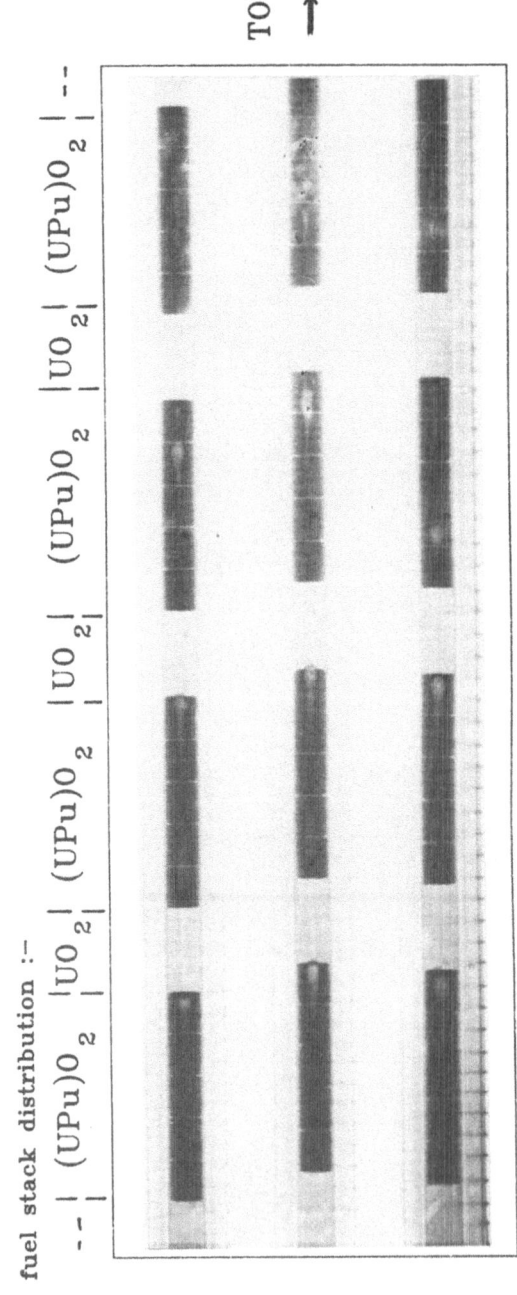

Fig 5 : Epi-thermal Neutron radiograph of 3 LMFBR Fuel pins after 1 hour irradiation at 700 W/cm maximum fissile power

is schematically shown in Fig. 3. On completion of the experiment, the 3 pins are withdrawn and placed into a special holder which is then positioned in the object holder of the neutron radiography installation.

The essential results are achieved using the process described above, in utilising epithermal neutrons. As a comparison, to indicate the clarity and advantages of carrying out epithermal neutron radiographs, a radiograph using thermal neutrons is shown in Fig. 4. Whilst this radiograph indicates that melting or even a central void may have formed in the breeder (UO_2) pellets only, it does not indicate the extent of melting within the fuel itself.

7. REFERENCES

1. "High Flux Materials Testing Reactor HFR Petten, Characteristics of facilities and standard irradiation devices", EUR 5700 EN, 1986

2. Bordo, J., Leeflang, H.P. and Veenema, J.J., "Neutron Radiography Installation in the HFR Reactor Petten", Proc. of the First World Conf. on Neutron Radiography, San Diego, USA, 1981.

IN-POOL NEUTRON RADIOGRAPHY OF BWR-CONTROL-RODS

WITHIN THE SCOPE OF A ROUTINE REFUELLING OPERATION

W. SCHULZ

PREUSSENELEKTRA AKTIENGESELLSCHAFT

TRESCKOWSTR. 5, 3000 HANNOVER 91

Users of BWR-Reactors know, that control rods of this reactor type
are a great problem in view of their livetime. Some authors already
reported about this point in a very detailed way (1,2). The most
common General Electric standard design shows a cruciform cross-section consisting of 4 metal-sheathed wings. In every wing you can find
up to 21 absorber tubes filled with boron carbide (B_4C) of a theoretical density (TD) by 70%.
By design the nuclear live limit of these control rods (in the following named CR) is reached when the reactivity worth at any significant axial section is reduced by 10%. During operation the following mechanical mechanism of failure was shown to be B_4C swelling and
stress corrosion cracking of the absorber tubes followed by B_4C
washout (s. 1).
The testing method presented in the following is able to detect and
lokalise these washouts and in addition to discern ^{10}B depletions.
For further analysis and valuation of the results a method is presented which allows to determinate the nuclear live limit of a BWR-CR
Using the radiographical instrument showed schematically in figure
one (page 2) we conducted examinations during the routine refuelling
operations in our nuclear power plant Würgassen upon 1984. So we were
able to reinsert the CR or to exchange them for new rods in case of
substantial boron losses.

The instrument works with a 500 µg californium-252 source
($Q_o = 1,2 \cdot 10^9 S^{-1}$) and is conceived as a dry system. Keeping a minimum
of distance below the waterlevel of the fuel storage pool water, the
CR to be radiographed is standing in a holding device. A cover,
looking like the CR is placed over and flooded with pressured air.
A collimation system located nearly at the bottom of the cover and
consisting of two divergent collimators extract the neutron flux.

At the front part of the collimator system a holding device is fixed
allowing to move and hold the neutron source. Opposite the collimator
in position of the testplane you can fix two plates at the cover loaded with a cellulose acetate film covered by $LiBO_4$ (CA 80 15 B). To
keep the watergap between cover and plate as small as possible you
must fill the plates with pressured air. So you have less influence
of scattering neutrons. Moving the cover over the CR you are able to
inspect 8 different CR-sections at a distance of about 25 (cm). In
the objectplane two segmented tubes are installed filled with B_4C
of theoretical densities by 50, 60 and 70%, air and water. The geometrical dimensions and the material of the two tubes are equal to
the absorber tubes. This is the way to value the radiographs in comparing the object with the reference rods. The exposed foils show
recognizable pictures after etching in a 10% (2,5N) solution of cau-
stic soda for about 3 hours at 58°C.

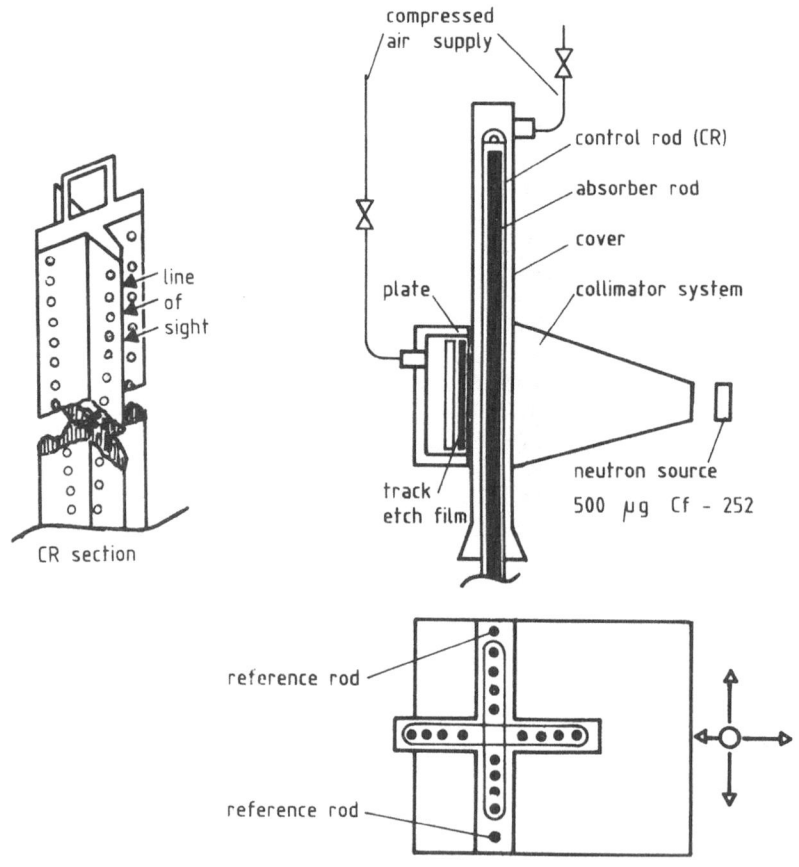

FIG. 1: SCHEMATIC REPRESENTATION OF THE NEUTRON RADIOGRAPHICAL SYSTEM

We startet our first tests 1983. At this time the neutron radiographical system had aggravating defects in physical respect, which became apparent in a long irradiation time, small resolution and contrast of detail. So we decided to reconstruct the radiographical system in a way that we were able to vary the important parameters with regard to the neutron radiography. In addition we replaced the 200 µg Cf-252 source by a 500 µg Cf-252 source and began to study the influences of collimator dimensions, quality of different moderators in connection to different source positions and shielding conditions. At the moment the best compromise in respect to the quality of the pictures and in connection to the irradiation time is shown in picture 2 and 3. With an irradiation time of about 30 minutes (*) you are able to produce radiographs giving a clear message in respect to show different boron contents. Referring to the dimensions of the radiographs (14x25 cm) you can get a relative irradiation time of 5.14 sec per cm^2 of the inspected area. In regarding picture 2, 3 and 4 you can see the following facts:

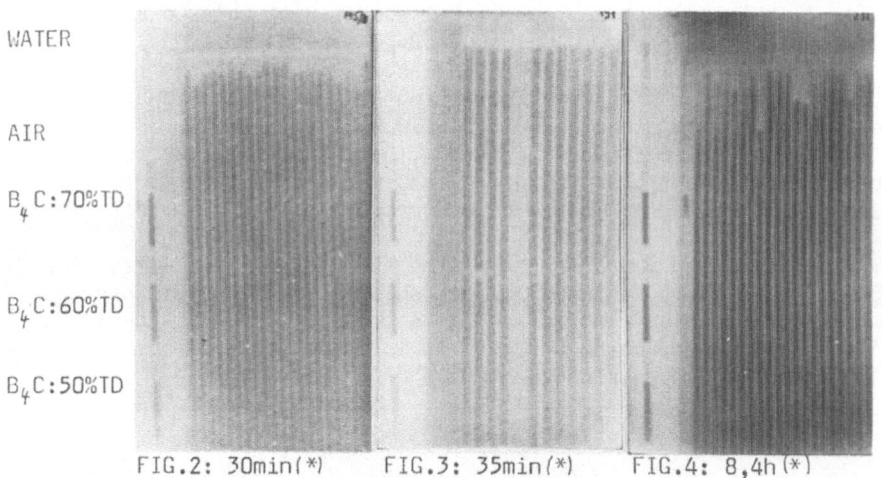

WATER

AIR

B_4C:70%TD

B_4C:60%TD

B_4C:50%TD

FIG.2: 30min(*)   FIG.3: 35min(*)   FIG.4: 8,4h(*)

RADIOGRAPHS OF ABSORBER WINGS WITH DIFFERENT RADIOGRAPHICAL IRRADIATION TIME

FIGURE 2 shows an unirradiated CR-segment in the upper part (GE standard design). On the left side of the picture you can see the reference rod containing B_4C of theoretical densities by 50, 60 and 70%, recognizable by different blackening of the relevant segments. However, the segments containing air and water are recognizable in a moderate way provided that the irradiation time is 30 minutes. The

(*) Every value of an irradiation time refers to a source strength of $Q_o = 1,2 \times 10^9 sec^{-1}$

CR section irradiated together with the reference rod shows all absorber material, the cover sheath (nearby the outermost tube) and shadowy the endplugs of the absorber tubes and finally the structure above the absorber section.

FIGURE 3 shows the top section of a prototypical CR wing (irradiation time: 35 min (*)). This CR, which is now on operation again, had a fluence of $0,7 \cdot 10^{21} \mathrm{cm}^{-2}$ (**)) during the inspection period. The periphery of this CR contains a border of hafnium and absorber tubes -as you can see- having a cross-section, which is a little bit bigger than a standard absorber tube. In the middle of FIGURE 2 you can see a border made of steel. The blackening of the absorber sections are in accordance to the designed reactivity effectiveness.

Making radiographs of CR the visibility of different concentrations in B_4C obviously depends on the fluence (irradiation time) by which the film was irradiated:

FIGURE 4 shows a CR section $(1,8 \cdot 10^{21} \mathrm{cm}^{-2}$ (**)) being exemplary irradiated for 8,4 hours. The structure of the reference rod and especially its sections filled with air and water are better to localize as for example in picture 2 or 3. But you are not able to see the differences of B_4C with TD between 50-70% any more.

In other words it is impossible to discern [10]B depletions in a range of about 0-30%. At a further increasing of the irradiation time you are even unable to discern [10]B depletions between 0-50%.

To reduce the effectiveness of the CR, some absorber tubes with [10]B depletions of about 50% are sufficient. Because of this a valuation, which is only based on losses of B_4C cannot lead to a correct result.

For evaluation and reproduction of the track-etch foils we developed a device (FIGURE 5), which allows to show the smallest differences in density in a high contrast and in a detailed way. This device is called "Film Reproduction Device". It works in principle by edge light illumination and a dark background.

FIGURE 5: FILM REPRODUCTION DEVICE

**) Thermal neutron fluence is the mathematical middled value of the concerned control rod quarter

Every radiograph is photographed to make a row of pictures, showing the radiographed CR section on an original scale.
In FIGURE 6 (page 6) the upper 90 cm of a high burn up CR section $(2,46 \cdot 10^{21} n/cm^2$ (**) are shown. You can recognize clearly the zones of boron losses. They are different to the zones of boron depletion due to the local burn up.The real valuation of every original radiograph however is made directly on the Film Reproduction Device. The boron content is valuated in a visual way by the help of the reference rod, measured and documented in a table. Making the valuation of the original radiographs it is helpful to include the row of photographs having a general view of the whole CR section.

The visual valuation schedule of the boron content shows FIGURE 7:

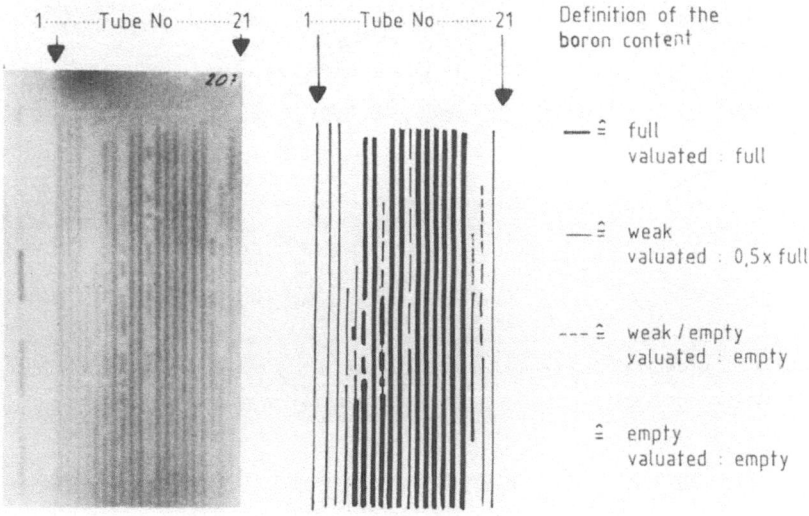

FIGURE 7: VISUAL VALUATION SCHEDULE OF THE BORON CONTENT

The tabled values of the CR valuation are given into a program, which works with the informations, put them in a graphical form and calculate the B_4C losses respectively the ^{10}B depletion with regard to the axial radiographical distortions and overlapped zones.
FIGURE 8 (page 7) shows the result of the described proceeding: On the left side of the respective CR wing (A,B,C,D) you can see the B_4C losses and ^{10}B depletions middled over the width of the absorber wing in (%) and in addition its mathematical burn up condition. As you can see the same CR is concerned, which is already partly shown in FIGURE 6. The middled boron content of this CR was determinated in the upper quarter by 75% of the original boron content. Per definition (see FIG. 6) the removal of the boron content by 25% is composed of the following 3 different parts:

A C D B

FIGURE 6: NEUTRON RADIOGRAPH OF A HIGH BURN UP CONTROL ROD

KWW/CR-R065/1985/FLUENCE : 2,5 10^{21}cm^{-2} (≈ ✳)

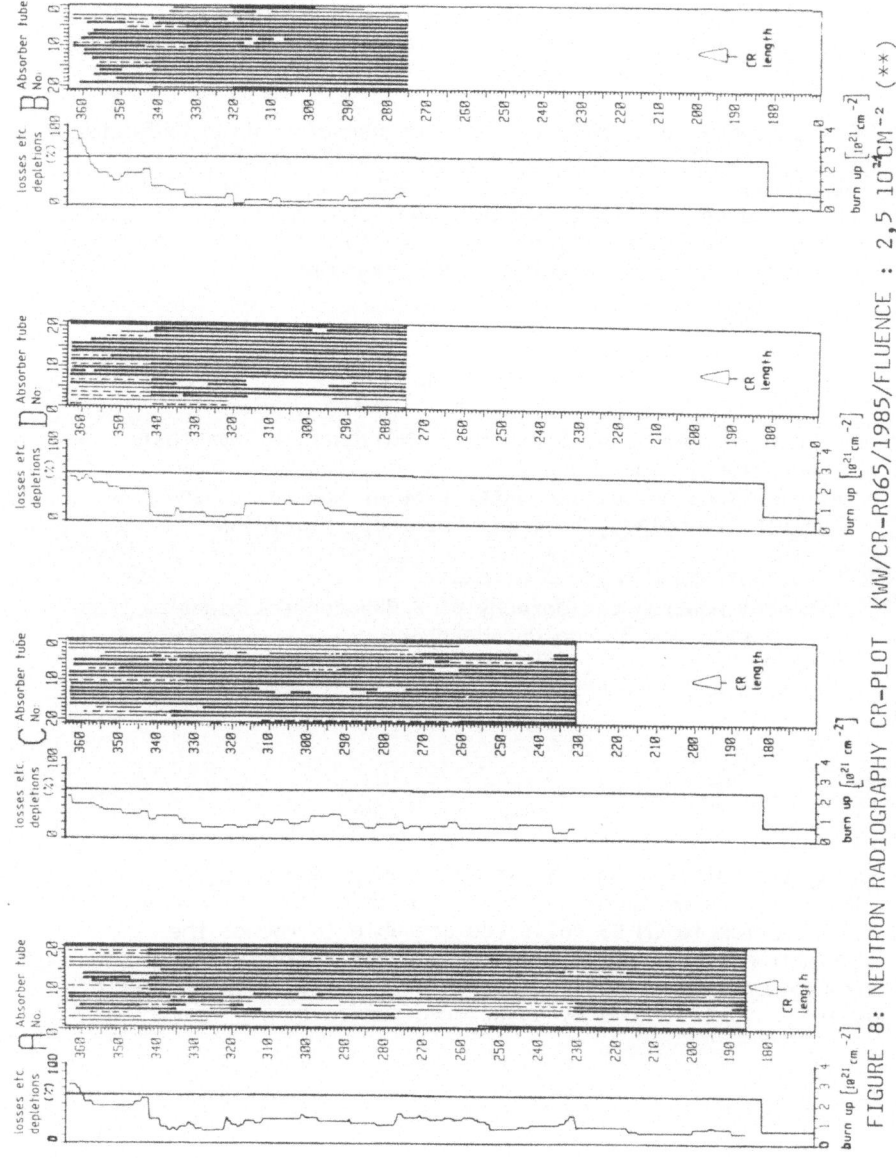

FIGURE 8: NEUTRON RADIOGRAPHY CR-PLOT KWW/CR-RO65/1985/FLUENCE : 2,5 10²²CM⁻² (**)

365

1. Boron losses (washouts)	- defined as "empty"		
	- valuated empty	-	12,5%
2. Boron losses /depletions	- defined as "empty/weak"		
	- valuated empty	-	5,5%
3. Boron depletions	- defined as "weak"		
	- valuated 0,5xfull	- 0,5 x 12,5%	

Considering the local effectiveness of the CR and the defined
nuclear live limit of these CR, we came to the result to exchange
all CR with this burn up history for new CR.

Finally it should be mentioned that the use of neutron radiography
of CR has security aspects as well as economic effects:
We found out, that we were able to save the equivalent costs of
9x3 CR cycles and the costs of the final storage of the defective
CR within the following three years, provided that we include the
neutron radiography in an optimal CR management.

REFERENCES

(1) N. Eickelpasch, R.W. Seepolt, J. Müllauer, W. Spalthoff
    "Operational experience with and postirradiation
    examinations on boiling water reactor control rods"
    Nuclear Technology, Vol. 60 Mar.1983 p 362-366

(2) W.J. Oosterkamp, F.J. v.d. Kaa,
    "In pool neutron radiography of a BWR control blade"
    J.P. Barton and P. von der Hardt (eds.),
    Neutron Radiography, 447-452
    Brussels and Luxembourgh

P.S.
During a first experimental study in July 1986 with
^{10}B Converters we found out that with the help of ^{10}B Converters
in connection to CN 85 foils you are able to reduce the
irradiation time from 30 minutes to about 8 minutes for producing
one exposure. So it will be possible to increase the number of
inspected control rods by a factor of up to 4.

NEUTRON RADIOGRAPHY FACILITY IN A STORAGE POND OF A NUCLEAR POWER STATION EQUIPPED WITH ANTIMONY-BERYLLIUM NEUTRON SOURCE.

L. Greim, F. Borchers, M. Greim, G. W. Schumacher, H.-W. Schmitz, G. Rudholzer*

GKSS Research Center, Geesthacht FRG
*Bayernwerk AG, Munich, FRG

ABSTRACT

A neutron radiography facility is described, which has been installed in the storage pond of the reactor power station Isar 1 near Munich (KKI). The facility is designed for the inspection of BWR control elements. Some of the control elements contain the neutron absorbing material boron carbide, filled in tubes. Neutron radiography can be applied to detect absorber losses by nondestructive testing.

The design of the facility corresponds to the principle of the diving bell. A tube, with the upper side closed, is submerged at the edge of the pond and the water is pressed out by air. The set up contains an antimony-beryllium neutron source. For imaging the track track etch method is used.

In an inspection program the blades of nine control elements were examined. The evaluations of the images yield the amount of boron losses in relation to neutron fluence exposure of the control elements.

INTRODUCTION

Nondestructive testing by neutron radiography is an important inspection method for control elements of boiling water power reactors. These control elements (fig. 1.) contain the neutron

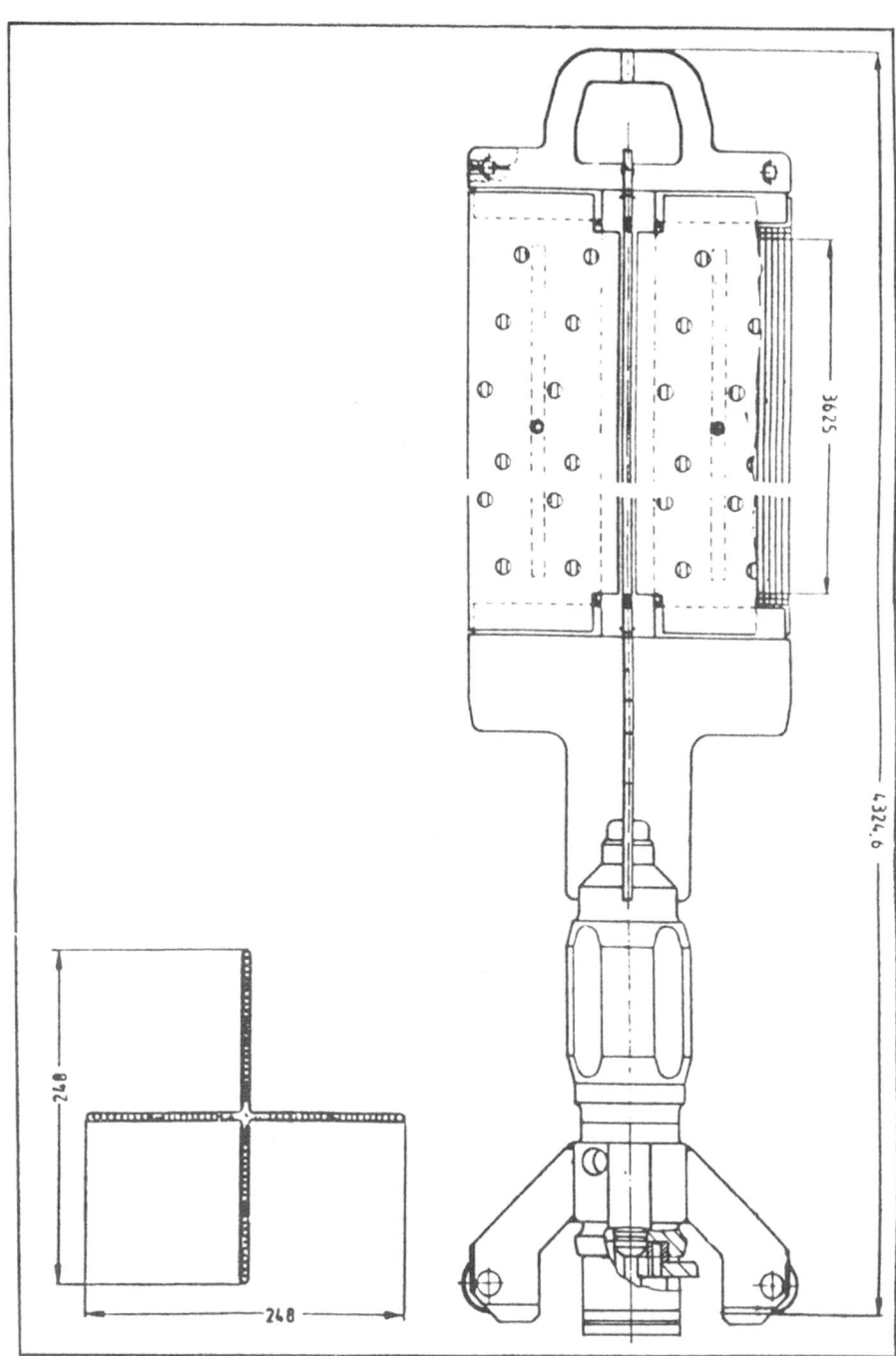

Figure 1. BWR Control Element

absorber boron carbide, which is filled in stainless steel tubes. These tubes are arrayed in the blades of the control element, forming a cruciform cross-section.

Losses of boron absorber had been suspected to occur before the end of the originally scheduled life-time of those control elements (1) and had been detected by neutron radiography (2,3) in regions where boron 10 burn up was more than 30%. Obviously the reasons of this losses are cracks in the tubes caused by boron carbide swelling, radiation induced, and the subsequent washout.

Extensive informations about the irradiation behavior of control elements are necessary. The easiest method is to perform the inspection in the storage pond of the power station. The control elements are too big for most neutron radiography facilities at research reactors and transportation of many high radioactive control elements is expensive.

For this reason an underwater neutron radiography facility had been developped. It had been installed and operated in the storage pond of the reactor power station Isar 1 (KKI). Design and construction of the set up are described and procedures and results of an inspection program are reported. A particularity of the facility is the utilisation of an antimony-beryllium neutron source.

Antimony-Beryllium Neutron Source

This neutron source had been designed for underwater radiography applications (fig. 2). It is watertight and has a cylindric shape with a diameter of about 17 cm and a total length of 27.1 cm

Figure 2.   Antimony-Beryllium Neutron Source

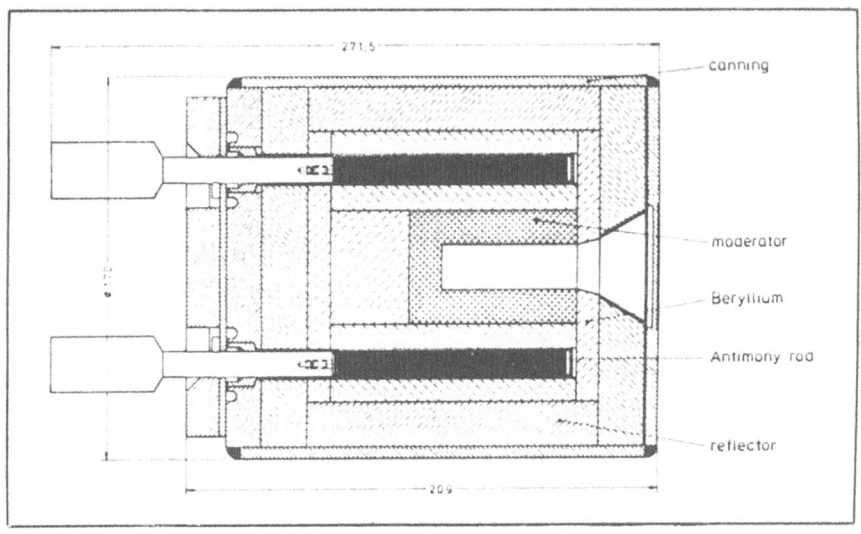

Figure 3. Design of the Sb-Be Neutron Source

The concentric structure (fig. 3) consists of a nickel reflector, covering a beryllium body with drillings for canned rods of antimony. The inner moderator (polyethene) has an outlet diaphragm of 2 cm diameter.

The neutrons are produced by $^9Be(\gamma,n)$ reaction. With a source of 37TBq ^{124}Sb a thermal neutron flux of $1.0 \cdot 10^8$ cm^{-2}s^{-1} in the central hole is obtained. This flux corresponds to that of a 10 mg ^{252}Cf source.

The radioactive antimony 124 is produced by neutron activation in a research reactor. The rods are delivered in a lead container to the storage pond and are inserted underwater with the aid of long tools.

NEUTRON RADIOGRAPHY FACILITY AND PERFORMANCE

The design principle of the facility is that of the diving bell (fig. 4). A tube, with the upper end closed, is submerged at the edge of the storage pond. A layer of more than 2 m water shields the radiation from the source and from the control element. The water is pressed out of the tube by pressurized air. The neutron source is mounted in a chamber at the lower part of the tube.

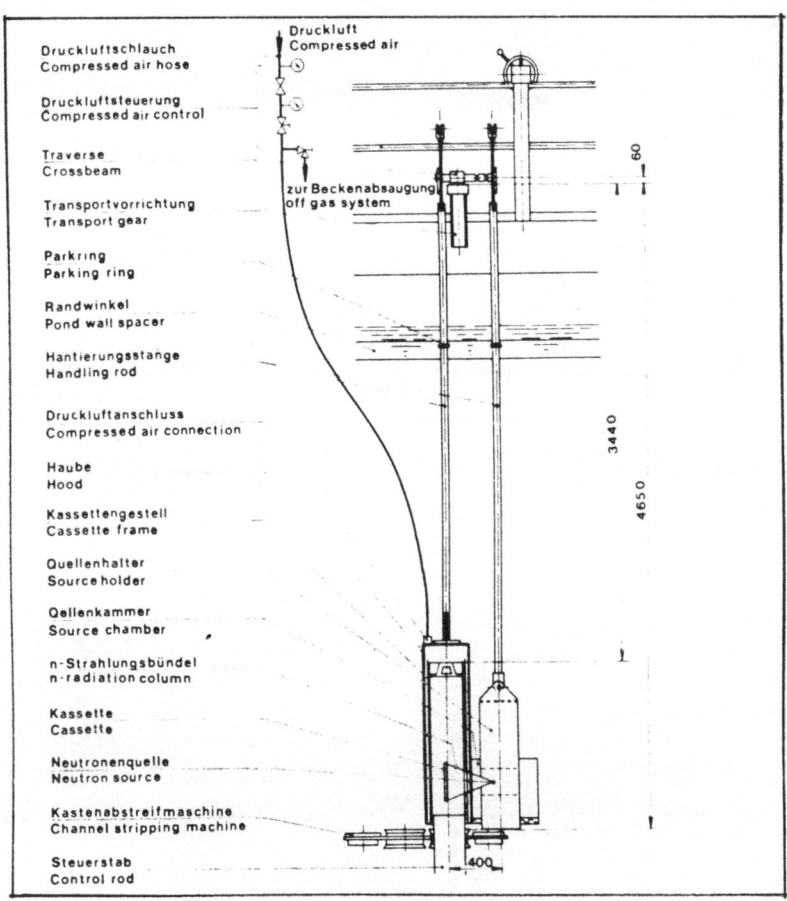

Druckluftschlauch	Compressed air hose
Druckluftsteuerung	Compressed air control
Traverse	Crossbeam
Transportvorrichtung	Transport gear
Parkring	Parking ring
Randwinkel	Pond wall spacer
Hantierungsstange	Handling rod
Druckluftanschluss	Compressed air connection
Haube	Hood
Kassettengestell	Cassette frame
Quellenhalter	Source holder
Qellenkammer	Source chamber
n-Strahlungsbündel	n-radiation column
Kassette	Cassette
Neutronenquelle	Neutron source
Kastenabstreifmaschine	Channel stripping machine
Steuerstab	Control rod

Druckluft
Compressed air

zur Beckenabsaugung
off gas system

Figure 4.   Underwater Neutron Radiography Facility

For neutron radiography the control element is placed in a lift system below the tube. A frame device is fixed at the control element to adjust foil-cassettes at the blades to be inspected (fig. 5). Two opposite blades can be radiographed simultaneously at a length of 27 cm. Track etch foil technique is used for imaging (CN 85-B). The neutron flux at the imaging plane is about $6 \cdot 10^4$ cm^{-2}s^{-1}, the collimation ratio L/D is 12.5. Generally an irradiation time of three hours was used. For exposure the control element is moved into the bell to the source position.

After the end of an inspection period the facility can be dismounted. The components, taken out of the pond, must be decontaminated by washing. The radioactive rods of antimony are kept in a storage position in the pond.

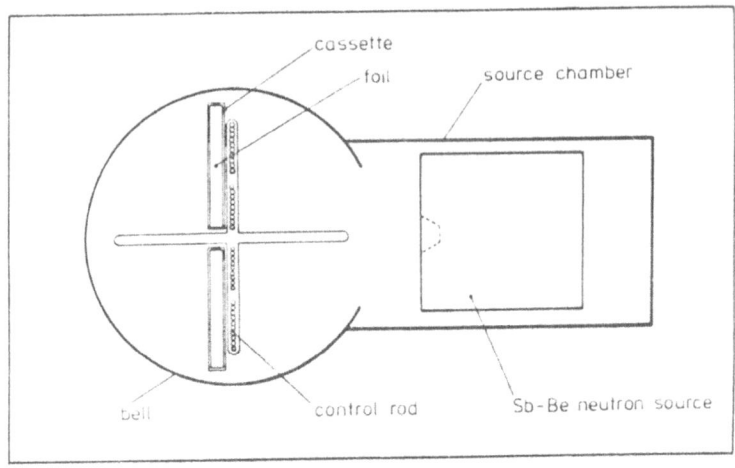

Figure 5. Neutron Radiography Set Up

The parts of the facility (fig. 6) consist of stainless steel resp. aluminium. They have to fullfill the regulations for storage ponds of the licensing authorities.

Figure 6. Components of the Facility

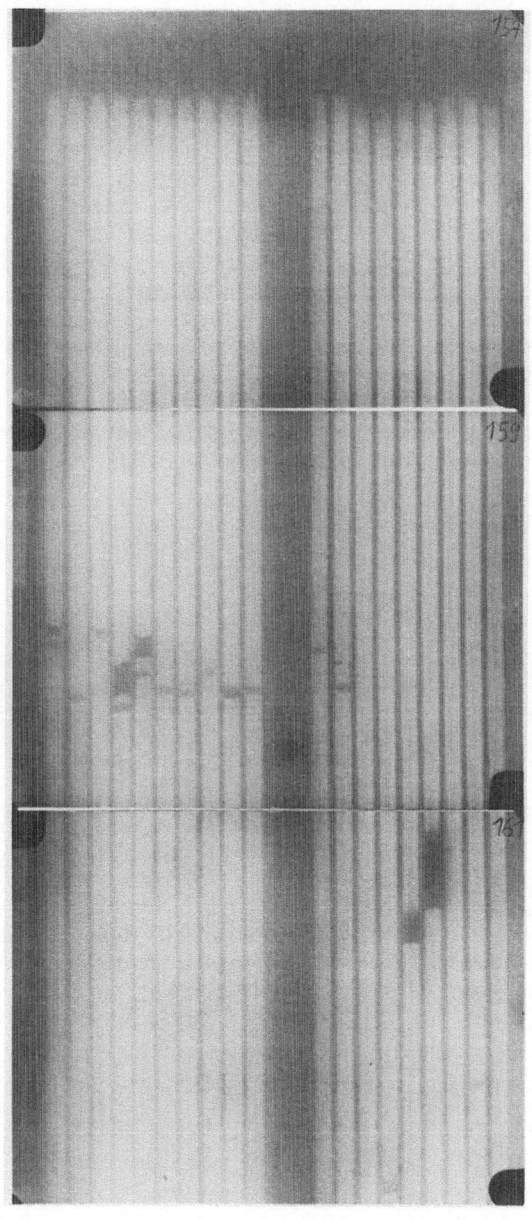

Figure 7. Neutron Radiograph of an Absorber Blade

INSPECTIONS AND RESULTS

Several control elements were selected for inspection. They had been exposed at different fluences in different places in the power reactor core.

Examinations with neutron radiography were performed at regions of the control elements with high fluence exposure. In these positions all blades were radiographed. Totally 52 radiographs have been made within a few weeks. One of the images is shown in fig. 7. It shows the boron containing tubes, gaps in the filling, and the steel balls which are inserted in the boron carbide column. Boron losses can be determined quantitatively.

The inspections will be continued in order to obtain more information about the behavior of BWR control elements.

The facility has proved to be very useful for those examinations. According to the design with a wide tube the radiography is not restricted to control elements. Other objects can be radiographed if necessary.

REFERENCES

1.   N. Eickelpasch, R.W. Seepolt, J. Müllauer, W. Spalthoff, "Operational experience with and postirradiation examinations on boiling water reactor control rods", Nucl. Tech. 60 pp. 362 - 366 (1983).

2.   W.J. Oosterkamp, F.J. v.d. Kaa, R. Tanke, "In Pool Neutron Radiography of a BWR Control Blade", p. 447 Neutron Radiography, Proc. of the 1st World Conf., San Diego, Cal. USA, Dec 7 - 10, 1981.

3.   L. Greim, W. Spalthoff, "Inspection of Reactor Fuel- and Absorber-Elements by Means of Neutron Radiography with a Small Size Antimony-Beryllium Neutron Source, p. 453 Neutron Radiography, Proc. of the 1st World Conf., San Diego, Cal., USA, Dec 7 - 10, 1981.

REFERENCE NEUTRON RADIOGRAPHS
OF NUCLEAR REACTOR FUEL

J. C. Domanus
editor

Risø National Laboratory
DK-4000 Roskilde, Denmark

Abstract

Reference neutron radiographs of nuclear reactor fuel were produced by
the Euratom Neutron Radiography Working Group and published in 1984 by the
Reidel Publishing Company.

In this collection a classification is given of the various neutron
radiographic findings, that can occur in different parts of pelletized,
annular and vibro-compacted nuclear fuel pins. Those parts of the pins are
shown where changes of appearance differ from those for the parts as fabricated. Also radiographs of those as fabricated parts are included.

The collection contains 158 neutron radiographs, reproduced on photographic paper (twice enlarged) and on duplicating film (original size).

1. INTRODUCTION

For weldings and castings many standard reference radiographs were published as ASNI/ASTM standards. Also the IIW has issued similar collections of reference radiographs of welds. Therefore it was felt that a similar collection of standard reference radiographs will be needed in the field of neutron radiography of nuclear fuel. Thus the assessment of neutron radiographs of nuclear fuel elements can be faster and simpler if reference can be made to typical defects that can be revealed by neutron radiography.

One of the first tasks of the Euratom Neutron Radiography Working Group (NRWG), constituted in 1979, was to start standardization work by establishing a classification of defects revealed by neutron radiography of nuclear fuel and to collect adequate examples of corresponding neutron radiographs. Such a classification together with a collection of 36 neutron radiographs illustrating those defects was published in 1979 by Risø National Laboratory |1| and was accepted by the NRWG as a first step in its standardization activity.

Just recently, a new edition was published of a collection of reference neutron radiographs of neutron reactor fuel |2|. It contains 158 examples of defects in nuclear fuel as well as examples of its different parts as fabricated, assembled from different neutron radiography centers of the European Community, participating in the activities of the NRWG.

2. WHAT THE REFERENCE RADIOGRAPHS SHOW

It must be mentioned that in the reference radiographs published by ANSI/ASTM the term "discontinuity" is used, instead of "defect". The illustrations of those discontinuities are graded or ungraded and each graded discontinuity type has several severity levels.

In the IlW collection of reference radiographs the term "defect" is used, and the radiographs have been divided into grades, graded in accordance with the relative importance of the different types of defects.

In the present collection of reference neutron radiographs the term "defect" is used to designate a change in appearance shown on an original radiograph of a particular part of the fuel as fabricated, to that shown on a subsequent radiograph, usually post-irradiation.

3. FUEL PIN COMPONENTS

In fig. 1 typical examples of nuclear fuel pins are given, containing pelletized, annular and vibro-compacted fuel.

The components of those pins are:

A.   FUEL
A.a. Pellets
A.b. Annular fuel
A.c. Pellet-to-pellet-gap
A.d. Dishing
A.e. Vibro-compacted fuel
A.f. Fuel-to-clad-gap

D.   PLUGS
D.a. Bottom plug
D.b. Top plug

E.   INSTRUMENTATION
E.a. Thermocouple
E.b. Pressure transducer

A.g. Fuel column
A.h. Fuel composition

E.c. Diameter gauge
E.d. Length gauge
E.e. Other instrumentation

B. CLADDING

C. PLENUM
C.a. Spring
C.b. Spring sleeve
C.c. Insulating disc
C.d. Spacer
C.e. Fuel column to plug distance

4. CLASSIFICATION OF FINDINGS

All of the nuclear pin components listed above are shown on the right side of fig. 2, whereas various differences in appearance from the fuel component as fabricated are listed at the top of fig. 2. They are the following:

0. (Fuel pin part) AS FABRICATED	3. CHANGE OF SHAPE OR LOCATION
1. CRACKS	3.1 Enlarged or swollen
1.1 Random	3.2 Contracted
1.2 Longitudinal	3.3 Filled-up or closed
1.3 Transverse	3.4 Deformed
1.4 Annular	3.5 Broken
1.5 Stratified	3.6 Dislocated
	3.7 Extended
2. CHIPS	3.8 Accumulated
2.1 Corner	3.9 Restructured
2.2 Other	3.10 Melted
2.3 In pellet-to-pellet gap	3.11 Disintegrated
2.4 Missing	3.12 Migrated
4. VOIDAGE	6. CORROSION
4.1 In one pellet	6.1 Hydrides
4.2 Through several pellets	6.2 Oxides
4.3 Through whole fuel column	6.3 Other
5. INCLUSIONS	7. NUCLEAR PROPERTIES
5.1 Of plutonium	7.1 Different enrichment
5.2 Of poison	7.2 Different burnup
5.3 Other	
	8. COOLANT
	8.1 Present
	8.2 Absent

As can be seen column "0" contains neutron radiographs of as fabricated fuel pin parts.

The neutron radiographic findings were selected from radiographs of light water (L) and fast (F) reactor fuel. If L of F is marked in fig 2 it means that in the collection |2| an example is given. It does not, however, mean that there cannot be such a finding if neither L nor F marking occurs. It means only that none of the participants of the NRWG has found such an example among his radiographs.

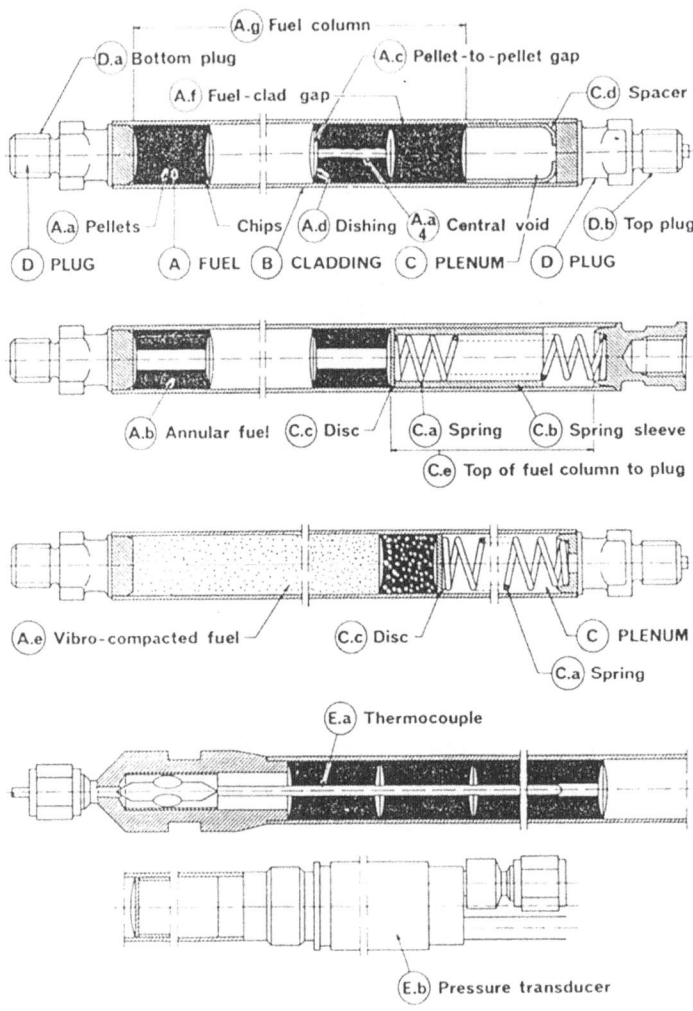

Fig. 1. Nuclear fuel pins components

5. CONTENTS OF THE COLLECTION

The collection of neutron radiographs of nuclear reactor fuel |2| contains besides examples of nuclear pins components and a classification of neutron radiographic findings (described in 3 and 4 above) 158 copies and duplicating film (in original size) and on photographic paper (twice enlarged) of neutron radiographs taken on silver halide or nitrocellulose film.

A list of contents of the collection describes in detail the type of defect illustrated on the radiographs, as well as the type of nuclear fuel and the origin of the radiograph.

6. USE OF THE COLLECTION

The copies of the neutron radiographs on film can be viewed without removing them by illuminating the blank page which follows with a shaded desk lamp.

The reference radiograph may also be removed from the collection and be viewed on an illuminator together with the actual radiograph under assessment.

7. TERMINOLOGY

The text of this collection is produced both in England and French. Special terms used throughout the collection, as well as some useful ones in the field of neutron radiography, are reproduced in Danish, Dutch, English, French, German and Italian.

8. INSTALLATIONS IN THE EUROPEAN COMMUNITY

A survey on the main technical data and addresses of the neutron radiography installations in the European Community applicable for examination of nuclear reactor fuel is given at the end of the collection.
More technical details can be found in reference |3|.

REFERENCES

|1| J. C. Domanus. Neutron radiographic findings in light water reactor fuel. Risø National Laboratory, Metallurgy Department, June 1979.

|2| J. C. Domanus (editor). Reference neutron radiographs of nuclear reactor fuel. D. Reider Publishing Co., 1984. EUR 8916 EN EP ISBN 90-277-1717-6.

|3| P. von der Hardt, H. Röttger (editors). Neutron radiography handbook. D. Reidel Publishing Company, 1981, EUR 7622e, ISBN 90-277-1378-2.

NEUTRON RADIOGRAPHY OF LIGHT WATER REACTOR FUEL RODS

AT THE HIGH FLUX REACTOR PETTEN

J. Bakker, A. Baritello, J. Bordo, J.F.W. Markgraf
Joint Research Centre of the Commission of the European
Communties, Petten Establishment
Postbus 2
NL 1755 ZG Petten, The Netherlands

H.P. Leeflang
Netherlands Energy Research Foundation
Postbus 1
NL 1755 ZG Petten, The Netherlands

I. Ruyter
Kernforschungsanlage Jülich GmbH
Postfach 1913
D 5170 Jülich 1, Federal Republic of Germany

ABSTRACT

Within the LWR fuel rod testing programmes at HFR
- where more than 200 individual fuel rod tests have been
performed - neutron radiography is an important and unique
non-destructive test technique.
Equipment, methods and typical results related to neutron
radiography, of LWR fuel rods at HFR are presented.
The importance of neutron radiography is shown for some
representative cases, addressing the condition of the fuel
rod and fuel stack.

1.0 INTRODUCTION

For approx. 15 years Neutron Radiography (NR) is being used
on a routine basis at the High Flux Reactor (HFR) Petten /1/ for
non-destructive examination of Light Water Reactor (LWR) fuel rods.
About 1000 images have been taken with the HFR underwater NR camera
at various irradiation stages on more than 200 LWR fuel rods /2/.

Together with pool inspection systems such as Eddy Current
Integrity Test and Fuel Rod Profilometry, NR has become an
important and versatile tool for the investigation of LWR fuel rod
behaviour. Its capability to provide an image of the interior of
the LWR fuel rod makes NR unique for the examination of fuel and
internal components behaviour. The special features of NR for LWR
fuel rod investigation are shown for selected examples.

During the HFR vessel exchange /3/ period (1984/85), the 13 years
old underwater NR camera /4/ was replaced by a new, redesigned
underwater NR camera. This new NR camera, Fig. 1 and related
techniques are presented hereafter.

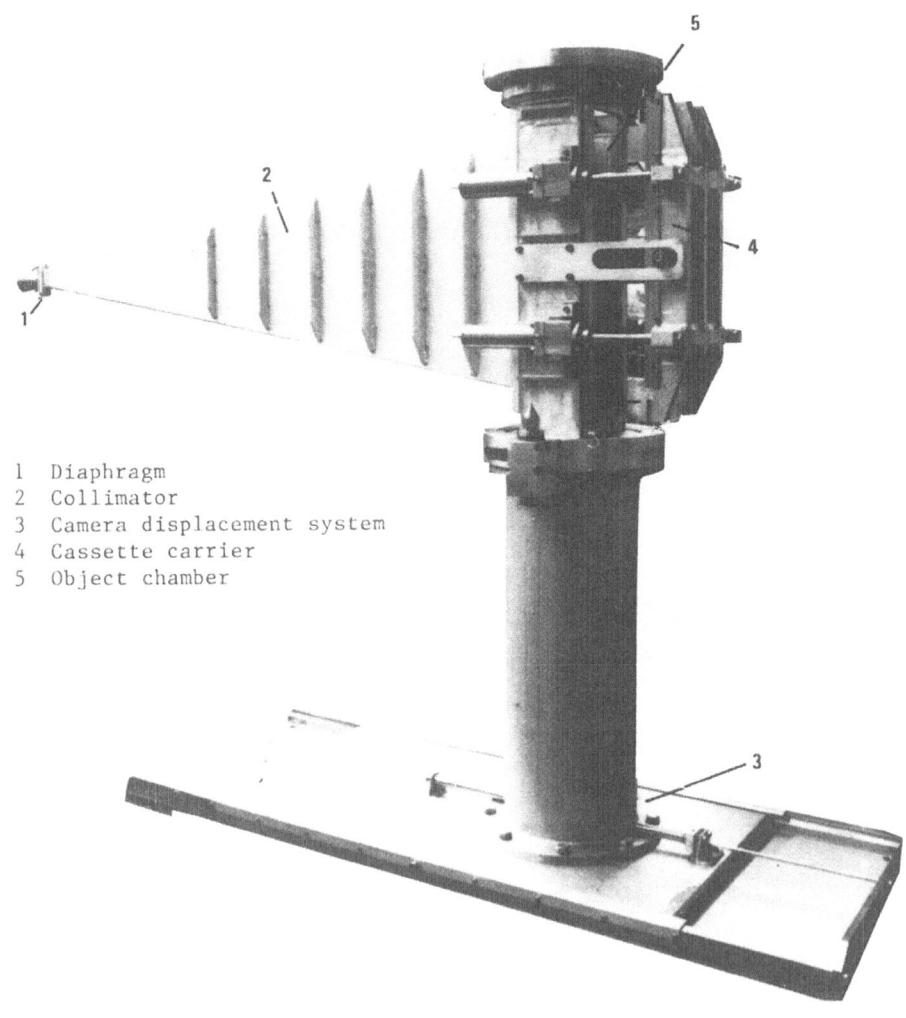

1 Diaphragm
2 Collimator
3 Camera displacement system
4 Cassette carrier
5 Object chamber

Fig. 1 New underwater neutron radiography camera

2.0 NEUTRON RADIOGRAPHY TECHNIQUES

2.1 Underwater NR Camera at HFR

Based on the operational experience with the first underwater NR camera and the user's requirements for this facility, a new underwater NR camera was designed and installed at HFR.

The main characteristics of the NR camera were maintained and are now :

o   Collimator
    - Material : Main structures - Aluminium,   8 mm thick
                 Liner   - Boral (B_4C), 6.35 mm thick
                 Gas filling   - Helium
    - Collimation ratio (L/D) : 220
    - Diaphragm, mm   : 8, exchangeable
    - Beam dimensions
      in object plane, mm   : 564 x 86
    - Thermal neutron fluence
      at object plane, $m^{-2}s^{-1}$: 2 x 10^{11}

o   Object chamber, mm : 150 x 200 x 1558

o   Cassette carrier hydraulically operated

o   Cassette floating on cassette carrier providing close contact between object and NR film/foil

o   Image generation by remote displacement of the NR camera into neutron flux at the Pool Side Facility

o   Loading and unloading of objects by remote technique (objects are nearly always highly radioactive)

o   Removal of water in object chamber and object holder by pressurized air

On the following points the new NR camera differs from the old one :

o Modular design to ease maintenance, repair and replacement of components
o Collimator and object chamber including the cassette carrier can be remotely unloaded from and loaded into the camera displacement system during HFR power operation
o All components are made from materials providing smooth cleanable surfaces (no cast structures as used previously for e.g. the collimator), thereby improving the decontamination procedure
o Simplified, but neutronically improved diaphragm; diaphragm easily exchangeable; see Fig. 2

o Locking and positioning of object holder on object chamber
  facilitated by the introduction of a new remotely operated
  mechanical locking system
o Positioning and guidance of the cassette camera is improved by
  an additional guiding and support device, see Fig. 3.

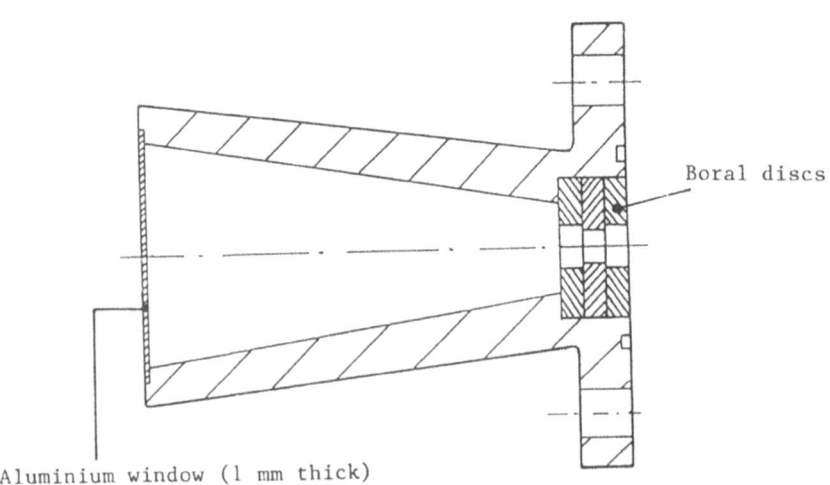

Boral discs

Aluminium window (1 mm thick)

Fig. 2 Inlet diaphragm of the neutron radiography camera

1 Floating cassette carrier
2 Object chamber

Fig. 3 Detail of cassette carrier system

2.2 Procedures, image techniques and evaluation equipment in use for NR of LWR fuel rods

LWR fuel rods are always put into aluminium containers prior to NR. Irradiated fuel rods are loaded und unloaded remotely in the HFR pool in aluminium containers which are drained by pressurized air.

The fuel rod containers are placed in special object holders (flanges) providing

o parallel positioning of the fuel rod relative to the NR film or foil for both radial image planes (0 and 90 degree) and

o a sealed object chamber.

The object chamber is drained and dried after loading of the object and cassette by pressurized air.

For NR of LWR fuel rods, at present nearly only the direct imaging technique using nitrocellulose film is employed at HFR. Currently the Kodak-Paté CN 85 nitrocellulose film with separate converter screen type BN1 is in routine use. Images are taken on 600 x 80 mm film and copied to Kodalith contact film type 2571. The exposure time for an image on nitrocellulose film of a LWR fuel rod is typically 7 minutes.

When the indirect imaging technique is applied, Dysprosium foil 0.1 x 600 with transfer to Kodak SR X-ray film or Kodak Industrex M X-ray film is employed. Typical exposure time for an image on Dysprosium foil of a LWR fuel rod is approx. 20 minutes. The utilization of the direct image technique is preferred due to its better resolution, shorter exposure times and handling ease.

Evaluation of NR images from LWR fuel rods is primarily oriented towards structural integrity checking, proper functioning of internal components, fuel stack length changes, dishing closure, fuel pellet crack pattern and water ingress. For the visual examination, a light box with adjustable viewing area and light intensity is generally used. Images on nitrocellulose film are viewed through contrast enhancing polarizing screens. For dimensional evaluations the original film is used. A modified Nikon 6C-2 profile projector with provisions for fuel stack, fuel rod length and dishing size determination is available for dimensional and detailed image evaluation. A computerized travelling microdensitometer /5/ for more precise dimensional investigations of e.g. fuel clad gap size is at present being developed.

3.0 NEUTRON RADIOGRAPHY APPLICATION FOR LWR FUEL RODS

The information obtainable from non-destructive inspection systems for irradiated LWR fuel rods in the HFR pool is summarized in the table below.

Content of information from NDT at HFR	NDT-method		
	Neutron Radiography	Eddy current check	Profilo-metry
Cladding condition	X	X	X
Clad tube perforation	(X)	X	X
Total length of fuel rod	X		(X)
Outer diameter changes	(X)		X
Ovality of clad tube			X
Ridges in clad tube		X	X
Length and condition of fuel column	X		
Pellet-to-pellet interface positions	X		
Pellet height	X		
Pellet diameter	X		
Pellet cracking pattern	X		
Change in dishing	X		
Gap between clad tube and fuel	(X)		
Condition of spring or support tube	X		
Condition of insulation pellets	X		
Presence of water in fuel rod and plenum	X		

In the following chapters examples of some typical NR work related to LWR fuel rod inspection will be shown.

3.1 Check of LWR fuel rod condition

The fuel rod condition is mainly determined by its power history. Therefore, comparison of NR images from different stages of an irradiation programme yields the most information about the fuel and fuel rod component condition including its behaviour.

Figs. 4 and 5 show the same section of a PWR fuel before and after a transient. Dishing closure as a result of the power increase during the transient is very pronounced in this case.

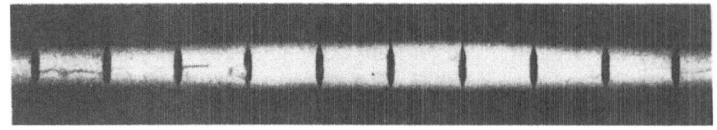

Fig. 4 PWR fuel rod before transient test

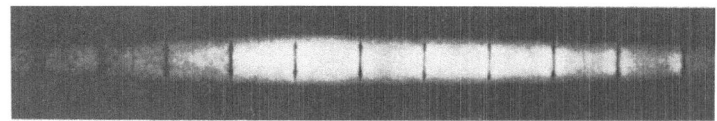

Fig. 5 PWR fuel rod after transient test

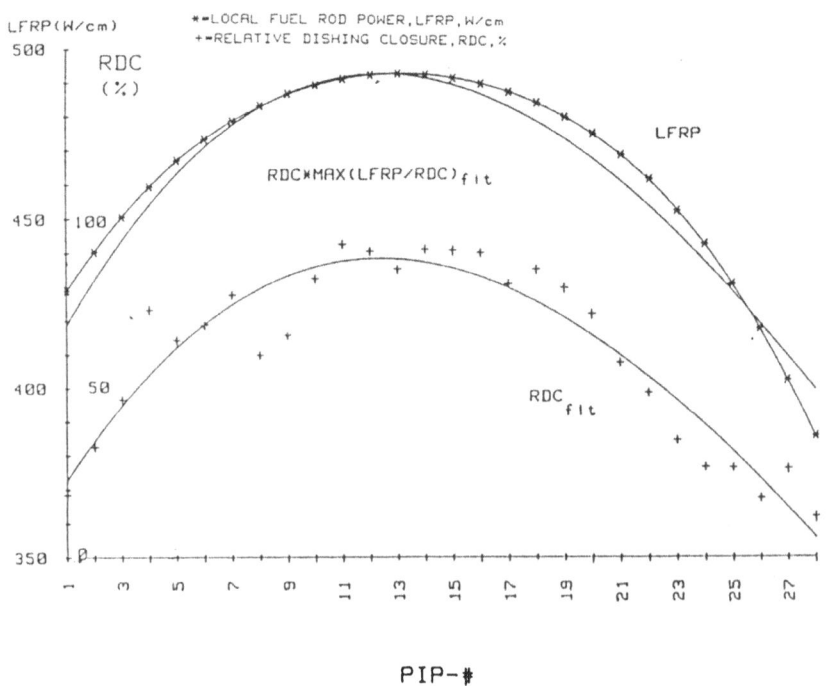

PIP-#

Fig. 6 Determination of dishing closure as a function of axial
power level and pellet interface position (PIP)

Dishing closure is a function of the local power that the various pellets have been operating at. Fig. 6 compares the results from dimensional measurements of the dishing closure derived from NR images with the axial power level and pellet interface positions of a LWR fuel rod. Comparison of the axial local fuel rod power distribution (LFRP) with the fit function of the relative dishing closure (RDC Norm) normalized to the same peak power confirms the previously mentioned dishing observation.

Water ingress in defect fuel rods is easily detectable due to the neutron attenuation of hydrogen. Fig. 7 gives a typical example of a partially water filled plenum, traces of water in the pellet cracks and other voids inside the fuel rod.

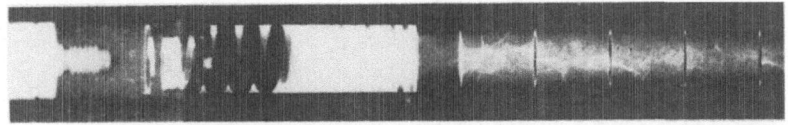

Fig. 7   Partially water filled plenum of a defective BWR fuel rod

3.2. Check of fuel stack condition

Visual assessment of NR images from LWR fuel rods yields that pellet crack propagation in axial direction, as shown in Fig. 8, is typical for local power levels below approx. 350 W/cm (for PWR only) when radial interaction forces are not yet dominant.

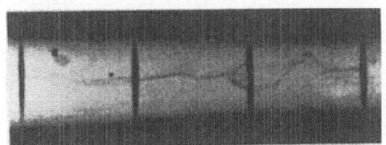

Fig. 8   Consecutive axial crack propagation in pellets

For quality assurance, LWR fuel rods are checked at HFR prior to any transient test for fuel stack separation occurence. Fig. 9 gives an example of a fuel stack separation due to rough hot cell handling during the preparatory work on irradiated fuel rods.

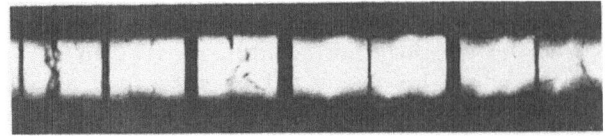

Fig. 9   Fuel stack separation

Central void formation (Fig. 10) is mainly observed in low density
pellet fuel and vibrocompacted fuel. The reason for central void
formation is the higher center temperature due to lower thermal
conductivity of low density fuel.

Fig. 10   Central void formation (vibrocompacted fuel)

3.3   Check of fuel rod components

The main objectives of NR images checks with regard to fuel
rod components are typically status control of these components
and specifications conformance.

After the pre-irradiation of LWR fuel rod segments in
commercial power reactors, the segments are prepared in hot cells
for further testing at HFR. Fuel rods are assembled with stainless
steel support and centering pieces. NR, as shown in Fig. 11, is
used to check the proper assembly.

Fig. 11   Check of assembly of BWR fuel rod and lower centering
          pieces

The fuel rod plenum contains either a spring or support tube in order to keep the fuel stack axially in place. Figs. 12 and 13 give typical examples of the plenum with a spring and a support tube.

Fig. 12 Plenum with support tube (at bottom of fuel rod)

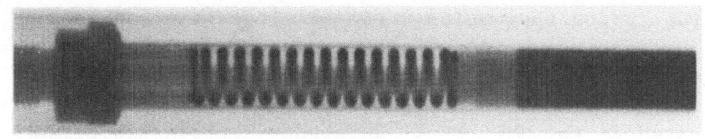

Fig. 13 Plenum with spring (at top of fuel rod)

The condition of the insulation pellet can also be checked by NR. Fig. 14 shows a broken Al_2O_3 insulation pellet.

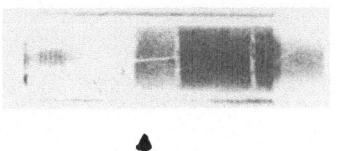

Fig. 14 Broken Al_2O_3 insulation pellet

3.4 Reference Neutron Radiographs

Within the framework of the Neutron Radiography Working Group (NRWG) /6/, typical examples of fuel rod appearance, fuel stack condition and components of fuel rods have been collected and are presented in the NRWG-book titled "Reference Neutron Radiographs" /7/.

Approximately one third of these reference neutron radiographs originated from exposures performed at HFR.

4.0 CONCLUSIONS

- Neutron radiography is an important and unique tool for LWR fuel rod investigation.
- No other non-destructive technique yields a comparable amount of integral information about the fuel rod state.
- Neutron radiography using nitrocellulose film provides a basis for qualitative and quantitative evaluation of LWR fuel rod images.
- Typical examples of details of LWR fuel rod appearance are presented in the NRWG-book "Reference Neutron Radiographs".

REFERENCES

/1/ H. Röttger, A. Tas, P. von der Hardt, W.P. Voorbraak
High flux materials testing reactor HFR Petten.
Characteristics of facilities and standard irradiation
devices, EUR 5700 EN, 1986

/2/ J. Markgraf
HFR irradiation testing of light water reactor (LWR) fuel
EUR 9654 EN, 1985

/3/ N.G. Chrysochoides, M.R. Cundy, P. von der Hardt, K. Husmann,
R.J. Swanenburg de Veye, A. Tas
High Flux testing reactor Petten. Replacement of the reactor
vessel and connected components. Overall Report.
EUR 10194 EN, 1985

/4/ J. Bordo, H.P. Leeflang, J.J. Veenema
Neutron Radiography Installation in the HFR Pool.
Proceedings of the First World Conference on NR, San Diego,
Dec. 1981, EUR 8296 EN

/5/ K.H. van Otterdijk, B. Shapiro
Computer Controlled Travelling Microdensitometer.
This conference

/6/ J. Markgraf
Neutron Radiography Working Group (NRWG). Summary of
activities and publications.
This conference

/7/ R. Barbalat, J.C. Domanus, J. Markgraf, F. Michel,
D.J. Taylor
Reference Neutron Radiographs of nuclear reactor fuel.
D. Reidel Publishing Company
ISBN 90-277-1717-6, EUR 8916 EN EP, 1984

NEUTRON RADIOGRAPHY OF IRRADIATION DEVICES

AT THE HIGH FLUX REACTOR PETTEN

J. Bakker, A. Baritello, M. Beers, J. Bordo, R. Conrad,
R. Lölgen, J.F.W. Markgraf, P. Zeisser

Joint Research Centre of the Commission of the European
Communities, Petten Establishment

Postbus 2

NL 1755 ZG Petten, The Netherlands

ABSTRACT

Within the irradiation testing programmes at the High Flux
Reactor (HFR) Petten, neutron radiography is applied for
quality assurance and inspection of irradiation devices.

The equipment, methods and typical results related to neutron
radiography of irradiation devices at HFR are presented.

The unique non-destructive test capability and importance
of neutron radiography is shown for some representative
applications.

1.0 INTRODUCTION

Neutron Radiography (NR) is performed at HFR Petten on a
routine basis for quality control, functional and dimensional
evaluation and inspection of irradiation devices. An underwater NR
camera /1/ in the HFR pool is used for this task.

Special object holders and flanges are available for
reproducible positioning and guidance of the components or
irradiation devices to be investigated by NR. An example of a
special object holder for NR-inspection of fast breeder fuel rods,
contained in sodium filled irradiation devices, is presented in
Fig. 1.

395

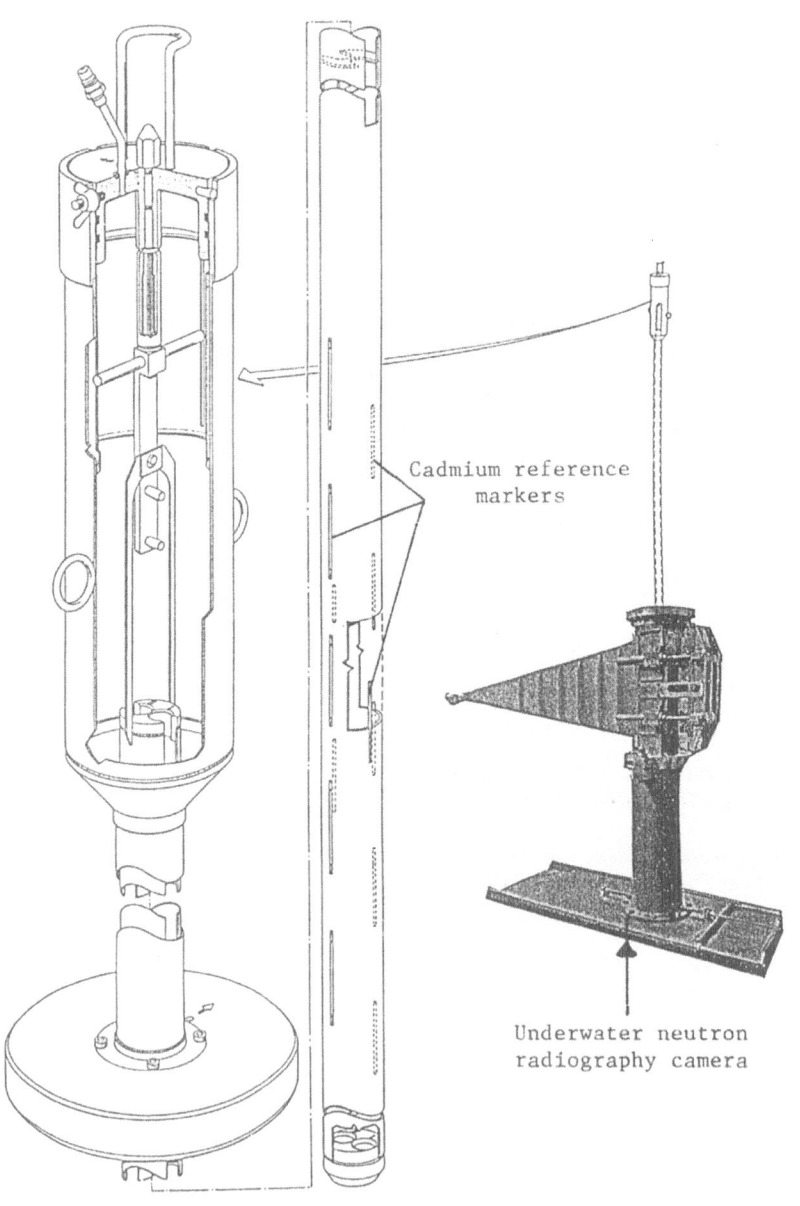

Cadmium reference markers

Underwater neutron radiography camera

Fig. 1   Special sample holder for simultaneous inspection of two
          LMBFR fuel rods by neutron radiography

From four typical irradiation research areas at the HFR, some
representative applications of NR are shown hereafter.

2.0 TYPICAL APPLICATIONS OF NEUTRON RADIOGRAPHY

2.1 Light Water Reactor Fuel Rods with Instrumentation and
     attendant Irradiation Devices

     NR inspection of LWR fuel rods at various irradiation stages
is one of the major tasks of the NR services at the HFR /2/.

     Instrumented LWR fuel rods and their irradiation devices are
examined by NR in order to check specification conformity,
function, position of instruments, device and fuel rods status
after e.g. severe fuel rod defects.

     Fig. 2 shows a NR image of an instrumented LWR fuel rod with
an integral void volume measurement system and pressure
transducer. The NR-image is used primarily for functional check of
the void volume measurement system.

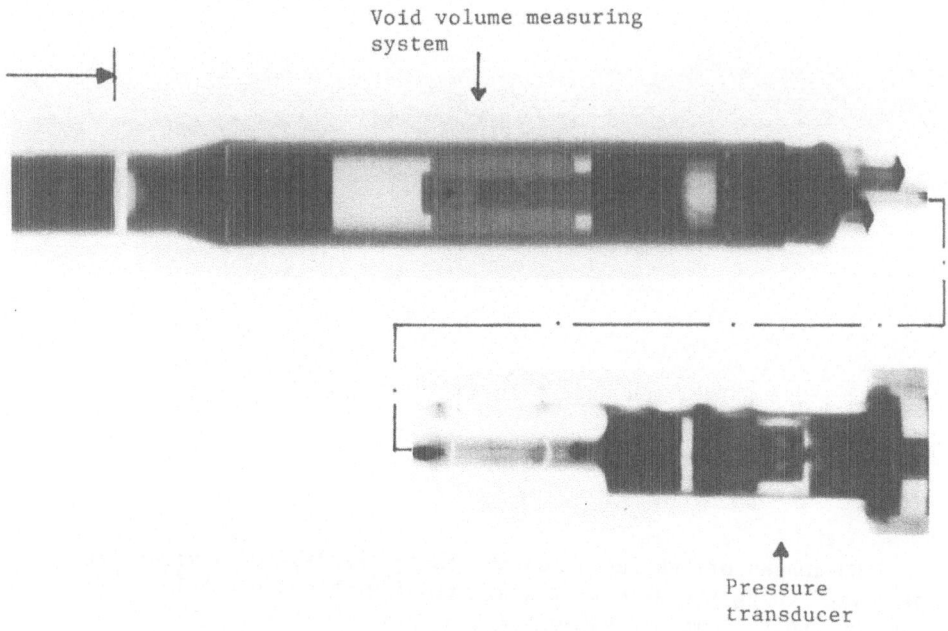

Fig. 2 Instrumented LWR fuel rod

Fig. 3 shows NR and X-ray images of a severe fuel rod defect
which occurred in a LWR irradiation device. The NR-images give an
overview of the location and form of the fuel rod debris and
status of the intact section of the fuel rod, whereas the X-ray
image taken in a hot cell provides the main information of the
condition of the irradiation device.

Neutron radiography

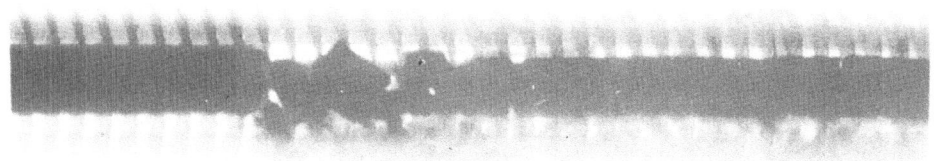

X-ray

Fig. 3   Severe fuel rod defect in an irradiation capsule

NR-images of LWR fuel rods and their irradiation devices are
performed using the direct imaging method with nitrocellulose film
and the indirect method employing a Dysprosium foil.

2.2 Irradiation Devices for Creep Strain Measurements on Stainless Steel and Graphite

Stainless steel is one of the candidate materials for future Fusion Reactors /3/. Characterization of the creep behaviour of this material under various combinations of uni-axial tensile load and temperature as a function of fluence is a major objective of present irradiation tests at the HFR. Creep strain measurements are performed intermittently during reactor shut-down periods. As an alternative to the conventional technique with hot cell sample unloading, strain measurement and sample reloading, intermittent dimensional measurements with reference to the marker positions on NR-images of· sample stacks are proposed. Fig. 4 indicates schematically the components of a sample stack. Fig. 5 shows a NR-image of a sample stack with Dysprosium markers on the tensile sample measuring shoulders and at the surrounding reference half-shells. Dimensional evaluation of the NR-images is intended to be performed by the computerized microdensitometer /4/ now being developed at the HFR.

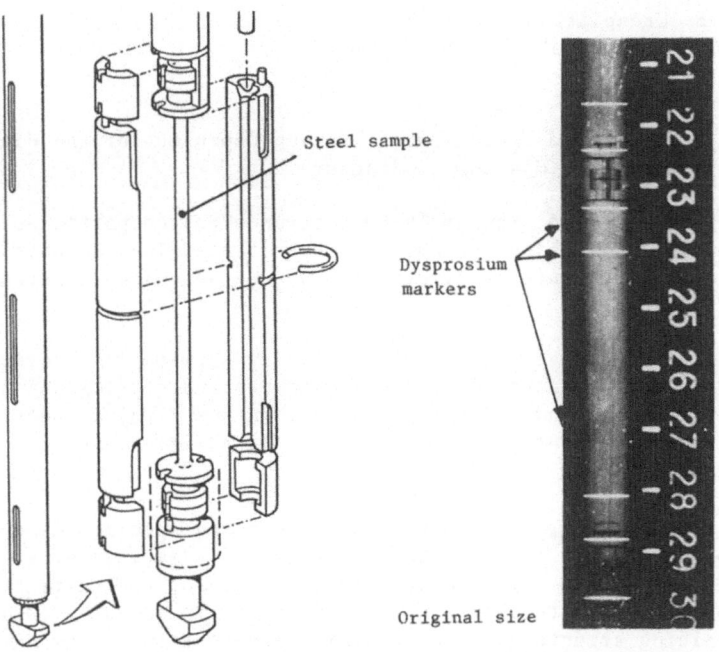

Steel sample

Dysprosium markers

Original size

Fig. 4 Steel creep irradiation device assembly of the sample stack

Fig. 5 Neutron radiography of the steel sample stack with Dysprosium markers

399

NR is often the only method applicable at the HFR pool for investigation of the status of a radioactive irradiation device. Fig. 6 shows an example of a broken tensile graphite sample in an irradiation device for creep measurement. Although graphite is not an easy material to check by NR, this example indicates that this technique gives a satisfying result when failures can be visualized through variations in material distribution.

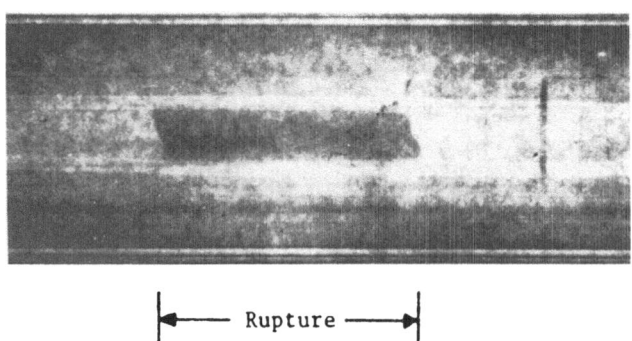

|←——— Rupture ———→|

Fig. 6 Creep irradiation device
        Rupture of a graphite sample

2.3 Instrumented Fuel Element Testing Capsules for the High Temperature Gas Cooled Reactor

Low enriched uranium TRISO reference coated particles embedded in graphite coupons have been tested in-pile under defined temperature gradients (150-300 K) and high temperatures (1400-1760 K).

The objective was to quantitatively examine interactions between kernel material, fission products and coating which can lead to amoeba and SiC corrosion /5/. NR and X-ray inspection is applied to check the integrity of the irradiation device and specimens.

In addition, the graphite structure and specimen dimensions before and after the irradiation are also investigated. Fig. 7 gives an example of a pre-irradiation X-ray of the testing capsule. Due to the low contrast of the NR-images of graphite . containing structures, photographic presentation of the NR-images is not possible. However, the expected information was obtained by evaluation of the original NR-images generated on nitrocellulose film.

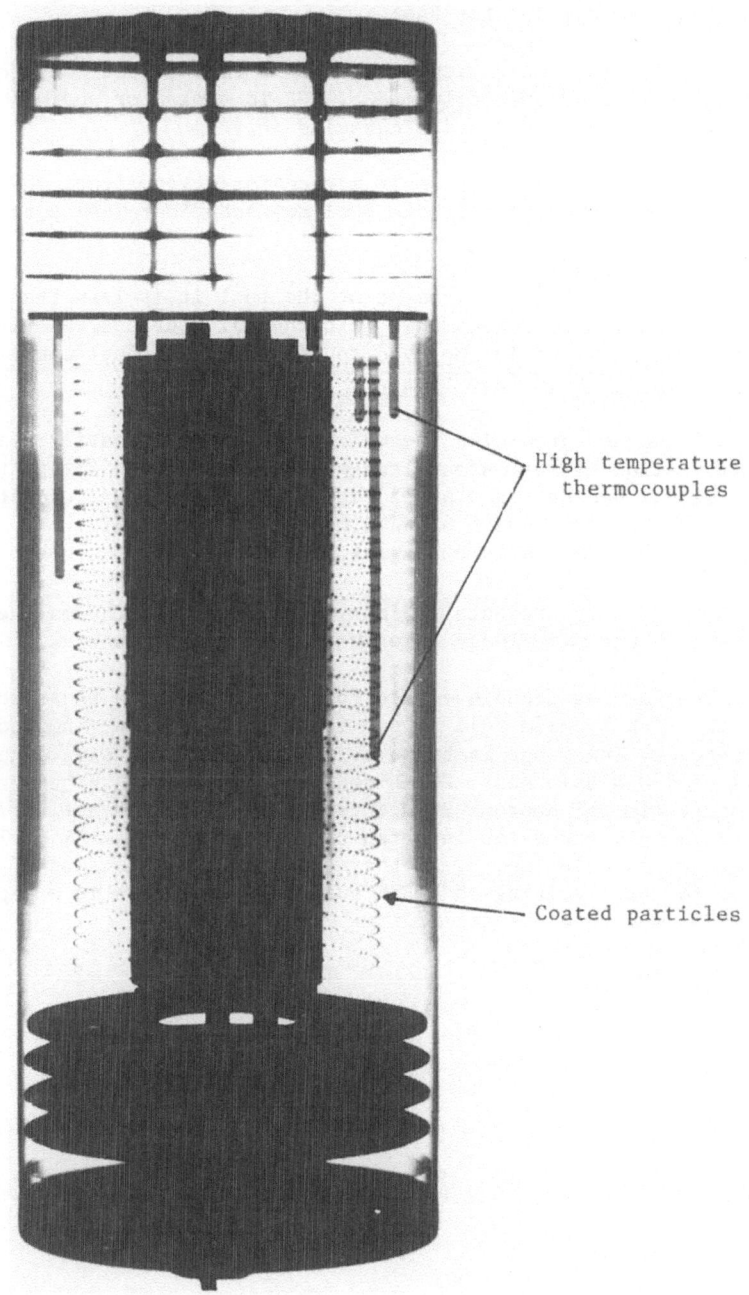

High temperature
thermocouples

Coated particles

Fig. 7   Pre-irradiation X-ray image of an irradiated device
containing HTR-LEU coated particles in graphite coupons

2.4 Fast Breeder Reactor Fuel

Irradiation tests with Liquid Metal Fast Breeder Reactor
(LMFBR) fuel rods are performed at HFR in sodium or sodium/
potassium-filled irradiation capsules /6/.

NR is practically the only method for intermittent and non-
destructive examination of the fuel rod behaviour inside these
irradiation capsules.

Due to the high enrichment of the fuel it is less transparent,
so, in addition to standard NR with thermal neutrons, NR with
epithermal neutrons is employed to visualize the fuel condition,
e.g. crack and void formation.

NR images taken with thermal neutrons are primarily used for
determination of fuel stack length. In order to facilitate the
dimensional evaluation, the object holder for these irradiation
devices is equipped with a Cadmium marker system (Fig. 1).
Fig. 8 shows a NR-image of a part of this system.

Fig. 8 also presents the NR-image taken with thermal neutrons
and Fig. 9 the same image taken with epithermal neutrons.

The imaging technique with epithermal neutrons is as follows :
A 2 mm thick Cadmium foil is put in the neutron beam in front of
the object. The image is taken by indirect method using a gold
foil of 100 µ thickness. For LMFBR fuel rods as shown before an
exposure time of approx. 90 minutes is applied. The activation of
the gold foil generates in a ten day period on Kodak SR5 film an
image as shown in Fig. 9. Another example, the detection of the
extent of fuel melting in LMFBR fuel rods, is given in a separate
paper at this conference /7/.

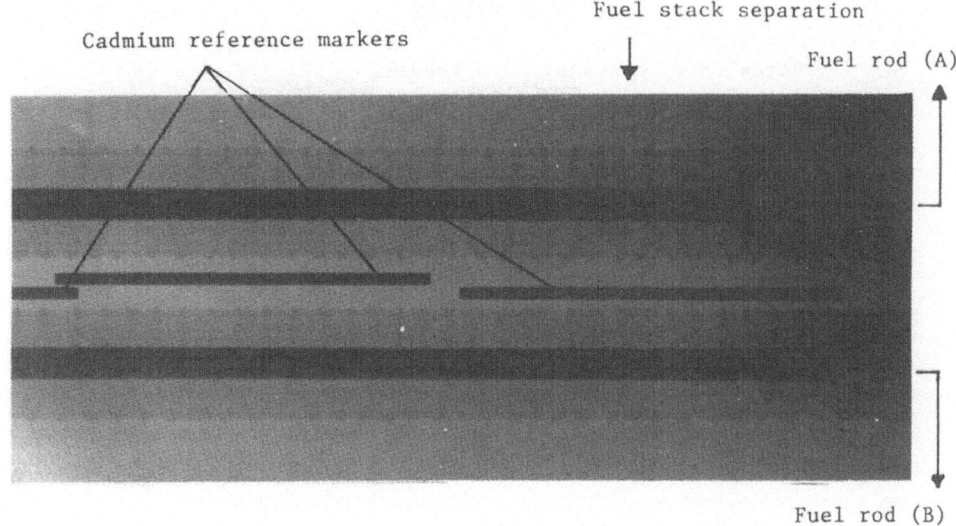

Fuel stack separation

Cadmium reference markers

Fuel rod (A)

Fuel rod (B)

Fig. 8 Neutron radiography with thermal neutrons of LMFBR fuel
rods

Top

Fuel rod (B)

Central void formation

Fuel stack separation   Fuel rod (A)

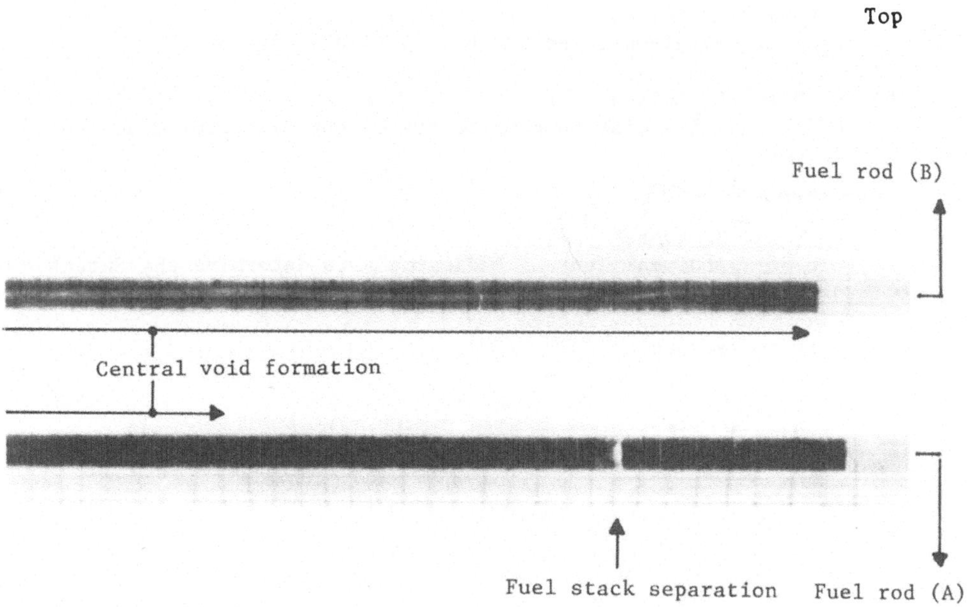

Fig. 9 Neutron radiography with epithermal neutrons of LMFBR
fuel rods

REFERENCES

/1/ J. Bordo, H.P. Leeflang, J.J Veenema
Neutron Radiography Installation in the HFR Pool
Proceedings of First World Conference on NR, San Diego,
Dec. 1981, EUR 8296 EN

/2/ J. Bakker, A. Baritello, J. Bordo, J. Markgraf, H.P. Leeflang,
I. Ruyter
Neutron Radiography of Light Water Reactor Fuel Rods at the
HFR Petten
This conference

/3/ R. Conrad, P. von der Hardt, R. Lölgen, H. Scheurer,
P. Zeisser
HFR Irradiation Testing of Fusion Materials
Topical report EUR 9515 EN

/4/ K.H. van Otterdijk, B. Shapiro
Computer Controlled Travelling Microdensitometer
This Conference

/5/ R. Conrad
Irradiation device for HTR fuel testing under abnormal
thermal conditions
Atomkernenergie-Kerntechnik Bd. 40 (1982) Lfg. 3

/6/ R. Moss, P. Zeisser
LMFBR Fuel Irradiation experiments in the High Flux Reactor
at Petten
ENC4-Conference, Geneva, June 1986

/7/ R. Moss, M. Beers
Use of Epithermal Neutron Radiography to determine the Extent
of Melting in Mixed Oxide LMFBR Fuel Pins Irradiated in the
High Flux Reactor, Petten
This conference

PART VI

CORROSION APPLICATIONS

EXPERIENCE WITH AN ON-OFF MOBILE NEUTRON RADIOGRAPHY SYSTEM

J. J. ANTAL
Army Materials Technology Laboratory
Watertown MA 02172-0001 U.S.A.
W. E. DANCE
LTV Aerospace and Defense Company
P.O. Box 650003, Dallas TX 75265 U.S.A.
J. D. MORAVEC
U. S. Army Proving Ground
Yuma AZ 85365 U. S. A.
S. F. CAROLLO
LTV Aerospace and Defense Company
P.O. Box 650003, Dallas TX 75265, U.S.A.

ABSTRACT

In September of 1979 the U. S. Army and the Vought
Corporation demonstrated an unique mobile neutron radiography
system designed as a prototype for the examination of light-
weight aircraft structures. Its source of neutrons is a
small (d,t) accelerator available commercially. The system
has produced hundreds of useful radiographs under a variety
of field maintenance conditions and was transported over
12,000 km with ease. The system's physical size, weight,
shielding, and operational requirements appear to be optimal
for acceptance by NDT personnel for routine survey examinations. We summarize here our experience with the system's
performance, reliability and accommodation to a variety of
needs. The success of the system has produced a revival of
interest in field applications of neutron radiography.

INTRODUCTION

The ability of neutrons to image hydrogen-containing materials
such as plastics, rubbers, explosives, water and corrosion products
has made neutron radiography an interesting nondestructive evaluation tool for use in defense-related activities. Older aircraft
must be examined for hidden corrosion damage and fuel and hydraulic
fluid leakage, while newer aircraft are making increasing use of

polymeric adhesive bonding and composite structural materials. The United States Army, much aware of these requirements, joined with the Vought Corporation* of Dallas, Texas, U.S.A., in exploring the possibility of developing a small, transportable neutron radiography system suited to field maintenance procedures. Characteristics sought for the system were on-off radiation control, simple operator handling, good reliability and a configuration suitable for helicopter aircraft examination. It was hoped that off-the-shelf, commercially available components could be used to manufacture even this first prototype of the system. The system was to use standard film imaging techniques; the use of electronic imaging was not a part of this development program.

THE NEUTRON SOURCE

The primary neutron source chosen was the A-711 accelerator manufactured by Kaman Sciences Corporation of Colorado Springs, Colorado, U.S.A. We had used this accelerator system in our analytical chemistry laboratory for many years and were impressed with its relatively high operational reliability and small size. The high energy of the neutrons produced (14 MeV) is disadvantageous, but the ready availability of the accelerator, its small size, and its known reliability were overwhelming factors in choosing it. The accelerator system is delivered by the manufacturer in three units: the accelerator head (250mm Dia. x 550mm L), the power supply (1000mm Dia. x 1800mm L) and the cooling unit (450mm H x 550mm W x 425mm D).

The accelerator head was mounted into an iron sphere which was filled with oil which acts as both moderator and shielding. The head arrangement is sketched in Figure 1. The manufacturer normally locates the tritium target just outside the hemispherical accelerator head enclosure. For this application, it was mounted a few centimeters away from the head surface to allow moderating material to be placed over a much larger solid angle around the target in this critical moderation zone. Convection currents reduce the degree of radiation damage to the moderator oil, although it is easily replaced if that is necessary.

THE CARRIAGE

The sphere is mounted on a specially-designed carriage which provides it with rotational, lifting, and lowering motions for positioning it to an aircraft structure. The carriage is completely maneuverable in the horizontal plane via motor-driven wheels. Power for all motions is provided by batteries mounted on the carriage and a single pendant provides operator control. The large high voltage

---

*Presently the Vought Missiles and Advanced programs Division of LTV Aerospace and Defense Company, Dallas, Texas, U.S.A.

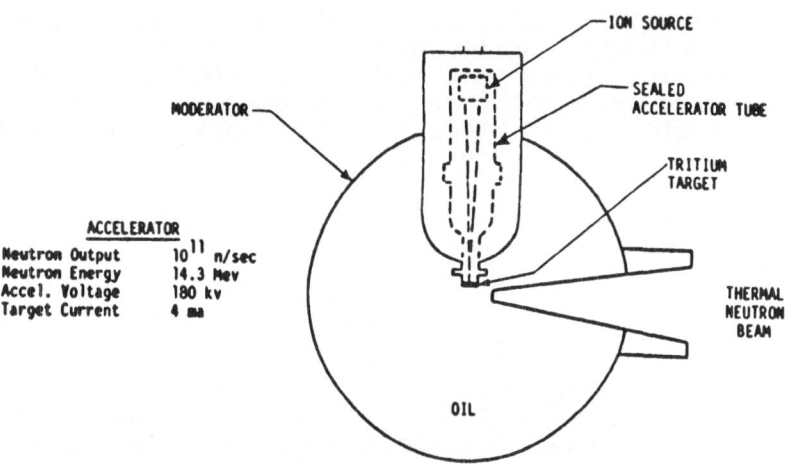

Fig. 1. Radiography system head, showing the position of the accelerator
head in the sphere. The sphere diameter is 910mm (36in).

Fig. 2. The transportable neutron radiography system demonstrating a
field application at a desert test area.

power supply is also mounted on the carriage. Power to the generator is supplied by a line cord. An auxilliary generator for this
power would make the system completely portable. The carriage is
4.6m Lx1.3m Wx1.3m H (15ftx4.4ftx4.3ft) and weighs 2205kg (4860lbs).

THE COMPLETE SYSTEM

Figure 2 is a photograph of the completed system. The cooling
unit rests on its own wheels nearby the carriage. It might have been

simpler to mount the cooling unit on the carriage, but a requirement that it pass through a reactor airlock did not allow it. The control console is placed at a location remote from the radiation area during operation and is connected to the carriage by a single cable. A more complete description of the system is given in References (1) and (2).The system began producing useful radiographs immediately after completion and has required no alterations since that time. In this sense it has behaved from the beginning more like a finished product than a prototype.

SYSTEM TESTING

After its construction the unit was tested in several work environments. Our aim in this testing phase was to evaluate its operation and acceptance as a nondestructive testing tool in realistic workplaces settings. Neutron radiography is a method unfamiliar to most nondestructive testing centers outside the nuclear industry.

The system was first tested in the indoor environment of the development laboratory and next at an Air Force aircraft maintenance center (1). At this center it was operated in conjunction with the x-ray radiography staff in their normal workplace after the last workshift of the day. Figure 3 shows the system in use at this site and a radiograph taken at that point. It was later transported to a desert test area where again the x-ray radiography personnel operated the system, radiographing a variety of items common to their daily activities as an x-ray facility (2). Finally it was transported to the Army Materials Technology Laboratory where it now serves as a test-bed for further developments of similar small systems, and provides radiography on a routine basis (3).

The system was shipped between field stations by ordinary truck transport with the carriage supported upon and restrained by a wooden skid. Figure 4 shows a common loading procedure. The system has sustained over 12,000km (7700 miles) of travel in this manner with only minor damage attributable to the rigors of shipping. At one station it arrived in non-working condition and the problem was traced to a broken lead on a capacitor in the accelerator control unit. The capacitor was given better mechanical support after its replacement.

During operation in the field, a down-time of about 1% of operating time was sustained by the system. Some of the downtime was caused by one-time failures in high voltage cables. Otherwise, maintenance of the system has been routine.

QUALITY OF THE RADIOGRAPHS

Some early experimentation at the Vought Corporation showed that a system utilizing this accelerator would be feasible even though the thermal neutron output would be low. The collimation of the

Fig. 3. The radiography system in operation at an aircraft maintenance site. (Inset) An engine nacelle is being examined for internal corrosion. (Background) The resulting neutron radiograph.

Fig. 4. The radiography system being loaded into a truck for transport to a field site.

411

system was designed therefore to allow a minimum L/D ratio of 13:1 in order to maximize the neutron flux at the film plane. This turned out to be an important element in the success of the machine. The thermal neutron flux measured at the film plane is the order of 3×10^4 neutrons/cm^2. With this flux density and low resolution film systems, exposure times are about 30 minutes. Resolution was sacrificed to gain intensity. However, for the types of artifacts expected to be imaged by the system, areas of corrosion and entrained moisture, and adhesive bond voids, all in rather large structural assemblies, the highest resolution is not necessary. Thus this is a satisfactory trade-off.

It was common for us to find that x-ray personnel judged the neutron radiographs often as poor in quality. Their reference for quality however was their experience with x-ray radiography wherein high resolution and high contrast are routinely attained with ease. Once personnel understood that the neutron radiographs, regardless of their quality, could solve many of their examination problems, they accepted the method as a valuable adjunct to their x-ray work.

Actually, the neutron radiographs are not so poor in quality but simply not of the high quality one is trained to strive for at all times. A radiograph of some electronic components in a cable amplifier of 6.4mm dia. (3/8 in dia.) taken by the system at a 21:1 L/D ratio, showed the undesirable migration of potting compound to a metal bellows section of the cable and allowed a solution to a problem of flexibility. In this case it was necessary to resolve a 0.5mm (0.02in)-wide artifact. An ASTM Type A sensitivity indicator on this radiograph showed 5 holes resolved.

IMAGING SYSTEMS

A wide variety of screen-film combinations were experimented with during the field testing as a means of surveying the available options for small systems. Routinely we now use a gadolinium oxysulfide screen (Kodak Lanex Fine) with Kodak SB film for rapid radiographs and an evaporated gadolinium screen with Kodak AA film for high resolution work.

The choice of screen-film combinations is an interesting exercise when one gives due consideration to the size and visibility of the artifacts to be imaged. Radiographing the same object with a variety of screen-film combinations to obtain the same average film density gave the results shown in Fig. 5. It is clear that if one is interested only in determining the presence of regions of internal corrosion (dark, ill-defined regions in Fig. 5), then one would be strongly tempted to consider only the screen-film combination which allows a minimum exposure time, since it would appear to satisfy the need most rapidly. This is another way to see how useful radiographs can be obtained from low-intensity neutron sources for specific radiography purposes.

While in the field, a small amount of time was spent evaluating the use of an electronic imaging system supplied by the Vought Corp-

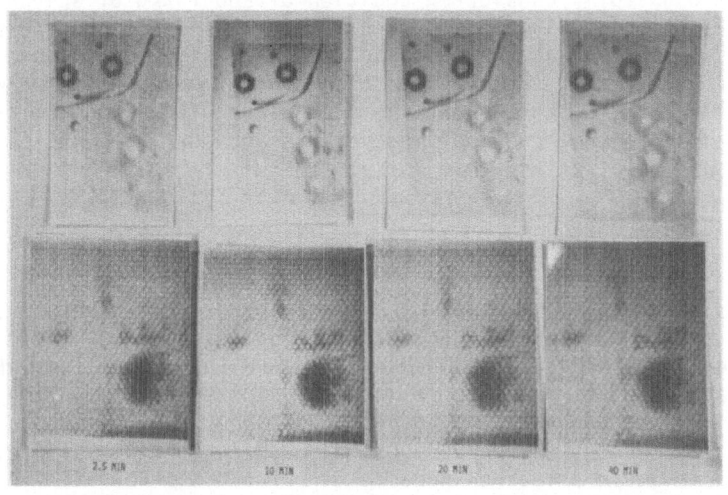

Fig. 5. Pairs of radiographs showing corrosion in two different areas of an aluminum aircraft skin, taken at 2-1/2, 10, 20, and 40 minutes of exposure time using various screen-film combinations to attain approximately the same average density level. Note that most all of the artifacts of interest are visible in all radiographs. (Left to Right [screen/film]: Vought DC/DuPont NDT 75, Vought DC/DuPont NDT 65, Kodak Lanex Fine/Kodak SB, and Vought DC/DuPont NDT 45)

oration. With this system, satisfactory radiographic images were obtained with less than five minutes of integration (exposure) time. Electronic imaging provides an immediate view of the radiograph and is ideal where archiving of every radiograph is not necessary, such as in maintenance work. We believe that in the future electronic imaging will be the recording system of choice for small, field-deployable systems.

FUTURE ENHANCEMENTS

The program which produced this transportable system was concerned with proving the concept of an on-off system which could be moved to the worksite. Details of optimization of system performance were not considered at tht time, but have been since. Shorter exposure times are always desirable, and we are aware of three manufacturers who are presently considering neutron generator designs with flux increases of a factor of two to ten envelope that does not become overly cumbersome. The avenue of consideration we have taken however, is to be content with the present generator output and attempt to optimize the performance of the system components. Such improvements would not increase the safety problems associated with the machine nor change its weight and size significantly.

A thoughtfully considered addition of about 2kg of shielding
internal to the radiography head has been applied and has resulted
in a 60% reduction in the radius of the exclusion area around the
operating machine. An improved collimator is being designed which
will reduce the background at the film plane without adding significant weight to the head and improvements in moderation efficiency
and image detection are also forseen.

SUMMARY

During the past six years a small transportable neutron radiography system with on-off radiation capability has been field tested
in actual work areas. The machine has more than fulfilled our
expectations with regard to reliability and versatility. The good
reliability is probably the result of using proven, commercially-
available components wherever possible and introducing simple,
rugged design in all other areas.
Those who work daily with x-ray radiography readily accepted the
machine since its size, weight and safety requirements were not too
different from high voltage x-ray units within their experience.
The presence of a familiar kilovoltmeter and milliameter on a fairly
simple control panel also played a role in this acceptance. The
quality of the radiographs produced was not accepted so readily.
Retraining to place emphasis on problem solution rather than attain-
ing perfection in radiographic quality was necessary to overcome
this lack of acceptance.
Our experience with this machine indicates that it is possible
to utilize today's technology to create neutron radiography systems
which can be transported to the work site. However, the design must
take full advantage of any relaxed system requirements allowed by
the type of examinations to be performed. We have been gratified to
see a renewed interest in taking neutron radiography into the workplace since this project began and hope that it indicates a time of
many new developments.

REFERENCES

1. W. E. Dance, S. F. Carollo and H. Bumgardner, "Mobile Acceler-
    ator Neutron Radiography System", AMMRC TR 84-39, Army Materials
    and Mechanics Research Center, Watertown, Mass., Oct 1984.

2. J. D. Moravec, Sr., "Final Report - Mobil N-Ray Technical Review
    and Evaluation", YPG Report #488, U. S. Army Proving Ground,
    Yuma Ariz., Dec. 1983.

3. J. J. Antal, "A Renaissance in Neutron Radiography via Acceler-
    ator Neutron Sources", in Materials Characterization for Systems
    Performance and Reliability, Plenum Press, New York, 1986, pp.
    385-401.

HIGH SENSITIVITY ELECTRONIC IMAGING SYSTEM FOR

REACTOR OR NON-REACTOR NEUTRON RADIOGRAPHY

W. E. Dance and S. F. Carollo

LTV Aerospace and Defense Company

Dallas, Texas, USA

ABSTRACT

A filmless neutron imaging system for either small mobile neutron radiography sources or high flux reactor sources has been developed and evaluated. In reactor neutron beams this system provides true real time images, while in non-reactor beams good quality electronic images are provided in fields as low as 10^3 n/cm^2-s . In the imaging head a high sensitivity neutron-to-light converter is viewed by an intensified camera. Digital image processing provides summing and enhancement operations on the image data. In the normal mode of operation, a 645 cm^2 field of view (25.4 cm x 25.4 cm) presents a 1:1 image of the object. The system can also be operated in an x-ray imaging mode without changing its configuration. Evaluation of the imaging system has been performed in both laboratory and field environments. Typical results using non-reactor neutron radiography systems are presented which illustrate practical applications·in structures inspection.

INTRODUCTION

Historically, attempts to achieve practical electronic imaging in neutron radiography with low intensity sources using low light level television cameras have been seriously limited and eventually abandoned. This is due mainly to the poor particle statistics and low intensity of the light emitted from converter screens below approximately 10^5 n/cm^2-s. These limitations have been overcome by an imaging system developed in the LTV laboratories for nondestructive inspection of aerospace structures. High quality images from the full range of neutron sources in current use for neutron radiography (10^3 to 10^8 n/cm^2–sec) are provided by the system. The chief design objective was to provide an electronic imager sensitive to low flux thermal neutron radiation from which the gamma rays and background neutron radiation have been minimized, as in the LTV-developed mobile "on-off" neutron radiography system (Ref 1). Utilizing readily available sub-components for ease in maintenance and parts replacement was an important consideration in the design. In addition to a brief description of the system, this paper presents typical radiographic results.

SYSTEM DESCRIPTION

In the LTV imaging system, a high output radioluminescent converter screen is combined with a low light level television camera and digital image processor to acquire neutron radiographic images. The major components are: (1) an imaging module, (2) the digital processing and storage system, and (3) the display and recording system. A schematic diagram of the basic neutron imaging system is given in Figure 1.

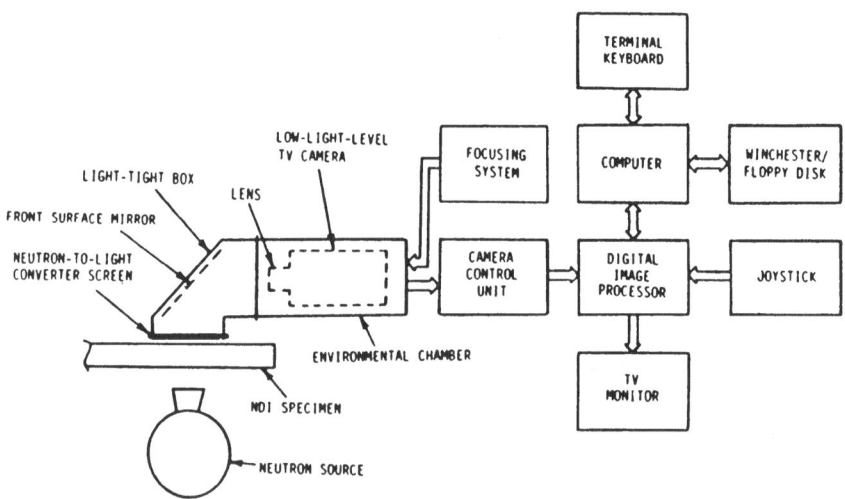

Figure 1. Diagram of Neutron Imaging System

Imaging Module

Comprising the imaging module are a demountable neutron-to-light converter screen, an intensified camera with optics system, a proprietary noise reduction system, and a remote camera focusing system. Figure 2 is a photograph of the module.

416

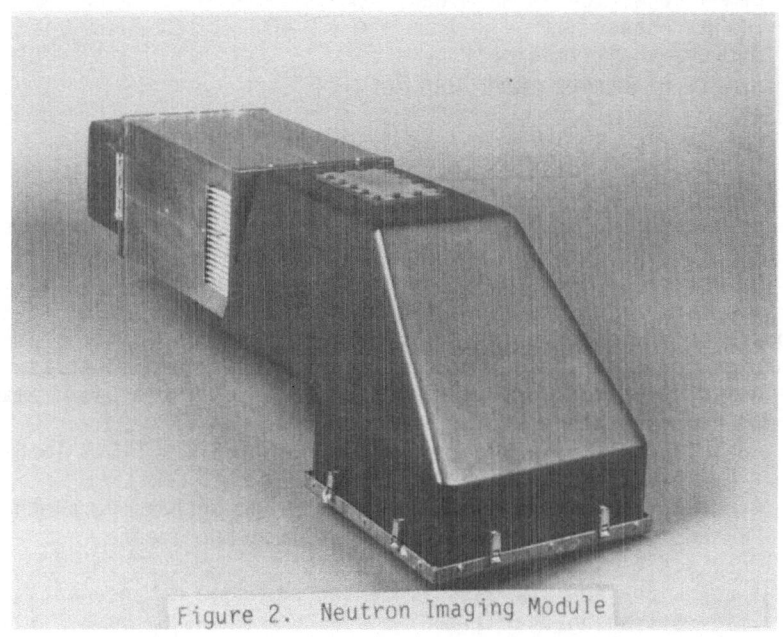

Figure 2.   Neutron Imaging Module

The converter screen provides a 10-in x 10-in square image or 100 in^2 (645 cm^2) active viewing area. The screen is responsive to x-rays, making it suitable also for electronic x-ray imaging. For x-rays the sensitivity is approximately equal to that of commercially available gadolinium oxysulfide (GOS) screens. For thermal neutrons, however, the screen is five to ten times as sensitive as commercial GOS screens. Thus when used in the neutron imaging mode, the screen is quite insensitive to the gamma background present, as neutron radiography beams typically have n-to-γ ratios in the range 10^4 to 10^7 n/cm^2-mr.

In the development of the imaging module, proprietary techniques (patents pending) were applied which substantially increase the video signal to noise ratio when used in low intensity neutron beams. For example, in a beam of 4 x 10^4 n/cm^2-s, the S/N was measured to be within 20 percent of the calculated value 19. By way of comparison with other systems, the calculated S/N for the video signal from one direct imaging tube (converter on inside of tube face) when imaging in a flux of 4 x 10^4 n/cm^2-sec is approximately 1 (unity).

Salient characteristics of the imaging module are listed below:

a.   Minimum flux required for good quality image: <10^3 n/cm^2-s
b.   Field of view: 645 cm^2 (100 in^2) square format, 1:1 size
c.   Dual neutron/x-ray imaging capability

417

d. Gamma rejection in NR mode: demonstrated @ 10^3 n/cm^2-mR
e. Field tested
f. Off-the-shelf imaging tube
g. Remote electronic and lens focus.

Digital Image Processing

The development of compact sophisticated digital image processing equipment in the 1970's also provided a tool which was previously missing for practical electronic neutron imaging from low intensity sources. In current use in the LTV imaging system is a Gould Model FD 5000 processor. Automatic image acquisition and enhancement is provided through frame/sum integration, shading correction, contrast stretching and other menu-driven operations. Some of the basic image enhancement capabilities are:
a. Capacity to handle two digital images of 512 x 512 x 16 bits
b. Viewing of the image as it accumulates
c. Accumulating one image while viewing and processing another
d. Pan and scroll of image in single pixel increments
e. Gray scale expansion and compression
f. Image high- and low-pass filtering and edge enhancement
g. Image averaging over time
h. Image subtraction for background elimination
i. Pseudocolor enhancement.

Special software interfaces the processor with the imaging module and allows "single command" automated image acquisition, processing and display.

Image Display and Recording

Display of the radiographic image is through a standard resolution 19-inch RGB color monitor; a hard copier provides 8-1/2" x 11" prints in a copy time of approximately 20 seconds. For storage of image data and software, the host computer is equipped with 8-inch floppy disk and 80 Mbyte hard disk. Standard videorecorders may also be used for image storage.

RESULTS

Typical radiographic results are best illustrated by photographing the screen of the television monitor used to display the electronic images. Many types of specimens were used to evaluate the performance of the system and repeatability of the results. One such specimen is an ASTM Type A sensitivity indicator, modified by this laboratory in 1978 for evaluating electronic images. The LTV modification of this device, shown in Figure 3, is currently being utilized in the USA as an interim IQI while an ASTM standard is being developed for electronic neutron imaging. Another test article used at LTV for

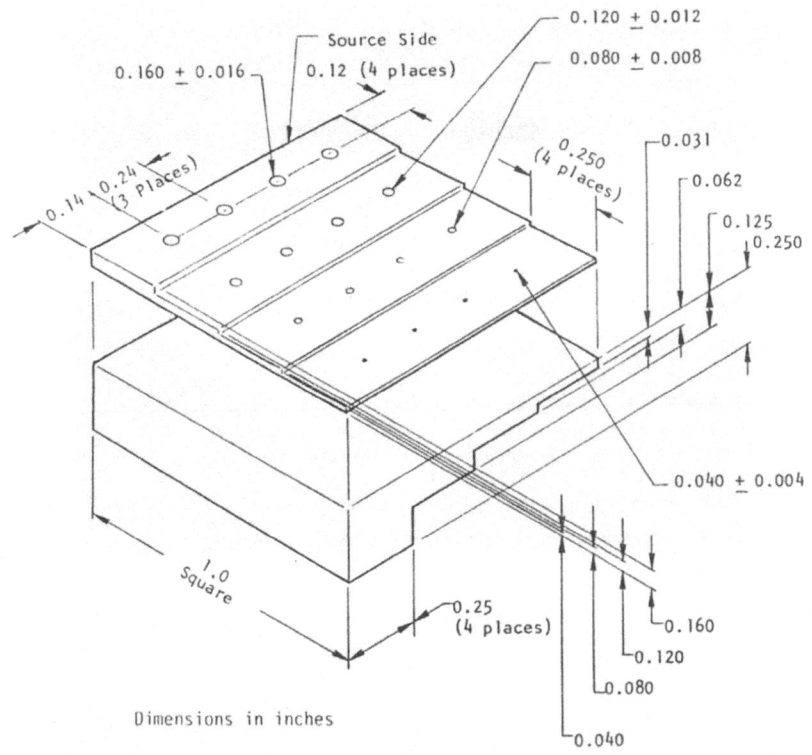

Figure 3.   LTV-Modified ASTM Type A Sensitivity Indicator

evaluating image quality and repeatability is a reference plate containing the following specimens:
  a. 4" x 4" x 3/4" aluminum honeycomb panel containing corrosion
  b. 3" x 8" x 1/2" aluminum honeycomb helicopter panel with multi-
      ple adhesively bonded steel inserts, with adhesive defects.
  c. 3" x 3-3/4" x 1/4" aluminum aircraft structural joint with
      intergranular corrosion present around fastener holes
  d. 1" x 3" x 1/4" aluminum slab with 5/8" diameter patches of
      transparent tape, .004", .006" and .008" thick.

    The photograph of Figure 4 shows these specimens as they were
typically mounted for radiography. Figure 5 is a Polaroid photograph
of the electronic neutron image of the specimens. This digitized
image is background and shading corrected. As seen, the system provides a high quality image of the corrosion in the honeycomb and the
structural joint, the adhesive deficiencies in the steel inserts, all
patches of transparent tape, and the holes in the modified ASTM sensitivity indicator. An x-ray image of the reference plate using the
same imaging system and processing is presented in Figure 6, which
illustrates the system capability in this mode and the complementarity
of the two modes of inspection.

419

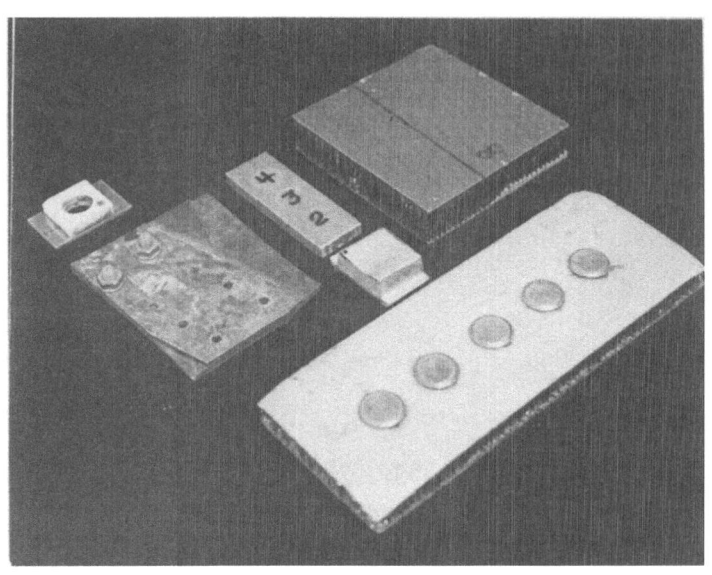

Figure 4.   Photograph of Neutron Radiography Test Specimens

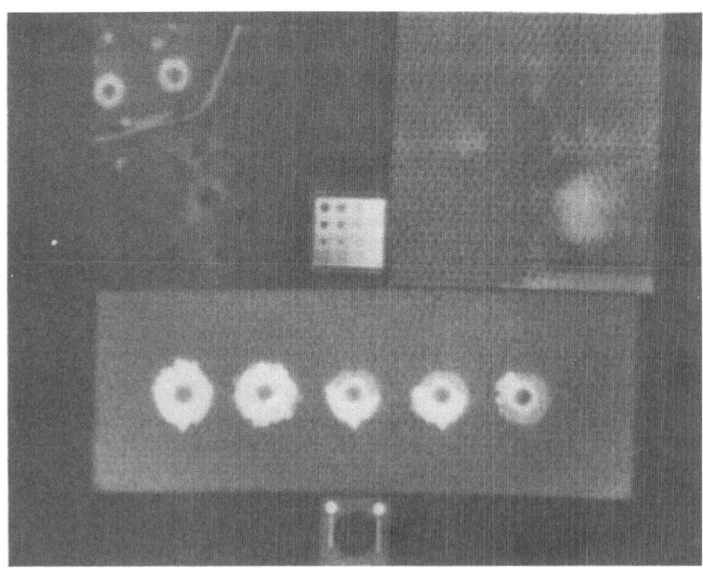

Figure 5.   Electronic Neutron Image of Test Specimens - Corrected for
Background and Shading

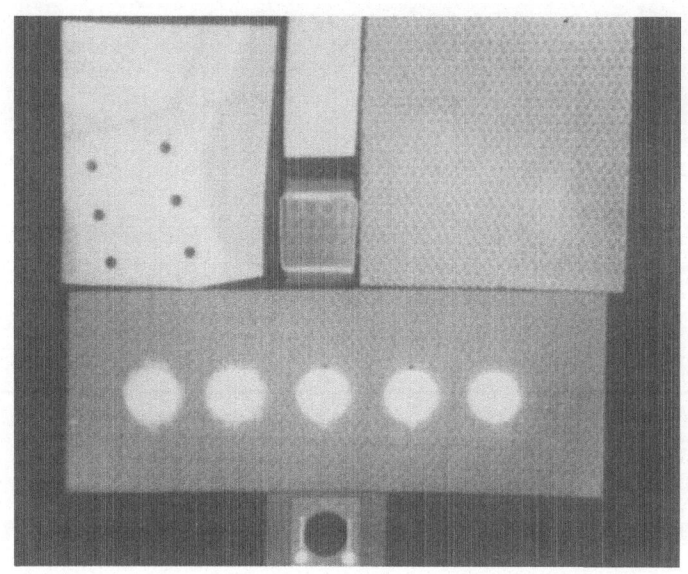

Figure 6.   Electronic X-Ray Image of Test Specimens - Corrected for
Background and Shading

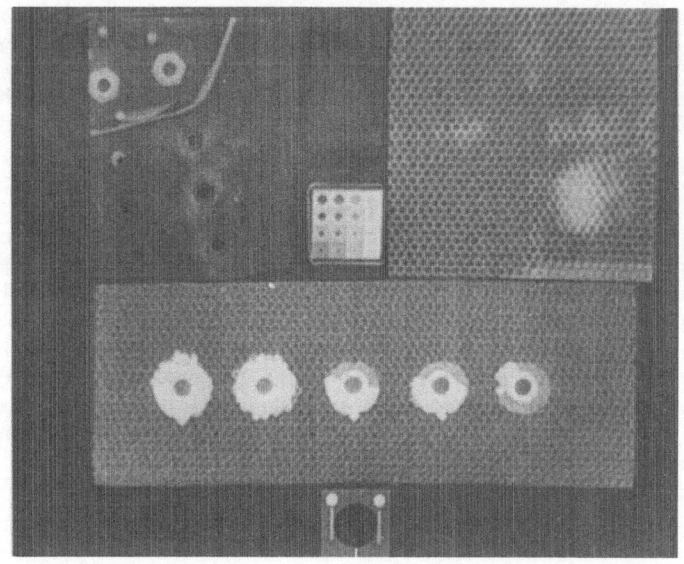

Figure 7.   Electronic Neutron Image of Test Specimens Corrected for
Background and Shading, and Enhanced by Contrast and
Sharpening Routines

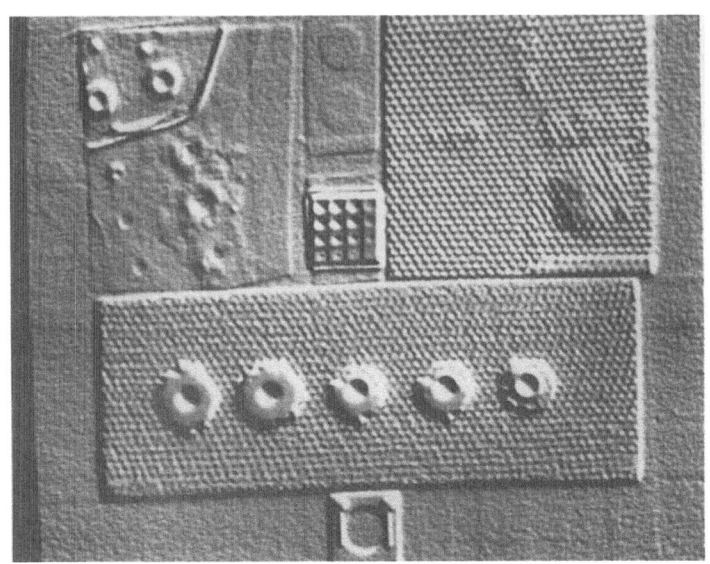

Figure 8. Electronic Neutron Image of Test Specimens - Corrected for
Background and Shading and Contrast Enhanced, with Additional
Two-Dimensional Gradient Enhancement

Additional routines have been developed with the digital image
processing unit for further enhancement of radiographic images. One
such routine is image sharpening. Figure 7 shows an example of this
routine applied to the neutron radiograph of Figure 5. With this
technique some of the finer image details are easier to read. For
example, the 15th consecutive hole in the modified ASTM sensitivity
indicator is easily read in this figure. Another routine is a
gradient enhancement. Application of this operation to the neutron
radiographic image of Figure 5 results in the image shown in Figure 8.
The image exhibits a three-dimensional effect on certain distinctive
features, presenting them in bold relief. In many cases this technique enables one to visualize certain details more easily, such as
the .004"-thick disc of tape, and the 16th hole in the modified ASTM
sensitivity indicator. Other techniques such as histogram equalization, Laplacian edge enhancement and pseudocolor have been applied
routinely in this laboratory for enhancing the visualization of
radiographic images.

REFERENCES

1.   W. E. Dance, S. F. Carollo and H. M. Bumgardner, Mobile
Accelerator Neutron Radiography System. AMMRC TR 84-39, Contract
Final Report Prepared for Army Materials and Mechanics Research
Center (October, 1984).

ECONOMICS OF NEUTRON RADIOGRAPHY
FOR COMMERCIAL AVIATION

Douglas A. Froom
McClellan AFB CA   95652
and
John P. Barton
N-Ray Engineering Company
5709 Waverly Ave
La Jolla CA   92037

ABSTRACT

The objective of this paper is to consider
whether neutron radiography used for corrosion and
moisture detection has a role to play in maintenance
of commercial aircraft. If so, what are the economic
benefits? And how might this be implemented
internationally?

Entrapped moisture and hidden aluminum corrosion
can cause serious problems in critical parts of
certain aircraft. Neutron radiography can provide
unique information that, when used in conjunction with
other methods, will provide the needed inspection.
Studies published by several independent groups have
confirmed the feasibility of using neutron radiography
for aircraft inspection. Plans have been announced
for three custom designed neutron radiography systems
for application to military aircraft. Benefits relate
to maintenance costs, extended operational fleet life,
and aircraft safety. Implementation for commercial
airfleets appears to be most practical in nations with
major government owned airlines and major government
sponsored neutron radiography facilities that can be
used on a shared basis.

Feasibility Studies

This discussion is focused on possible future neutron radiographic inspection of commercial aircraft for detection of moisture and corrosion. Neutron radiography is already being used for inspection of aircraft engine turbine blades, and for numerous special applications such as inspection of emergency escape actuators.

At least four separate US companies have independently documented studies showing that neutron radiography has unique capabilities for aircraft inspection. Between 1974 and 1978, IRT Corporation published reports on the use of the isotopic source CF-252 (1). During the same time period, Vought Corporation reached equally positive conclusions with emphasis primarily on use of an accelerator source (2). Science Applications International Corporation and GA Technologies Inc. are two other companies that have studied the subject using a range of sources with advanced electronic imaging capabilities (3,4).

The studies have related to military aircraft, but the similarities in military and civilian aircraft construction indicate that neutron radiography could also play an important role in commercial aircraft maintenance programs.

Two areas of interest are aluminum honeycomb and graphite composites. Commercial aircraft such as the Boeing 737, 747, 757, and 767 use such structures in flaps, spoilers, trailing edges, elevators, rudders, etc.

Detection of hidden moisture and early stage corrosion of aluminum are serious problems. Neutron radiography is the only known method of nondestructive inspection for those types of defect. It would be used most effectively in conjunction with other technologies such as microfocus x-ray and dual energy x-ray (5).

Precedent

Several custom designed neutron radiography systems have been installed in the USA for applications such as reactor fuel inspection and pyrotechnics inspection (6). These have used various types of neutron source: reactor, accelerator and isotopic. For military aircraft inspection, three custom designed neutron radiography systems, valued in excess of 15 million dollars, have been announced.

The Transportable Neutron Radiography System (TNRS)
which uses a low yield accelerator source, is designed
to be used without an enclosure. The Maneuverable
Neutron Radiography System (MNRS) housed in a
specially equipped aircraft hangar, will be a full
time application to meet aircraft maintenance needs.
Both the TNRS and the MNRS are designed to provide for
complete scanning of intact aircraft. Because these
systems use low intensity neutron sources, the
sensitivity will be relatively low. The Stationary
Neutron Radiography System (SNRS), will use a small
nuclear reactor, and is designed for high sensitivity
inspection of components removed from aircraft.

The contract for the TNRS was placed in 1982 and
equipment is now ready for acceptance tests. .The
contract for the MNRS was placed in 1985. The request
for proposals for the SNRS was issued in Jan 1986 and
a contract award is imminent.

Economic Benefits

With the precedent established for custom
designed neutron radiography systems, some dedicated
to military aircraft inspection, the discussion can be
focussed on economic benefits for civilian aircraft
application.

Outage Time: The processing of an aircraft
through its major maintenance cycle will typically
take 100 days every 1,000 days. For a commercial
fleet of 4,000 aircraft, about 400 will be out of
service at any one time. If use of a Transportable or
Maneuverable Neutron Radiography System could reduce
the outage time for maintenance by say just one
percent by reducing unnecessary dismantling of the
aircraft, this would equate to the value of four
aircraft retained in service. The 1984 average
replacement cost of commercial jet aircraft was $30M.
Reduction in maintenance outage time by one percent is
therefore worth $120M.

Efficiency of Repair: The introduction of a high
sensitivity, stationary neutron rediography system
will have two separate results for detached aircraft
panels. It will identify early stage corrosion that
could not previously be detected, and it will provide
higher precision for those types of repairs previously
conducted. The first capability will meet needs for
improved maintenance. The second capability will lead
to cost savings in the present inspection/repair
cycle.

Without neutron radiography, a typical
inspection/repair cycle for a detached panel is as
follows: In the first series of steps, the panel is
inspected using all available methods. If corrosion
is found, the panel is next heated in the autoclave.
This initial heating causes moisture and other gases
to generate high pressure which ruptures honeycomb
cells (blown core). The panel is then inspected with
x-ray to show the blown core, and, identify the area
to be repaired. In the repair cycle the panel is
heated to cure the new adhesive, and then inspected to
detect voids in the sealant. Any voids are
subsequently filled by injecting hot sealant and the
inspection cycle is repeated.

Neutron radiography, when used in conjunction
with x-ray and other methods will provide more precise
locations where repair is needed, thereby permitting
local repair and reducing the need for autoclave
recycle, blown core, and unnecessary repair. In some
circumstances cost savings in the repair cycle can
repay the cost of a neutron radiography system in a
period of one to two years.

Extended Fleet Life: Aircraft typically remain
in service for twenty years or more. Economic fleet
life depends on the most rigorous inspection and
repair procedures (7). There are many factors which
ultimately cause an aircraft to be withdrawn from
service. Corrosion is one important factor, and
specific examples can be cited where the inability to
adequately repair corrosion has led directly to
withdrawal of aircraft from service.

A hypothetical analysis indicates very large
potential savings. There are 4,000 major aircraft in
the US commercial fleet, of which 200 were purchased
in 1984 at an average price of $30M each. The fleet
replacement value is therefore approximately 4,000 x
$30M = $120 billion. Assume that one tenth of this
fleet (400 aircraft) are maintained at one center.
If, for this small fraction of the total fleet,
neutron radiography inspection help extend the average
useful fleet life from say 20 years to just 21 years
(5% increase), the savings in replacement costs would
be approximately five percent of $12 billion or $600M.
Because the average value of an aircraft is
approximately half the new replacement cost the
savings should be reduced by a factor of two ($300M).
This one time savings of $300M equates to an annual
savings of $15M over the 20 year life of the facility.

426

Thus, even a small contribution to fleet life extension could have substantial economic benefits.

Crash Risk Reduction: Corrosion, inadequately inspected and repaired, can lead to inflight structural failure or fire. Improvements in inspection capability could reduce crash risk at any given aircraft age.

A single crash, attributable to inadequate inspection, would cost far more than the replacement value of the aircraft . Other factors to be considered include legal liabilities, and the large awards that have been made in courts following accidents.

Implementation Probability

A few years ago, the US government took steps to encourage free competition between commercial airlines (deregulation). Results have included the formation of many new airline companies, cheaper airfares, the bankruptcy of older airlines, and an apparent reduction in aircraft inspection and maintenance. Table 1 shows that the number of reported structural problems (Service Difficulty Reports) for most aircraft dropped between 1980 and 1984 in spite of an ageing fleet: an indication of significantly reduced aircraft inspection.

Implementation of neutron radiography for commercial aircraft is not likely at the present time in the USA, because of the unusual nature of the technology, and a capital cost of equipment that is high relative to other inspection methods. Other nations may have different circumstances. Some of the following steps appear necessary for implementation of neutron radiography:

1. Successful demonstration of the TNRS, MNRS, or SNRS.

2. A detailed cost-benefits evaluation for a specific commercial aircraft maintenance program.

3. A lead by a nation in which the government provides highly sophisticated neutron radiographic facilities for general industry application, or in which there is a government owned airline .

4. Initiation of aircraft neutron radiography on

a small scale, possibly with a neutron
radiography facility shared with other
applications - such as nuclear fuel and
pyrotechnics inspection.

5. A situation in which the responsibility for
aircraft inspection is separated from ownership
or promotion responsibilities.

6. One scenario considers that civilian airliner
crash were to occur, and the inquiry proves this
to be due to corrosion that would have been
prevented using known neutron radiography
technology. Very expensive legal claims follow.

Conclusion

Neutron radiography could be considered for
civilian aircraft maintenance to help detect entrapped
moisture and early stage aluminum corrosion. Factors
favoring such studies are:

1. positive feasibility studies have been
documented by a variety of users, suppliers, and
independent researchers.

2. in the USA plans exist for construction of
neutron radiography systems for maintenance of
military aircraft.

3. it is possible to argue that the economic
benefits could be large.

Factors against early implementation by commercial
airlines are:

1. the cost of custom built neutron radiography
systems which is likely to be in the range of one
to ten million dollars.

2. the specialized nature of the technology.

3. the long term or intangeable nature of the
economic benefits.

It is recommended that cost-benefits studies be
undertaken on selected civilian aircraft maintenance
programs and that these studies be run in parallel
with the utilization of the production systems
described elsewhere in these preceedings.

428

References

1. IRT Corporation, San Diego, CA (unpublished reports)

2. Vought Corporation, Dallas, TX (unpublished reports)

3. Science Applications, Incorporated, San Diego, CA (unpublished reports)

4. G.A. Technologies, Incorporated, San Diego, CA (unpublished reports)

5. A New Radiographic Corrosion Inspection Capability. AFWAL TR-85-4130. Wright Patterson Air Force Base, OH

6. Proceedings - First World Conference on Neutron Radiography, J.P. Barton and P. Van Der Hardt, Editors, D. Reidel, 1982

7. McDonnell Douglas Corporation, Aircraft Corrosion and Detection Methods. Douglas Paper 7483, 1984.

8. McDonnell Douglas Corporation, Supplemental Inspection of Aging Aircraft. Douglas Paper 7610, 1985

FIGURE 1

REPORTED REDUCTIONS IN AIRCRAFT INSPECTION FINDINGS

Aircraft	Average Fleet Age 1980	Average Fleet Age 1984	Structural Inspection Findings Reported Per Aircraft 1980	Structural Inspection Findings Reported Per Aircraft 1984
727	9.0	12.2	0.76	0.5
DC 9	10.7	11.7	1.8	0.6
DC 10	6.3	918	0.76	0.3

FIGURE 2

SCENARIO - NR APPLICATION FOR COMMERCIAL AIRCRAFT

1. U.S. systems demonstrated for military aircraft.

2. French Neutron Radiography Center (government funded) inspects critical components for Air France.

3. Crash in U.S.A.. Corrosion is identified as cause. Families of deceased sue airline for failure to perform adequate inspection.

4. Commercial airlines worldwide increase use of neutron radiography using shared facilities.

PLANS FOR AIRCRAFT MAINTENANCE NEUTRON RADIOGRAPHY SYSTEMS

Douglas A. Froom
McClellan AFB CA   95652
and
John P. Barton
N-Ray Engineering Company
5709 Waverly Ave
La Jolla CA   92037

ABSTRACT

Plans have been underway since 1982 to acquire a Maneuverable Neutron Radiography System for scans of intact aircraft, and a Stationary Neutron Radiography System for high sensitivity inspection of removed components. This paper outlines the key steps taken in the acquisition process, including the conceptual design, preparation of specifications, and procurement plans.

Two separate sources of californium-252 are included in the conceptual design of the manueverable system. One for scans of wings and tail section, will be coupled primarily with step motion electronic imaging. The other, for scans of the empty engine bay regions, will be used with film imaging. Major sub-systems include the programmable overhead positioner, the programmable underside positioner, the source shield and transfer system, the radiation protection systems, and the image interpretation systems.

The Stationary Neutron Radiography System will feature a TRIGA reactor in a below grade configuration. Four beam tubes will be extracted, of which two will be optimized for electronic imaging of corrosion products in honeycomb panels. The shield and containment system will be designed as an integral part of the system, and will conform with a detailed environmental assessment.

INTRODUCTION

Improved methods of inspection for early stage corrosion will lead to major economic advantages in aircraft maintenance. Benefits will include reduced repair costs, reduced crash risk, and extended fleet life. At present a combination of non-destructive inspection methods are used for corrosion detection including acoustic emission and regular x-radiography. However, the existing methhods are inadequate. Neutron radiography can detect small amounts of hydrogen. Hydrogen exists both in moisture, and in aluminum corrosion products caused by moisture. Neutron radiography can therefore add important additional information, that, taken in conjunction with other existing methods, will significantly improve the detection capabilities.

A typical aircraft maintenance program calls for major inspection and repair every four years. At McClellan AFB new aircraft arrive for inspection at average intervals of three days. With all planned systems installed, the initial inspection sequence will include the following steps:

1. Maneuverable Neutron Radiography System - Inspect intact aircraft.

2. Maneuverable X-Radiography System - Inspect intact aircraft.

3. Stationary Neutron Radiography System - Inspect detached panels.

4. Microfocus Real Time X-Ray System - Inspect suspect regions.

5. Accoustic Emission Inspection to determine active corrosion.

6. Automatic Ultra-Sonic Water Squirter - Inspect for debonds.

After an initial inspection the cycle of repair steps becomes more involved, including use of the autoclave for adhesive curing. Extensive use of the autoclave, which causes blown core, and extensive de-skinning is prohibitively expensive. Effective initial inspection for hidden moisture or corrosion is critical for performance of adequate maintenance, and reduction of repair costs.

MANEUVERABLE NEUTRON RADIOGRAPHY SYSTEM

Basic Design Parameters

The first neutron radiography system for scanning intact aircraft will be located at North Island, NARF, San Diego. The following basic design parameters differentiate the McClellan Air Force Base maneuverable system from the North Island-NARF system which is designed to operate outdoors, be mobile, and use an accelerator source.

1. The McClellan MNRS must be suitable for production application in all weather, 52 weeks per year. It must therefore be inside a suitable building. The building will be part of an integral non-destructive inspection complex. Personnel radiation protection will be provided by both thick walls and shielding around the source, rather than just by distance and exclusion areas, as used in the NARF system.

2. The MNRS need not be mobile (i.e., transportable to another site of operation). Rather it will be dedicated to a particular building and workload.

3. The MNRS will use two separate neutron sources and robotic systems, one designed for general aircraft surfaces using programmable overhead positioner (POP), the other designed for aircraft engine bay inspection using a programmable underside positioner (PUP).

4. The MNRS will use 50 mg of californium-252 for each of the neutron sources. This provides a thermal neutron flux significantly more intense than the accelerator source used in earlier studies.

5. The MNRS will be designed to form one part of broader inspection capabilities including a suitable image storage and retrieval system.

Acquisition Approach

The acquisition consists of two parts: custom designed building and the custom designed equipment (the MNRS). The two parts are processed concurrrently, but are treated quite differently.

The user, the McClellan AFB Directorate of Maintenance, provided input to the design of the building, but the detailed

433

drawings were prepared by an outside architect/engineer. The building was bid competitively for construction in agreement with the detailed drawings. The equipment, like other custom designed systems, has been procured under a competitive bid process for a turnkey system (i.e. a system which is complete and ready to use). In this process the user provides detailed performance specifications and invites technical proposals from the bidders. The performance specifications are based on a conceptual design which is outlined below.

Conceptual Design

The neutron source is surrounded by a sphere, about one meter (approximately 39 inches) in diameter, which serves to moderate the neutrons, collimate a divergent beam on one side, and provide the necessary shielding for gamma rays and fast neutrons. A large yoke is used to hold this Moderator, Collimator, and Shield (MCS) unit on one side of a wing, or other aircraft section, while holding real-time imaging equipment on the opposite side. The 50 mg source of Cf-252 will produce a thermal neutron flux of about 5×10^8 n/cm^2-s at the collimator input, or about 5×10^4 n/cm^2-s when collimated at L:D=30:1. Exposure times in the range 10 seconds to 100 seconds will be needed.

Programmable Overhead Positioner (POP). The positioner is essentially a highly accurate bridge crane providing movement of the N-ray Overhead System (NOS) as follows: X-axis (rails), Y-axis (trolley), Z-axis (telescope tubes and cable hoist). In addition, motor drives between the telescope tube and the yoke provide pitch, yaw, and roll. The system can be programmed to scan the NOS in continuous motion or in-step motion along the aircraft parts. Programming is performed by removing the source to a storage position and then operating the POP in the teach mode using a teach control pendant inside the aircraft bay. The bridge span is approximately 93 feet and the work envelope of the system is approximately 77 feet (x), 72 feet (y), and 16 feet (z). The system was designed for an estimated load of 9000 pounds so that an accelerator can be substituted for the Cf-252 source if required later.

N-ray Underside System (NUS). The NUS is designed for film imaging of the aircraft engine bay panels, with the source and MCS positioned on the inside of the engine bay, and film in flexible cassettes on the relatively inaccessible tank side of the structures. The system is designed such that several film cassettes can be set in place during a single down-time, and exposed one at a time by remote control, during simultaneous operation of the overhead system.

Programmable Underside Positioner (PUP). The PUP is a

relatively simple, ground-based robotic system designed to move the N-ray Underside System (NUS) from a safe shielded position to various pre-selected exposure positions along the axis of the engine bay.

Collision Avoidance System (CAS). Sufficient precautions must be taken to prevent the massive POP and NOS from damaging the aircraft through impact. The CAS is designed to provide three levels of protection. The first will use software methods. The second and third redundant CAS levels will use physical methods. Special consideration is needed to allow the neutron imaging surface to be placed very close to the surface being inspected. A two-level CAS will be provided for the NUS positioner.

Source Shield and Transfer Systems (SSTS). Three levels of SSTS must be provided. The first level, to be used most frequently, will provide sufficient shielding for operators to enter the N-ray inspection bay. This can be achieved simply by placing the entire source and yoke behind a concrete shield. The second level of SSTS permits operation of the PUP and NOS in teach mode. This level therefore removes the source from the MCS unit. The third level of SSTS must provide for source replenishment at approximately annual intervals. A backup system will be provided to minimize the risk that the source could get stuck in a position such that entry to the inspection bay would be hazardous.

Radiation Protection System (RPS). The equipment and building will be designed to meet applicable radiation protection codes during all routine and emergency operations. The building will have thick concrete walls (thickness 4 feet at rear, 3 feet at side, 1 foot at large doors). Additional administrative controls will be provided through the operational procedures to include limits on radiation exposure directions and times. In addition the RPS will provide necessary area monitors and interlocks to ensure a safe personnel radiation system.

Image Interpretation System (IIS). This is perhaps the most important, and most demanding component of the total MNRS. Whereas N-rays can show corrosion in test samples, the presence of corrosion is not easy to interpret under most actual inspection conditions. Sophisticated, custom-designed interpretation methods will be supplied. For example, a frequent occurence is a fairly uniform layer of corrosion product on the inside skin. This is not distinguishable on viewing the neutron radiograph by eye but would require other methods such as quantitative penetration standards and histogram plots. Similarly thin coatings of corrosion products on honeycomb cell walls are only visible when neutron radiographed at certain angles. Finally, corrosion can be hidden by other hydrogenous materials found in aircraft components, such as bonding adhesives. Comparison of images taken with different methods (N-ray and X-ray), and at different times can be an additional help to image interpretation. The MNRS will also have an elementary comparison capability through video cassette recorders (VCR) on the N-ray and the neighboring x-ray system.

Specifications

The specifications control the design and the acceptability of the finished system. They are performance specifications, and are written in four (4) levels of detail. The first level concerns the turnkey requirement. It states, in effect, that the system must be complete and operable in all respects. The second level of detail specifies some of the approaches to be used to meet the first requirement. The third level of detail gives the degree of sensitivity or accuracy to be met by various sub-system or components. The fourth level of detail provides the minimum acceptance tests. Acceptance tests are required, but are not necessarily sufficient. The controlling requirement is that the overall turnkey system must function to meet the specified overall requirements.

STATIONARY NEUTRON RADIOGRAPHY SYSTEM

Basic Design

The main requirements for the Stationary Neutron Radiography System are (1) very high sensitivity for detection of low level corrosion in aluminum aircraft panels (typically 0.001 inch of corrosion products visible in a 0.5 inch thick aluminum test block); (2) very high throughput capability to meet the maintenance program needs (500 square feet per day of inspected area including initial scan and detailed investigation. An analysis of all available sources focussed attention on the cyclotron and various nuclear reactors. The selected source is a new TRIGA reactor of power 1 MW dedicated to neutron radiography. The complete turnkey system consists of three major sub-components: (i) the nuclear reactor and associated equipment, (ii) the shield and containment system (i.e., all building work), and (iii) the robotics and radiography systems necessary to scan the parts through the beams.

Acquisition Approach

The following key steps were part of the necessary preparation for this project:

1.   Contract for conceptual design and planning assistance.

2.   Approval of conceptual design up to senior levels.

436

3.   Formation of safety review processes and safety review
        committees, including independent outside experts.

4.   Approval of license approach using section 91(b) of the
        Atomic Energy Act.

5.   Completion and approval of a detailed environmental
        assessment.

6.   Approval of a staffing plan and initial hiring of reactor
        experts.

7.   Completion of specification and solicitation of proposals.

Conceptual Design

Nuclear Reactor System.   The reactor is a standard TRIGA
reactor positioned near the bottom of an open tank of diameter
about 7 feet and height of about 26 feet. Pulsing capability and
isotope irradiation rack are two normal features that are not
required.

General Layout.   The reactor core is below ground level.
Four beams are extracted tangentially from the reflector region,
and pass into four radiography bays that, by use of beam
shutters, can be operated simultaneously.   The shield and
containment system is to include all necessary control rooms,
parts staging areas etc., and is to meet all supply requirements
of the environmental assessment.

Positioning Systems.   Three componenet positioners each with
6 axes are required to move the aircraft panels through the
neutron beams.   One is large enough to move an entire wing over
30 feet long.   The positioners are programmable, and have
sufficiently free motion to permit true real-time radiography.

Neutron Image System.   The imaging systems include both
electronic imaging, and film radiography. Each electronic imager
is coupled to an image intepretation system capable of field
flattening, contrast stretching, and frame averaging.   A
sophisticated image storage and retrieval system will be an
important capability, permitting comparison of images taken at
different times with both X-ray and neutron radiography.

SUMMARY

Two neutron radiography systems have been planned for
application to aircraft maintenance at McClellan AFB.   The
Maneuverable System uses an isotopic source.   The Stationary
System uses a small reactor.   A third type - The Transportable
Neutron Radiography System - is under consideration so that

intact aircraft can be inspected outside the hangar. This model would use an advanced acelerator, preferably with a neutron yield at least ten times that of the KAMAN-711. Information on such advanced sources, that may be developed during the next few years, is solicited.

NEUTRON IMAGING WITH LOW-INTENSITY NEUTRON SOURCES

D. Kedem*, R. Polichar, V. Orphan, D. Shreve

Science Applications International Corporation

San Diego, CA, 92121

ABSTRACT

The performance of low intensity thermal neutron field radiography depends on the quality of the imaging system and the intensity and quality of the neutron beam. Therefore, the configuration and composition of the neutron moderator/collimator assembly were analyzed and optimized. The imaging system was designed to achieve maximum performance for the thermal neutron flux obtained.

INTRODUCTION

Thermal neutron radiography has been recently recognized in the military and industrial communities as a necessary and essential non-destructive testing method. In many cases it is the only applicable technique. Field applications of thermal neutron radiography, however, require extensive work in the areas of thermal neutron fluxes and imaging. The performance of such a system depends on the four following components:

o High flux neutron sources
o Moderator/collimator design for an optimal source of high thermal neutron flux
o Imaging system which matches the thermal neutron beam quality
o High technology manipulation equipment which will enable optimum performance in field radiography

The neutron sources available for radiography are divided into three classes: radioisotopes, accelerators and nuclear reactors. M. R. Hawkesworth, in his review (1), summarizes

* now at Kedem Technologies, Rohovot, Israel

439

different neutron sources and their properties. Since we are dealing with field applications, nuclear reactors and heavy-weight accelerators are not applicable. It is also obvious that radioisotope neutron sources with the neutron-producing reaction, (α,n), have a low neutron yield, high mean-- neutron-energy (4-6 MeV), and intense high energy gamma radiation from the encapsulated source. The (γ, n) reaction has the best mean-neutron-energy, considering moderation, but the neutron yield is too low and the gamma flux too high to consider it as an applicable neutron source for field uses.

Some of the applicable neutron sources are:

- ^{252}Cf spontaneous fission isotopic source
- T(d,n) accelerator, operating at 160 KV, with a sealed separate tube head
- A light-weight small-dimension accelerator, using the reaction Be(d,n)

MODERATOR/COLLIMATOR DESIGN

The total neutron yield of a reaction as a function of the bombarding ion energy (2) is given in Fig. 1. In addition to the yield, one has to take into account the neutron energy spectrum of the source, since the higher the neutron energy, the greater the average distance travelled by the neutron into the moderator until thermalized. As a result, the thermal flux distribution will be more dispersed than that of a low energy source, and will have a higher "thermalization factor", which is a measure of the ratio of the neutron source yield to the peak thermal flux produced in a given moderator. The thermalization factor and the H atom number density of the moderator are inversely proportional (3). MORSE calculations performed for the Be(d,n) 2.8 MeV accelerator source in high density polyethylene ($\rho=0.96$) show a thermalization factor of K=80. In addition, it was shown, Fig. 2, that a 12 cm. radius of high-density polyethylene is enough to support that thermalization factor. A decision to use 18 cm. radius is supported by the ratio of of the thermal-to-total neutron flux which peaked at ~ 80%.

SAIC, for the "Mobile Neutron Radiography System" (MNRS), uses a neutron source based on the reaction T(d,n). It is a Kaman A-711 with a sealed accelerating tube of 10 in. diameter and 25 in. long. The measured yield is about 10^{11} n/sec of 14 MeV energy. Because of the high neutron energy, the moderator/collimator design has to be optimized in order to achieve maximum thermal flux and minimum weight. The complexity of the moderator/collimator configuration dictated the use of the three-dimensional MORSE transport code.

The final configuration consisted of a natural uranium booster surrounded by high-density polyethylene moderator, Figure 3. No perturbations were included, other than the accelerator tube itself. The reaction (n,2n) and (n,3n) of the 14 MeV

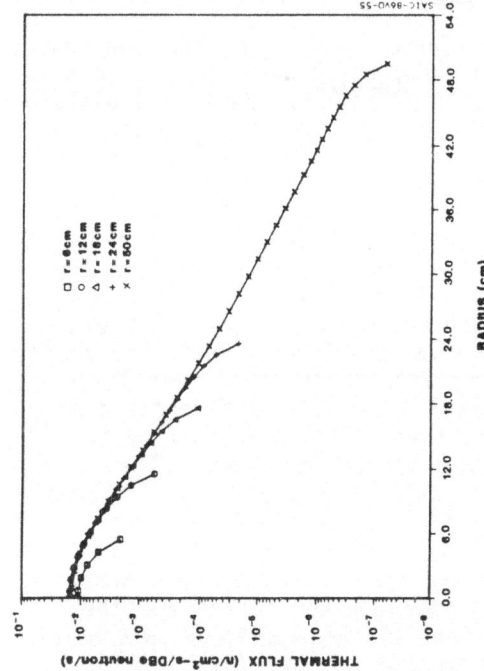

Figure 2. Thermal Neutron Flux vs. Radius for Polyethylene Moderators of Various Radii.

This illustration is printed from the best available original

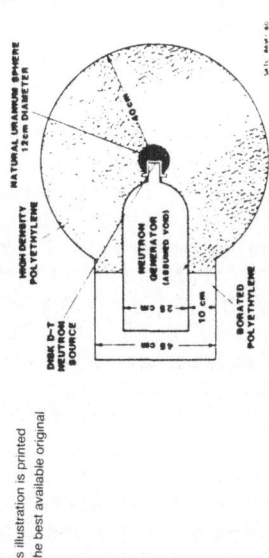

Figure 3. Geometry/Material used for 3-D "Flux Mapping" Calculations.

Figure 1. The Thick-Target Neutron Yield as a Function of Bombarding Ion Energy for Various Neutron sources of Interest for Radiography. The T(d,n) Yield refers to a Titanium Target Coating of 2.2 mg cm-2 and an H:Ti Atom Ratio of 1:2.

neutrons reduces the source energy spectrum towards that of a fission neutron spectrum to give a lower thermalization factor.

MORSE calculations were performed: (1) to determine the appropriate placement of the collimator hole and shape in order to maximize the thermal neutron flux output, (2) to calculate the high energy (non-thermal) neutron and the 2.2 MeV capture gamma-ray flux at the radiographic detector position and (3) to optimize the final moderator/collimator configuration.

Fig. 4 shows 3-D flux map results. Fig. 4(a) shows the energy group containing the 14 MeV source neutrons, Figure 4(b) the sum of all non-thermal neutron energy groups, and Figure 4(c) the thermal neutron group. These are fluxes in the x - z plane at one cm intervals ($\Delta x = \Delta z = 1$ cm, $\Delta y = 5$ cm). To reduce some of the statistical fluctuations, results averaged over large volume bins ($\Delta x = \Delta z = 3$ cm, $\Delta y = 5$ cm) were performed and are shown in Fig. 5. It is evident that a dip in thermal neutron flux exists at the site of the uranium booster. The thermalization factor achieved is K=160, and the ratio of thermal neutron flux to the fast neutron flux is above 0.55.

In order to obtain a high-quality beam from the thermal neutrons, it is best to contain the neutron source and the moderator within a neutron absorbing shield, and to allow some of the neutrons to stream through a hole in this shield. The angular spread of the emerging beam will be confined by the length-to-diameter ratio of the collimator hole, L/D. The divergent collimator is one of the most widely used for neutron radiography. Fig. 6 shows two divergent collimators with the same L/D. When estimating the neutron flux emerging from the collimator tube, one must assume that not all the neutrons in the collimator originate at the entrance aperture, but in practice, some are emerging from the collimator wall adjacent to the aperture. It was shown by Hawkesworth that the total flux is given by:

$$\phi_0 = 1/16 \ (D/L)^2 \ (1+2l/L) \ \phi_i$$

where l is the length of the collimator wall which emits neutrons. The MORSE calculations demonstrate that when using pre-collimation with enough absorption media, the leakage contribution is less than 6%.

A comparison between the divergent collimators of Fig. 6(a) and (b) demonstrate that, due to the entrance collimator geometry, configuration (b) is superior to configuration (a), due to the neutron field flatness at the detector site. Since the neutron source (after boosting) is about 1.5×10^{11} n/sec, and K=160, the thermal neutron flux at the collimator entrance is about 4.5×10^8 n/cm^2/sec. Therefore, the calculated flux at the detector site is

$$\phi_0 = (4.5/16) \ (1/24)^2 \ 1.05 \times 10^8 = 5 \times 10^4 \ \text{n/cm}^2/\text{sec.}$$

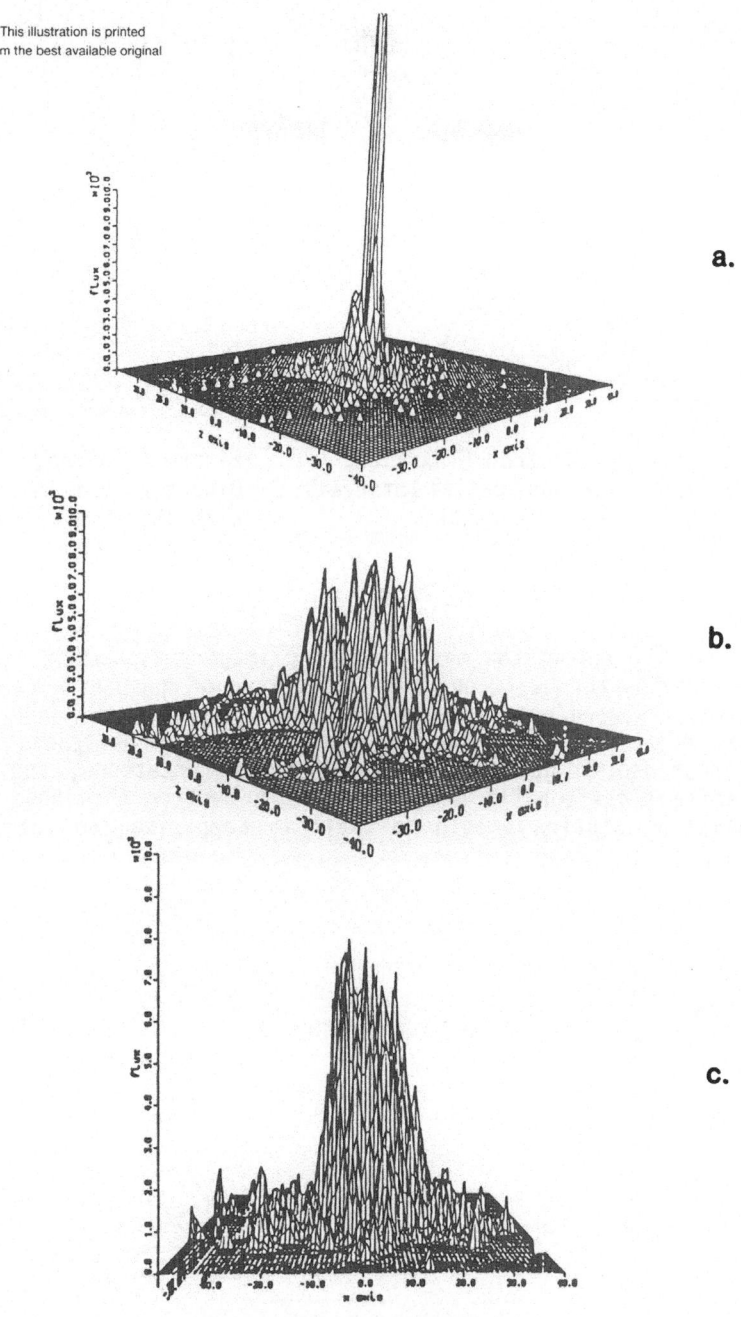

Figure 4. Flux Map for (a). Fast Neutrons, (b). Non-Thermal
Neutrons and (c). Thermal Neutrons.

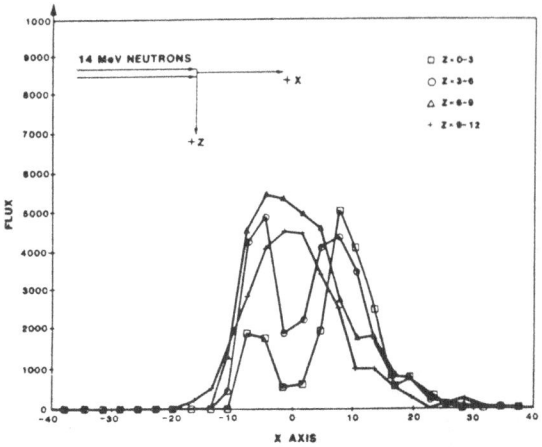

Figure 5. Thermal Neutron Flux along a Beam Direction in Various Spatial Intervals Perpendicular to Beam Direction.

Figure 6 (a) and (b). Divergent Collimators with the same L/D.

NEAR REAL TIME/REAL TIME IMAGING

Before discussing the most promising imaging system, it is useful to review the basic quantum limitations of neutron imaging with an ideal screen/recording system. Such a system will have a detection efficiency of 100% for incoming thermal neutrons, minimal response to gamma rays and non-thermal neutrons, and system noise significantly below the single neutron threshold. The spatial resolution should be small in comparison to the

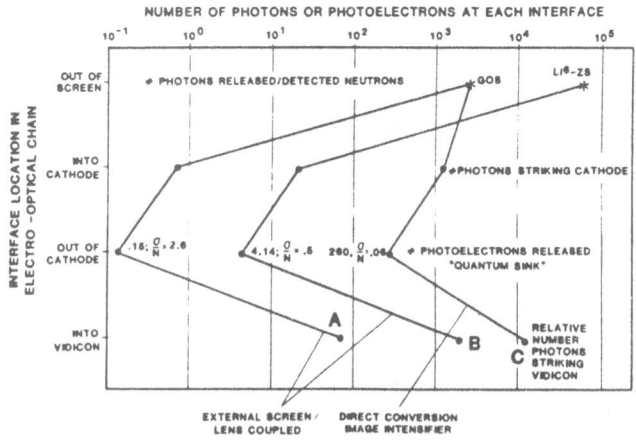

Figure 7. Photo Optical Signal through Neutron Radiographic Imaging System. In A and B, the Screen and Cathode are Lens-Coupled Systems, while C is a Direct Conversion Image Amplifier.

geometric blur caused by the L/D ratio of the source. One can illustrate a scene composed of an array of square picture elements, in which one picture element is assumed to have a higher luminance level than the surrounding background elements:

$$n = \frac{k^2}{c^2 d^2 t}$$

where:

n is the density of detected neutrons per unit area per unit time

k is the certainty coefficient

d is the dimension of the resolution element

t is the integration time of one observation

c is the contrast

Let us assume a contrast of 3%, a certainty of 68% (k=1) and pixel size d=0.05 cm, which fits a scene of 25x25 cm (10"x10") into a 512x512 pixel memory, then

$$n = \frac{1}{0.03^2 x 0.5^2 xt} = \frac{4.4x10^5}{t}$$

which yields a thermal neutron flux of $4.4x10^5$ n/cm^2/sec. In cases where the thermal flux is of the order of $5x10^4$ n/cm^2/sec, such an image quality will be achieved in 9 sec.

The real time imaging systems fall broadly into two categories, those which use an external converter screen optically coupled to a low light level TV (LLLTV) camera, and those which use a direct conversion image amplifier (using an internal converter screen) coupled to a more conventional TV image tube. The main difference between the two approaches is the way in which the light from the screen is converted into photoelectrons in the intensifier tube and in the effect that the relative efficiency of this process has on the final recording of the signal of each neutron.

In an optically coupled system the main drawback is in the light gathering ability of the lens. From basic optics theory, when using a single lens, the fraction of light captured from the screen is:

$$\frac{\Omega}{4\pi} = \frac{m^2}{1+m} \times \frac{1}{16 f^2}$$

where:

$\frac{\Omega}{4\pi}$ is the solid angle viewed by the lens

m is the optical system magnification (m<1 in our case)

f is the ratio of the focal length to the diameter.

The above formula assumes that the light is emitted isotropically from the screen.

445

The standard silicon-intensified target (SIT) tube has an image diagonal of 16 mm. The fastest practical lens has an f number of 0.65. Using these figures, the optical capture efficiency for such a lens coupled system is $\Omega/4\pi = 2.8 \times 10^{-4}$, and for a 40 mm image diagonal, $\Omega/4\pi = 3.2 \times 10^{-3}$. These numbers must be compared to the direct conversion system, where the cathode is deposited directly on the converter material. In this case, 50% of the photons are expected to be emitted into a 2π solid angle, and thus about 20% of them interact with the photocathode to produce electrons. The image is reformed on an output phosphor, using an electrostatic lens with efficiency nearly 100%. The gain of the image converter tube is about 10000, in comparison to a SIT tube which has a gain of 1500. The coupling of the SIT to the image converter tube is via lenses, which causes a loss of light. This can be overcome using a fiber-optic coupling.

Fig. 7 shows the quantum gain in the two systems. It is assumed that the fluorescence screens A and C have identical luminous efficiencies, and the final images are of the same size. In the system C (direct conversion image amplifier), the luminous flux on the SIT target is 280-300 times higher than in system A. System B has a different entrance screen, with a photon efficiency higher than C by a factor of nine. The luminous flux on the SIT target is 5-6 times higher is system C than in system B.

CONCLUSIONS

Results obtained with the Direct Conversion Image Amplifier (Type C in Fig. 7) fit the preliminary calculations. For very low thermal neutron fluxes ($<5 \times 10^3$ n/cm^2/sec) the image obtained was very grainy due to the low statistics and low signal-to-noise ratio, even after long integration (<30 sec). For thermal neutron fluxes of 2×10^4 to 2×10^5 n/cm^2/sec, integration times of 8 sec to 2 sec were needed in order to get good quality images (hole #11 in the modified A type penetrameter), and no enhancement was needed. When field flattening and other mathematical enhancements were applied, the quality of the images were dramatically improved. For thermal neutron fluxes above 5×10^5 n/cm^2/sec, no integration, or very little (2-4 frames) was needed to obtain very good quality images. At fluxes $<10^6$ n/cm^2/sec and integration times of 8-10 frames, very good image quality was obtained with a plumbicon tube.

BIBLIOGRAPHY

1. M.R. Hawkesworth, Atomic Energy Review 152, 169 (1977)

2. M.R. Hawkesworth and J. Walker, Basic Principles of Thermal Neutron Radiography, J.P. Barton and P. Von der Hardt (eds), Neutron Radiography, pp. 5-21 (1983)

3. L. Holland and M.R. Hawkesworth, Low Voltage Particle Accelerator for Neutron Generation, Non Destructive Testing 4,5 (1971)

MOBILE NEUTRON RADIOGRAPHY SYSTEM FOR AIRCRAFT INSPECTION

V. Orphan, D. Kedem, F. Johansen

SCIENCE APPLICATIONS INTERNATIONAL CORPORATION

SAN DIEGO, CA 92121

ABSTRACT

A Mobile Neutron Radiography System (MNRS) has been designed and fabricated for the inspection of naval aircraft for hidden corrosion in aluminum structures. The MNRS is comprised of an accelerator-based thermal neutron source, near real-time neutron imaging system, robotic positioner and manipulator for the source/imager, control trailer housing system control electronics and digital image processing system, mobile dark room for film processing, a self contained electrical power source and a radiation safety system. For insitu aircraft inspection, the Robotic Scanner is programmed (in a teach-learn mode) to scan a region of the components, using a control pendant.

The thermal neutron source consists of a Kaman A-711 neutron generator using the deuterium-tritium (D,T) reaction and an optimized moderator/collimator assembly. The sealed-tube neutron generator produces about 1×10^{11} 14 MeV neutron/sec which is projected to produce a thermal neutron flux of about 5×10^4 thermal $n/cm^2/sec$ at the image plane for an L/D = 24. The neutron imaging system uses a 9-inch diameter neutron-sensitive Thomson CSF Image Intensifier tube coupled to a television camera and a user-oriented digital image processing system. With a thermal flux of only 6×10^3 $n/cm^2/sec$ and L/D = 30, the system easily resolves 11 holes (Modified Type A Penetrameter) and detects 6 layers of Scotch "Magic Tape" (0.012 inches) with 8 sec integration.

MNRS, when delivered to the North Island Naval Air Rework Facility (San Diego) in late 1986, will provide the first opportunity to demonstrate the effectiveness of neutron radiography for in situ inspection of aircraft in a realistic aircraft overhaul environment.

INTRODUCTION

Thermal neutron radiography has been demonstrated both in laboratory and limited field demonstrations (1,2) to be capable of detecting hidden corrosion in aluminum aircraft structures (wings, horizontal and vertical stabilizers, etc.) as well as trapped moisture in honeycomb materials, inadequate adhesive bonds, improper lubrication and flaws in composite materials. The presence of hydrogen in the corrosion products of aluminum, either as aluminum hydroxide or hydrated aluminum oxide, allows thermal neutron imaging to detect incipient corrosion layers on the order of 0.03mm (0.001 in.), in relatively thick, up to equivalent thickness of 100mm (about 4.0 in.), aluminum structures. The spatial resolution afforded by neutron radiography facilitates discrimination of interfering hydrogen-containing materials (such as adhesives, sealants, paints, etc.) with easily recognizable shapes. The irregular shape of a corrosion area can usually be readily identified even in the presence of other hydrogen-containing materials.

Although neutron radiography has been demonstrated to be uniquely capable of detecting hidden corrosion in aluminum aircraft structures, its use has been restricted because practical neutron radiography equipment for insitu inspection of aircraft has not been generally available. Alternate NDT techniques, such as X-radiography, ultrasonics, eddy current and acoustic emission testing have been used (3) for hidden aircraft corrosion inspection despite serious limitations in performance; for instance, an inability to inspect complex structures without requiring costly and time-consuming "de-skinning" of the aircraft and inability to detect corrosion initiation or small areas of corrosion. These shortcomings of "conventional" NDT techniques can be overcome by a practical implementation of insitu thermal neutron radiography. This implementation requires a mobile thermal neutron source, electronic neutron imager and robotic source/imager manipulator capable of scanning aircraft to produce neutron radiographs of adequate quality for corrosion detection in practical exposure and overall inspection times. The candidate neutron sources, Cf-252 radioisotopic source and accelerator sources, each have advantages and disadvantages; however, the ability of an accelerator source to be turned off between exposures makes it the logical choice for a mobile source with optimum radiation safety features.

In this paper we describe an accelerator-based Mobile Neutron Radiography System (MNRS) developed by Science

448

Applications International Corporation (SAIC) for the Naval Air Rework Facility, North Island, California. The MNRS, when deployed, will demonstrate in a realistic aircraft overhaul environment the application of insitu neutron radiography for detection of corrosion and other structural defects. The goal of the MNRS project is to demonstrate and quantify the productivity enhancement made possible at a Naval Air Rework Facility by the use of neutron radiography.

MOBILE NEUTRON RADIOGRAPHY SYSTEM DESCRIPTION

Overview

The Mobile Neutron Radiography System (MNRS) is comprised of an accelerator-based thermal neutron source, near real-time neutron imaging system, robotic positioner and manipulator for the source/imager, control trailer housing system control electronics and digital image processing system, mobile dark room for film processing, a self contained electrical power source and a radiation safety system. Figure 1 shows a typical deployment of the MNRS components around an aircraft during inspection. The operator prepares for an inspection by "driving" the Robotic Manipulator Cart; using a small portable control panel (pendant) attached by cable to the cart; to position the source/imager close to the aircraft component to be inspected. Next, the Robotic Scanner is programmed (in a teach-learn mode) to scan a region of the components, also using a control pendant. Finally, personnel are excluded from the radiation safety zone around the aircraft, the operator enters the shielded control trailer and turns on the neutron source to commence the neutron radiography which is carried out according to the preprogrammed sequence of the Robotic Scanner. The principal subsystems of the MNRS are next described.

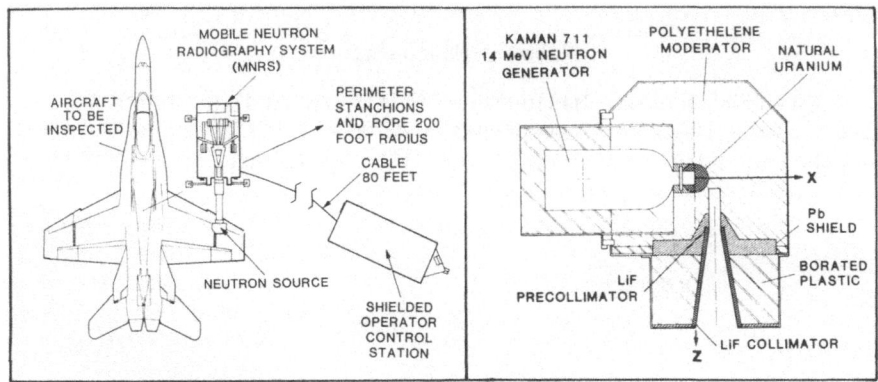

Figure 1. Overview of Aircraft Corrosion Inspection by Thermal Neutron Radiography.

Figure 2. Mobile Neutron Radiography System-Neutron Source/Moderator/Collimator.

Thermal Neutron Source

The thermal neutron source, sketched in Figure 2, consists of a Kaman A-711 neutron generator using the deuterium-tritium (D,T) reaction and an SAIC optimized moderator/collimator assembly. The sealed-tube neutron generator, operating at approximately 160KV, produces about 1×10^{11} 14 MeV neutron/sec. The moderator assembly, which uses a natural uranium shell around the 14 MeV neutron source and a high-density polyethylene moderator to optimize the resulting thermal neutron flux, produces approximately 5×10^4 thermal n/cm^2/sec at the image plane for an L/D = 24. The combined weight of the thermal neutron source assembly is about 700 kg (1540 lb).

Near Real-Time Imager

Two general approaches for the design of an electronics near real-time imager are illustrated in Figure 3. The External Screen/LLLTV approach is significantly less sensitive than the Direct Neutron Imaging Tube approach because a large fraction (~98%) of the light from the neutron-to-light converter screen is lost in the relatively inefficient lens-coupled optical system. On the other hand, the Direct Image Tube, which converts the visible light from the converter screen to photoelectrons in the adjacent photocathode, focuses the photoelectrons by means of an electrostatic lens with little loss. The MNRS neutron imaging system uses a 9 in. diameter Thomson CSF image intensifier tube. The intensifier tube is optically coupled to a high quality television image tube and camera. The imaging subsystem also contains a sophisticated, user-oriented computer-based digital image processor capable of providing all necessary functions, including image grab, image integration and averaging, contrast stretching, image filtering, convolution by NxN kernel, edge enhancements, pan/zoom, alphanumeric image labeling and image archiving.

Robotic Manipulator

The MNRS Robotic Manipulator is illustrated in Figure 4 as a sketch with the major components indicated, and as a photograph of the manipulator taken shortly after assembly. The Robotic

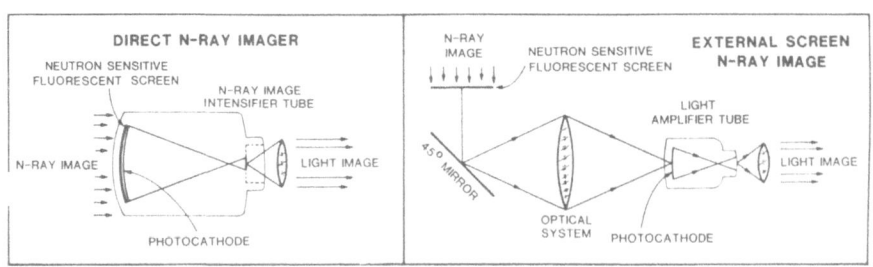

Figure 3.   Real-Time N-Ray Approaches.

Manipulator performs many precise and automated inspection movements required to carry out the following functions:

- Reach difficult access areas such as at the wing/fuselage interface.
- Manipulate the imager close to the complex curvature of each target structure while avoiding external obstacles on the surface.
- Radiograph around internal obstacles on the surface, and align the source/imager with honeycomb structures.
- Continuously maintain neutron source/imager alignment and spacing.

General design features of the Robotic Manipulator are as follows:

- Imager may be positioned on either side of the object
- Nominal scan pattern - 7-ft deep by 4-ft wide
- Wing height - maximum: Grumman E-2 - Minimum: McDonnell Douglas F-18

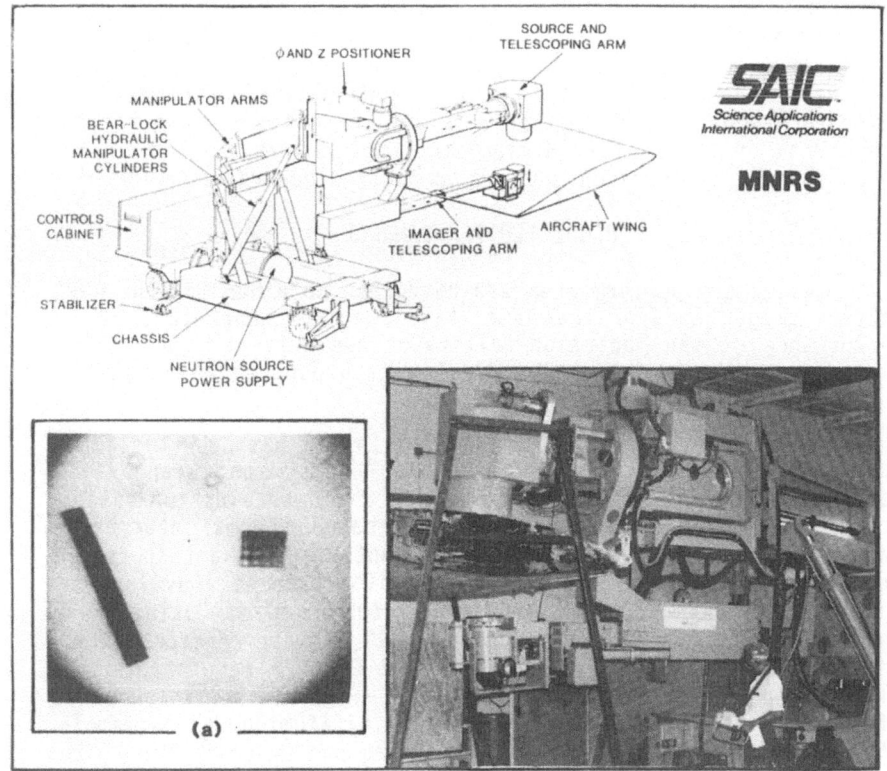

Figure 4. Science Applications International Corporation Mobile Neutron Radiography System (MNRS). 4(a). Thermal Neutron Image taken with MNRS Imager at a Flux of 6 x 10^3 n/cm^2 / sec and Recursive Average of 8 sec (Kaman A-711 Neutron Source).

451

- Vertical airfoil up to 16-ft above ground level
- Digital readout of inspection position within scan pattern
- Re-inspect capability (repositionability) $\pm$ .25 inch
- Redundant collision-avoidance systems

The MNRS Robotic Manipulator was successfully demonstrated and passed a Factory Acceptance Test in November 1985. Final Acceptance Testing is scheduled for the summer 1986 pending receipt of a radiation safety operating license which will allow the neutron source to be turned on.

The Control Trailer, approximately 8 ft. wide x 8 ft. high x 15 ft. long is shielded (12 in. polyethylene for the wall facing the source) to provide a dose level of less than 2 mR/hr inside. A shielded window with a vision field 20^0 x 45^0 horizontal allows the operator to view the manipulator from the control console.

Radiation safety is assured by a moveable stanchion and rope barrier with a radius of about 200 ft which is placed around the aircraft. The barrier is equipped with "Radiation Area" signs and flashing red warning lights. In addition, an infra-red intrusion system is located inside the barrier so that any intrusion will automatically interupt the high voltage to the neutron generator tube. Other radiation safety devices include a rotating beacon on the Robotic Manipulator Cart and an audible alarm which sounds for 15 seconds prior to neutron generation.

PROJECTED PERFORMANCE

Preliminary performance data have been obtained for the MNRS Neutron Imager and for the Kaman A-711 Neutron Generator. The performance of the Moderator/Collimator assembly has been predicted based on a detailed radiation transport calculational model.

The Near Real-Time Neutron Imaging System has been tested at several nuclear reactor thermal neutron beams with a range of neutron flux levels typical of those from an accelerator-based source and at an existing accelerator-based thermal neutron radiography source. Figure 5 shows a photograph of a test sample and thermal neutron images of this sample taken at flux levels ranging from 1 x 10^4 n/cm^2/sec to 2 x 10^5 n/cm^2/sec using the MNRS imager at the GA Technologies TRIGA nuclear reactor. The test sample consisted of ASTM Standard E545-75 penetrameters (Type A and Modified Type A), 0.010 in. and 0.024 in. thicknesses of polyethylene on a 0.25 in. thickness of aluminum and a sample of aluminum aircraft structure containing hidden corrosion. The neutron images in Figure 5 taken at an L/D = 40 and with an integration time of 4 sec (128 frames), show that the two thicknesses of polyethylene can be clearly seen in all cases. Furthermore, about 11 holes can be seen in the Modified Type A Penetrameter for the image obtained at a flux of 1 x 10^4 n/cm^2/sec while more

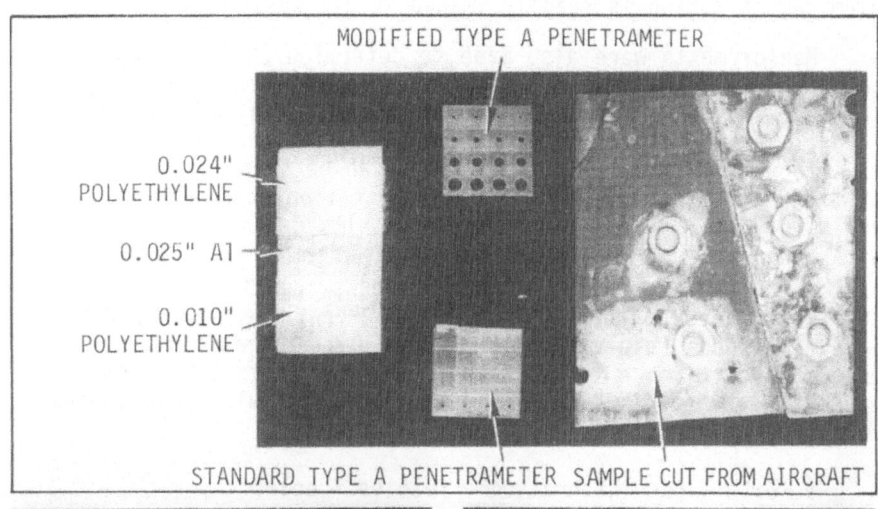

MODIFIED TYPE A PENETRAMETER

0.024" POLYETHYLENE

0.025" Al

0.010" POLYETHYLENE

STANDARD TYPE A PENETRAMETER   SAMPLE CUT FROM AIRCRAFT

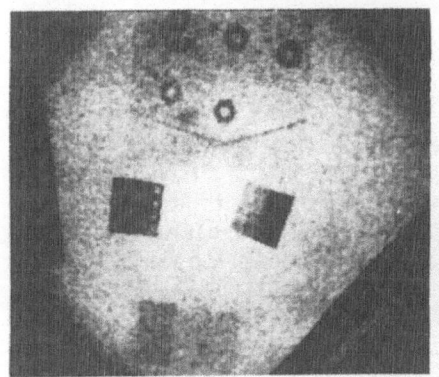

1×10^4 n/cm^2/sec 128 Frames

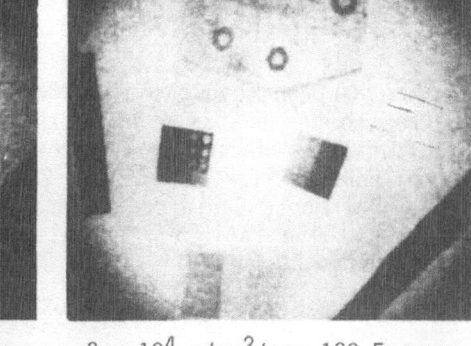

2×10^4 n/cm^2/sec 128 Frames

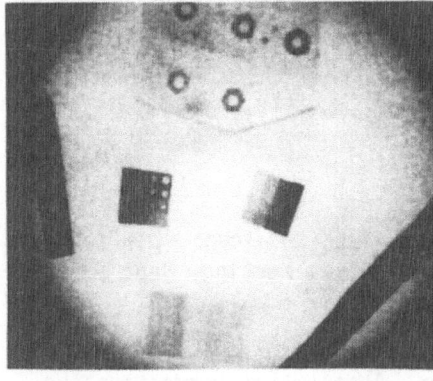

5×10^4 n/cm^2/sec 128 Frames

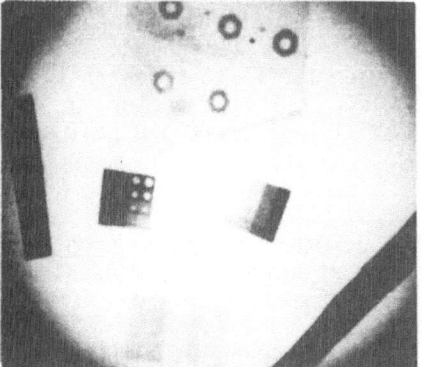

2×10^5 n/cm^2/sec 128 Frames

Figure 5. Photograph and Thermal Neutron Images
of a Test Sample using the MNRS Imager.

holes can be seen for the higher flux levels. Note that the aluminum corrosion is readily imaged in all cases.

Measurements were also made to determine the performance of the imager with a transportable thermal neutron source using the Kaman 14 MeV neutron generator at the U.S. Army Materials Technology Laboratory, Watertown, MA. Typical results are shown in Fig. 4(a) which shows that 11 holes can be seen in the Modified type A Penetrameter with an 8 sec integration at an L/D = 30 for a flux of 6×10^5 $n/cm^2/sec$. These results demonstrate that even with a non-optimum moderator/collimator assembly, the MNRS imager still exceeds the performance requirements specified by the Navy: resolve 7 holes on the Modified A Type Penetrameter with 60 sec integration and resolve 11 holes with 300 sec integration. The enhanced performance of the Direct Neutron Image Tube will far exceed the original MNRS performance goal and will achieve the following performance:

- With 4 sec integration - resolve 7 holes

- With 8 sec integration - resolve 11 holes and image 6 layers of Scotch "Magic Transparent Tape" (0.012 in.) on 1/4 in. aluminum.

The neutron output of the MNRS neutron generator has been measured to be about 9×10^{10} n/sec. For this 14 MeV neutron output in the moderator/collimator assembly (see figure 2), the thermal neutron flux at an L/D = 24 is calculated (using a three-dimensional Monte Carlo radiation transport code) to be 5×10^4 $n/cm^2/sec$. The ratio of thermal-to-total neutrons (thermal = fast) is about 0.5.

CONCLUSIONS

The MNRS projected performance far exceeds the original goals established by the Navy. When delivered to the North Island Naval Air Rework Facility in late 1986, the MNRS will provide the first opportunity to demonstrate the effectiveness of neutron radiography for insitu inspection of aircraft in a realistic aircraft overhaul environment.

REFERENCES

1. J. John, Mobile Neutron Radiograph System, 1973. Air Transport Association Nondestructive Testing Meeting, Sept 11-13, 1973.

2. E.P. Roeser, et al, Neutron Radiographic Systems for Aircraft Maintenance, 1976 Air Transport Association Nondestructive Testing Meeting, Sept 14-16, 1976.

3. D. J. Hagemaier, et al, Aircraft Corrosion and Detection Methods, Materials Evaluation, 43, March 1985.

THE SENSITIVITY OF NEUTRON RADIOGRAPHY FOR DETECTION OF ALUMINIUM

CORROSION PRODUCTS

J. Rant[1], R. Ilić[1], G. Pregl[1,2], P. Leskovar[3], B. Žnidar[4]

[1] "J.Stefan" Institute, "E. Kardelj" University of Ljubljana, Ljubljana, Yugoslavia
[2] Faculty of Engineering, University of Maribor, Maribor, Yugoslavia
[3] Faculty of Mechanical Engineering, "E. Kardelj" University of Ljubljana, Ljubljana, Yugoslavia
[4] Adria Airways, Brnik, Yugoslavia

A B S T R A C T

Within a feasibility study concerning the introduction of neutron radiography as a nondestructive method into the domestic aircraft industry and for the maintenance of domestic commercial civil aircraft, the sensitivity of neutron radiography for detection of Al corrosion products was studied experimentally. The mass thickness of a large area surface corrosion deposit which could be easily detected through thick (1 - 2 cm) Al slabs was found to be at least 0.02 g/cm^2. The minimal detectable mass thickness of corrosion products depends on the relative amount of $Al(OH)_3$ and $AlO(OH)$, moisture and possible organic materials in their composition and should be in the range of 0.01 - 0.02 g/cm^2.

INTRODUCTION

Since the fundamental work of John (1) and Dance (2), neutronographic detection of various types of corrosion on aluminium structures has found broad applications, especially in the maintenance of commercial and military aircraft (3). In order to promote the use of this efficient nondestructive technique in the maintenance of domestic commercial aircraft and generally for applications in the aircraft industry, a feasibility study and a demonstration programme was performed.

In this communication we shall describe experiments we performed in order to assess the possibilities of the method for quantitative determination of the depth of corrosion products on aluminium slabs. In addition to the detection and localization of the corrosion, neutron radiography would also help to determine the extent of the corrosion damage and the loss of aluminium metal. Further, we were interested in the minimal thickness of corrosion products detectable with our technique on aluminium objects.

ATTENUATION PROPERTIES OF ALUMINIUM AND ITS CORROSION PRODUCTS FOR THERMAL NEUTRONS AND SOFT X-RAYS

In many common cases, e.g. in the presence of halides in the marine environment, the corrosion process on aluminium can be considered as the following transformation:

Al (density 2.7 g/cm^3) $\longrightarrow$ various Al trihydrates (Al(OH)$_3$),

Al monohydrates (AlO(OH)) and Al salts

In practice Al(OH)$_3$ (hydrargilite or bayerite) constitutes 60-90 % of the corrosion products. In addition to trihydrates, monohydrates (boehmite, diaspore), oxides, various aluminium salts and organic compounds can also be present. Moisture and water droplets often accompany corrosion nests.

Attenuation properties of Al corrosion products can thus substantially differ from case to case depending on the actual composition. Therefore for each typical case of inspection the expected actual composition of corrosion products has to be determined. Only as a rough approximation can one assume Al(OH)$_3$ as the only constituent of Al corrosion products. Nevertheless, for illustrative purposes it is interesting to compare attenuation of base aluminium metal, Al(OH)$_3$, AlO(OH) and water for thermal neutrons and soft X-rays (30 kV), which are typically used for radiographic inspection of Al structures. The attenuation coefficients of the relevant elements and materials are presented in Table I.

Table I: Linear attenuation coefficients of thermal neutrons (Σ) and soft (30 kV) X-rays (μ) for major constituents of Al-corrosion products.

Material	Density (g cm^{-3})	Linear attenuation coefficient (cm^{-1})	
		X-rays (30 kV)	Neutrons (0.025 eV)
Al	2.7	3.0	0.0861
Al(OH)$_3$	2.53	1.501	2.40
AlO(OH)	3.014	2.16	1.50
H$_2$O	1.0	0.368	2.706

On the basis of the above attenuation coefficients we can draw the following conclusions about the radiographic properties of neutrons and soft X-rays for the detection of corrosion on aluminium:

- aluminium is almost transparent to neutrons and it is possible neutronographically to inspect even thick (several cm) Al objects. This is increasingly difficult and costly with soft X-rays. In this case, thick Al parts can easily mask structural features and loss of material due to corrosion,

- due to the high neutron scattering cross section of hydrogen, corrosion products are highly opaque to neutrons and thus can be detected even through thick Al slabs. In the case of soft X-rays the difference in the attenuation coefficients of aluminium and its corrosion products is smaller than with thermal neutrons and hence one can expect a smaller detection sensitivity.

In the assessment of the attenuation properties of aluminium corrosion products it is also important to consider the relation between the depth of corrosion (loss of Al metal) and the thickness of corrosion products as 1 cm^3 of corroded Al yields 3.22 cm^3 of Al(OH)$_3$. Therefore in the case of Al(OH)$_3$ the thickness of the layer of corrosion products is 3 times the depth of corroded base aluminium. Knowing the composition of corrosion products and measuring their thickness by a suitable nondestructive method, e.g. from neutron radiographs, it is possible to estimate the loss of the base aluminium metal. This has been first proven by John (1).

CONTRAST SENSITIVITY OF RADIOGRAPHIC DETECTION OF CORROSION PRODUCTS FOR THERMAL NEUTRONS AND SOFT X-RAYS

The relative figure of merit of neutron radiography against soft X-ray radiography for the detection of corrosion products on aluminium can be obtained by comparing the relative contrast K_{cp} of the image of an Al(OH)$_3$ deposit imbedded in aluminium base metal:

$$K_{cp} = \frac{D_{cp} - D_{Al}}{D_{Al}} \tag{1}$$

where D_{cp} is the film density of the image of corrosion products and D_{Al} is the film density of the image of uncorroded surrounding aluminium.

Assuming a linear relationship between film density and film exposure, and for the geometry of the radiographic inspection presented in Fig. 1 and for thin layers ($x_{cp} \ll 1/\Sigma_{cp}$) of corrosion products, one obtains:

$$K_{cp} \text{ (neutrons)} = 1 - e^{-\Sigma_{cp} x_{cp} + \Sigma_{Al} x_1} \approx 3 \Sigma_{cp} x_1 \tag{2}$$

$$K_{cp} \text{ (X-rays)} = 1 - e^{-\mu_{cp} x_{cp} + \mu_{Al} x_1} \approx (3 \mu_{cp} - \mu_{Al}) x_1$$

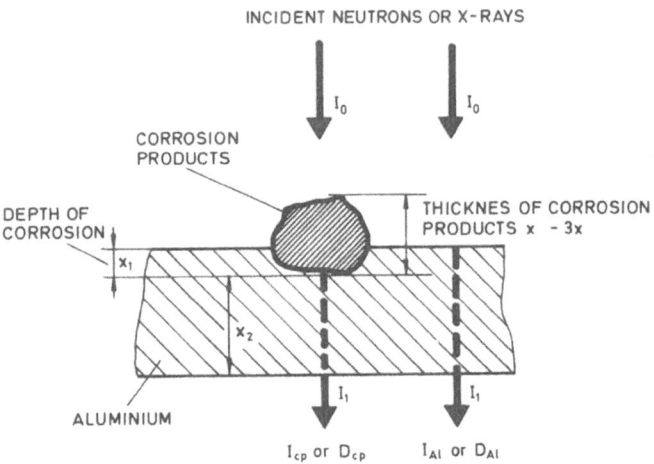

INCIDENT NEUTRONS OR X-RAYS

I_0 I_0

CORROSION
PRODUCTS

DEPTH OF
CORROSION

THICKNES OF CORROSION
PRODUCTS $x - 3x$

x_1

x_2

ALUMINIUM

I_1 I_1

I_{cp} or D_{cp} I_{Al} or D_{Al}

Fig. 1: Geometry of radiographic inspection of surface corrosion

Taking into account the approximate relation $x_{cp} = 3 x_1$ between the thickness of corrosion products and the depth of corrosion (x_1), and for the case of inspection of very thin layers, we obtain:

$$\frac{K_{cp} \text{ (neutrons)}}{K_{cp} \text{ (X-rays)}} = \frac{3 \Sigma_{cp}}{3 \mu_{cp} - \mu_{Al}} \approx 4.8 \qquad (3)$$

The relative image contrast and therefore also the contrast sensitivity for the detection of surface corrosion on aluminium is almost 5 times greater with thermal neutron radiography than with soft X-ray radiography.

Knowing the minimal visually detectable relative image contrast K_{cp}^{min} we can estimate the minimal detectable thickness X_{cp}^{min} of corrosion products:

$$X_{cp}^{min} \text{ (neutrons)} = \frac{K_{cm}^{min}}{\Sigma_{cp}}$$

$$X_{cp}^{min} \text{ (X-rays)} = \frac{K_{cp}^{min}}{\mu_{cp} - \frac{1}{3}\mu_{Al}} \tag{4}$$

Assuming the minimal visually detectable relative image contrast K_{cp}^{min} conservatively to be 0.01, the minimal neutronographically detectable thickness of pure $Al(OH)_3$ is about 0.04 mm, and for soft X-rays amounts to about 0.2 mm. In practice these figures can be somewhat greater depending on the actual composition of the corrosion products.

DESCRIPTION OF THE EXPERIMENT

The relation between the film density and the thickness of corrosion products was experimentally determined from neutron radiographs of aluminium test objects. In each of a series of 1 cm thick Al blocks two holes of equal diameter (sets of 1.0, 0.4, 0.2 and 0.1 cm diameter) and depth have been drilled. For each set of holes, the hole depth ranged from 0.01 to 0.8 cm. One of the holes was filled with corrosion products, the other left open and the Al block covered with thin (0.02 mm) Al foil. The corrosion products were obtained by scraping Al pieces corroded in salt (marine) water, drying in the open air, grinding and sieving.

An X-ray difraction study of the powder confirmed the presence of its both major constituents, bayerite and boehmite. By chemical analysis their relative weight fractions were determined as 63 and 37 %, respectively. The mass thickness of corrosion products in each hole was determined by precision weighting and the tap density (g/cm^3) of filled powder ranged from 1.1 to 2.0 g/cm^3, significantly lower than the calculated value of 2.69 g/cm^3 for powder or the value of 2.53 g/cm^3 for bayerite. Each set of Al blocks was simultaneously neutron radiographed in the thermal column NR facility of the Ljubljana TRIGA Mark II reactor, described previously (4, 5). A direct neutron radiographic method with a Gd (0.025 mm) screen and Kodak M and Fotokemika FIR 10 films (both single coated) were used.

On the neutron radiographs all the holes (diameter 1.0 and 0.4 cm) were visible, including the hole with tap mass thickness of about 0.02 g/cm^3 (depth 0.11 mm), as shown in our previous work (6). In the case of smaller holes (0.2 and 0.1 cm in diameter), holes with a tap mass thickness of 0.03 g/cm^3 could be detected.

Calibration curves relating the net density of neutronographs and the thickness of corrosion products were measured. The densitometric readings were taken in the middle of the image of the filled hole (D_{cp}), in the middle of the empty hole (D_e) and in the middle of Al block (D_{Al}). The precision of densitometric readings was only 5-10 %. The contribution of γ-rays to the optical density of neutron radiographs was subtracted.

It is easy to show that the graph of ($D_e - D_{cp}$) is a linear function for small values of the mass thickness σ_{cp} of corrosion products. The experimentally obtained graph of ($D_e - D_{cp}$)/D_e for 1.0 cm D holes is presented in Fig. 2. From the slope of the linear portion of this graph, an estimate of the mass attenuation coef-

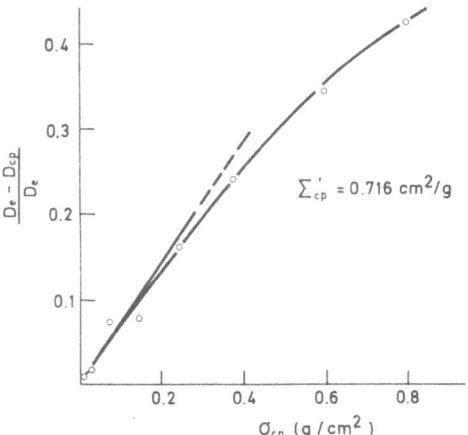

Fig. 2: Calibration curves relating $(D_e - D_{cp})/D_e$ to the mass thickness of the layer of corrosion products, σ_{cp}.

ficient, Σ'_{cp}, can be obtained. In our case the estimated value for Σ'_{cp} (≈ 0.72 cm²/g) is significantly lower than the value (0.95 cm²/g) of pure $Al(OH)_3$. This again indicates the presence of monohydrate and other aluminium salts in addition to $Al(OH)_3$ in our test corrosion products. For our particular case, with accurate densitometry of neutron radiographs, the minimal thickness which could be reliably determined is at least 0.1 mm (mass thickness 0.02 g/cm²). The minimal detectable thickness x_{cp}^{min} calculated according to eq. (4), using the value of 0.72 cm²/g for Σ'_{cp} amounts to 0.05 mm, which is perhaps the limiting value of the thickness of corrosion products we can detect by thermal neutron radiography. This is in agreement with the value of 0.051 mm previously reported by Dance (2). Hence, knowing the composition of corrosion product it is possible to determine the thining of Al-base metal in the range of about 0.02 - 0.03 mm.

APPLICATIONS

The neutron radiographic detection of surface corrosion on aluminium structures has been found useful in the maintenance of aircraft, especially those exposed to marine environments. As an example we present in Fig. 3 a neutron radiograph of an Al clad honeycomb structure, artificially aged to contain moisture and corrosion nests. Simultaneously it is possible to observe surface corrosion, air bubbles in the glue and the distribution of glue. With neutron radiography the sensitivity of detecting surface corrosion, moisture or defects in the glue distribution should not be impaired by the varying thickness of the Al structure.

460

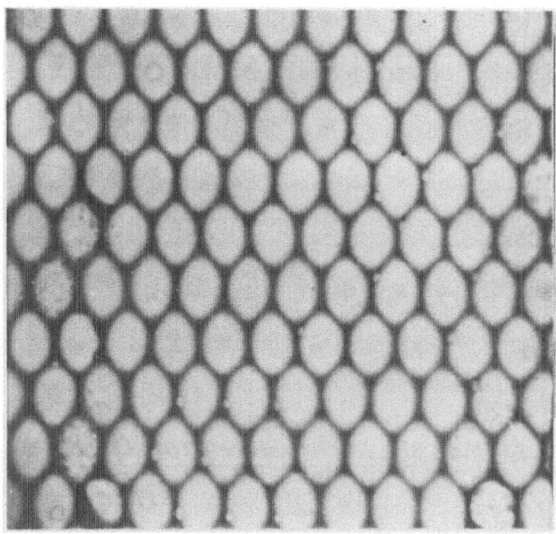

Fig. 3: Neutron radiograph of artificially aged aluminium clad honeycomb
structure

CONCLUSIONS

Results of our study can be summarized as follows:

i)   the minimal thickness of corrosion products one can neutronographically detect
       on aluminium depends on the amount of $Al(OH)_3$ and $AlO(OH)$ in their composition
       and in our case was found to be about 0.022 g/cm^2,

ii)   using previously calibrated measurements it is possible on neutron radiographs
       to estimate the thickness of corrosion products,

iii)   the inspection capability of neutron radiography to detect surface corrosion
       is practically almost insensitive to the variations in the thickness of Al
       base metal. Several cm thick Al structures can easily be inspected.

REFERENCES

1. J. John, "Californium Based Neutron Radiography for Corrosion Detection in Air-craft"; Practical Applications of Neutron Radiography and Gaging, ASTM STP 586, American Society for Testing and Materials, 1970, 168

2. W.E. Dance, "Neutron Radiographic Nondestructive Evaluation of Aerospace Struc-tures", Ibid., 137

3. J.John, "Neutron Radiography for Nondestructive Testing", Proc. Conf. "Nondes-tructive Evaluation of Materials", (Burke, J.J., Weiss V., Eds.), Plenum Press, New York, (1979), 151

4. Neutron Radiography Newsletters, "Neutron Radiography and Autoradiography at the J.Stefan Institute in Ljubljana", Brit. J. of NDT, 28, (2), (1986), 93

5. J. Rant et al., Proc. Nat. Conf. on Utilization of Nuclear Research Reactors in Yugoslavia, Belgrade (1978), 505–13

6. J. Rant, R. Ilić, G. Pregl, S. Demirović, Neutron Radiographic Detection of Sur-face Corrosion on Aluminium, Proc. 14th Symp. on Autoradiography, Reinhardsbrunn (GDR), Ed. H.C. Treutler, ZfI – Mitteilungen No. 100, (1985), 173

NEUTRON RADIOGRAPHY OF ALUMINIUM ALLOY CORROSION DAMAGE

A. Ridal
Australian Atomic Energy
 Commission
PMB, Sutherland NSW 2232
Australia

N.E. Ryan
Aeronautical Research
 Laboratories
Melbourne Vic 3001
Australia

ABSTRACT

The potential for neutron radiography to detect
hydrogenous corrosion products in aircraft structures
was examined using a series of naturally and artificially
corroded aluminium alloy samples. The sensitivity of
this technique for detecting corrosion damage was
intially assessed using the Moata reactor facility and
commercial neutron radiography services. Neutron
radiographs of the corroded samples were obtained using
a portable accelerator and a californium-252 isotope
source. Reactor, accelerator and isotope neutron
radiographs have been compared to assess the sensitivity
of portable neutron sources.

INTRODUCTION

The results of a collaborative program between the Australian
Atomic Energy Commission (AAEC) and the Aeronautical Research
Laboratories (ARL) to evaluate the use of neutron radiography as a
non destructive technique for the detection and assessment of
corroded aluminium alloys in aircraft structures are reported. A
series of corroded aluminium alloy specimens was supplied by ARL
for neutron radiography at the AAEC's Moata facility a 100 kW
Argonaut type reactor(1). The specimens represented various
aircraft structural configurations damaged by different types and
levels of corrosion.

An initial assessment of the sensitivity of neutron radiography for corrosion detection was performed using the Moata reactor. This evaluation was subsequently broadened to include assessments of neutron radiographs from thermal neutron reactor facilities and mobile sources in the USA.

CORRODED SPECIMENS

The corroded aluminium alloy specimens supplied by ARL consisted of simple laboratory specimens containing surface damage and specially assembled specimens in which significant corrosion damage was hidden. The latter specimens were designed to simulate typical aircraft structure damage. Apart from the hidden corrosion, the paint film and sealant could have neutron absorbing features likely to cause changes in image definition during the neutron radiography of aircraft structures. Details of the specimens are given in Table 1.

TABLE 1
Corrosion Specimen Details

SPECIMEN	TYPE OF CORROSION
Standard	Contained 4 recesses filled with Al(OH)$_3$ gel. The gel was obtained from corrosion products formed on 7075 aluminium alloy in a sodium chloride environment. Recesses were 12.5 mm diameter and 0.62, 0.38, 0.25 and 0.12 mm deep. The assembly was sealed to retain the moisture.
A1, A2, A3	Artificially induced surface corrosion, stimulated using standard "EXCO" solution in accordance with ASTM G-34-72.
N1, N2, N3	Naturally occurring exfoliation corrosion, the corrosion product being similar to that formed by "EXCO" solution.
Exfoliation	Hidden example of significant naturally occurring exfoliation corrosion.
Surface	Hidden example of surface corrosion, induced by "EXCO" solution.

FACILITIES

Thermal Neutron Reactors

The operation of two research reactor facilities and two organisations offering commercial neutron radiography participated

464

in the evaluation program. Table 2 lists the reactor facilities,
together with their preferred neutron radiographic technique for
the supplied corrosion specimens. Gadolinium conversion screens
were used by all the facilities.

TABLE 2
Thermal Reactor Facilities & Optimum Radiographic Technique

FACILITY	POWER LEVEL (kW)	L/D RATIO	FILM	EXPOSURE (mins)	FLUX STRENGTH (n $cm^{-2}s^{-1}$)
Moata	10	100	Agfa-D4	8	1.4×10^6
Georgia Tech	15	N/A	Fuji M1 (Sc)	1	5×10^7
General Electric Co	100	179	R Kodak (Sc)	4	4×10^6
Aerotest	250	N/A	SR5	2.9	9.9×10^6

The quality of the neutron radiographs, especially those
provided by General Electric Co., PO Box 460, Pleasanton, CA 94566,
USA and Aerotest Operations, F&C, San Roman, CA 94583, USA, was
excellent, an example being shown in Fig 1.

Californium Multiplier

The californium multiplier (CFX) is a sub-critical assembly of
enriched uranium surrounding a californium-252 neutron source; this
enabled the californium-252 thermal neutron flux to be increased by
a factor of 30 (2,3). The neutron radiographs were produced at the
Mound Facility CFX, courtesy of the Monsanto Corp., USA, using a
californium source strength of 19.6 mg (1.5×10^5 n $cm^{-2}s^{-1}$).
Neutron radiographs using different films, L/D ratios and exposure
times were evaluated; all the radiographs were comparable to those
obtained from the thermal reactor facilities. The optimum
radiographic technique was L/D ratio 103, film SR, exposure time
20 h; the radiograph is shown in Fig. 2.

Mobile Sources

Neutron radiographs of the corroded specimens were taken using
a mobile californium-252 radiography facility and the Vought mobile
accelerator neutron radiography system (4), both provided by the US
Army Materials and Mechanics Research Center. Table 3 gives the
supplied data.

TABLE 3
Mobile Neutron Radiographic Details

FACILITY	L/D	FILM	CONVERSION SCREEN	EXPOSURE min	n cm^{-2}
^{252}Cf. (10.8 mg)	25	Kodak AA	Evaporated Gd	539	2×10^8
Accelerator	27	Kodak	Evaporated Gd	150	1.5×10^8
Accelerator	27	Kodak SB-S	G.O.S.	30	3.05×10^7

The quality of the californium-252 source radiograph was slightly inferior to those from the thermal neutron reactor and CFX facilities. Despite poor resolution and contrast, corrosion was detected in all the test specimens (Fig. 3). The radiographs taken by the accelerator source were inferior to those of the californium-252 source.

RESULTS

Neutron radiographs were obtained from thermal neutron reactor, californium-252 multiplier, isotope and accelerator sources of all the corroded aluminium alloy specimens. All the radiographs were able to image surface and subsurface corrosion in the specimens, but the thermal neutron reactor and californium multiplier sources provided better resolution and contrast.

A standard or control specimen was used to determine the sensitivity for imaging corrosion products. Neutron radiographs from all facilities showed that the corrosion product absorbed neutrons to a depth of 0.12 mm. Preliminary experiments on surface corrosion test specimens yielded imaged corrosion products between 0.015 and 0.050 mm in depth.

To establish sensitivity for the detection of exfoliation corrosion, specimens N1 and N4 were examined metallographically through a cross section of the corroded area, (Fig. 4); N4 was examined at the minimum level of detection and N1 at intermediate levels. In each case, the neutron radiographs detected the full extent of the exfoliation corrosion damage and, as indicated in specimen N4, it was relatively low, the detected area being 2.5 mm wide and 0.15 mm deep. Areas of corrosion in specimen N1 were 10 mm wide by 1.1 mm deep and 12.5 mm wide by 0.55 mm deep, respectively. The only neutron radiograph which did not image the corrosion was obtained with the mobile accelerator source.

DISCUSSION

Comparison of the neutron radiographs from the different facilities confirmed, as was expected, that the best resolutions were obtained from fixed facilities, e.g. thermal neutron reactors and californium-252 multipliers. In these cases, the collimation ratios (L/D) were considerably higher than those of the mobile sources (≥ 100 compared to ≤ 27). Cutforth(5), stated that the collimation ratio (which affects geometric resolution) should be 10 to give useful resolution, but collimator L/D ratios of 50 or greater are recommended for most practical applications.

The most important result of this evaluation program was that neutron radiography again demonstrated its ability to detect corrosion in aluminium alloys. Although the fixed facilities gave the best results, the mobile systems are of greatest interest for detecting aircraft corrosion damage since it is not feasible to bring the aircraft to the reactor and very few structural components in an aircraft are immediately detachable. It was established that both the radioisotope and the accelerator sources can detect low levels of corrosion.

Apart from health, safety and legal constraints on the use of radioactive materials, the use of californium-252 as a neutron source has a number of disadvantages, a major one being its high cost (>$10,000 per mg), Hilditch and Chrimes(6), estimated the cost of a mobile californium-252 source for operation in Australia at $US $\sim 10^6$. Furthermore, the high level of operating costs is augmented by the need to replace the californium-252 (2.64 years half-life) source at frequent intervals. A standard neutron radiography system using californium-252 is too heavy to be used as a convenient portable source, but a truly portable system consisting of the source, moderator and thermal neutron collimator and weighing one tenth of a fixed system has been shown to be feasible when remotely operated(7).

Although the neutron radiographs from the on-off mobile accelerator were slightly disappointing, the attractiveness to the system is that the neutron emission can be generated and halted at will. Which makes it more suitable in Australian working conditions where the work load is not sufficient for an isotope source which, to be cost-effective, needs to be used 24 hours a day. Second, the cost of an accelerator unit is approximately half that of a californium-252 system, i.e. $US550,000(8).

The evaluation tests were performed over two years. In assessment of these results, it is important to note that subsequent improvements, particularly in mobile neutron sources (9, 10), may modify the conclusions of similar experiments.

ACKNOWLEDGEMENTS

The authors are indebted to Aerotest Operations Inc., General Electric Co., Neely Nuclear Research Center and the US Army Materials and Mechanics Research Center for their participation in the program.

REFERENCES

1. P.A. Gillespie, P.A. and T. Wall, "Neutron Radiography at Lucas Heights", Neutron Radiography, Proceedings 1st World Conf. San Diego, California USA, Dec 7-10, 1981, Ed. J.P. Barton, pp. 85-92.

2. K.L. Crosbie, C.A. Prestith, J. John, and J.D. Hastings, "Californium Multiplier Part I, Design for Neutron Radiography", Materials Evaluation, Vol 40 April 1982, pp. 579-583.

3. J.D. Hastings, K.L. Crosbie, C.A. Preskith, and J. John, "Californium Multiplier Part II: Performance of the Mound System", Materials Evaluation Vol. 40, April 1982 pp. 584-589.

4. W.E. Dance, and S.F. Carollo, "AMMRC Mobile Accelerator Neutron Radiography System Operations at US Army Yuma Proving Ground" Interim Technical Report No 3-41000/4R-110, 15 April 1984.

5. D.C. Cutforth, "Neutron Sources for Radiography and Gaging" ASTM Special Technical Report STP 586, 1976 pp. 20-34.

6. R.J. Hilditch, and N.W.D. Chrimes, "AAEC Private Communication, 1983".

7. J. John, "Mobile Neutron Radiography System for Aircraft Inspection", IRT Corp. Technical Report RT-TB-151, 1973.

8. J.J. Antal, "US Army Materials and Mechanical Research Center, Private Communication, March 1985".

9. W.E. Dance, S.F. Carollo, J.J. Antal, & J.P. Moravec, "Experience with an off-on Mobile Neutron Radiography System", 2nd World Conference on Neutron Radiography, 16-20 June 1986, Paris France.

10. V. Orphan, D. Redem and F. Johansen, "Mobile Neutron Radiography System for Aircraft Inspection", 2nd World Conference on Neutron Radiography, 16-20 June 1986, Paris, France.

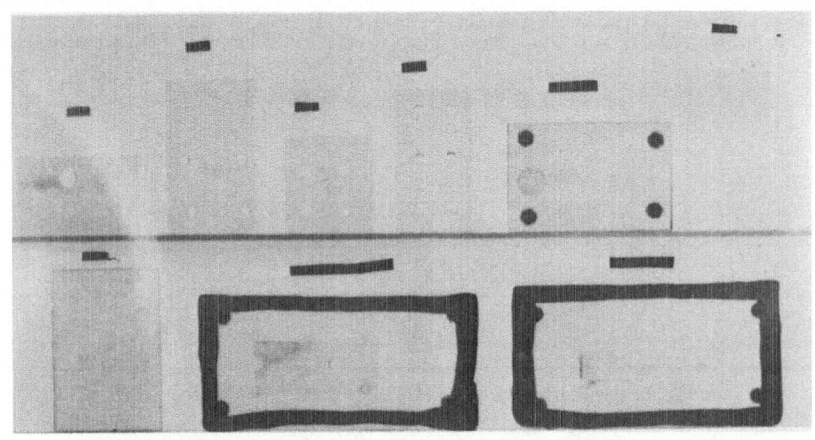

Figure 1

Thermal Reactor Neutron Radiograph

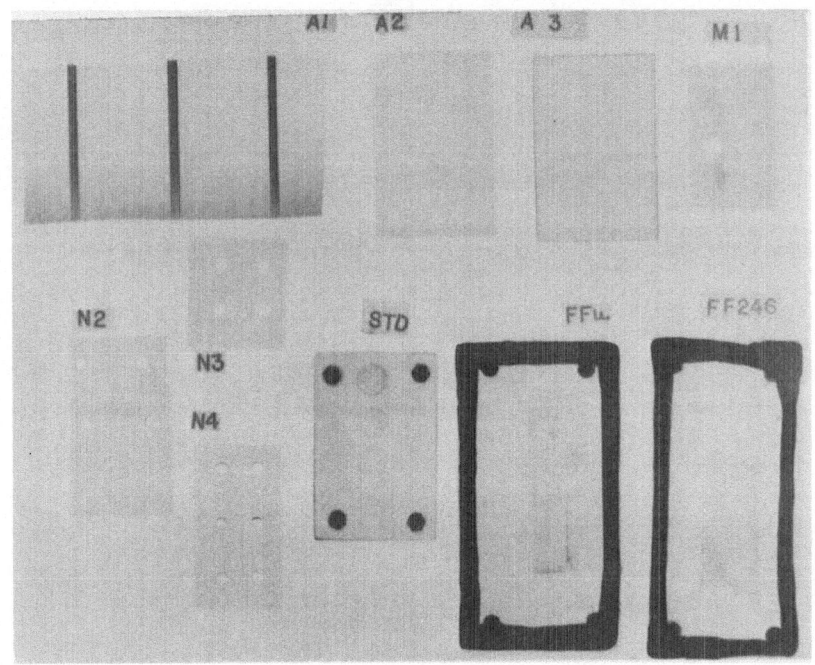

Figure 2

Californium-252 Multiplier (CFX) Neutron Radiograph

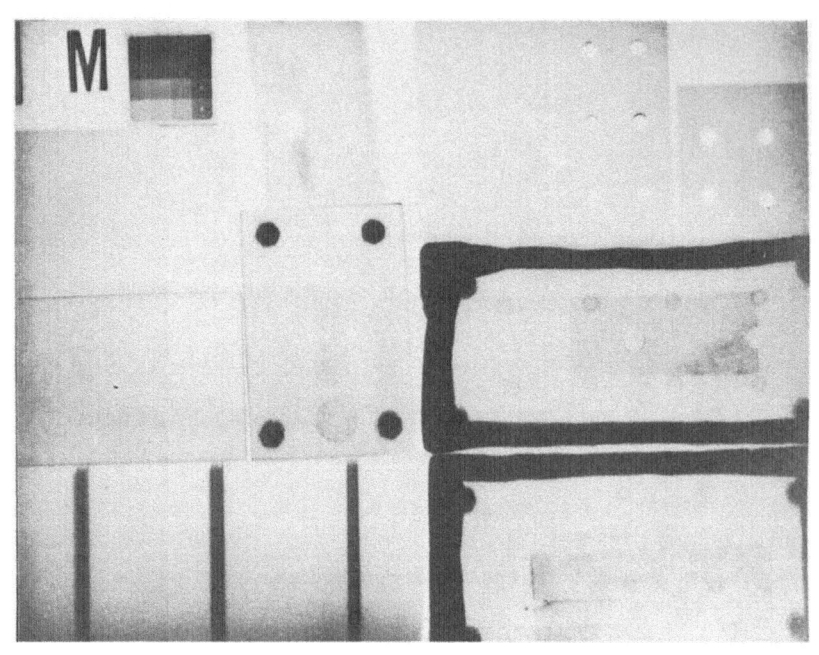

Figure 3

Californium-252 Isotópe Neutron Radiograph

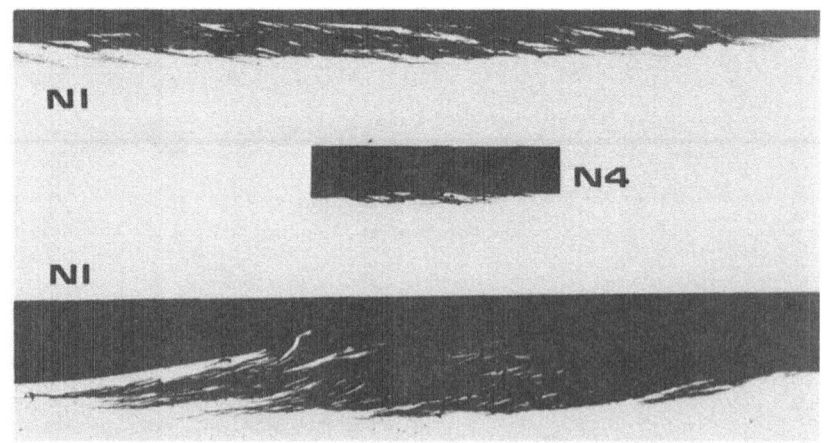

Figure 4

Cross Sections specimens N1 and N4, mag. x 12

PART VII

SPECIAL APPLICATIONS

NEUTRON RADIOGRAPHIC INSPECTION OF METAL-TRITIUM-HELIUM-SYSTEMS[*]

T.Buchberger, H.Rauch, E.Seidl

Atominstitut der Österreichischen Universitäten

A-1020 Wien, Austria

ABSTRACT

In this investigation neutron radiography is used as a method which is highly sensitive to even small quantities of ^{3}He. A tantalum sample, charged with 12.74 at% tritium has been observed for more than 2 years. Now the helium concentration, resulting from radioactive decay of tritium is at about 1.5 at%. The time dependent beam attenuation has been measured and compared with values derived from theoretical considerations. A new conical collimator for high resolution neutron radiographic measurements has been installed to observe the enrichment of helium and so-called helium bubbles at or near to grain boundaries.

INTRODUCTION

With the progress of nuclear technology a comprehensive knowledge about the behaviour of He in solids has become increasingly important. Significant quantities of He will be produced in spallation sources by an enhanced number of (p,α)-reactions and in structural components or in fuel of future fusion reactors and fast breeder reactors by (n,α)-reactions or by direct α-injection.

[*] This work has been supported by Bundesministerium für Wissenschaft und Forschung (project "Tritium in Metals").

^{3}He will be produced in tritium storage and breeder materials by tritium decay ($^3_1T \rightarrow ^3_2He + e^- + \bar{\nu}_e$).

As He is practically insoluble in solids it strongly tends to precipitate into bubbles. Especially the formation of these bubbles and their enrichment at grain boundaries are considered to be the live-time limiting factor for many materials (1-3).

For investigations the specimens have to be charged with He in a controlled quantity. As it is not soluble in solids it has to be introduced into a metal either by α-implantation or by the "tritium trick". The latter utilizies the fact, that most metals easily absorb high quantities of hydrogen and its isotopes. So a sample can be charged with T which decays into ^{3}He with a half-life period of 12.3 y. Using this method He can be introduced into metals without significant damage to the host lattice. Due to the relatively long half life period one has to wait considerable time until sufficient amounts of ^{3}He are produced. The tritium trick has the additional advantage, that ^{3}He has the extremely high absorption cross section for thermal neutrons of $\sigma_a = 5333$ b, which gives neutron radiography a fair chance to detect very small quantities of ^{3}He.

The present work is a continuation of previous neutron radiographic investigations of metal-hydrogen- and metal-deuterium-systems (4-6) and supplements other neutronic inspection methods of metal-hydrogen systems (7,8).

PROPERTIES OF HELIUM IN METALS

He bubbles are presumed to be formed athermally by a self-trapping mechanism of interstitial He atoms. When small numbers of He atoms (5 in a fcc metal, 6 in a bcc metal) cluster together, the resulting force is sufficient to displace a lattice atom spontaneously, thereby creating a Frenkel pair and a deeply bound He cluster. Prefered nucleation centers are solid precipitates, dislocations, grain boundaries, grain boundary junctions and surfaces. During further growth of the bubble matter may be transported away either by emission of self interstitials or by the emission of prismatic dislocation loops. The morphology of such bubbles ranges from disks or platalets to spherical shapes (9). The resulting pressures in the bubbles are extremely high and reach values of several GPa. Under such pressures He can be considered as liquid or even solid. It has been observed that upon reaching a critical pressure bubbles and platalets divide into a cluster of several bubbles (10). The diameters of He bubbles range from 1 to 100 nm, but typical values are 1 to 10 nm. In general the bubble size increases at grain boundaries or near surfaces, which was found in a small angle neutron scattering (SANS) and an

Figure 1: Tantalum sample showing the position of the grain-
boundaries.

electron microscopy (TEM) study with a nickel sample implanted
with He (11). For metals having a high degree of crystalline perfection within the grains the formation of gas bubble super-
lattices has been observed (12).

SAMPLE PREPARATION

To study phenomena at grain boundaries with neutron radiography a sample with macroscopic grains was needed. This was
achieved by secondary recrystallisation of polycrystalline Ta
(MARZ - grade, purity 99.996%) under UHV and temperatures above
2000 $^\circ$C. By cutting, grinding and polishing we got two samples
with the dimension of 40x8x1 mm^3 with very similar grain-structure.
One of these samples was designated to be charged with tritium,
the other serves as constant reference. Figure 1 presents one of
these samples. Tritium loading was performed by NUKEM in Hanau,
BRD. A tritium concentration of 12.74 at% (at%=$N_{Tritium}/N_{Ta}$) was
reached. At this concentration the metal-tritium system remains in
the α-phase at 300 K, that means that T is distributed uniformly
in the specimen, which is important for the following considerations.

NEUTRON RADIOGRAPHIC FACILITIES AT THE VIENNA TRIGA REACTOR

At the Vienna TRIGA MARK II reactor, which has a power of 250
kW (thermal), two different facilities for neutron radiography
have been constructed. Both use collimators of conical geometry
for beam collimation. One is designed mainly for industrial applications and, therefore, has the relative large beam diameter of
40 cm, thus allowing the inspection of very large objects. The
other facility is intended as a high resolution system for micro-
neutronogaphy. Figure 2 shows the general arrangement of the facility which is used for the neutron radiographic work described
in this paper. Table 1 presents some characteristic data of these
two facilities.

Facility	Thermal Column	Experim. Tank
Neutron fux ($cm^{-2}s^{-1}$)	1.5×10^5	3.0×10^5
Cd-ratio	20	3
L/D-ratio	125	45
γ-background (r/h)	2	10
Beam diameter (cm)	9	40
Typical exposure time		
Structurix D7 (min)	20	10
Structurix D4 (min)	80	40
γ-filter	4 cm Bi	10 cm Bi
	polycrystalline	single-crystal
Inlet aperture		
diameter (cm)	2	5
material	3x5 mm Pb	2x40 mm Pb
	3x5 mm B_4C	4x5 mm B_4C
		6x.5 mm Cd
Collimator lining	B_4C	B_4C

Table 1: Characteristic data of the neutron radiographic facilities at the Vienna TRIGA reactor.

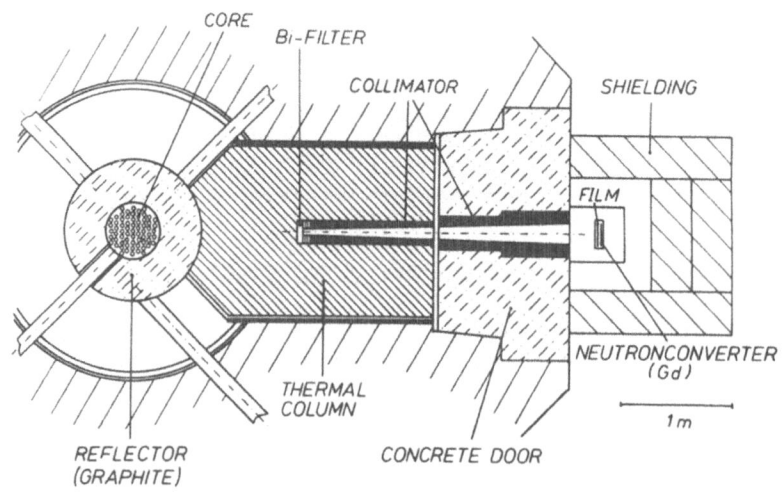

Figure 2: Conical collimator for neutron radiographic investigations at the thermal column of the TRIGA reactor.

EXPERIMENTAL PROCEDURE

All neutronographs presented were taken at the facility mentioned above (Figure 2). The usable beam diameter at the position of the specimen was reduced to 6 cm to avoid any additional increase of scattered neutrons. The homogeneity of the neutron beam is better than 2% and this effect has been corrected by means of an appropriate computer procedure. The neutron flux at the center of the beam is 1.5×10^5 $cm^{-2}s^{-1}$. The geometrical resolution of our experimental setup is about 10 μm.

As neutron converter gadolinium foils with a thickness of 25 μm are used. The films used are Strukturix D7 and Strukturix D4 from Agfa Gevaert, a conventional double sided X-ray film. The necessary exposure time to achieve a film density of 1.5 is about 5 min for D7 and 1 hour for D4. Films are developed in a half automatic processing system at constant temperature ($\Delta T = 0.1$ $^{\circ}C$). Film densities are measured with a scanning microdensitometer from Gamma Scientific (model 2800) contolled by a Micro PDP-11. Recently the neutronographs were also digitized on a Perkin-Elmer microdensitometer (model PDS 1010A), which was connected to a VAX via a microcomputer.

RESULTS

Due to tritium decay the concentration of ^3He in the specimen increases with time corresponding to

$$c_{He}(t) = c_T(0) - c_T(t) = c_T(0) (1 - e^{-\lambda t}) \qquad (1)$$

where $\quad c_{He}$ = ^3He concentration
$c_T(0)$ = T concentration immediately after T loading
$c_T(t)$ = T concentration at the time t after T loading
λ = decay constant of T ($\lambda = 0.1779 \times 10^{-8} s^{-1}$)

Assuming as a first approximation that ^3He is distributed uniformly in the specimen (single He bubbles are too small to be resolved), the beam attenuation in the sample can be calculated :

$$A(t) = 1 - e^{-(n_{Ta}\sigma_{Ta} + n_{He}(t)\sigma_{He} + n_T(t)\sigma_T)d} \qquad (2)$$

with

$$n_{He}(t) = n_{Ta}c_{He}(t) \quad \text{and} \quad n_T(t) = n_{Ta}(c_T(0) - c_{He}(t)) \qquad (3)$$

where $\quad$ n = atomic density of Ta, ^3He or T respectively (cm^{-3})
σ = microscopic cross-section of Ta, ^3He or T (cm^2)
d = thickness of specimen (cm)

The visibility of ^{3}He results from its large attenuation cross-section (σ = 5333 b) whereas the attenuation cross-sections for Ta (σ = 26.6 b) and for T (σ = 3.1 b) are rather small.

For the reference sample, which does not contain ^{3}He, equation (2) simplifies to

$$A = 1 - e^{-n_{Ta}\sigma_{Ta}d} \tag{4}$$

where A is no longer time-dependent.

Figure 3 presents the measured values of the absorptipon A(%) of the ^{3}He containing sample and the pure sample as a function of the aging time t as well as the functions derived from theoretical considerations (continuous lines).

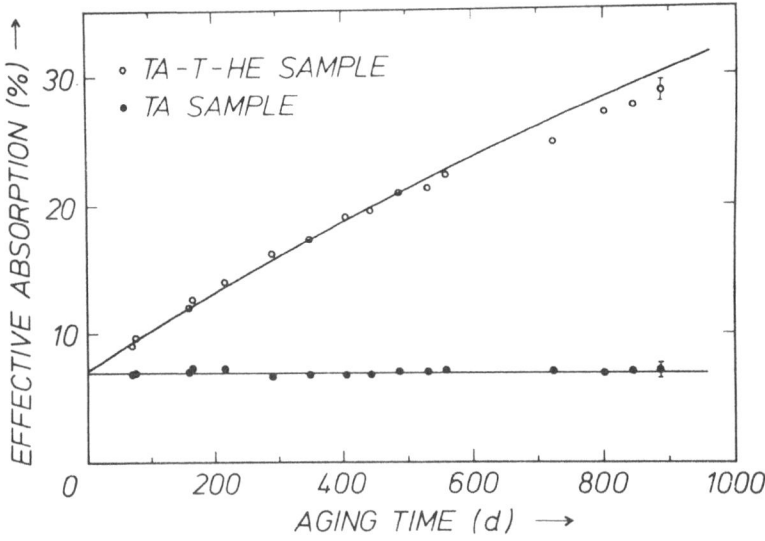

Figure 3: Effective absorption A (%) of a ^{3}He containing Ta sample and a pure Ta sample as a function of the aging time t.

Figure 4: Neutronograph of a 1.56 at% ^{3}He containing Ta sample
(above), a pure Ta sample (below) and a silver stair
(middle).

Figure 4 shows a neutronograph of the two samples at about
850 days after T charging. The corresponding ^{3}He concentration is
1.56 at%. The object in the middle is an Ag stair used to check
film quality and processing conditions. The sensitivity of ^{3}He
detection in our 1 mm thick samples is better than 400 ppm.

The expected enrichment of He bubbles at grain-boundaries has
not been proven at this time. Recently an analysis of the radiographs was performed with a high precision scanning microdensitometer, which has a resolution of $1.\mu m$. The noise of the film
density values is reduced by the application of appropriate image
processing techniques (Fourier filter, spline filter, maximum entropy method). The results of this method have not been analysed
completely and, therefore, no final decision can be made whether
an accumulation of He-bubbles exists or whether it is not detectable by neutron radiographic methods. Therefore, parallel neutron
small angle scattering experiments are in progress.

ACKNOWLEDGEMENTS

The authors wish to thank Doz.W.Weiss and Dr.J.Hron from the
Institute for Astronomy in Vienna for their assistence in the
densitometric measurements.

REFERENCES

1. H.Ullmaier, "Helium in Metals", Rad.Effects 78 (1983) 1-10.

2. H.Schröder, "High Temperatur Embrittlement of Metals by
    Helium", Rad.Effects 78 (1983) 297-314.

479

3. H.Ullmaier, "The Influence of Helium on the Bulk Properties of Fusion Reactor Structural Materials", Nucl.Fusion 24 (1984) 1039-1083.

4. K.Chountas, H.Rauch, "Neutronenradiographie von Metallklebungen, Legierungen, aktiven Brennelementen, Diffusion von H in Zr und Diffusion $H_2O - D_2O$", Atomkernenergie 13 (1968) 444.

5. A.Zeilinger, W.Pochmann, "A New Method for the Measurement of Hydrogen Diffusion in Metals", J.Appl.Phys 47 (1976) 5478.

6. H.Rauch, A.Zeilinger, "Hydrogen Transport Studies Using Neutron Radiography", Atomic Energy Review 15 (1977) 249.

7. H.Rauch, E.Seidl, A.Zeilinger, W.Bauspiess, U.Bonse, "Hydrogen Detection in Metals by Neutron Interferometry", J.Appl.Phys.49 (1978) 2731.

8. A.Miksovsky, H.Rauch, E.Seidl, W.Wachter, "Order-Disorder Transition in the Vanadium-Hydrogen System Measured by Perfect-Crystal Neutron Small-Angle Scattering", Z.Naturforschung 39a (1984) 1077.

9. J.H.Evans, A.van Veen, L.M.Caspers, "In-situ TEM Observations of Loop Punching from Helium Platalet Cavities in Molybdenum", Scripta Met. 17 (1983) 549-553.

10. M.W.Finnis, A.van Veen, L.M.Caspers, "The Energy of Helium Platalets and Bubbles in Molybdenum", Rad.Effects 78 (1983) 121-132.

11. W.Kesternich, D.Schwahn, H.Ullmaier, "Size of He Bubbles in Bulk, Grain Boundaries and Regions Near the Surface of Nickel" Scripta Met. 18 (1984) 1011.

12. P.B.Johnson, D.J.Mazey, J.H.Evans, "Bubble Structures in He [+] Irradiated Metals", Rad.Effects 78 (1983) 147-156.

SONIC ROOT CANAL INSTRUMENTS EFFECTIVENESS

STUDY WITH NEUTRON-RADIOGRAPHY AND MICRODENSITOMETRY

J.M. VULCAIN[*], A. PILVEN[*], J.H. ESPIE[*], G. BAYON[**]

[*] Laboratoire de biologie buccale, université Rennes I

[**] Laboratoire du C.N.D., C.E.N., Saclay, C.E.A.

ABSTRACT

This experiment shows the use of neutron-radiographic and microdensitometric analysis for evaluation sonic instruments efficiency (for recall sonic waves are 1500 Hz).

Neutron-radiography is first a use for viewing marked predentin with Gd-Cit. and secondly for controlling in a preservative and reproducible way the instrument action on parietal material of the endodont.

Microdensitometric analysis from neutron-radiography evaluates the instrumental action by quantifying the neutron binding change which corresponds to predentin tracing.

INTRODUCTION

As in high technic industry, probing materials is a need (3) in medecine it is the same. Pulpectomy is a surgery act which consists in the removal of the pulp and the parietal predentin along side the root canal with specialized instruments. So it seems important that operators know in which way these instruments act (4).

In 1981, in San Diego, we introduced neutron-radiography and microdensitometry for analysing instrumental behaviour during endodontics (8).

Then we improved our technics and by now our target is a standardized method which role is the systematic and reproducible control of efficiency of root canal instruments (9).

We propose this method through the study of three sonic instruments, RISPI-SONIC, SHAPER, HELI-SONIC. During their use, these instruments receive waves from 1500 to 3000 Hz with a pneumatic endodontic contra angle from a new type MM 3000 Sonic-air (5-7).

MATERIAL and METHODS

- 30 human teeth, avulsed for orthodontics are displayed in 3 batches.
- Each sample is refered and an endodontic treatment with a nerve broache from a smaller diameter than the mesio-distal diameter of root canal is done.
- Each sample is then conditionned with Isomet Buerler III for spotting each time in a reproducible way in a positionning ruler. This preparation is made under physiologic serum irrigation to prevent heating (8) (fig. 1).

First manipulation

Each sample in a batch is conduct through an initial neutron-radiography refered N 0. This one is get by a neutron flux of $3 \times 10^{14} n.cm^{-2}s^{-1}$ from $< 0,025$ ev energy given with the "Orphee" nuclear reactor which power provides 14 MW (3).

Neutronic image is capted on an industrex R Kodak film fitted under vacuum in a closed neutron-radiographic tape.

This one contains a neutron convector (fig. 1).

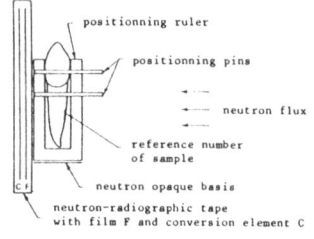

fig. 1

N 0 neutron-radiographies are analysed with JOYCE LOEBL MARK III CS microdensitometer using on XY axe apical 1/3 or medium 1/3 root canal.

Parameters are : lens x 10 ; microdensitometric spot C 565 ; additionnal filter of D.O. x 1,5 D.O. ; analysis ray 250 x 40 μm ; scale of D.O. 0,007u D.O./mm ; recording scale x 10.

So each N 0 neutron-radiography corresponds to a M 0 microdensitometric analysis.

Second manipulation

Each predentin sample is marked with a rare earth from

lanthans serie. These elements can fuse through non mineralized tissues and complex with glyco-amino-glycans (G.A.G.) (1-6). They can be adsorbed on mineralized tissues and those in the process of mineralisation (6). An other property is to stop very easily neutrons (2).

After tracing, for each sample we have an other neutron-radiography N 1 and a new microdensitometric analysis M 1. These are done on the same axe XY.

Third manipulation

The three batches of samples are treated with Rispi-Sonic, Shaper and Heli-Sonic during 30, 60, 90 and 180 sec. The sampling half is treated with physiologic serum simultaneous irrigation.

This is done in respect of clinical conditions. After each instrument period, neutron-radiographies N 30, N 60, N 90 and N 180, microdensitometric analysis M 30, M 60, M 90 and M 180 are done.

ANALYSIS

N 0 and N 1 give the tracing quality.

Superposition of M 0 and M 1 gives a S 1 surface in mm^2 which indicates neutron-binding of tracing. And then it gives a volumetric evaluation of predentin on XY axe.

N 1-N 30, N 1-N 60, N 1-N 90, N 1- N 180 give informations on tracing changes after 30, 60, 90 and 180 sec. of instrument use.

M 1-M 30 superposition induces a S 30 surface in mm^2 which shows the decrease or the disparition of neutrono-binding after 30 sec. of instrument use. This quantifying corresponds to the volume change of predentin following XY axe.

It is the same for M 1-M 60, M 1- M 90, M 1-M 180, for 60, 90 and 180 sec. of instrument use.

Superpositions M 0-M 30, M 0- M 60, M 0-M 90, M 0-M 180 show if parietal removal is homothetic all around canal pit.

Removal surfaces report S 30, S 60, S 90, S 180 on S 1 surface gives efficiency coefficient of improved instrument with time. So for...

$$30 \text{ sec.} \longrightarrow C\ 30 = \Sigma S\ 30\ /\ \Sigma S\ 1$$
$$60 \text{ sec.} \longrightarrow C\ 60 = \Sigma S\ 60\ /\ \Sigma S\ 1$$

idem for 90 and 180 sec.

Σ S is the surface sum determined for a batch of samples.

A graph establishes the instrument behaviour on XY axe as these different coefficients involve.

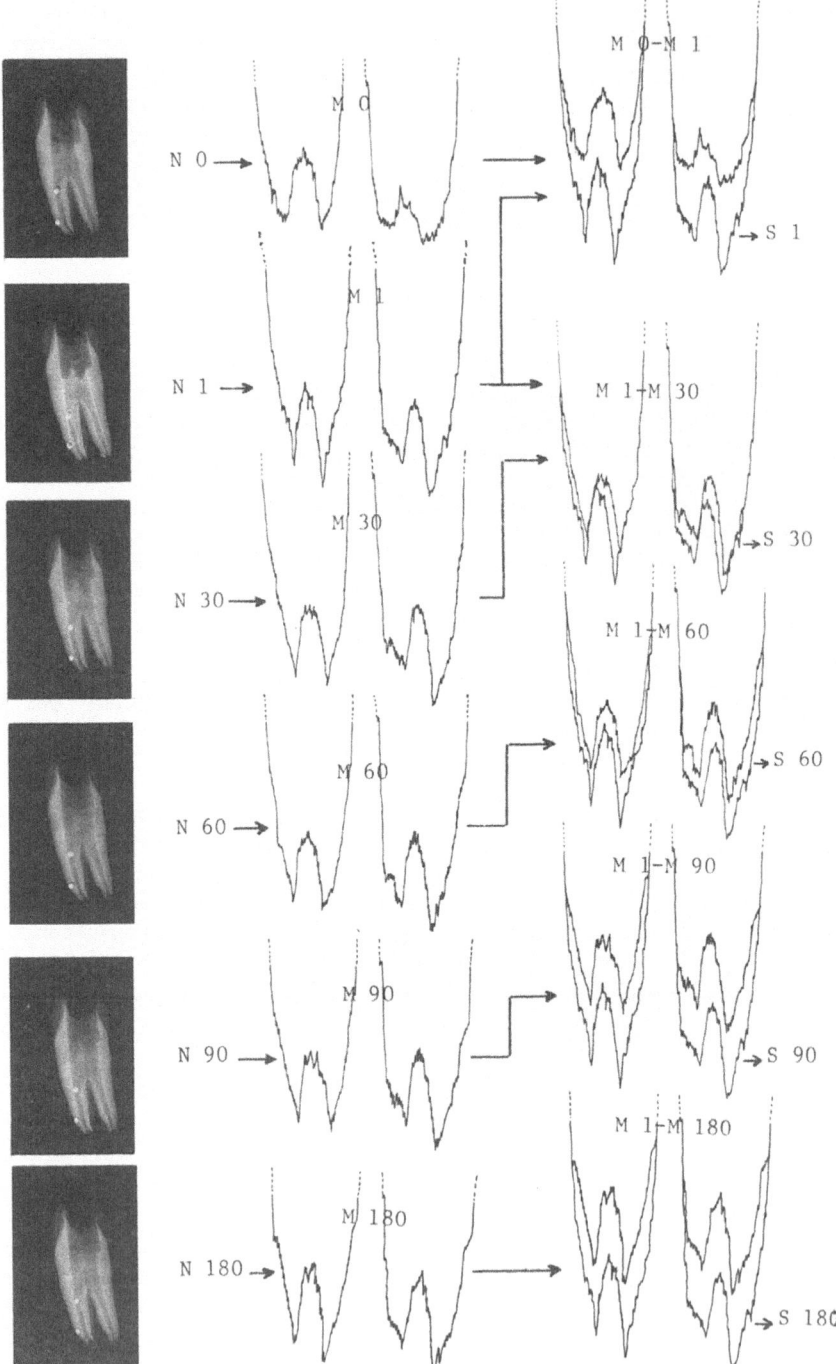

This illustration is printed from the best available original

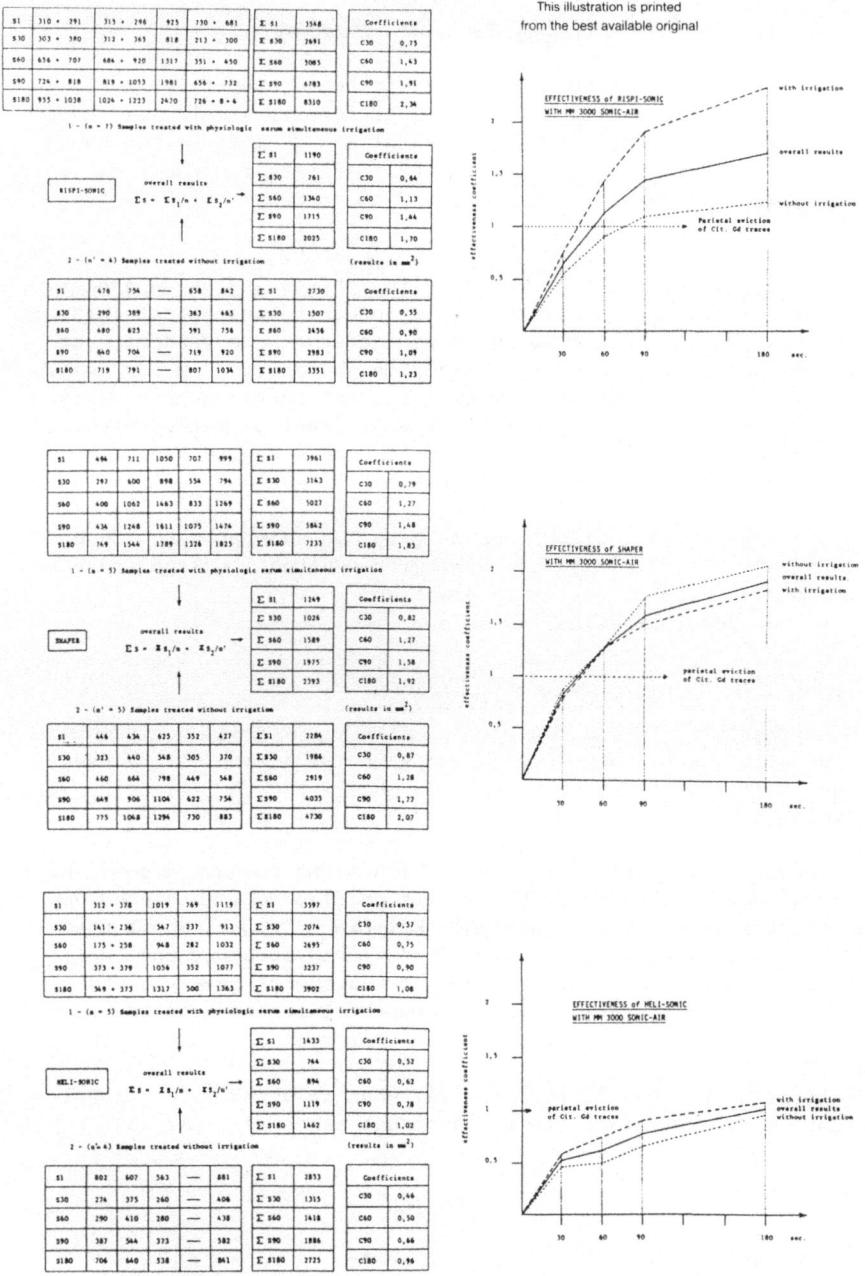

RISPI-SONIC

S1	310 + 291	315 + 296	925	730 + 681		Σ S1	3548		Coefficients	
S30	303 + 390	313 + 363	818	213 + 300		Σ S30	2691		C30	0,75
S60	656 + 707	684 + 920	1317	351 + 450		Σ S60	5085		C60	1,43
S90	724 + 818	819 + 1053	1981	656 + 732		Σ S90	6783		C90	1,91
S180	955 + 1038	1024 + 1223	2470	726 + 8 + 4		Σ S180	8310		C180	2,34

1 - (n = 7) Samples treated with physiologic serum simultaneous irrigation

overall results
$$\Sigma s = \Sigma s_1/n + \Sigma s_2/n'$$

	Σ S1	1190		Coefficients	
	Σ S30	761		C30	0,64
	Σ S60	1340		C60	1,13
	Σ S90	1715		C90	1,44
	Σ S180	2025		C180	1,70

2 - (n' = 4) Samples treated without irrigation   (results in mm²)

S1	476	754	—	658	842		Σ S1	2730		Coefficients	
S30	290	389	—	363	465		Σ S30	1507		C30	0,55
S60	480	625	—	591	756		Σ S60	2456		C60	0,90
S90	640	704	—	719	920		Σ S90	2983		C90	1,09
S180	719	791	—	807	1034		Σ S180	3351		C180	1,23

EFFECTIVENESS of RISPI-SONIC WITH MM 3000 SONIC-AIR
— with irrigation
— overall results
— without irrigation
Parietal eviction of Cit. Gd traces

SHAPER

S1	494	711	1050	707	999		Σ S1	3961		Coefficients	
S30	297	600	898	554	794		Σ S30	3143		C30	0,79
S60	400	1062	1463	833	1269		Σ S60	5027		C60	1,27
S90	434	1248	1611	1075	1474		Σ S90	5842		C90	1,48
S180	749	1544	1789	1326	1825		Σ S180	7233		C180	1,83

1 - (n = 5) Samples treated with physiologic serum simultaneous irrigation

overall results
$$\Sigma s = \Sigma s_1/n + \Sigma s_2/n'$$

	Σ S1	1249		Coefficients	
	Σ S30	1026		C30	0,82
	Σ S60	1589		C60	1,27
	Σ S90	1975		C90	1,58
	Σ S180	2393		C180	1,92

2 - (n' = 5) Samples treated without irrigation   (results in mm²)

S1	446	434	625	352	427		Σ S1	2284		Coefficients	
S30	323	440	548	305	370		Σ S30	1986		C30	0,87
S60	460	664	798	449	548		Σ S60	2919		C60	1,28
S90	649	906	1104	622	754		Σ S90	4035		C90	1,77
S180	775	1068	1294	730	883		Σ S180	4730		C180	2,07

EFFECTIVENESS of SHAPER WITH MM 3000 SONIC-AIR
— without irrigation
— overall results
— with irrigation
parietal eviction of Cit. Gd traces

HELI-SONIC

S1	312 + 378	1019	769	1119		Σ S1	3597		Coefficients	
S30	141 + 236	547	237	913		Σ S30	2074		C30	0,57
S60	175 + 258	948	282	1032		Σ S60	2695		C60	0,75
S90	373 + 379	1056	352	1077		Σ S90	3237		C90	0,90
S180	549 + 373	1317	500	1363		Σ S180	3902		C180	1,08

1 - (n = 5) Samples treated with physiologic serum simultaneous irrigation

overall results
$$\Sigma s = \Sigma s_1/n + \Sigma s_2/n'$$

	Σ S1	1433		Coefficients	
	Σ S30	744		C30	0,52
	Σ S60	894		C60	0,62
	Σ S90	1119		C90	0,78
	Σ S180	1462		C180	1,02

2 - (n'= 4) Samples treated without irrigation   (results in mm²)

S1	802	607	563	—	881		Σ S1	2853		Coefficients	
S30	274	375	260	—	406		Σ S30	1315		C30	0,46
S60	290	410	280	—	438		Σ S60	1418		C60	0,50
S90	387	544	373	—	582		Σ S90	1886		C90	0,66
S180	706	640	538	—	841		Σ S180	2725		C180	0,96

EFFECTIVENESS of HELI-SONIC WITH MM 3000 SONIC-AIR
parietal eviction of Cit. Gd traces
— with irrigation
— overall results
— without irrigation

485

DISCUSSION

Methods

1 - Tracing : Although sampling has been conduct in an homogeneous way, tracing quality was not equal for all samples and some of them have been refused. The old samplings seem to be the cause of these poor tracings. For getting a standardized method, the very first target is the optimistic tracing of predentin with neutron marker.

2 - Microdensitometric analysis is interesting on an arbitrary axe XY after tracing on a level of second third and apical third of the sample. May be this probing seems inadequate to readers, but microdensitometric care of each analysis shows very similar response for a special batch. We shall view a numeral densitometric analysis of the whole parietal tracing of the root canal as marked with neutron-radiography. This will allow us to understand in a better way how acts the instrument during intra root surgery.

3 - Neutron-radiographic exams give us same answers. Much more investigation in exposure time with neutron flux and frame screen should give us better information.
By now development has to go on.

Instruments

Results show a lot of differences in the action of the experimented instruments. This behaviour is modified with irrigation with physiologic serum.

Shaper is the only instrument which seems to have a good shape for sonic waves (1500 to 3000 Hz). Its profile cuts in a good way the parietal layer of the endodont. Efficiency graphs shows that Gd-Cit. marked predentin is eliminated within 60 sec., and that there is an activity decrease when the instrument goes through highly mineralized dentin. This instrument takes care of the previous anatomy of the root canal which corresponds to actuel therapeutic needs.
Rispi-Sonic goes in the same way when it works with a good irrigation but Heli-Sonic is an inadequat instrument for sonics.

CONCLUSION

Neutron-radiography (tracing of predentin with Gadolinium citrate) and microdensitometric analysis are the positive way for getting a good information about the efficacity of root canal instruments.

This experiment shows the interest of neutron-radiography in biology.

REFERENCES

1 - T.B. Engefeld, A. Hjerpe, "glycoaminoglycans of dentin and pre-dentin" calc. Tissue. Res., 152-159, 10, 1972.

2 - P. Hardt Von der, H. Rottger, "Neutron radiography hand book" nuclear science and technology, D. Reidel Publishing company, 1981.

3 - A. Laporte, "Neutron radiography associated with the Orphee reactor" C.E.N. Saclay C.E.A., 1982.

4 - Y. Launay, A. Claisse, J.M. Laurichesse, "Les instruments à canaux spécificité et intégration dans les séquences opératoires" R.F.E., vol. 2, n° 2, 1983.

5 - J.M. Laurichesse, "La technique de l'appui pariétal (T.A.P.)" R.F.E., vol. 4, n° 3, 1985.

6 - R. Masse, H. Métivier, R. Guillaumont "Fixation osseuse des terres rares et des éléments transuraniens. Physico-chimie et cristallographie des apatites d'intérêt biologique", colloques internationaux du C.N.R.S., n° 230, 1973.

7 - F. Riitano "Manuale di tecnica endosonica", grafiche abramo S.p.A., sept. 1984.

8 - J.M. Vulcain, J. Tamisier, J.H. Espié, R. Masse, A. Laporte, "The use of neutron radiography and microdensitometry in the study of instrumental effectiveness on dental tissue in the process of mineralization", proceedings of the first World conference, San Diego, dec. 1981.

9 - J.M. Vulcain, J.H. Espié, G. Bayon, A. Laporte, "contribution à l'amélioration d'une méthodologie permettant l'étude de l'efficacité ampliative des instruments endo-canalaires", A.I.F.R.O. 1984.

NEUTRON RADIOGRAPHY APPLICATION TO ANCIENT ARTS

Fumitake MASUZAWA
Gangoji Institute for Research of Cultural Property
Nara 630, Japan

Munehiko SAKATA and Yoshiharu INOKUCHI
Nara National Museum
Nara 630, Japan

Eiichi HIRAOKA, Yukio TSUJII and Shuichi OKUDA
Radiation Center of Osaka Prefecture
Sakai, Osaka 593, Japan

Kosuke KATSURAYAMA, Tadashi TSUJIMOTO
and Ken-ichi OKAMOTO
Research Reactor Institute, Kyoto University
Sennan-gun, Osaka 590-04, Japan

Shuichi TAZAWA* and Takehiko NAKANII†
Sumitomo Heavy Industries, Ltd.
*Niihama, Ehime 792; †Tōyo, Ehime 799-13, Japan

and

Megumu UJITANI
National Museum of Ethnology
Suita, Osaka 565, Japan

Neutron radiography testing (NRT) has been applied
to several ancient relics at Kyoto University Research
Reactor and the cyclotron-based facility of Sumitomo
Heavy Industries, Ltd. Some of the results are
presented with comments on the usage and applicability
of NRT. The objects studied are an excavated bronze
mirror with four bells (late 5c.A.D. to early 6c.A.D.),
an unearthed corroded bronze mirror with cloth

489

(4c.A.D.), unearthed Buddhist sutras in a bronze case
(10c.A.D.), a Tibetan unglazed Buddhist statue cushioned
with cloth in a copper box with a window and an
excavated iron sword conserved and restored (4-5c.A.D.).
From comparisons with X-ray radiography testing, it has
been found that NRT can well identify the stones located
inside the bells and suggest the casting procedure.
Moreover, NRT has been found to be remarkably effective
for elucidating unobserved organic objects in bronze
fine arts.

---

1. INTRODUCTION

X-ray radiography is indispensable for the study and
restoration of archaeological and ethnological objects. We
presently use the method for the following purposes:

(i) Identification of the shape of an unearthed object covered
     with clay and/or corrosion.
(ii) Observation of the inner structure of an object.
(iii) Searching inlay or gilding made in an object covered with
     rust.
(iv) Inspection of cracks, corrosion and deterioration of an
     object.
(v) Recording data on an object immediately after conserving
     and restoring.

Some ancient art objects are, for instance, composed of gold,
silver, bronze and/or iron, as well as wood, Japanese paper, silk,
linen and so on. X-ray photographs can reveal the existence of
these metals. However, it is very difficult to see the organic
materials listed above when they are obscured by the presence of
metals, even if the exposure to X-rays is permissible for the
organic material because of little irradiation effects.

Until Hilling (1) reported on neutron radiography applied to
art objects, solution to this problem had not been found.
Realizing the importance of this problem, the present authors
devised an application of neutron radiography as a suitable
method. They have tried to use this method on several ancient art
objects and ethnological objects. The present paper describes the
satisfying results obtained.

2. SAMPLES

The samples studied are as follows:

(a) An excavated bronze mirror with four bells (late 5c.A.D. to early 6c.A.D.): diameter of the mirror = 12.6 cm, diameter of the bells = 1.6 cm.

(b) An unearthed corroded bronze mirror with cloth (4c.A.D.): diameter of the mirror = 13.8 cm. Green rust covers the bronze mirror. Particles of the cloth can be seen in the corrosion on the mirror.

(c) Several unearthed Buddhist sutras in a bronze case (10c. A.D.) (see Fig. 1): dimensions of the case; diameter = 12.1 cm, height = 29.5 cm, thickness of the wall and the base = 1-2 mm. The bronze sutra case and the Buddhist sutras inserted were excavated from a sutra mound. The sutras made of Japanese paper had deteriorated under the earth for about 1000 years, and had changed into a gray lump similar in appearance to excrement. After excavation, they were reinforced by impregnation with acrylic resin.

(d) A Tibetan unglazed Buddhist statue cushioned with cloth in a copper box with a window (see Fig. 2): dimensions of the box; height = 12.7 cm, width = 10.5 cm, depth = 4.8 cm.

(e) An excavated iron sword conserved and restored (4-5 c.A.D.). It has acquired magnetite and reddish brown corrosion on its surface. Fragments of a sheath remain on it, and it is permeated with iron oxides. For conservation, all parts of the sword were impregnated with a non-aqueous acrylic emulsion in a vacuum, and were dried. The broken parts were joined by epoxy adhesives. Two of them were reinforced with a stainless steel wire. The missing parts were filled with a mixture of epoxy resin and phenolic microballoon.

3. EXPERIMENTAL

Neutron radiography testing (NRT) of the aforementioned samples were compared with X-ray radiography testing (XRT) of them. Table 1 gives the characteristics of the neutron radiography facilities used of the Kyoto University Research Reactor (KUR E-2) and Model 370 Cyclotron, Tokyo Works of Sumitomo Heavy Industries, Ltd. (SHI-CT). The X-ray equipment used for comparison was of type Radioflex 160EG (XRF), Rigaku Electric Industries, Ltd.

Table 2 gives the radiographic conditions used for the samples.

Table 1. Characteristics of the radiography facilities.

No.	Facility	Power	Collimator (L/D)	Neutron flux (n/cm²/sec)	Field size
1	Kyoto University Research Reactor (KUR E-2)	5 MW	50 100 170	4.5×10^2 1.2×10^2 0.55×10^2	6"ϕ 6"ϕ 6"ϕ
2	Model 370 Cyclotron Tokyo Works of Sumitomo Heavy Industries, Ltd. (SHI-CT)	Protons 18 MeV 50 μA	30-85	1.0×10^6*	14"×17"

* At L/D = 30.

Table 2. Conditions for radiography of the samples.

Sample	No.	Facility	Tube voltage /Tube current	Focus-film distance or L/D	Screen or Converter	Film
a	a-1	XRF	...	...	LF 0.03	I×100*
	a-2	SHI-CT	...	70	Gd	SR†
	a-3	SHI-CT	...	70	Gd	SR
b	b-1	XRF	120 kVp/5m A	100 cm	LF 0.03	I×100
	b-2	KUR-E2	...	100	Gd	SR
	b-3	SHI-CT	...	30	Gd	SR
c	c-1	XRF	160 kVp/5 mA	100 cm	LF 0.03	I×100
	c-2	KUR-E2	...	100	Gd	FG§
	c-3	SHI-CT	...	30	Gd	SR
	c-4	SHI-CT	...	70	Gd	SR
	c-5	SHI-CT	...	70	LiF	SR
d	d-1	XRF	100 kVp/5 mA	100 cm	LF 0.03	I×100
	d-2	KUR-E2	...	100	Gd	FG
e	e-1	XRF	70 kVp/5 mA	100 cm	LF 0.03	I×100
	e-2	KUR-E2	...	100	Gd	FG

* Fuji Industrial X-ray Film 100.
† Kodak Industrial X-ray Film SR.
§ Fuji Industrial Softex Film FG.

4. RESULTS

The comparisons of NRT of the samples with their XRT are as follows:

(a) The excavated bronze mirror with four bells

XRT: The image of the pattern is very clear. However, the insides of the bells and the traces of molding are invisible.

NRT: The image of the pattern has not come out clear. However, the cubic stones in the bells has appeared. In the convex center (chū), a mold cavity is recognized.

(b) The unearthed corroded bronze mirror covered with cloth

XRT: Invisible patterns and pitting corrosion are imaged distinctly. It reveals that the mirror was designed with four metamorphosed animals. However, the cloth is not radiographed.

NRT: The eyelet of chū is clear, but the patterns of animals are not clear except one of them. The condition of covering cloth is recognized, but the texture cannot be recognized.

(c) The unearthed Buddhist sutras in the bronze case

XRT: Part of the sutras are visible but unclear. It cannot be recognized what they are. The corrosion of the sutra case and its tinkered part are very clear.

NRT: The sutras in the bronze case are imaged most clearly as if the bronze case were transparent. Even rolled pieces of paper of sutras can be seen in a film photographed with KUR-E2 when L/D is 100. The less L/D is, the less clear the image is. When a scintillator screen converter (LiF) is used, the contrast of the image is higher, but the quality of the image is worse compared with the case of Gd metal screen.

(d) The Tibetan unglazed Buddhist statue cushioned with cloth in the copper box with a window

XRT: The patterns of the copper box are plainly seen. Buddhist statue can be recognized vaguely. The cloth is not imaged.

NRT: The outline of the bronze case is very clear, and the thickness of the copper plate can be recognized. The Buddhist statue is imaged only faintly, but folds of cloth is vary clear.

(e) The excavated iron sword conserved and restored

XRT: Details of the sword blade and corrosion can be observed. The image of the stainless steel is brightest and clear. Adhesives and resin are not imaged.

NRT: On the same film, we can see the condition of corrosion of the sword, adhesives of pure epoxy resin, fillers of mixture of epoxy resin and microballoon and a partial wooden sheath. The adhesives made of resin was imaged white; the fillers of resin and microballoon, gray; the stainless steel, dark.

5. CONCLUSION

From the facts described above, we can conclude as follows:

(i) As was reported previously (1), NRT shows conditions of casting bronze and stones more clearly than XRT.

(ii) Organic objects in a metallic case, not shown in XRT, are revealed.

(iii) NRT gives very clear images when L/D is 100, and especially can show structures and sections of cultural properties.

(iv) NRT shows the condition of organic adhesives and fillers unobservable with XRT as well as metallic materials in metallic arts repaired, so that this technique is expected as a new inspection method for restoring arts. In future, NRT will be used effectively in tracing the change in arts and materials used for repairing and restoring.

(v) NRT with cyclotrons is useful as well as the one with reactors. In the former, however, images should be made clear by increasing L/D.

(vi) The use of LiF instead of Gd as the converter reduces the neutron exposure time by an order of magnitude. Therefore, radioactivity induced in cultural properties, an undesirable effect on the use of activation analysis, is also reduced. In this case, however, a method to obtain a higher quality of the image should be developed.

ACKNOWLEDGMENT

The authors thank Mr. M. Yamamura for supplying the unearthed bronze mirror with four bells, and the Board of Education of Kakegawa City for supplying the excavated iron sword. The research was supported in part by Grant-in-Aid for Co-operative Research (A) (1984) from the Ministry of Education, Science and Culture.

REFERENCE

1. O. R. Hilling, Neutron Radiographic Enhancement Using Doping Materials and Neutron Radiography Applied to Museum Art Objects, p. 268 in Practical Applications of Neutron Radiography and Gauging, ASM Special Technical Publication 586, H. Berger, Editor, ASTM, 1975.

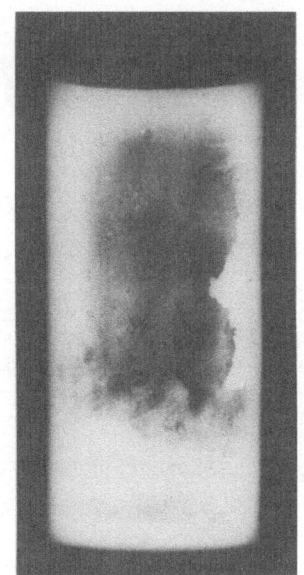

Photograph   XRT

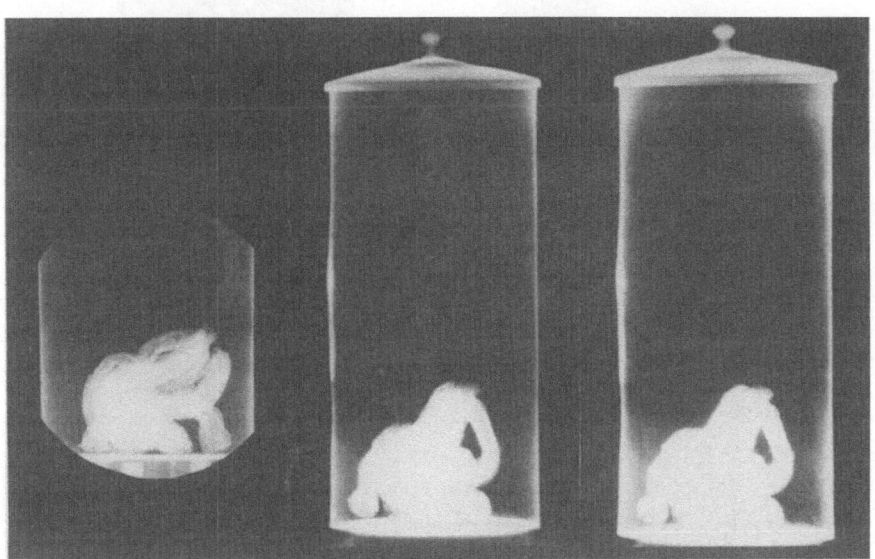

NRT (KUR E-2)   NRT (SHI-CT,   NRT (SHI-CT,
                     Gd converter)   LiF converter)

Fig. 1. Photograph and radiographs of unearthed Buddhist sutras
         in the bronze case.

Photograph   XRT   NRT

Fig. 2. Photograph and radiographs of Tibetan unglazed Buddhist statue.

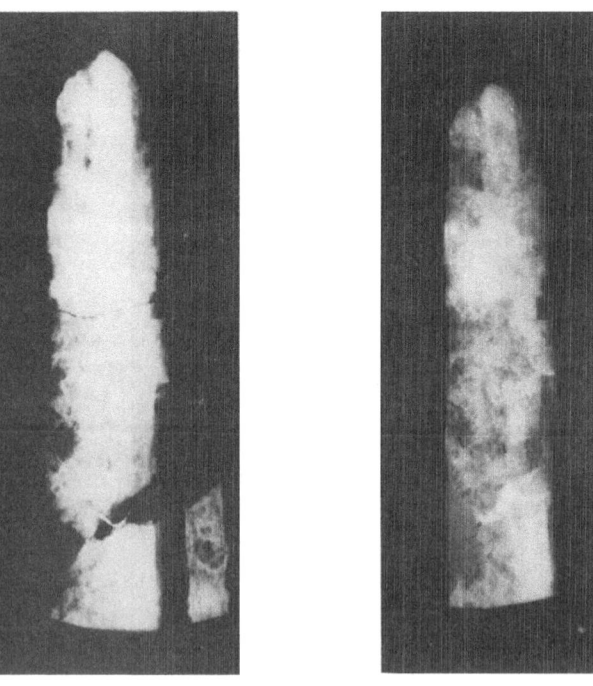

XRT   NRT

Fig. 3. Radiographs of excavated iron sword conserved and restored.

NEUTRON MICRORADIOGRAPHY FOR CELL-SEEKING BORON COMPOUNDS

B. Larsson, O. Sornsuntisook and G. Eriksson.

The Gustaf Werner Institute, Uppsala University

Box 531, S-751 21 UPPSALA, Sweden

E. Johansson, K. Sköld and B. Nilsson

Studsvik Energiteknik AB

S- 611 82 NYKÖPING, Sweden

M. Fantini

Kodak-Pathé, Direction des Recherches

B.P. 60, F-94302 Vincennes Cedex, France

A facility for the irradiation of biological samples for the
purpose of thermal neutron microradiography at the R2-0 reactor in
Studsvik, Sweden, is presented. Analysis of boron was carried out
in cultivated cells and thin histological sections using cellulose
nitrate film detectors. The aim was to determine concentrations of
boron in the interval 0.01-10 ppm with or without spatial representation at microscopic resolution. Examples of the results are
presented from current studies of tumour-seeking boron compounds
developed for their potential usefulness in neutron capture therapy
in malignant disease. Special emphasis is on problems encountered
at low boron concentration and the experimental requirements for
cellular localization of boron. Current developments are presented
and discussed.

INTRODUCTION

Neutron capture radiography (NCR) for boron compunds is based on the large nuclear absorption cross section for the boron isotope ^{10}B, which has an isotopic abundance of 19.8 atom per cent. Upon neutron capture in ^{10}B, isotropic emission of two ions, ^{7}Li and ^{4}He, occurs. The near-linear trajectories of these particles are anti-parallel and their range is less than 10 micrometer in condensed organic materials. The particle traces can be made visible by chemical etching in the solid-state detectors used. The tracks thus formed can be studied, visually, in a microscope or scored by automatic counting techniques. Such methods for visualization of boron in biological systems has signigicant potential for application in experimental and clinical physiology. The present study is part of a search for boron compounds suitable for the treatment of tumours, in the context of neutron therapy (cf. Hatanaka 1985), where targeting of boron the neoplastic cells is a critical problem. The methods and apparatus described are applicable also to the detection of natural boron in its role as a biogenic element, artificial contaminant in plants (Martini and Thellier 1980) or animals, as well as artificial and enriched ^{10}B as tracer for natural boron compounds (Thellier, Duval and Demarty 1979) or boron-labelled pharmaceuticals such as monoclonal antibodies and other cell-seeking drugs (Larsson et al. 1984). In most biological studies of the types referred to above, the boron content is low. The neutron fluence must therefore be high and parasitic tracks must be supressed. For those reasons unwanted nuclear reactions causing increased background and radiation damage in the detector materials must be avoided. These requirements impose rigorous demands on the quality of the neutron field, the handling of the detector system, the preparation of the biological specimen and the data analysis. These aspects are treated separately below. First, however, the prerequisites for the use of boron compounds as labels in tracer studies of macromolecules are presented. The account is, to a great extent, relevant also to other biological or organic-chemical situations, where minute amounts of boron specimens are analysed.

PREREQUISITES FOR DETECTION OF ^{10}B IN BIOLOGICAL MICROSPECIMENS

Among nuclides with a large cross section for neutron capture, ^{10}B is unique, in the sense that it can be covalently bound to organic molecules, In the chemical combinations considered, it also seems relatively harmless from the toxicological point of view (Kliegel 1980). Physical prerequisites for the use of NCR for topographic analysis of ^{10}B in cells or tissue specimens are proper detection techniques and a high and pure fluence of slow neutrons. In this paragraph we deduce the prerequisites that

determine the signal to background ratio, and the energy transferred to the detector by the ionic fragments that appear upon
neutron capture in the specimen. In order to evaluate the potential of NCR, we consider a theroretical model sample apporpriate
for the experimental situation. It consists of 1 ng of tissue
$H(C_5 H_{40} O_{18} N)n$ with 0.1 $\mu g.g^{-1}$ natural boron and 0.01 $\mu g.g^{-1}$
natural lithium (the "background"), charged with a superimposed
^{10}B load of 10 $\mu g.g^{-1}$ (the "label").

One ng of tissue correspond to a typical mammalian cell (1000 μm^3)
of density 1g cm^{-3} and may also represent a mass element in an
organic solid-state detector in contact with the specimen. The
model sample is exposed to 10^{12} thermal neutrons per cm^2, a
typical value for the experimental situation referred to (cf.
Larsson et al 1984).

The chosen background levels of natural boron and lithium are
representative for human soft tissue specimens (Iyengar et at 1978).
The level of labelling with enriched ^{10}B corresponds to a typical
experiment, in which macromolecular carriers are being used. With
markers such as aminophenylborate-substituted carboxymethyl-
cellulose or decachlorocarborane (Gabel and Walczyna 1982), local
concentrations of boron of approximately 10 $\mu g.g^{-1}$ have been
obtained in the cellular or tissue specimens under study.

Analysis of boron by NCR is based on the tracing of the produced
^7Li and α particles. In the detector material radiochemical
changes occur so as to form latent tracks. These latent tracks can
be developed into visible tracks which can be scored in a microscope either by eye or by automatic techniques. The total energy
released as kinetic energy of the reaction products (including,
in 95% of the events, a 0.48 MeV capture gamma ray) is 2.8 MeV and
on the average (2.33 $\pm$ 0.02) MeV is transferred to the film
material per reaction. The equivalent ^4He ion energy is 1.48 MeV,
the ^7Li ion energy is 0,85 MeV and the corresponding ranges in
water are about 8.9 μm and 4.8 μm, respectively. In freeze-dried
soft tissue the tracks are 5-10 times longer. With a thermal cross
section for the reaction ^{10}B (n,α) ^7Li of 3837 barns, a ^{10}B-
concentration in the tissue of 10μg cm^{-3}, and a thermal capture
fluence of 10^{12} neutrons cm^{-2}, there is, on the average, about
2.4 events in a volume element of 1000 μm^3, the model chosen for a
mammalian cell (Table I).

The calculated types and numbers of significant neutron capture
events obtained under the same conditons in ^1H, ^{14}N, ^{17}O, ^{10}B and
^6Li, i.e. the significant background events, in the absence of
fast neutrons, are given in Table I. In terms of $^4He(\alpha)$ tracks,
the signal to background ratio calculated would be 2.4/0.0068=350.
Since discrimination would be possible against all reactions
except neutron capture in natural ^{10}B (as well as ^6Li), the

theoretically possible ratio is even higher, namely $2.4/0.00434 =$
550. In reality the ratio would be inversely related to the
natural boron content of the specimen. It has been possible to
obtain useful experimental information at signal levels one or
two orders of magnitude lower than the values assumed in the above
model, i.e., down to about 0.01 µg ^{10}B per g of tissue in cases
where background correction can be applied on the basis of proper
counting statistics (Gabel et al. 1986).

The only biogenic nuclides in animal tissues that have thermal
cross sections of importance in the present contect, are ^{1}H, ^{14}N
and ^{17}O (Table I). The hydrogen capture reaction $H(n, \gamma)^{2}H$
produces very short recoils of deuterium ions which are harmless
from the point of view of absorbed radiation dose and easily
discriminated in the etching procedure. Capture in ^{14}N, however,
contributes a significant amount of parasitic tracks through the
reaction $^{14}N(n,p)^{14}C$. In natural as well as freeze-dried soft
tissues ^{14}N is present at a concentration of about 1.5×10^{12} atoms
per 1000 µm^{3}. Its cross section for thermal neutron capture, 1.82
barns, leads to 2.7 events per 1000 µm^{3} at a fluence of 10^{12} n
cm^{-2}, a value similar to that for the boron label in our model
sample.

In this paper, NCR based on solid-state, alpha track detectors
and a slow neutron beam, practically free from contaminating
radiations (Figs 1 and 2), is presented. In this facility,
tracing of macromolecules by neutron capture techniques is now
becoming possible, thanks also to the development of boron
compounds suitable for conjugation. Immunoglobulins, for example,
have been labelled with about 5% (by weight) of boron with
retainde solubility (Gabel and Walczyna 1982). At a sufficiently
high concentration of the ^{10}B label in the tissue, the tracks of
α and ^{7}Li ions dominate over parasitic tracks, including those
ascribes to naturally present ^{10}B in the tissue.

With the beam of slow neutrons now available at Studsvik, there
is little deterioration of the detector material attributed to
fast beutrons or photons, even at the high neutron fluences
necessary to achieve high detection sensitivity for ^{10}B. For a
given solid-state detector material, the limits of sensitivity
are set by nuclear capture reactions in the specimen and in the
detector, not by the nuclear recoils and nuclear reactions
produced by contaminating fast neutrons. That this is an important
aspect has been demonstrated in experiments aiming at analysis of
environmental actinides with solid-state fission-track detectors
(Sohnius and Dehnschlag 1982).

NEUTRON RADIATION

The developments described here were inspired by the positive
results of explorative experiments performed with a cold neutron
beam from the high-fluc reactor at Institut-Laue Langevin in
Grenoble. By use of this unique facility Larsson et al (1984)
presented evidence that such slow neutrons, practically free from
contaminating fast neutrons, could be used for high-sensitivity
NCR. On the other hand, the beam area available at the window of
the neutron guide was too small to permit practical work on a
large scale. For these reasons a thermal beam facility has now
been constructed at the R2-0 reactor in Studsvik (Fig.1.) It
employs an approximately 1 mater thick heavy water moderator and
gives a near-pure flux of thermal neutrons. The size of the
useful field is 30x30 cm^2 , the thermal flux available being more
than 10^{10} n. cm^{-2}. s^{-1} over this entire area.

Flux calculations and design of the neutron facility. For
guidance in designing the D_2O thermal neutron irradiation facility
at the R2-0 reactor, we have calculated the neutron flux as a
function of space and energy. These calculations were performed
with the two-dimensional diffusion theory code DIXY. The model
configuration, representing the reactor with the H_2O reflector
replaced on one side by the D_2O tank with the surrounding
graphite blocks, was cylindrical and divided into 55 (axial) x 17
(radial) = 935 microregions. The irradiation compartment was
represented by graphite as a void in the structure would lead to
numerical difficulties. The use of graphite in this region is an
acceptable approximation for the present purpose. The energy was
divided into four groups as follows:

i = 1	10MeV - 0.821 MeV
i = 2	0.821 MeV - 5530 eV
i = 3	5530 eV - 0.625 eV
i = 4	0.625 eV - 0 (themal group)

The neutron cross sections and diffusion constants needed in DIXY
for the various materials involved were obtained from calculations
with the 69-group transport theory code CASMO, including condensation to the four-group structure used in DIXY.

The results from the DIXY calculations for the configuration
which sas finally shosen are presented in Figure 2. The diagram
shows the flux for each group averaged over a 900 cm^2 large area
perpendicular to and centered on the axis of the tank. In the
experimental position the variation of the thermal flux over this
area is less than 5%.

The calculation predicts a thermal flux in the experimental position of 1.5×10^{11} n. cm^{-2} s^{-1}. Subsequent measurements have given a lower value, 4×10^{10}. Even this value is, indeed, large enough for the NCR experiments. Furhermore, the flux of fast neutrons (groups 1 and 2) is less than 10^{-5} of the thermal flux. The flux in group 3 (5530 to 0.625 eV), is about one percent of the thermal flux. However. due to the 1/E-shape of spectrum, the distribution is heavily weighted towards lower energies within this group. It should also be observed that only neutrons of energy >10 keV are likely to produce recoils that leave detectable traces in a track detector.

CHOICE OF DETECTOR

The cellulose-nitrate films are practically insensitive to the γ and β^- radiation but they can register protons of energy less than 400 keV (Fantini, unpublished) and α - tracks of energy less than 5MeV (Baroni et al 1974). They are thus well suited for detecting neutron capture in ^{10}B. The Kodak LR 115 (type 1) detector film is red-coloured, 6 µm thick, and stuck on a 100 µm thick polyester support. A special variation of the Kodak LR 115, non-coloured and only 4 µm thick, was also used. The α-tracks as well as the Li-tracks do easily perforate these latter detectors and the contrast is good enough for automatic track-counting. The latent tracks are intesified by means of precessing in a 10% solution (2.5 N) of NaOH of analytical purity, at 60^0 C. The etching time ranges from 15 to 40 min for the LR 115 (standard type 1) and from 10 to 30 min for the uncoloured, thin detector.

The cellulose nitrate films were manufactured by Kodak-Pathé. For this study, we could also use their special modifications of the above-mentioned films that can be removed from the plastic base. Although fairly well known, the study of the effect of etching conditions continues, the prupose being to obtain the best use of various types of detectors, viz. by optimization of the thickness of cellulose nitrate and the other physical para-meters involved. It was also our ambition to optimize the visi-bility of the nuclear tracks in the detector film after etching by use of fluorescent stain in order to facilitate the counting of the tracks in the biological materials under study.

In preliminary experiments, various detector materials were tested, including polycarbonate and CR-39. So far, the best results have been obtained with cellulose nitrate films, and for the purpose of illustrating the technique, the examples given in this presenta-tion are based on the use of this type of detector (Kodak-Pathé LR 115). During irradiation, the detector and the freeze-dried tissue section were pressed tightly together in an evacuated bag of 0.1 mm thick polyethylene.

PREPARATION OF BIOSPECIMENS FOR IRRADIATION

The biological problems under study require supra-cellular, cellular or sub-cellular localisation of ^{10}B, with progressively increasing demands for spatial resolution, detection sensitivity and high signal background ratio. Special attention has been paid to localisation of ^{10}B markers in cultivated aggregated human tumour cells , and in healthy or malignant human or murine tissues.

After incubation with, or injection of, chosen boron conjugates, for varying lengths of time in vitro or in vivo , respectively specimens were secured by deep-freezing of the samples to be studied and a corresponding control material. Cell spheroids and whole rats were rapidly frozen in cool propane, freeze-mounted and freeze-sectioned. Most of the radiographic studies so far have been made on freeze-dried 5-20μm thick sagittal sections prepared from the skull or the abdomen of adult rats. The detector film and the freeze-dried tissue section were pressed tightly together in an evacuated 0,1 mm thick polyethylene bag and irradiated at a   fluence of 10^{10}–10^{14} n.cm^{-2}.

PREPARATION OF THE IRRADIATED DETECTOR FOR TRACK ANALYSIS

The normal etching process entailed etching the irradiated films for 20 minutes in a bath filled with 2.5 M NaOH at a controlled temperature of 60 ^{0}C. The NaOH solution was kept in circulation at a defined rate by a pump and by a slowly flowing stream of nitrogen bubbles. After etching the film was washed for 30 minutes in running tap water.

To facilitate the microscopic study of the tracks in non-colored LR-115 films we have developed a fluorescent stain technique demonstrating the location of boron in tissue. Optical brightener, a stilbene derivative which binds to the cellulose layer was used as a fluorescent stain to fill the etching pitch. The fluorescens-spectrum emitted from the optical brightener at the excitation wavelength of 325 nm was measured by Zeiss NM3 fluorescensspectro-meter. The maximum intensity of the emitted light was in the blue-green range. To maintain the fluorescent stain in the etched pits of the stripping film, agarose gel (melting point 65^{0}C) was melted in the optical brightener. The optimum concentration of the optical brightener (CDM-AB) was found to be 1/8 of the original concentra-tion of the solution supplied by the factory.

TRACK ANALYSIS

The etching of the latent, submicroscopic traces of the alpha and lithium particles in the solid state detectors produces visible "tracks" that could be studied macroscopically (Fig.3) or in the microscope (Fig. 4). The information can then be used to evaluate, qualitatively or quantitatively, the variation of the boron content in the specimen. As the full exploitation of such methods would require automatized evaluation of the autoradiograph, we here present the result of a pilot study made with a computer-based image processing system (Diascanner, IMTEC, Box 1633, S-751 46 UPPSALA, Sweden).

The Diascanner system consists of a computer-controlled TV-microscope that permits automatic analysis and recording of the track structures at a spatial resolution of ca. 1 micrometer, i.e. the size of the individual pixel elements. The pilot study comprised the analysis of four fields, each corresponding to an area of 128 x 128 micrometers. After recording of the primary information the pictures were analysed by means of the image analysis language ILIAD (IMTEC, Uppsala). The analysis comprised the setting of a threshold, the filtering of the picture, the counting and size measurement of detected tracks. The results were presented by histograms (Fig. 4).
The number of tracks recorded by visual inspection over areas representative of a typical freeze-dried mammalian cell is shown in Fig. 3. Values are given for different concentration of boron.

CONCLUSIONS

In our field of experimental biological research, NCR for boron-labelled compounds is now an alternative to the corresponding radioactive methods currently in use for the demonstration of macromolecular bahaviour in biological systems. For example, there is a potential for the study of monoclonal human antibodies in clinical investigations. In such situations the use of secondary antibody ("sandwich") techniques cannot be used and a practically useful alternative to radioactive methods would be of great interest.

From a more general point of view, the present method of NCR for boron seems useful for several other applications (see above). Since we are interested in the further development of such techniques we would welcome contacts with other groups with similar interests. The capacity of the present system of irradiation is very large and there is a place for enlargement of this pilot study.

The present features of the neutron capture radiography (NCR) technique, as it appears in the present version in Studsvik, may be summarized as follows:

(i) Single neutron capture events and their localization may be studied.

(ii) The sensitivity of the method allows pictorial representation of ^{10}B at concentrations of 1 p.p.m.

(iii) The background situation is such that meaningful quantative studies of local boron concentrations could be made at levels of 0.01 p.p.m.

(iv) The capacity of the slow neutron facility presented in this paper allows irradiation of up to ca 1000 specimens per day, for analysis of boron at the p.p.m. level or below.

(v) Automatic image analysis in a computer-controlled TV-microscope would allow detailed, quantitative analysis of boron in biological specimens.

ACKNOWLEDGEMENTS

This work was supported by the Swedish Natural Science Council and the Swedixh Cancer Society.

REFERENCES

Baroni G., Di Liberto S., Petrera S., Romano G. and Sgarbi C.,
    Track discrimination of low-energy nuclei in plastics.
    Nuclear Instruments and Methods. 115, (1974) 545-552.

Gabel D., Holstein H., Larsson B., Gille L., Eriksson G., Sacker D.,
    Som P. and Fairchild R.G. Pers. comm. 1986.

Gabel D., and Walczyna R.B. Z. Naturforsch. 37C (1982) 1038-9.

Hatanaka H., In "Neutron Capture Therapy" (Hiroshima H., Ed.),
    Nishimura Ltd, Niigata, Japan (1986) p. 447

Iyengar G.V., Kollmer W.E. and Bowen H.J.M.,
    The Elemental Composition of Human Tissues and Body Fluids
    (Weinheim, New York: Verlag Chemie -1978).

Kliegel W.,
    Bor in Biologie, Medizin und Pharmazie, Springer (1980)

Larsson B., Gabel D. and Börner H.G.,
    Phys. Med. Biol. 29, (1984) 361-370.

Martini f. and Thellier M., Planta, 150 (1980) 197.

Sohnius B. and Denschlag H.O.
    Nucl. Instrum. Methods, 197 (1982) 449-52.

Thellier M., Duval Y. and Demarty M., Plant Physiol., 63
    (1979) 283.

Table 1. Neutron capture reactions induced by a thermal capture fluence equivalent of 10^{12} neutrons cm^{-2} in 1 ng of $(C_5H_{40}O_{18}N)_n$ containing 0.1 μgg^{-1} natural boron and 0.01 μgg^{-1} natural lithium. (From Larsson et al. 1984). Types and calculated number of capture events and nuclear fragments or recoils (particles) are given for reactions of significance in studies of ^{10}B-labelled compounds by neutron capture radiography.

Reaction	^{1}H(n,γ)^{2}H	^{14}N(n,p)^{14}C	^{17}O(n,α)^{14}C	^{10}B(n,α)^{7}Li	^{10}B(n,$\alpha\gamma$)^{7}Li	^{6}Li(n,α)^{3}H
Isotopic abundance of target nuclide (atom per cent)	99.985	99.63	0.038	19.8	19.8	7.5
Mass of target element (ng)	0.100	0.0348	0.715	1.00×10^{-7}	1.00×10^{-7}	1.00×10^{-8}
Number of target atoms	6.0×10^{13}	1.5×10^{12}	1.02×10^{10}	1.10×10^{6}	1.10×10^{6}	6.5×10^{4}
Microscopic cross-section(b)	0.332	1.82	0.235	242	3595	942
Number of events *	20.0	2.7	2.5×10^{-3}	2.7×10^{-4}	4.0×10^{-3}	6.6×10^{-5}
Particle, kinetic energy(MeV)	^{2}H,0.0013	p,0.59 C,0.04	α,1.42 ^{14}C,0.40	α,1.78 ^{7}Li,1.01	α,1.48 ^{7}Li,0.84	α,2.05 ^{3}H,2.73
Total energy carried (MeV)	0.026	1.70	0.0045	0.0008	0.0093	0.0003
Particle, range in water(μm)	^{2}H,<1	p,10 ^{14}C,<1	α,8 ^{14}C,<1	α,10 ^{7}Li,6	α,8 ^{7}Li,5	α,11 ^{3}H,45
Particle **, approximate linear energy transfer (keVμm^{-1})	-	p,50	α,220	α,220 ^{7}Li,340	α,230 ^{7}Li,300	α,210 ^{3}H,20

* In the text, these figures (background) are compared with the figure 2.4, i.e. the number of capture reactions in a 1 ng specimen considered to contain 10μg ^{10}B g^{-1}, a typical concentration in a tracer experiment (signal).

** Here only particles of >1μm range are given.

R2-0 THERMAL BEAM FACILITY

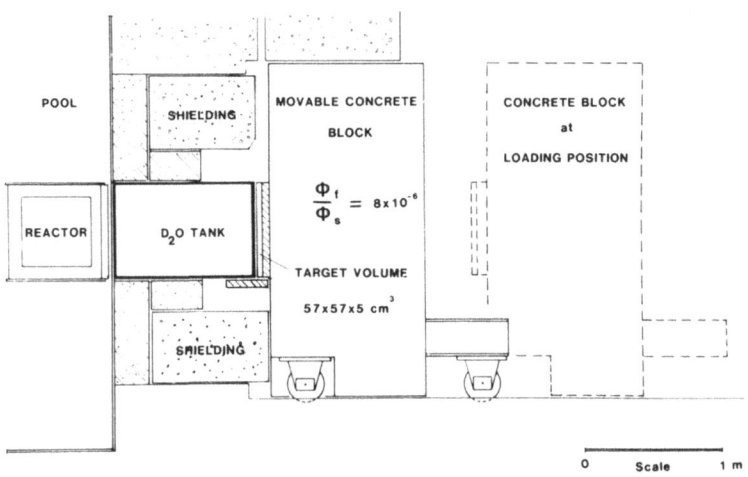

Figure 1   The thermal neutron irradiation facility at the R2-0
reactor at Studsvik, Sweden.

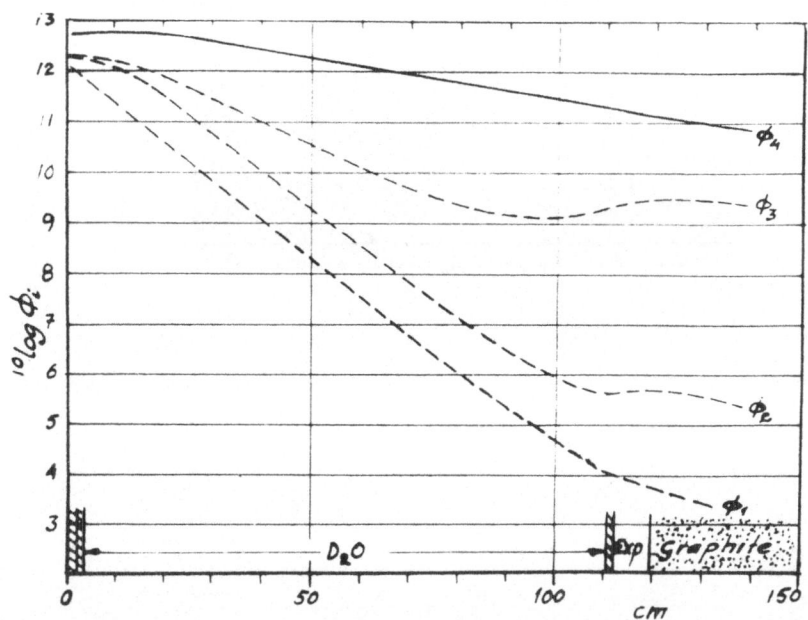

Figure 2 Calculated flux, normalized to full-power operation for
the R2-0 reactor, i.e. 1 MW. The flux is in neutrons/cm² sec.
The calculations predict a thermal flux in the experimental
position of 1.5×10^{11}. Subsequent measurements have given a lower
value, 4×10^{10}. However, even this value is large enough for the
experiments planned which require doses of approximately 10^{12} n
cm⁻² (see text). Furthermore, the flux of fast neutrons (groups 1
and 2) is less than 10^{-5} of the thermal flux. The flux in group 3
(5530 to 0.625 eV), is about one percent of the thermal flux. Due
to the 1/E-shape of the spectrum, the distribution is heavily
weighted towards lower energies within this group. Thus, both the
thermal neutron flux and the level of fast neutron contamination
are adequate for the applications described above.

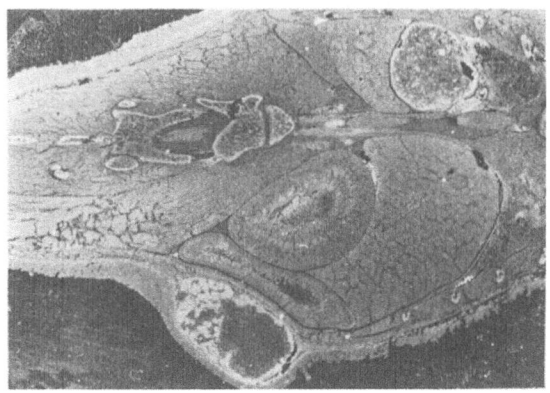

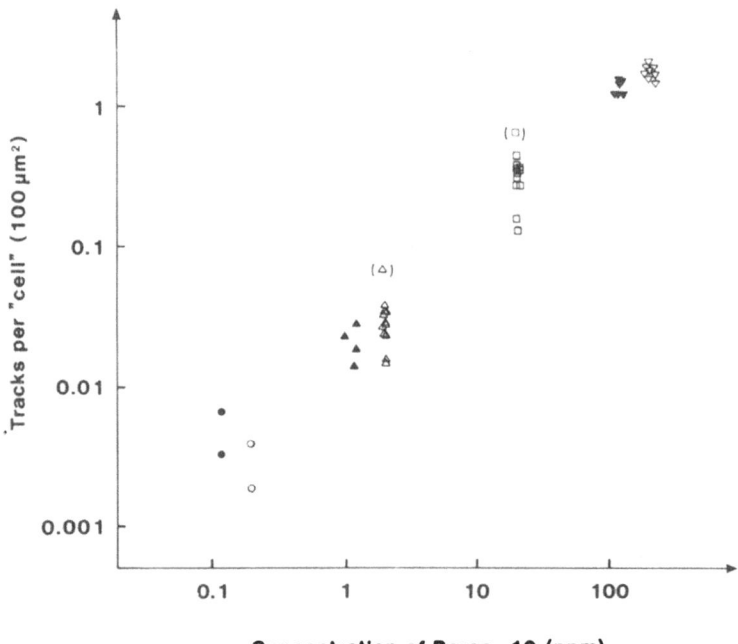

Figure 3 Autoradiograph of a section containing a boron compound (boronated chloroquine) with affinity for melanoma cells (above) and the number of tracks expected for various concentrations of boron-10 (below). The neutron capture radiograph is from an unpublished study (Bengt Larsson et coll., Uppsala university). The calibration diagram summarizes the result of measurements based on standard boronated plastic foils with known amounts of ^{10}B.

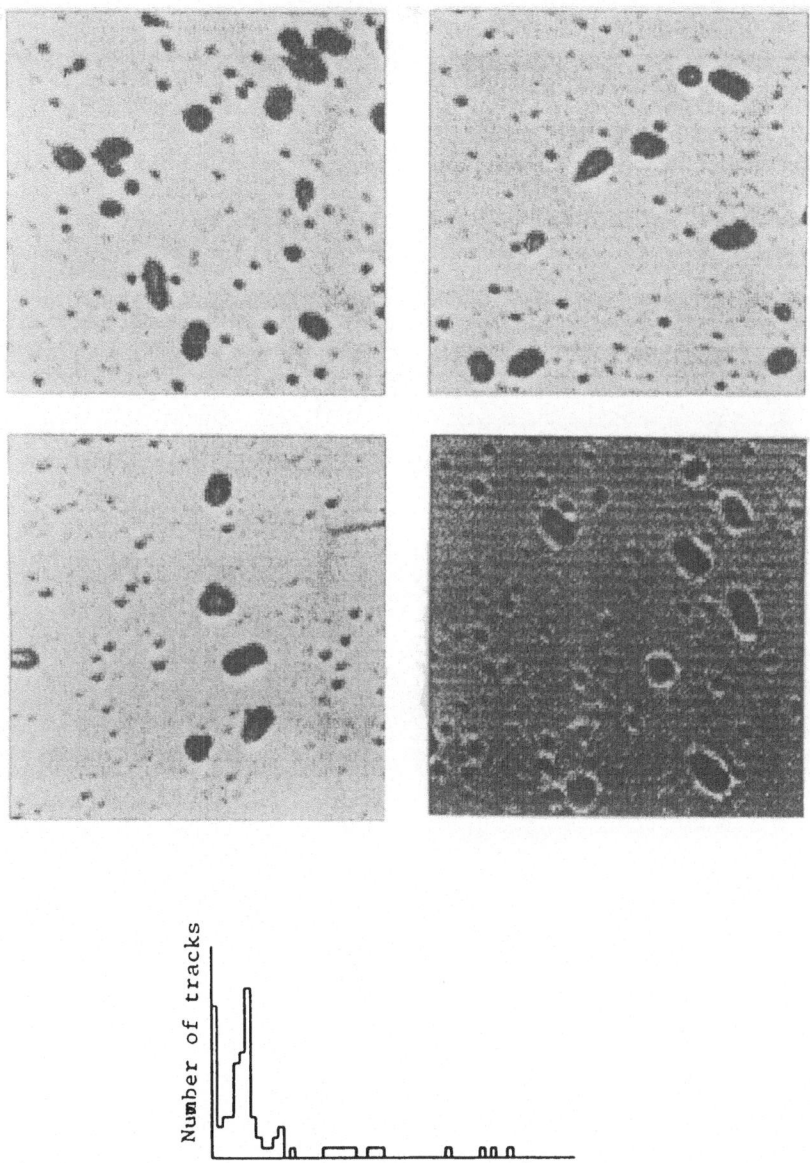

Figure 4 Tracks studied on the screen of a computer-controlled TV-microscope (Diascanner), in four different fields 128 x 128μm². The diagram shows the number of tracks vs the area of the individual tracks.

DELINEATION OF PATHOLOGIC INTRAOSSEOUS

LESIONS BY NEUTRON RADIOGRAPHIC IMAGING

By

P.J. BOYNE, D.M.D.
Loma Linda University
Loma Linda, California
and
W.L. WHITTEMORE, Ph.D.
GA Technologies
San Diego, California

ABSTRACT

Neutron radiography provides a significant diagnostic tool for use in medical research and surgical pathology as an adjunct to the usual microscopic methods of tissue examination. Recent application of the NR method has identified the presence of developing cystic lesions or daughter cysts within the marrow vascular spaces. These were not detectable by X-radiographic examination. Further, NR imaging has provided a powerful method to examine nondestructively the interface beteen metal implants and the host mandible. The delicate tissue layers can be examined closely by this technique without the destructive resection required by alternative diagnostic techniques.

DISCUSSION

Neutron radiographic (NR) imaging of bony tumors has been previously described by Boyne and Whittemore (1, 2, 3) with respect to solid bony tumors and aberrant masses in the facial bones. The neutron radiographic source was a nuclear reactor producing a thermal neutron beam which interacts predominantly with the hydrogen in tissue cells of the bone specimen. It was found in this

513

previous work (1, 2, 3) that the enhanced hydrogen atom concentration in aberrant bony masses, particularly neoplastic disease, could be imaged by the neutron radiographic technic.

While some cystic involvement in tumor masses was detectable in earlier work, it was felt desirable to apply the neutron radiographyic imaging technic to inherently cystic bone lesions to determine specifically:

A. if the neutron radiographic imaging technic could delineate the true extent of the cystic mass as compared with the apparent cystic margins as outlined by X-ray radiographic technics which are dependent upon considerable bone destruction to be detectable;

B. if the neutron radiographic technic could give an indication of cystic breakdown in the bone prior to the actual histologic appearance of the cystic lesion; that is, the precystic or the pre-tumor invasive phase;

C. whether the technic offers any advantage over routine radiographic imaging in determining cystic characteristics and aggression of the destructive abilities of the cystic lesion.

To compare NR imaging with histologic evaluation of the cystic mass and X-radiographic imaging of the cystic mass, a series of patients who underwent osseous surgery to irradicate cysts of the jaws were treated by block resection of the cyst to hemi-resection of the mandible. The treatment received was considered to be routine surgical procedures for the lesions. The surgical pathologic specimen was radiographed by a standard neutron imaging technic as described earlier (3, 4). The same specimen was subjected to X-radiograph imaging. Further, the specimen was sectioned in toto and histologic sections were made of the entire mass of the cyst and with the surrounding bone intact, thus providing comparison of the histologic specimen directly with the radiographic images.

RESULTS

Two patients in the study were afflicted with keratocysts of the mandible involving the entire ramus and posterior body of the ramus. Each had a long history with repetitive surgical procedures to irradicate the disease. It was found that the NR image (Figure 1) was able to detect the presence of developing small cystic lesions or daughter cysts within the marrow vascular spaces of the area surrounding the main portion of the keratocystic mass. These daughter cysts were not detectable by routine X-radiographic examination of the specimen (Figure 2).

514

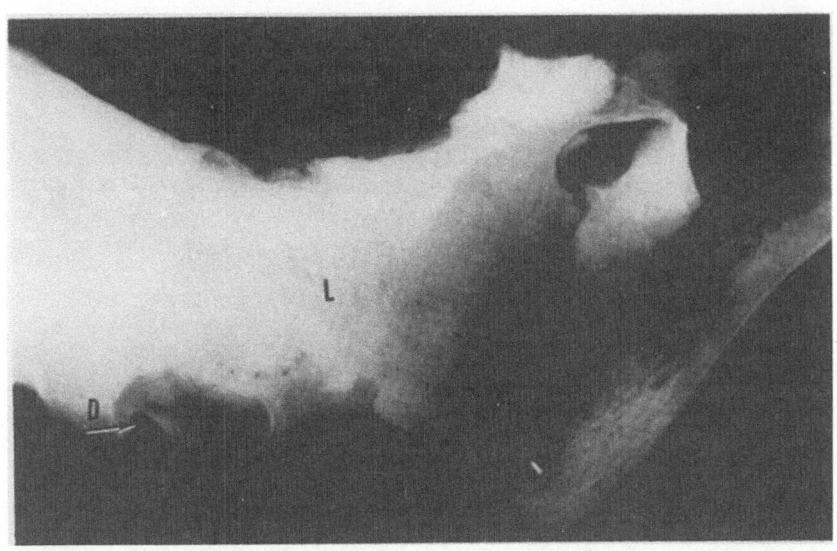

Fig. 1. Neutron radiograph of a keratocyst within a mandible. An impacted tooth is located on the upper right. "L" indicates the main lesion of the cyst showing an increase of hydrogen atom activity. "D" is a small daughter cyst within the larger daughter cyst at the inferior border of the mandible.

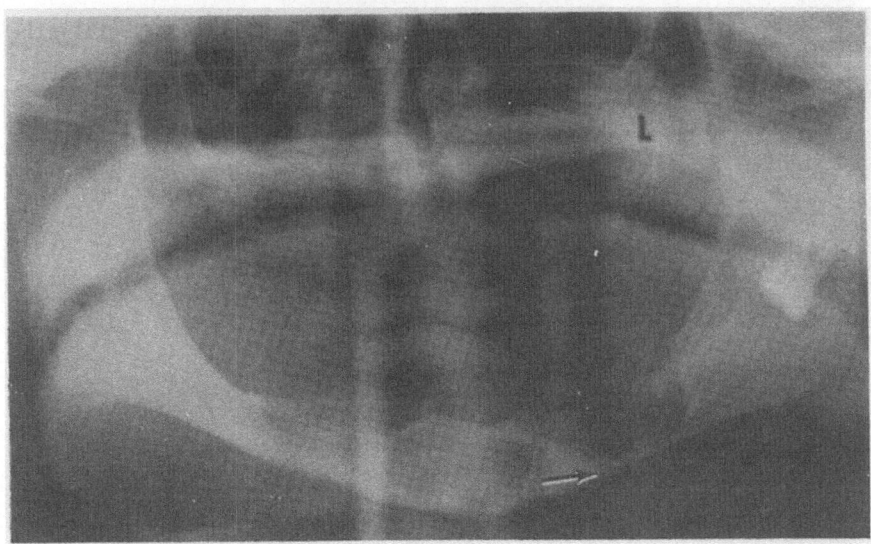

Fig. 2. X-ray of the lesion of the mandible showing the impacted tooth to the left. "L" indicates the main cystic region and the arrow indicates the position of the two daughter cysts which are not visable in the X-ray.

Histologically, the areas indicated by the NR images were found to contain epithelial cells and epithelial lining of beginning deatocysts.

In addition, it was found that the neutron radiograph could determine the presence of beginning cystic invasion by direct proliferation of epithelial cells from the main mass of the cyst itself. Thus, the cystic mass was growing and expanding by two histologic methods:

A. by direction extension of the epithelial mass from the main cystic portion of the lesion, and

B. by the formation of peripheral daughter cysts, not necessarily associated with the main cystic mass, but capable of growing and later coalescing with the main lesion.

Additionally, degenerative areas in the main cystic mass composed of masses of keratin could be distinguished from growing cells or proliferating cell masses. Thus it appears that the accumulation of keratin cellular debris does not have the same imaging potential as growing cells with the nuclear material of a

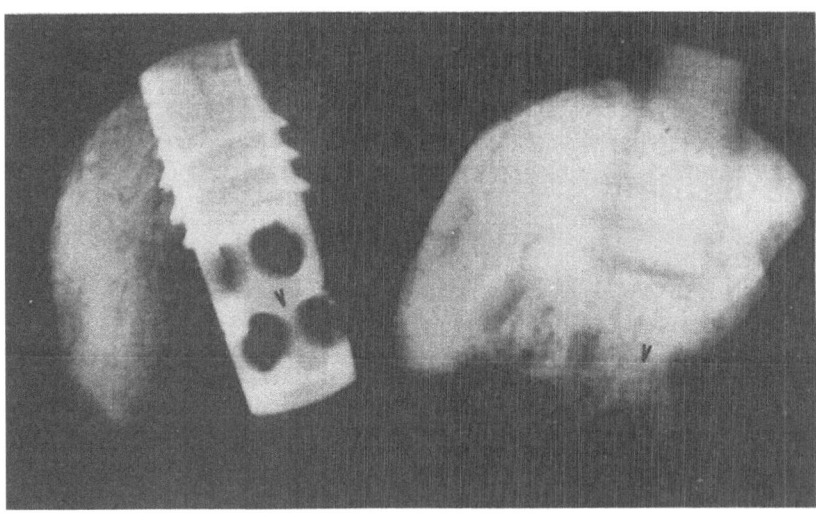

Fig. 3. An X-ray (left) of a intra-osseous metal implant with the superior portion being at the level of the alveolus process of the mandible. The vented portion "V" is situated in a deeper portion of the mandible, with no evidence of bone or soft tissue formation within the core or vented portion of the implant. The Neutron Radiograph (right) of the same specimen shows bone and soft tissue in the vented portion of the implant "V" and considerable soft tissue in the surrounding bone, which appears radiopaque.

viable nature. This may have some future potential in delineating neoplastic disease in the specimens as distinguised from inflammation of other types of tissue reaction. Further, the lesion continues to show proliferation of actual cystic tumor tissue into the lining of the cyst as previously described by us (1, 2). These neoplastic masses are distinguishable from the cystic lining itself and normal epithelial proliferations.

We believe that the neutron radiographic imaging technique has much potential particularly in view of the recent proliferation in the use of bone implants and permucosal and percutaneous metallic implants placed orthopaedically in bone through the body skin or mucosa of the oral cavity. In an earlier publication (1) we showed two examples of metal implants and the benefits provided by NR imaging. These types of implants can now be studied in toto without removal of the metal bone fixture from the bone mass. Thus, it will be possible to determine the interface responses at the level of the implant itself. A recent example of this is given in Figure 3 where the neutron radiographic image is compared with the X-Radiograph.

CONCLUSION

It is concluded from the cases examined under this ongoing pilot study that further work can profitably be undertaken. Additional histological identification of the different types of tissue imaged by NR will continue to be rewarding as it was the in the earlier phases of this pilot study. The work reported herein on surgically removed specimens continues to confirm the value of the NR technique to examine tissue with diseased masses. In addition, NR offers the valuable possibility to examine tissue immediately adjacent to metal implants without removing the metal implant as is required for histologic examination.

REFERENCES

1. P.J. Boyne, W.L. Whittemore, A.M. Harvey, Oral Surgery, Oral Medicine, Oral Pathology, Vol. 37, January 1974, pp. 124-130.

2. P.J. Boyne, W.L. Whittemore, Oral Surgery, Oral Medicine, Oral Pathology, Vol. 31, February 1971, pp. 152-156.

3. P.J. Boyne, W.L. Whittemore, "Application of Neutron Radiography to Histopathology," NEUTRON RADIOGRAPHY AND GAGING, H. Berger, editor, ASTM Special Technical Publication 586, Philadelphia, Pa, 1976, pp. 77-86.

4. W.L. Whittemore, J.E. Larsen, J.R. Shoptaugh, "A Flexible Neutron Radiographic Facility Using a TRIGA Reactor Source," Materials Evaluation Vol. 29, No. 5, May 1971, pp. 93-98, 104.

AUTORADIOGRAPHY OF OIL PAINTINGS AT THE

BERLIN EXPERIMENTAL REACTOR (BER II)

C.-O. Fischer*, Claudia Laurenze**,
W. Leuther*, K. Slusallek***

*Hahn-Meitner-Institut, Berlin
**Gemäldegalerie SMPK, Berlin
***Rathgen-Forschungslabor SMPK, Berlin

ABSTRACT

As part of the international research of art-historians about Rembrandt and his time, 15 paintings of the Gemäldegalerie in Berlin have been investigated by neutron activation autoradiography. The improved experimental technique is explained and results are described for a painting by Jan Vermeer van Delft.

INTRODUCTION

The application of physical methods is getting increasing importance for the investigation of old painting by art historians. Besides the infrared-reflectography, the X-ray-radiography was until recently the only method to look into the underlying paint layers. But the sensitivity of these methods are restricted. Infrared-reflectrography shows only black under-paintings and X-rays make visible only the distribution of lead-based pigments.

The method of neutron activation autoradiography was developed between 1964 and 1966 in the Brookhaven National Laboratory [1, 2]. By irradiation with neutrons in a research reactor an artificial radioactivity is induced. Up to ten different isotopes within various pigments decay with very different half lives. Autoradiographs are generated by the ß-rays when a photographic film is brought to close contact to the surface of the painting. A series of exposures is obtained which show different images of the painting. That depends on the relative aboundances of the different isotopes which vary due to their different half-lives. The autoradiographs make visible underlying structures like preparatory sketches, conceptional changes, signatures and the individual characteristics of the artist's brush-work. These are valuable informations for the art-historians and restaurators.

The method of neutron activation autoradiography was first applied for a systematic study by the Metropolitan Museum of Art in New York. A total of 39 paintings, among them 28 by or formerly attributed to Rembrandt, was investigated between 1976 and 1980. The impressive results focussed the interest of the Berlin art-historians on this new method for their own Rembrandt-research.

EXPERIMENTAL TECHNIQUE

The experimental procedure of the New York investigation [3] could not be applied directly at the Berlin reactor, because the neutron flux density at the thermal column was originally only $1 \times 10^7 cm^{-2} s^{-1}$. This was far less than the flux densities of $3 \times 10^9 cm^{-2} s^{-1}$ at the Brookhaven Graphite Reactor and $1 \times 10^{10} cm^{-2} s^{-1}$ at the Brookhaven Medical Research Reactor.

The first step to reduce this disadvantage factor of 1000 was the modification of the thermal column by removing 80 cm of graphite over an area of 40 cm x 70 cm. A bismuth shield of 5 cm thickness was installed to reduce the γ-Doserate to 200 rem/h. So the flux density was increased to $6 \times 10^7 cm^{-2} s^{-1}$. With an irradiation time of 72 h we achieved a neutron fluence of $1,5 \times 10^{13} cm^{-2}$ in comparison with $5,4 \times 10^{13} cm^{-2}$ at the Brookhaven experiments [3]. This residual disadvantage could be compensated by the application of fluorescent foils for a four-fold amplification of film blackening. Further details, different from the Brookhaven techniques, are described in [4]. We like to point out, that we performed a very detailed γ-spectroscopy of numerous selected areas of 1,2 or 5 cm diameter with a portable germanium detector. The area for this activitation analysis were proposed by the restaurator or determined immediately after estimation of the developed film.

RESULTS

The evaluation of 180 autoradiographs and 303 γ-spectra is in progress and will be published together with a careful analysis by the art-historians.

One of the investigated paintings, the famous "The Man with the Golden Helmet" has been presented to the public in March, 1986 after the restauration was completed. Due to the diagnosis of the responsible art-historian it is definitly stated, that this most popular painting of the Berlin galleries is no authentic work of Rembrandt, but seems to be painted by one of his most familiar assistants.

The autoradiographs of paintings by Rembrandt, Rubens, Frans Hals, de Gelder, Vermeer, v.d. Velde and others show many interesting details. As an example for an autoradiograph which supports without any doubt the authenticity of a painting we present here the result for the "Young Lady with a Pearl Necklace" of the Berlin gallery.

The painting (fig. 1) shows a conceptional simplicity which is in general not typical for Vermeer.

The X-ray radiography indicates very poor structure of the underlying layers (fig. 2).

The third autoradiograph (fig. 3), exposed between 28,5 and 47,75 h after reactor shut-down shows the dominant radiation of Cu-64 in the azurit pigment of the blue drapery and indicates only slightly some fine structures on the right-hand part of the painting.

The structure of the fifth exposition (fig. 4), exposed between 171 and 1659 h after reactor shut-down is much more detailed. The Cu-64 no longer contributes to the film-blackening, which now is due to P-32 of the underlying sketch. The fifth autoradiograph shows very clearly two typical attributes of paintings by Vermeer: paintings or landscape hanging at the wall of the room and a floor made of squared tiles. These typical composition of Vermeer can be seen very impressive in the paintings "Lady and Gentlemen Drinking Wine" in Berlin and "The Artist in his Atelier" in Vienna.

522

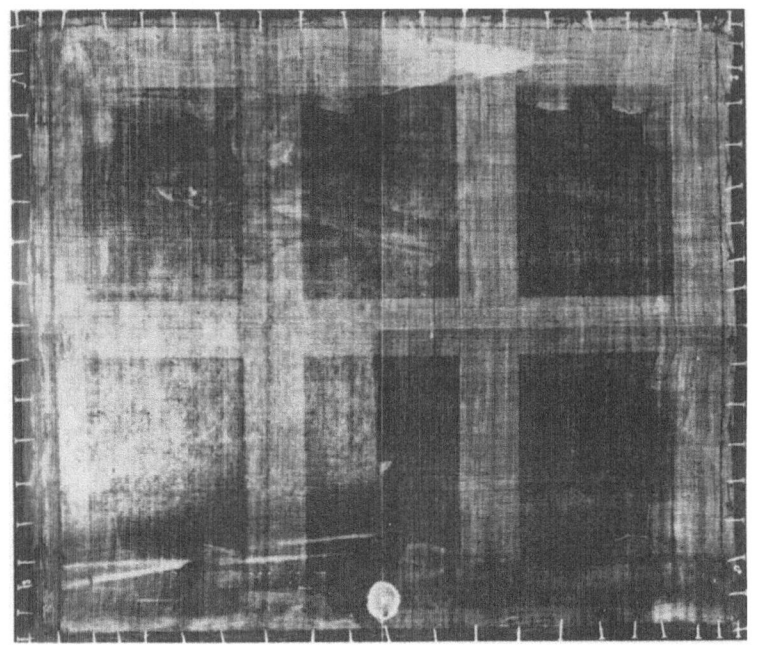

Fig. 1:
Jan Vermeer van Delft:
"The Girl with the Pearl Necklace",
Gemäldegalerie SMPK Berlin

Fig. 2:
X-Ray Radiography

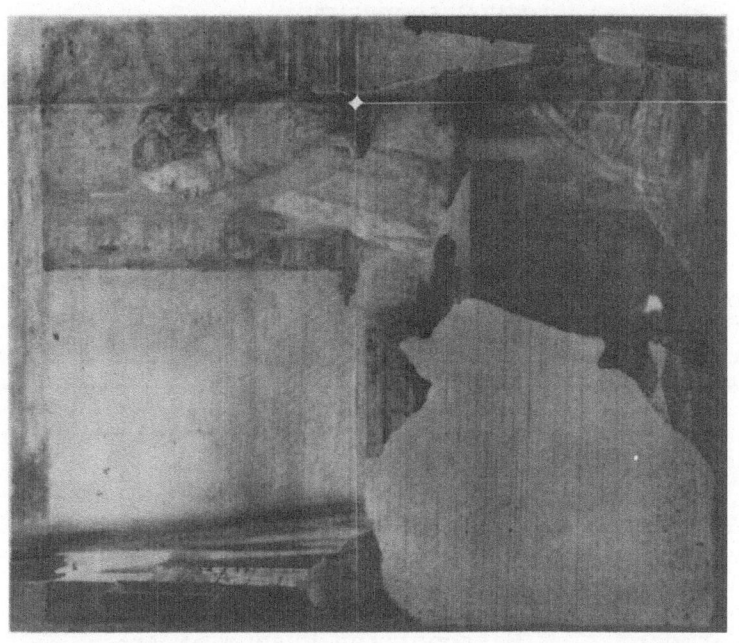
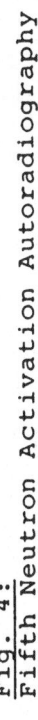

Fig. 3:
Third Neutron Activation Autoradiography

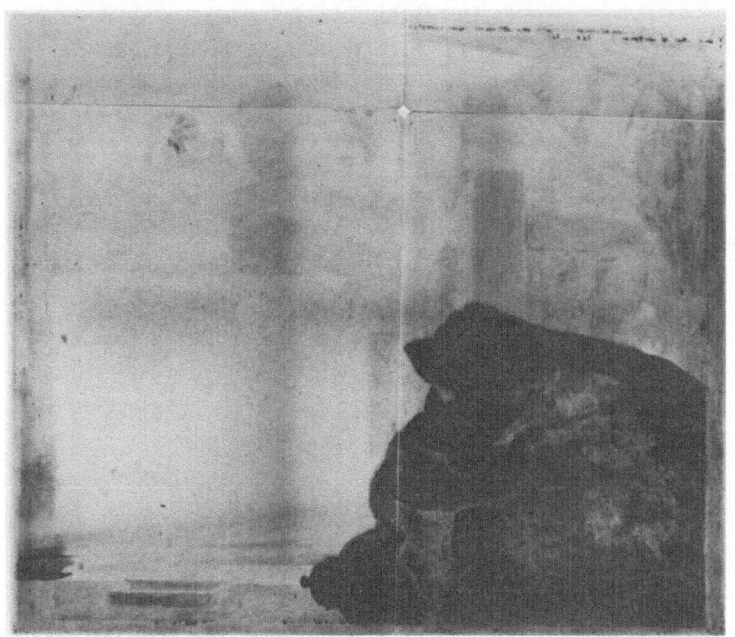

Fig. 4:
Fifth Neutron Activation Autoradiography

ACKNOWLEDGEMENT

The authors are indepted to Dr. Jan Kelch, the coordinator of the Rembrandt-research at the Gemäldegalerie, for many discussions and his support of the investigations.

REFERENCES

[1] Heather N. Lechtman
Neutron Activation Autoradiography of Oil
Paintings, Master of Arts Thesis, Institute
of Fine Arts, New York University (June 1966).

[2] E.V. Sayre, Heather N. Lechtman
Neutron Activation Autoradiography of Oil
Paintings, Studies in Conservation 13 (1968) 161.

[3] P. Meyers, M.J. Cotter, L. v. Zelst, E.V. Sayre
p.105 in Maryan W. Ainsworth (ed.): "Art and Radiography: Genesis of Paintings by Rembrandt, Van Dyck
and Vermeer", The Metropolitan Museum of Art,
New York (1982).

[4] C.O. Fischer, Claudia Laurenze, W. Leuther,
K. Slusallek
p.38 in J. Kelch (ed.): "Bilder im Blickpunkt: Der
Mann mit dem Goldhelm", Gemäldegalerie Staatliche
Museen Preußischer Kulturbesitz, Berlin (1986)

PART VIII

ELECTRONIC IMAGING

ROBUST EQUIPMENT FOR DYNAMIC NEUTRON FLUOROSCOPY

S.J. Cocking* & D.H.C. Harris

Materials Physics & Metallurgy Division, AERE Harwell

Didcot, Oxon, OX11 ORA, U.K.

ABSTRACT

Compact and reliable equipment for dynamic neutron fluoroscopy has been developed and used in inspection tasks. The equipment is based on a zinc sulphide-lithium fluoride composite neutron fluorescent screen viewed by a silicon intensifier target television camera. The real time image can be recorded with a standard video recorder and definition enhanced using a digital image processor.

During development, two fluorescent screens, ZnS/LiF and gadolinium oxysulphide, with additional image intensification have been tested.

Discussion of the performance of the imaging system is based on analysis of the mean luminance of the fluorescent screens. Achievable resolution is limited by the statistics of the neutrons forming each picture element within the time span imposed by the rate of change of the observed phenomenon.

Using a cold neutron beam at the DIDO reactor, well contrasted image features of 1mm dimension moving at 3 cm/sec can be readily observed.

---

*Now with Courtenay Technical Services, Oxon, U.K.

INTRODUCTION

Neutron radiography using film recording of the image was demonstrated at Harwell in 1955 and during the next decade became widely established as an inspection technique. Real time neutron radiography was demonstrated by Berger[1] in 1966. He used an image intensifier which included a neutron absorbing fluorescent screen (ZnS:Ag & Li^{6}F mix) coated with a photo-emitting layer.

Subsequently alternative assemblies for real time neutron fluorography have been demonstrated. For example, a fluorescent screens image can be focussed by a lens onto a low light television imaging tube. A commercially available medical X-ray fluoroscopy system (the Delcalix by Oude Delfte, Holland) was adapted for neutron fluoroscopy and used by Haskins[2], Bracher and Garrett[3] and Stewart[4]. A silicon intensifier target (SIT) imaging tube was used by Cocking[5] for neutron fluoroscopy following its use in a X-ray fluoroscopy system. A gadolinium oxysulphide based screen developed for X-ray imaging, quite independent of its potential use as a neutron detector, offered greater X-ray detection efficiency than the traditional zinc/cadmium sulphide or calcium tungstate based screens, allowing a marked improvement in X-ray fluorography (Burch and Cocking[6]. Gadolinium absorbs neutrons very efficiently (σ_a = 46,000 barns at 0.025 ev) exciting fluorescence in the oxysulphide by emission of an internal conversion electron (mean energy, 71 Kev). The spatial resolution of commercially available gadolinium oxysulphide screens is appreciably higher than that of available ZnS(Ag)/Li^{6}F composite screens. An intensifier incorporating gadolinium oxysulphide coated with a photo emitter is commercially available (from Thomson CSF, France).

Thus a variety of assemblies and components have been demonstrated. This paper aims to emphasise the major considerations governing the choice of screen and of the sensitivity of the associated imaging equipment for real time neutron fluoroscopy. In particular, the luminance under neutron bombardment of readily available neutron detecting screens namely the ZnS(Ag) and Li^{6}F composite screen manufactured by Nuclear Enterprises Ltd. (reference no. NE 426) and a thin gadolinium oxysulphide screen, Trimax 2 available from 3M's Co. Ltd., are compared. From these results the sensitivity requirement of the optical imaging system using these screens is derived. The statistical nature of the image formation process is discussed. As fluorographic image quality is normally limited by the number of detected neutrons which form the image, the design requirement is to ensure that each neutron detected is registered at the output device (usually a television monitor). Once this condition is met, further optical gain is not beneficial.

LUMINANCE OF NEUTRON ABSORBING FLUORESCENT SCREENS

The light energy output $= I \; \varepsilon \; C \; \eta \; E_a$ eV m^{-2}s^{-1}

with

I is the incident neutron flux n m^{-2} s^{-1}

ε is the fraction absorbed by the screen

C is the fraction of those absorbed which give light output

E_a is the energy emitted by each absorption

η is the energy conversion efficiency (from Stevels[7])

Best available values for NE 426 (ZnS(Ag) & LiF) and Trimax 2 (Gd$_2$O$_2$S:Tb) screens are given in table 1. It is notable that the major difference lies in the value of E_a, neutron capture by the lithium-6 nucleus resulting in release of 4.79 Mev recoil energy of the (n,α) reaction while capture by gadolinium results in energy release in the screen of only 71 kev by an internal conversion electron.

Using a beam flux $I = 5.10^9$ n m^{-2}s^{-1} (measured in the 30cm diameter cold neutron beam at DIDO) we can calculate the screen irradiance in watt m^{-2} (table 2).

Stevels[7] gives the spectral output from ZnS(Ag) & Gd$_2$O$_2$S fluorescence and also the relative luminous efficiency, $\bar{D}$ of several detectors (imaging tubes, film, the human eye) to these emitted spectra. For an imaging device having an S-20 photo emitting surface as most commonly used on low light imaging devices, Stevels gives $\bar{D} = 0.77$ & 0.55 for ZnS(Ag) & Gd$_2$O$_2$S emission. These figures are used in table 2 to convert to candela m^{-2} (1 candela = 1 lumen per steradian).

Table 1

Screen	ε	C	E_a eV	η	$(\varepsilon C \, E_a \eta)$ eV
NE 426	0.40	(0.5)	4.79 10^6	0.18	1.72 10^5
Trimax 2	0.95	(0.5)	7.1 10^4	0.15	5.06 10^3

Table 2

Screen	watt m^{-2}	cd m^{-2}
NE 426	1.38 10^{-4}	0.574 10^{-2}
Trimax 2	4.05 10^{-6}	1.19 10^{-4}

CHOICE OF IMAGING TUBE AND SCREEN

The optical image from the fluorescent screen is converted in an electro-optic device to an equivalent electron image. In the case of a fluorescent screen enclosed as a component inside an image intensifier, a photo-emitting layer is deposited directly onto the screen. The probability of emitted photons striking the photo-emitting surface is then high. However, since such intensifiers can only be fabricated with screen area of up to about 20 cms. diameter, the maximum image size available with these devices is sometimes not acceptable for imaging large specimens.

In order to image a large area fluorescent screen on a television tube, the screen can be focussed on the face plate of a low light imaging tube, using a wide aperture lens. The face plate illumination (in lux) is then,

$$\frac{B\pi T}{4f^2(1+M)^2}$$

B is the screen luminance in candela m^{-2}, T the lens transmittance
f the lens aperture ratio (aperture diameter/lens focal length)
M the magnification of the screen image projected on the tube.

For large area screens M <1, T = 0.9 and f = 1.4 are practicable values. Using these figures with screen luminance values of table 2, gives face plate illuminances in table 3.

Figure 1 reproduces manufacturers data on performance of a SIT (silicon intensifier target) tube. With the higher illuminance resulting from an NE 426 screen a SIT imaging tube would yield, for a 100% contrast image, a resolution of about 500 TV lines, that is close to its peak performance. The Trimax 2 screen, with 50 times lower illuminance, should achieve a resolution of about 260 TV lines, for a 100% contrast image. In practice information is required from images of far lower contrast, when resolution could be unacceptably low. Experiments confirm that the SIT tube yields bright images with the NE 426 screen; while barely detectable images with the gadolinium

Table 3

Screen	Face plate illuminance
NE 426	$2.0 \; 10^{-3}$ lux
Trimax 2	$4.2 \; 10^{-5}$ lux

530

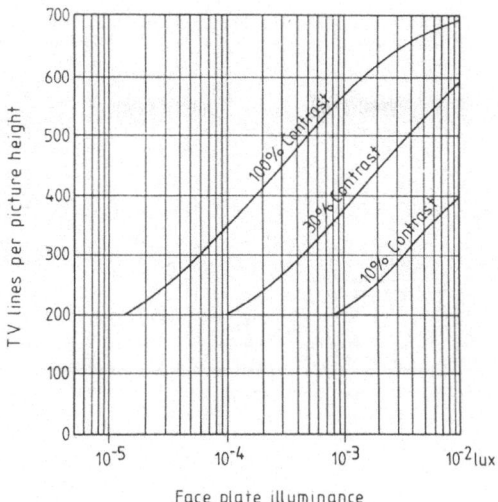

Figure 1 - Limiting resolution (T.V. lines per picture height as a function of face plate luminance for 25mm SIT tube (RCA inc. data).

Figure 2 - 'Real time' image of a coleus plant using cold neutrons (in a 30cm dia. beam position) and NE 426 screen (25 x 25cms) viewed by SIT camera, photographed direct from the TV monitor with 1/8 sec. exposure.

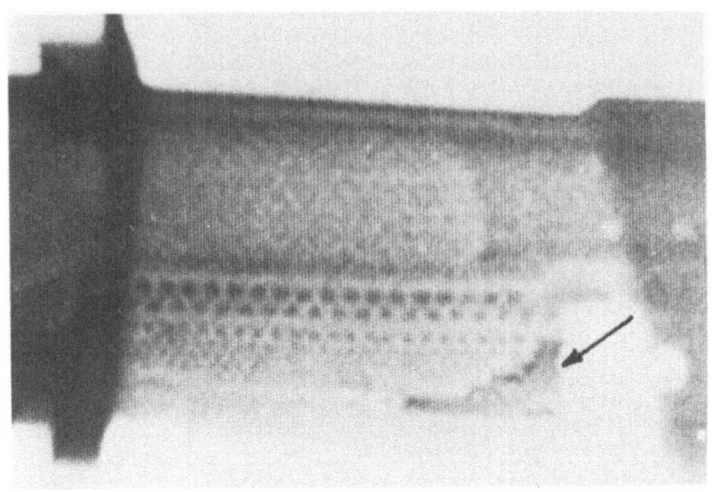

Figure 3 - Neutron fluorograph of aeroengine turbine blade
showing some residual core material in an airway (arrowed). A
frame store allowed exposure of 20 secs. to reduce mottle due to
neutron statistics.

oxysulphide screen are obtained with the same camera and lens
system. Figure 2 shows that, viewed in real time, a noisy image
is seen, while Figure 3 shows that a useful image for many
inspection tasks can be obtained when a frame store was used to
extend the exposure to 20 seconds

In earlier work at Harwell by Anderson[8], a two-stage image
intensifier (light gain 6000) was used with lenses coupling
screen to intensifier and also intensifier to SIT camera. This
equipment used with Trimax 2 showed poor optical resolution and
appreciable geometrical distortion. With an NE 426 screen the
optical gain can be reduced (a factor of 64) by closing the first
lens aperture; the image resolution was in no way deficient to
that obtained with Trimax 2 and a flatter field of focus was
obtained. However, still better resolution and image flatness was
achieved with an NE 426 screen viewed directly by lens on the SIT
camera.

One variant on the use of the 2-stage intensifier to achieve
higher resolution, albeit over a limited area (39mm diameter), was
tested. The Trimax screen was pressed into contact with the
fibre-optic input face of the image intensifiers. Spacial
resolution of the order 0.2mm was measured. However, a factor 1.8
improvement in resolution was found using the NE 426 and SIT
camera under the same conditions.

We conclude that the higher resolution capability of Trimax 2 screen compared with NE 426, which we have observed using film recording, is lost in the image intensifier needed with the gadolinium oxysulphide based screen.

In association with Rolls-Royce Ltd. we have made preliminary comparisons of the performance of our equipment against that of the Thomson CSF intensifier tube with enclosed gadolinium based screen and photo emitter. The resolution of the two methods was comparable, being 0.1 to 0.14mm.

NEUTRON STATISTICS

The discussion of section 2, while relevant to the selection of an imaging system of adequate optical sensitivity, can nevertheless be misleading. The calculated luminance of the fluorescent screen is proportional to the incident neutron flux. This could be used to infer that a more sensitive optical imaging device viewing the screen would allow similar imaging with a lower neutron flux. This however ignores the discrete nature of neutron detection and the statistical variation about the mean detected number which finally determines the accuracy with which each image element is determined.

Thus taking the beam flux already used namely $5.10^9 n \, m^{-2} s^{-1}$ and using the fraction 0.4 absorbed by the screen and a time $\frac{1}{25}$ sec for each TV frame, each element of say 1mm x 1mm area of the image on the screen absorbs 80 neutrons during each TV frame. With normal counting statistics this number is determined to an accuracy $\frac{100}{\sqrt{80}} = 11\%$. If the flux were lower by say an order of magnitude the accuracy of determination becomes $100/\sqrt{8} = 35\%$.

Thus the accuracy of detection of detail in an image, that is of small variations of neutron flux over a small area can have severe and fundamental limitation. While the accuracy can be raised by increasing either the neutron flux or the time over which the neutrons are detected (exposure), this is not possible by increasing the optical sensitivity of the screen imaging equipment. Once each neutron interaction at the screen results in a signal at the output device (usually a TV monitor), further optical sensitivity does not aid the statistically limited determination of the neutron generated image.

The energy each neutron interaction contributes to fluorescent emission is $E_a \eta$. With $\bar{\lambda}$ as the mean wave-length of the fluorescent emission, $E_a \eta \bar{\lambda} C / hc$ photons leave the screen. If a fraction L of these photons are directed onto the sensor input and there converted to photoelectrons with efficiency e, the resulting number of electrons per neutron interaction is $E_a \eta C L e \bar{\lambda} / hc$.

Table 4

	Mean Photon Energy ev	Photons per Absorbed Neutron	Photo Electrons per absorbed Neutron
NE 426	2.0	$2.1 \; 10^5$	590
Trimax	2.1	$2.5 \; 10^3$	7

Table 4 gives values for these parameters, using $f = 1.4$, magnification $M = \frac{1}{20}$ and transmittance $T = 0.9$, $L = 2.6 \times 10^{-2}$ and for a trialkali photo emitter of S-20 type, $e = 0.108$.

Thus for either screen each neutron absorbed results in more than one photo-electron in the imaging tube. The probability that a photo electron in an imaging tube produces an output signal has apparently not been studied in detail but examination of the lower illuminance limit of operation suggest that the probability is higher than 0.1. Thus for the NE 426 screen, 590 photoelectrons from each neutron event will reliably register an output signal The situation is less clear for the Trimax screen (7 electrons per neutron event). The use of a higher optical collection efficiency by at least four times is clearly to be recommended. Such an increase can be achieved, using faster collecting optics (f/0.7). Far higher collection efficiency can be achieved with direct fibre optic coupling or inclusion of the screen and photo electron converter in an intensifier unit. The image size limitations of these solutions have already been noted.

CONCLUSIONS

Calculations of the luminance of NE 426 and Trimax 2 under irradiation in the 30 cm cold neutron (6H) beam from DIDO shows that a 25 mm S.I.T. TV camera tube can provide well resolved fluorographic images with the NE 426 screen. Images of up to 25 x 25cm area have been demonstrated using such equipment. Well contrasted features of 1mm dimensions, moving at 3cm/sec, can be readily observed on a TV monitor; frame summing allows static features of ~0.25mm to be resolved. The use of this video system is reported in other papers at this conference (Attwood & Swift[9] Harris & Seymour[10]).

ACKNOWLEDGEMENTS

The authors wish to express their thanks to Messrs. W.A.J. Seymour and N.K. Bealing for their support and assistance with the tests and development of the imaging systems described in this report.

REFERENCES

1. H. Berger "Characteristics of a Thermal Neutron Television
   Imaging System" Mater. Eval. 24.9 (1966) pp 475 - 81.

2. J.J. Haskins "Evaluation of a real time imaging system for
   Neutron Radiography" General Electric reportNEDC 12512 29
   May (1973).

3. D.A. Bracher & D.A. Garrett (Abstract only) Mater. Eval. 33
   (1975) pp 47A.

4. P.A.E. Stewart Patent Specification 1 542 860 filed 23 Dec.
   1975.

5. S.J. Cocking "Resolution in Neutron Radiography using
   television & film imaging systems" AERE internal memorandom
   (1982).

6. S.F. Burch & S.J. Cocking "Evaluation of a Computer aided
   X-Ray fluorographic System: Part I - System Analysis", AERE
   Harwell Report R-10409 available from H.M.S.O.

7. A.L.N. Stevels " New Phosphors for X-ray Screens"
   Mecicamundi 20, 1 pp 12 - 22.

8. M.R. Anderson "Dynamic Neutron Radiography - a Research Tool
   in Automotive Tribology" Proc. of I. Mech. E. Conference on
   Tribology, 19th Jan. 1982, pp 23 - 30.

9. P.A. Attwood & P. Swift "Cold Neutron Radiography of
   Mechanical Connectors" GB.10, this conference.

10. D.H.C. Harris & W.A.J. Seymour "Applications of Real-Time
    Neutron Radiography at Harwell" GB.3, this conference.

DIGITAL IMAGE PROCESSING FOR REAL-TIME NEUTRON RADIOGRAPHY

Shigenori FUJINE, Kenji YONEDA and Keiji KANDA

Research Reactor Institute, Kyoto University

Kumatori-cho, Sennan-gun, Osaka 590-04, Japan

ABSTRACT

Real-time neutron radiography (i.e. neutron television - NTV) system of the Kyoto University Research Reactor Institute (KURRI) has been practically applied to identifying the location of boron burnable poison in the side plates of MTR type reactor fuel, to investigation of moving objects and to neutron computed tomography (NCT).

At present, however, a direct image from the TV system is still somehow in low-contrast and poor-resolution, image integration is effective in increasing image quality in neutron radiography.

The present paper describes several digital image processing approaches, such as image integration, adaptive smoothing and image enhancement, which have beneficial effects on image improvements. Details invisible in direct images of NTV can be revealed by digital image processing, such as reversed image, gray level correction, gray scale transformation, contoured image, subtraction technique, pseudo color display and so on. For real-time application a contouring operation and an averaging approach can also be utilized effectively.

INTRODUCTION

For neutron radiography (NR), photographic techniques have been mainly used for many years. To observe a dynamic event and to test many samples, however, it is desirable to be introduced the real-

537

time NR system. At KURRI, both a direct film and a dynamic imaging methods have been supplementally utilized earch other for these several years. In dynamic method, neutron radiographic images can be directly taken from a neutron converter screen with an image orthicon television camera (525 lines, 30 frames/sec) without image intensifier at the thermal neutron intensity of 1.2×10^6 n/cm^2.sec at the E-2 hole of the KUR.

The system demonstrated a spatial resolution of 0.5 mm for small holes in a thin Cd plate. The use of a digital image integrator has been employed with the NTV system to acquire high quality images. The image integration technique will widen the variety of application for the NTV. High quality images taken at the neutron flux of 10^6 n/cm^2.sec or less, make it possible to use the NTV system with neutron sources of lower intensity such as accelarators.

In addition to image integration, digital image processing is very useful for neutron radiography as described in the present paper.

For real-time application, an integration technique cannot be used because of processing time, however, contouring, averaging, gamma correction and/or enhanced operations can be effectively introduced.

SYSTEM CONFIGURATION

A block diagram of the KUR NTV system is illustrated in Fig. 1. A neutron radiographic image is directly taken from a neutron sensitive screen with an image orthicon camera without image intensifier. As a neutron converter, besides NE-426 or BN+ZnS(Ag) converter, KH (made by Konishiroku Photo. Industries Co.), G4 or G8 (made by Fuji Film Co.) screen (based on $Gd_2O_2S(Tb)$) can be used for the moment.

The radiographic images from the TV system are digitized through a high speed video analog-digital converter (ADC) (1/30 sec/frame). A digitized image with pseudo color has resolution of 640x480 pixels, 256 gray levels (IMAGE Σ). The device can digitally sum TV frames and feed resulting image to a CRT monitor and a TV image processor system (TVIP-2000) through a video D/A converter. A high performance personal computer (NEC-9801) is connected for image processing with a DMA interface to the TVIP-2000.

In the near future, an optical memory disc file system will be provided for image storage and image data base system.

EFFECTIVE NEUTRON EXPOSURE IN THE NTV SYSTEM

The effective exposure is equivalent to 4×10^4 n/cm^2 per television frame, as the thermal neutron beam intensity at the irradiation field is 1.2×10^6 n/cm^2.sec at the KUR power of

538

5000kW. Therefore, an integrated frame by summing 400 frames collects a total exposure of 1.6×10^7 n/cm^2.

At the reactor power of 500kW, the values are 1.2×10^5 n/cm^2 in a direct image (30 frames) and 4.8×10^6 n/cm^2 in an integrated image of summing 1200 frames (Fig. 3i).

The image with the best quality at 5000kW was obtained at the exposure of about 10^7 n/cm^2. An effective exposure of this NTV system becomes two orders in magnitude smaller than the value with a direct film method using a Cd metal screen with a neutron exposure of 15 min. duration at 5000kW (equivalent to 1.1×10^9 n/cm^2, Figs. 1i and 6i).

For moving events, it is required to obtain a clear image because of real-time application (4×10^4 n/cm^2).

DIGITAL IMAGE PROCESSING

Image quality can be simply improved by integrating over many images for a stationary event [4], and digital image processing for high quality image widened the variety of application for the NTV system [3,4,5] as shown in Table 1.

Figure 2 shows the neutron radiographic images of the ASTM standard indicators, and reversed image has usually supplemental effects in these images (Figs. 2iv and 4ii).

In Fig. 3, images at 500kW and 5000kW are presented. Figure 3i suggests that the NTV system will be applicable to use in a low intensity field like as accelerators. In comparison with Fig. 3ii taken at a high intensity field, contrast in an image (Fig. 3i) is still a little lower.

Where a neutron beam intensity of about 10^5 n/cm^2 sec or less is available, an image intensifier must help to make an image clearly.

Figure 4iii shows an image after a 3x3 matrix filter operation with noise reduction and contoured effects. Figure 4iv gives an enhanced image after gray level slide with intermediate level enhancement.

Figure 5 presents some other examples of useful processing, namely, gamma correction (Fig. 5ii), histgram equalization (Fig. 5iii), gray level transformation (Fig. 5iv), enlarged image (Fig. 5v) and contoured processing by a Laplacian filter.

APPLICATIONS

(1) The system has been used to identify the location of · burnable poison in the side plates of MTR type reactor fuel manufactured by CERCA in France [1,2]. From the photos in Fig. 6, the irregular locations of burnable poison in ten side plates are clearly observed, but they nevertheless satisfy the manufacturing specification.

(2) Figure 7ii gives a subtracted image, which shows a nylon string behind a lead block of 5 cm in thickness. The string was not so clear in an integrated image (Fig. 7i), and was obviously enhanced in a subtraction method [4].

(3) For real-time applications, filter operation, averaging and contoured images, are very effective for studying moving objects. A contoured image is used for bubble-motion investigations in water as shown in Fig. 8ii, where the air bubble and water level are clearly revealed [4].

(4) Image integration in NCT is particularly valuable for the measurement of projection data. The high quality images (for example Fig. 3ii) make it easy to discern the small density differences between stainless steel and copper [3]. Figure 9 illustrates a dot-image printing of an 81x81 reconstructed image produced by the Shepp and Logan method [6].

RESULTS AND DISCUSSION

The poor-resolution and low-contrast image from the NTV system has limited the application of the NTV system in the past. At KURRI, however, images of high contrast and resolution were obtained through image integration, and the system has provided an experimental spatial resolution of about 0.5 mm for an integrated exposure of 1.6×10^7 n/cm^2.

The primary image could be taken at a flux level of 10^5 n/cm^2·sec and integrated as presented in Fig. 3i. In comparison with an image obtained at high power (Fig. 3ii), the contrast is still a little lower, nevertheless, the NTV system will be possibly used to practical applications at a lower intensity field.

This paper described about image improvements with image restoration such as integration, smoothing and median or another matrix filter, and with image enhancement such as gray level correction (gamma correction), gray scale transformation (histgram modification), sharpening (Laplacian filter and contouring) and pseudo color display, as summarized in Table 1.

It will be important to be used these digital processing approaches as the case may be reasonable.

In application to identification of the location of boron burnable poison in the side plates of MTR type reactor fuel, boron parts were more evident in reversed and contoured images (Fig. 6iii and Reference [2]) than in normal image, and pseudo color display was noteworthy (Fig. 6iv).

A material in a low contrast image, for example the nylon string in Fig. 7i, was also distinctly enhanced in a subtracted image as Fig. 7ii.

The contoured image was particularly effective when used to observe moving objects in real time. Compared with Fig. 8i, the moving air bubble and water level are more clearly presented in Fig. 8ii.

540

For NCT, projection images of high quality could generate a clear reconstructed image capable of revealing small holes filled with water and of distinguishing the slight density difference between stainless steel and copper as shown in Fig. 9.

The image processing approaches will widen the variety of applications for real-time neutron radiography.

To obtain a higher-quality radiographic image, (1) some advanced neutron converters, (2) a high sensitive and resolution TV camera, and (3) a high performance image processing system are under development at KURRI.

This work was partially supported by a Grant-in-Aid from the Ministry of Education, Science and Culture (1984). The authors would like to express their thanks to Prof. G. Matsumoto of the Nagoya University for suggestions and discussion and also to members of Nuclear Reactor Division of their institute for their assistance in carrying out the experiments.

REFERENCES

1. K. Kanda, K. Yoneda and S. Fujine; "Development of an Online Neutron Radiography System of High Resolution for Nuclear Materials", pp. 219-225 in Neutron Radiography, J.P. Barton and J. von der Hardt, Editors, D. Reidel Publ. Co., Dordrecht, Boston (1983).

2. S. Fujine, K. Yoneda and K. Kanda; "An On-line Video Image Processing System for Real-time Neutron Radiography", Nucl. Instr. Meth., 215, pp. 277-289 (1983).

3. S. Fujine, K. Yoneda and K. Kanda; "An Application of the Neutron Television Fluoroscopic System to Neutron Computed Tomography", ibid., 226, pp. 475-481 (1984).

4. S. Fujine, K. Yoneda and K. Kanda; "Digital Processing to Imorove Image Quality in Real-time Neutron Radiography", ibid., 228, pp. 541-548 (1985).

5. S. Fujine and K. Yoneda; "Digital Image Processing for Neutron Television Fluoroscopic System and Its Application to Neutron Computed Tomography", J. At. Energy Soc. Japan, Vol. 26, No. 9, pp. 793-801 (1984) (in Japanese).

6. L.A. Shepp and B.F. Logan; "The Fourier Reconstruction in Computed Tomography", IEEE Trans. Nucl. Sci., NS-21, pp. 21-43 (1974).

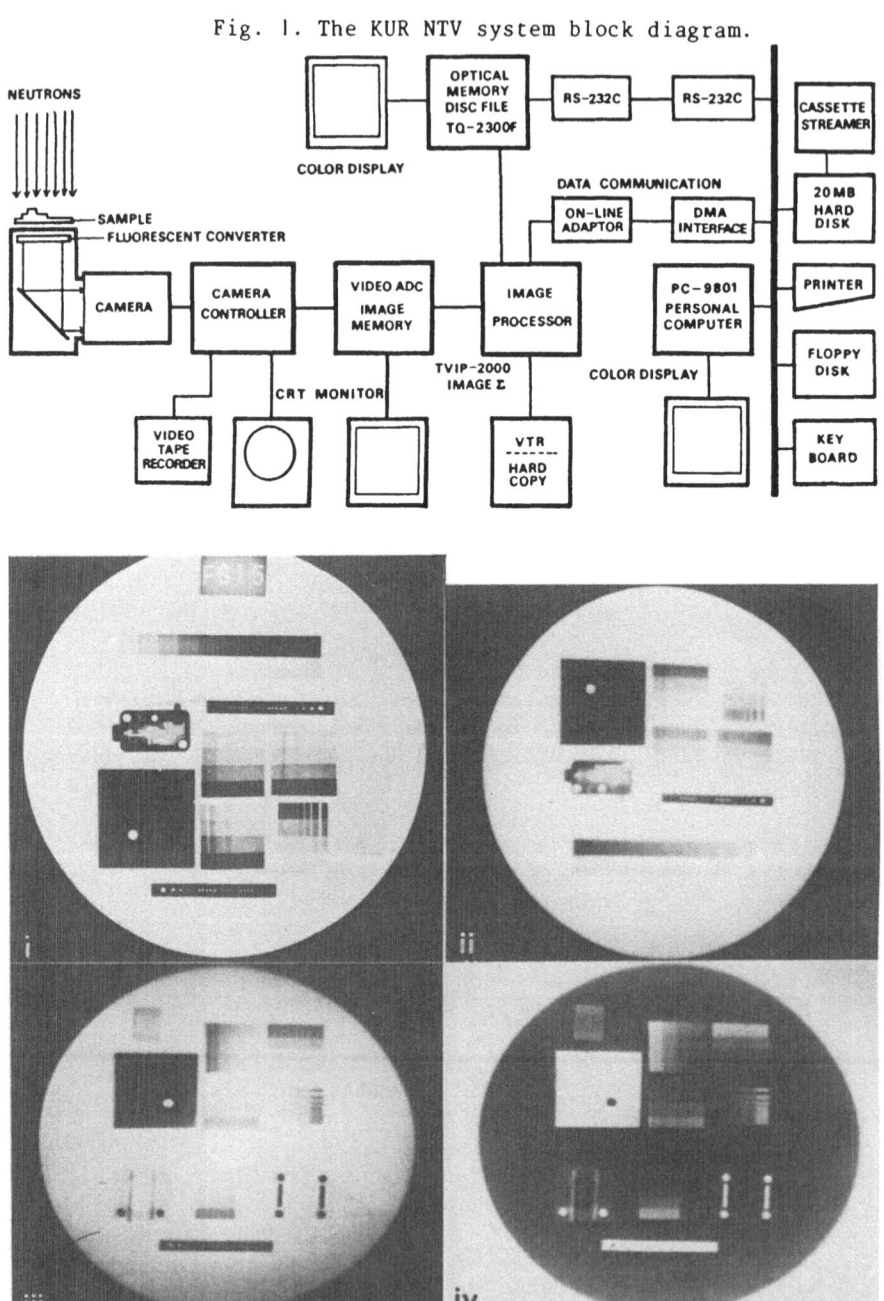

Fig. 1. The KUR NTV system block diagram.

Fig. 2. ASTM IQI images. i: The film method using vacuum cassette with Gd converter and single emulsion FG film [$1.1 \times 10^9 n/cm^2$]. ii: Integrated image at 5MW [$1.6 \times 10^7 n/cm^2$]($BN+ZnS(Ag)$). iii: Integrated image at 5MW ($KH, Gd_2O_2S(Tb)$). iv: Reversed image of fig. 2iii.

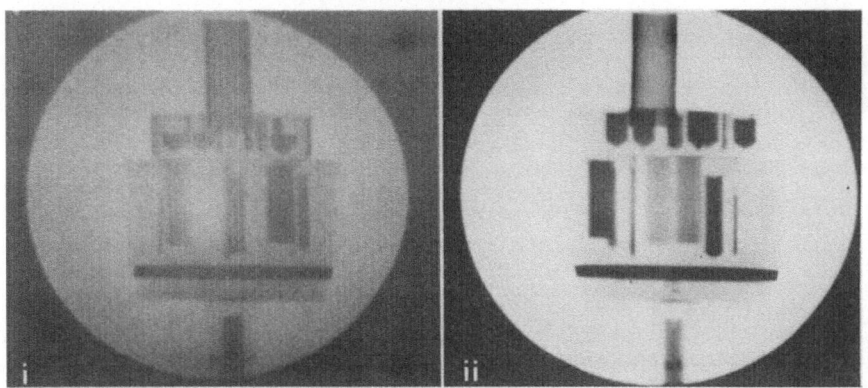

Fig. 3. Images at 500kW and 5000kW. i: Integrated image by summing 1200 frames at 500kW [$4.8 \times 10^6 n/cm^2$]. ii: Integrated image by summing 400 frames at 5MW [$1.6 \times 10^7 n/cm^2$].

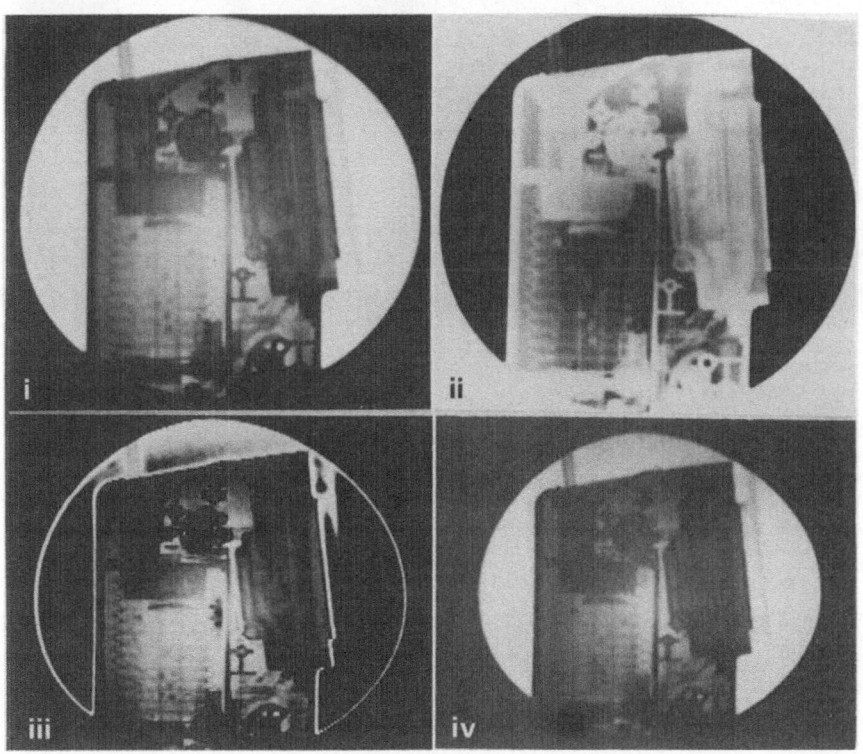

Fig. 4. Images of small steam iron taken at 5MW. i: Integrated image (400 frames). ii: Reversed image of fig. 4i. iii: Enhanced image after 3x3 matrix filter operation and gamma correction. iv: Enhanced image after saw level slide operation.

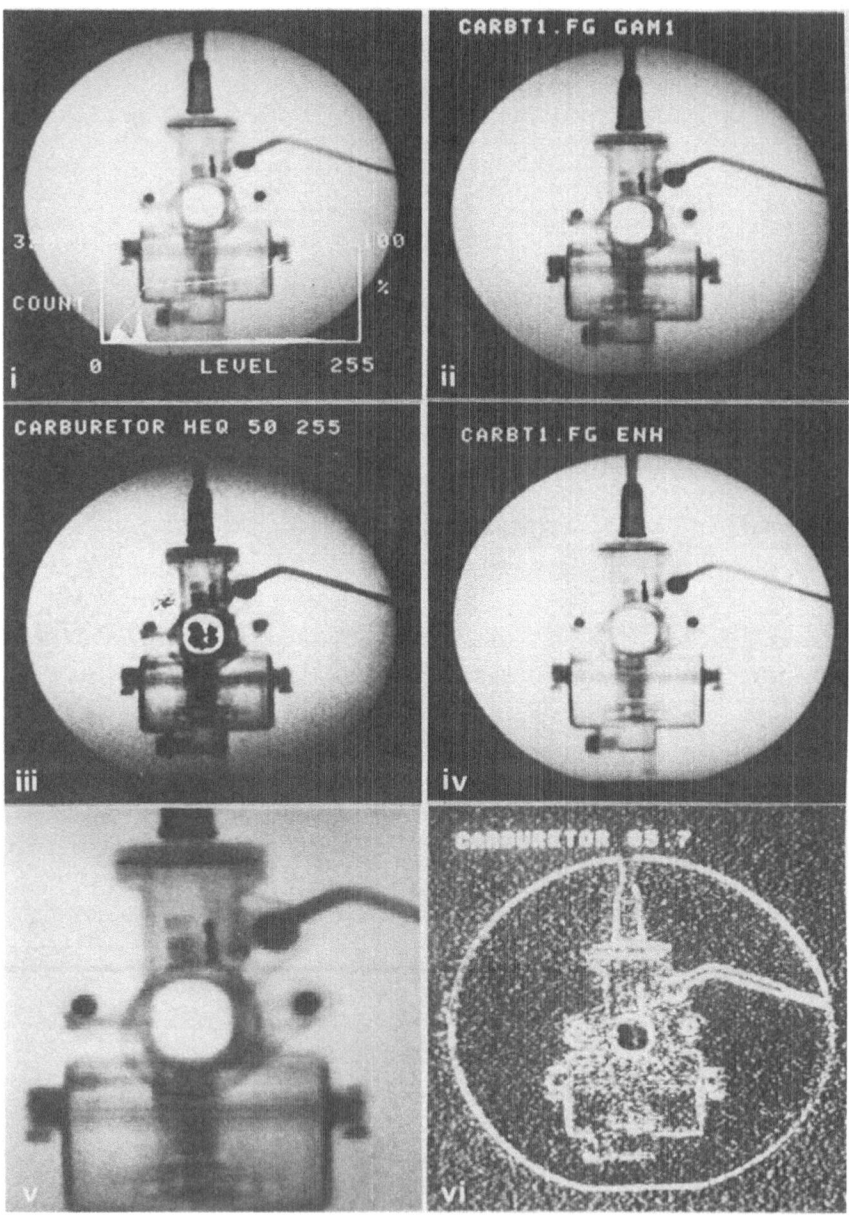

Fig. 5. Images of a carburetor for a motor cycle (KH converter).
i: Integrated image. ii: Enhanced image by gamma correction. iii:
Enhanced image after histogram equalization. iv: Enhanced image
after gray scale trasformation. v: Enlarged image by zooming
operation. vi: Contoured image by Laplacian filter operation.

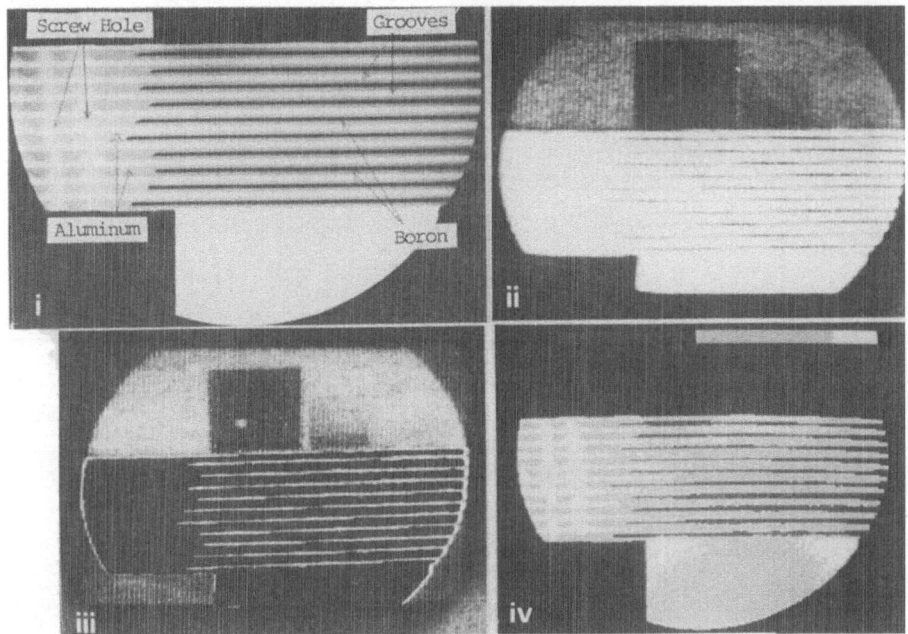

Fig. 6. Location-identification of burnable poison in side plates.
i: Image of Gd-direct method [1∶1x10⁹n/cm²]. ii: Direct TV image
at 2MW [4.8x10⁵n/cm²](NE-426). iii: Contoured image of Fig. 6ii.
iv: Pseudo color image enhanced by gray scale transformation.

Table 1. Digital Processing for NTV Images

DIRECT IMAGE ------------ Low Contrast and Poor Resolution
     Direct   Image   ---- *Real-Time Application   (Figs. 6ii and 8i)
     Reversed Image   ---- *Identification of the Location
     Contoured Image   ---- ⎰ of Burnable Poison in the Side   (Fig. 6iii)
     Pseudo Color Image --- ⎱ Plates of the MTR Type Fuel   (Fig. 6iv)
     Enhanced Image   ---- *Real-Time Application   (Fig. 8ii)
     Average Image   ---- *Real-Time Application   (Fig. 8ii)

INTEGRATED IMAGE -------- High Sensitivity and High Resolution
     High Sensitivity   ---- *Capability at Low Intensity   (Fig. 3i)
     High Quality   ---- *High Resolution   (Figs. 2ii, 2iii, 3ii and 4i)
                        ---- *Neutron Computed Tomography   (Fig. 3ii)
     Reversed Image   ---- *Complemental Use   (Figs. 2iv and 4ii)
     Subtracted Image   ---- *Low Contrast Image   (Fig. 7ii)
     Zooming Image   ---- *Enlarged Image   (Fig. 5v)

ENHANCEMENT
     Gamma Correction   ---- *Gray Level Transformation (Figs. 4iii and 5ii)
                        ---- *Real-Time Application
     Matrix Filter   ---- *Noise Reduction and Contouring (4iii and 5vi)
     Histgram Modification ------ *Enhancement of Intermediate (Fig. 5iii)
     Gray Level Transformation -- ⎰   Level   (Figs. 4iv and 5iv)

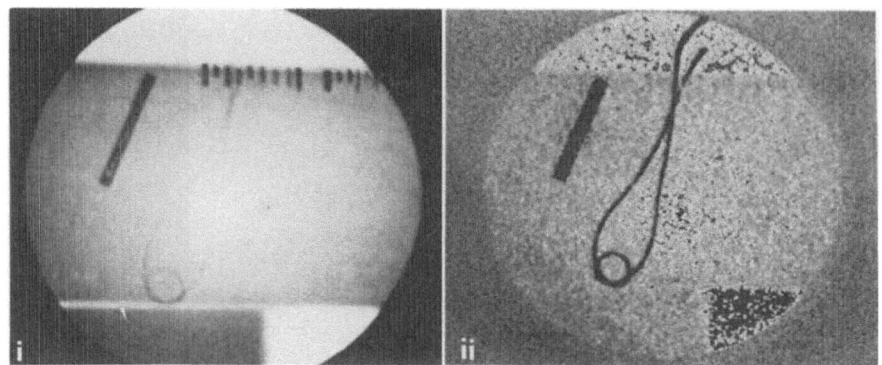

Fig. 7. Images of a nylon string behind a lead block (5 cm in thickness). i: Integrated image (400 frames). ii: Subtracted image.

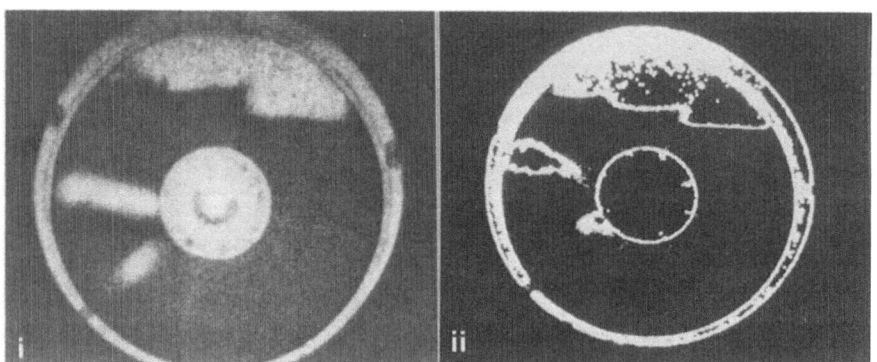

Fig. 8. Moving images. i:Direct image of a rotating water vane at 5MW [$1.5 \times 10^5 n/cm^2$]. ii: Contoured image by enhancing air bubble and water level.

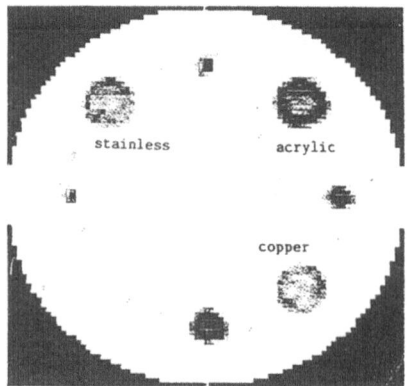

Fig. 9. Reconstructed image with 16 gray levels.

REAL-TIME IMAGING WITH LESS INTENSE NEUTRON BEAM

G. Matsumoto, K. Ohkubo and Y. Ikeda

Department of Nuclear engineering,
Faculty of Engineering, Nagoya university

Furo-cho, Chikusa-ku, Nagoya 464 Japan

ABSTRACT

The real-time imaging was tried with small and weak neutron sources. A separate type imaging system using an image orthicon tube was assembled for the tests. The neutron fluxs available at the exposure plane for these sources were $3x10^3$ to $3x10^5$ n/cm^2 .s. Dynamic motions of samples could not be obtained at the flux less than 10^4 n/cm^2.s. But at 10^5 n/cm^2.s, fairly good NTV images of dynamic motions were obtained. Image processing by electronic computers was very effective to improve their image quality.

INTRODUCTION

To realize a portable neutron television (NTV) facility for in-situ non-destructive tests, less intense neutron beam from a sealed-tube neutron generator as well as other weak sources, such as a small accelerator (Baby-cyclotron) or a radionuclide isotope should be utilized. The neutron beam intensities at the exposure positions in these NR facilities are from $3x10^3$ to $3x10^5$n/cm^2 .s. The Kinki University Research Reactor, whose thermal output is just one watt, can be used with a simple collimator to change the beam intensity in a fairly low range by selecting the thermal output. In order to study the feasibility of NTV at these rather low neutron intensities, some NTV imaging experiments were carried out in this work.

The sensitivity of the imaging system used with such tansportable NR equipments should be as high as possible. For usual much intense reactor sources, various converters and TV cameras are possible to use. However, for the weaker sources the usable converters and TV systems may be very restricted. There are two types of NTV systems, one being an image-intensifier type, which is ued in DIDO facility, and another being a separate type, whose converter and TV camera are arranged separately. In spite of fairly lower cost, the latter system can takes various variations of selecting the converter and the TV camera employed. It is anothor advantage to be easily able to change the parts if they are damaged. Therefore, the latter separate system was used in this experiment. For the sake of testing, several scintillation converters including a comercially avalable NE-426 and an image orthicon camera with or without an image-intensifier were prepared and employed case by case. A SIT camera was also tested instead of an image orthicon.

At the beam intensity of 3×10^5 n/cm^2.s, fairly clear images of dynamic motion of samples were obtained without an image intensifier. But with neutron fluxs of 3×10^3- 1×10^4 n/cm^2.s, only static or nearly static samples could be imaged. In the latter case, a handy-type image processer was used and gave an appreciable improvement in image quality.

A movable neutron TV system in the Hot Laboratory Water Pool at Nagoya University was designed and constructed. A plan to convert it to a portable one by addition of radiation shieldings is in progress now.

TEST FACILITIES

The neutron sources used for the test are the following.
(1) Kaman sealed-tube neutron generator (Nagoya University,:NU),
(2) Baby-cyclotron (Japan Steel Works, Co. :JSW),
(3) UTR (Kinki University Reactor, Higashi-Ohsaka).
The characteristics of these facilities are shown in Table 1, together with the Kyoto University Reactor (KUR) E-2 facility used as the reference in this experiment.

The schematic diagram of the NU NTV system is shown in Fig.1*1). Its imaging system consists from a NE-426 fluorescent converter followed by a TV camera. The latter is C-1679-P1 type 3-inch image orthicon tube with F-1217 image intensifier purchased from HAMAMATSU PHOTONICS Co.. A NIKON optical lens (F1.2) was set at the entrance of the camera. The output signal of it was sent to a monitor display and an image processer (Image Σ purchased from Japan AVIONICS Co.) simultaneously to record the image on a video tape. In the case of JSW, a BN converter *2, which

Table 1 Characteristics of utilized NR facilities

Facility	KUR E-2	JSW	NU	UTR
Thermal Neutron Flux (n/cm^2.s)	1.2×10^6	3.0×10^5	1.0×10^4	3×10^4
Cd ratio	400	5	5	3
n/ ratio(n/cm^2.mr)	1.0×10^6	3.0×10^5	1.9×10^4	1×10^4
L/D ratio	100	76	28	21
Beam Dimension	160mm dia.	360x435mm^2	260x260mm^2	200mm dia.
Source type	Research Reactor	Be(p,n)B	T(d,n)He	Research Reactor

was developed by the coordination of SHOWA-DENkOH Co. with us, was used in some caes. In this work, the NU NTV system was transported to JSW or UTR for imaging test. In addition to these, a image processer, TOSPIX 2 (TOSHIBA Co.), was intermitently used for image processing.

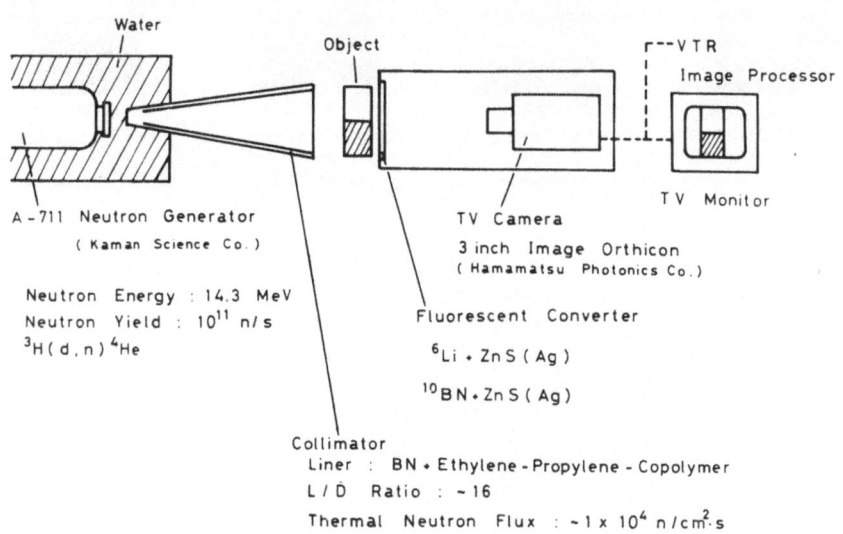

Water
Object
VTR
Image Processor
TV Monitor

A-711 Neutron Generator
(Kaman Science Co.)

Neutron Energy : 14.3 MeV
Neutron Yield : 10^{11} n/s
^{3}H(d,n)^{4}He

TV Camera
3 inch Image Orthicon
(Hamamatsu Photonics Co.)

Fluorescent Converter
^{6}Li・ZnS(Ag)
^{10}BN・ZnS(Ag)

Collimator
Liner : BN・Ethylene-Propylene-Copolymer
L/D Ratio : ~16
Thermal Neutron Flux : ~1 x 10^4 n/cm^2.s

Fig.1 Schematic diagram of the movable NTV system

EXPERIMENTS

<JSW>

The main subjects of this experiment were the real-time observations of the dynamic behaviors of hydrogeneous materials inside steel or aluminium vessels. The tests we carried out were the followings,

    (a) Water behavior in a water thermosiphon*3,
    (b) Oil evaporation in a vacuum diffusion pump,
    (c) Motion of water and air bubbles in a steel pipe*3,
    (d) Water behavior in carbureters of an automobil and an autobicycle,
    (e) Water behavior in a two stage water heat pipe,
    (f) Lithium melting in a simulation model of a liquid metal heat pipe,
    (g) Behavior of working fluid in a Na-Li heat pipe.

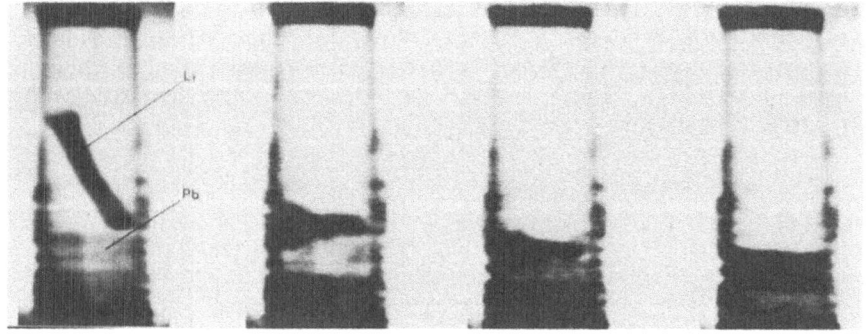

PRE-MELTING   1 SEC AFTER   40 SEC. AFTER   60 SEC. AFTER

Li MELTING   Li MELTING   Li MELTING

Fig.2   Real-time NTV images of Pb-Li co-melting
(By JSW Baby-cyclotron, In tegration of
8 frames)

These imaging were performed at KUR E2 facility in collaboration with the KUR investigators. Afterward the similar experiments were repeated at JSW facility. The image quality of the latter case was a little less than that with the KUR E2, but for the sake of industrial applications, the NTV at JSW was fully acceptable. Figure 2 shows the photograph of the (f) by the JSW facility.

<NU>

Up to now, the sample objects imaged by this movable NTV facility have been all static, not dynamic, because of its insufficient capability. They were Cd plates with holes of various sizes and a steel rectangular box of 4x4x12 cm whose wall was 3 mm thick (Fig.3a). Figure 3b is the NTV image by the NU facility. The photograph of 3b is obtained after image processing by TOSPIX 2.

Fig.3a
Samples for NTV (Cd plate with holes and Steel vessel, 40x40x 125 mm, wallthickness of 3 mm, containing water in it)

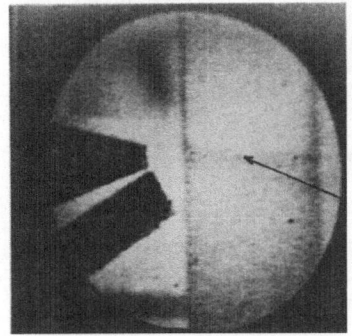

Fig.3b
NTV image of the samples of Fig. 3a with the movable NTV system in the Pool of the Hot Laboratory at Nagoya University
(Integration of 900 frames)

Water level

<UTR>

Figure 4b is the NTV image of Cd plates with holes shown in Fig.4a by the UTR facility. The neutron beam intensity was 3×10^4 n/cm^2.s at the exposure position at 1 watt thermal output operation. And Fig.4c is the NTV image at 0.1 watt, so that the neutron beam intensity was estimated only about 3×10^3 n/cm^2.s.

UNIT: mm

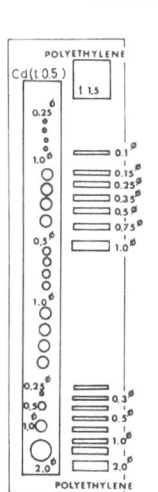

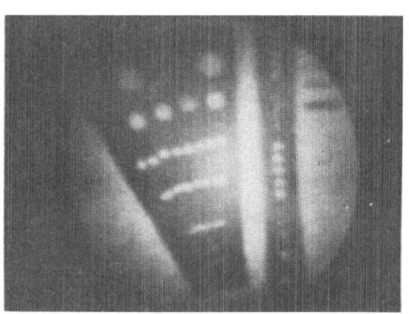

Fig.4b NTV image of the samples of Fig. 4a at the thermal output of 1 watt of UTR (Neuron flux: 3×10^4 n/cm^2.s, Integration of 900 frames)

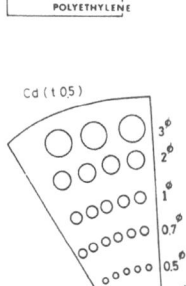

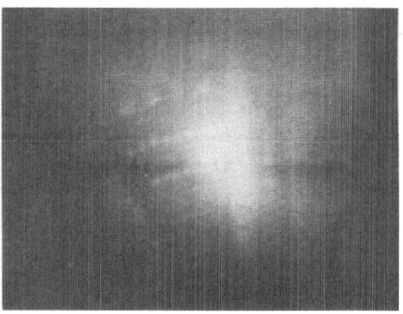

Fig.4a
Samples for UTR NTV experiment

Fig.4c NTV image of the samples at the thermal output of 0.1 watt of UTR (Neutron flux: 3×10^3 n/cm^2.s, Integration of 900 frames)

SUMMARY OF THE RESULTS

Using the neutron beam intensity of 3×10^5 $n/cm^2.s$ at JSW, the real-time imaging of dynamic behavior of various samples could be obtained with good image quality for industrial applications. But at NU and UTR, insufficient beam intensity of $3 \times 10^3 - 3 \times 10^4$ $n/cm^2.s$ was not enough to get image of dynamic motion of samples.

The frame-number of this NTV camera is 30/s. Therefore, one frame interval is 1/30 sec. During this time interval the number of thermal neutrons which pass through the $1mm^2$ area perpendicular to the neutron beam's derection is calculated to be

$$3 \times 10^4 \ (n/cm^2.s) \times 1/100 \ (cm^2/mm^2) \times 1/30 \ (s) = 10 \ (n/mm^2)$$

Because of the quantum effects of the netron sources, the pertubation of the neutron beam intensity may be large, that is, the standard deviation of the neutron coming in one frame interval is about 3.3. Furthermore, the quantum effects originating in NTV imaging system components superpose on them. Therefore, rather noisy images are given in these cases. In order to cancel those noises, a sort of image integration has to be innevitable. In one case of this work, 900 frames were integrated to give fairly good images, which are corresponding to 30 sec exposure. These image processing gave us some hopes to materialize a portable NTV facility.

References

(1) G.Matsumoto et al., Nucl.tecnol., 68(1),pp94-101(1985)
(2) G.Matsumoto et al., Material evaluations, 42(11), pp1379-1383,1388 (1984)
(3) G.Matsumoto et al., "Real-time Imaging of Working Fluids in Heat Pipes by Neutron Television" The Proc. Part 2, Fifth Int. Heat pipe Conf., May 14-18 (1984), Tsukuba Science City, Japan, pp 133-138

STATISTICAL PROPERTIES OF REAL-TIME NEUTRON RADIOGRAPHIC IMAGE

Ryoichi TANIGUCHI and Eiichi HIRAOKA
Radiation Center of Osaka Prefecture, Sakai, 593 Japan

Atsuo ONO and Koichi SONODA
College of Liberal Arts, Kobe University, Nada-ku, Kobe, 657 Japan

Shuichi TAZAWA and Takehiko NAKANII
Sumitomo Heavy Industries, Ltd., Niihama, 792 Japan

Abstract

Actually statistical quality of raw images produced by neutron radiography is very poor. We analysed the statistical properties of neutron images obtained from the cyclotron-based real-time neutron radiography system of Sumitomo Heavy Industries, Ltd., which uses a set of LiF/ZnS(Ag) screen and high sensitive TV camera.

The distribution of image intensity at a constant neutron flux domain is represented well by a Poisson distribution and ripple noises appearing in the real-time image can be attributed to the fluctuation of the distribution of neutron flux. These suggest that some digital image processing techniques based on a local statistical method would be effective for a ruffled image like real-time neutron radiography. Some of the examples using contrast stretching, median filter etc. are also presented.

Introduction

In recent years, real-time neutron radiography (NRG) has been developed for some industrial applications. It has the following significant characteristics. First, sensitivity is high so that the requisite neutron fluence per one neutron image can be reduced by three or four order magnitude compared with the normal NRG. This would have wide applicability to various objects, and release us from the problem of the radio-activation with neutron irradiation. Second, the real-time NRG would be capable of observing dynamic behaviour of some objects. At the last point, the output data, generally TV video signals, can be processed by the digital image processing technique which has been remarkably advanced [1].

Up to the present, all the attempts of real-time NRG have
attained by utilizing the high sensitive TV camera in place of the
photo-emulsion. The neutron fluence per one frame, generally 33
msec in the TV system, is regretably insufficient. For instance,
neutron flux of 10^5 n/cm^{2}sec, even if it may be enough for the
conventional NRG, corresponds to only 33 n/mm^2 per one TV frame
for real-time image. Fig. 1 shows the neutron signal flow of our
real-time NRG, using a sub-compact cyclotron of Sumitomo Heavy
Industries, Ltd. A neutron signal glows to nearly one hundred
photon at the SIT (Silicon Intensifier Target) camera. A gamma
ray's response, as a comparison, was also shown by a broken line
in the same figure. Neutron signal, though amplified to pertinent
signal level, has insufficient statistical confidence. Actually
raw images of real-time NRG are noiseful. Fig. 2 shows a typical
sample of one frame image. These noises are usually reduced by
the image-integration, as shown in Fig. 3 and ít suggests that the
origin of these noises is some random phenomena. In this paper,
we discuss about the property of the noise by using statistical
techniques, and about the origin of it. The limit of the
resolution and suitable image processing method for the real-time
NRG are also discussed.

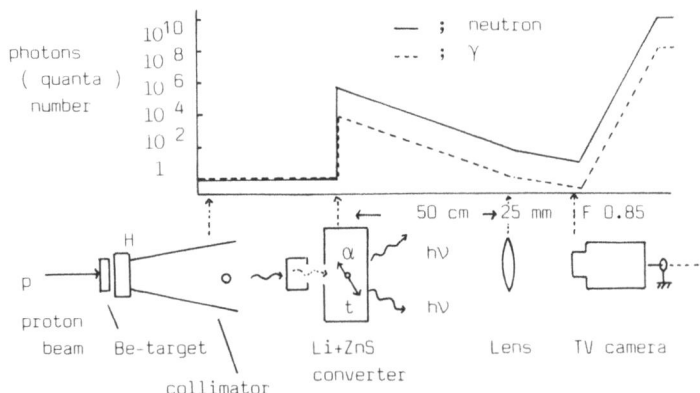

Fig.1   Schematic signal flow of the real-time NRG system and the
        numbers of quanta (photons, neutrons, electrons) per one
        event of neutron bombardment in the conversion screen

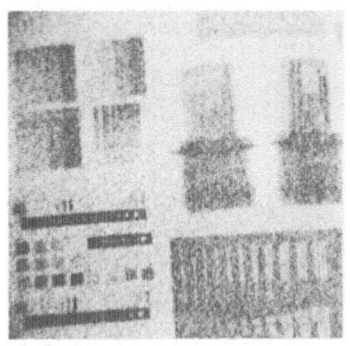 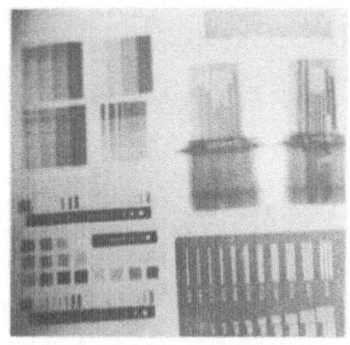

Fig.2 Typical real-time picture of
one frame for hollow blade

Fig.3 Image integrated picture
for 256 frames

Noise analysis

We adopt here a 'local statistical analysis method' [2] which
is based on the assumption that the mean value and variance for
the samples at the same picture element (pixel) are equal to the
respective local mean and variance arround there. First,
attention will be focussed on redefinition of local mean and
variance.

In digital image processing, the image is represented by a
finite array of brightness values, each pixel being quantized to a
finite number of bits (8 in our case). Let $x_{i,j}$ be the brightness
of the (i,j)-th pixel in a two-dimensional image. Then the local
mean and variance are defined as

$$f_{i,j} = \frac{1}{(2n+1)(2m+1)} \sum_{k=i-n}^{i+n} \sum_{l=j-m}^{j+m} x_{k,l} \qquad (2,1)$$

$$s^2_{i,j} = \frac{1}{(2n+1)(2m+1)} \sum_{k=i-n}^{i+n} \sum_{l=j-m}^{j+m} (x_{k,l} - f_{i,j})^2 \qquad (2,2)$$

Now, we consider a response model of the neutron image in
which the brightness $x_{i,j}$ is linear to the neutron fluence $n_{i,j}$ in
the same region, that is

$$x_{i,j} = an_{ij} + b \qquad (2,3)$$

Here, b is a response part independent of neutron fluence.

The fluctuation of neutron flux would be represented by a

557

Poisson distribution. Thus, the variance of n is represented by n
and equation (2,3) makes

$$f_{i,j} = a \, n_{i,j} + b \tag{2,4}$$

$$s^2_{i,j} = a^2 n_{i,j} + s^2_b \tag{2,5}$$

where s_b is a variance of b.

From equations (2,4) and (2,5), n or a can be eliminated and

or
$$s^2_{i,j} = (1/n_{i,j})(f_{i,j} - b)^2 + s^2_b \tag{2,6}$$

$$s^2_{i,j} = a(f_{i,j} - b) + s^2_b \tag{2,7}$$

Thus the relation of local mean and variance is second order in
the region of constant neutron fluence, and linear in the region
of the same a value.

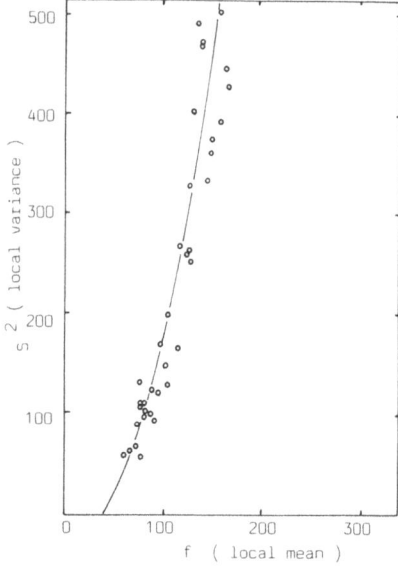

Fig.4 Relation of local mean value and
variance at a constant flux domain

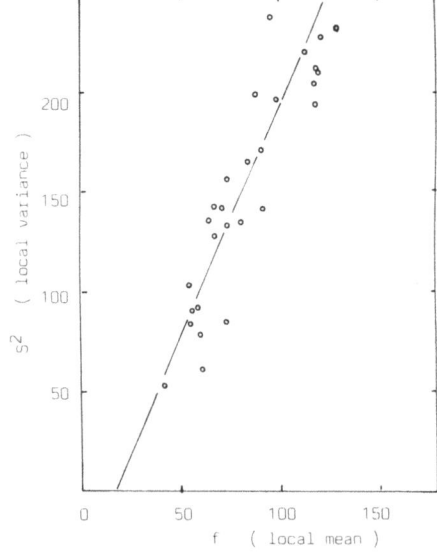

Fig.5 Relation of local mean value and
variance in a region affected with
constant conversion gain

558

On the real-time NRG image, incident neutron flux on the screen is considered to be nearly constant in the absent region of any specimen. Thus the validity of equation (2,6) would be checked with samples in these regions. The picture, shown in Fig. 2 was sampled (m = 4). The obtained relation is shown in Fig. 4. The curve of second order in this figure was obtained by applying the method of least-squares to these data. As is shown, these data well conform to our supposition, to be second order function. In addition, the neutron fluence $n_{i,j}$ could be calculated with the coefficient of secondary part of equation (2,6). In the case of Fig. 4, $n_{i,j}$ was 45. Considering one frame interval (33 msec) and an area of one pixel (0.33 x 0.38 mm^2), the neutron flux on this picture was estimated to be about 1.07×10^6 $n/cm^2 sec$. This value shows good agreement with the neutron flux measured by other method, foil activation method etc.

On the other hand, similar calculation was operated for the samples in the regions where the conversion gain, from neutron to video signal, would be same in the same figure. Fig. 5 shows the obtained relation. This distribution can conform to linear function, agreeable to equation (2,7). These results seem to support the validity of the response model given by equation (2,3), described above.

Limit of contrast resolution

The contrast resolution of real-time NRG image, similarly to the spatial resolution, depends on the neutron flux. Therefore, by means of quantum statistics, the relationship between spatial and contrast resolution's limits can be estimated.

Now, we define the still-recognisable contrast as s/f, the ratio of noise-to-signal level. In equation (2,3), a is assumed to be local function, then

$$f_{i,j} = a_{i,j} n_{i,j} N + bN \qquad (3,1)$$

$$s_{i,j}^2 = a^2 n_{i,j} N + s_a^2 (n_{i,j} N)^2 + s_b^2 N \qquad (3,2)$$

where N is the integration times.

Thus, the resolution between contrast resolution and integration times is given by

$$\frac{s_{i,j}}{f_{i,j}} = \frac{\sqrt{s_a^2 n_{i,j}^2 + (s_b^2 + a^2 n_{i,j})/N}}{a n_{i,j} + b} \qquad (3,3)$$

When the variance of a is small, first part of (3,3) is negligible. Then

$$\frac{s}{f} = (1 + \frac{b}{a n_{i,j}})^{-1} \frac{1}{n_{i,j} N} \qquad (3,4)$$

559

Equation (3,4) predicts that the contrast resolution increases linearly to the square root of N.

In the experiment, integrated images, whose integration times were different each other, were prepared. Fig. 6 shows the result deduced from them. It seems that the contrast resolution is proportional to the reciprocal of the square root of N. At the limit of N, however, the contrast resolution appeared to have a finite value. From equation (3,3), the limit would be represented as follows.

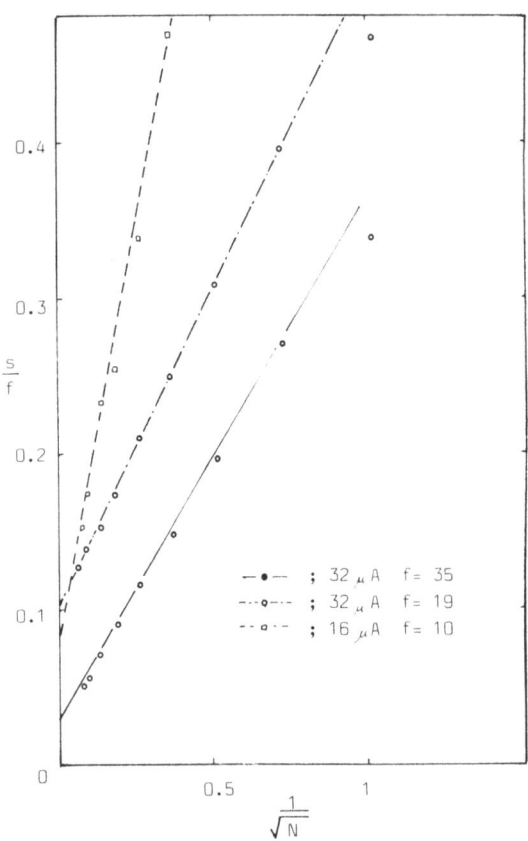

Fig.6 Relation between contrast resolution and the number of image integration times with the parameter of the beam current of the cyclotron and the mean value of the image

$$\lim_{N\to\infty}\frac{s}{f} = \frac{\sqrt{s_a^2 n_{i,j}^2}}{an+b} = \frac{s_a}{a} \qquad (3,5)$$

It means that the limit of contrast resolution, or image restoration by image integration, is restricted by the fluctuation of a.

Image processing for the real-time NRG

The following has been clarified with the above discussion.
(1) The ripple noises, frequently appeared in real-time NRG images, are mainly due to the fluctuation of neutron response.
(2) The limit of image restoration would be determined with the fluctuation of the conversion gain, neutron to video signal.

For the improvement of the quality of real-time NRG image, neutron flux should be increased and brightness of the optical system also improved.

With the view to develop the image processing for real-time NRG, emphasis should be placed on image enhancement. In addition, the conventional image processing method, premised a continuous data, may be unsuitable for the ruffled image as real-time NRG. Image processing methods suitable for discontinuous data are expected to be developed. Local statistical image processing [2], similar method was used in the previous section as the analytical technique, would be effective. Table 1 shows these algorithms and Fig. 7 shows the results of the processing. Another characteristic of these algorithms is the computational economy, in contrast with the spatial filtering and others. It would be favourable for parallel processing and real-time processing. Future research is to extend the method of the real-time and dynamic inspection, with the use of the local statistical processor in the hardware.

Table 1. The list of the image processing method by use of local statistics to discuss in our presentation

	Algorithm	Function
1	noise filter	noise reduction and image restoration
2	contrast stretching	contrast enhancement for the detail of the image
3	median filter	noise suppression

This work was supported in part by a Grant-in-Aid from the Ministry of Science and Technology Agency.

<u>References</u>

[1] R.E.Twogood, F.G.Sommer, IEEE Trns. Nucl. Sci. NS-29 No. 3
    (1982) 1076.

[2] Jong-Sen Lee, Computer graphics and image processing 15 (1981)
    380.

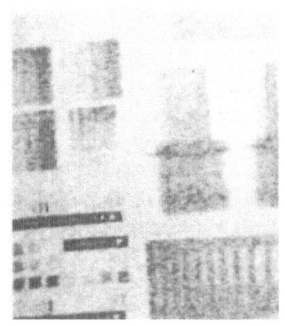

(1) noise filter   (2) contrast   (3) median
                            stretching   filter

Fig.7 The results of the image processing, applied on the raw image (1),(3)
      and the integrated image (2).

CORROSION DETECTION BY REAL-TIME NEUTRON IMAGING*

Harold Berger
Industrial Quality,
Inc.
Gaithersburg, MD USA

Raulf Polichar
Science Applications
International Corp.
San Diego, CA USA

W.J. Rowe
Lockheed Georgia
Company
Marietta, GA USA

ABSTRACT

Neutron radiographic imaging offers advantages
for the detection of corrosion in metallic assemblies
because of the excellent neutron sensitivity for the
hydrogen in the corrosion product. Corrosion products usually are mixtures of hydroxides and often contain extra water from the moisture in the associated
environment. The relatively high slow-neutron attenuation for hydrogen coupled with the relative neutron
transparency for many metals leads to the prospect for
early detection of hidden corrosion in assemblies.
Neutron detection of corrosion depends on the hydrogen
content of the corrosion product buildup whereas other
nondestructive methods for corrosion detection, such
as x-ray imaging, ultrasonics or eddy currents,
depend on metal loss as the principal means for detection. These nondestructive testing approaches
are compared to real-time neutron radiographic methods for the detection of corrosion. Real-time imaging methods offer advantages in prompt response at
a remote location, capability for scanning and the
presentation of an electronic signal that lends itself to image processing. Enhancement techniques
such as frame averaging, beam flattening, and

* Work supported in part through Air Force Contract F33615-84-C-
5021, Wright-Patterson Air Force Base; Lockheed Georgia Company,
Prime Contractor.

gray-level stretching are all shown to be effective
in bringing out neutron radiographic corrosion
images. Detection capability in aluminum is shown
to be at least 0.18 mg H/cm^2, representing an aluminum metal loss of about 25 μm.

INTRODUCTION

Many nondestructive testing (NDT) methods are used to detect
corrosion in structures. Where inspection surfaces are accessible,
visual inspection can be very effective. This surface inspection
is sometimes supplemented by the use of borescopes or television
cameras to gain improved access. The other conventional NDT methods
for corrosion detection normally depend on the fact that some of
the base metal has been lost. In effect, the techniques depend on
detection of a reduced thickness, pitting or cracking as caused by
the corrosion process [1].

Neutron radiography, on the other hand, depends on the accumulation of the hydrogen-bearing corrosion product as the basis for
detection. The high neutron sensitivity for hydrogen and the high
probability that the corrosion product will be rich in hydrogen
(hydroxides, entrapped water, etc.) provide the basis for improved
corrosion detection by neutron methods. In this report, some comparisons are made for neutron radiography versus other NDT methods.
The primary neutron radiographic technique [2] discussed in this
report is the real-time, electronic detection method used with slow
neutrons. The discussion is limited to corrosion in metals.

NDT METHODS TO DETECT CORROSION

Ultrasonics [3-6], x-radiography [7,8] and eddy currents [9]
are used in field inspections to detect corrosion. Acoustic emission (AE) techniques [10,11] can also be applied but the use of AE
to detect corrosion by itself is limited. For ultrasonic (UT),
radiography (RT) and eddy current (ET) testing, the basis is loss
of metal (thinned structure) or signals from pits or cracks. The
basis for AE as applied to the direct detection of corrosion is
evolution of hydrogen; water or moisture must normally be present.
In addition, AE is effective for detecting cracking associated with
corrosion [11].

In ultrasonic testing, it is possible to detect scattering from
pits or cracks as indicators of corrosion. In addition a thickness
test approach is used either by causing the material to resonate or
by reflecting an ultrasonic pulse off the back surface. Both the
resonant and pulse-echo methods are sensitive, especially when
applied to a surface with two flat faces. Resonance methods are
capable of detecting thickness changes in the order of 0.1 percent.

Users of pulse-echo methods claim sensitivities as small as 0.025 mm in materials 1 to 250 mm thick. However, in a typical corrosion situation where one side is not flat or there is intervening structure, ultrasound is scattered and accuracy is reduced.

The eddy current technique is sensitive both to the change in thickness caused by corrosion thinning and also to the different electrical and magnetic properties of the corrosion product. Typical sensitivities are of the order of 1-2 percent thickness changes; Bond [9] claims sensitivities as small as 0.2 percent of the wall thickness for the eddy current technique. Phase sensitive eddy current methods offer potential for deeper penetration and far-side detection of corrosion.

Acoustic emission is being applied to the detection of corrosion in aluminum honeycomb aircraft parts at McClellan Air Force Base [12]. The technique has also been studied for large structures [13] and turbine blades [14].

Conventional radiography [15] offers the advantage that large areas can be inspected at one time. Differences in thickness, including small corrosion pits, on the order of 1 to 2 percent, can be visualized under good conditions (for example, when overlying structures do not mask detail). Real-time detection methods can be used [8].

NEUTRON RADIOGRAPHY

Neutron radiographic methods [16-23] differ from these other NDT approaches in that the primary mechanism for detection involves not the change in thickness of the basic structure but the accumulation of the corrosion product. The corrosion product contains hydrogen and is typically in the form of a hydroxide deposit. Slow neutrons are highly attenuated by hydrogen and yet are transmitted relatively easily through most metals - a situation directly opposite to that for x-rays. Neutron radiography has been applied to the detection of many hydrogenous materials [24] - rubber, water and adhesives as well as corrosion [25-29].

Some quantitative work has been reported for corrosion detection. Dance [25] investigated flat bottom holes filled with aluminum hydroxide. Such filled holes inside aluminum structure could be detected with neutron radiography to hole depths as small as 0.05 mm. In the NBS study, Garrett [28] used standardized approaches to create corrosion in aircraft grade aluminum. Corrosion resulting from a 6.5 hour super-saturated salt spray could be detected. These investigations were useful in demonstrating what could be achieved with neutron radiography. However, the key to the detection, the actual hydrogen content as determined by thickness and density (or weight per unit area) was not available.

Other investigations have been directed toward determining the amount of hydrogen that can be detected. Work has been reported using layers of cellulose [30] and hydrogen in zirconium alloys [31]; in the latter case, the limiting value reported was 0.66 mg H/cm^2. A 30 to 40X improvement in neutron sensitivity to hydrogen has been demonstrated using a scattering and filtering technique [32] but this low spatial resolution method offers only limited applicability for practical corrosion work because of sensitivity to any other hydrogen present and difficulty of practical field application.

Concerning the corrosion products a review of previous work reported in the literature showed very little experimental data on chemical composition or physical density of naturally occurring aluminum corrosion products. Where chemistry is discussed, aluminum hydroxide $Al(OH)_3$ and hydrated aluminum oxide $Al_2O_3.nH_2O$ are identified as the corrosion product compounds which contain hydrogen. This previous experimental work using neutron imaging can be used to derive a hydrogen density of 30 mg/cm^3 in the natural corrosion products studied.

Laboratory procedures were used in this program at Lockheed-Georgia to characterize samples of naturally occurring aluminum corrosion products in terms of hydrogen content. The primary method used on these dry powder samples was thermal analysis, in which the water vapor mass lost as the samples were heated was measured. The mean hydrogen density derived from several measurements was 27 mg/cm^3, a value close to that calculated from previous work.

In the present investigation, detection tests have been made with aluminum aircraft structure that included, (1) polymer layers with known hydrogen content, (2) flat bottom holes containing actual corrosion products from aircraft structure and (3) deliberately induced corrosion by exposure in environmental chambers. Tests with single and multiple layers of thin polyethylene film, 12.5 µm in thickness, have yielded quantitative data. Enhanced, real-time neutron radiographic images have shown good contrast even for a single layer of polyetheylene as described above; the detected hydrogen content was 0.18 mg H/cm^2. Figure 1 shows an enhanced real-time neutron radiographic image of single and multiple layers of polyethylene. The enhancement methods used included frame averaging (256 frames), beam flattening and gray-level stretching. The detection method is described in a companion paper at this Conference [33].

CONCLUSIONS

If one assumes a conservative value for hydrogen content of 0.2 mg H/cm^2 that can be detected by practical real-time neutron radiographic methods, then that value can be compared to the values

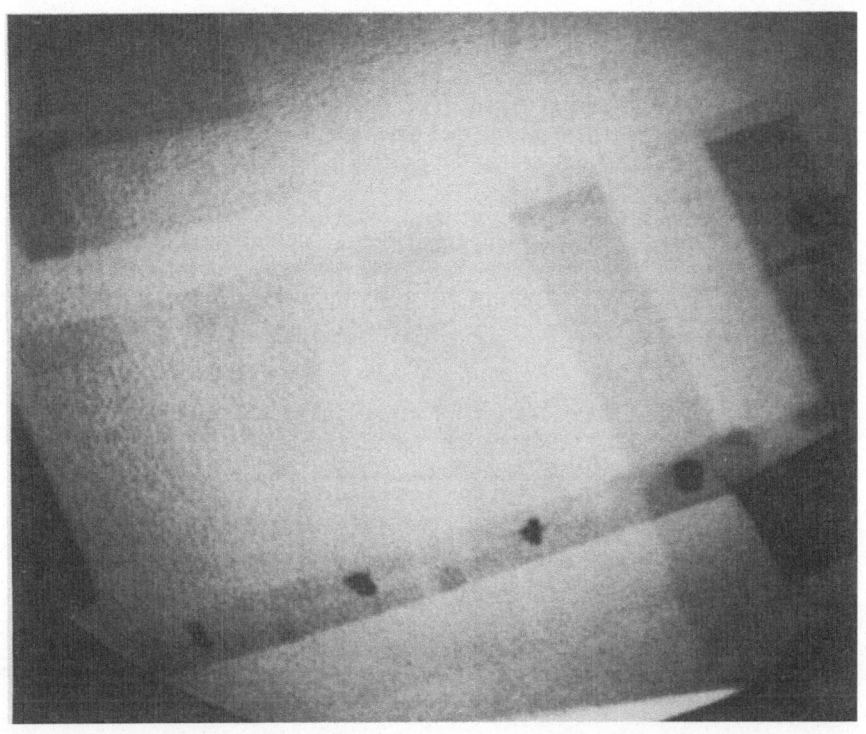

Figure 1. Enhanced, real-time neutron radiographic image of layers of polyethylene (12.5 μm = 0.0005 in. thick, 0.18 mg H/cm^2). The large rectangular images are 1, 2, 4, 8 and 16 layers thick, left to right. There is also the image of a small polyethylene stepped wedge in the upper left of the photo, 16, 8, 4, 2 and 1 layers thick, respectively, left to right. The thermal neutron beam at the time this image was detected had an intensity of 2 X 10^6 thermal n/cm^2-sec and an L/D ratio of 45 (taken with a TRIGA reactor at GA Technologies). The image was averaged over 256 frames. Detection was with thermal neutron image intensifier and a plumbicon TV camera.

previously determined for the hydrogen density of aluminum corrosion products. This leads to the conclusion that the thickness of corrosion product that can be detected in practice by real-time thermal neutron radiography is in the order of 70 μm (0.003 inch). Since metal loss will be less than the build-up thickness of the corrosion product (usually considered a proportion of 1:3), this implies that real-time thermal neutron radiography will permit detection of corrosion product build-up in aluminum resulting from a metal loss as small as 25 μm (0.001 inch).

ACKNOWLEDGMENTS

This work was supported in part through a contract with the U.S. Air Force, Wright Patterson AFB, Ohio, Lockheed-Georgia Co., prime contractor. This support and the cooperation of the Air Force monitor for this program, Captain J. Tarnacki, are gratefully acknowledged.

REFERENCES

1. D.J. Hagemaier, A.H. Wendelbo and Y. Bar-Cohen, "Aircraft Corrosion and Detection Methods," Materials Evaluation, 43, No. 4, 426-437 (1985).

2. H. Berger, "Real-Time Neutron Radiographic Observations of Fluid Motion," ASNT 1982 Paper Summaries, 487-489, Am. Soc. for Nondestructive Testing, Columbus, Ohio, 1982.

3. A. Singh, R. McClintock, T.W. Rudwick and R.L. Brackett, "Automated Inspection of Corroded Steel Structures," Materials Evaluation, 41, No. 5, 568-570 (April, 1983).

4. A.D. Cordellos, R.O. Bell and S.B. Brummer, "Use of Rayleigh Waves for the Detection of Stress-Corrosion Cracking in Aluminum Alloys," Materials Evaluation, 27, 85-90 (1969).

5. B.L. Weil, "Stress Corrosion Crack Detection and Characterization Using Ultrasound," Materials Evaluation, 27, 135-139 (1969).

6. D. Erdman, "Ultrasonic Pulse-Echo Techniques for Evaluating Thickness, Bonding and Corrosion," Nondestructive Testing, 18, 408-410 (1960).

7. B.F. Peters, "Radiography for Corrosion Evaluation," Materials Evaluation, 23, 129-135 (1965).

8. P.W. Lott and J. Lott, "Near Real-Time Radiologic Corrosion Monitoring of Arctic Petroleum Gathering Lines," Materials Evaluation, 43, No. 4, 408-412 (1985).

9. A.R. Bond, "Corrosion Detection and Evaluation by NDT," Brit. J. Non-Destructive Testing, 17, No. 2, 46-52 (1975).

10. J. Rodgers, "Acoustic Emission in Aircraft Structural Programs," J. Acoustic Emission, 2, No. 2 (1982).

11. C. Thaulow and T. Berge, "Acoustic Emission Monitoring of Cor-

rosion Fatigue Crack Growth in Offshore Steel," NDT International, 17, No. 3, 147-153 (1984).

12. D.R. Johnson, "Nondestructive Inspection of Honeycomb Panels
    at McClellan Air Force Base," DOD NDI Conference, Seattle,
    151-166 (Nov., 1976).

13. M.N. Bassim and D.L. Piron, "Acoustic Emission Monitoring of
    Large Structures Under Corrosive Environment," Brit. J. Non-
    Destructive Testing, 24, No. 5, 259-262 (1982).

14. W.D. Feist, "Acoustic Emission Inspection of Aircraft Engine
    Turbine Blades for Intergranular Corrosion," NDT Int., 15, No.
    4, 197-200 (1982).

15. H. Berger, "Radiographic Nondestructive Testing," Standardiza-
    tion News, 3, No. 3, 21-29 (March, 1975).

16. H. Berger, "Neutron Radiography," Elsevier, Amsterdam, 1965.

17. M.R. Hawkesworth, ed., "Radiography With Neutrons," British
    Nuclear Energy Society, London, 1975.

18. H. Berger, ed., "Practical Applications of Neutron Radiography
    and Gaging," ASTM STP 586, Am. Soc. for Testing and Materials,
    Philadelphia, 1976.

19. Neutron Radiography Issue, Atomic Energy Review, 15, No. 2,
    123-364, International Atomic Energy Agency, Vienna, June, 1977.

20. N.D. Tyufyakov and A.S. Shtan, "Principles of Neutron Radio-
    graphy," translation from 1975 book published in Russian, 1975,
    Amerind Publ. Co., New Delhi, 1979.

21. P. Von der Hardt and H. Röttger, "Neutron Radiography Handbook,"
    D. Reidel Publ. Co., Dordrecht, Holland, 1981.

22. J.P. Barton and P. Von der Hardt, eds., "Neutron Radiography,"
    D. Reidel Publ. Co., Dordrecht, Holland, 1983.

23. L.E. Bryant, ed., "Nondestructive Testing Handbook, Radiography
    and Radiation Testing," Vol. 3, Sections 12 and 13, Am. Soc.
    for Nondestructive Testing, Columbus, Ohio, 1985.

24. R.L. Tomlinson, "Industrial Neutron Radiography in the United
    States of America," ref. 19, pp 291-326 (1977).

25. W.E. Dance, "Neutron Radiographic Nondestructive Evaluation
    of Aerospace Structures," ref. 18, pp 137-151 (1976).

26. J. John, "Californium-Based Neutron Radiography for Corrosion Detection in Aircraft," ref. 18, pp 168-180 (1976).

27. J. John, J.E. Larsen, F. Patricelli, M.J. Devine and A.J. Koury, "Neutron Radiography for Maintenance Inspection of Military and Civilian Aircraft," Californium-252 Source Technology, Scientific and Industrial Applications, R.L. Berger and W.R. Cornman, eds., Report CONF-760436, E.I. du Pont de Nemours & Co., Savanah River Laboratory, Aiken, NC, pp V-47 to V-75, 1976.

28. D.A. Garrett, "The Microscopic Detection of Corrosion in Aluminum Aircraft Structures with Thermal Neutron Beams and Film Imaging Methods," Report NBSIR 78-1434, National Bureau of Standards, Washington, DC, De., 1977.

29. H. Berger, "Neutron Radiographic Detection of Corrosion," ASTM Symp. NDT and Electrochemical Methods of Monitoring Corrosion in Industrial Plants, Am. Soc. for Testing and Materials, Philadelphia, in press.

30. N.D. Tyufyakov, "Determination of Hydrogen Content in Materials and Products by Neutron Radiography Method," ref. 22, pp 303-307 (1983).

31. H.H. Klepfer, H.D. Kosanke and E.L. Esch, "Neutrographic Hydrogen Determination in Zirconium Alloys," ASTM STP 458, "Applications-Related Phenomena for Zirconium and Its Alloys," pp 372-385, Am. Soc. for Testing and Materials, Philadelphia, 1969.

32. H.D. Kosanke, "Hydrogen Sensitive Neutron Radiography," Trans. Am. Nuclear Soc., 14, No. 2, 533 (1971).

33. R. Polichar and D. Shreve, "Image Processing of Real-Time Images for Quantitative Neutron Radiography," Second World Conf. Neutron Radiography, Paris, June 17-19, 1986.

AN ADVANCED VIDEO SYSTEM FOR REAL-TIME NEUTRON RADIOGRAPHY

J. S. Brenizer[1], B. Hosticka[1], R. W. Jenkins, Jr.[2] and
D. D. McRae[2]

[1]University of Virginia, Charlottesville, VA   22901
[2]Philip Morris USA, Richmond, VA   23261

ABSTRACT

An advanced video system has been assembled from
commercially available equipment to support the real-time
neutron radiography facility established jointly by the
University of Virginia Reactor Facility and the Philip Morris
Research Center. The system includes a neutron sensitive
image intensifier with a modified video camera, a video
timer, a special effects generator, a time base corrector, a
high resolution video tape recorder and a digital image
processor. The digital image processor permits quantitative
analysis of the image with non-commercial as well as
commercially supplied processing software. Each of the
system components and other video accessories are discussed
along with examples of their use.

INTRODUCTION

The video system for the support of neutron radiography at the
University of Virginia has been assembled from commercially
available video equipment in a project conducted jointly by the
Department of Nuclear Engineering and Engineering Physics and by
the Philip Morris Research Center (1). It must be noted that the
products named in this paper are those at the University of
Virginia and do not represent endorsements of these products.
Other components that perform the same functions are available and
may be more appropriate for any particular application. The signal
path of the video system is shown as a block diagram in Figure 1

and originates with a neutron sensitive image intensifier and a modified video camera (Precise Optics) mounted on a table with two degrees of translational motion and one degree of rotational motion. The signal from the camera is sent to a video laboratory, removed from the shielded room surrounding the beam port, where it is passed through a video timer (Panasonic model WJ-810) which places the time and date on each video field and a special effects generator (Panasonic model WJ-4600B) which permits the image from a second camera to be inserted into the neutron radiographic image. The image can be stored as an analog signal on either a high resolution 3/4-inch video cassette recorder (Panasonic model NV-9420) or one of two 1/2-inch video cassette recorders (Panasonic VHS or Sony Beta). The heart of the quantitative analysis capabilities is a digital image processor (Micro Consultants Intellect 100 utilizing a Digital Equipment Corporation LSI 11/23 computer) that can perform many processing functions under software control. The processor can also be used to store single frames of an image in digital form on either hard or soft magnetic disks. Each of these components along with examples of the applications to quantitative real-time neutron radiography will be discussed below.

MATERIALS AND METHODS

Neutron Camera

A neutron camera utilizing an electrostatic image intensifier with a neutron-sensitive gadolinium oxysulfide screen and a monochrome video camera viewing the anodic phosphor screen is used to collect a neutron image and produce a standard RS-170 composite video signal (2,3). With the exception of the neutron sensitive screen, this system is similar to those used in medical X-ray radiology. This type of camera system is typically designed to automatically create an image that has subjectively normal contrast and brightness. This automatic analog signal processing in the camera makes quantitative comparisons of the neutron optical thicknesses based on signal levels of two objects or portions of objects difficult. When qualitative studies are done on the neutron image, the automatic controls produce a picture that is easily interpreted. Before routine quantitative analysis could be performed on the video image, the automatic target control, the automatic gain control, the anti-vignette control and the automatic black level control in the video camera had to be defeated. In order to produce an image of the desired brightness and contrast range, manual controls for both the black level and the gain in the video camera were substituted for the defeated automatic controls. These manual controls along with switches to invoke their automatic counterparts are located outside of the shielded room to allow adjustments to be made while the object is in the neutron beam.

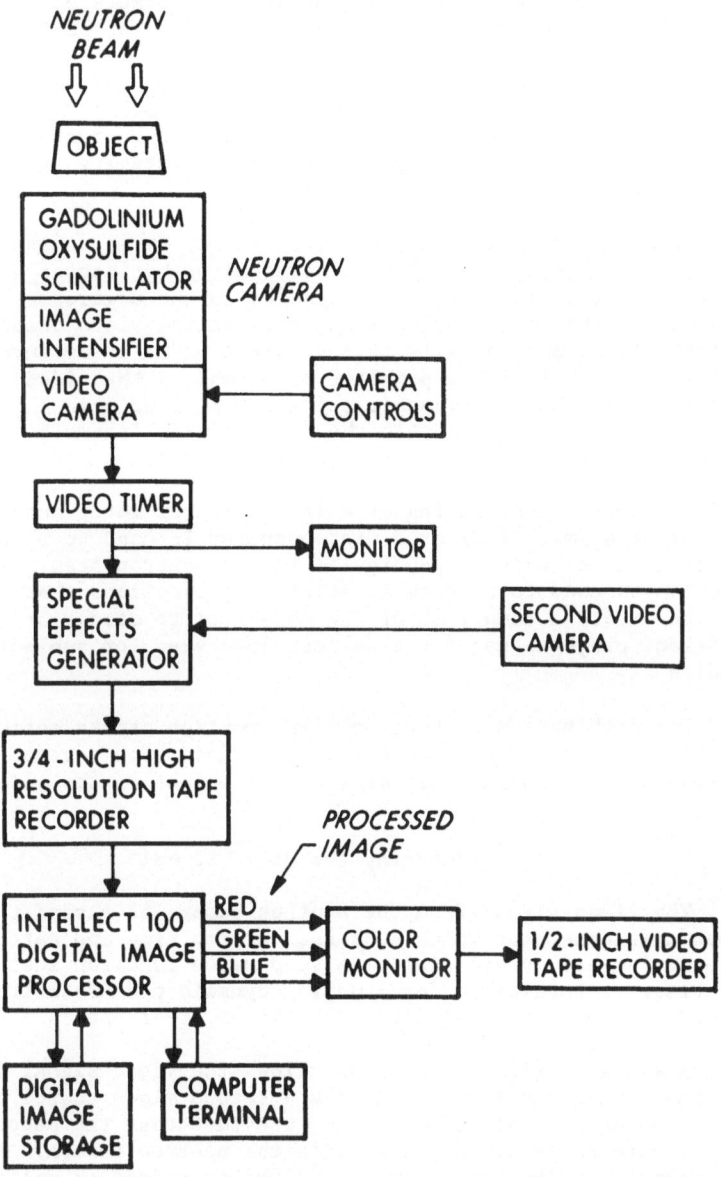

Figure 1. Schematic diagram showing the path of the video signal.
The wires to and from every piece of video equipment are
routed through a patch panel so that the system can be
easily rearranged as desired for different experiments.

The neutron camera is mounted on a table that is in turn mounted on two pairs of mutually perpendicular, horizontal rails. Each pair of rails has a small motor attached to a pinion and rack to move the table under remote control. The camera can be moved parallel to the neutron beam to remove it from the path of the vertically traveling stage used to move objects into and out of the shielded room. Once the object is in position, the camera can be moved forward to a location where it almost touches the object. The camera can also be moved from side to side to center the object horizontally in the video image. Vertical centering is achieved by raising or lowering the traveling stage.

When analyzing a digitized video image, it is convenient to work with the object in either a vertical or horizontal orientation relative to the video raster. This led to the development of a remotely controlled rotating mount between the video camera and the intensifier to move details of the object under study into a position which is either parallel or normal to the video scan. By changing just the relationship between the video camera and the intensifier, the need to tilt the much heavier intensifier to achieve the desired image rotation was avoided.

A turntable consisting of a large stepper motor under the control of a small dedicated microcomputer is used to rotate an object about an axis perpendicular to the neutron beam. This rotation is used to produce a series of projections that can be used to create a tomograph of the object or to merely bring different projections of the subject into view for subjective studies (4).

The combination of stage motion, neutron camera motion, video camera rotation and turntable rotation gives five degrees of freedom in positioning the image of the object in the video field.

Processing the Video Signal

The video signal from the neutron camera is passed through a video time generator which adds the current time and date, uniquely identifying each video field. A stop watch function included in the timer is invaluable for studying dynamic processes from video recordings running at either real-time or other speeds.

A special effects generator allows the video signal from the neutron camera and the signal from a second video source to be displayed on a single video field, provided that the second video can be externally synchronized with the neutron camera. A time base corrector (Microtime model S-230D) is needed to split the screen between a live image and an image recorded on magnetic tape

574

since the synchronization of the tape varies as the tape physically stretches or the heads move. The time base corrector adjusts the timing of the signal from the tape recorder to be consistently in step with an external synchronizing signal which allows the recorded image to share a field with a live image.

Image Recording

The images formed using the equipment in the laboratory can be stored in several different formats. For high resolution archival storage, a 3/4-inch video cassette recorder that has been modified by the manufacturer to achieve a very high monochrome resolution at the expense of color recording capabilities is used. The high resolution information on the tapes cannot be realized by standard 3/4-inch recorders because the high frequency response of the modified recorder is used to improve resolution whereas it is used for color information on unmodified recorders. Standard half-inch video cassette recorders in two different formats (VHS and Beta) are used to make recordings of images that must be edited or are to be viewed outside of the facility. The image processor can digitally store single frames of an image on either soft or hard magnetic disks but this technique suffers from format compatibility problems when extended beyond the facility. On those occasions when it has been found necessary to distribute the image of a single frame and a static film radiograph could not be made owing to motion in the object, good quality photographs of the frozen image on the video monitor screen have been made. The photographs are not as good as either film radiographs or the digitally stored image from which they were derived but the technique represents a low cost solution to the problem of reproducing single frames of a dynamic process.

Image Processing

For most quantitative work with the neutron camera, a commercially available digital image processor that can digitize and store an image as an array of 512 by 512 picture elements each with 256 levels of gray tones is used. The digitization takes place at the frame speed of the live-time video image (30 frames per second) so that some functions of the processor such as contrast enhancement, noise reduction, and frame addition can be done on a dynamic image. Once an image has been frozen in the processor's digital memory, any mathematical manipulation or analysis of the image can be done by treating the image as an array of integer numbers from 0 to 255 in a 512 by 512 square matrix (5). Either software controlled or fixed function image processing routines are available. A software controlled unit was chosen for this laboratory to permit tailoring the processing functions to specific projects utilizing the neutron radiography system. The fixed function processors are typically faster than the software controlled ones and may be more appropriate in applications where speed is important.

The manufacturer of the image processor provided compiled routines to perform several enhancement functions as well as manipulation of the data in the frame storage memory. These routines can be divided into two broad categories. The first category includes the routines that either modify the image by routing the data through look-up tables at real-time speed, or add and subtract the luminance data of subsequent frames into the frame storage memory at acquisition speed. These manipulations are useful when studying dynamic processes. The second category includes the routines that manipulate the data in the frame storage memory at a rate quite a bit lower than real-time video speeds. These manipulations can be performed only on one frame which has been frozen in time.

Examples of routines in the first category include contrast stretching or compression, noise reduction by partial updating of sequential frames and imaging only the parts of a picture that are different from a reference picture. Examples of routines in the second category include storing and retrieving an image from a magnetic disk, and the highlighting of contours of equal luminance. Extracting contrast features by applying various bit mask filters on the data can be done in real time only if the processor has the capability to store multiple video frames in memory at one time. Otherwise such feature extractions must be done at a slow speed on a frozen frame.

Monitors

Several video monitors are used in this system to allow viewing the image at several stages in the signal stream. One monitor is used exclusively to display the raw image from the neutron camera while others display recorded images, split screen images and the processed image from the digital image processor. The monitors have a high frequency response and some include a feature that permits the signal to be displayed slightly undersized on the monitor screen. This is useful because monitors often have a great deal of spherical distortion at their edges or sometimes fail to display the edges of an image at all. One monitor connected to the output of the image processor is a red-blue-green color monitor which allows false coloration of the image by assigning different colors to different luminance values via the look up tables in the processor.

DISCUSSION

The video laboratory described above has been used for a wide range of studies utilizing the imaging characteristics of neutron radiography. Each piece of equipment was included to meet specific needs encountered in these studies. The real-time neutron camera, for example, has been used to image the flow of water in a soil column (6). When the timing of specific events were needed to study the dynamics of the flow, the video timer allowed the recordings of the flow to be analyzed on a frame by frame basis without losing track of the time. The special effects generator has proved useful to observe the visual image of an object on the upper half of the monitor screen along with the neutron image on the lower half of the screen. This enables the external appearance of the objects under study to be correlated with the internal processes revealed by the neutron image. The special effects generator has also been used to insert the image of a meter monitoring the temperature of the object in the neutron beam into an otherwise unused corner of video field. This was used in studies of the characteristics of neutron contrast agents at elevated temperatures.

It was necessary to write in-house programs to perform quantitative analyses based on the variation of the neutron optical thickness of samples under study. One such program compares the luminance values along a line plotted through the sample to the luminances along the same line when the sample is removed from the neutron image. By making this comparison to a standard or reference of known density, it is possible to determine the variation in the density of the material along the line. This was used to measure the profile of the quantity of gadolinium chloride aerosol deposited along the length of a filter bed (7). Other in-house programs have been used to produce neutron tomograms and to study the modulation transfer function of the system while viewing a sharp contrast edge (4). For these and other studies, accurate positioning of the object in the video field is very important. The camera mount and stage combination which yields five degrees of freedom in orienting the object in the video field has been invaluable.

Once the picture is treated as numbers, any manipulation of the data can be performed depending on the information needed from the picture. However, no amount of manipulation will reveal information about an object that is not already contained in the digitized picture.

CONCLUSIONS

By taking advantage of the wide range of commercially available, relatively low cost video hardware now on the market, an advanced video laboratory can be added to an existing real-time neutron or other form of radiographic facility with little in-house development of equipment. The equipment as provided by the manufacturers will often be designed to automatically yield a pleasing picture. If quantitative results are needed, slight modifications to the circuitry of the video equipment will be necessary to give signals that are a fixed function of intensity. The radiation fields in the vicinity of neutron beam will usually preclude making adjustments to the equipment electrical settings or position while observing the neutron image; therefore, remote controls for all such functions must be provided.

REFERENCES

1. J. S. Brenizer and M. F. Sulcoski, "Real-Time Neutron Radiography at the University of Virginia," in Use and Development of Low and Medium Flux Research Reactors, p.958, O. K. Harling, L. Clark and P. von der Hardt, Eds., Karl Thiemig Graphische, W. Germany, 1984.

2. M. Verat, H. Rougeot and B. Driard, "Neutron Image Intensifier Tubes," in: Neutron Radiography, p. 601, J. P. Barton and P. von der Hardt, Eds., D. Reidel Publishing Company, Dordrecht, Holland, 1983.

3. P. Grivet, Electron Optics, Pergamon Press, New York, 1965.

4. M. F. Sulcoski, Neutron Computed Tomography Using Real-Time Neutron Radiography, Ph.D. Dissertation, University of Virginia, Charlottesville, VA, U.S.A., 1985.

5. R. C. Gonzalez and P. Wintz, Digital Image Processing, Addison-Wesley Publishing Company, Reading, MA, 1977.

6. H. E. Gilpin, An Investigation of Water Flow in Porous Unsaturated Media Using Neutron Radiography, M.S. Thesis, University of Virginia, Charlottesville, VA, U.S.A., 1985.

7. D. D. McRae, R. W. Jenkins, Jr., J. S. Brenizer, M. F. Sulcoski and T. G. Williamson, "Real-Time Observation of Aerosol Deposition by Neutron Radiography," in Aerosols, p. 241, B. Y. H. Liu, D. Y. H. Pui and H. J. Fissan, Eds., Elsevier, New York, 1984.

REAL TIME NEUTRON RADIOGRAPHY AND ITS' APPLICATION TO THE STUDY OF INTERNAL COMBUSTION ENGINES AND FLUID FLOW

J.T. Lindsay', J.D. Jones, and C.W. Kauffman

Phoenix Memorial Laboratory and Aerospace Engineering

University of Michigan, Ann Arbor, MI. 48109, U.S.A

ABSTRACT

Real time neutron radiography(RTNR) is now proving to be a valuable research tool in the study of hydrogenous fluid flow. Whereas neutron radiography compliments other forms of radiography by making it possible to image many objects and substances that otherwise can not be imaged, RTNR adds the important dimension of motion which greatly enhances the value of the technique. Further, video processing techniques used in real time radiography provide many ways in which the image can be enhanced and studied. The Phoenix Memorial Laboratory(PML) at the University of Michigan has recently developed and installed a facility dedicated to RTNR. The work at PML has shown that RTNR of dynamic events can provide information enabling the researcher to follow dynamic events that were previously impossible or impractical.

INTRODUCTION

Many of the materials used in mechanical systems are transparent to thermal neutrons when compared to the fluids used. This provides a method of radiography that not only compliments x-ray radiography but often results in the ability to observe phenomena that x-ray radiography is incapable of imaging(1). One of the most significant advantages of neutron radiography is the ability to image hydrogenous substances (such as lubricants, coolants, and fuels) inside metallic materials such as aluminum engines(2,3). By using RTNR, one can then study dynamic events such as the movement of liquids inside of these solids(4,5).

The RTNR facility at PML is located in the research reactor facility on the North Campus of the University of Michigan. The thermal neutron source is a two Megawatt, Materials Testing Reactor(MTR) type reactor reflected on one side with D_2O. The divergent beam of neutrons is then passed through the object to be radiographed and on to the imaging system. Currently, two imaging systems are in use at PML. One consists of an 8" x 8" gadolinium oxysulfide screen mounted in a light tight box, a front surface mirror to reflect the image at right angle to the screen, a f/ 0.8 lens, and an EMI magnetically focused image intensifier tube. The other imaging system is a recently acquired LIXI Neutron Imaging Device(NID) manufactured by LIXI Inc.,Downers Grove, Illinois(6,7). This device uses an input phosphor that is high in gadolinium to generate a light image outside the vacuum envelope of a high gain visible light microchannel plate image intensifier tube. In order to avoid lateral light spread and degradation of resolution, both the input and output face plates of the intensifier are made of fiber optics which provide intimate physical contact with the gadolinium phosphor. Because the LIXI NID is completely portable and relatively small in size (51 mm input diameter), it is easily placed strategically in an area where internal dynamic motion is to be observed. Either system can be viewed by eye, video camera, or standard 35 mm camera. When viewed by video camera, the video signal from either imaging device is sent to a Quantex QX-9200 image processing system. This is an IBM 9002 laboratory computer based real time image processing system which has a library of several pre-programed image processing routines as well as several custom routines written by the PML staff.

SPRAY STUDIES

In the summer of 1984, feasibility studies were begun applying RTNR to the study of lubricant and spray systems. These studies resulted in demonstrating the ability of RTNR to image sprays inside metallic structures. Figure 1 shows two different spray profile images

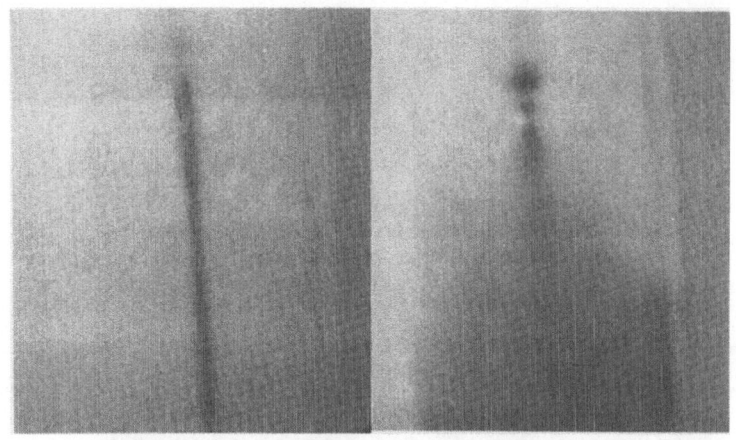

Figure 1. RTNR of two sprays using the EMI system.

using RTNR for two different nozzles with the EMI system.
Later, the LIXI NID mentioned previously was found to
greatly increase the sensitivity of spray detection and
the resolution achievable. Figure 2 shows the same
sprays as in Figure 1 using the LIXI NID.

The next stage of development in spray imaging at
PML was to develop the capability to do tomographic,
three dimensional, imaging of sprays. Two reconstruction algorithms were implemented on the IBM CS-9002
computer at PML and applied to the conical spray shown
in Figure 1. Figure 3 shows the spray used and three

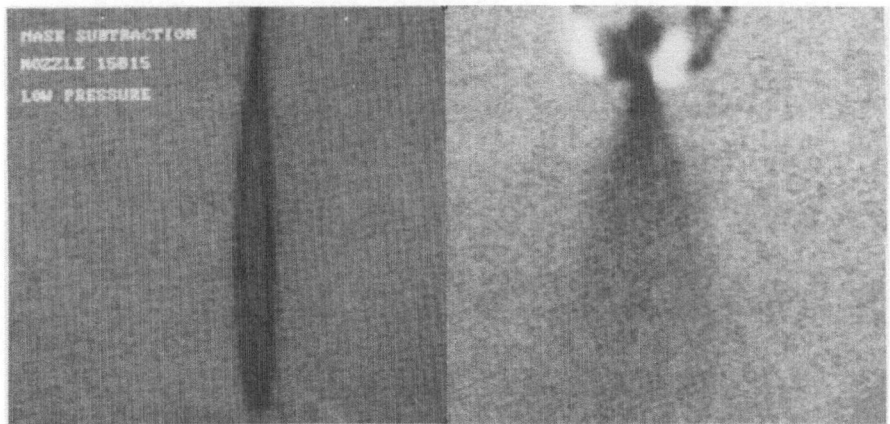

Figure 2. RTNR of two sprays using the LIXI NID system.

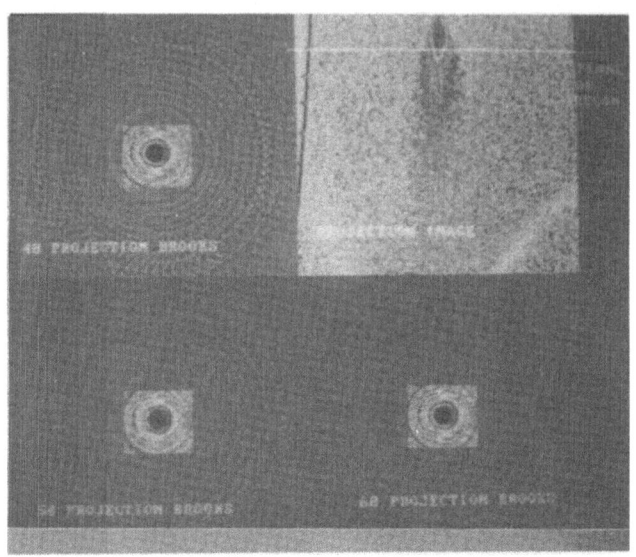

Figure 3. Tomographic spray reconstructions.

reconstructed images of the spray 1.0 cm from the tip of the nozzle using 40, 50, and 60 projections. As can be seen the greater the number of projections used, the better the image. Figure 4 shows an enlargement and enhancement of the reconstructed image using 60 projections. This is a reconstruction using the EMI imaging system. Use of the LIXI system and development of a reconstruction algorithm designed for sprays should greatly improve the reconstruction images obtainable.

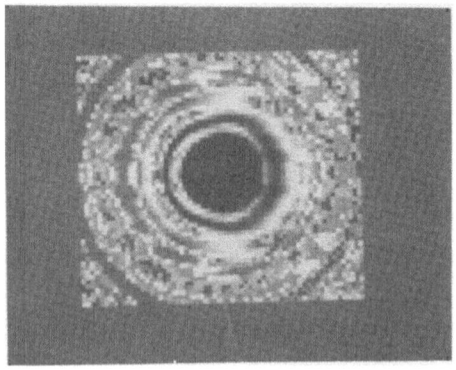

Figure 4. Enlargement of 60 projection image.

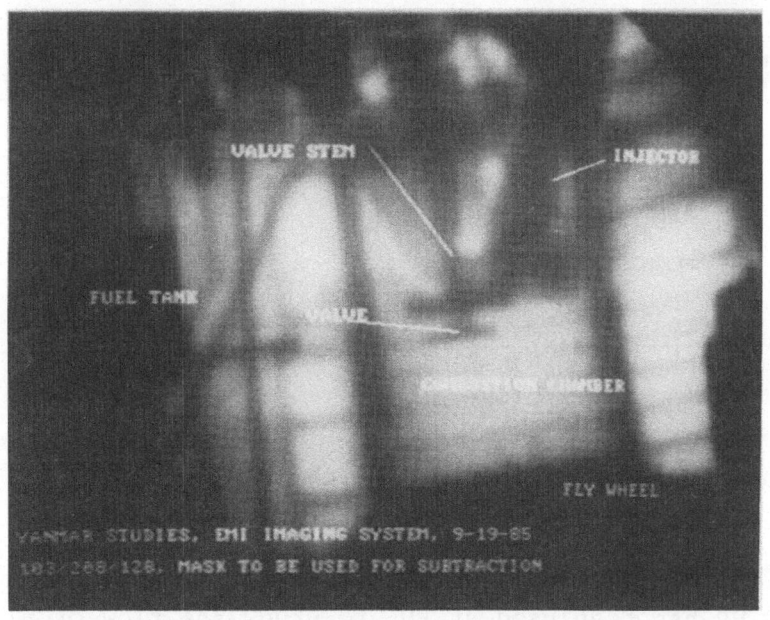

Figure 5.   EMI image of Yanmar diesel engine.

ENGINE AND TRANSMISSION STUDIES

Use of RTNR enables the engineer to image engine
phenomena without the necessity of introducing material
changes which may influence the phenomena. These phenomena include fuel spray density and spray pattern of the
injector, fuel spray density profile and location in the
combustion chamber, the presence or absence of wall wet-
ting by the fuel spray, changing dynamics of phenomena as
a function of "time" in combustion cycle(i.e. top dead
center), lubrication of the piston walls, performance of
oil pumps, lubrication of the valve chain, and any other
fluidic dynamic functions which involve hydrogenous
fluids inside operating engines. Figure 5 shows an image
of the Yanmar single-cylinder, diesel engine obtained
with the EMI system. This image was used as a mask to
subtract out undesired detail from the real time images.
This results in real time images showing only the movement of the fluid of interest(oil or fuel). Various
improved techniques are currently being used, including
ensemble frame averaging and new computer enhancement
techniques, to study lubrication and fuel phenomena in
the Yanmar engine.

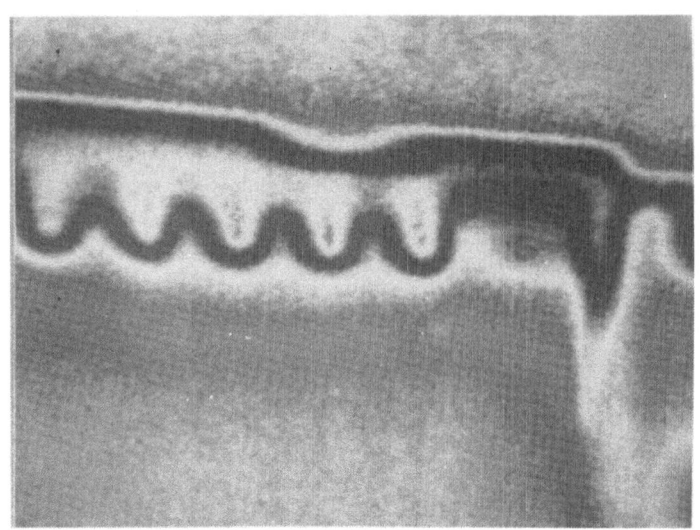

Figure 6. Fluid flow in lower reverse clutches, 1000 RPM.

RTNR has been used at PML to study the flow of fluid in a front-wheel drive transmission. Figures 6 and 7 show the difference in fluid flow in the lower reverse clutches at different RPM's. RTNR was used to image the fluid flow throughout the transmission and to quantify the change in fluid flow with internal modifications made to improve the fluid flow in areas of concern.

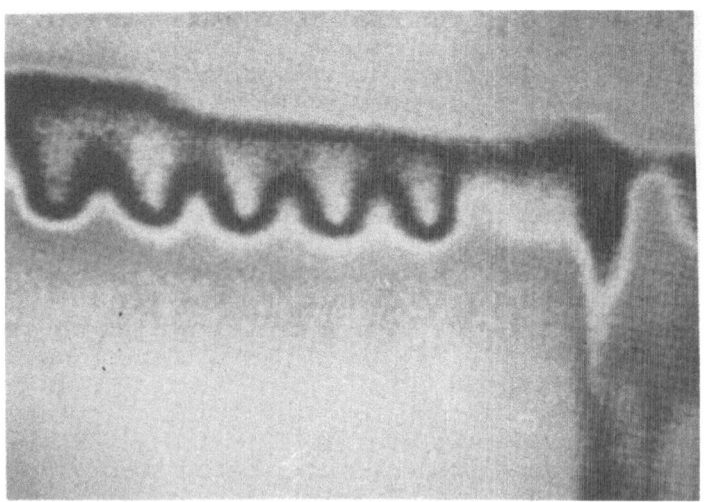

Figure 7. Fluid flow in lower reverse clutches, 5000 RPM.

Figure 8. Jet turbine fuel nozzle.

MISCELLANEOUS USES OF RTNR

Figure 8 shows a fuel injector used in a jet turbine engine with internal parts and passages easily visible. RTNR will be used at PML to study cavitation and clogging in this fuel nozzel. Figure 9 shows RTNR being used to study the transport of water through soil. This is currently an area of great interest in predicting the transport of waste and toxic materials through soil into the water table.

Figure 9. Water transport through soil.

CONCLUSIONS

RTNR has a variety of applications in a wide diversity of areas. RTNR is growing and as general awareness of RTNR and its' unique qualifications increases, RTNR will become a routine research and quality assurance tool.

REFERENCES

1. H. Berger(1976), <u>Practical Applications of Neutron Radiography and Gaging</u>, ASTM STP 586.

2. J.T. Lindsay(1983), <u>Development and Characterization of a Real Time Neutron Radiographic Imaging System</u>, Ph.D. Disertation, University of Missouri.

3. M. Kuriyama, W.J. Boettinger, and H.E. Burdette (1980) "Basic Limits in Real-Time Industrial Radiographic Systems", <u>Real-Time Radiologic Imaging: Medical and Industrial Applications</u>, ASTM STP 716, D.A. Garret and D.A. Bracher, Eds.

4. G.S. Okawara and A.A. Harms(1976), "Neutron Radiography of Fast Transient Processes", Nuclear Technology, Vol. 31.

5. J.D. Jones, J.T. Lindsay, C.W. Kauffman, A.T. Vulpeti, and B. Peters(1985), "Real Time Neutron Imaging Applied to Internal Combustion Engine Behavior", SAE Technical Paper Series No. 850560, International Congress and Exposition.

6. J.T. Lindsay, J.D. Jones, C.W. Kauffman, and B. VanPelt(1985), "Real Time Neutron Radiography Using a LIXI Neutron Imaging Device", Nuclear Instruments and Methods in Physics Research,(Jan. 1986), Volume a242, No. 3, pp. 525-530.

7. I. Yin, J.I. Trombka, and S.M. Seltzer(1979), "Portable X-ray Imaging System for Small-Format Applications", Nucl. Inst. and Methods 158 pp. 175-180.

ACKNOWLEDGEMENTS

This work has been supported in part by the following agencies and companies: U.S. Army, Department of Defense, General Motors Corp., and Ford Motor Company.

PROCESSING OF REAL-TIME IMAGES FOR

QUANTITATIVE NEUTRON RADIOGRAPHY

R. Polichar and D. Shreve

Science Applications International Corporation

San Diego, CA, 92121

ABSTRACT

Modern real-time imaging systems may provide an excellent sensor for quantitative neutron radiographic measurements. When properly designed, these sensors are much less sensitive than film to reciprocity failure and may be used over a very wide range of incident fluxes. Image processing can be used to correct the effects of system response and nonlinearities, permiting the quantitative measurement of neutron removal from all portions of the field of view. This is particularly useful for corrosion detection. The systems sensitivity to very small quantities of hydrogenous material and its uniformity of response are critical. The additional benefit of providing an accurate direct measurement of the degree of absorbtion is a great aid in interpreting the image.

The goal of this work is to develop an image processing approach that provides a quantitative relationship between grey scale and absorption over the entire image. The system removes the instrumental response of the sensor and the beam distribution and provides a direct contrast scale in units of μx, the effective absorption constant. The study has also been extended to develop a non-subjective approach to the evaluation of system quality and a means of predicting the level of hydrogen detectability as a function of total integrated fluence. A summary of data taken between 10^6 to 10^7 neutrons/cm^2 sec is presented for varying integration times. Total system performance is summarized in a single figure of merit that predicts spatial and contrast sensitivity.

INTRODUCTION

This work was initiated to develop computer aides for the interpretation of real-time neutron radiographic images. One of the primary uses of such images are to detect corrosion on aluminum structures. Such corrosion products generally consist of Hydroxides and Hydrated Oxides of Aluminum and are generally entrained in inaccessible places within the structure. The goal of the current image processing effort is to provide images which produce a reliable and constant relationship between the grey level and total areal density of hydrogen present. As a secondary goal, we wished to determine a relationship that would quantitatively predict the detectability of specific amount of corrosion products at a variety of useable neutron fluxes.

Our technical objectives in this work are to develop procedures to:

o   Remove the effects of imager response and beam non-uniformity
o   Correct for other instrumental effects such as non-linearity and additive noise
o   Correlate the effects of neutron fluence and the systems detection efficiency on the minimum detectable defect area and contrast

All data used in the study were taken on a Thomson CSF neutron sensitive image amplifier viewed by a camera employing a diode gun lead oxide image tube. The images were processed on a Recognition Concepts Inc. TRAPIX 55128 image processor with a Digital Equipment Co. LSI11/73 computer as host. Data were stored digitally on IOMEGA removable 10 Mbyte "Bernouli" disks. Analog display was available either on an RGB pseudo color monitor or could be converted to NTSC composite video where it could be recorded on VCR or sent to a video hardcopy printer.

Imager Response Removal

Clearly the simplest way to remove the variations in local response of the imager and the intensity of the beam is to divide the image of a test object by the image without the object. There are, however, a number of details that must be accounted for before such a procedure can be used. If the beam has some distribution E_{ij} over the face of the imager, then the camera will produce a signal I_{ij} at each point that is related by the expression

$$I^0_{ij} = Z + G_{ij}E_{ij}^{\gamma}$$

where: Z is measured zero offset
         G_{ij} is the gain of the system which may vary over the
               image
and γ is the power law non-linear response of the image
               tube.

If we put an object with some degree of absorption to neutrons
then the new intensity becomes

$$I_{jj} = Z + G_{ij} \ (e^{-\mu x_{ij}} E_{ij})^{\gamma}$$

where: μx_{ij} is the absorption coeffecient.

Before one can derive useful information from the point-by-
point division of local image values, the constant offset Z must
be removed. (The camera must also preserve a constant Z level
regardless of the actual intensity on the screen).

In order to do a division at t.v. frame rates, we use a log
look up table and do a subtraction of the logarithmetic grey
scales. Done properly, this procedure removes the dependence of
gain and the incident flux yielding a log difference D_{ij}

where: $D_{ij} = \log(I^0_{ij} - Z) - \log(I_{ji} - Z) = \gamma \mu x_{ij}$

In our imaging system γ is unity so that the log difference
is simply a reflection of the actual hydrogen density at each
point of the field of view.

This procedure must be carried out with some care as taking
the log of an 8-bit number requires at least 12-bit accuracy to

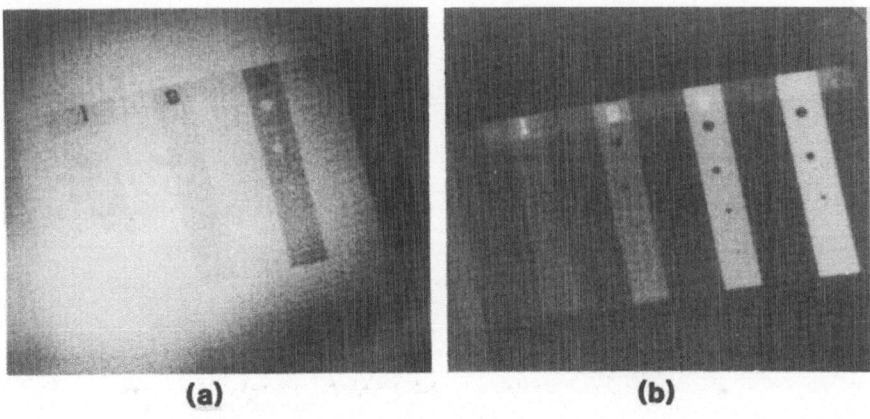

                    (a)   (b)

Figure 1.   Effects of Response Removal (Flattening) as a Function
of Contrast Enhancement: (a). Contrast Enhancement only, and (b).
Field Flattened and Contrast Enhanced.

ensure that every initial value is mapped into a unique integer value. Typically, we use an initial multiplier and then divide the result after subtraction. The actual values in our operation are equivalent to

$$D_{ij} = 200 \,\mu x_{ij} + 62$$

where the constant is added arbitrarily to move the integer valves of D_{ij} above zero. Typically, the absorption due to corrosion is quite small and considerable contrast enhancement is necessary to visualize the effects. Figure 1a shows what happens when the unprocessed image is simply contrast stretched. The fixed pattern response of the images is much larger than the effect being observed and much of the edge regions are lost. Figure 1b shows the contrast stretched image after response removal is applied. The contrast scale is now reversed and the brightness is directly proportional to the local density of neutron absorber within the field of view.

Measurement of System Performance

In order to develop an image quality indicator (IQI) that would reflect the systems sensitivity to small controlled levels of hydrogen, a gauge was constructed consisting of varying levels of commerical cellophane tape approximately .05 mm in thickness. A thick sample of the tape material was weighed and measured on a neutron backscatter gauge. Using cellulose and polyethelene as standards, the tape was determined to have a density of 0.686 mg/cm^2 and a Hydrogen fraction of .097 by weight. The original gauge was to have a sequence of 1, 2, 4 and 8 layers of tape with precision holes of 0.5, 1, 2 and 4 mm diameter drilled in all of them as shown in Figure 2. When the neutron radiographic data

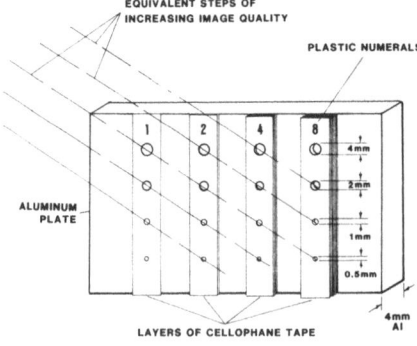

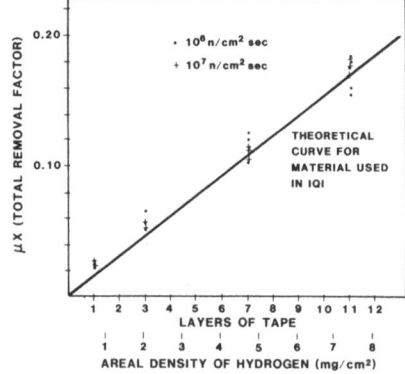

Figure 2. Thin Cellophane IQI for System Evaluation Minimum Layer: 0.68 mg/cm^2 Hydrogen.

Figure 3. Measured Absorption of Tape IQI using Quantitative Image Processing.

590

were analyzed, the measured hydrogen thicknesses were inconsistent with the labled number of tape layers. The gauge was checked and was found to consist of 1, 3, 7 and 11 layers of tape. The absorption of each group was measured on the image and plotted against the true amount of tape as shown in Figure 3. The values were obtained from a number of different fluence levels and taken at both 10^6 n/cm^{2}sec and 10^7n/cm^2 sec fluxes. The straight line is a theoretical absorption calculated from the published hydrogen removal cross section and the independently measured properties of the tape. The agreement shows convincing evidence of the capability of the process to derive absolute quantitative absorption data from a real-time image over a wide range of fluences.

Defect Detection as a Function of Fluence

It can be shown from recognition statistics that the observed contrast must be several times larger than the signal-to noise ratio. If the only source of statistical fluctuation is due to that of the incident neutrons, then these should follow the inverse square root of the fluence. The data shown in Figure 4 are the standard deviation per pixel obtained from processed images over a very wide range of integration times and flux rates. The data appear to be totally dominated by statistical effects up to total fluences of about 10^7 n/cm^2. Beyond this level, the measured standard deviations seem to flatten out indicating small instrumental effects of the order of 1%. These data are consistent with the predictions of simple counting statistics shown by the dashed curves. The lower curve shows the statistical limit where the imager response is normalized at 256 frames at 10^7 n/cm^2/sec while the upper curve is for normalization statistics of 1/10 that total fluence.

The relative sensitivity to a circular flaw of diameter d has been shown to be proportional to the product of diameter,

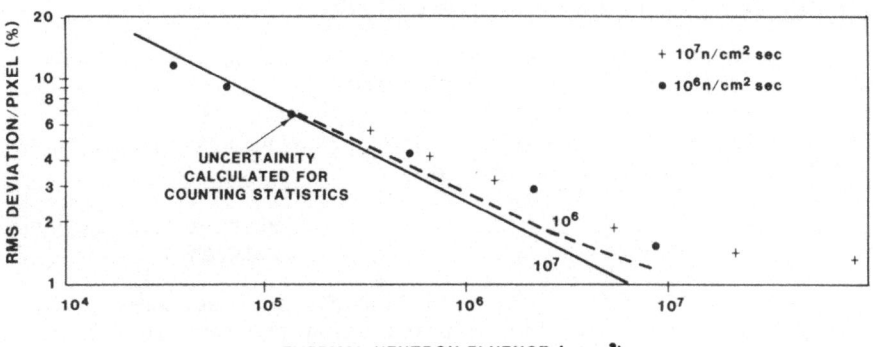

Figure 4. RMS per Pixel Fluctuations as a Function of Total Fluence.

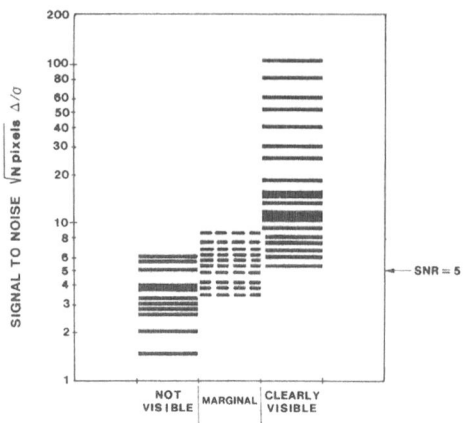

Figure 5.   Scatter Diagram of Detectability of Holes
on IQI as a Function of Measured Signal to Noise Ratio.

signal-to-noise ratio and a function that is related to the
system modulation transfer function and the object diameter. An
emperical measure was undertaken using the IQI described above to
determine which holes were visable for a given fluence. The data
were recorded as a scatter diagram shown in Figure 5. Each hole
was classified as to being clearly visable, marginally visable or
not visable as a function of measured signal-to-noise. The
quantity measured was Δ or the difference between the average μx
in the hole and in the surrounding tape region. This is a direct
measure of contrast ratio. The standard deviation was obtained
by taking histograms of the two regions. We have neglected the
effect of MTF in this case since it will seriously affect only
the smallest holes. The plotted data of Figure 5 overlap, but it
is clear from the graph that the quantity $\sqrt{N}\ \Delta\ /\sigma$ must be
greater than 4 or 5 to ensure detectability. (N is the number of
pixels within the defect diameter). This threshold number has
some human factors such as apriori knowledge included but is
consistent with other work on recognition.

The results can be related to the actual quantities known
about the experiment such as μx, the area of the hole A, the flux
ϕ, integration time t and the detector efficiency ϵ . The
relationship can be expressed as the fluence necessary to detect
a defect of area A and contrast μ as

$$\phi t > \frac{k^2}{\epsilon A (\mu x)^2}$$ where k is somewhere
between 4 - 5

For smaller objects, the relationship must be modified by
the MTF but for comparative studies using the same IQI, this
becomes a valid expression of detectability. An example of a
series of experiments taken from 1 frame to 256 frames integration at $\phi = 10^7$ n/cm^2 sec is shown in Figure 6.

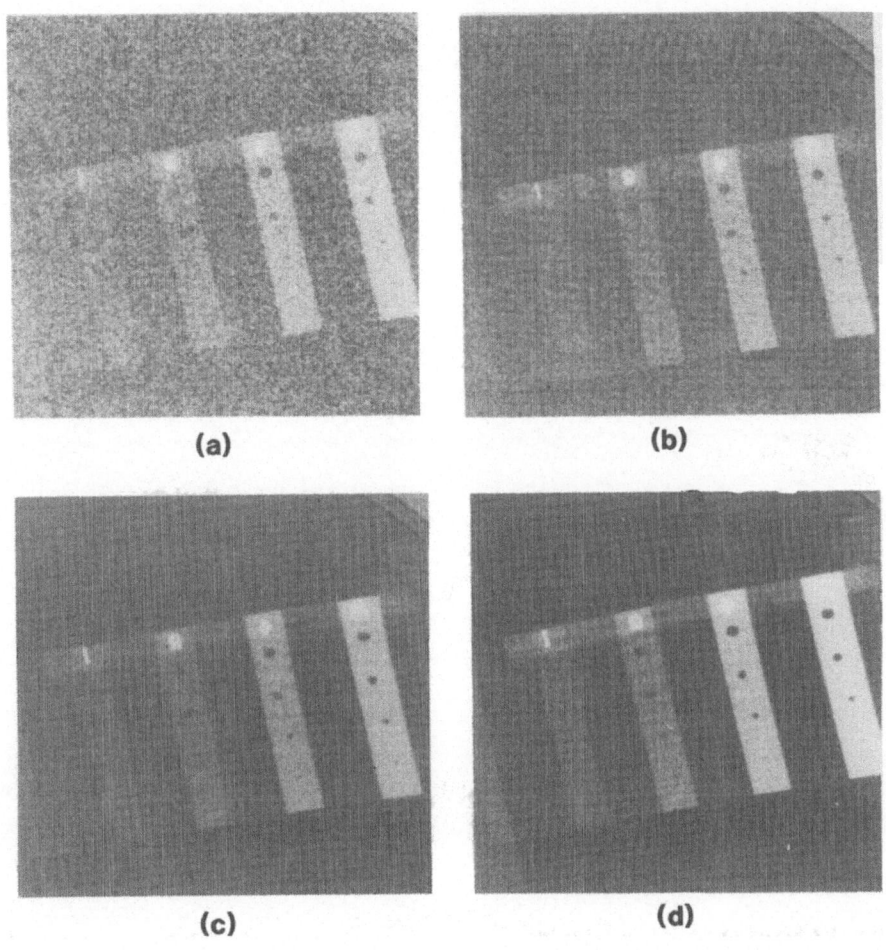

Figure 6. Demonstration of the Effect of Increasing Integration on Resolution and Contrast Sensitivity, Test Object T4 at 10^7 n/cm^2-s: (a) 1 Frame, (b) 4 Frames, (c) 16 Frames, (d) 256 Frames of Integration.

Conclusions

From our results we conclude that it is feasible to process real-time data to provide a quantitative measure of hydrogenous material. Moreover, we have set criteria for the detection of such indications based on their areal density and area and the number of detected neutrons. The use of this information provides a powerful interpretive tool in reading images suspected of containing corrosion products to determine the actual extent of material present.

APPLICATIONS OF REAL TIME NEUTRON RADIOGRAPHY AT HARWELL

D.H.C. Harris and W.A.J. Seymour

Materials Physics and Metallurgy Division, AERE Harwell,

Didcot, Oxon, OX11 ORA, U.K.

ABSTRACT

There are two areas of application of real time neutron radiography at Harwell. In the first area the technique is used to observe dynamic events such as liquid or gas flow in operating systems. Secondly real time imaging is used to inspect objects under examination and to select the best view for subsequent screen film radiography. The dynamic work is illustrated by showing studies of the flow patterns which occur when a preheated stainless steel tube is quenched by water. The results obtained have been used to improve film boiling theories used in reactor safety analysis codes.

The real time inspection work shown is of an archaeological investigation of the contents of a Saxon container from <u>ca.</u> 650. The examples are. presented on video tape as well as in 'still' form in the text below.

INTRODUCTION

The two Neutron Radiography facilities at Harwell[1] are installed on the DIDO reactor. Figure 1 shows the layout of both sets of apparatus. The thermal neutron radiography facility (6HGR9) is used extensively for non-destructive testing of industrial and nuclear reactor components using conventional gadolinium foil and X-ray film exposures as well as the image transfer technique with indium foil. Only occasional real time

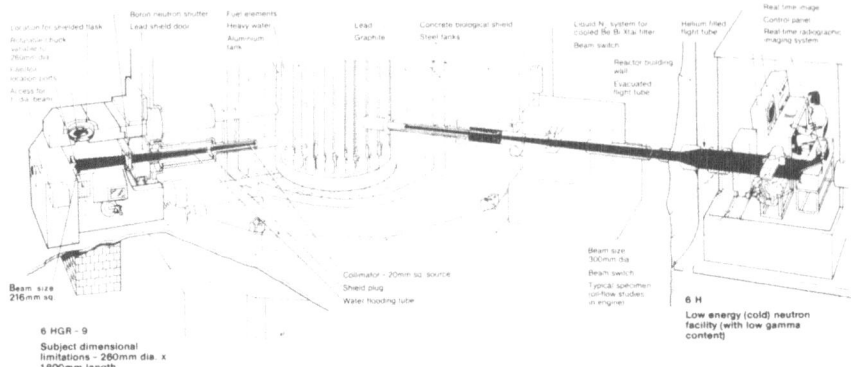

Figure 1   Layout of the two neutron radiography facilities on the DIDO reactor at Harwell.

observations are made with scintillator screens due to the high gamma ray content of the beam (6.5 Svh^{-1}). The 6H low energy (cold) neutron facility has a low gamma ray contamination (15mSvh^{-1} maximum) due to the presence of a 215mm long bismuth single crystal filter. This feature together with a flux of 7 x 10^6 ncm^{-2}s^{-1} produces very good real time images using the Harwell neutron video system, a 25mm S.I.T. camera with a LiFZnS(Ag) scintillator N.E.426, described in another paper submitted to this conference[2]. The applications of real time imaging shown here were obtained on the 6H cold neutron facility and were recorded on video tape at normal TV frame repetition rate.

REAL TIME DYNAMIC IMAGING

Studies of the consequences of loss of coolant accidents in water cooled reactors have stimulated research into the mechanisms of heat transfer in the vicinity of a quench front. Real time dynamic neutron radiography has been used at Harwell, by G. Costigan[3,4] to discover what actually occurs inside a red hot tube as it is slowly quenched by water. The assembly of the apparatus is shown in figure 2. The stainless steel tube, 12.2mm O.D and 9.25mm I.D., is heated to 600°C by passing a high electrical current at low voltage through its 600mm length. Water/steam flow inside the tube is observed when cold neutrons are attenuated by the H_2O and the LiF2nS(Ag) screen fluoresces when absorbing neutrons. A mirror mounted at 45° to the beam reflects the images formed to the S.I.T. low light video camera. The screen, mirror and TV camera are mounted inside an aluminium light tight box.

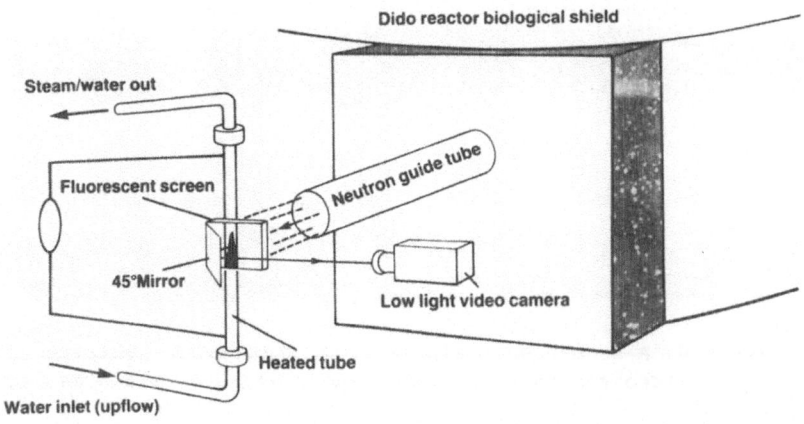

Figure 2   The apparatus used to study the flow of water inside
a stainless steel tube at 600°C

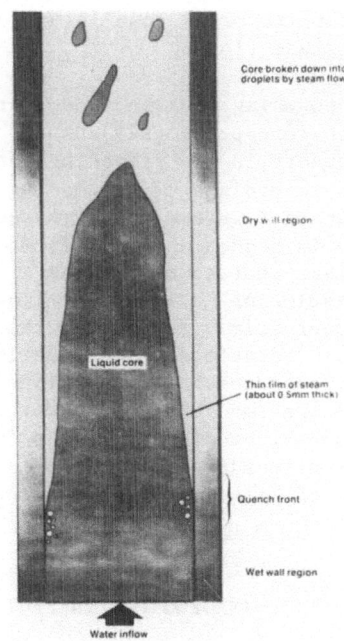

Figure 3   Diagram showing theoretical behaviour of water
inside a tube at 600°C.

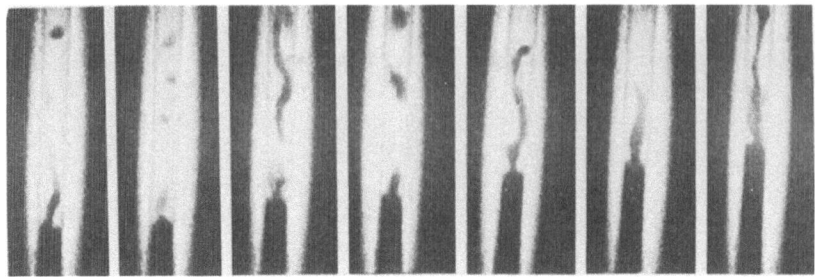

Figure 4 A sequence of single frame photographs obtained from the neutron video recording of water advancing up a stainless steel tube at 600°C.

Previously it was thought that the coolant behaviour would be as illustrated in figure 3. It was believed that water flows up the tube until it reaches a narrow region known as the quench front. The quench front moves relatively slowly and water moves past it to form a liquid core. This core is separated from the hot dry wall by a thin film of steam. Eventually enough steam is formed to cause the core to break down into droplets which are carried along in the steam flow.

Sequences obtained using neutron video were able to illustrate some phenenomena which were previously unknown and were not consistent with the theory in the preceding paragraph. Frozen frames from the video recordings of reflooding with water at a flow rate $25mms^{-1}$ are shown in figure 4 where the quench front is seen advancing up the tube in sequence from left to right. At the quench front it is clear that so much steam is being generated that it entrains large globules of water almost immediately and there is no sign of a thin vapour film. Further up the tube the liquid breaks down into smaller droplets which sometimes coalesce and fall back. TV video recordings at 7 positions were obtained covering the whole of the tube under both downflow and upflow conditions with reflooding veocity of $25mm\ s^{-1}$ and $75mm\ s^{-1}$. These and other observations are being used to improve predictions of a the consequences of loss of coolant accidents in water cooled reactors.

REAL TIME IMAGING FOR INSPECTION

During excavations by Dr. D. Miles, University Museum, Oxford of a Saxon cemetery (circa. 650 A.D.) at Lechlade, Gloucestershire, England a casket was unearthed in one of the graves, fig. 5. Similar caskets known as Thread Boxes have been found in high

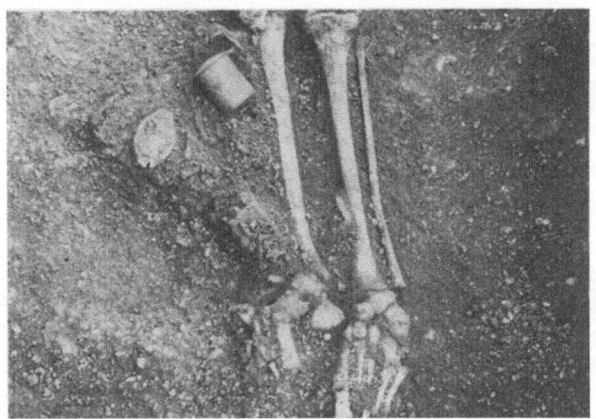

Figure 5   Excavation of the Saxon grave showing the
cylindrical casket (nearest to the skeleton).

status female graves from other Saxon sites. Several examples
have been found to contain textiles wrapped around medicinal
plants or seeds. It is likely that these boxes were worn as
amulets and have early Christian rather than pagan
associations.

The archaeologists wished to investigate the contents of the
thread box prior to opening it since the act of opening can often
lead to a collapsing of the contents and subsequent loss of
information. A neutron radiographic examination was chosen since
the metal box walls of copper alloy were transparent to neutrons
whereas textile and plant residues would be reasonably opaque.

The thread box was examined through a 360° rotation using the
Harwell Real Time Imaging System and the Arlunya TF4000 TV frame
storage system capable of storing up to 4000 TV frames. From this
examination the archaeologist was able to select the most suitable
orientations for a hard copy standard radiograph. Considerable
savings in time and film were thus achieved.

A comparison of the picture qualities achievable is shown in
fig. 6, where we compare photographs obtained from the TV screen
of firstly a single frame and then stored frames of information
with a standard film radiograph. The quality of the stored image
can be seen to compare quite favourably with the radiograph.

Objects observed inside the box were identified on opening as
small balls of flax thread and a number of carcasses of millipedes
which surprisingly had been able to penetrate the casket at some
time and consumed the vegetable matter.

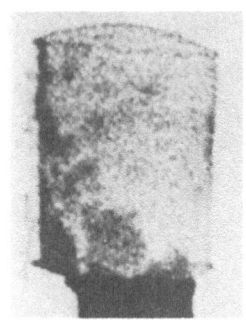  

Figure 6   Comparison of single frame and frame summed video
with a standard film radiograph.

The ability to quickly inspect objects in real time has been
applied to a variety of investigations for commercial customers at
Harwell and has enabled radiography to be more comprehensive and
more cheaply targeted on a specific area of the industrial
component than is achievable using film techniques alone.

ACKNOWLEDGEMENTS

The authors wish to express their thanks to Mr. G. Costigan,
Thermal Hydraulics Division, Harwell and Dr. D. Miles of the
Archaeological Unit, University of Oxford for their kind permission
to publish their results.

REFERENCES

1.   D.H.C. Harris & W.A.J. Seymour, Neutron Radiography
     Facilities, DIDO Reactor, Harwell, England, INR Newslett No.
     8, The British Journal of Non-Desructive Testing, Vol.27,
     No.4, July 1985.

2.   S.J. Cocking & D.H.C. Harris, Robust Equipment for Dynamic
     Neutron Fluoroscopy, GB2, this conference.

3.   G. Costigan & C.D. Wade, Visualisation of the Reflooding of a
     Vertical Tube by Dynamic Neutron Radiography, International
     Workshop on Post-Dryout Heat Transfer, Salt Lake City, Utah,
     April 1984.

4.   G. Costigan, J.C. Ralph & C.D. Wade, Heat Transfer above the
     Quench Front in Single Tubes, European Two-Phase Flow Group
     Meeting, Marchwood Engineering Laboratories, June 1985.

KUR NEUTRON TELEVISION SYSTEM

Keiji Kanda, Shigenori Fujine and Kenji Yoneda

Research Reactor Institute, Kyoto University

Kumatori-cho, Sennan-gun, Osaka 590-04, Japan

ABSTRACT

A neutron television system has been installed at
the E-2 experimental hole of the KUR (Kyoto University
Reactor, 5 MW), including an online video processing
system. The image of the KH on a CRT of a super high
quality TV camera can be observed directly and visually.
The video image signals from the TV camera are digitized
through a video A/V converter and can be stored in the
image buffer of a microcomputer system. The image can
be treated in various ways according to the purposes of
experiments. Some applications using this system and
further studies in progress are described in this paper.

INTRODUCTION

In order to obtain a direct and real time image of neutron
radiography samples, and online neutron radiography system has been
developed in KURRI (Kyoto University Research Reactor Institute).
In the previous papers (1-4), The neutron television system, the
online video image processing system and the neutron computed
tomography are reported.

In this paper, The KUR neutron radiography system and some
recent improvements of the KUR television system are described.
In addition, some applications of this system and further studies
in progress, such as application for the analysis of gas-liquid
two-phase flow, are mentioned.

KUR TELEVISION SYSTEM

Figure 1 shows the Horizontal view of the KUR experimental holes. The neutron radiography facility is installed in the E-2 experimental hole as shown in Fig. 2. The characteristics of the KUR neutron radiography system are shown in Table 1. The view of the sample irradiation port is shown in Photo. 1. The block diagram of the revised KUR television system is depicted in Fig. 3, and its photography is shown in Photo. 2.

The digital processing for NTV images is summarized in Table 2.

COMPARISON OF SCINTILLTORS

Besides an NE-426 or BN + ZnS(Ag), three kinds of intensifier screens for X-ray used as converters, KH of Sakura, G4 and G8 of Fiji were compared, and it was found that a luminescence scintillator has a better sensitivity than neutron scintillators. Figure 4 shows the experimental results on relations of film density and exposure time. Photo. 3 shows the comparison of images on a CRT taken at the KUR operation power of 500 kW. Photo. 4 , in which images were taken at the closest camera position and treated by the contour technique, demonstrates that even a small hole of 0.25 mm in diameter can be clearly observed by using G4 with so called a low sensitivity.

APPLICATIONS

In the past several years, the system has been applied in various ways, as follows of which details are described in JP.2 in this conference:

(1) The system has been used to identify the location of burnable poison in the side plates of MTR type reactor fuel manufactured by CERCA in France.

(2) A nylon string behind a lead block of 5 cm in thickness was observed clearly by the subtraction method.

(3) In the application for moving objects a bubble motion in water was studied. The air bubble and water level are clearly revealed by the contoured technique.

(4) The high quality images were obtained by the image integration in NCT to distinguish a small density differences between stinless steel and copper.

The subsequent works are in progress:

(1) The construction of the data base of images accumulated in the past using an optical memory disc file system.

(2) The examination of the neutron image indicators has been done, then a new indicator systems will be proposed (6). Some results are to be presented in this conference. A direct image and an integrated image of indicators are shown in Photo. 5, which was taken in the arrangement shown in Photo. 6.

(3) A new project has just started, namely the gas-liquid two-phase flow is analyzed with a ultra high speed camera and neutron radiography technique.

ACKNOWLEDGEMENT

The authors express their thanks to Prof. G. Matsumoto of Nagoya University and Prof. K. Katsurayama for their valuable suggestion and discussions. This work was supported by a Grant-in-Aid of the Ministry of Education, Science and Culture.

REFERENCES

(1) K. Kanda, Y. Yoneda and S. Fujine, Development of an Online Neutron Radiography System of High Resolution for Nuclear Materials, pp.219-225 in Neutron Radiography (Edit. J.P. Barton and J. von der Hardt), D. Reidel Publ. Co., Dordrecht, (1983).

(2) K. Kanda, S. Fujine and K. Yoneda, Development of an Online Neutron Radiography System with High Resolution, pp. 946-971 in Use and Development of Low and Medium Flux Research Reactors (Edit. O. K. Harling, L. Clark, JR. and J. von der Hardt) Karl Thiemig, (1984).

(3) S. Fujine, K. Yoneda and K. Kanda, An online Video Image Processing System for Real -time Neutron Radiography, Nucl, Instr. Meth., 215, 277-289 (1983).

(4) S. Fujine, K. Yoneda and K. Kanda, An Application of the Neutron Television Fluoroscopic System to Neutron Computed Tomography, ibid, 226, 475-481 (1984).

(5) S. Fujine, K. Yoneda and K. Kanda, Digital Processing to Improve Image Quality in Real-time Neutron Radiography, ibid, 228, 541-548 (1985).

(6) H. Yamagata, K. Yoneda, S, Fujine, K. Kanda and K. Katsurayama, Standardization for Determing Image Quality in Thermal Neutron Radiographic Testing, in press in Annu. Rep. Res. Reactor Inst. Kyoto Univ.

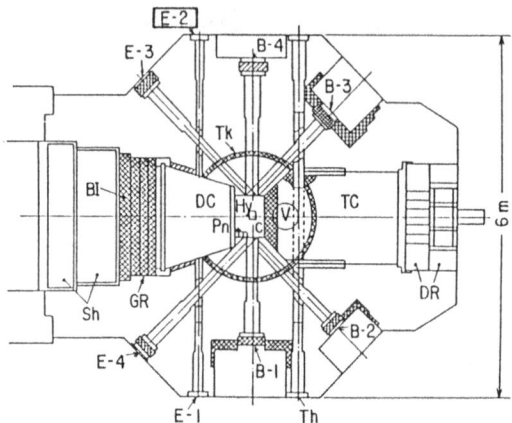

Fig. 1. Horizontal view of the KUR experimental holes.

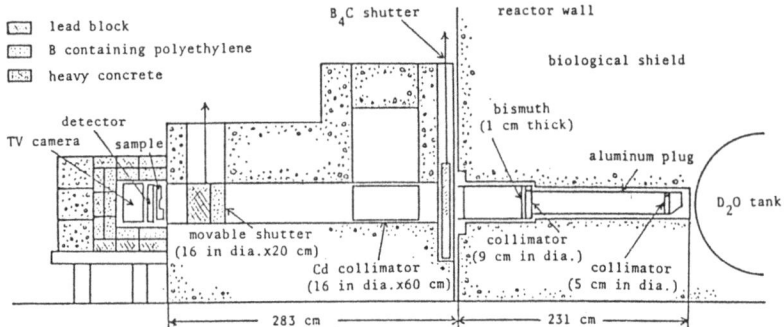

Fig. 2. KUR neutron radiography facility.

Table 1. Characteristics of the KUR neutron radiography facility.

1. Reactor / Power	KUR / 5000 [KW]
2. Peak Φ_{th} in core	6×10^{13} [n/cm^2.sec]
3. Range of L	500 [cm]
4. Standard L / D	100
5. Φ_{th} at film	1.2×10^6 [n/cm^2.sec]
6. Gamma dose rate	4.2 [R/h]
7. Cadmium ratio	400
8. Neutron/gamma ray ratio	1.1×10^6 [/cm^2.mR]
9. Film size available	16 [cm in dia]
10. Beam uniformity	+3.5 [%]
11. ASTM-75 specification	85-12-11
12. ASTM-81 (NC-H-G)	79- 7- 7 (Categorie I)

604

Photo. 1. Sample irradiation port.

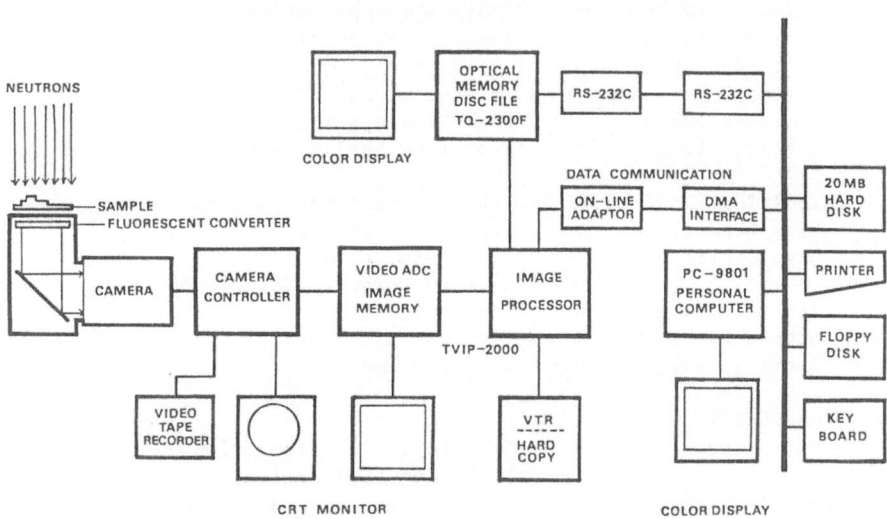

Fig. 3. Block diagram of KUR television system.

Photo. 2. KUR television system.

Table 2. Digital processing for NTV images.

```
DIRECT IMAGE  ------------  Low Contrast and Poor Resolution
   Direct   Image     ---- *Real-Time Application
   Reversed Image     ---- *Identification of the Location
   Contoured Image    ---- { of Burnable Poison in the Side
   Pseudo Color Image ---  { Plates of the MTR Type Fuel
   Enhanced Image     ---- *Real-Time Application
   Average  Image     ---- *Real-Time Application

INTEGRATED IMAGE  --------  High Sensitivity and High Resolution
   High Sensitivity  ---- *Capability at Low Intensity
   High Quality      ---- *High Resolution
                     ---- *Neutron Computed Tomography
   Reversed Image    ---- *Complemental Use
   Subtracted Image  ---- *Low Contrast Image
   Zooming  Image    ---- *Enlarged Image

ENHANCEMENT
   Gamma Correction   ---- *Gray Level Transformation
                      ---- *Real-Time Application
   Matrix Filter      ---- *Noise Reduction and Contouring
   Histgram Modification ------ *Enhancement of Intermediate
   Gray Level Transformation -- {    Level
```

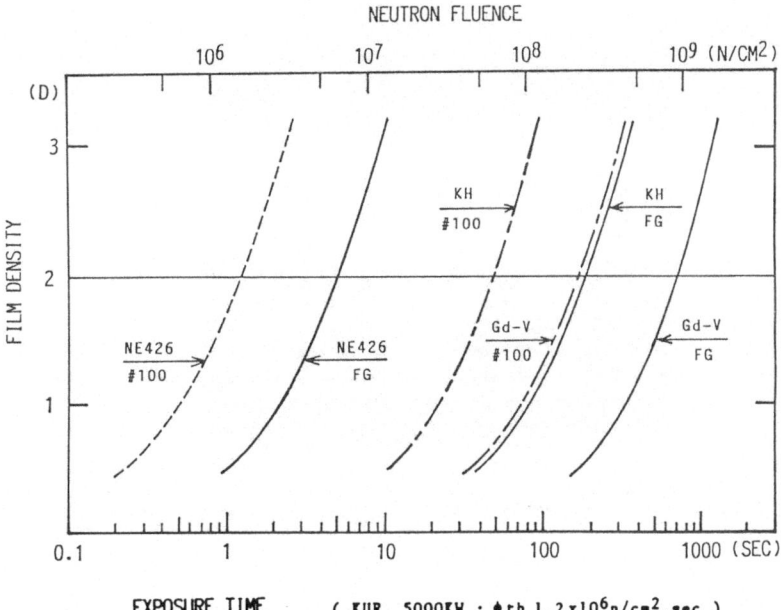

Fig. 4. Relations of film density and exposure time.

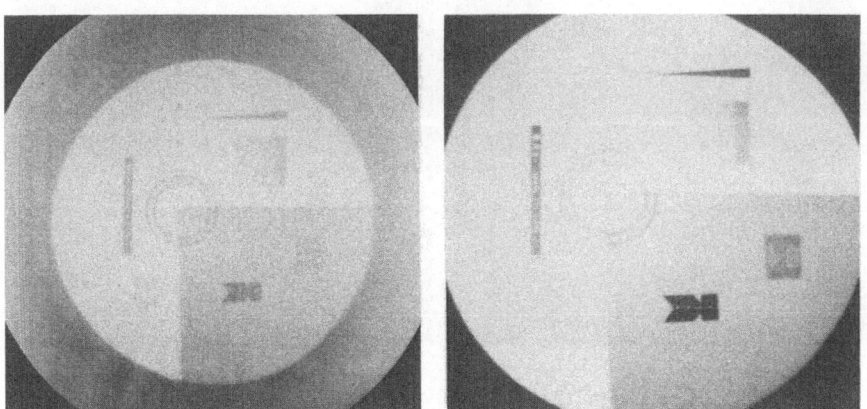

Photo. 3. Comparison of images on CRT taken by three intensifier screens for X-rays. (Operation power of KUR: 500kW; left: Sakura KH, right top: Fuji G8, and right bottom: Fuji G4.

Photo. 4. Contoured image of Photo. 3.

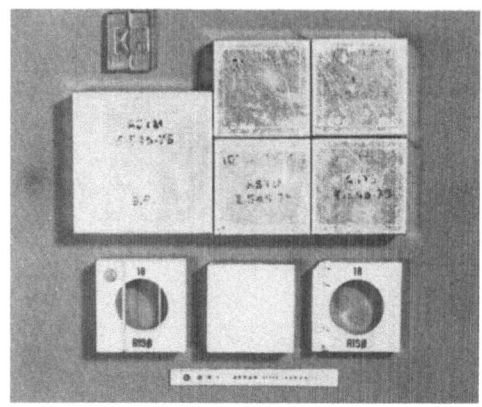

Photo. 5. Intercomparison of indicators.

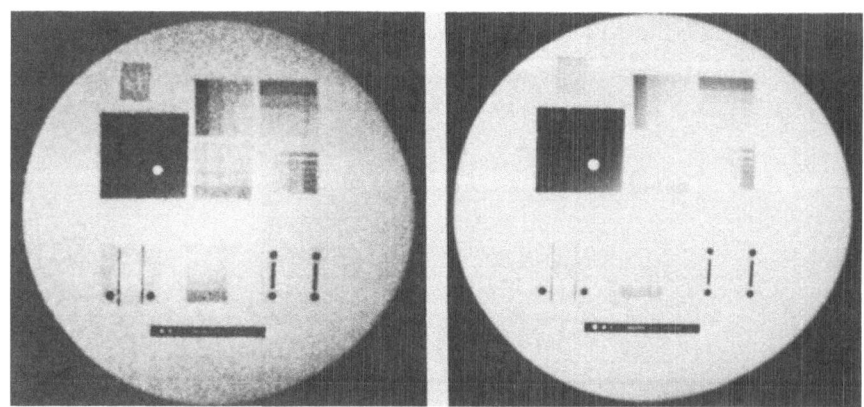

Photo. 6. Experimental Arrangement of Intercomparison.

ANALYSIS OF TWO PHASE COUNTER FLOW IN HEAT PIPE

BY NEUTRON RADIOGRAPHY

M. Tamaki, K. Ohkubo, Y. Ikeda and G. Matsumoto

Department of Nuclear Engineering,
Faculty of Engineering, Nagoya University
Furo-cho, Chikusa-ku, Nagoya 464 Japan

ABSTRACT

The heat pipe operates by the four processes composed
of evaporation, vapor flow, condensation and reflux of the
working fluid. Neutron radiography and neutron television
were applied to investigate the thermal-hydraulic behavior
of the working fluids(boiling, vapor flow, condensation
and liquid flow) in steel heat pipes. Light and heavy
waters, propanol, isooctane, sodium and 10%lithium-sodium
were used as working fluids. All the bulk condensates were
easily visualized. The films of light water, propernol,
isooctane and 10%lithium-sodium contained in the wick were
also detectable. Dimension of the liquid films and the
spatial distribution of the working fluids were evaluated
by image processing. The interface between vapor and non-
condensable gas in the 10%Li-Na heat pipe was speculated
from the condensing position of the vapor.

INTRODUCTION

Heat pipe is a simple structure device having high thermal
conductance. It is composed of a closed container and a working
fluid. A wick having capillary action attaches to the inner wall
of the container. The working fluid is a volatile liquid. The
heat pipe operates by the four sequential processes; evaporation,
vapor flow, condensation and reflux of the working fluid in the
closed system. By the evaporation-condensation process, the

latent heat of the phase change of the working fluid is transferred from the evaporator to the condenser. The operational characteristics of the heat pipe were influenced by boiling and condensation of the working fluid, interaction between liquid and vapor in the counter-current two-phase flow, capillary capability of the wick and the existing non-condensable gas. These are able to be investigated by visualization of the working fluid on the stationary and dynamic conditions of the heat pipe(1). In the present study, by the neutron radiography(NR) and the neutron television(NTV) techniques, the spatial distribution and the dynamic flow of the working fluids in the small heat pipes were visualized. In addition, the qualitative analysis was attempted by image processing of the NR and NTV data.

EXPERIMENTALS

Some kinds of heat pipes were prepared for imaging by NR and NTV. Fig.1 shows the schematic of the heat pipe. The heat pipe container and the wick(2 layers of 400 mesh) were made of stainless steel. Six kinds of working fluids were used(Table 1). Sodium and lithium were for the investigation of the capability of the visualization by NR. Light water(H_2O) was the reference material. The others were used to compare with the operational characteristic of the light water. They have the boiling point of about $100°C$. These six heat pipes were simultaneouly operated by the special heating and cooling system and were observed by NR and NTV. The NR and NTV systems had been mentioned in elsewhere (2,3). NR photos and NTV video were analyzed by the image processor and the computer.

RESULTS AND DISCUSSIONS

Fig.2a shows a NR photo of the light water heat pipe(LW). The black part(bottom) of the heat pipe was corresponded to the bulk water. The dark part(midst) indicated the thin film water contained in the wick. The bright part(upper) was the unwetted wick. The thickness of the thin water film was evaluated to be about 0.4mm by the image processing. Fig.2b shows a NR photo of the sodium(N) and 10%lithium-sodium(LN) heat pipes. Bulk sodium and 10%lithium-sodium were detectable. Moreover, thin lithium-sodium film was also detectable. But sodium in the wick was difficult to be detected. Fig.3 shows three different types of photos of optical camera (3a), NR(3b) and NTV(3c) for the six heat pipes. Number on the heat pipes indicated the species of the working fluid(see Table 1). The NR photo(3b) showed the spatial distribution and the wettability of the working fluids. The degree of the darkness of the bottom of the heat pipes depended on the integral cross section of the working fluids. Sodium and heavy water in the wick were undetectable. Fig.3c was the photo from the NTV monitor. Spatial resolution was evaluated to be about 1mm in NTV and 0.1mm in NR. Image quality of NTV is considerably coarse compared to NR. However, NTV is applicable to the dynamic observation of the working fluids. Fig.4 shows the instantaneous images of the heat pipes. Fig.4a was the original image of the single frame of the NTV monitor (1/30 sec) operated at 70-75°C. Fig.4b was the reference image of the heat pipes before operation. Fig.4c was the image of the differential between 4a) and 4b) images, which was obtained by the image processing. This image in Fig.4c represented only the moving working fluids. Bright parts reveals the instantaneous spatial distribution of the working fluids in the operating heat pipes. In the water heat pipes(No.3 and 4), working fluids moved violently in the mode of the two-phase counter-current flow. The violence of the movement of the working fluids decreased in sequence of water, propanol and isooctane. It seems to be related to the latent heat of the vaporization of the working fluids. Fig.5 shows the photos of the 10%lithium-sodium heat pipe obtained from the NTV monitors(3). Fig.5a was corresponding to before operation. The black part(bottom) was the bulk 10%Li-Na solid. Fig.5b and 5c show the operating ones at 510 and 570°C, respectively. The dark region(midst) revealed the condensing film of the working fluid. It may be speculated that the interface between the dark and bright regions(✳) corresponded to the diffusing front of the vapor to the existing non-condensable gas. Furthermore, by the image processing technique, the velocity of the splashing liquid and the density of the vapor phase were qualitatively evaluated.

Conclusionally, stationary and dynamic working fluids in the heat pipes were visualized by the neutron radiography and the neutron television techniques. Qualitative analysis of the thermal-hydraulic behaviors of the heat pipes was enabled by the image processing of the NR and NTV.

REFERENCES

1. Ed. by K.Ohshima, Research and Development of Heat Pipe Technology(Japan Technology and Economics Center Inc.,Tokyo 1984).

2. G.Matsumoto, M.Tamaki,K.Ohkubo and Y.Ikeda," Real-time Imaging of Working Fluids in Heat pipe by Neutron Television." Proc. of 5th Inter. Heat Pipe Conf.,Tsukuba Sci. City,Japan, ed.by K.Ohshima(Japan Technology and Economics Center Inc. Tokyo 1984),vol.II,pp.133-138.

3. G. Matsumoto, " Development and Application of Neutron Radiography." Doctor Thesis of Nagoya University(1985).

Table 1. Specifications of the working fluids in the heat pipes

No.	1	2	3	4	5	6
working fluid	sodium (Na)	sodium-10%lithium (Na-10%Li)	heavy water (D_2O)	light water (H_2O)	propanol (C_3H_7OH)	isooctane ($i-C_8H_{18}$)
m.p.(K)	371	628	277	273	147	166
b.p.(K)	1156	>1000	375	373	370	372
density (g/cm^3)	0.97	0.90	1.10	1.00	0.80	0.69
molecular weight	23.0	18.7	20.0	18.0	60.1	114.2
$\Sigma_t(cm^{-1})$	0.095	0.4	0.50	3.72	3.3	3.4
latent heat(J/g)	3873	3000	2077	2257	787	272

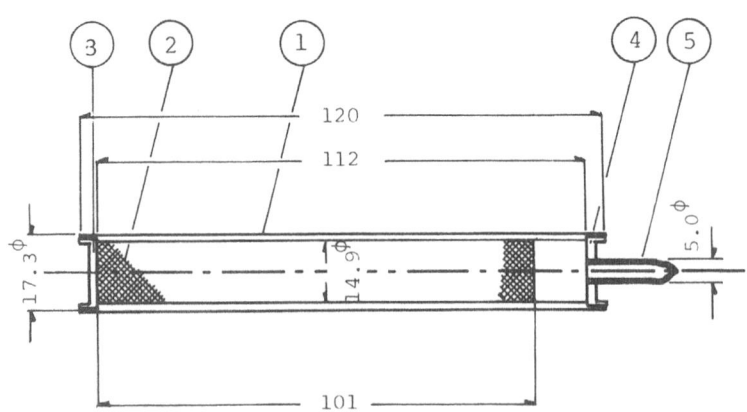

Fig.1 Schematic of Heat pipe for Imaging Analysis by Neutron Radiography and Neutron Television
(1:envelope pipe, 2:400 mesh wick(2 layer), 3:end cap 4:end cap, 5:fill tube(sealed))

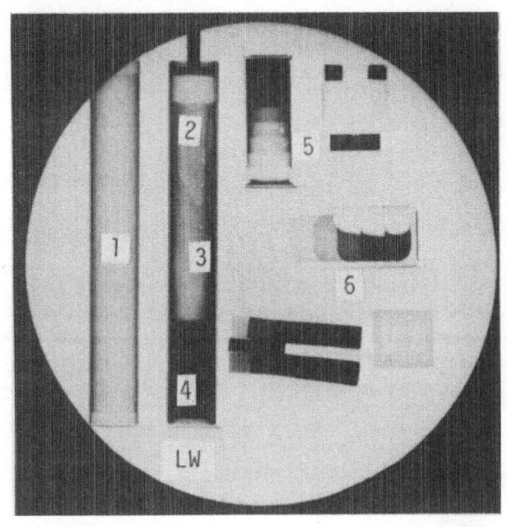

1: reference steel pipe

2: unwetted wick

3: wetted wick

4: bulk water

5: H_2O step wedge (cylindrical)

6: H_2O step wedge (flat)

a) Light water(H_2O) Heat Pipe

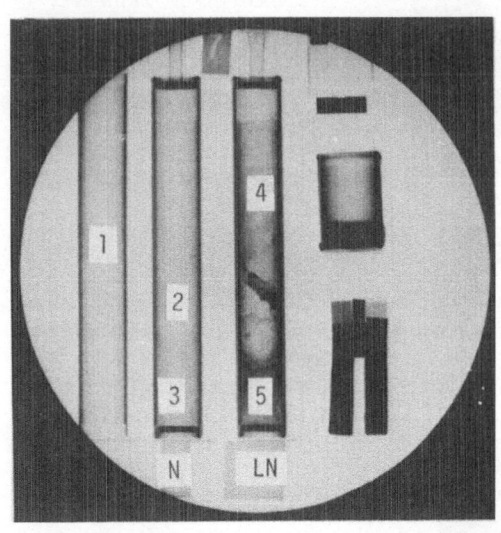

1: reference steel pipe

2: hollow hole in bulk solid sodium

3: bulk sodium(solid)

4: Li-Na in wick

5: bulk 10%Li-Na(solid)

b) Sodium(N) and 10%Lithium-sodium(LN) Heat Pipes

Fig.2 Neutron Radiographs of Heat Pipes

(Gd/Fuji FG/$1.2 \times 10^9 n/cm^2$/KUR E-2)

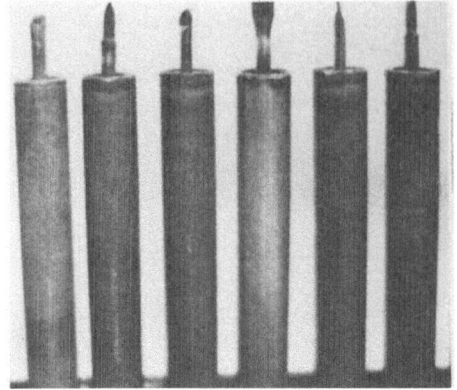

a) Image by Optical
Camera

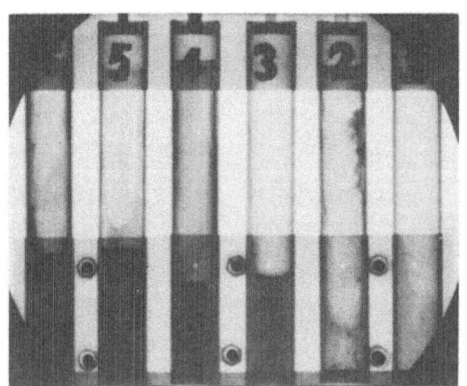

b) Image by NR
(before Operation)
(Sakura$_8$KH/Fuji FG/
2.3x10^8 n/cm^2/KUR E-2)

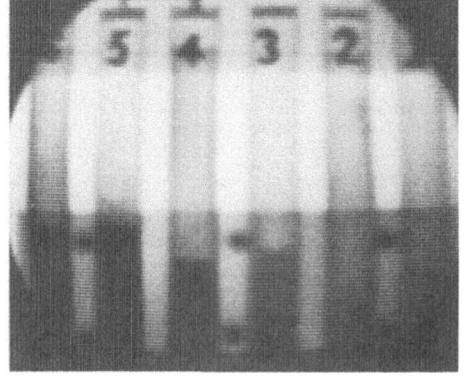

C) Image by NTV
(before operation)
(NTV/30$_6$frames(1 sec),
1.2x10^6 n/cm^2/KUR E-2)

Fig.3 Images of Heat Pipes by Optical Camera, NR and NTV

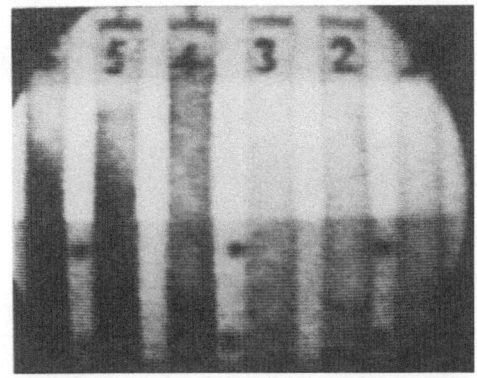

a) On Operation at 70-75°C

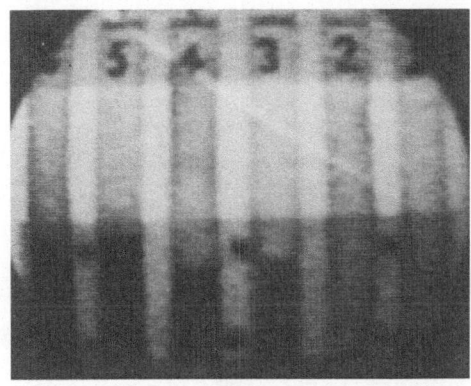

b) Pre-operation

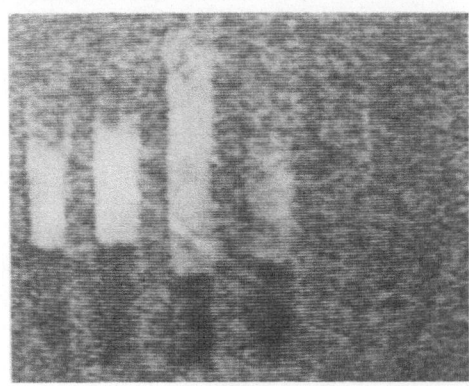

c) Processed Image
   (differential: a-c)
   (Movement of Working
    fluids : white regions)

Fig.4 Instantaneous Images from NTV monitor
   (NTV/single frame(1/30 sec), $4 \times 10^4 \text{n/cm}^2$/KUR)

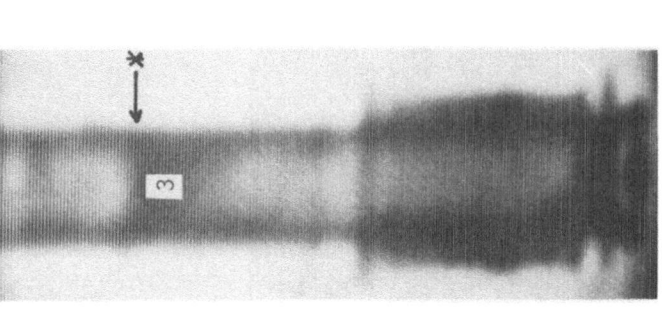

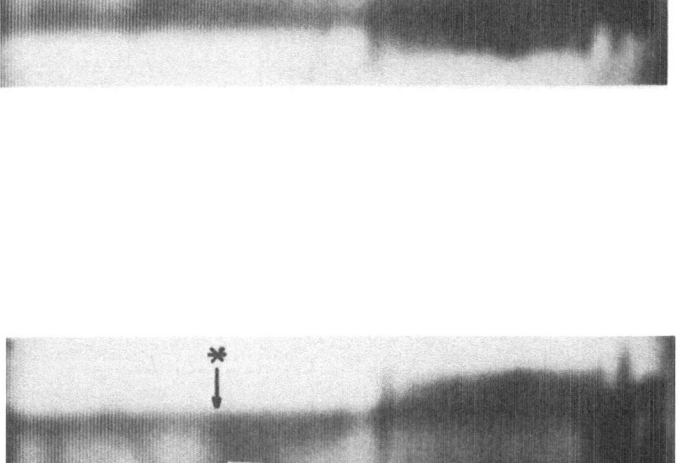

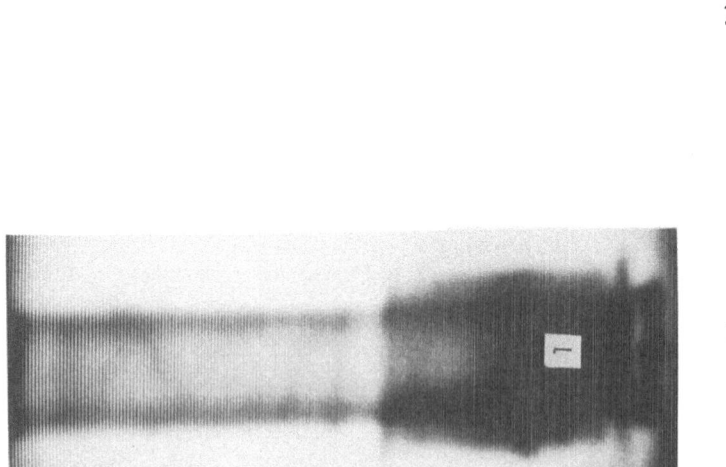

a) Pre-operation
(1:bulk 10%Li-Na)

b) On Operation at 510°C
(2:condensing film)

c) On Operation at 570°C
(3:condensing film)

Fig.5 10%lithium-sodium Heat Pipe images by Neutron Television
(NTV / 120 frames(4 sec), $1.2 \times 10^6 \text{n/cm}^2$ / JSW) (Ref.3)

APPLICATION OF CYCLOTRON-BASED REAL-TIME

NEUTRON RADIOGRAPHY SYSTEM FOR SOME INDUSTRIAL COMPONENTS

Atsuo ONO and Koichi SONODA
College of Liberal Arts, Kobe University,
Nada-ku, Kobe, 657 Japan

Ryoichi TANIGUCHI and Eiichi HIRAOKA
Radiation Center of Osaka Prefecture,
Sakai, 593 Japan

Shuichi Tazawa, Takehiko NAKANII, Yorihisa ASADA and Munenori YANO
Sumitomo Heavy Industries, Ltd.,
Niihama, 792 Japan

Kazuo HAYASHI
Central Laboratory, Nissan Motors,
Yokosuka, 237 Japan

Takehiko HOSOKAWA
Aerospace Division, Ishikawajima Harima Industries, Ltd.,
Tanashi, 188 Japan

Abstract

Reliable facility of cyclotron-based real-time neutron radiography system has been developed and applied to some industrial components. The equipment for neutron fluoroscopy is based on a sub-compact cyclotron and a LiF/ZnS(Ag) fluorescent screen viewed by a silicon intensifier target TV camera. The real-time image is monitored on a CRT, recorded with a standard video recorder and processed by a digital image processor.

The effectiveness of our real-time neutron radiograph has been demonstrated to be applicable to not only the dynamic observation but also the magnifying and stereoscopic observation of fluoroscopic images. The results are presented on photos and video tape.

Introduction

A real-time neutron radiography facility was installed at the Toyo Works of Sumitomo Heavy Industries, Ltd., where model 370 cyclotron had been equipped themselves. A schematic diagram of the facility is shown in Fig. 1. Although more detailed descriptions about the facility were given elsewhere in this proceedings [1,2], the outlines are as follows:

Fig. 1 Schematic diagram of the real-time neutron radiography system using the sub-compact cyclotron.

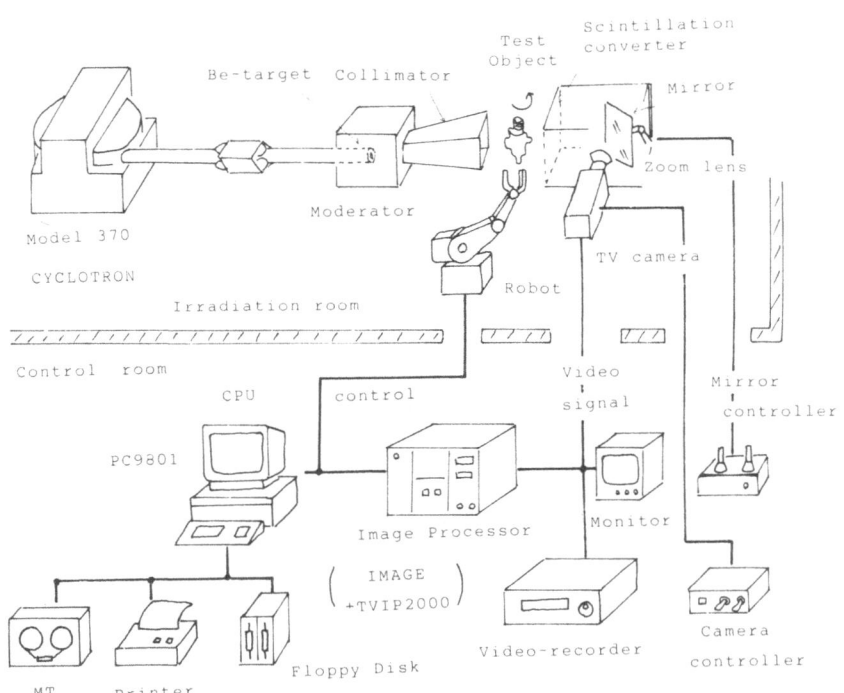

Extracted 18 MeV protons from the sub-compact cyclotron are converted into neutrons by beryllium target and created neutrons are thermalized by moderator. Three-stage collimator which is capable to change the L/D ratio to 15, 20 and 32.2 is used. At the end of the longest collimator, nominal flux is 1.1 x 10^6 neutrons/cm^2/sec. Since the outgoing neutron density per frame is low, high sensive LiF/ZnS(Ag) fluorescent screen of type A4 (Kasei Optonix, Ltd.) is applied to the real-time imaging though it has slightly poor spatial resolution. Normally 14" x 17" large scaled screen is used.

The image producing on the fluorescent screen is reflected by a mirror into a zooming lens attached to a TV camera VC 7000 (Tokyo Electronics Industies Co., Ltd.) in which a silicon intensifier target (SIT) tube of RCA 4804H is used. Visual field is limited, so the inclination angle of the mirror is varied by a motor-drive to make observation of the wide range of the screen.

The original video images obtained from TV camera are statistically poor, and has low contrast and low space resolution (which is at least limited by the number of scan lines) [3]. Use of image enhancement techniques such as noise reduction, image subtraction and integration increases the capabilities of radiographic system and is accomplished with the digital image processor. We used a combination of Image Sigma and TVIP-2000 of Nippon Avionics Co., Ltd.

The main purpose of real-time radiography is to observe the image of the hydrogen containing substances inside metal vessel instantly or dynamically. The experiments were performed to check the capabilities of the technique to provide visual information in real time.

Observation of dynamic events

As a first step, a rotating blower fan in an oil bath was observed dynamically. One frame image is shown in Fig. 2(a), where oil layer flows along the edge of the fan, was clearly observed.

As experimental tests to observe the oil movements in real-time in actual automobile components, a fuel injection pump and a shock absorber were radiographed. Fig. 2(b) and Fig. 3 show the oil flows in a fuel injection pump and a shock absorber. We present the oil movements within a shock absorber by giving shocks on the video tape. Local variation of oil density was observed with the movement of the piston.

Fig. 2 (a) Dynamic raw image of rotating blower fan in oil
        bath.
        (b) Dynamic image of automobile fuel injection pump.
          (Averaging 5 frames).
          (a)   (b)

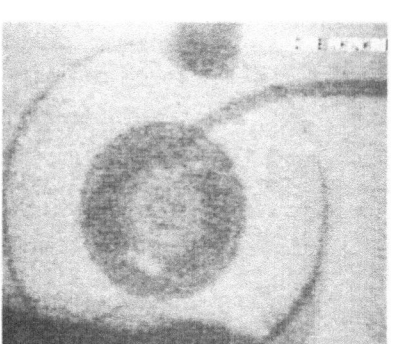

Fig. 3 Dynamic raw images of shock absorber.

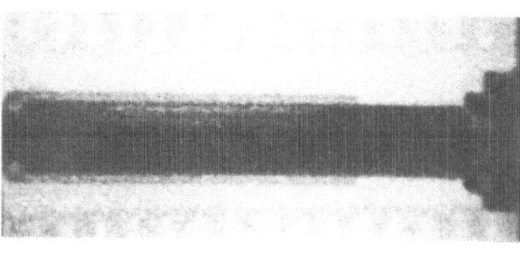

A plasticating extruder has been constructed to study the melting mechanism which depends on operating conditions such as screw speed and production rate. Fig. 4(a) shows a schematic picture of the extruder. To make contrast with pellets to be fed, cylinder of 2 mm thickness and brass screw were examined. Fig. 4(b) shows a fluoroscopic image of the extruder when it is out of operation. Packed styrole pellets are clearly identified in the integrated image. The outer diameter and pitch of the screw are 20 and 18 mm, respectively.

Fig. 4 Plasticating extruder.
    (a) Schematic picture.
    (b) Static image. Styrole pellets are packed.
    (c)-(d) Dynamic images. Screw speed is 6 rpm.
      (Averaging 8 frames).

(a)   (b)

(c)   (d)

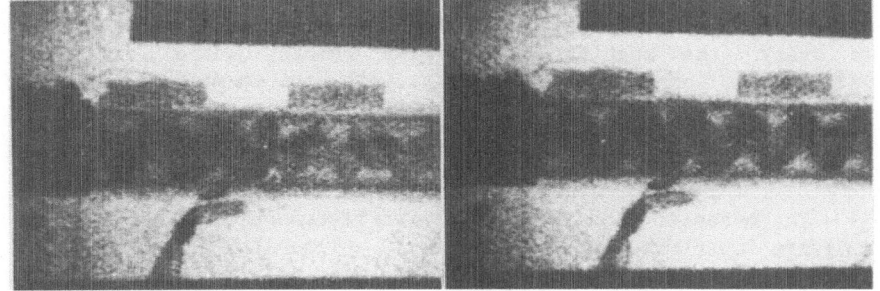

The screw was rotated in the speed range of 0-200 rpm in operation. We show on a video tape the dynamic behavior of the styrole pellets within an extruder. Stills of averaging images are shown in Fig. 4(c)-(d). Fusing styrole twines round the screw to the conveying zone and its width grows wide. We also show a dynamic flow of extruding styrole within a metallic mold. Fig. 5 shows the stills of this experiment.

Fig. 5 Dynamical flow of extruding styrole within a metallic mold. (raw image).

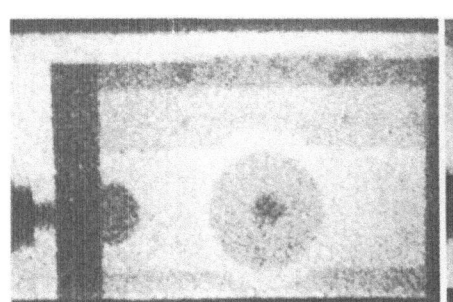 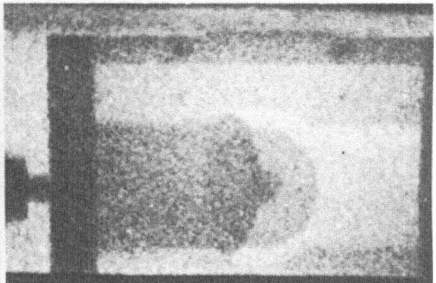

Magnifying observation of fluoroscopic image

Compared with the film method, visual field and spatial resolution are subjects to restriction in the video imaging. At the object field size of 200 x 250 mm^2, obtained resolution is 1.0 line pairs/mm. Usually we uses a zooming lens which varies focal distance from 18 to 108 mm. Though a zooming lens is easily able to vary the magnifying power by remote control, obtained radiographic image grows gloomy with the magnification and resolving power obtained is at most 1.25 line pairs/mm in life size. This value is far from the expected one from the film method for the screen used.

In order to observe objects with higher resolution, we place a large aperture lens additionally close to the fluorescent screen. In this optics, we examined the minimum imaging capabilities of the real-time neutron radiographic system. By observing the line slits on a cadmium plate of 100 microns thickness, the resolving power is at most 4 line pairs/mm in the neutron fluoroscopic image though we can easily resolve 5 line pairs/mm in a direct image for dark light.

ASTM sensitivity indicators of E545-75 and -81 were radiographed. The obtained images are shown in Fig. 6. The largest consecutive value of R, which is visible in the fluorescent image of the E545-75 type A indicator, is 4. This value is comparable to obtained one with the film method by using the same fluorescent screen, type A4. In addition to the examples mentioned above, we also show magnifying fluoroscopic images of an encased explosive and an igniter.

Fig. 6 Magnifying images of ASTM sensitivity indicators.
   (a) E545-75.   (b) E545-81.

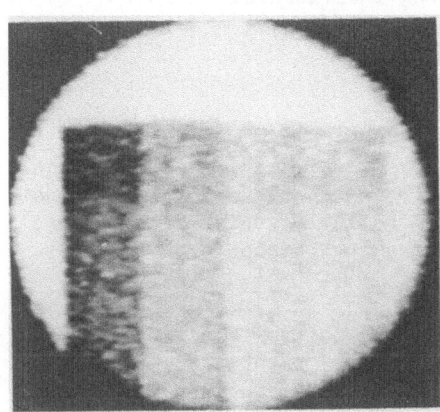 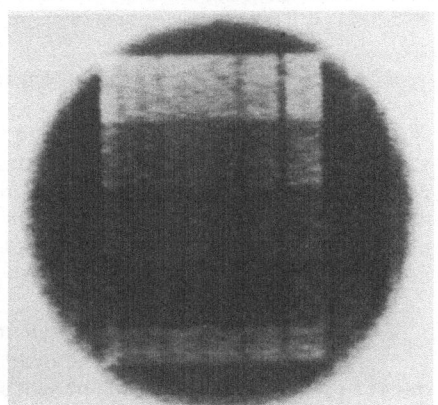

Stereoscopic observation

To provide three-dimensional information about the internal condition inside the object, neutron computed tomography has been developed elsewhere, but it seems to be expensive and time consuming to process in real-time with high resolution.

Stereoscopic displays of the object are of interest because they produce true three-dimensional visual sensations. In neutron fluoroscopy, however, the obtained image is a shadow of the object and it is difficult to prepare two similar neutron radiography systems simultaneously. We tried to test the possibility of the stereoscopic observation by using one neutron source.

By rotating a object 5 or 10 degrees on a fixed axis with robotic manipulator, two integrated fluoroscopic images are easily recorded. Stereoscopic images are produced by merging them into one frame displayed once each image in a refresh cycle. The processing time is very small. The composite image is viewed through the electro-optic shutters in the stereoscopic viewer which are operated synchronously with the CRT vertical retrace synchronized pulse. As shutters, both lead lanthanum zirconate titanate (PLZA) ceramics and liquid crystal were tested.

Representative applications of stereoscopic display techniques are given for a turbine blade and a vacuum flange to which BNC connectors and an o-ring are attached. Flatter objects are unsuitable for stereoscopy though they are appropriate to fluoroscopy. In cases of the above samples, stereoscopic effect is clear. But in general, the effect is ambiguous sample by sample. From our experiments, we realized that there is difference in stereoscopic sense among individuals.

Conclusions

Many samples were sequentially observed for the purpose of industrial application. Preliminary work has demonstrated the feasibility of viewing oil and fusing plastic fluids inside metallic materials. Thus, it is reliable to observe dynamic phenomena in real time by utilizing a sub-compact cyclotron as a neutron source.

Flexible optical system is necessary to vary the visual range and magnification. We demonstrated the capability of making detailed observation of objects. As a result, we can resolve at most 4 line pairs/mm. This limit causes from the fluorescent screen used.

Simple techniques to realize stereoscopic observation of fluoroscopic images have been developed and checked the availability. The application will be the following problem.

This work was supported in part by a Grant-in-Aid from the Ministry of Science and Technology Agency.

References

[1] S. Tazawa et al., this conference.
[2] E. Hiraoka et al., this conference.
[3] R. Taniguchi et al., this conference.

DEDICATED IMAGING SYSTEM FOR REAL-TIME NEUTRON RADIOGRAPHY

Vijay Alreja and Leonard Corso

Precise Optics/PME, Inc.

Bayshore, New York

1. Abstract

This paper describes the principles of an image intensifier based real-time system, as well as the basis for selecting components to produce a state of the art imaging system capable of high contrast and resolution standards. Various components or sub-systems are discussed, and advantages and limitations of the system are enumerated.

2. Introduction

Real-time radiographic imaging systems depend on the fact that neutron scintillations yield light when irradiated with thermal neutrons. This light can be amplified by image intensifiers and/or detected by television cameras. The television display can provide real-time viewing of the neutron images at locations removed from the radiation area. The real-time image intensifying systems can be used with any neutron source, whether reactor, accelerator or isotope based. A typical real-time radiographic imaging system includes an image intensifier and a CCTV camera. The input screen is made from gadolinium oxysulfide which is very sensitive to neutrons.

Consideration to high contrast spatial resolution in the system is dependent to a large degree on neutron density. Tri-field tubes, which increase field size resolution and contrast instantly, by changing magnified modes provide the flexibility needed for real-time radiography. Imaging system resolution also depends on the thickness of the radiographic object and the geometry of the system. The design of a real-time imaging system must have the versatility to view different objects of various densities to provide the best picture quality needed for critical interpretation. We have recommended the criteria for selecting imaging system components to provide the best system with the least compromise.

3. Description

Figure 1 shows a typical real-time neutron image intensifying system, consisting of the image intensifier tube and closed circuit TV camera.

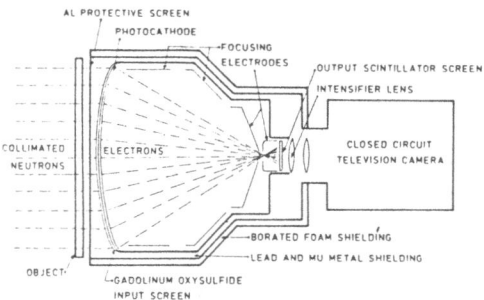

REAL TIME NEUTRON IMAGE INTENSIFYING SYSTEM

Figure 1

3.1 Image Intensifier Tube

The image intensifier tube has a gadolinium oxysulfide input screen providing high neutron sensitivity, 9/6/4-inch tri-field tubes instantly increase field size resolution and contrast. The tri-field tube allows continuous focussing of the

image between 9 and 4.5 inches. Figures 2 and 3 show
photo cathode current and measured light output versus
energy of operation with the image intensifier system
in use at the Breazeale Nuclear Reactor at
Pennsylvania State University. The illustrations
show the linear response and non saturation of the
image intensifier tube in spite of very high flux
rates.

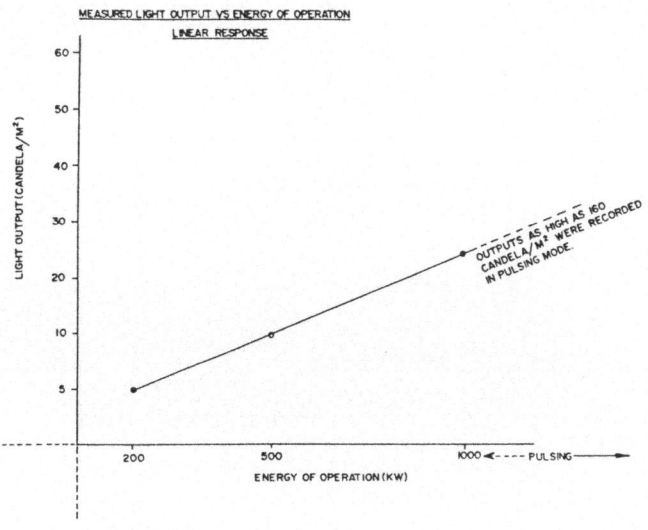

Figure 2

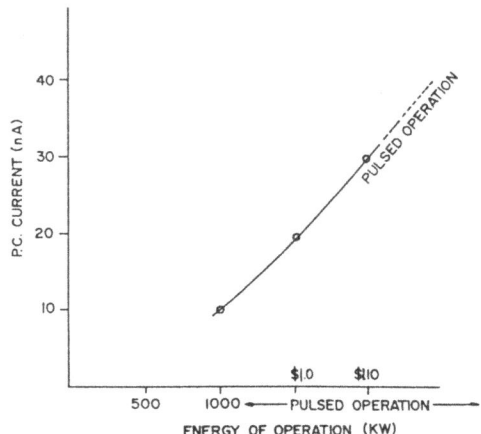

Figure 3

3.2 Closed Circuit TV Camera

The CCTV camera should provide high resolution, high sensitivity at low light levels, and low lag characteristics. The camera should have the ability to display areas of divergent brightness, such as raw n-ray next to highly attenuated n-rays without picture blooming.

3.3 Neutron Sources

Use of reactor is the best source of high neutron flux for greatest resolution, contrast and short exposure time in an imaging system. Refer to Table 1 Characteristics of Thermal Neutron Radiographic sources.

Table 1

Type of Source	Typical radiographic intensity $(cm^{-2} s^{-1})$	Radiographic resolution	Average exposure time	Characteristics
Radio Isotope	10^1 to 10^4	Poor to medium	Long	Stable operation, medium investment cost, possibly portable.
Accelerator	10^3 to 10^6	Medium	Average	On-off position, medium cost, possibly portable.
Subcritical	10^4 to 10^6	Good	Average	Stable operation, medium to high investment cost, portability difficult.
Nuclear	10^5 to 10^8	Excellent	Short	Stable operation, medium to high investment cost, portability difficult.

629

Figures 4 and 5 show a typical imaging system setup using the Breazeale Reactor at Pennsylvania State University.

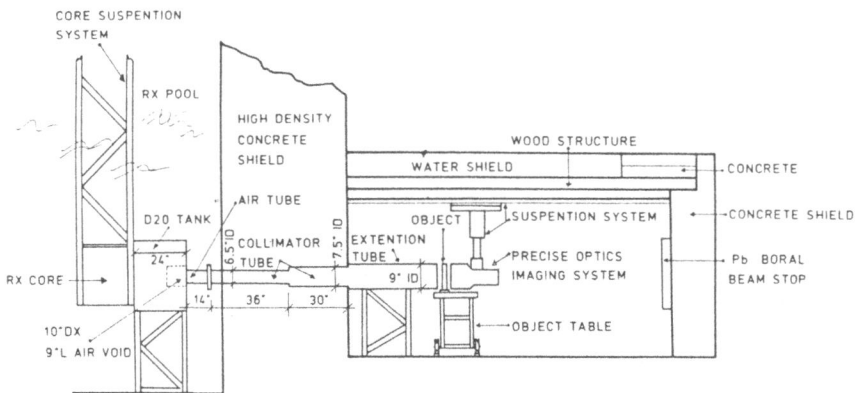

SIDE VIEW OF THE BEAM HOLE FACILITY SECTIONED THROUGH THE CL OF THE CORE
THE REACTOR CORE IS IN POSITION FOR RADIOGRAPHY

Figure 4

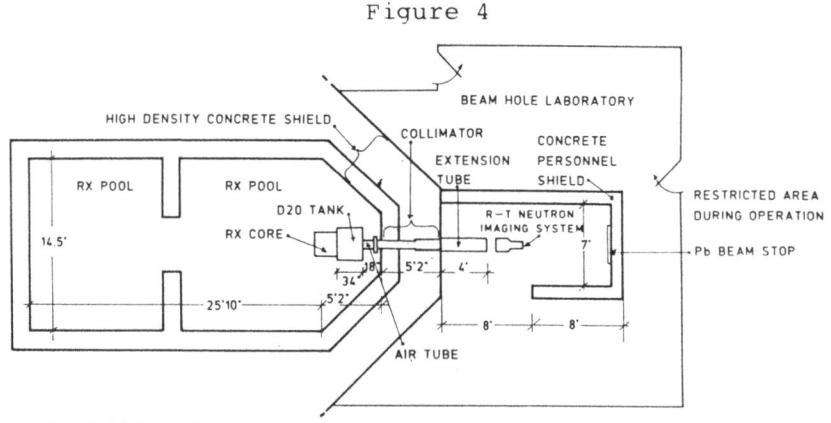

REACTOR POOL AND BEAM HOLE LABORATORY
WITH THE REACTOR CORE IN POSITION FOR NEUTRON RADIOGRAPHY

Figure 5

Figure 6 shows the cross section of a neutron collimator.

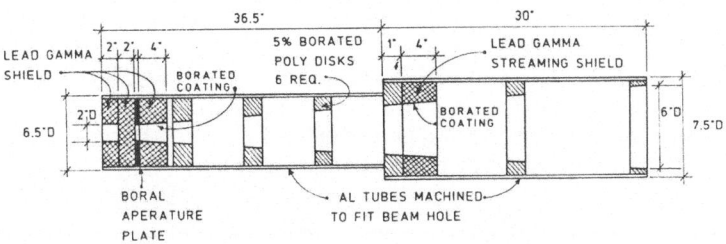

NEUTRON COLLIMATOR CROSS SECTION

Figure 6

4. Image Intensifier System

4.1 Advantages

The obvious advantage of a real-time imaging system over x-ray film is that object motion can be followed. In an industrial setting, parts could be continuously inspected. A real-time system is ideal for inspection of large objects such as aircraft components (engines, wings, etc.) because of the speed and ease of application. Electronic control over brightness, contrast and image manipulation achieves exceptional spatial resolution over other systems. Objects in real-time can be positioned so as to obtain the best view.

4.2 Limitations

The high neutron flux necessary to generate high quality radiographs is usually only obtained from a reactor, often making portability difficult. New systems using a portable neutron source with an emission of 10^{12}n/cm^2/sec is now feasible so that this problem may be overcome. The use of radio-isotopes or accelerators does not produce the quality of images that can be obtained using a reactor, but today, image processing systems reduce much of this gap.

5. Recommendations and Conclusions

Real-time radiography has been aided by the progress of a tube design which has been developed by the medical industry. This has improved the performance of image intensifier based systems to an extent that the quality is almost comparable to that of film. Along with the other advantages of being able to use digital image processors and the flexibility of operation, real-time imaging demands attention from anybody considering neutron radiography.

REFERENCES

Berger, Harold, "Detection Systems for Neutron Radiography". Practical Applications of Neutron Radiography and Gaging. ASTM STP 586, American Society for Testing and Materials, 1976 p. 43-44.

Berger, Cutforth, Garrett, Haskins, Iddings, Newacheck, "Non-Destructive Testing Handbook". Neutron Radiography American Society for Non-Destructive Testing p. 20-21.

Berger, Harold, "The Technological Development of Neutron Radiography, Atomic Energy Review 152 (1977) p. 132, 135, 137.

PART IX

TRACK-ETCH IMAGING

ADVANTAGES AND POSSIBILITIES OF CELLULOSE NITRATE DETECTORS

IN CORPUSCULAR IMAGERY

M. FANTINI

Centre de Recherches Kodak-Pathé

BP 60   94300 Vincennes France
Personal address: 69 rue de Fontenay 94130 Nogent sur Marne F

ABSTRACT

The relatively low, but adjustable contrast given by
cellulose nitrate films permits, in neutron radiography
to record very fine details in images contrary to the
high contrasted silver halide images. The resolution is
excellent. The simplicity of use is remarkable: all
operations are made in full light, the insensivity to
gamma rays permits use in direct, non filtered neutron
beams and the radiography of gamma active objects.
The converter screens used with cellulose nitrate are
non-activable. This is not the case for converters
used with X rays films.
Cellulose nitrate films are excellent for alpha or
proton autoradiographs, visualising Boron, Lithium,
Nitrogen, etc.;extremely interesting for Medicine,
Biology, Industry, etc.

INTRODUCTION

Out of all the polymers tested over the past fifteen years,
the most adapted for nuclear particle imaging is still cellulose
nitrate. Developed originally for neutron radiography of
irradiated fuel rods, its use is now extended to other types of
radiography; in particular alpha and proton autoradiography
induced by neutrons, and radiography using any heavy nuclides.

QUALITY OF NR IMAGES ON CELLULOSE NITRATE

Often people think a high contrast is the major quality
of a radiographic image. This, true for certain routine controls
of defects coming from missing matter, for example, is not
true for recording highly detailed images. The cellulose nitrate
film with its low, but adjustable contraste (by variation of
etching parameters) gives extremely detailed images, inaccessible
with silver halide emulsions.

As everybody knows, images are not produced directly by
neutrons, but by ionizing particles or rays of their conversion.
In the case of cellulose nitrate images are made by alpha
particles. Being heavy, they almost do not diffuse and give a
high image resolution. In the case of silver halide emulsion
images are made by gamma rays or electrons (very light particles)
Then, the diffusion phenomenon is important giving less sharp
images.

EASY TO USE AND INEXPENSIVE

Storage, irradiation, process and all handling of cellulose
nitrate are made in full light. The possible storage-time at
room temperature is at least three years, enormous advantage in
regard of silver halide films.

Not at all breetle, cellulose nitrate films, with
incorporated converter, can be used directly in contact with
curved objects, simply applied with small adhesive tapes.

Insensitive to gamma rays, cellulose nitrate films do not
require expansive devices for filtering them in the neutron beam.
Moreover, they can be used for direct examination in contact
with active objects. In such a case, the use of conventional
X ray plates requires indirect method with a primary activation
of a converter which is a long and delicate handling operation.
The simplicity of cellulose nitrate use saves time and money.

Another unique advantage is the possibility to stop the
etching operation at several steps of image formation. In this
way it is possible to observe the equivalent of several
exposure level, successively, using only one irradiation. This
operation is cheaper than several irradations, as needed by
conventional neutron imaging.

Another simplification increasing the speed of the
procedure and lowering the cost, is the fact that Boron-made
Neutron-Alpha converters are not active after irradiation. So,
the film can be removed for etching immediately after exposure.
Even the etching process to get images after exposure is very

simple and cheap: only one "home-made" bath of pure solution of
sodium or potassium hydroxide.

SENSITIVITY

As the cellulose nitrate film response to alpha particles
is superior to 95%, the highest sensitivity is obtained using
the converter of highest Neutron-Alpha ratio: the Boron 10.
For example, using the "$_2$BE 10 " Kodak-Pathe screen, a fluence
of 5.10^8 neutrons per cm^2 is sufficient to obtain good
radiographs. This sensitivity is competitive with regard to
silver halide plates.

ION RADIOGRAPHY

Thermal and epithermal induced alpha and proton radiography
is practically possible only with plastic films and especially
cellulose nitrate ones, because of their flexibility and the
possibility to be used in micro-thin layers.

This technique permits Boron image visualization in the
Semi-conductor Industry, in biological Research (both vegetal and
animal).

Recently, in medicine, Boron seeked antibodies permit the
visualization of cancerous cells (1).Lithium, Nitrogen 14 are
visualized too, in Biology with Cold, thermal and epithermal
neutrons. So study of the metabolism of these elements is
possible (2) .

FAST NEUTRON RADIOGRAPHY

Cellulose nitrate films are quite well adapted for imaging
with Fast Neutrons of one MeV and above. In this case the
radiograph is recorded directly in the film without any converter.
Etchable tracks are induced by recoil nuclide of the detector
itself: Hydrogen, Carbon, Nitrogen, Oxygen. This technique can
be precious when a deep penetration is needed in highly
hydrogenated materials, like very soft tissues for medical
purposes.

CONCLUSION

This is a small list of corpuscular imaging utilisations
in full expansion. We suspect the existence of a large number
of non initiate potential users of these type of imaging .

But unfortunately, they are difficult to inform without a large diffusion of information.

REFERENCES

1. B. Larsson, O. Sornsuntisook, E. Johansson, K. Skhld, M. Fantini, Neutron Microradiography for cell-seeking Boron Compounds. 2nd. World Conference on Neutron Radiography, Paris June 1986.

2. A. Hartmann, J.C Wissocq, M. Thellier, H. Börner, W. Mampe , A. Fourcy, M. Fantini, and B. Larsson, Use of Cold Neutrons to increase the sensitivity of th Detection of Stable Isotopes with a (n,α) Nuclear Reaction for Tracer Experiments, Physiol. Vég. 1984 22 (6), 887-895 Gauthier-Villars Editor.

FAST NEUTRON RADIOGRAPHY (FNR) WITH TRACK ETCH

TECHNIQUE

E. Dühmke[1], L. Greim[2], and H.-W. Schmitz[2]

[1]Dept. of Radiotherapy, University of Göttingen, FRG

[2]GKSS Research-Center, Geesthacht, FRG

ABSTRACT

Neutron radiography of extended hydrogen containing objects is feasible only with fast neutrons, which guarantee sufficient radiation transmission. The imaging neutrons can be detected most reliably using plastic foils, by developing the tracks of locally corresponding recoil nuclei with an appropriate etching technique. This detection method is insensitive to photons, electrons, and thermal neutrons. There are defined correlations between etch pit and optical density, which allow optical recordings of foil images by the help of parallel light. The dependence of optical density on fast neutron dose as well as on various etching conditions is reported. The relatively low image contrasts of single foils can be enhanced considerably by a multifoil stack detector. By this, the image quality is improved, and the neutron exposure can be lowered by at least one order of magnitude. Radiographic examples show advantages and possible applications of FNR with respect to technical materials as well as biological specimens and in vivo radiography.

INTRODUCTION

In comparison to thermal and epithermal neutrons, fast neutrons have a considerably higher penetration capability for hydrogen containing matter (1). For this reason, optimal neutron radiographs of extended biological and comparable technical objects can only be obtained with fast neutrons: Fig. 1 shows the details of an animal below 1 cm thickness better with thermal, beyond 2 cm better with fast neutrons. Although attenuation differences decrease with increasing neutron energy, fast neutron radiography is of fundamental interest in addition to x- and γ-ray inspections, especially with respect to heavier materials (2, 3).

Fast neutrons can be generated by fission reactions (^{235}U, ^{239}Pu, ^{252}Cf: 1-2 MeV), by radionuclide sources using (α,n) reactions (e.g. ^{226}Ra: 4-6 MeV), or by accelerators of the ^{2}H (d,n)-type (14 MeV).

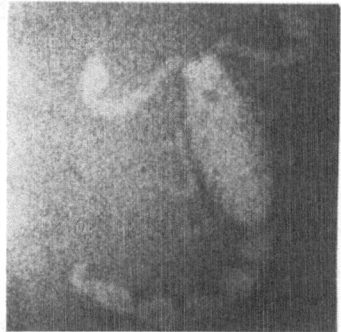

Fig.1 Neutron radiographs of a scorpion using thermal neutrons ($3 \cdot 10^{9}$ n/cm^{2}, boron coated cellulose nitrate foil, CN 85$_{10}$B, Kodak Pathé, Paris, left side) and fast neutrons ($3.6 \cdot 10^{10}$ n/cm^{2}, E>0.1 MeV, cellulose nitrate foil, CN 85, Kodak Pathé, Paris, multifoil detector, right side).

IMAGING METHOD

Among all fast neutron detection methods, applied to broad beam radiography (4,5), plastic foils developed by track etching combine the advantages of specific sensitivity and good resolution with a high, but reducible neutron exposure. These foils, especially cellulose nitrate (CN) foils, are insensitive to photons, electrons and thermal as well as epithermal neutrons. Fast neutrons, however, generate recoil nuclei via (n,p)- and (n,α)-reactions in the hydrogen, carbon, and oxygen containing material. The primarily invisible tracks of those secondary particles can be developed by lye etching (6). Fig. 2 shows the magnified surfaces of etched CN-foils after different fast neutron exposures. The positive correlation between etch pit density and neutron dose can be determined quantitatively by measuring the optical net density of the foil with a microphotometer (Fig.3), as etch pits scatter part of traversing parallel light. So, within the limits of linearity, the local optical densities of the developed plastic foil detector correspond to the applied fast neutron dose distribution.

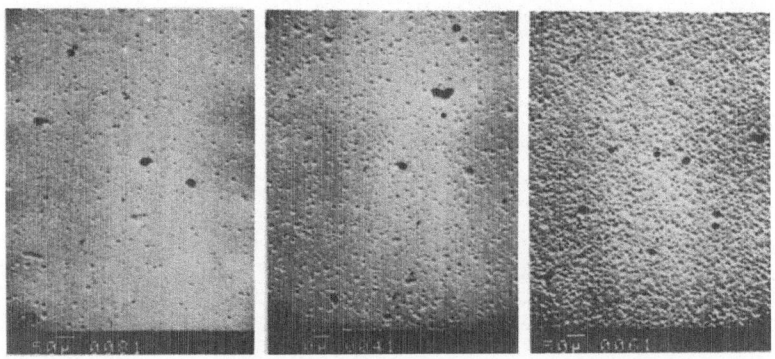

Fig.2 Etch pit densities for $0.3 \cdot 10^{10}$ n/cm^2 (a), $1.2 \cdot 10^{10}$ n/cm^2 (b), and $3.6 \cdot 10^{10}$ n/cm^2 (c). CN-foils (CA 80-15, Kodak Pathé, Paris), etching conditions: 4h, 50°C, 6n NaOH (7).

As the optical net density of the irradiated foil does not only increase with the etch pit density as given by the neutron dose, but also with the mean pit diameter, it is clear, that the foil density is also positively correlated with lye concentration, etching temperature and etching time. Fig. 4 shows the mutual dependence of these etching conditions for the highest net densities, which are achievable with single CN-foils.

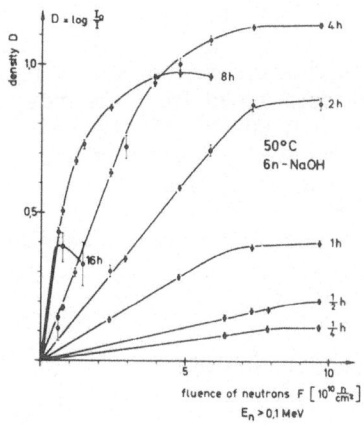

Fig.3 Optical net density of CN-foils (CA 80-15) depending on neutron dose and etching time (7).

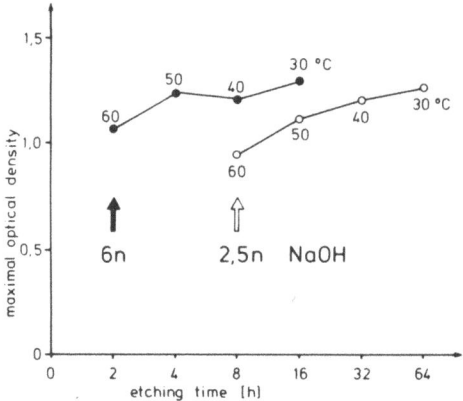

Fig.4 Highest optical net densities of a single CN (CA 80-15) foil, which are achievable after irradiation with $1.6 \cdot 10^{11}$ n/cm^2 with different lye concentrations and etching temperatures, depending on the available etching time.

IMAGING QUALITY

According to the small maximal densities of the plastic detector, compared to photographic films, the expected contrasts of fast neutron radiographs are rather small, too. In principle, contrast can be enhanced up to a certain degree by photographic methods or electronic device, but these methods are limited by foil artifacts, which are enhanced as well. Effective contrast augmentation, however, is possible by the use of several plastic foils in a stack (Fig. 5).

In this case a gain in the conversion rate K (tracks per neutron) is obtained, which can be determined from the measured conversion rates $K_1 = 10^{-5}$ of a single CN (CA 80-15) foil, taking into account both sides of the foil. The effect of the fluence attenuation in the stack is described by the half value layer of plastic HVL$_p$, which has been determined for fission neutron with 3.4 cm (5). So, the total conversion rate K_t of a multifoil package of n foils can be expected theoretically:

$$K_t = K_1 \frac{1 - 0.5^{n \cdot \delta}}{1 - 0.5^{\delta}}$$

where δ is single foil thickness d/HVL_p.

So, it is possible to enlarge the density range of the detector and enhance the image contrasts as well as to improve the image quality to a high degree by adding up numerous mosaic like images, while at the same time the foil artifacts are reduced (Figs. 7-10).

642

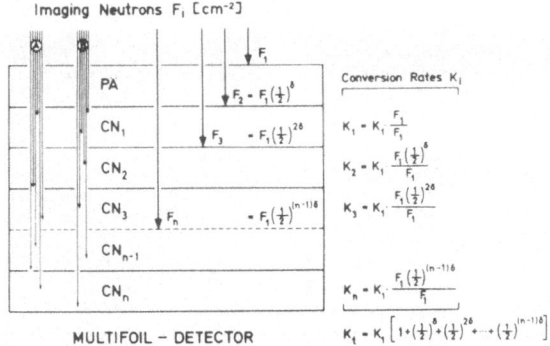

Imaging Neutrons F_i [cm^{-2}]

F_1

PA $\quad F_2 = F_1(\frac{1}{2})^\delta$

Conversion Rates K_i

CN$_1$ $\quad F_3 = F_1(\frac{1}{2})^{2\delta}$ $\quad K_1 = K_1 \frac{F_1}{F_1}$

CN$_2$ $\quad K_2 = K_1 \frac{F_1(\frac{1}{2})^\delta}{F_1}$

CN$_3$ $\quad F_n = F_1(\frac{1}{2})^{(n-1)\delta}$ $\quad K_3 = K_1 \frac{F_1(\frac{1}{2})^{2\delta}}{F_1}$

CN$_{n-1}$

CN$_n$ $\quad K_n = K_1 \frac{F_1(\frac{1}{2})^{(n-1)\delta}}{F_1}$

MULTIFOIL — DETECTOR $\qquad K_t = K_1 \left[1+(\frac{1}{2})^\delta+(\frac{1}{2})^{2\delta}+\cdots+(\frac{1}{2})^{(n-1)\delta}\right]$

Fig. 5 Multifoil detector system. Stack of n CN-foils with polyethene (PÄ) converter in front, F_i neutron fluence, K_i conversion rate at foil i. A and B scheme of formation of image points. δ ratio of single foil thickness d and half value layer of plastic HVL$_p$ (7).

In addition, neutron exposures can be reduced with this detector system by at least one order of magnitude, so that parts of the living human body can be radiographed during a fast neutron radiotherapy fraction (Fig. 10).

Regarding imaging quality, it may be mentioned, that the internal unsharpness of the CN-foil is about 35 µm (8). The maximal geometrical unsharpness of the multifoil system, on the other hand, depends on the thickness of the stack.

FINDINGS WITH FAST NEUTRON RADIOGRAPHY (FNR)

Technical applications of FNR may consist of imaging the shape of active radiation sources (Fig. 6) and testing hydrogen containing materials. Compared to x-ray examinations, FNR can give additional details, such as showing exsiccation areas with wooden specimens (Fig. 7).

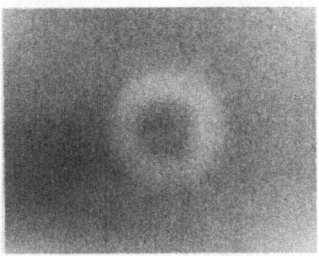

Fig. 6 Pin hole imaging of the radiating cylindrical target of a a ^{3}H (d,n) accelerator ("Korona", GKSS-Research Center, $2.5 \cdot 10^{12}$ n/s)

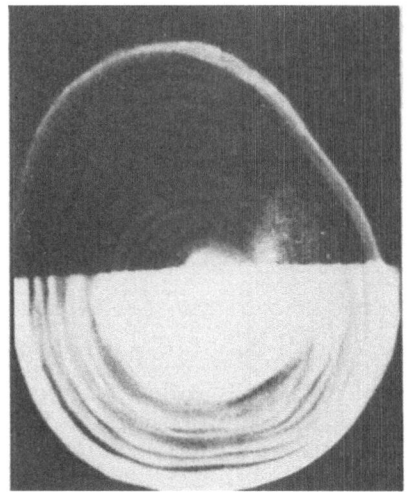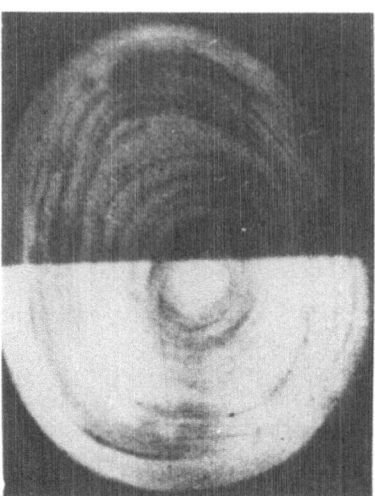

Fig. 7 Axial x-ray (left) and fast neutron image (right) of a spruce trunk. Exsiccation areas are visible only with FNR, while annual rings are to be seen with either technique (3).

Furthermore, FNR is advantageous with the inspection of biological details, which, on x-ray radiographs, are covered by calcium, such as soft tissues of animals coated by a calcified shell (Fig. 8) or tumor tissue in bone structures (Fig. 9).

Finally, in vivo-FNR of fast neutron radiotherapy volumes occasionally show tumor extension, which may be hidden on x-ray or photographic verification films (Fig. 10).

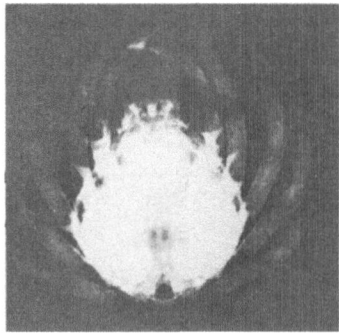

Fig. 8 Maja squinado by x-ray (left) and FNR (right), showing the calcified coat and the soft tissue of the animal, respectively.

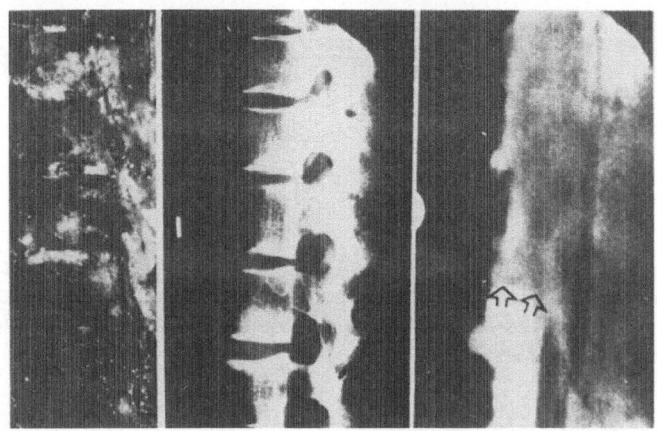

Fig. 9 Human vertebral column with manifestations of Hodgkin's sarcoma, to be seen on frontal anatomical section (left) as greyish spots, invisible on lateral x-ray film (middle), clearly to be distinguished on the lateral FNR (right) as dark regions of higher transmission compared to normal marrow containing bone structures (3).

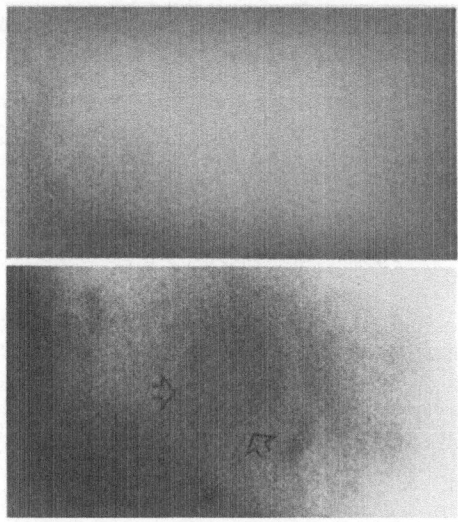

Fig. 10 Lateral FNR of lower jaw region as generated by a single fraction of fast neutron radiotherapy, given for tumor of the base of the tongue. The upper image recorded from one, the lower one from four foils, showing more details, esp. the tumor infiltration of the tongue.

REFERENCES

1. E. Dühmke, L. Greim, Fast Neutron Radiography in Biology and Medicine, p. 573-580 in Neutron Radiography, J.P. Barton and P. von der Hardt, Editors, ECSC, EEC, EAEC: Brussels and Luxembourg, 1983

2. H. Berger, Some Experiments in Fast Neutron Radiography, Mater. Eval. 27, 245-253, 1969

3. E. Dühmke, L. Greim, Biomedical Applications of Fast Neutron Radiography, p. 559-565 in Use and Development of Low and Medium Flux Research Reactors, Supplement Atomkernenergie-Kerntechnik 44, 1984

4. H. Berger, Practical Applications of Neutron Radiography and Gaging, STP 586, Am. Soc. Testing and Materials: Philadelphia, 1976

5. E. Dühmke, Medizinische Radiographie mit schnellen Neutronen, Thiemig Taschenbuch Nr. 86, München, 1980

6. H. Berger, Track etch radiography: alpha, proton and neutron, Nucl. Technol. 19, 188-198, 1973

7. E. Dühmke, L. Greim, Fast Neutron Imaging by Cellulose Nitrate foils, p. 565-571, in Neutron Radiography, J.P. Barton and P. von der Hardt, Editors, ECSC, EEC, EAEC: Brussels and Luxembourg, 1983

8. M. Čopič, D. Horvat, R. Ilić, M. Najžer, J. Rant, Über die Unschärfe einiger radiographischer und neutronenradiographischer Prüfverfahren für aktive Spaltstoffelemente, Materialprüfung 8, 171-175, 1976

ACKNOWLEDGEMENTS

The authors are indepted to Prof. H.-D. Franke and Dr. Hess, Dept. of Radiotherapy, University of Hamburg, FRG, for the opportunity to take FNR of a therapy patient (Fig. 10) and to Mr. O Goemann, Biologische Anstalt, Helgoland, FRG, as well as to Dr. R. König, Zoologisches Institut, Universität Kiel, for the animals (Figs. 1 and 8).

EXPERIMENTAL STUDY OF YIELDS OF DIFFERENT (n, alpha) CONVERTERS

USED WITH NITROCELLULOSE FILMS

R. BARBALAT and J.L. PERSON
IRDI/DERPE/SPS
CEN/Saclay
91191 GIF-SUR-YVETTE   Cedex

I - INTRODUCTION

In connection with a neutron source mainly characterized by its intensity, it becomes essential to choose a suitable converter type (or couple of types).

Using the ISIS reactor neutron fluxes, we have been able to observe the experimental result of a good image quality obtained by use of a couple of converters realized either with natural Boron or with enriched Lithium Fluoride (95 %) simultaneously with a fluence close to 5.10^9 n/cm^2.

At its nominal power of 700 kW, the characteristics of ISIS reactor (CEN/Saclay) generates on the image plane a thermal flux of about $1,4.10^7$ $n/cm^2/s$, able to produce the referenced fluence during a 6 mn irradiation.

Those parameters are taken as reference for the comparative following study in which we investigate the yields of 6 kinds of converters in terms of exposure time in the first part, and processing time in the second part.

In case of a double converter, we have also measured the absorption effect of the converter in front position.

II - CHARACTERISTICS OF UTILIZED MATERIALS

II.1 - Converters

Type (i)	Symbol	Reference
Natural Boron	B	KODAK BN 1
Lithium Borate	$Li_2B_4O_7$	KODAK CN 85B
Lithium Fluoride	LiF	CEA patend n° 6913104
Enriched Boron	^{10}B	KODAK BE 10
Enriched Boron Carbide	$^{10}B_4C$	Experimental KODAK product
Enriched Lithium Fluoride	^6LiF	CEA patend n° 6913104

II.2 - Support
Polyester thickness : 100μ

For each referenced converter and close to the film, we have positionned :
a) a Single converter (back position) . . . . S
b) a Double converter (back and front position) . . . . D
c) a Single converter (back) + a reverse converter (front) . S+d
d) a Single converter (back) + a support (front) . . . . . S+s

II.3 - Film
The Nitrocellulose film is supplied by KODAK-PATHE, ref. CN 85 (thickness : 100μ).

III - EXPERIMENTAL CONDITIONS

III.1 - Exposure
The reactor being at stable power close to 700 kW, a 4-series irradiation has been operated :

a) a 2-series exposures with duration times increasing from 0.5 mm to 30 mm,

b) a 2-series of 4 exposures with the same duration time of 3 and 6 mn.

III.2 - Processing

Both first series have been processed during 30 mn at 42°C in a thermostated bath including a 150 g/l KOH solution.

Both second series have been processed in the same solution at 42°C, but in increasing duration times from 30 mn to 90 mn.

III.3 - Measurements

The film densities have been evaluated using :
- a travelling microdensitometer JOYCE-LOEBL (Mark III CF),
- a densitometer MACBETH (TD 100 A) (for a series of increasing exposure times) in which the film was positionned between a couple of polarizing filters (ref. HN 32).

IV - RESULTS

IV.1 - Numerical

As indicated in paragraph I, we have therefore chosen to standardize all the measurements d_i with the film density obtained with a double natural Boron $[\ d_B\ (D,6)\ ]$ converter irradiated during 6 mn (fluence $\sim 5.10^9$ n/cm^2). Thus, we achieved table 1 in which the yields are calculated as following :

$$\rho_i = \frac{d_i - d_f}{d_B\ (D,6) - d_f}$$

d_i = rough density of the film in front of converter i

d_f = background density of the film

where :

a) microdensitometer measurement series :

$$d_f = 0 \qquad\qquad d_B\ (D,6) = 0,399$$

b) densitometer measurement series (with polarizing filters) :

$$d_f = 3,42 \qquad\qquad d_B\ (D,6) = 1,90$$

IV.2 - Curves

We have plotted the comparative yields of the most usual converters : natural Boron, Lithium Borate and enriched Boron because in fact natural Lithium Fluoride converter shows a too low yield and moreover enriched Lithium Fluoride is not commercialized.

Graph nb.1 shows the yield variations of those elements ; we propose, on graph nb.2, to compare the yield of a natural Boron double converter to the calculated yield which would be obtained with one enriched Boron converter in back position and enriched Lithium Fluoride converter in front position. We intend to check the above estimated result by experimental measurements.

Lastly, the graph nb.3 compares the yields changes with different processing times.

V - CONCLUSION

A double converter yield is not the double of a single converter, but it appears clearly that utilization of double converters allows shorter exposure times. Thus, table 1 shows that a double natural Boron converter requires an irradiation 0.6-time shorter than a single converter to get a similar yield, respectively with the fluences located between 7.10^8 to $1.5 \ 10^{10}$ n/cm^2.

This is even more verified in the case of Lithium Borate but this converter can not be recommended because of its poor efficiency which is 5 to 6 times weaker than a natural Boron converter.

An enriched Boron element shows a very interesting yield : in case of a double converter and with a 10 mn-exposure time, its yield is 1.7 higher than natural Boron (with a single converter, the yield is even 2.3 higher).

We have therefore to point out the importance of the front converter absorption which could be responsible for image alteration. This is due to a probable inhomogeneity of the converter and suggests even more to prefer an enriched Lithium Fluoride in front position, because of its low absorption.

The graph nb.1 shows also that for more than a 5 mn-exposure time, its yield is probably increasing while enriched Boron yield reaches its maximum with a 10 mn-exposure time. Moreover, we can point out that the yield of an enriched Boron carbide is decreasing at the same time.

On another hand, the graph nb.3 shows how it is possible to modulate the image contrast : this observation is also justified for enriched Boron in the case of processing times shorter than 30 mn.

Δt(mn) $\phi.t$(n.cm^{-2})	El. Geom.	B	Li$_2$B$_4$O$_7$	LiF	^{10}B	^{10}B$_4$C	^{6}LiF
0.5 3.6 10^8	S D	0.05 0.09	0.00 0.00	0.00 0.00	0.19 0.25	0.19 0.18	0.05 0.09
1.0 7.2 10^8	S D	0.13 0.18	0.00 0.00	0.00 0.00	0.40 0.55	0.38 0.35	0.09 0.20
1.5 1.1 10^9	S D	0.14 0.24	0.00 0.00	0.00 0.00	0.52 0.73	0.52 0.48	0.13 0.24
2.0 1.5 10^9	S D	0.18 0.32	0.04 0.07	0.00 0.00	0.67 0.94	0.67 0.62	0.18 0.32
2.5 1.8 10^9	S D	0.23 0.41	0.05 0.10	0.00 0.00	0.87 1.25	0.87 0.80	0.25 0.44
3.0 2.2 10^9	S D	0.29 0.48	0.07 0.13	0.00 0.00	0.98 1.31	0.99 0.92	0.29 0.50
5.0 3.6 10^9	S D	0.47 0.82	0.11 0.22	0.03 0.06	1.50 1.98	1.49 1.43	0.51 0.84
6.0 4.4 10^9	S D	0.63 1.00	0.16 0.30	0.07 0.11	1.72 2.23	1.64 1.75	0.61 1.07
10.0 7.2 10^9	S D	0.92 1.50	0.25 0.43	0.08 0.14	2.16 2.54	2.15 2.27	0.97 1.68
15.0 1.1 10^{10}	S D	1.29 1.76	0.34 0.62	0.09 0.20	2.32 2.63	2.32 2.51	1.42 2.14
20.0 1.5 10^{10}	S D	1.61 2.29	0.45 0.80	0.15 0.28	2.37 2.67	2.39 2.62	1.77 2.49
30.0 2.2 10^{10}	S D	2.02 2.52	0.69 1.24	0.20 0.43	2.40 2.66	2.40 2.67	2.25 2.70

Converters Yields
(Microdensitometer Measurements)

Table 1

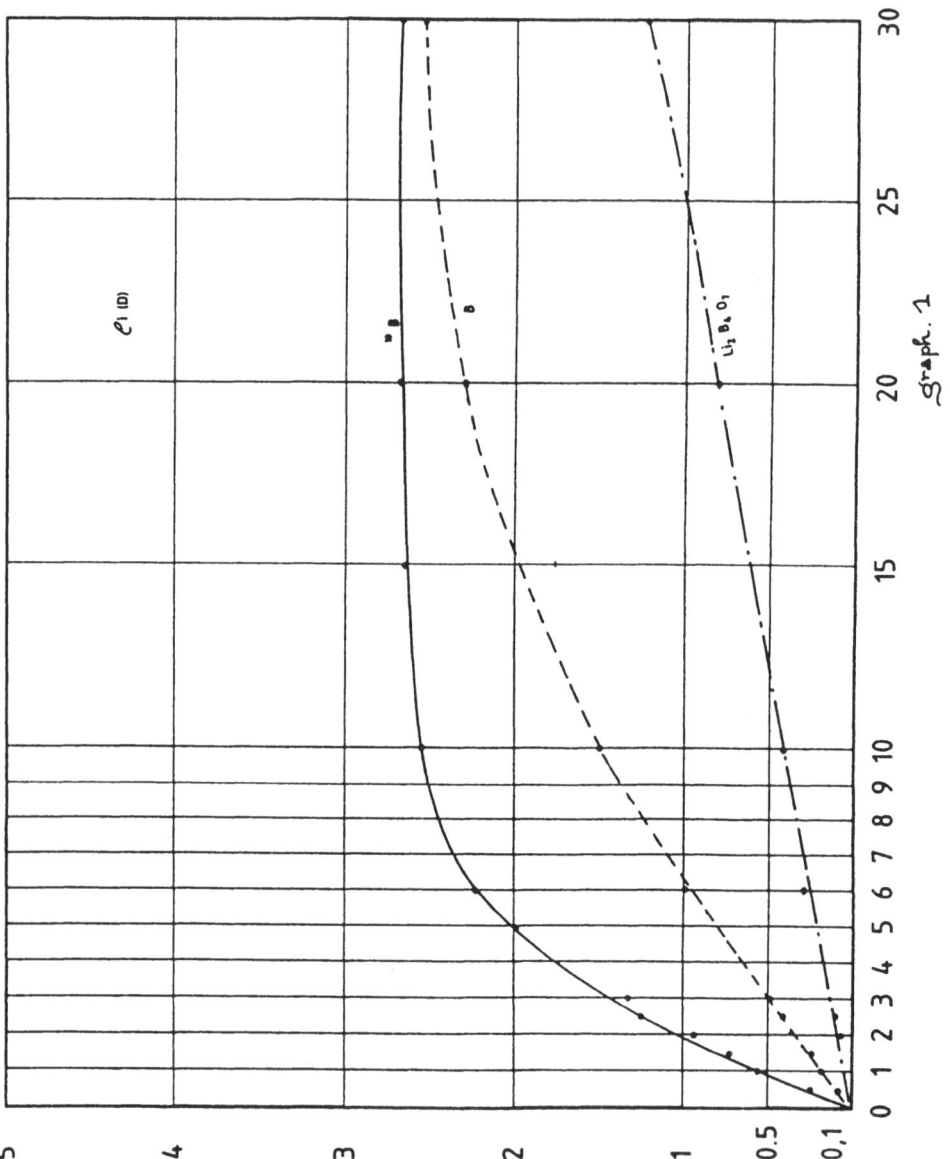

graph. 1

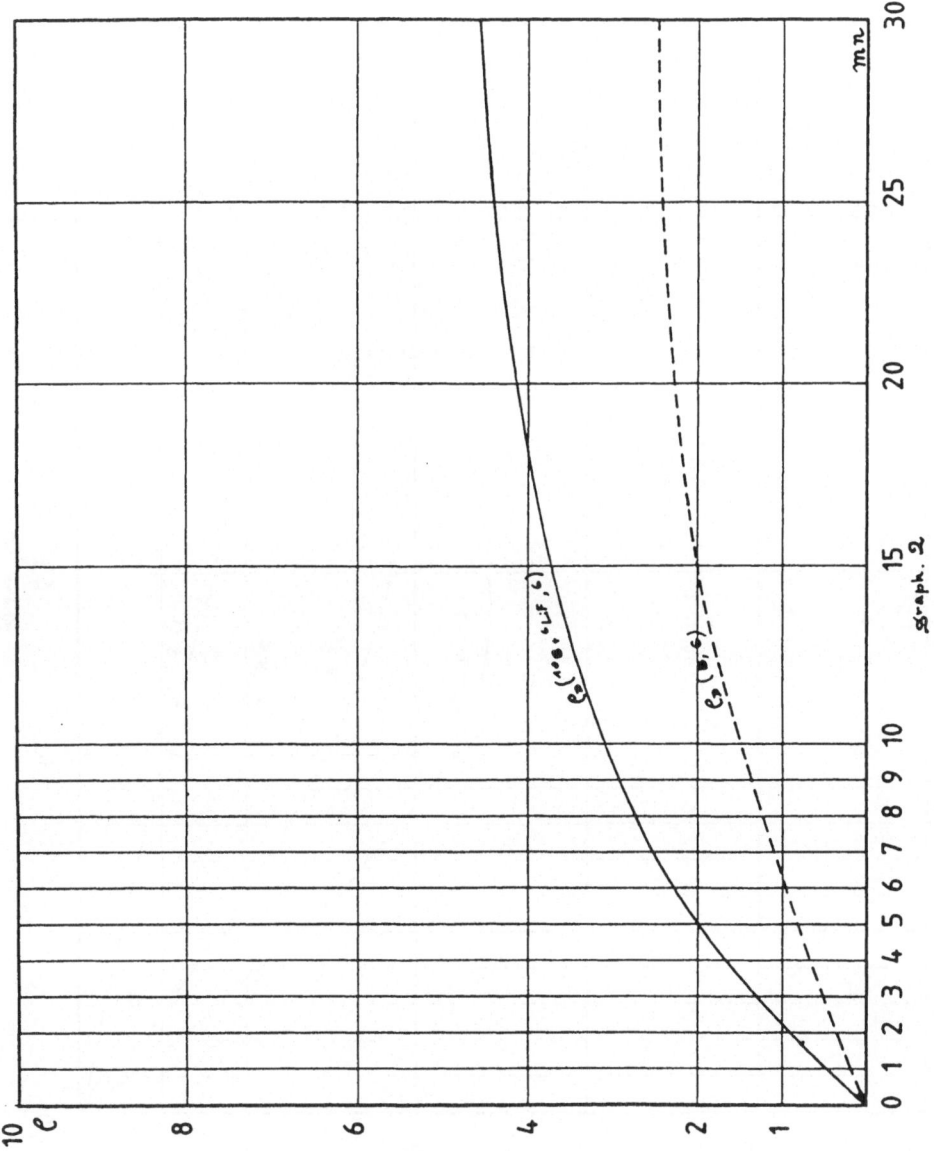

graph. 2

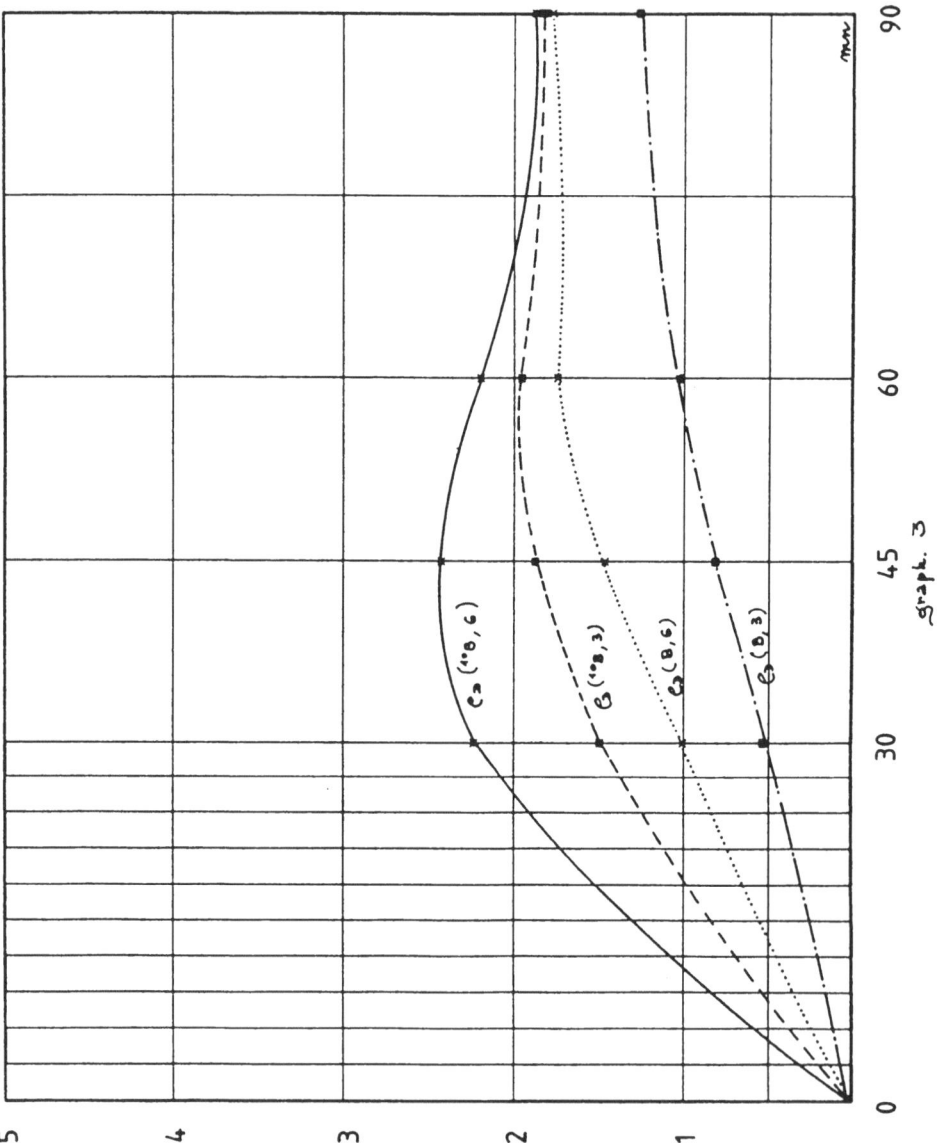

graph. 3

PART X

DIMENSIONAL MEASUREMENTS

PHYSICAL BASIS AND TECHNICAL REQUIREMENTS FOR

DIMENSIONAL MEASUREMENTS IN NEUTRON RADIOGRAPHY

A.A. Harms

McMaster University

Hamilton, Ontario, Canada

ABSTRACT

Selected aspects of neutron radiographic dimensioning are examined. A descriptive quantitative assessment is established identifying high and low dimensional quality.

INTRODUCTION

The extraction of accurate dimensional information from neutron radiographs is emerging as a timely and challenging objective. Commonly encountered applications of interest involve fuel-pin diameter determination and gap/crack/inclusion assessment.

While the general mathematical-physical basis for dimensional analysis may well be identified by heuristic approaches[1], there remains the problem of quantitative assessment of system components for their role in object dimensioning. These considerations are essential to the implementation and routine application of dimensional measurements in neutron radiography. It is the objective of this paper to identify the critical features of this problem domain.

SYSTEM-PROCESS CHARACERIZATION

Neutron radiography is characterized by the use of a penetrating beam of neutrons serving as a tool of diagnosis. The transmitted beam carries information about the object which is then displayed for visual inspection[2]. The processes affecting the neutron beam, the transport of the information carried by the neutron beam, and the several radiation conversion processes all contribute to a visual product

657

useful for analysis. Figure 1 provides a schematic depiction of the important process and also displays some notation to be used in the subsequent analysis.

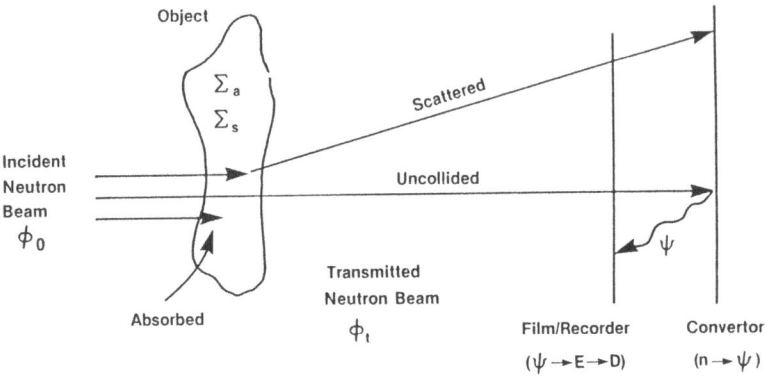

Fig. 1: Schematic of neutron radiographic system and processes.

The primary feature to recognize is that high resolution neutron radiography and accurate dimensional radiographic applications demand a high degree of correlation between the material heterogenity of the object and the associated optical density pattern in the film/recorder plane. Processes such as incident beam divergence, neutron scattering in the object, secondary radiation spreading from the converter, nonlinearities of the film/recorder and other factors, all combine to degrade this one-to-one point correlation between the object and its image

COMPONENT ANALYSIS

A quantitative analysis of the system features and radiographic conversion processes as they relate to the extraction of high-quality dimensional information is suggested in Table I. We consider these in sequence emphasizing their physical-technical basis.

Incident Beam:

1.   Rectangular objects demand high collimation with ϕ_0 = constant.

2.   Directionally uniform anisotropy of $\phi_0(\theta)$ = constant is acceptable for radially symmetric cylindrical objects providing diverging correction for object-film/converter distance is introduced.

Object:

1. Rectangular objects with high neutron absorbing cross section $\Sigma_a >> \Sigma_s$, provide for simple theory and ease of analysis.

2. Less absorbing and highly scattering rectangular objects, $\Sigma_s >> \Sigma_a$, provide for scattering-edge enhancement[3].

3. Cylindrical and other irregular objects introduce image unfolding complications; these can be decomposed into

$$\psi(x) = \sum_i \psi_i(x).$$

Converter Response:

1. Desired narrow line-spread function ψ with FWHM $< 50\,\mu$m; can be achieved with thermal neutrons in highly absorbing converter with short range secondary radiation.

Film/Converter:

1. Steep D-vs-E desired for high ΔD discrimination for small ΔE.

2. Linear response for $\Delta D \sim 1$ very useful.

3. Undesirable film/recorder noise effects δD can be minimized by "narrow-wide" apperture opening in scanning apparatus; dimension of $15\,\mu$m $\times\, 250\,\mu$m (with $15\,\mu$m in the radial directions) found to be a good compromise[4].

Information Extraction:

1. Stable, data-smoothed differentiation procedure is a powerful method for dimensional information extraction.

2. Visual identification of scanning domain may aid in rapid dimensional assessments.

3. Advanced techniques of differentiation may further aid in "signature" identification of various faults.

TABLE 1

Listing of component/processes and their correlation to
low/high dimensional information extraction

Component/Processes		Dimensional Information	
		Low	High
Incident	Rect. Obj.	Isotropy	Collimation
Beam	Cyl. Obj.	Geometry Translation	
Object		$\Sigma_s >> \Sigma_a$	$\Sigma_a >> \Sigma_s$
Converter Response		Wide LSF FWHM $> 50\ \mu m$	Narrow LSF FWHM $< 50\ \mu m$
Film and/or Recorder	Range	Nonlinear for $\Delta D \sim 0.1$	Nonlinear for $\Delta D \sim 1$
	Noise	High for $\delta D > 0.05$	Low for $\delta D < .01$
Information Extraction		Visual Bias	Mathematical Processing

CONCLUDING COMMENT

Recent investigations in dimensional neutron radiography -- theoretical, experimental, and calculational -- suggest that existing tools and techniques are adequate to provide quantitative dimensional information of broadly based scientific-technical utility.

ACKNOWLEDGEMENT

Support for the research reported here has been provided by Natural Sciences and Engineering Research Council of Canada.

REFERENCES

1. A.A. Harms and D.R. Wyman, <u>Mathematics and Physics of Neutron Radiography</u>, D. Reidel Publ. Co., Dordrecht (1986).

2. P. van der Hardt and H. Roettger, <u>Neutron Radiography Handbook</u>, D. Reidel Publ. Co., Dordrecht (1981).

3. D.R. Wyman and A.A. Harms, "The Radiographic Edge Scattering Distortion", J. Nondest. Eval., $\underline{4}$ (2), 75 (1984).

4. A.A. Harms and T.G. Blake, "Densitometer-Beam Effects in High Resolution Neutron Radiography", Trans. Am. Nucl. Soc., $\underline{15}$, 710 (1972).

EFFECT OF FILM EXPOSURE ON FUEL PIN DIMENSIONAL

MEASUREMENTS

A.A. Harms

McMaster University

Hamilton, Ontario, Canada

G. Shani

Ben Gurion University

Beer Sheva, Israel

ABSTRACT

The effect of the nonlinear $D-vs-E$ film relation on precision diameter measurement of fuel pins is investigated. It is found that this effect can be significant but nevertheless provides a linear plateau enabling adequately exact measurements.

INTRODUCTION

The extracton of sufficiently precise dimensional information from neutron radiographs continues to be of broad technical and industrial relevance[1]. Of particular contemporary interest is the determination of the diameter of radioactive fuel pins from their radiographs.

The underlying difficulty of such quantitative radiographic applications is attributable to several image blurring phenomena intrinsic to the radiographic system[1,2]. It is a common objective of neutron radiographic research and development activity to examine the various blurring contributions and attempt means for their accounting or ellimination.

We report here on an experiment designed to assess specifically the effect of the nonlinear Density-vs-Exposure relation of a film on the "apparent" fuel pin diameter. For this purpose we used a consistent objective methodology of extracting fuel pin diameters from a set of neutron radiographs which differed only by their exposure time.

EXPERIMENTAL BASIS

The experiment performed is suggested in Fig. 1. A commercial pressure cassette with a gadolinium converter foil and industrial R-type X-ray film was used. The object was the RISO standard fuel pin[3] and the neutron radiographic installation of the McMaster University Nuclear Reactor was used. A series of neutron radiographic exposures were obtained with all system parameters held constant except for the exposure time which was varied from less than 30 s to about 3 hrs; the associated optical density range varied from 0.26 to 4.7. A Leitz-Wetzlar travelling-stage microdensitometer was used to obtain optical density scans in a direction perpendicular to the fuel-pin axis.

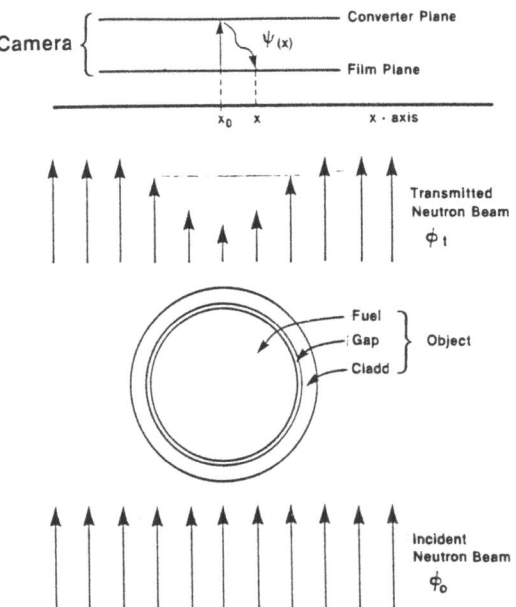

Fig. 1: Schematic of experimental set-up.

MATHEMATICAL-CONCEPTUAL BASIS

A primary quantity of interest is the converter response function, $\psi(x)$ of Fig. 1, for the particular object and facility of interest. For an edge of about 1 mm of length or longer, this converter response is well represented as a convolution integral[2]

$$\psi(x) = N \int_{-\infty}^{\infty} L(x, x_0) \phi_t(x_0) dx_0 \tag{1}$$

Here N is a normalization constant, $L(x,x_0)$ is the line-spread function for the radiographic system and $\phi_t(x_0)$ is the transmitted neutron current appropriate to the object.

It is important to emphasize that $\psi(x)$ is not the observed optical density $D(x)$ at the coordinate x. Indeed, the product of $\psi(x)$ times the exposure time τ defines the exposure $E(x)$ at x

$$E(x) = \psi(x)\tau$$

$$= \tau N \int_{-\infty}^{\infty} L(x, x_0) \phi_t(x_0) dx_0 \tag{2}$$

and the optical density $D(x)$ then follows as

$$D(x) = fc[E(x)]$$

$$= fc[\psi(x) \tau] \tag{3}$$

where fc[ ] represent the characteristic nonlinear Density-vs-Exposure relation for a given film. Notice, therefore that $D(x)$ is thus dependent upon the exposure time τ, i.e.

$$D(x) \rightarrow D(x,\tau) \tag{4}$$

In our experiment $\psi(x)$ is a constant -- since no changes in object or the radiographic system were introduced -- with x as the spatial microdensitometer coordinate along which the "apparent" location of the fuel pin edge might be found.

DATA ANALYSIS

An enlarged geometrical correlation between a typical optical density scan and the cross section of the fuel pin is suggested in Fig. 2. Here, for clarity of exposition and as a test of methodologies, we identify two "apparent" fuel-pin edge coordinates: one is labelled the Extremum Slope Method and determines x_{es} while the other is the Intersecting Slope Method and determines x_{is}. Both of these methodological criteria possess a theoretical-physical basis[2] and refer to the inherent difficulty of incorporating all relevant physical processes in the integrand of Eq. (1). Evidently x_{ex} and x_{is} can be determined by visual inspection or mathematical analysis of suitable portions of the optical density scans; we found adequate accuracy by a visual curve fitting of a suitably scaled microdensitometer scan using independent data extractions by two examiners. Figure 3 provides the results of this analysis.

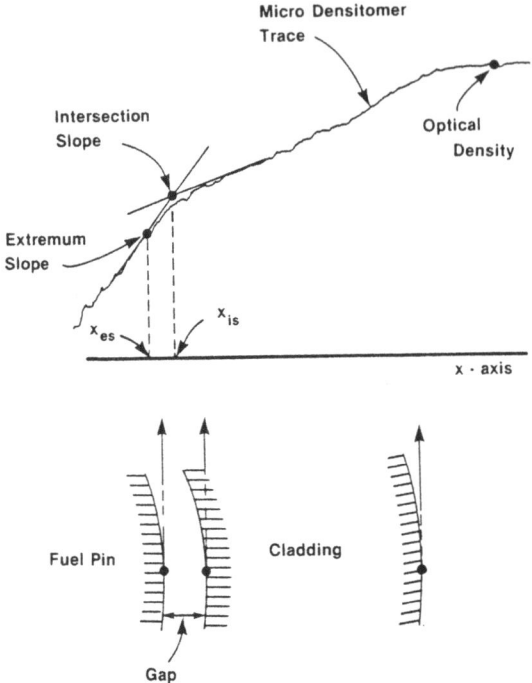

Fig. 2:   Correlation between object and microdensitometer scan showing also two
           criteria for the determination of the "apparent" edge location.

DISCUSSION AND CONCLUSIONS

The effect of film nonlinearity on the dimensional data obtained is illustrated
in Fig. 3 and suggests several features of general interest. First, we note the severe
effect of the "heel" and "toe" of the nonlinear Density-vs-Exposure characteristic of
the film. Second a significant plateau exists in which the linearity domain of the film
response dominates. Thirdly, we find a systematic error in the Extremum Slope
methodology for this kind of object. Finally, for this case, the Intersecting Slope
methods appear to provide diameter estimates to a satisfactory degree of accuracy.

We conclude therefore, that the nonlinear film response does provide a
sufficiently broad linear range justifying this assumption in theory and also provides
an adequate plateau to enable sufficiently accurate dimensional measurement in
routine practice.

ACKNOWLEDGEMENTS

Support of the research reported upon here has been provided by the Natural
Sciences and Engineering Research Council.

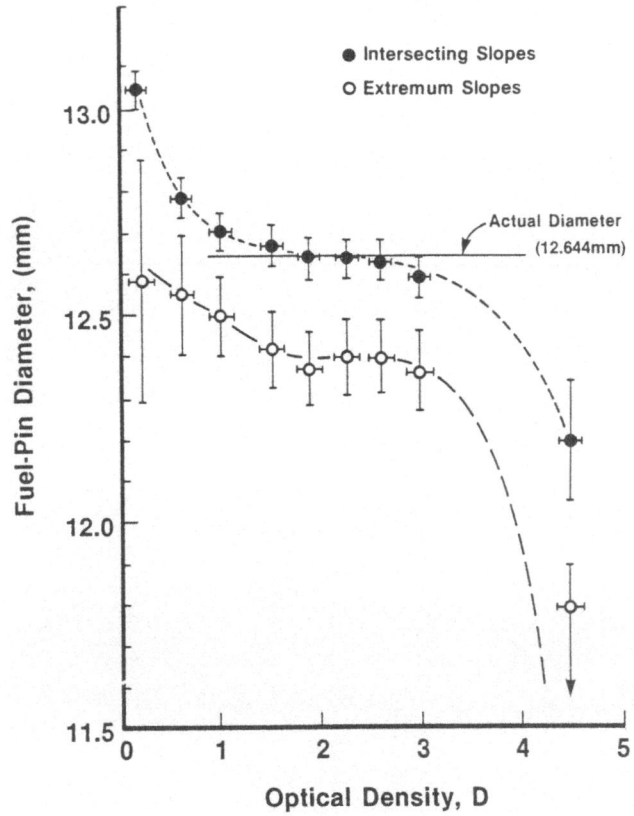

Fig. 3: Fuel pin diameter determined by two criteria.

REFERENCES

We refer to the several sessions on the subject at the First World Conference on Neutron Radiography, San Diego, USA, 7-10 December 1981 (J.P. Barton and P. von der Hardt, E., <u>Neutron Radiography</u>, Reidel Pub. Co., Dordrecht, 1983) and also to the activities of the EURATOM Neutron Radiography Working Group (NRWG).

A.A. Harms and D.R. Wyman, <u>Mathematics and Physics of Neutron Radiography</u>, Reidel Pub. Co., Dordrecht (1986). This monograph examines in some detail image sharpness processes and contains an extensive bibiliography on the subject.

J.C. Domanus, Riso National Laboratory, Denmark, Private Communication.

COMPUTER CONTROLLED MICRODENSITOMETER

AND SOME APPLICATIONS

L. Greim, M. Greim, H.-W. Schmitz, G. W. Schumacher

GKSS Research Center, Geesthacht, FRG

ABSTRACT

Microdensitometry on neutron radiographs enables quantitative evaluations of images.

The measuring set up presented consists of the Jenoptic Schnellphotometer G III with several essential alterations and extensions. The densitometer was equipped with a photodiode which has a fast and linear response over a measuring range of 6 powers of ten. A travelling table can be moved by step motors in x- and y-directions over an aera of 150 x 250 mm, step length 10 µm. The scanning program is controlled by minicomputer, with a speed of about 5 measurements per second. Data processing a n d evaluation is done on-and off-line.

Some applications are presented: the examination of several resolution functions, as well as the scanning and fitting for the evaluation of radiographs of the Euratom calibration fuel pin and of the AFNOR IQI and of a proposed L/D standard.

INTRODUCTION

Quantitative evaluation of radiographic films is interesting for several purposes:
- Determination of the quality of an image utilizing standard objects.
- Determinations of the properties of a radiography facility

669

- Measurements of lengths, gaps etc on radiographs
- Measurements of the thickness or of the structure of an object or of parts inside the object. An example is the determination of the content of neutron absorbers in a material.

Such measurements are done with densitometry. Per definition the optical density D of a film is:

$$D = \log_{10}(I_0/I)$$

with I_0 intensity of impinging parallel light
    I   intensity of penetrating light, scattered fraction included

Generally, microdensitometry makes use of narrow beam geometry. The scattered fraction of the light is not recorded in this case. The resulting density Dm is by a factor of about 1.3 greater (Callier coefficient) than the conventional density with respect to silver halide film.

It should be emphasized that the optical density of track etch foils can be measured using such a microdensitometer. The etch pits on these foils scatter the light, only a small fraction is absorbed. For this reason the conventional density is low. In narrow beam geometry, however, densities can be measured up to saturation values of 1.2 - 1.5. Therefore such images can be evaluated without copying on photographic film.

The light beam is defined by narrow slits. These slits should not be set smaller than necessary:
- light intensity should be measurable for a sufficient density range
- noise and fluctuations of density should be neglectable. They are caused by statistical distribution of the grains in the films resp. of the etch pits on a foil.
- disturbances from light reflections etc. grow with smaller slit width.

In scanning measurements, however, the slit width should be smaller than the inherent unsharpnes of the film.

With a travelling microdensitometer the density of a film is scanned along interesting lines. The obtained density distributions are the basis for further evaluations, graphically by plotting, or by mathematical fitting procedures.

670

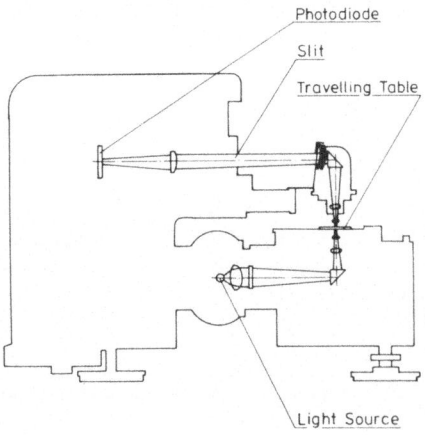

Figure 1. Schematic Drawing of the Jenoptic Schnellphotometer G III

Figure 2. Densitometry Set Up

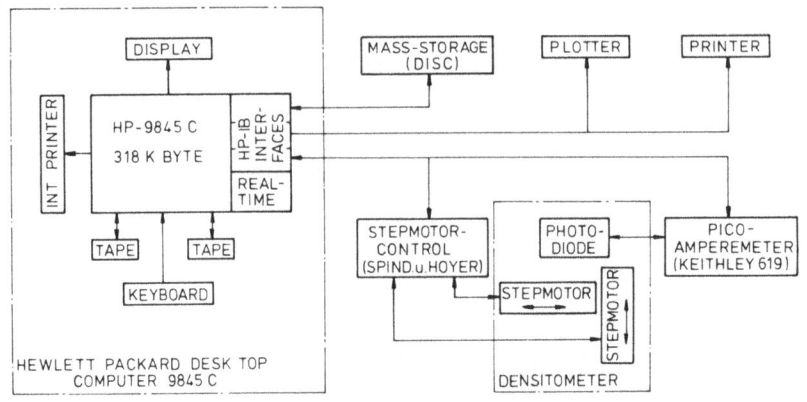

Figure 3. Diagram of Travelling Microdensitometer

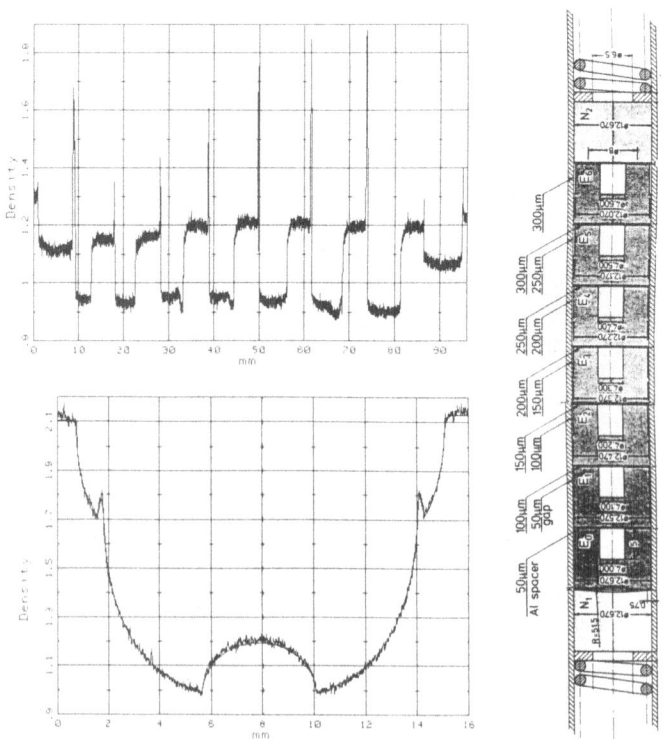

Figure 4. Calibration Fuel Pin CFP-E1 (3), Axial and Radial Scans on Neutron Radiograph

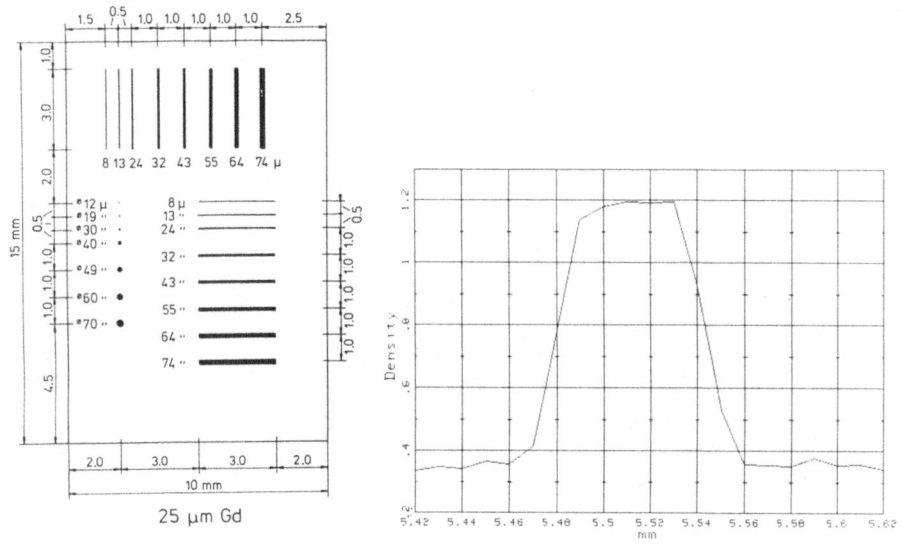

Figure 5. AFNOR IQI and Scan on Neutron Radiograph (Track Etch
Foil) across 64 μm slit

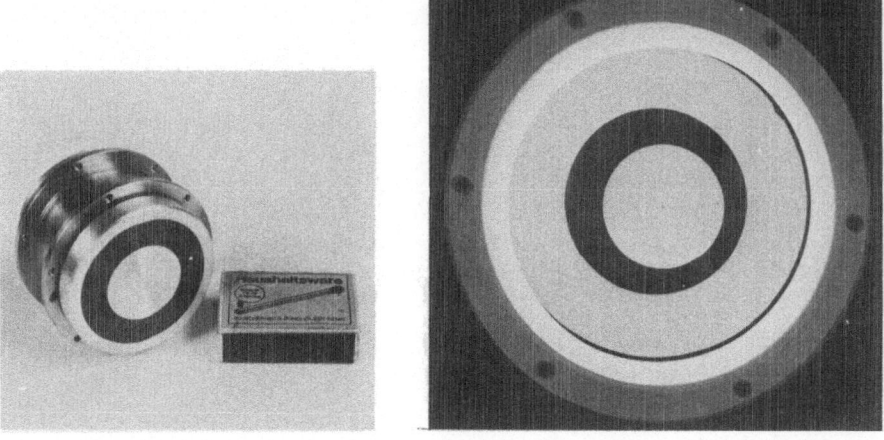

Figure 6. Photo and Neutron Radiograph of L/D Standard

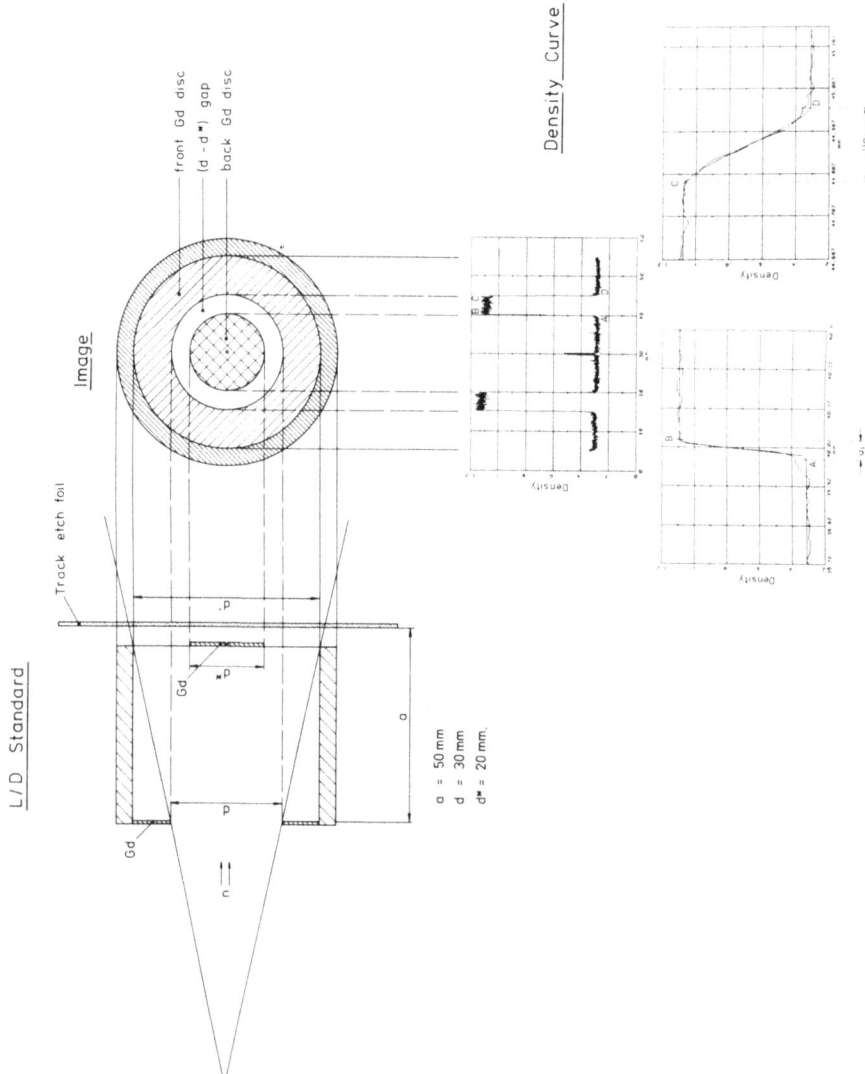

Figure 7. Diagram of L/D Standard Arrangement and Evaluation of Image

TRAVELLING MICRODENSITOMETER AT GEESTHACHT

The microdensitometer presented here is basically the Jenoptic Schnellphotometer G III (fig. 1). It has been equipped with a travelling table, which enables densitometry on an aera of 150 x 250 mm. The table is moved with computer controlled step motors in x- and y- directions. The length of one step is 10 μm. Light measurement was improved by installation of a photodiode (BPW20) on which the beam is focussed by extra lenses. The diode has a fast response. The linear measuring range extends over 6 powers of ten. This was achieved by compensating the temperature depending dark current with a circuit containing a reference photodiode. The lamp is supplied with stabilized DC.

The measuring aera on the film is defined by a slit which can be adjusted with a micrometer screw. For scanning with high resolution the slit width is set to 10 μm and the slit heigh to 250 - 500 μm. A smaller slit width is not recommended as in this case the grains of the film resp. the etch pits on a track etch foil will produce too much fluctuations of the signal.

The measuring set up is connected to a minicomputer (HP 9845C). By program it controls the scanning course and the positions of density measurements. The readings of the photodiode are taken from a picoamperemeter and calculated into densities. The speed of the procedure is about 5 measurements per second.

Figure 2 is a photograph of the measuring set up. It is installed in a case which is closed during operation to avoid disturbances from daylight or lamps. The diagram in fig. 3 shows the components of the systen.The collected data can be displayed on a screen, can be plotted, and can be stored on discs.

The minicomputer is also used for off-line evaluations and calculations.

EXAMPLES OF APPLICATIONS

The posibilities of microdensitometry are shown by some examples :
- use of resolution functions for determination of edges.
- evaluations of neutron radiographs in the Euratom test program for a calibration fuel pin.
- scanning of the radiograph of the AFNOR image quality indicator
- evaluations of the image of a proposed L/D standard object

Following the investigations of Harms (1,2) several resolution functions were tested. These functions characterized by width parameter B are listed in the following table:

Rectangle	width 2B height 1/2B
Triangle	width 2B height 1/B
Parabola	$0.75/B \left[1 - (X/B)^2\right]$
Lorentzian	$1/(\Pi \cdot B) \cdot 1/\left[1+(X/B)^2\right]$
"Circular diaphragm"	$2/(\Pi \cdot B) \cdot SQR \left[1-(X/B)^2\right]$

The radiograph of an edge ("knife edge") was scanned. Idealized it should be a step. The measured densities across the edge were fitted to the function which is obtained by convolution of the resolution function with the step. The fitting parameters were densities on both sides of the edge, the position of the edge, and the width of the resolution function. Good results for the step position were obtained with all listed functions. The best fits for the steps were achieved by the Lorentzian function (proposed by Harms) and by the function "circular diaphragm". The latter function was used for further applications. It is the mathematical description of the geometrical unsharpness in a beam which has circular inlet diaphragm.

With the evaluations of the neutron radiographs of the calibration fuel pin CFP-E1 (3) (fig. 4) fitting procedures were used for radial and axial scans across the image. The pellet lengths and the gaps between the pellets were determined . The parameter of the resolution function is greater as there is no sharp edge. Fitting the radial scans a function was used which describes the neutron attenuation by the thickness of materials . Besides other dimensions the built in gaps between cladding and fuel were obtained.

The AFNOR image quality indicator (4,5) consists of a thin gadolinium foil with engraved slits of different widths . A density scan across the image of the 64 μm slit is displayed in fig. 5.

Scanning and evaluation by fitting procedures is used in tests of a proposed L/D standard. The L/D ratio of a neutron radiography facility describes the beam divergence and determines the geometrical unsharpness of an image. The ratio is given by design of inlet diaphragm and collimator length, but there may be deteriorations caused by scattering and other effects. The measurement of the actual L/D ratio can be done radiographing a standard object designed by Newacheck (6). In this object absorbing wires are fixed in different distances of the image plane. The direct Gd imaging method is to be applied. In a mixed radiation field with γ-contribution, track etch technique should be the approbiate method. The evaluation is difficult because the track etch technique is sensitive to epithermal neutrons. For this reasons effective diameters of the wires are not well defined.

The proposed L/D standard object is suitable in this case (fig. 6). It consists of an annular and of a circular Gd-disc, each fixed on one side of a 50 mm long tube (fig. 7). The unsharpness (Ui) of the edge close to the image plane gives the inherent unsharpness, the unsharpness (Ug) of the edge of the remote disc is related to the collimation ratio. In this scanning an effective collimation ratio of 280 was determined. This is to compare to the design ratio of 375. The inherent unsharpness of the foil is about 35 μm. It can be expected that collimation ratios from 10 to 500 can be determined with this device. It should be emphasized that the evaluations can give the collimation ratios for different directions across the beam.

The radiograph is also an indicator of the beam direction. The images of the two Gd-discs are not concentric if the standard is adjusted incorrectly (fig. 6). Furthermore the determination of effective collimator length can be obtained from image enlargement of the front disc.

The usefulness of a travelling microdensitometer has been proved for many purposes. A growing interest in quantitative image evaluation can be expected. Microdensitometry may be an important method for such work.

REFERENCES

1.   A. A. Harms, Physical Processes and Mathematical Methods in Neutron Radiography, p. 143 in Atomic Energy Review Vol. 15, No.2 IAEA (1977)

2.   J. C. Osuwa, A. A. Harms, The Extremun-Slope Criterion for Precise Dimensional Measurements in Neutron Radiography, p. 859 in Neutron Radiography, J. P. Barton, P. von der Hardt (eds) D. Reidel Publishing Comp. Dordrecht (1983)

3.   J. C. Domanus, Revised Test Program for Testing of CFP-E1; ASTM (revised) BPI and SI and BPI-E, Report B-512 Riso Nationallaboratory, August 1981

4.   J. C. Domanus, Test Program of the AFNOR A09-220 IQI, Riso National Laboratory, December 1984

5.   G. Bayon, Utilisation d'un Indicateur de Qualité d'Image recommande par l'Association Francaise de Normalisation, Service des Piles de Saclay IRDI/DERPE/SPS/EAR/85/067 April 1985

6.   ASTM Designation E 803-81, Standard Method for Determining the L/D Ratio of Neutron Radiography Beams.

STRUCTURAL DATA DETERMINATION

AND IMAGE PROCESSING OF DIGITIZED NEUTRON RADIOGRAPHS

L. Steinbock, D. Piel

Institut fuer
Material- und Festkoerperforschung
Kernforschungszentrum Karlsruhe
Postfach 3640
D-75-Karlsruhe
West-Germany

Abstract

A series of neutron radiographs from fuel
pins were digitized with a video camera and
the image data were processed in order to
enhance details and to extract structural
data from the radiographs. These data are
compared to actual cuts from the destructive
examination of one sibling pin.

1. Introduction

The CABRI experimental program examines the behaviour of LMFBR
pins under severe accident conditions. A number of test pins had
been preirradiated in the PFR and PHENIX reactors and their pre
test state was documented by a series of non destructive
examinations. The neutron radiography in especial should show
the central channel and the gap geometry of these preirradiated
test pins. These data are necessary in order to provide input
data for the precalculations of the CABRI experiments with
several codes.

For twelve pins of the series with 5% burn-up it was decided to
analyze the radiographs digitally and to compare the results with
five cuts executed at two hot cell facilities. As already
demonstrated with X-Ray radiographs [1] and neutron radiographs

of other experiments the image analysis methods allow to extract
quantitative data about the axial and diametral mass distribution
inside a fuel pin or a test channel.

The neutron radiographs of rig 3 pins were generated at LDAC in
Cadarache with a small uranyl reactor and the transparencies were
later digitized with a video camera and the image data stored in
a computer. These data were analyzed on an IBM-PC equipped with
a graphic controller and a image hardcopy unit. Contrast
enhancement and spatial filtering was applied in order to improve
the visibility of the central channel and an algorithm was
developped which allows to determine the central channel diameter
automatically.

2. Contrast enhancement and spatial filtering techniques

The neutron radiographs had been copied on silver film and their
scale was about 1:1. The visual analysis of a radiograph is
limited by the film graininess and the resolving powers of the
human visual system. Microscopic enlargement does not change
this situation very much as can be verified by regarding fig. 1.
This is a section from the upper part of the pin 52 enlarged by
about a factor of three. The central channel is visible but its
edge is blurred.

A very simple method for improving the image is to use a point
transformation as described in fig. 2. The frequency
distribution or histogram of all density levels in the image is
plotted versus the intensity scale. In this case the scale
extends from 0 to 255 because the digitizer generates 8-bit
values. This is already more than the 50 grey levels which the
human eye is able to discern. The histogramm shows that the low
and high intensity levels are populated only sparsely. These
intensities do not contain relevant information and could be
suppressed by transforming them to black and white. The rest of
the intensity scale may be transformed linearly into the grey
scale of the output medium. 32 grey values were found sufficient
in order to generate acceptable paper prints on a Polaroid
Palette hard copy system. The linear contrast enhancement was
already applied to fig. 1 and allows to use the entire grey
scale of the image hardcopy system.

However the brightness of the image still depends on the
frequency distribution. There exists a transformation which
makes this distribution a constant or in ordinary terms corrects
badly exposed images. This optimum contrast transformation is
the partial sum series of the frequency distribution (fig. 2) and
results in an image (fig. 3) where the central channel is much
better visible but the film grain structure too.

The histogramm shows also peaks at modulo-16 values which indicates that the digitizer is defective and generates noise. This noise and the grain structure may be suppressed partially by convolution with a spatial filter. The effect of a very simple averaging filter may be seen in fig. 4 where the grain structure is less pronounced but another defect of the digitizing system appears. The sensitivity of the camera tube is not constant and produces artificial line shadows.

3. Central Channel Diameter Determination

The measurement of the central channel diameter therefor needs a method which is immune to local errors and noise. Fig. 5 shows ten average intensity scans from consecutive locations at the lower end of the pin 52 fissile column. The lower scans have parabolic shape not indicating a central channel. The upper five curves show u or v shaped central depressions with small asymmetries.

Fig. 6 shows the main constituents of a projectional distribution of a fuel pin. The first curve is a distribution of solid fuel with a central channel of one quarter the fuel diameter. The second curve represents mass inside the central channel. The two other curves are a constant for the film background and a curve which serves the centering of the scans.

The n measured points $s(x_i)$ of the actual scan are now considered as a linear superposition of these four basic density functions $f_j(x_i)$ with m_j as their relative contribution to the scan:

$$s(x_i) = \sum_{j=1}^{4} f_j(x_i) * m_j \qquad (3.1)$$

This linear equation system is overdefined and the inversion yields a best fit of the four function coefficients m_j. The first two coefficients represent the contribution of solid fuel and central channel to the measured density pattern. The inverse matrix of the four function table at the n transverse imaging points is a matrix with 4 lines and n points. These 4 lines constitute a spatial filters which extract the four constituents m_j from the measured projection.

The transverse scan at 37 cm of pin 52 was analyzed in that manner and compared to the best fit in fig. 7. Except at the central channel edge the measured curve is well approximated. If the scatter in the object and film blurring is accounted for by a gaussian blurring function the approximation is good at the central channel edge too.

At 37 cm the actual central channel diameter was the same as assumed in the four function system of fig. 6 and consequently the coefficients of the first two functions were equal. At other locations where the diameter is smaller the same function system may be used for the computation of the central channel diameter. The actual central channel diameter is then calculated from the ratio of the central channel and the solid fuel coefficient :

$$d_{cc} = d_{fuel}/4 * \sqrt{max(0, m_{cc} / m_{fuel})} \qquad (3.2)$$

Earlier image analysis of CABRI X-Ray radiographs showed that this computational simplification does not lead to serious errors. The use of the filling ratio eliminates errors which are introduced by fuel density changes or imaging system deficiencies like the striping in fig. 4.

A basic assumption in this analysis was the linear superposition of densities. However behind dense objects the neutron density varies exponentially with thickness. A simple way to avoid such nonlinearities is to take a measured distribution from a section with no central channel as the reference and to proceed as before. In this case d_{fuel} must be replaced by d_{pin} in the formula (3.2).

The fig. 8 shows the axial distributions of background, fuel and central channel void density along the radiograph of pin 52. The background density shows a sharp decline at 42 cm. This is caused by exposure differences between the lower and the upper neutron radiograph. Because only the ratios enter into formula 3.2 this difference does not cause a big step of the central channel diameter as shown in fig. 9.

This figure also shows the cuts made in two hot cell laboratories and the channel diameters derived from these cuts. The agreement is good except for the upper cut at 713 mm height. But the hot cell cuts stand for only a very small axial portion whereas the analysis represents an integration over a length of about 3 mm. When the analysis was repeated with only 0.3 mm long samples the calculated central channel at this location showed a sharp minimum. This explains the differences between the two hot cell cuts itself and the calculation.

The error margin for small central channels is rather high because cracks lying in the direction of view cannot be discriminated from a real central channel. Thus the strong oscillations at the beginning of the fissile column may be radial

682

cracks in the fuel. This problem arises because only one projection is available.

The other pins show central channel distributions similar to pin 52. The channel begins to form at about 5 cm BFC and oscillates between 1 and 1.5 mm above 10 cm of the fissile column. At the upper end the central channel decreases but not in all pins and not always to zero.

4. Conclusions

The quantitative analysis of neutron radiographs yields central channel data with sufficient accuracy along the entire fuel column. The algorithm developped is independent from the human analyst and relies only on general assumptions about projections. It is rather independent on the quality of the radiographs.

The central channel volume is a global characteristic of a fuel pin. If finer details like cracks or gaps should be determined it is necessary that the radiographic density should be digitized directly in the image plane in order to avoid errors inherent to secondary media like films, illumination platforms, cameras and digitizers.

Literature

[1] Steinbock, Deckers, Piel
    Quantitative Roentgenradiographie an CABRI-Testeinsaetzen
    KFK-Nachrichten, 3/84

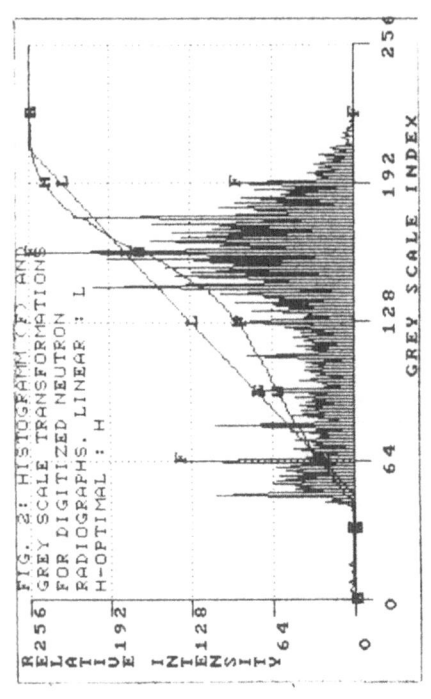

FIG. 1 :
PART OF DIGITIZED NEUTRON RADIOGRAPH
OF PIN 52 AT 90 % OF FISSILE HEIGHT
EXPANDED LINEAR GREY SCALE (32 STEPS)
NO FILTER
BOTTOM ←   → TOP

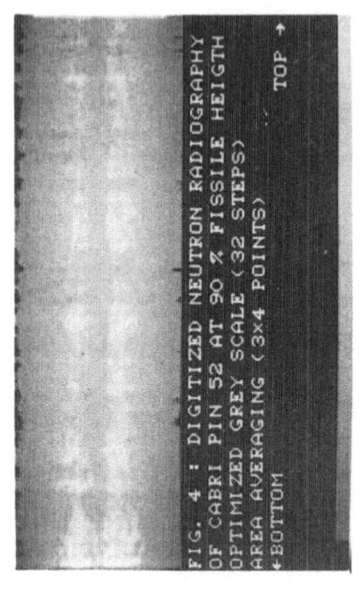

FIG. 2: HISTOGRAMM (F) AND
GREY SCALE TRANSFORMATIONS
FOR DIGITIZED NEUTRON
RADIOGRAPHS. LINEAR : L
H-OPTIMAL : H

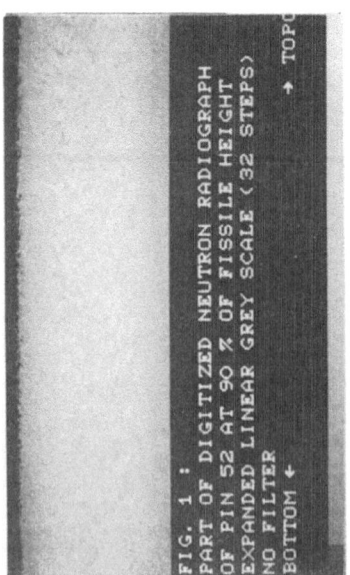

FIG. 3 : DIGITIZED NEUTRON RADIOGRAPHY
OF CABRI PIN 52 AT 90% FISSILE HEIGHT
OPTIMIZED GREY SCALE (32 STEPS)
NO AXIAL FILTER
←BOTTOM   TOP→

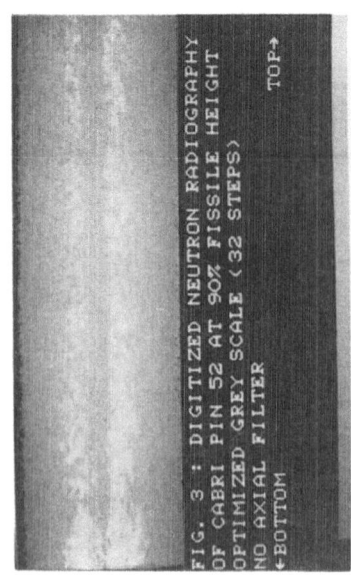

FIG. 4 : DIGITIZED NEUTRON RADIOGRAPHY
OF CABRI PIN 52 AT 90 % FISSILE HEIGTH
OPTIMIZED GREY SCALE (32 STEPS)
AREA AVERAGING (3x4 POINTS)
←BOTTOM   TOP →

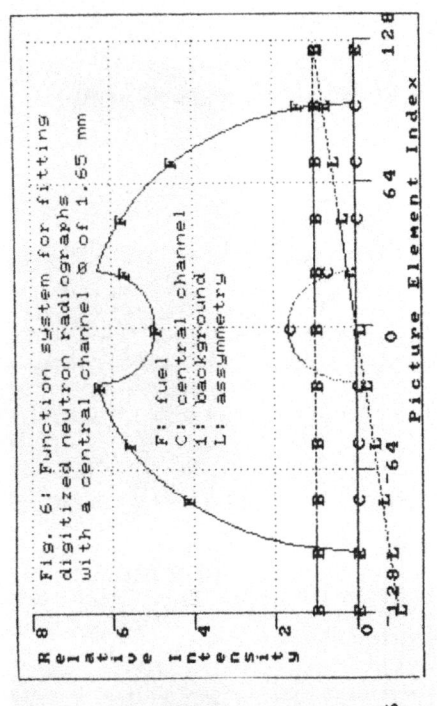

FIG. 5: DIGITIZED NEUTRON RADIOGRAPH OF CABRI PIN 52 BETWEEN 1.5 AND 9.5 CM CURVES SEPARATED; INDEX ~ HEIGHT

TRANSVERSE PIXEL INDEX

RELATIVE INTENSITY

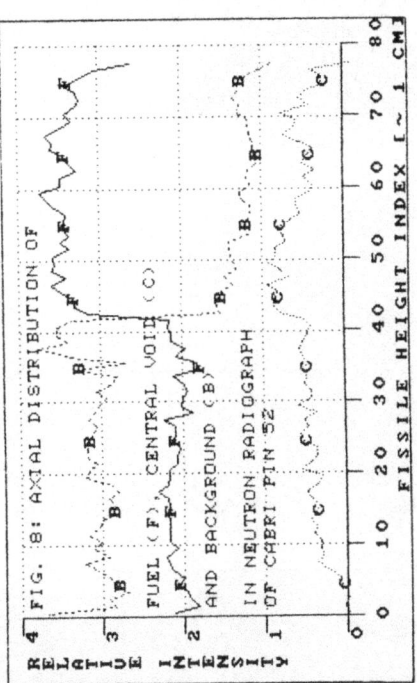

Fig. 6: Function system for fitting digitized neutron radiographs with a central channel @ of 1.65 mm

F: fuel
C: central channel
I: background
L: assymmetry

Picture Element Index

Relative Intensity

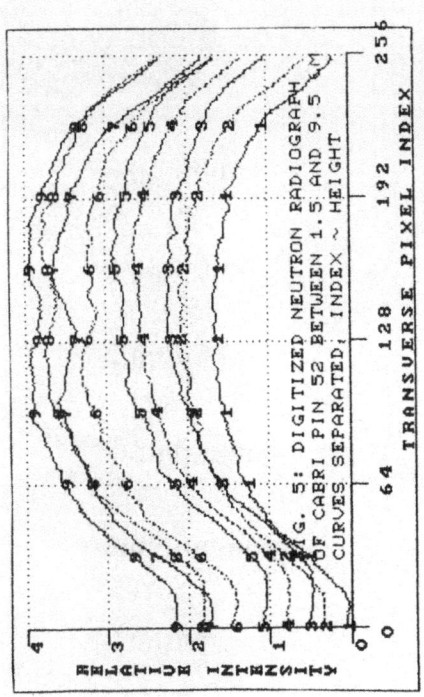

FIG. 7: COMPARISON OF MEASURED AND FITTED FILM DENSITY IN NEUTRON RADIOGRAPH

M: AVERAGE OF 256 SCANS FROM CABRI PIN 52 AT 37 CM
C: FIT WITH 1.65 MM CENTRAL VOID
G: FIT WITH SCATTER AND FILM BLUR BLUR FUNCTION :

TRANSVERSE PIXEL INDEX

RELATIVE INTENSITY

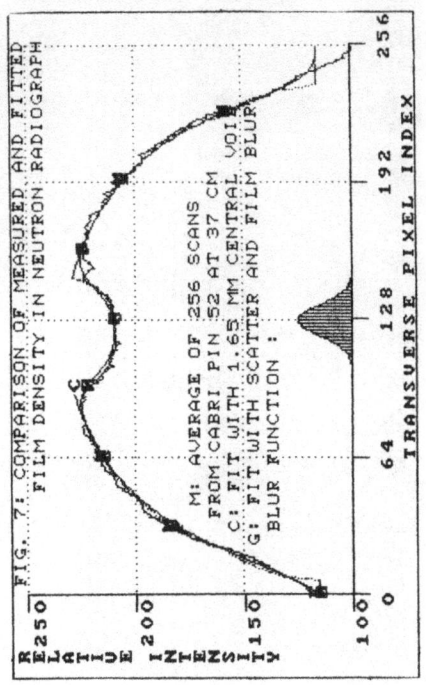

FIG. 8: AXIAL DISTRIBUTION OF FUEL (F) CENTRAL VOID (C) AND BACKGROUND (B) IN NEUTRON RADIOGRAPH OF CABRI PIN 52

FISSILE HEIGHT INDEX [~ 1 CM]

RELATIVE INTENSITY

685

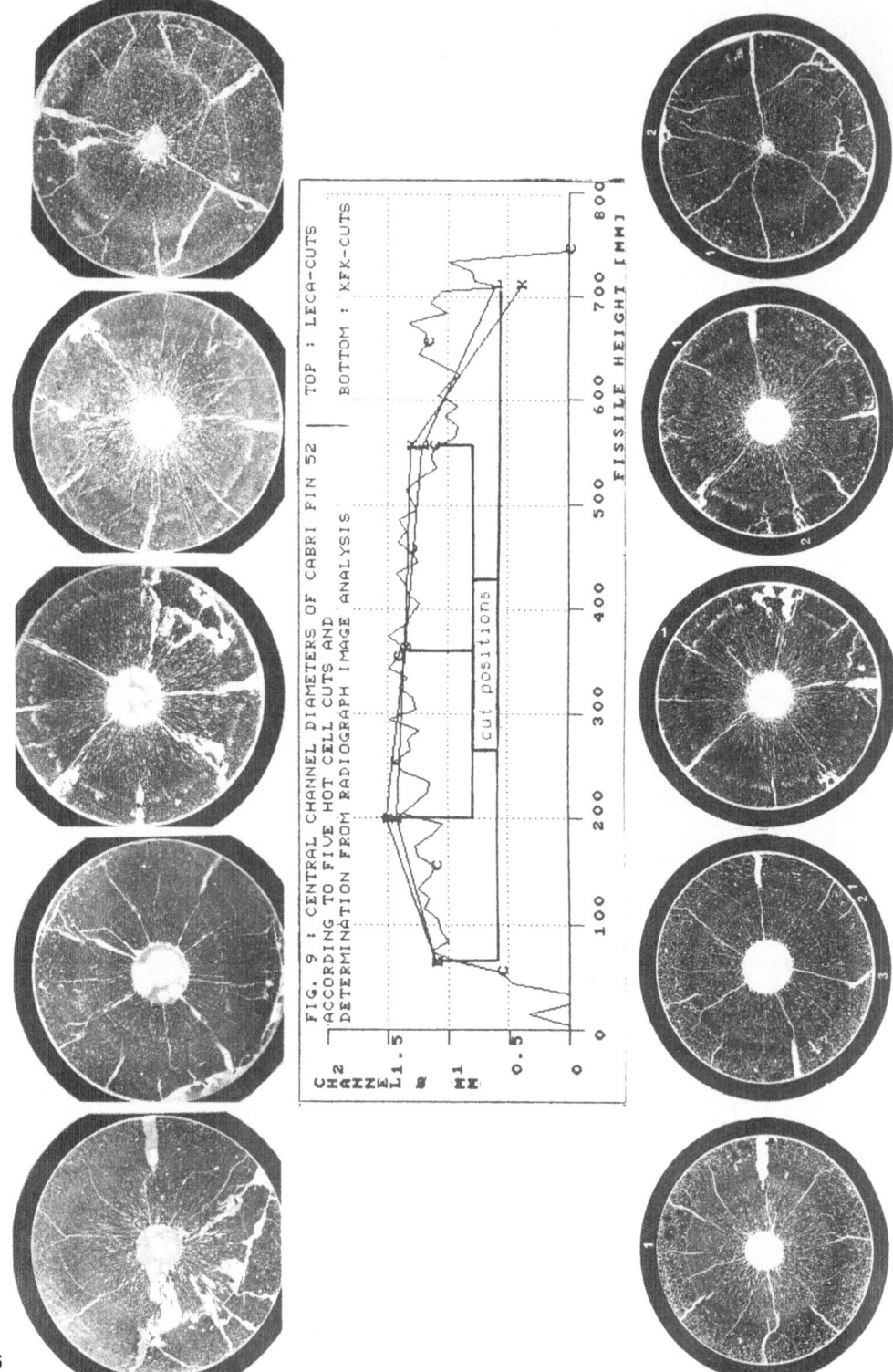

FIG. 9 : CENTRAL CHANNEL DIAMETERS OF CABRI PIN 52 | TOP : LECA-CUTS
ACCORDING TO FIVE HOT CELL CUTS AND
DETERMINATION FROM RADIOGRAPH IMAGE ANALYSIS | BOTTOM : KFK-CUTS

COMPUTER CONTROLLED TRAVELLING MICRODENSITOMETER

K.H. van Otterdijk, B. Shapiro, H.P. Leeflang
Netherlands Energy Research Foundation
Postbus 1
NL 1755 ZG Petten, The Netherlands

J.F.W. Markgraf
Joint Research Centre of the Commission of the
European Communities, Petten Establishment
Postbus 2
NL 1755 ZG Petten, The Netherlands

ABSTRACT

A short description of a computer controlled microdensitometer is presented. The system performance was assessed and
evaluated with respect to the feasibility of a dimensional
analysis to be carried out on the neutrographic film images
of irradiated fuel pins.

---

1. INTRODUCTION

To allow an accurate and reproducible dimensional analysis to
be carried out, an instrument, capable of adequate quantitative
assessment of the film-recorded image is required. The digital
image recording process should preferably not lead to any
information losses or noise induction so that the Signal-to-Noise
Ratio (SNR) remains unaffected. The required instrument
performance with regard to the image recording quality, shall
therefore be significantly better than that of the silver halide
film. Consequently, the following design targets have been laid
down :

>Spatial   Resolution : better than 20 micron
>Contrast   Resolution : better than 0.007 D
>Low Level Sensitivity : better than 3 D
>Useful Field of View : better than 4 x 48 cm

In view of the development costs involved, a concept for the
travelling microdensitometer, e.g. the point scanning, was chosen
to be implemented.

2.   APPARATUS

The development was based on the commercially available
photometer system (Schnellphotometer GIII, Jenoptik, Jena),
whereby an improvement of the light detection and measurement
circuit as well as the introduction of remote control and digital
data storage capabilities were carried out (Fig. 1).

Fig. 1 General view of microdensitometer

The mechanical subsystem provides a two-dimensional precision
scanning capability with a spatial resolution of
5 micron across the useful field of view (54.8 x 4.8 cm) situated
in the photometer object plane. The stepping motors drive, on
command, the travelling bench whilst the linear position encoders
continuously sense the absolute values of the X and Y coordinates.

The photometer light source and optics were applied in
combination with a selected photodiode and a high precision
current meter to provide the suitable measurement circuit
featuring the remote control, on-line dark current subtraction and
a digital readout (Fig. 2).

The remote control capability allows the stepping motors,
position encoders and the current meter to be commanded. The IEEE
interface accommodates both control functions and data flow. The
current and position data are collected and stored on the APPLE
IIe desk top microcomputer system, which feeds the controls back
into the system. The computer MOS memory is here configured by the
specially developed software to facilitate the execution of user-
specified scanning protocol. The step size, the scan length and
the ordinate can be selected. The acquisition software provides
the safeguards to avoid attempts to travel beyond the boundaries
posed by the field of view.

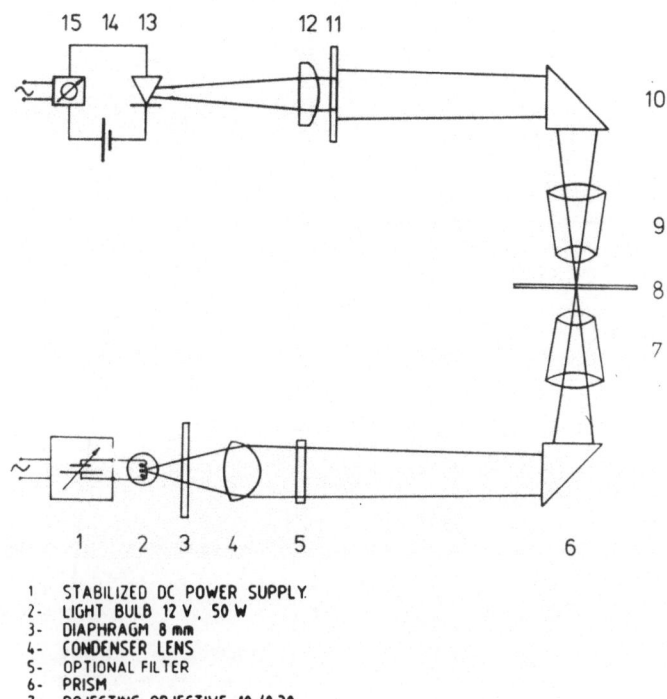

```
15  14  13              12 11
                                                          10

                                                           9

                                                           8

                                                           7

 1      2   3   4     5                        6
```

1   STABILIZED DC POWER SUPPLY.
2-   LIGHT BULB 12 V, 50 W
3-   DIAPHRAGM 8 mm
4-   CONDENSER LENS
5-   OPTIONAL FILTER
6-   PRISM
7-   POJECTING OBJECTIVE 10 /0.30
8-   FILM AND TRANSPARANT GLASS OBJECT HOLDER 2x 3.5 mm
9-   IMAGING OBJECTIVE 10 / 0.30
10- PRISM
11- ADJUSTABLE PRECISION DIAPHRAGM 0 - 3 mm, 0.01 mm 0 - 20 mm, 2 mm
12- FIELD LENS
13- PHOTODIODE 10^{-1} TO 10^{5} lx, 30 nA / lx (BPW20, AEG TELEFUNKEN)
14- BATTERY 4.5 V
15- AUTORANGING PICOAMMETER (MODEL H 85, KEITHLEY INSTRUMENTS)

Fig. 2 Light Measurement Circuit

3.   SYSTEM PERFORMANCE

The system performance was assessed with the help of a
standard film density calibration scale, by adjusting the light
beam diaphragm to yield the effective spot size of
10 x 100 micron, as under the actual measurement conditions.

The results of the sensitivity test demonstrate an almost
perfect linear response across the wide density range (Fig. 3).
The derived values of the contrast resolution of the low level
sensitivity amount to 0.003 D and 3.1 D respectively.

689

Proper experimental determination of the system spatial resolution assumes availability of an ideal "knife-edge" density transition test pattern to be imaged on the film. In an attempt to measure the 0.54 - 0.0 D density transition, a compromise procedure was applied when scanning was performed across the side boundary of the calibration scale mentioned above (Fig. 4). The value of 30 micron obtained according to the Klasens method, reflects not only the system performance, but also the optical imperfections of the test pattern and the object holder (two 3 mm thick plates of ordinary glass).

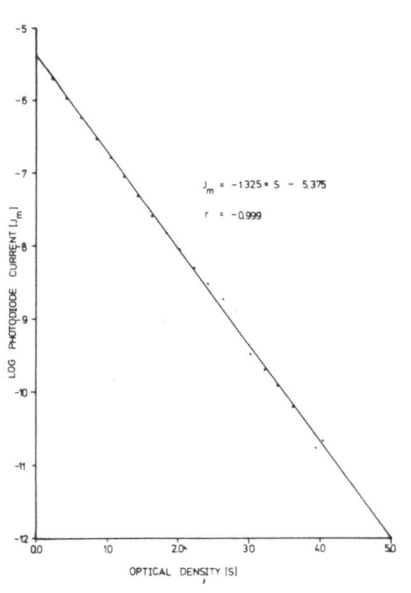

DIAPHRAGM : 10 × 100 micron

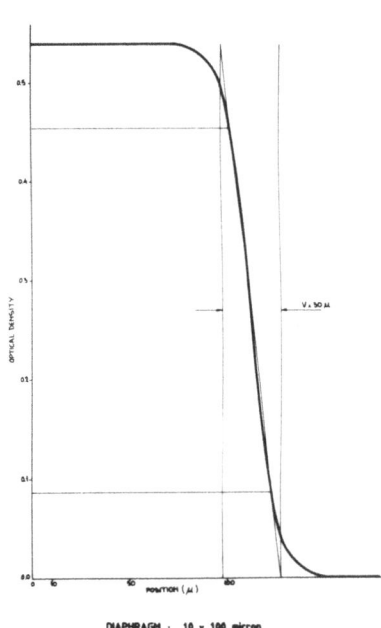

DIAPHRAGM : 10 × 100 micron

Fig. 3 Sensitivity Calibration   Fig. 4 Resolution Test

The 10 micron wide aperture implies a step size of 10 micron and a scanning speed of 5 micron/sec, so that 50 minutes will be required to complete the single transverse travel across the fuel pin image.

The total development costs, including the hardware as well as the software investments, amount to some $ 25.000,-.

Finally, the global assessment of the system performance was made by carrying out the transverse scan across the RISØ Calibration Fuel Pin (CFP-El) image, obtained with the help of the transfer technique on a silver halide film (Fig. 5). It should, be remembered however, that the density profile obtained reflects the whole imaging process and therefore may not be used for quality evaluation of digital image recording on its own.

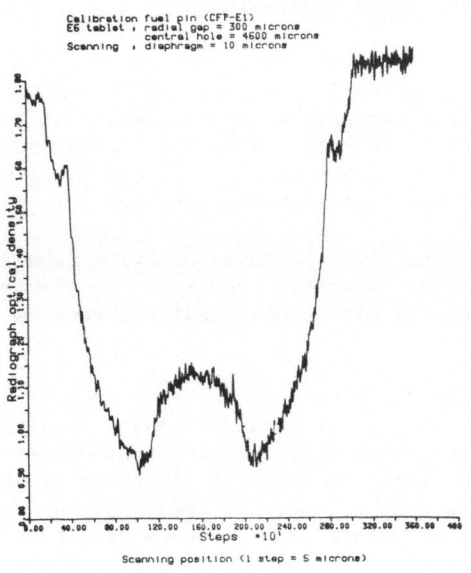

Fig. 5 Calibration Fuel Pin Scan

4. DISCUSSION AND CONCLUSIONS

As clearly indicated from the test results, the target specifications, with the exception of the spatial resolution, could be successfully achieved within the given technological and financial constraints.

The measured value of spatial resolution reflects not only the instrument performance but also the quality of the test pattern, the object holder and procedure being applied. Thus further improvement of their optical quality is required. The initial target value of 20 micron was based on the intrinsic nitrocellulose film resolution; nevertheless, the actual neutrographic image resolution, whilst being in excess of 60 micron, implies that the present digital recording quality is certainly adequate. In addition, a more fundamental approach involving quantification of SNR degradation due to image digitisation seems to be desirable.

The scanning speed appears to be rather low regarding the potential need for extensive detailed image quantification, so that only selective evaluation seems feasible at this stage. This limitation, however, is rather inherent to the concept of point scanning. Significant speed improvements could be achieved by stepping over to the line scanning approach.

The remote control and data acquisition capabilities of the system were found satisfactory. However, the adequacy of the desk top microcomputer for dimensional analysis remains to be proven.

In conclusion, the system costs could be kept within reasonably low limits, thus allowing gain in the initial experience to be made prior to any definite commitment with regard to the final concept. It should also be remembered that the optimal technique required for digital recording of a silver halide film image most probably differs from one meant for the nitrocellulose film. Therefore, it seems rather unrealistic to expect both types of films to be optimally digitized with the help of the same technique, which ultimately infers consequences on the instrument design.

692

PART XI

COMPUTED TOMOGRAPHY

TOMOGRAPHIE AUX NEUTRONS UTILISANT MART ET LE FILTRE WIENER

CRISPIM, V.R, LOPES, R.T, et BORGES, J.C.

Programa de Engenharia Nuclear - **COPPE/UFRJ**

C.P. 68509 - 21945 - Rio de Janeiro - RJ - Brasil

Dans le processus de reconstruction d'images à travers de projections par neutrongraphies, l'obtention de projections subissent l'influence de divers facteurs. Dans notre travail , nous avons fait l'usage du filtre de WIENER dans les projections obtenues aux neutrongraphies d'un objet déterminé, pour essayer de minimiser les effets de ces facteurs sur la qualité de l'image reconstruite. L'algorithme de reconstruction utilisé a été le MART (Technique de Reconstruction Algébrique Multiplicative). Des comparaisons qualitative et quantitative (en utilisant une fonction empirique) sont montrées entre l'image originale et celles reconstruites avec le MART, avec ou sans filtre.

I - INTRODUCTION

Le processus de reconstruction d'images à travers de projections sert à déterminer la distribution spatiale des densités dans une section transversale d'un corps d'essai. Une reconstruction d'image moyennant des neutrons peut être obtenue avec l'acquisition de neutrongraphies d'un objet qui, tournant sur son axe central, sera capable de réaliser un grand nombre de projections. Néanmoins, plusieurs facteurs peuvent avoir une influence sur cette reconstruction quand le processus d'enregistrement des données est une radiographie aux neutrons. Parmi ceux-ci nous pouvons citer: l'alignement des données de projection, le besoin de projections aussi coplanaires que possible, l'effet de divergence du faisceau neutronique, la dispersion des neutrons sur l'objet, la digitalisation des images radiographiques et les caractéristiques inhérentes au système du film convertisseur, le spectre de la radiation avec une contribution de neutrons de diverses energies et radiation γ , etc.

Dans ce travail nous avons fait l'application du Filtre de Wiener dans les projections obtenues aux neutrongraphies, dans la tentative de mininiser les effets des facteurs considérés ci-dessus dans la qualité´ de l'image reconstruite.

II - MÉTHODE

L'algorithme de reconstruction utilisé a été l'itératif ART (Technique de Reconstruction Algébrique) avec correction multiplicative[1], qui s'est avéré efficace dans le cas d'un nombre réduit de projections [2], cette limitation étant déterminée par les conditions expérimentales quand il s'agit de l'usage de neutrongraphies.

L'expression générale du MART pour le calcul de la fonction densité est donnée par:

$$f_{ij}^{q+1} = \frac{p(K, \theta)}{p^q(K, \theta)} \quad w_{ij}(k, \theta) \cdot f_{ij}^q \quad (1),$$

où $p(k, \theta)$ représente le rayon-somme, K étant l'ordre du rayon et la rotation angulaire; $p^q(k, \theta)$ représente le rayon-somme calculé après chaque itération et w_{ij} étant la fonction du poids.

Le critère de convergence accouplé à l'algorithme pour arrêter le processus itératif, a été celui de la fonction d'entropie maximale[3]

Dans la simulation, le corps d'essai utilisé a été cohérent avec celui utilisé dans l'essai expérimental, constitué de quatre lingots cylindriques disposés de façon assymétrique (3 étant en cuivre et un en acrylique). Les sections de choc considérées ($E_n = 0,0252$ eV) sont: cuivre 0,935 cm^{-1}, acrylique 2,56 cm^{-1} et nulle pour l'air.

Durant le processus d'enregistrement d'un signal, dans n'importe quelle méthode, un bruit est toujours présent. Le filtre de Wiener est excellent en ce qui concerne le critère de minimisation de l'erreur moyenne quadratique entre la sortie du filtre et le signal d'entrée. Ce filtre est utilisé pour récupérer un signal avec bruit additif dont le spectre de puissance est connu. Dans l'espace de fréquence spatiale, la fonction de transférence du filtre Wiener[4] est donnée par:

$$H(\omega) = \frac{S_s(\omega)}{S_s(\omega) + S_n(\omega)}$$ où $S_s(\omega)$ est l'auto spectre de puissance du signal et $S_n(\omega)$ est l'auto spectre de puissance du bruit.

Dans le calcul de la fonction de transference, le bruit simulé a été considéré comme une distribution Gaussienne de variation unitaire et valeur moyenne zéro, généré par la fonction RANDOM implicite dans l'ordinateur B-6800 de l'Université, avec des valeurs RMS("root mean square") égales à 1,0/0,5/0,25/0,1.

Dans les expériments, les projections p(K, θ) ont été obtenues de neutrongraphies réalisées à la sortie du canal J-9 du Réacteur Argonauta de l'Instituto de Engenharia Nuclear/CNEN, où le flux de neutrons thermiques est de l'ordre de 10^5 neutrons/cm²/s. Le temps d'exposition de l'objet au faisceau a été de 40 minutes. On a utilisé comme convertisseur une feuille métallique de Gadolinium-157 enrichie de 99,9% avec 125 μm d'épaisseur. Comme détecteur on a été utilisé le film Industrex-AA54, porduit par Kodak.La révélation a été réalisée dans des conditions standards pour le film.

La lecture des densités optiques du film a été faite par un microdensitomètre du type Jena, accouplé à un galvanomètre. La figure 1 montre la neutrongraphie du corps d'essai à 0º, où l'on peut remarquer des points pour l'alignement des données, nécessaires au processus correct de lecture. Le corps d'essai est constitué par un ensemble quadrangulaire assymétrique composé de lingots cylindriques de 1,1cm de diamètre, 3 lingots étant en cuivre et un en acrylique. La ligne de référence de la section transversale destinée à l'analyse est indiquée par des percettes sur de petites plaques de Cadmiun situées dans l'extremité de l'ensemble.

FIGURE 1: Neutrongraphie du corps d'essai.

III - RESULTATS ET COMMENTAIRES

L'importance de l'algorithme pour le corps d'essai a été évaluée qualitative et quantitativement par des mesures des distorsions entre les images originale et celles reconstruites. La mesure de la distorsion est donnée par:

$$\delta = \frac{1}{MN} \sum_{i=1}^{N} \sum_{j=j}^{N} | I_{ij} - J_{ij} |$$

où I est l'image originale, J est l'image reconstruite et i et j sont les coordonnées despixels et MN sont les dimensions de l'image.

Les reconstructions présentées ont été obtenues par le MART après 20 itérations(dimension de l'image 31 x 31 et 6 projections à $0°$, $30°$, $60°$,.....$180°$). Les figures 2, 3 et 4 montrent une comparaison qualitative entre les reconstructions simulées avec le corps d'essai. La figure 2 montre l'image originale et le comportement de l'algorithme MART dans les reconstructions.Pour le corps d'essai utilisé et avec peu de projections, la figure 2 montre que le MART peut fournir une bonne reconstruction. La figure 3 montre les résultats de reconstructions, quand on ajoute aux projections des bruits gaussiens avec des RMS différents. L'action du filtre de Wiener dans ces projections contaminées est montré dans la figure 3, où la fonction de transférence de chaque filtre a été calculée par l'utilisation de l'auto-spectre de puissance du bruit se référant à la RMS en question. Le tableau 1 montre la comparaison quantitative entre les reconstructions simulées, à travers la mesure de distorsion, en prennant par référence l'image originale montrée dans la figure 2a. Les résultats expérimentaux, montrés dans la figure 5 donnent,une évaluation qualitative du processus de reconstruction d'image aux neutrongraphies. Les reconstructions montrent une bonne image quant à la localisation des objets avec des densités différentes, ce que nous amène à conclure que le processus est valable pour la localisation de vides dans certains corps.

TABLEAU 1 MESURES DE DISTORSION

I M A G E	$\delta \times 100$	I M A G E	$\delta \times 100$
Original	0	MART + Bruit x 1 + Filtre	3,94
MART	1,82	MART + Bruit x 0,5 + Filtre	3,70
MART + Bruit x 1	10,88	MART + Bruit x 0,25 + Filtre	2,95
MART + Bruit x 0,5	5,05	MART + Bruit x 0,1 + Filtre	2,48
MART + Bruit x 0,25	3,13		
MART + Bruit x 0,1	2,57		

IV - REFERENCES BIBLIOGRAPHIQUES

1. GORDON, R.; BENDER, R. and HERMAN, G.T. J. Theor. Biol., 29
   471-481. 1970

2. FARIAS, V.S.O. Estudo do ART Multiplicativo na Reconstrução de
   de imagem obtidas em Neutrongrafias, Tese de M.Sc. COPPE/UFRJ.
   1983.

3. LENT.A. A Convergent Algorithm for Maximum Entropy Image Restoration, with a medical X-Ray Application , Image Analysis and E-
   valuation. R. Shaw. Ed., Society of Photography Scientists and
   Engineers, 249-257. 1977.

4. LATHI, b.P. An Introduction to Randon Signals and Communication
   Theory, Pensylvania, International Textbook Company, 251-262.
   1968.

Image originale   **Reconstruction par MART**

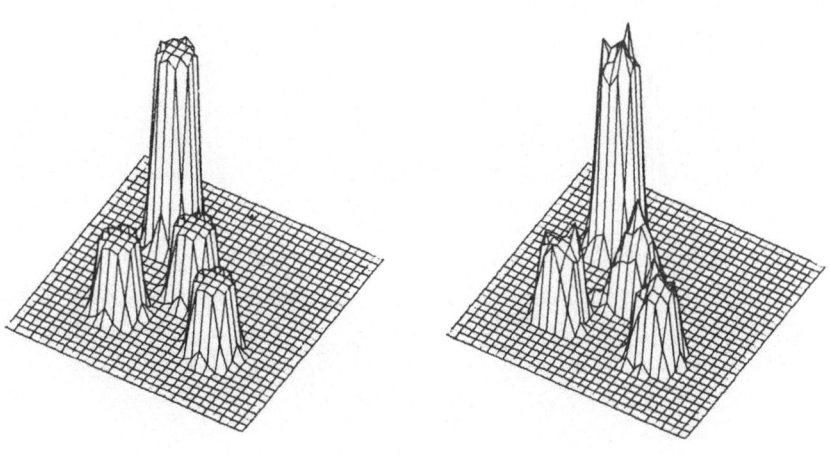

FIGURE 2: Reconstruction de l'image 31 x 31

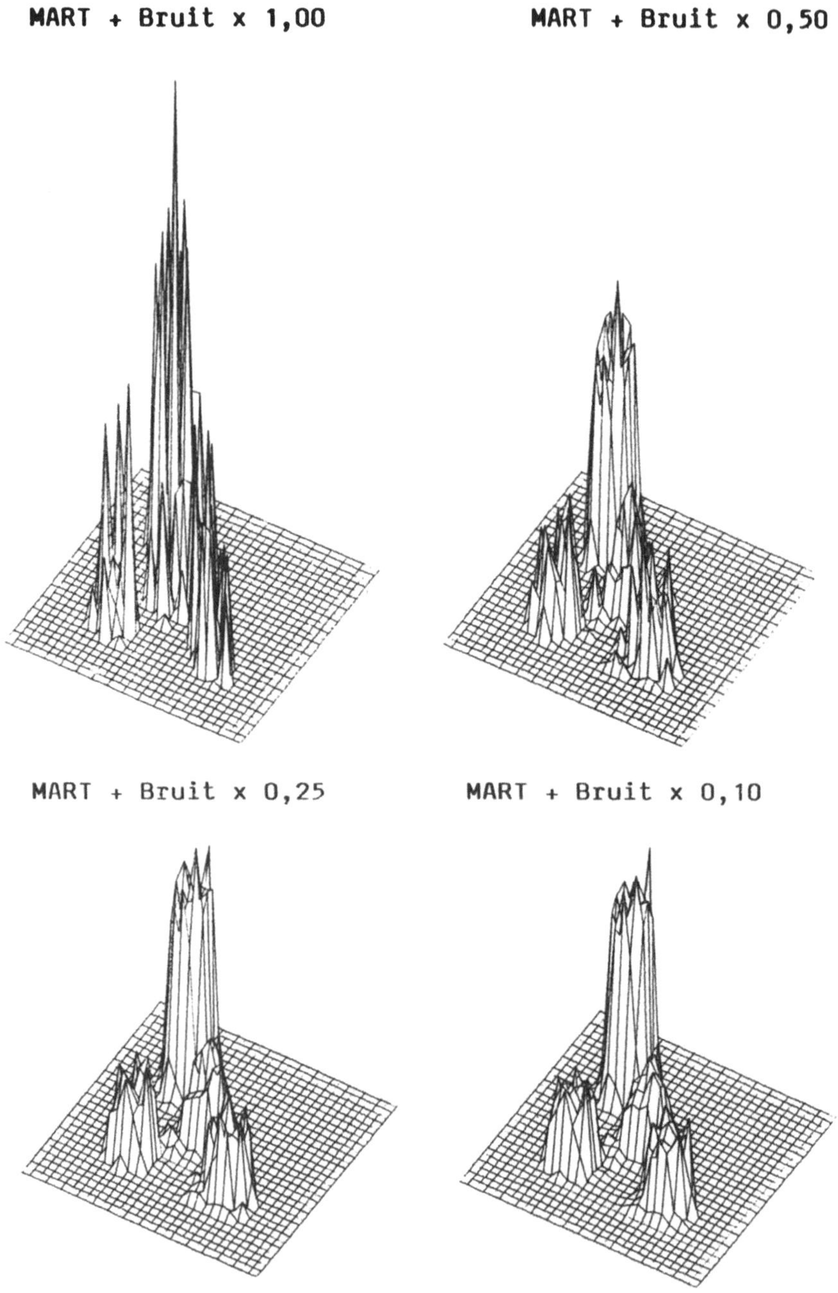

MART + Bruit x 1,00   MART + Bruit x 0,50

MART + Bruit x 0,25   MART + Bruit x 0,10

FIGURE 3: Reconstruction de l'image 31 x 31

MART + Bruit x 1,00   MART + Bruit x 0,50

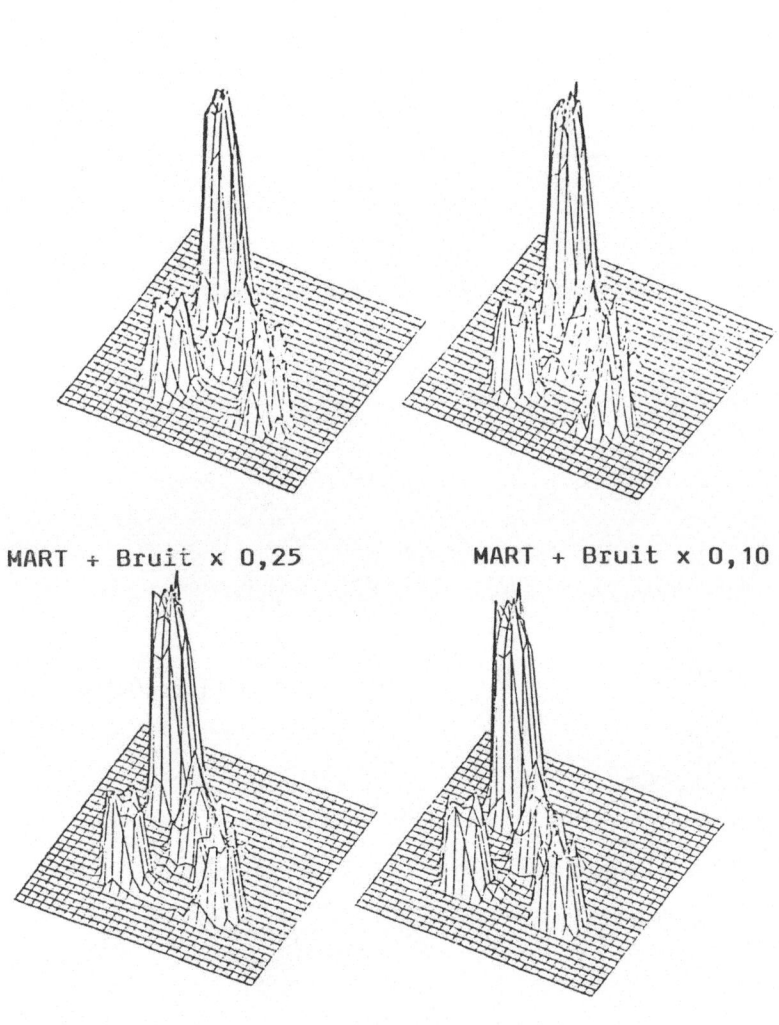

MART + Bruit x 0,25   MART + Bruit x 0,10

FIGURE 4: Reconstruction de l'image par filtre WIENER

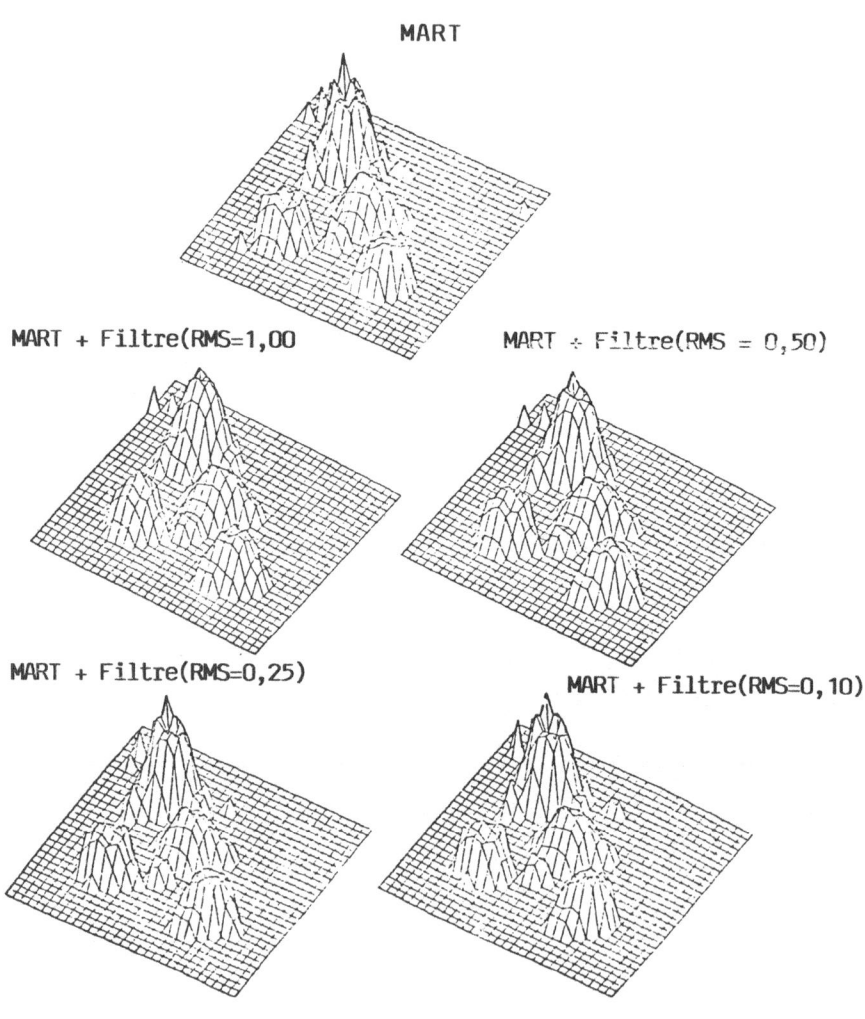

FIGURE 5: Reconstruction de l'image par neutrongraphies

TOMOGRAPHY USING NEUTRON RADIOGRAPHY PHOTOGRAPHS
FOR A BEAM OF SEVERAL FUEL PINS

P. RIZO

Centre d'Etudes Nucléaires de Grenoble
Avenue des Martyrs, 85 X Grenoble CEDEX, France

ABSTRACT

A method of tomography by neutron radiography has been developed
at the Centre d'Etudes Nucléaires de Grenoble (Grenoble Centre for
Nuclear Studies) with a view to examining beams of several fuel
pins during irradiation. During this study, the emphasis was made
principally on establishing a physical model of the neutron radiography measurements, so as to enable appropriate codes of reconstruction to be written. We also studied the conditions for use of
classical reconstruction codes after appropriate processing of the
readings.

INTRODUCTION

The evolution of safety experiments in the experimental reactors at
the Reactor Department in Grenoble leads to the irradiation of complex structures which may include several fuel pins.

It has become essential to develop techniques of non-destructive
testing which enable these pins to be visualized at various stages
of their irradiation.

To this end, we have studied a method of tomography by neutron
radiography which is adapted to our installations. This study was
carried out in 3 phases :

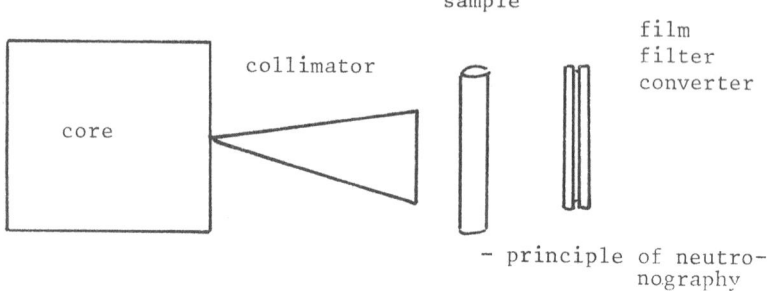

- principle of neutro-
  nography

- a study by simulation of the tomographic readings so as to
  evaluate the contribution of the various interactions of the
  neutrons

- establishment of a mathematical model of the neutron radiography
  readings

- research seeking appropriate methods of reconstruction

I STUDY OF THE NEUTRON RADIOGRAPHY READING

In order to compare the contribution of the various interactions
of neutrons between their emission by the collimator and
their reception by the converter, three simulations were performed using a Monte Carlo code.

First we considered that the neutrons were all parallel before
leaving the collimator and that they interact from this plane.
The following geometry was studied :

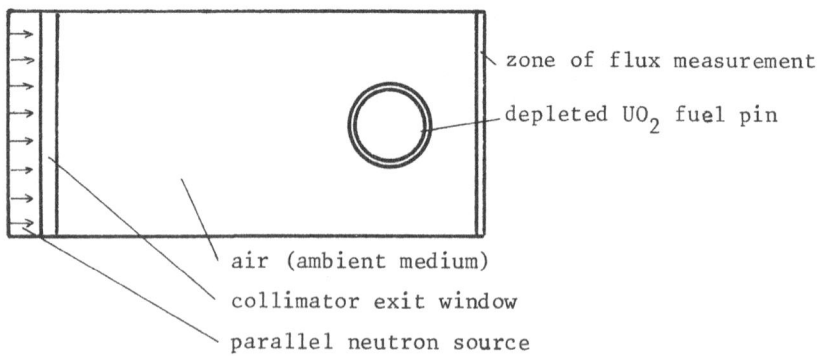

zone of flux measurement

depleted UO_2 fuel pin

air (ambient medium)

collimator exit window

parallel neutron source

1st simulation :   all the phenomena contributing to the flux
                   measurement are described as precisely as
                   possible;

2nd simulation :   we consider that the neutrons can only interact
                   in the object under investigation;

3rd simulation :   we consider that the neutrons can only interact
                   in the object, but that any interaction is ab-
                   sorption;

<u>Remark</u> : The third simulation describes the classic case of uni-
        directional attenuation for which most of the classical
        reconstruction codes by transmission are designed.

Results of
simulations

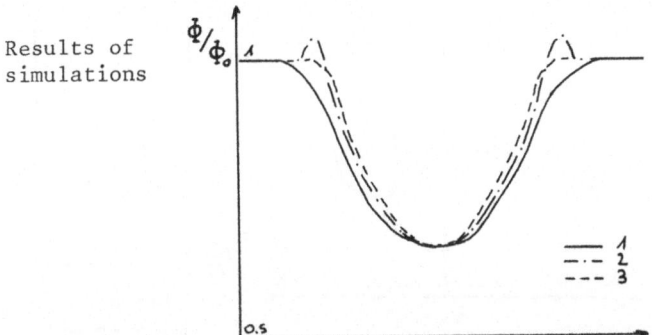

By comparing the previous curves, it can be seen that the interaction which determines the difference between unidirectionnal
attenuation modeling and the neutron radiography reading is the
scattering of neutrons in the ambient medium and when crossing
the collimator window. This is all the more true, the further
the object is from the converter. If indeed we consider that the
sample is a source of scattered neutrons, it can be seen that
their contribution to the reading varies with the square of the
distance from the object to the converter.

These three simulations show clearly that the neutron radiography
reading <u>cannot</u> be accepted to a reading obtained from neutrons
moving in a straight line within a collimator cone.

II ESTABLISHING A MATHEMATICAL MODEL

In order to write a mathematical model of the neutron radiography reading, scattering in the sample will be considered to a absorption. This is more or less true if the object is far from the converter.

To take into account scattering in the ambient medium, we consider that the neutrons interact between the collimator exit and a plane located just behind the sample (fig.   ), which we will call   the "emitting plane". We can then calculate that the angular distribution of the neutrons do not varies from the "emitting plane" to the converter. The advantage of such a model is that it can be considered as a superposition of readings for unidirectionnal attenuation, weighted by the angular distribution of the neutrons.

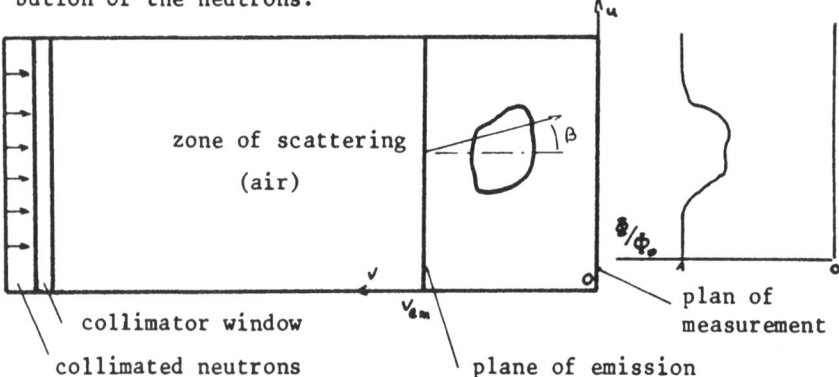

collimator window

collimated neutrons

plane of emission

plan of measurement

Let ϕ_0 be the flux emitted over $180°$ angular degrees by the emitting plane towards and ϕ (u) the flux measured in u.

$$\phi(us) = \phi_0 \int_{-\beta}^{+\beta} Rep(\alpha) \exp\left(-\int_0^v em\ \Sigma_T (u-\alpha v,v)dv\right) d\alpha$$

where β = maximal angular distance of the neutrons
$Rep(\alpha)$ = proportion of neutrons with direction α equalized by

$$\int_{-\beta}^{+\beta} Rep (\alpha) = 1$$

$\Sigma_T(u,v)$ = total apparent effective section at point (u,v)
calculated by

$$\Sigma_T (u,v) = \frac{\int \Sigma_T(u,v,E)\phi (E)\alpha E}{\int \phi(E)\ \alpha E}$$

in order to break free from the energy parameter.

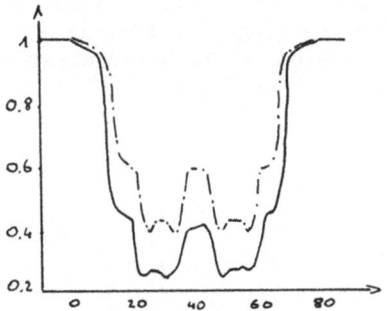

Comparison between
reading and the model

We note that for Rep(α) = δ (α) (δ = Dirac distribution), the
case of unidirectional attenuation reoccurs. Moreover, with this
model, the non homogeneity of the flux emitted by the collimator
can be taken into account by estimating a function Rep (α,u).

III RECONSTRUCTION

3.1. Reconstruction using the mathematical model directly

A method of reconstruction has been developed which takes
into account the previous model, based on an additive ART
method, as these methods are simple to apply in a nonlinear case. However, such methods require a long calculation time and can only be used for the purpose of vali-
dating the established model.

For the first phase, a beam of seven fuel pins with depleted
UO_2 were used as experimental matérial for the theoretical
investigations on the neutron radiography beam at
Melusine reactor, then with UO_2PUO_2 on the immersed installation at Siloe reactor.

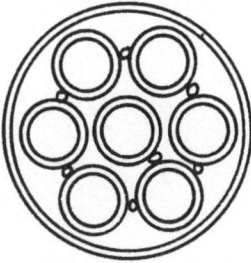

Fuel pin set—up observed
on the immersed installation

Fuel pin set—up used
on the external beam

The angular distribution of the neutrons Rep(α) were either calculated considering one single scattering of neutrons before the plane of emission, or measured directly using a collimator and a fission chamber.

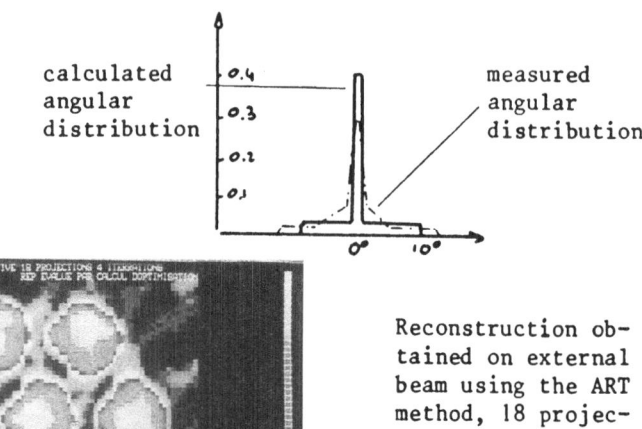

calculated angular distribution

measured angular distribution

Reconstruction obtained on external beam using the ART method, 18 projections over 180°

3.2 Reconstruction using the relation between unidirectional attenuation modeling and neutron radiography readings

This method consists of a transformation applied to the readings so as to make them compatible with a classic reconstruction code. Let us consider the logarithm of the neutron radiography reading.

$$\text{Log } \frac{\Phi(u)}{\Phi_o} = \text{Log } \left(\int_{-\beta}^{+\beta} \text{Rep}(\alpha) \exp\left(-\int_o^{v_{em}} \Sigma_T(u-\alpha v, v) dv \right) d\alpha \right)$$

$$\not\approx \text{Log } \left(\int_{-\beta}^{+\beta} \text{Rep}(\alpha) \int_o^{v_{em}} \Sigma_T(u+a\alpha, v) dv \, a\alpha + 1 \right)$$

$$\not\approx \int_{-\beta}^{+\beta} \text{Rep}(\alpha) \int_o^{v_{em}} \Sigma_T(u+a\alpha, v) dv \, d\alpha$$

with a ≠ 0.

It can be seen that the logarithm of the neutron radiography reading is the convolution product of the unidirectional attenuation and of the angular distribution of the neutrons. This approximation is only valid for small α ; these relations can then only be used when the neutrons are well collimated, and when they do not cross any medium that scatter too much before the emitting plane. We estimated Rep(α) for a simple sample made up of a single fuel pin, by deconvolution of the neutron radiography reading and by a simulation of that reading in unidirectional attenuation.

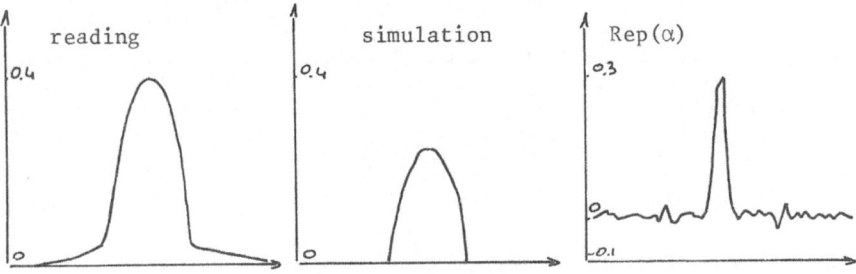

We then deconvoluted the readings by Rep (α), estimated previously, and reconstructed by two types of method : the ART method and optimal regularization.

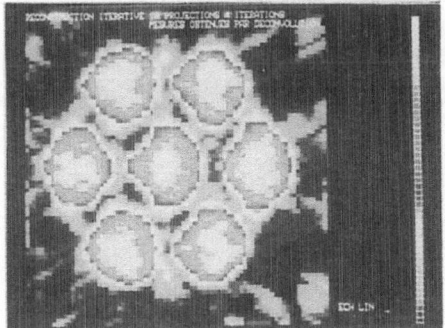
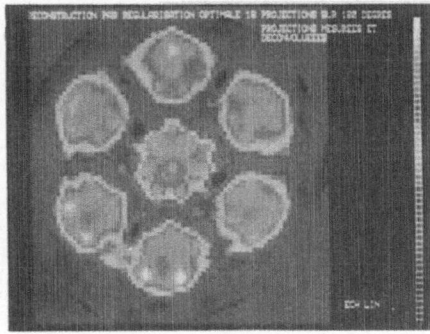

ART reconstruction   Reconstruction by optimal
                                      equalization

CONCLUSIONS

The most promising approach to this type of problem appears to be the application of deconvolution methods which allow rapid transformation of the readings, and the use of powerful algorithms of the optimal regularization type. Nevertheless, when the angular distribution of the neutrons is too broad, then the non-linear approxilation of the ART method remains a reliable and simple solution.

709

In order for tomography by neutron radiography to become a powerful method of non-destructive testing, the number of readings,
limited today by handling problems, must be increased. It may be
wagered that with an automatic system of photography and of fuel
pin assembly rotation, we could achieve quantitative reconstructions using approximately 100 angles of incidence.

REFERENCES

D.M. TOW 1st World Congress of Neutron Radiography 1981
Demoment - Segalen Constrained LMS adaptative algorithm elec-
          tronics letters 4th March 1982 Vol 18 N° 5 pp 226-227
Devolpi - Rode 1st World Congress of Neutron Radiography
Girard Doctorate Thesis Engineering INPG Oct. 1984
Rizo - Thesis   in preparation

NEUTRON TOMOGRAPHY OF

DAMAGED REACTOR FUEL ASSEMBLIES

Grant C. McClellan

Argonne National Laboratory
P.O. Box 2528
Idaho Falls, ID 83403-2528

David M. Tow*

Ilex Systems, Inc.
1423 South Milpitas Blvd
Milpitas, CA 95035

ABSTRACT

A neutron tomography capability has been developed at Argonne National Laboratory's Hot Fuel Examination Facility for use in the nondestructive evaluation of reactor fuel assemblies, including those deliberately damaged to simulate accident conditions. Such damaged fuel assemblies are typically irradiated under abnormally severe conditions, resulting in clad melting and fuel dispersion. Following in-reactor testing, the damaged fuel assemblies are transferred to the HFEF Neutron Radiography (NRAD) Facility, which was designed and equipped to perform high-quality thermal and epithermal neutron radiography of a wide variety of irradiated materials. The NRAD neutron source is a 250 kW TRIGA reactor. Indium-resonance radiographs are made of the damaged fuel assembly at many angles and over the full length of the fuel region. The resulting radiographs are then digitized, and the data are transmitted to the EG&G-Idaho computer laboratory for reconstruction into cross-sectional views of the fuel assembly. Although basic algorithms readily available from industry sources are used for the final reconstruction process, specialized data preparation is necessary to insure high quality results. Because the digital data set includes the entire fuel section, the reconstructions can be selected from any axial location along the assembly. Neutron tomographs provide prompt, reliable information about the internal condition of these damaged fuel assemblies and minimize the need for extensive sectioning which would disturb the as-irradiated condition of the assembly. To date, cross-sections of fuel assemblies containing up to 225 fuel pins have been successfully reconstructed.

---

*Formerly employed by EG&G-Idaho, Inc.

INTRODUCTION

Neutron tomography is a three-dimensional imaging technique developed at Argonne National Laboratory's Hot Fuel Examination Facility (HFEF) for the inspection of nuclear fuel assemblies irradiated under severe test conditions. This capability is the result of a cooperative effort between Argonne National Laboratory-West and EG&G-Idaho, Inc., which are both located at the Idaho National Engineering Laboratory (INEL).

Neutron tomography consists of three basic steps: (1) neutron radiographs are made of the assembly at many angles, (2) these radiographs are then digitized to convert the film density data to digital data, and (3) the digital data set is then used to produce computer-generated cross-sectional representations of the assembly. These cross-sectional images can be generated at any desired axial location and provide early assessment of fuel assembly damage and failure mechanisms without risking displacement of the internal components by cutting. In fact, neutron tomography reduces or even eliminates the need to conduct costly and time-consuming destructive examination of the assemblies.

Although X-ray tomography is used extensively in medical and industrial applications, X-rays are not suitable for the inspection of irradiated fuel assemblies having gamma intensities of up to 10^6 R/hr. This gamma radiation would fog X-ray film immediately. Neutron tomography is insensitive to the high gamma emission of the specimen and can image the materials used in these assemblies better than X-rays. Argonne's Neutron Radiography (NRAD) Facility, which is integrated with the HFEF hot-cell complex, is ideally suited for neutron tomography of these radioactive assemblies.

FACILITIES

The NRAD facility[1] is located within a large, modern hot-cell complex equipped with state-of-the-art remote handling and examination equipment. A 250 kW TRIGA reactor is located in the basement of HFEF and provides the source of neutrons for two separate beam tubes and neutron radiography stations. One station directly interfaces with the HFEF main hot cell and the other station is located in a separate, clean, shielded room outside the hot cell. Both beam collimators are equipped with remotely variable apertures that provide L/D ratios of from 50 to 300 and from 190 to 700, respectively, where L is the distance from the collimator inlet aperture to the imaging plane and D is the diameter of the inlet aperture.

The inlet of each beam tube is within 13 mm of the fuel region of the reactor. This results in a beam energy spectrum that is well suited for neutron radiography of even highly enriched reactor fuels because of the high epithermal-to-thermal neutron ratio of the beam.

Because it is impractical to enter the radiography rooms when radioactive specimens are present, an automated foil transport system is used to remotely position and retrieve cassettes containing the imaging foils. Foils up to 43x36 cm can be accommodated. Both radiography stations are equipped with specimen elevators that lower the fuel assembly into the beam and rotate it between exposures.

NEUTRON TOMOGRAPHY TECHNIQUES

Successful neutron tomography is dependent upon being able to penetrate the fuel assembly with neutrons. Indium-resonance energy (1.4 eV) neutrons will penetrate fairly large masses of fuel and are easily detected using indium foils shielded with cadmium thermal-neutron filters. The normal indirect neutron radiography process is used; that is, the activated indium foil is removed from the neutron beam immediately following the neutron exposure and placed against X-ray film to form the latent image. The film is processed using standard procedures.

Neutron radiographs are made at many (usually about 75) radial angles by rotating the fuel assembly about its long axis and collecting radiographs at evenly spaced orientations between 0 and 180 degrees. It is not essential that the views be evenly spaced or that the full 180 degree range of views be available; however, some computer reconstruction algorithms are easier to use if these conditions are met. In particular, the filtered back projection algorithm used for most of the inspections at the NRAD facility assumes that views are evenly spaced over 180 degrees.

The mathematical reconstruction algorithms require that the axis of rotation be identifiable in each radiograph of the fuel assembly. In the NRAD facility the axis of rotation of fuel bundles is assumed to be parallel to the force of gravity. Experience indicates this assumption to be valid as long as the bundles hang freely; i.e., they do not contact any part of the facility as they rotate. The axis of rotation is also assumed to undergo only negligible lateral translation. This assumption has likewise proven to be accurate. In the NRAD facility, a scribe line has been machined in an aluminum frame which is mounted in the neutron beam such that the scribe line is parallel to the axis of rotation. Additionally, a machinist's scale is mounted to the frame next to the scribe line. The scale is essential for correlating elevations in different views of the specimen. The scribe line and scale are treated with gadolinium-oxide and are not shielded by the cadmium thermal-neutron filter; therefore they appear in every radiograph alongside the image of the fuel assembly and are used as fiducial references in the radiographs.

The film density data on each of the multi-angle radiographs must be converted to digital information before cross-sectional images can be computed. The NRAD facility uses a computer-controlled Perkin-Elmer Model 1010A scanning microdensitometer for this purpose. The calculation of a single cross-sectional image requires one linear scan acquired at the same elevation, and perpendicular to the axis of rotation, in each view. By scanning the entire surface of each radiograph, cross-sectional images can be computed at any location on the fuel assembly. A three-dimensional "density" map of the entire specimen can be calculated by generating closely spaced cross-sectional images over the entire range of elevations. Slices at any orientation through this three-dimensional distribution can then be displayed (Fig. 1 and 2). Axial, rather than transverse, slices through rows of fuel pins have also proven valuable in some fuel assembly examinations performed at the NRAD facility (Fig. 3).

Reconstruction algorithms require projection data in the form of ray integrals. The following method is used to convert the digital film density data to the required form:

The attenuation of the neutron beam along a ray passing through a specimen is:

$$A(x, y) = \frac{F(x, y)}{F_o} = \exp[-d(x, y)]$$

where,

A (x, y) is the attenuation along a ray which inter-sects the plane of the film at (x, y),

F (x, y) is the neutron flux at (x, y),

F_o is the flux incident on the specimen,

d (x, y) is the ray integral through the specimen representing the magnitude of the total neutron attenuation.

The flux F (x, y) is proportional to the exposure at (x, y) of the neutron radiograph. The exposure is determined from measured film density using a plot of density versus exposure provided by the film's manufacturer. The attenuation is then:

$$A(x, y) = \frac{X(x, y)}{X_o}$$

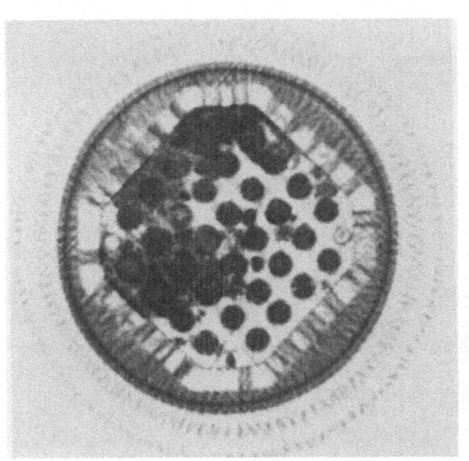

(a)

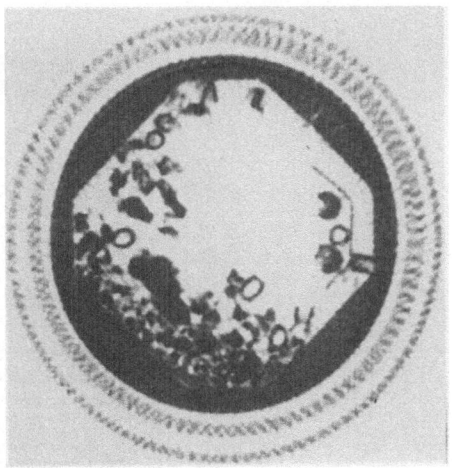

(b)

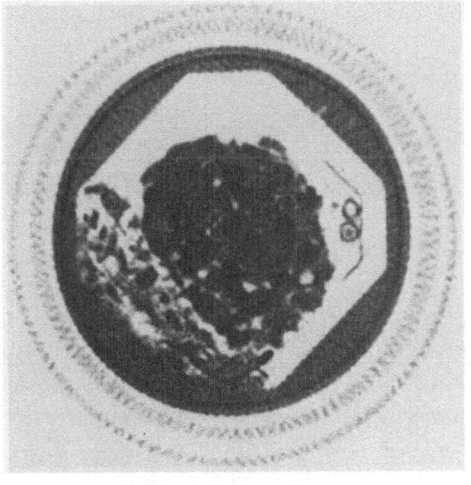

(c)

Fig. 1. This series of neutron tomographs shows three different
elevations of the same severely damaged fuel assembly following in-reactor safety tests. (a) near the bottom of
the fuel rods; intact rods visible with debris in the
cooling channels. (b) no intact rods; large void; most of
inner can and insulator region has disappeared. (c) mass
of fuel and melted structural material can be seen along
with some rubble.

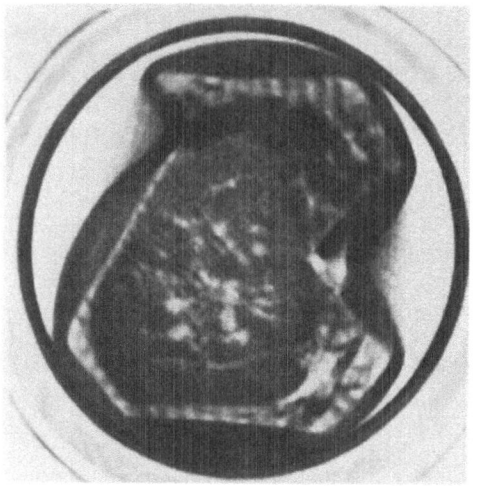

Fig. 2. Neutron tomograph of damaged fuel assembly in which the inner can has collapsed around fuel matrix remaining after the cladding has melted.

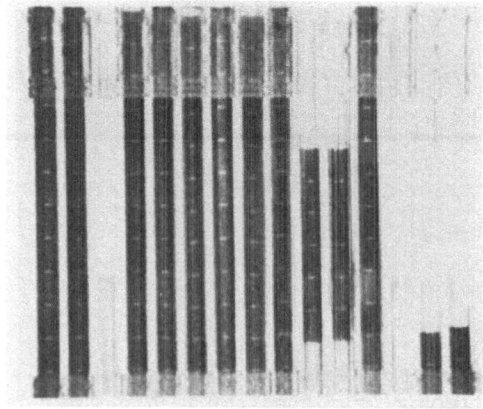

Fig. 3. This axial neutron tomograph of a mockup of a TMI-2 fuel assembly isolates a single row of fuel rods in a 15x15 array containing fully fueled rods, partially fueled rods, and empty rods.

where,

X (x, y) is the exposure at (x, y),

X_o is the exposure resulting from the unattenuated incident flux F.

The ray integral through the specimen is then:

$$d (x, y) = \ln \frac{X_o}{X (x, y)}$$

The reconstruction algorithm used at the NRAD facility is the filtered back projection algorithm.[2] After analysis of several algorithms, filtered back projection was chosen because it executes quickly, requires little computer memory, and provides superior spatial resolution. This algorithm is most applicable when a relatively large number of views (40-90) are used. A modified version of this algorithm has been used in some examinations. In the modified version, the number of views of the specimen is artificially increased by sinc function interpolation[3] between actual views. Interpolation results in smoother, less-noisy reconstructions. The resolution of the reconstruction remains essentially unchanged.

EXPERIENCE

Experimental neutron tomography was performed within a year after startup of the NRAD facility in 1978. However, neutron tomography did not become a standard capability until 1983 in response to special problems associated with the evaluation of severely damaged reactor safety experiments. Neutron tomography is a relatively expensive nondestructive evaluation technique that would probably not find general application for routine examination of fuel assemblies. However, the large expense of safety tests, and the valuable results expected, easily justify the cost of neutron tomography on damaged fuel assemblies. In these instances, the deliberately induced, simulated accident conditions have caused gross displacement of the fuel and components. Some of these features would surely be lost or disturbed during normal disassembly. Typically, features as small as 1 mm can be seen in the tomographs. To date, 12 irradiated assemblies have been evaluated using neutron tomography.

FUTURE APPLICATIONS

So far, the expense of neutron tomography has limited its application to a few well-funded projects. Because the cost of

717

neutron tomography is directly proportional to the reactor time required, anything that can reduce the incremental exposure time for each radiograph will have a significant effect on the cost. Recently, the cost of neutron tomography was reduced by nearly 50% just by using a more sensitive film. A proposed increase in the NRAD reactor power level will further reduce the cost of neutron tomography by reducing exposure times. Other improvements can be made that would make neutron tomography attractive to a wider range of applications, both nuclear and non-nuclear. These improvements might include the use of portable neutron sources, or eliminating the intermediate steps of processing and digitizing film by using a neutron detector system that would interface directly with a computer. The nondestructive evaluation of any materials or composites containing hydrogen, or other elements having a high neutron attenuation, could certainly benefit from neutron tomography if the cost can be lowered sufficiently.

REFERENCES

1. W. J. Richards and G. C. McClellan, "Hot Fuel examination Facility Neutron Radiography Reactor Design", Proc. First World Conf. on Neutron Radiography, San Diego, California, 257 (1981).

2. Avinash C. Kak, "Computerized Tomography with X-Ray Emmission and Ultrasound Sources,", Proc. IEEE, Vol. 67, No. 9, 1234 (1979).

3. Alan V. Oppenheim and Ronald W. Schafer, "Digital Signal Processing", Prentice-Hall, 29 (1975).

IMAGING WITH NEUTRONS IN TRANSMISSION TOMOGRAPHY

Kusminarto and Nicholas M. Spyrou

Department of Physics, University of Surrey

Guildford GU2 5XH, Surrey, UK

ABSTRACT

Reconstructive tomography was carried out using multiple neutron radiographs of an object as well as multiple projections of the transmitted neutron beam through the object using an He-3 proportional counter. Gamma-ray transmission tomography employing the 60 keV gamma-ray line of a well-collimated Am-241 source was also applied to the same object.

A 35mm camera was modified to incorporate a neutron converter so that sequential radiographs could be obtained without disturbing the geometrical arrangement between beam, object and recording medium. A conventional converter/film combination, in one single cassette, was also used. Results show that the technique using the collimated He-3 neutron detector provides tomographic images which are superior in terms of contrast and noise due to statistical fluctuation.

INTRODUCTION

In transmission radiography the image is a shadow of the object, usually formed on photographic emulsion or other detector placed behind the object. Tomography is concerned with the provision of three-dimensional information about the structure or the elemental distribution within the object, by formation of planes, or more accurately of thin slices, through the object, at various depths and orientations. Whatever ionising radiation is used as the probe, the resulting image is essentially dependent upon the quantity governing the energy loss (attenuation coefficient or stopping power) within the specimen to be imaged (1). Since the technique of tomography requires a computer for reconstructing the images, this has been termed Computerised Tomography (CT).

Transmission tomography has been well established in the medical field as a diagnostic technique, where the image of a section of a body under examination, employing X- and gamma-rays, is produced. More recently CT has been applied in industrial areas using gamma-rays (2,3) and neutrons (4,5,6,7,8,9,10,11,12).

The purpose of this study is to produce tomographic images using neutrons in transmission, where various modes of neutron detection and recording of data for reconstruction of images are compared and the effect of scattering on image quality is investigated.

EXPERIMENTAL METHOD

Experiments were carried out at the Institute Laue-Langevin, Grenoble, France employing a horizontal neutron beam with a thermal flux of 1.8×10^{12}n mm^{-2}s^{-1} covering an area of 12.5 mm by 20 mm. The experimental arrangement is shown in Figure 1.

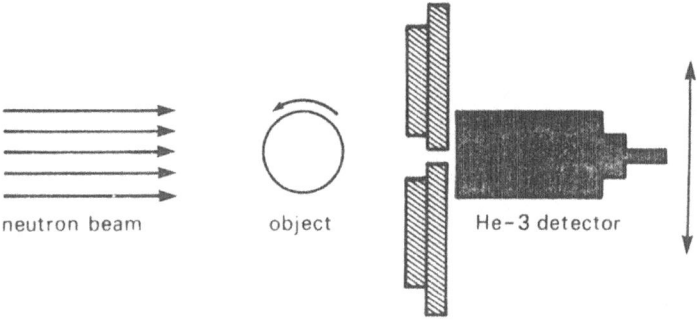

collimator

Fig. 1. Experimental arrangement for measurement of transmitted neutrons. In the case where film/converter combination was used, the collimated He-3 detector was replaced by a cassette or a 35 mm camera.

The test object, a cylinder 10 mm diameter was made of teflon. Two holes, 5mm diameter, were drilled perpendicularly to the axis of the cylinder. In one hole brass and copper rods were placed, separated at the centre by a cadmium foil and silver and indium foils fixed at each end, respectively. The other hole was filled with copper and perspex rods only. The cross section of the slice of interest is shown in Figure 2.

Projections of the test object were recorded in three ways. In the first, a well collimated He-3 proportional counter recorded the

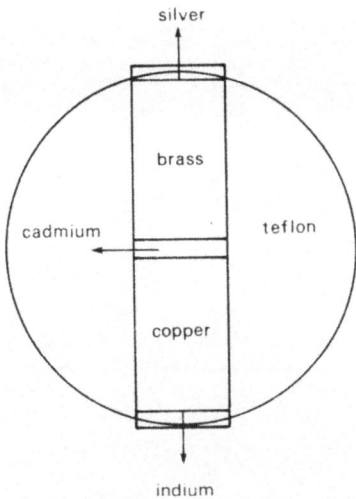

Fig. 2 The cross-section of the slice of interest of the test object.

neutrons transmitted through the object at each position and for each angle of projection for a counting time of 40 s. Fifteen projections of the section i.e. for every 12 degrees of angle of rotation were obtained, each projection being made up of nineteen raysums. The image, reconstructed using a filtered back projection algorithm, is shown in Figure 3.

In the second case, a conventional converter/film combination in a single cassette was employed. Fuji RX film and a gadolinium foil as the direct converter were used to obtain thirty projections for 10 second exposures per projection over 180 degrees of rotation of the object.

Since the single film/converter combination in the cassette technique is time consuming, a 35 mm camera was modified to incorporate a neutron converter so that sequential radiographs of the projections of the object could be obtained more rapidly without disturbing the geometrical arrangment between beam, object and recording medium. The modified camera is shown diagramatically in Figure 4. The neutron beam was incident on the back of the camera (the lens system was not used) and the gadolinium foil was placed in contact with the light sensitive part of the film. The exposed and unexposed film, not directly in the neutron beam area was shielded from background gamma-rays using lead blocks. Ilford HP5, ASA 600, 35 mm film was used and developed using a fast developer ('Microphen"). Thirty projections of the same object were obtained for a 10 second exposure per projection.

The radiographs obtained were then digitised using a computerised video camera-based microdensitometer with a spatial

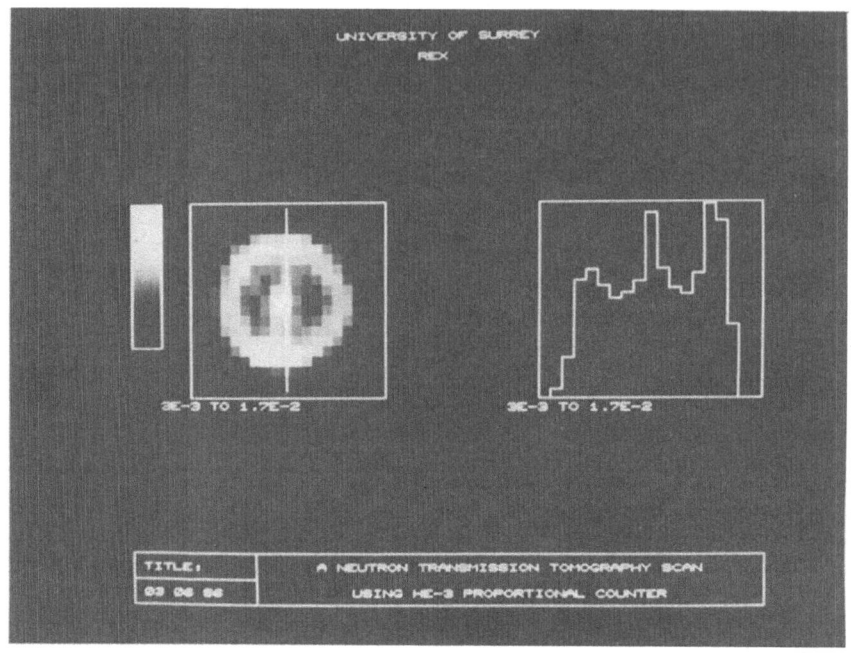

Fig. 3. The reconstructed image obtained using a He-3 detector. The right hand side represents a line scan taken from the top to the bottom of the image and shows the variation of the macroscopic cross-section along the line.

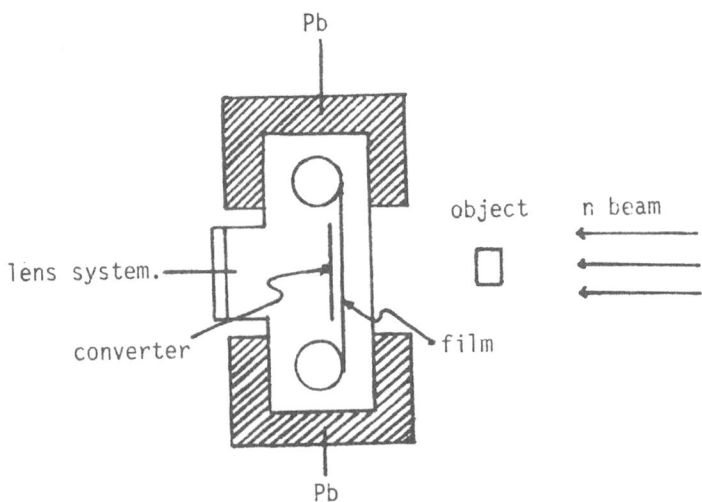

Fig. 4. The diagram of the modified 35mm camera

722

resolution of 12 pixels per millimeter and a 256 gray/colour scale. The slice of interest was then reconstructed using 44 raysums per projection. Figure 5 shows the projection of the object obtained at 0 degrees for the Fuji RX film a) and ILFORD HP5 film b) respectively.

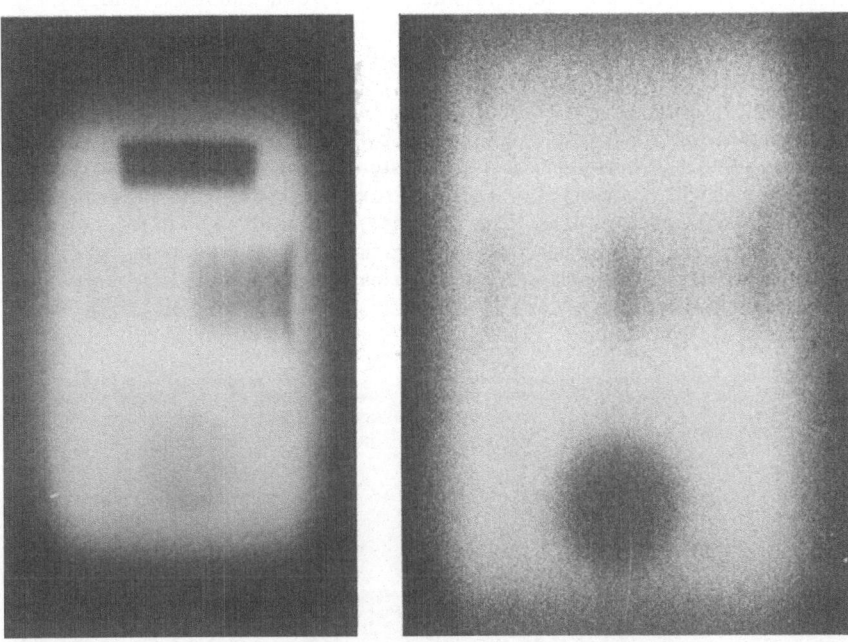

a.   b.

Fig. 5. The projection of the test object obtained at 0 degrees using (a) Fuji RX and (b) Ilford HP5 films, respectively.

Gamma-ray transmission tomography of the same object was also carried out using the 60 keV gamma line of a well collimated Am-241 source, for the same slice of interest. The transmitted photons were recorded using a NaI(TI) detector collimated to 1 mm diameter. Fifteen projections over 180 degrees were obtained comprising 19 raysums per projection.

RESULTS AND DISCUSSION

In Figure 3 the left hand side is the image representing the distribution of macroscopic cross-section in the slice obtained using the collimated He-3 proportional counter. It is displayed on a matrix of 19 by 19 pixels. The right hand side of Figure 3 represents a line scan taken from the top to the bottom of the image, and shows the variation of the macroscopic cross-section along the line. The three materials, indium, cadmium and silver can be distinguished in the image as well as in the line scan, however the values of the calculated macroscopic cross-sections obtained do not agree with the tabulated ones (13).. This is due to the partial volume effect since the foil thicknesses are 0.24, 0.1 and 0.05 mm for indium, cadmium and silver, respectively whereas the pixel size is 1 mm by 1 mm. Teflon, brass and copper could not be differentiated in the image. The system was not sensitive enough to detect differences between their total macroscopic cross-sections (0.3, 0.5 and 0.9 cm^{-1}) respectively. Furthermore, the solid angle subtended by the collimated He-3 detector was large (0.628 steradian) so that a significant fraction of scattered neutrons was also recorded. This in turn degrades the image. The effect becomes more apparent in the case where photographic emulsion is used since the converter/film combination acts as an open, bare and unshielded detector which receives information from both scattered and transmitted neutrons.

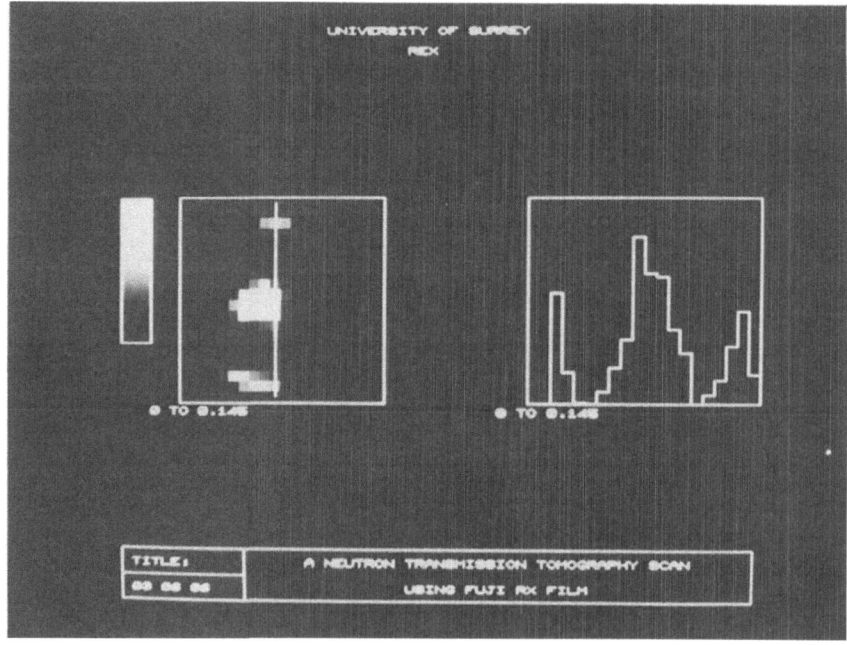

Fig. 6. The reconstructed image obtained using the FUJI RX film

The minimum number of projections, M, required to fully determine a density function i.e. the macroscopic cross-section in this case is given by (14)

$$M \stackrel{\sim}{=} n\pi/4$$

where n is the number of raysums per projection. Thirty projections were obtained experimentally using multiple radiographs, and considering the equation above, 44 raysums per projection were used for reconstruction. The reconstructed images are shown in Figure 6 and Figure 7 for the FUJI RX and ILFORD HP5 film, respectively. These images were affected by the contribution of scattered neutrons to the projection data, as expected. When the 35 mm camera was used, the reconstructed image was found to be even more degraded and blurred due to imperfect contact between film and converter.

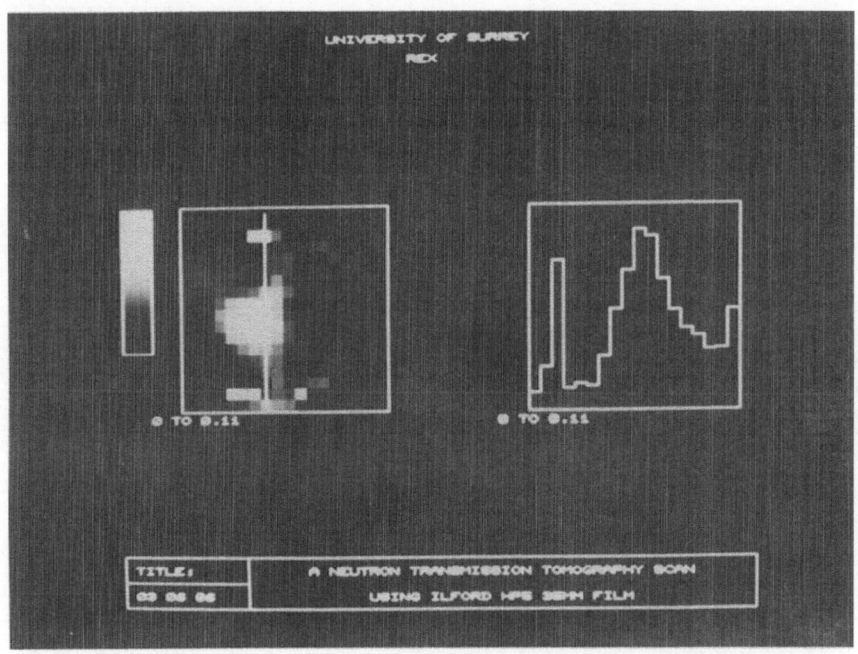

Fig. 7 The reconstructed image obtained using the ILFORD HP5 35mm film

A gamma-ray transmission tomography experiment was also performed on the same object for the same slice, for comparison. The reconstructed image is shown in Figure 8. This represents the distribution of the photon linear attenuation coefficient of the object in the slice. The result does not show the structure of the object. This can be explained as follows. The minimum detectable length l of a material with a linear attentuation coefficient μ_c in a matrix with alinear attenuation coefficient μ_r of length L is determined by equation (1).

$$\frac{\ell}{L} = \frac{\mu_r}{\mu_c - \mu_r} \, f$$

where f is the expected contrast or the expected fractional change in the raysum. In this experiment, the minimum detectable length of cadmium, indium and silver was found to be for all three elements ~ 2 mm for a ten percent change in the linear attenuation coefficient of the teflon matrix (i.e. f = 0.1) whereas the thicknesses of the foils used were considerably less than the minimum detectable length found.

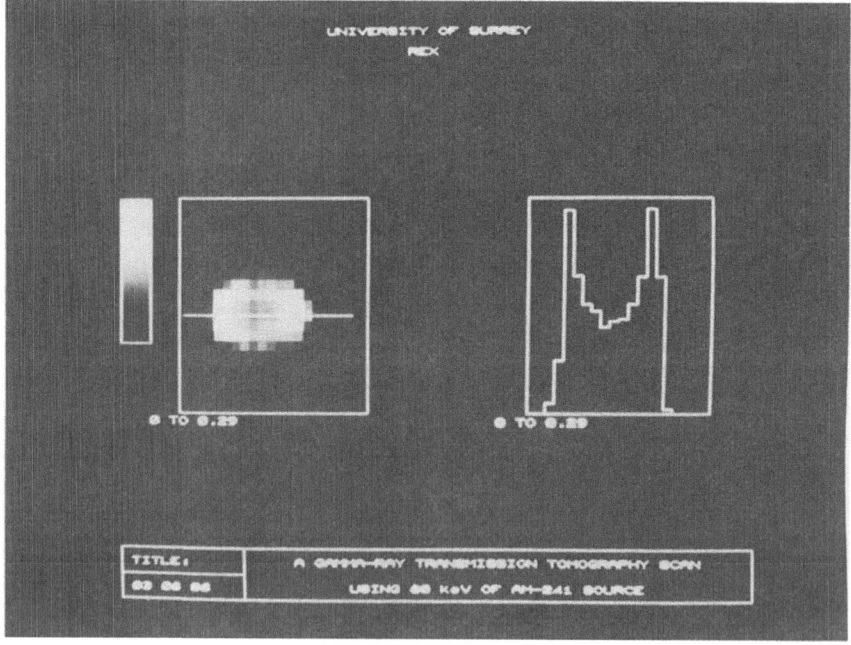

Fig. 8. A gamma-ray transmission tomography image of the same object using the 60 keV line of Am-241 source.

CONCLUSIONS

The results have shown that the collimated He-3 detector gives the best image. The advantages of using a well collimated He-3 detector are that the contribution of scattered neutrons can be suppressed and the statistical error can be controlled by adjusting the counting time but only one slice of interest can be examined at a time unless an array of detectors is employed. On the other hand the converter/film combination records the projection of the whole object at a time and subsequently several slices can be examined simultaneously. However it does suffer from statistical limitations and is not accurate for imaging objects which contain a significant amount of scattering material.

Redesign of the 35mm camera should give better contact between converter and film thus diminishing scattering of the emitted radiation and then the system would be preferable to the single cassette converter/film combination.

ACKNOWLEDGEMENTS

The authors would like to thank David Munro of the Department of Physics for setting-up the digitisation system. The Science and Engineering Research Council UK and the ILL-Grenoble have made these studies possible through financial support and the Indonesian government has provided a research scholarship to one of the authors (K).

REFERENCES

1. K. Kouris, N.M. Spyrou, D.F. Jackson, "Imaging with Ionizing Radiations" , Surrey Univ. Press/Blackie & Son, Glasgow, 1982

2. W.B. Gilboy et al, Nucl. Instr. Meth. 193 (1982) 209-214

3. N. McCuaig et al, Applied Optics, 24 (1985) 4083

4. C.F. Barton, Trans. Am. Nucl. Soc., 28, (1977) 212-213

5. G.D. Zakaib et al, Trans. Am. Nucl. Soc., 65, (1978) 145-154.

6. G.A. Schlapper et al, Materials Evaluation, 39 (1981), 1121-1125

7. R.A. Keoppe et al, J. Comp. Assist. Tomography, 5, 1 (1981) 79-88

8. W.J. Richards et al, Material Evaluation, 40, (1982), 1263-1267.

9. A de Volpi and E.A. Rhodes., Material Evaluation, 40, (1982), 1273-1279

10. J.C. Overley, IEEE Trans. Nucl. Sci., NS-30, (1983) 1677-1679.

11. D.M. Tow, Postirradiation exam using Neutron Tomography, in Barton J.P., and von der Hardt, P., Editors, Neutron Radiography, ECSC, EEC, EAEC, Brussels & Luxembourg, (1983), 425-435

12. G. Matsumoto and S. Krata, The Neutron Computer Tomography, in J.P. Barton and P. von der Hardt, Editors, Neutron Radiography, ECSC, EEC, EAEC, Brussels & Luxembourg (1983) 899-906.

13. P. von der Hardt and H. Röttger, Neutron Radiography Handbook, D. Reidel Publishing Co. 1981.

14. R.A. Brooks, G. Di Chiro, Phys. Med. Biol. 21 (1976) 689-732.

A COMPARISON BETWEEN COLD NEUTRON RADIOGRAPHY AND POSITRON EMISSION TOMOGRAPHY FOR LIQUID LOCATION

J. Walker and M.R. Hawkesworth

Department of Physics,

University of Birmingham, Birmingham B15 2TT.

ABSTRACT

In collaboration with industrial colleagues, we are studying the value of positron emission tomography in engineering. Our equipment is described briefly to illustrate the main features of the technique which is being used to trace fluids in various assemblies; there is currently a particular interest in lubricants. A number of applications are presented, and in the case of a model journal bearing, a direct comparison is made between positron tomography and cold neutron radiography. In a more general comparison, the differences and similarities arising from the basic properties of the radiations, sources and detectors, spatial resolutions, the extraction of three-dimensional information, and the inclusion of time dependence are all mentioned.

PRINCIPLE OF POSITRON EMISSION TOMOGRAPHY (PET)

The principle of PET is shown in figure 1. A positron emitter in suitable chemical form is used to label the fluid of interest in an engineering subject such as an engine or a bearing. The positrons are slowed down quickly in a short distance in the fluid and are then annihilated on encountering electrons. In each annihilation two 511keV photons are emitted in exactly opposite directions and detected simultaneously in two detectors, one on each side of the subject; simultaneity is the first measured indicator of a positron annihilation. Each detector registers the coordinates of a photon interaction it (figure 2) and the data system computes the collinear flight paths. The intersections of many such paths provide the spatial distribution of the positron annihilations and hence of the host fluid. Images can be viewed by selecting slices in the object/image space; figure 1 illustrates the concentration of intersections appearing in one slice from a point source while diffuse (defocused) information appears in other slices. The positron source can be mixed uniformly in the total volume of fluid for continuous observation or injected in bursts for time-dependent measurements. Time dependence can also be obtained by appropriate gating of detector signals; stroboscopic observation is an example.

RADIATION SOURCES AND DETECTORS

A positron emitter for engineering use needs a chemical form which is compatible with the fluids involved, an activity

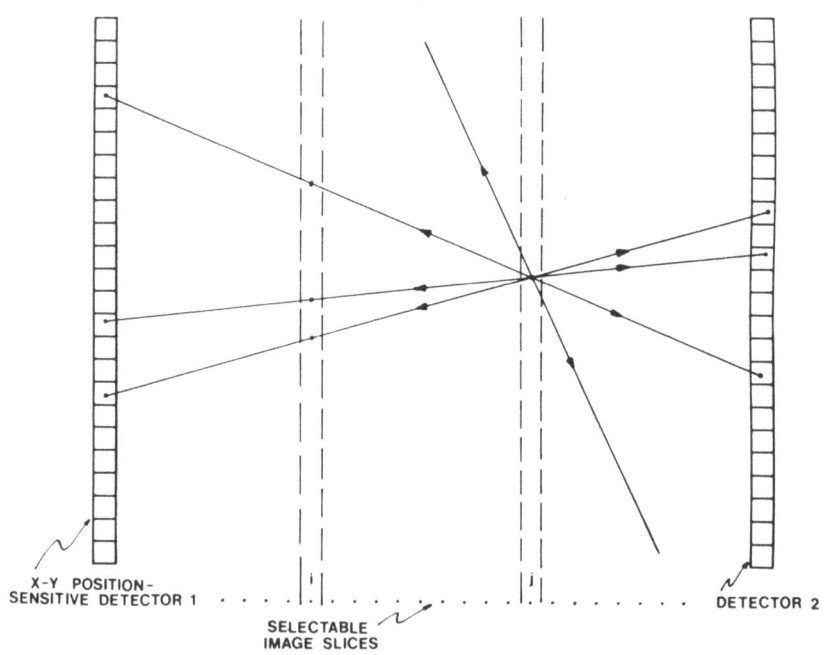

Figure 1 Illustration of tomograph construction with annihilation photons from a source of positrons. The tomograph of subject slice "i" shows only a wide-spread background whilst that of slice "j" shows the point source in good focus.

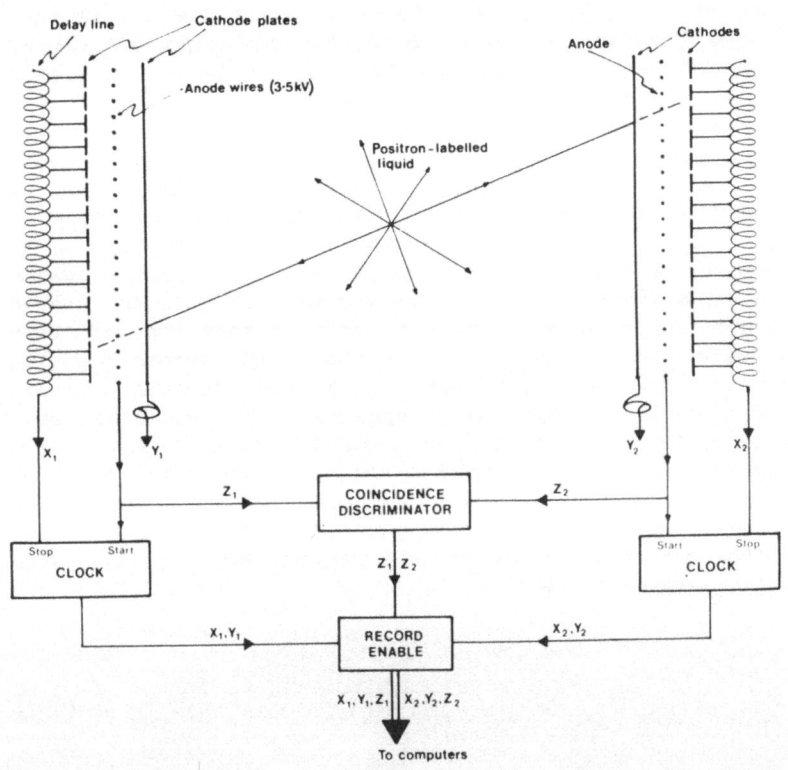

Figure 2 Illustration of the delay line method of coordinate readout used for the Birmingham positron camera. It should be noted that the "X" and "Y" cathodes and readout delay lines are identical in construction. The detectors were constructed by the SERC's Rutherford Appleton Laboratory[3].

sufficient to give good images in an acceptable time, but a half life long enough for satisfactory measurements to be made and short enough to avoid persistent radioactivity in the test objects[1]. Gallium—68 (half life: 68 minutes) is used for our work on lubricants; it is eluted as required from a generator carrying the parent nuclide germanium—68 (half life: 287 days). The generator technique, which can also be used with other nuclides, is well known in nuclear medicine and provides a simple form of positron source. The variety of positron emitters can be increased if a cyclotron or other accelerator is located near enough to the testing centre to allow the direct production and transport of the positron emitters.

Each detector (figure 3) is a multiwire proportional

731

counter with a sensitive area of 0.6m x 0.3m and 20 sensitive
planes to give a detection efficiency of about 11% for photons
of 511 keV. The overall efficiency for detecting a pair of
annihilation photons from a source between the detectors is thus
about 1% multiplied by an appropriate geometrical factor.   In
practice, the effective geometrical factor depends on whether
the data processing includes steps to equalise the response over
a large fraction of the sensitive area.   In addition to the
condition that a photon event in one detector must have an
associated event within 10ns in the other to be treated as
simultaneous, a number of additional tests are applied which
will not be elaborated in this short paper but will be covered
elsewhere[2,3]. It is sufficient to mention that the effective
maximum rate for recording data is about 3kHz before applying
response equalisation.   This rate would arise from an activity
of about 100µCi (3.7 MBq) in a specimen with negligible self
shielding 0.25m from each detector window, and an exposure time
of 500s would then give 1.5 million recorded positron
annihilations, sufficient for a good image. Actual test
specimens may well need an activity of 10—100 mCi (0.37-3.7 GBq)
between the detectors to provide the maximum data rate and hence
the minimum exposure time.

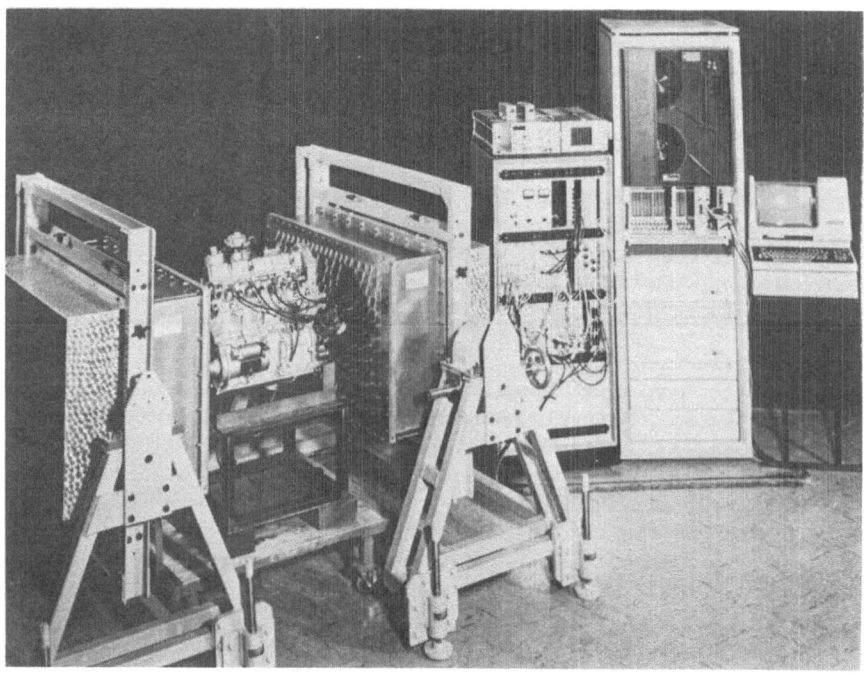

Figure 3   Photograph of the complete Birmingham positron camera
ready for measurements on a small (1000cc) automotive engine.

A COMPARISON WITH COLD NEUTRON RADIOGRAPHY AND OTHER EXAMPLES

The model journal bearing shown in figure 4 has been viewed by both PET and cold neutron radiography. Figure 5 is the two-dimensional radiograph obtained with a cold neutron beam at the DIDO reactor, Harwell; obviously, a second view would be necessary for three dimensional information, and is not always possible although it is in this case. The positron tomographs contain the full three dimensional information which can be viewed in any suitable way; in figure 6, 32 consecutive slices have been used as indicated in figure 4 and the activity in the central slice is shown in figure 8. The histogram shows not only the image width of a few millimetres (6mm FWHM) for a thin line source but also the much finer resolution made possible by using line intensities. The areas under the peaks on the histogram give an accurate comparison of oil film thicknesses. With the neutron radiograph a travelling micro-densitometer is used to provide the corresponding information.

Figure 9 shows how the interpretation of complex images is helped by the use of computer software to model the actual structure of the subject and to superimpose accurately the corresponding sectional drawing on any tomograph. This approach is also valuable in correcting observed image intensities for photon scattering and absorption by the materials traversed.

COMPARISON OF NEUTRONS AND POSITRONS IN NON-DESTRUCTIVE TESTING

Properties of Radiations and their Implications

Neutrons: The main features are penetration and a strong variation of nuclear interaction rates with neutron energy and isotopic composition of the subject. It is well known that these features give prominence to hydrogenous materials behind heavy ones and to the observation of strong neutron absorbers. **Positrons:** The annihilation radiation is also penetrating, and there is no restriction on materials to be penetrated or on the material to be visualized provided labelling with a positron emitter can be accomplished without affecting the equipment under test or the flow characteristics.

Radiation Sources

Neutrons: Low energies are most useful and moderation of neutrons from nuclear reactions is therefore required. The strongest sources from reactors or accelerators are not portable at present but a new generation of accelerators will improve the position.

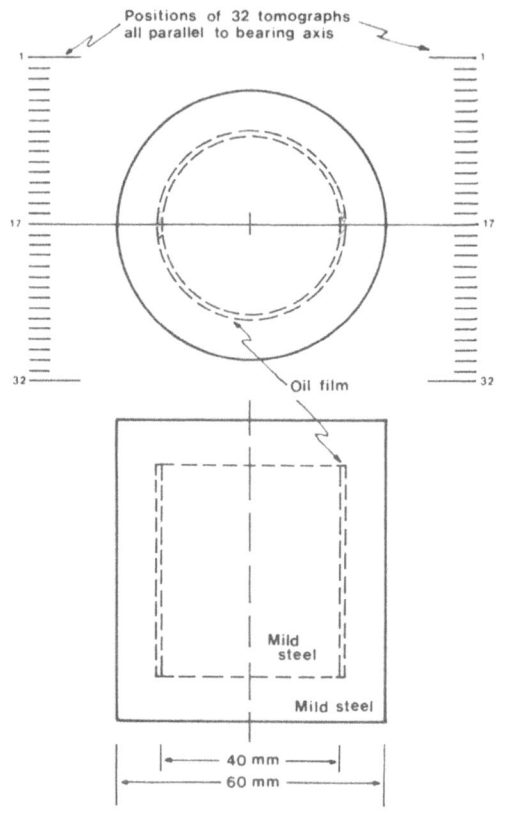

Figure 4 Drawing of a model journal bearing holding a cylindrical film of oil. The positions of the 32 tomographs constructed parallel to the bearing axis (figs. 6 and 7) and the volume of oil represented by tomograph 17 are indicated. Note the oil film thickness, which is a uniform 125μm when the journal is central is not shown to scale.

Figure 5 Cold neutron radiograph (positive print) of the model bearing illustrated in figure 4. The bearing surfaces are shown with an incomplete oil film to illustrate the capability of cold neutron radiography to delineate a 125μ film of oil through a full 60mm of steel in addition to showing it clearly, as expected, when seen tangentially. Note the edges of the outer steel cylinder are undercut and do not show on this print. (Exposure 1.5×10^{10}n cm^{-2}, 25μm Gd foil, Kodak CX film).

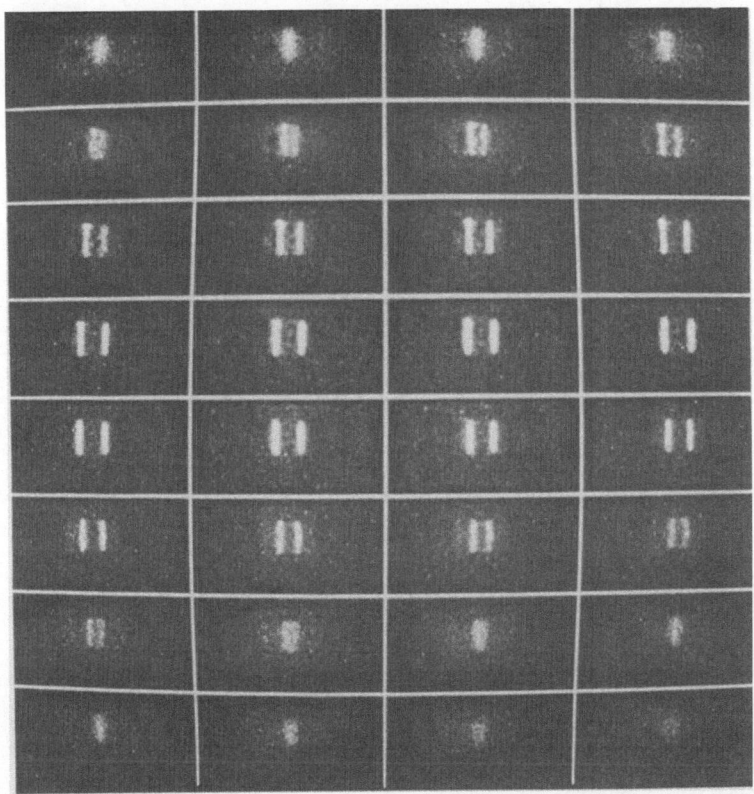

Figure 6 Array of positron tomographs of oil in 32 consecutive slices through the model bearing illustrated in fig. 4. In this case the journal (shaft) was almost central in the bearing.

Positrons: Cameras are movable and Portable sources of adequate intensity are available through the use of isotope generators. New accelerators will provide portability with a greater variety of lifetimes and source materials.

Radiation Detectors

Neutrons: Detection needs converters to produce photons or charged particles; they are readily available and the detection of the secondary radiation can be very simple with photographic film or plastic foils. More complex electrical detectors involving position sensitivity can also be used.

Positrons: Detected through annihilation radiation. An array of scintillation detectors is the main type in nuclear medicine.

Figure 7 Histogram across the tomograph through the central slice (tomograph 17, figure 6) of the bearing illustrated in figure 5. Clearly the journal is slightly off centre and, knowing the total bearing clearance is 250μm, it can be deduced that the thicker film is 137μm thick and the thinner 113μm.

The camera based on a pair of multi-wire proportional chambers discussed in this paper is simpler and less expensive.

Two or Three Dimensional Information and Spatial Resolution

Neutrons: At present the main uses need only two-dimensional information in each exposure but tomography and, more recently, holography have been used to provide three dimensional information. Spatial resolution depends on the converter and detector; it can be as good as about 20μm.

Positrons: Emission tomography inherently provides three-dimensional information which can be viewed in any convenient way. A choice of two-dimensional sections can obviously be made. The direct spatial resolution for the type of equipment described here is a few millimetres but much higher resolution, similar to the best in neutron radiography is possible when image intensities are used (figure 7).

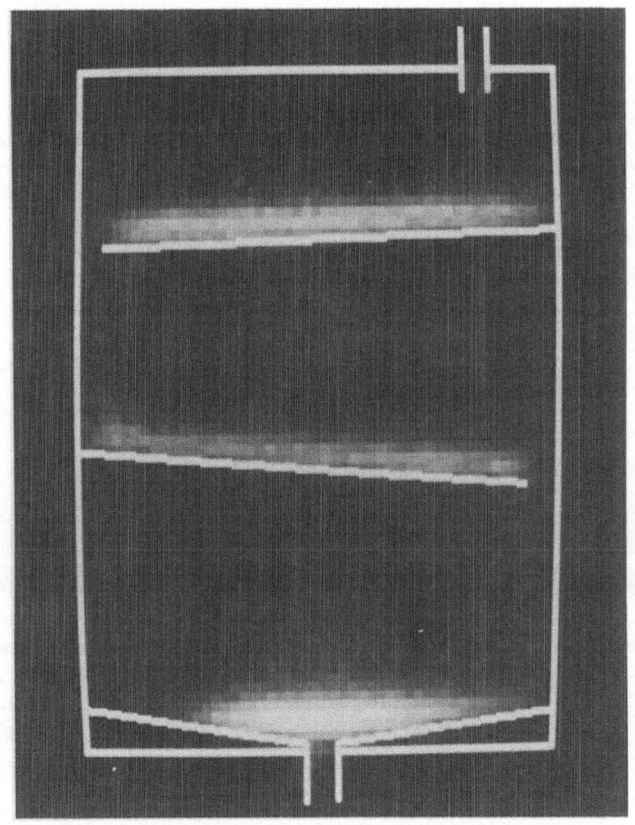

Figure 8 Positron tomograph showing oil flowing down apparatus containing inclined planes. The upper plane is nearly horizontal and clearly the oil film is thicker here than that on the middle plane which has a greater inclination. This figure also shows the value of overlaying an accurate appropriately scaled drawing of the subject on the positron image; normally the overlay is presented in a contrasting colour to avoid confusion with the positron image. The tomograph and overlay represent the central slice of the tower which is 160mm high and 120mm wide.

Time—dependent Applications

Neutrons: Dynamic neutron radiography involving scintillators and some form of closed—circuit television is now well-known but requires intense neutron beams.

Positrons: Pulsed injection of positron emitters into fluids and/or the gating of detector signals can be used.

Measurements in Strong Gamma Fields

Neutrons: Can be used by employing the transfer technique.
Positrons: Cannot be used.

Production Radioactivity in Subjects

Neutrons: Induced activity depends on materials used and can be a limitation.
Positrons: Activity has to be injected but its life-time can be short to avoid any problems.

ACKNOWLEDGEMENTS

We are pleased to acknowledge contributions by our colleagues at Rolls-Royce, Castrol and the Rutherford Appleton Laboratory, and financial support from the Science and Engineering Research Council and our industrial collaborators. We are grateful to the Atomic Energy Research Establishment, Harwell for the use of the DIDO cold neutron beam to obtain figure 5.

REFERENCES

1. J. Heritage et al, An imaging system for the radioactive tracing of lubricants in automotive components. Inst. Mech. Engineers (London) Conf. Series C71/85, 111-119.

2. M.R. Hawkesworth et al. A positron camera for industrial application. Submitted for publication in Nucl. Inst. and Meth.

3. J.E. Bateman et al. Development of the Rutherford Laboratory MWPC positron camera. Nucl. Inst. and Meth. **176** (1980) 83-88.

NEUTRON COMPUTED TOMOGRAPHY USING THE NEUTRON TELEVISION SYSTEM

Kenji Yoneda, Shigenori Fujine and Keiji Kanda

Reseach Reactor Institute, Kyoto University

Kumatori-cho, Sennan-gun, Osaka 590-04, Japan

ABSTRACT

At the Kyoto University Reactor(KUR), it is possible to apply the neutron television(NTV) to the neutron computed tomography(NCT).
The characteristics of the system can be acquired in a single measurement and simultaneously the projection image can be observed on a CRT monitor. The data was sampled from an optional horizontal scanning line at the video image. It was calculated by an on-line personal computer in a short time.
The Fourier-convolution technique is used to produce the reconstructed image and its image has a good enough quality for revealing water in a small hole of 1.5 mm in diameter.

INTRODUCTION

The KUR real-time neutron radiography system has been recently developed (1) and used for several practical applications as described in the previous papers (2,3) and presented in this conference (JP2 and JP3).
Recently the NTV system has been graded up by adopting some useful scintillators and an image integrated system to improve the image quality.
As another application, NCT using a NTV system has been made using the reconstructed image by the personal computer. This system is made only by adding a stepper motor control system to turn the sample by 180°.

NCT SYSTEM

The KUR-NCT system consists of the following four units as shown in Figs. 1 and 2: (i) a neutron TV system, (ii) a video image processing system, (iii) a personal computer system, and (iv) a stepper motor control and drive unit.

A 16-bit high performance personal computer (NEC PC-9801) system was introduced to the video image processing system. The personal computer system is provided for the arithmetic calculation of data in CT applications and receives the image data through a GP-IB interface from the video image system.

A stepper motor control and drive unit was operated by manual or computer controller as shown in Fig. 3. The stepper motor turned a sample table at an angle of 0.9 degrees/pulse.

In NCT using the NTV system, the transverse motion, which is commonly used in an X-ray CT, is not required, and the drive system can be made simple.

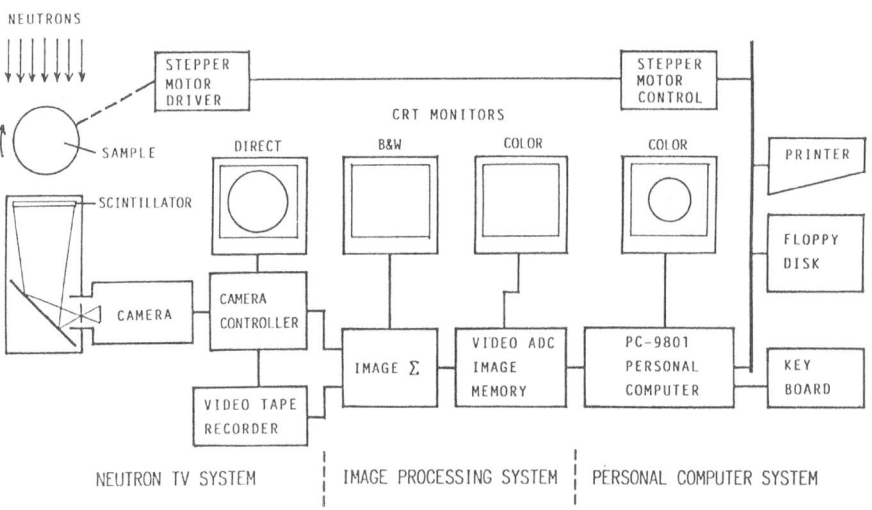

Fig. 1. KUR neutron television fluoroscopic system block diagram.

Fig. 2. Photo of KUR neutron TV system.

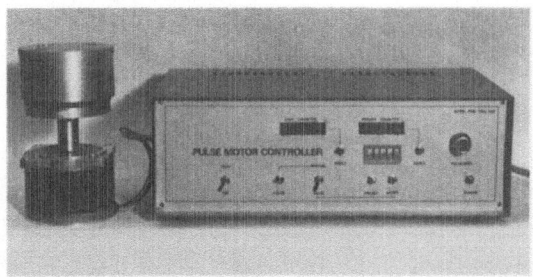

Fig. 3. Photo of the stepper motor control and drive unit.

SAMPLE

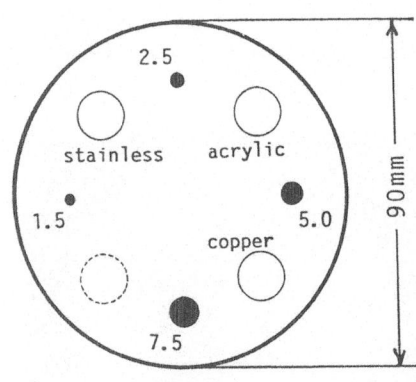

Fig. 4. A sample for NCT application.

A sample was made of an aluminum diecasting column of 9 cm in diameter and 4.5 cm in height. There were four small holes of 1.5, 2.5, 5.0 and 7.5 mm in daiameter and four large holes of 12 mm in daiameter as shown in Fig. 4.

Four small holes are filled up with water or water solution of gadolinum chloride, and large holes are filled with sticks of acrylic resins, stainless steel or copper, respectively.

For projection data, the image from the TV system was in low-contrast and mottled, because each 1/30 sec frame was divided into a 240 x 256 array of pixels in the NTV system. Therefore the digitized data cannot be used directly as the projection data as shown in Fig. 5(i). About 400 frames of data were summed to make an available image as shown in Fig. 5(ii). Figure 6 is shown as a degitized image for projection image and as a position of the projection data by dotted line.

In NCT using the KUR-NTVsystem, the projection data were collected in transverse increments of one pixel size, which is about 0.75-0.8 mm, with 9° rotation between projections similar to the Rotate/Rotate method. Twenty one projections covering 180° were taken to reconstruct the axial image.

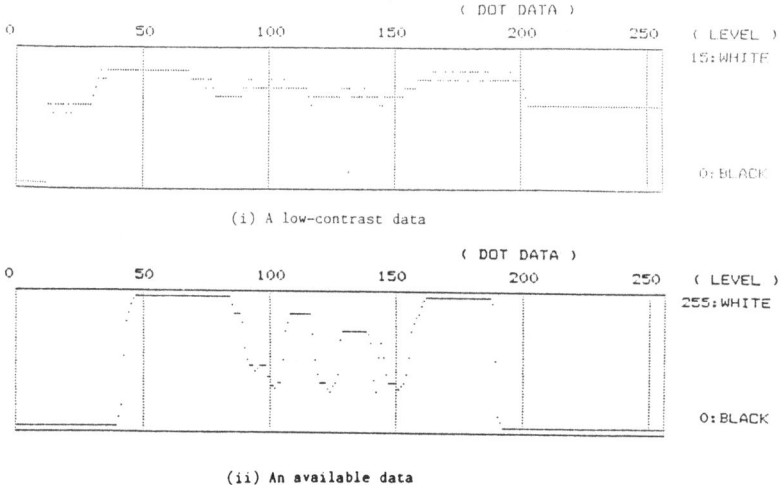

(i) A low-contrast data

(ii) An available data

Fig. 5. An example of projection data. (i) A low-contrast data from one image. (ii) An available data from 400 images.

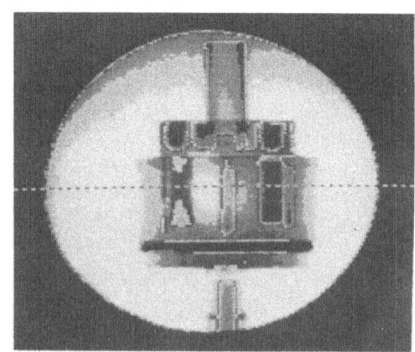

Fig. 6. Photo of the digitized
image and display
position.

For image reconstraction, the digital processing was executed with the language BASIC using a personal computer. The Fourier-convolution technique was applied to obtain a reconstructed image by using the Shepp and Logan method as the weighting function.

The reconstruction images are shown in Figs. 7 and 8. In Fig. 7, CT image is illustrated with a dot imaging technique for computer printer display with 16 gray levels. In Fig. 8 four images with 87 x 87 grids using different constants are displayed on a color CRT.

RESULT AND DISCUSSION

We have developed the NCT using the NTV system, which has the following distinct characteristics:

(1) Projection data can be acquired in a single measurement.
(2) Projection image can be observed on CRT during measurment.
(3) The transverse movement of the sample is unnecessary, therefore the rotating control and drive unit is simplified.
(4) The transverse increment is fixed and limited by the pixel-size.
(5) The artifacts depending on distortion of horizontal linearity in video image signals, make a reconstruction image quality inferior.

As to reconstracted images, Figs. 7 and 8 show a comparison of four examples to detect small holes in this sample.

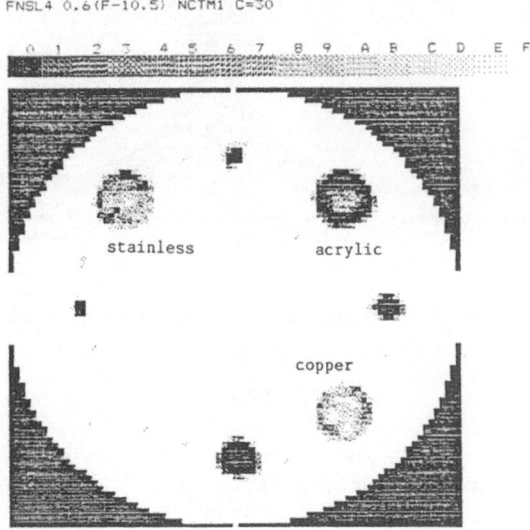

Fig. 7. A display of 81 x 81 image with 16 gray levels by printer.

The density difference between stainless steel and copper
is not so distinct, but Fig. 7 shows that a reconstructed image
with high quality which can differentiate these materials. The
ability to differentiate materials on NCT is enough to detect
water in a 1.5 mm diameter hole in a 9 cm diameter aluminum
diecasting column.

The processing time is approximately ten minutes to make an
87 x 87 image with BASIC program processing, however a machine
language software or an addition of the new type image processor
will provide much faster arithmetic operations.

The use of the minicomputer and the improvement of the
software will make possible to obtain a faster processing and
higher quality.

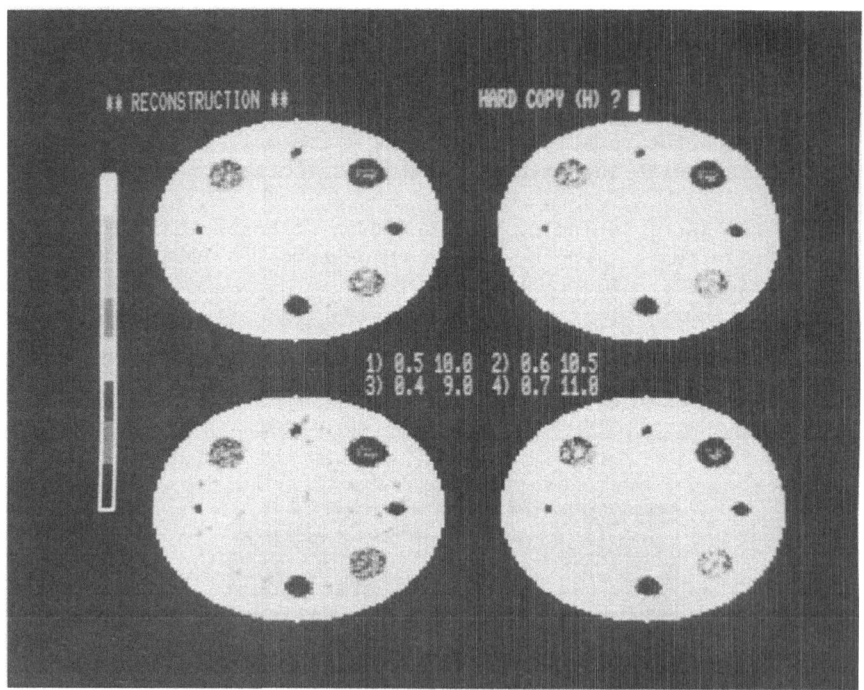

Fig. 8. Four reconstructed images using different constants on a
CRT. An 87 x 87 image is displayed with 8 levels of color.

REFERENCES

1. K. Kanda, K. Yoneda and S, Fujine, Neutron Radiography, (D.
   Reidel, Dordrecht, 1983) p.219.
2. S. Fujine, K. Yoneda and K. Kanda, Nucl. Instr. and Meth.
   215 (1983) 277.
3. S. Fujine, K. Yoneda and K. Kanda, Nucl. Instr. and Meth.
   226 (1984) 475.

IMPROVEMENT OF RECONSTRUCTED IMAGE QUALITY

IN NEUTRON CT

G. Matsumoto, T. Sendo, S. Honda, K. Ohkubo and Y. Ikeda

Department of Nuclear Engineering
Faculty of Engineering, Nagoya University

Furo-cho, Chikusa-ku, Nagoya 464 Japan

ABSTRACT

Improvement of Neutron CT technique were tried with an automatic CT camera and an image input equipment attached to a large scale computer. Neutron CT using Neutron TV was also carried out by a high speed image processor. The time needed for inputting optical density data of films for computational reconstruction of CT images was greatly reduced by 1/40 compared to the previous method done by using a microdensitometer. The image quality was improved appreciably. By using Neutron TV, the process time for CT was further shortened, and the obtained images were fairly good.

INTRODUCTION

This work is the extension of the one presented to the First World Conference on Neutron Radiography. To expand the application field of the Neutron CT (NCT), some improvements had been made in following directions,
1. reduction of time needed for imaging and computational processing
2. grade-up of CT-reconstructed image quality.
An automatic camera for NCT was designed and fabricated in this

laboratory to use roll films. Beforehand, a microdensitometer had been utilized to obtain the projection data from NR films, but at the present time, a visicon-type image input equipment at the large scale computer of Nagoya University was employed. The data acquisition time for NCT was much shortened. For this reason, the number of the projection angle for CT was able to increase up to 200 without difficulty. The increased projection data were able to contribute much to the grade-up of CT reconstructed image quality.

In the case of NCT with neutron television, the projected TV images for each angle were memorized in video-tapes, and were image-processed and digitized with a high speed image processor, TOSPIX-2. In this case, on-line processing for NCT could be potentially possible, but now the experiment was restricted to off-line processing. Although the resolution of TV images are generally not as good as those by films, the obtained image quality with the CT was fairly good because of image processing.

In both cases, mainly a convolution integral algorism was used for reconstruction of NCT images.

FACILITIES

The neutron exposure were done with the KUR E-2 NR Facility. Its characteristics are shown in Table 1.

Table 1   KUR E-2 Facility

Neutron beam (n/cm^2.s)	1.2×10^6
Cd ratio	400
n/γ ratio (n/cm^2.mR)	1.0×10^6
L/D ratio	100
Beam dimension	160 mm dia.

The main feature of the automatic camera for neutron CT is shown in Fig.1. In brief description, the camera is made from four components, that is, the neutron exposure part being made of a Gd metal foil converter (may be used with any others), a roll films and its carrier with a film length counter, radiation shielding for the film and its container painted partially with boron-containing paint. The converter is movable with the action

746

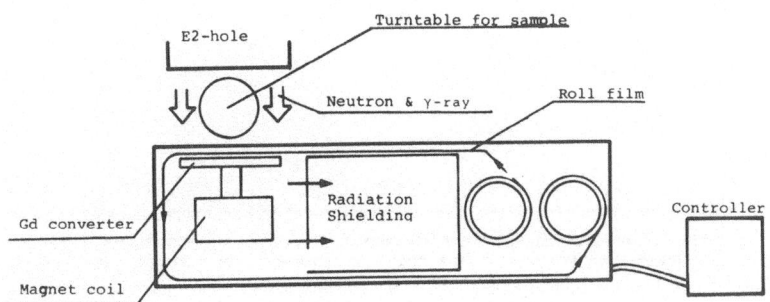

Fig.1 Automatic neutron CT camera

of two springs, being attached to the film when exposured, and being apart from it in the time of film winding. The sequence of neutron exposures and film windings is automatically controlled with a control unit which is placed apart from the NR exposure position. Therefore, the radiation exposure and labor of workers are extensively reduced. In front of the camera, the CT objects to be tested are placed on a turn-table moving by means of a stepping motor operated with the same camera controller. By using this controller, the exposure time, the film winding length and the number of projection angles are set freely.

The optical densities of the NR films were measured with the image input equipment of the Nagoya University Computer Center (the Nac Co. DVP-MTI) and inputted into the FACOM M-382 large scale computer of it. Fig.2 is the photograph of the image input

Fig.2

Image input equipment of Nagoya University Computer Center

equipment, which uses a visicon camera for sensing the optical density of the films. Table 2 shows the main features of it.

Table 2 Nac DVP-MTI image input equipment

Number of pixcels	480x640
Time for measurement	1 min
Range of density measurement	0.00-2.55
Minimum field of measurement	35 mm film

The computerized reconstruction of NCT images was carried out with the FACOM-M382. The reconstruction algorisms studied in this laboratory are the convolution integral, the FFT and the filtered back projection method. The CPU time and other characteristics of them are investigated.

In the case of NTV CT, the NTV equipment of the KUR E-2 was utilized, which is an image orthicon with a Gd oxy-sulphide converter and a bright optical lens of F0.7. The image output of the camera was memorized in a conventional video-tape recorder to transmit the images to the high speed image processor, TOSPIX-2 at the research center of Chubu Electric Co.. The high speed image processing unit bilt in and an auxiliary memory of 90 Mbite helped the CT computation greatly.

EXPERIMENTS

The experimental conditions at KUR E-2 are as follows.

Table 3 Experimental conditions at KUR E-2

No. of projection	200
Interval of projection angles	0.9 deg. for 180 full angle
Film	Fuji Softex HS or MI-FX
Converter	Gd metal foil (25 m thick)
Exposure time	4 min. for HS

Newly, the roll-type film, Fuji MI-FX, was used, the suitable exposure time being 7 min. When the combination of HS and Gd converter was used, the resulting resolving power of the NR image on the film was nearly 70 μm and that for the FX and Gd combination was not so inferior to the former. By means of the newly developed automatic CT camera, the imaging time needed for the latter case was allowable.

The measured optical density by the image input equipment was converted to the integlated total cross section of the neutron path through the sample by using experimental data for the relationship of density vs neutron exposure obtained by a 15 stage Cd step-edge.

When the convolution algoris was employed for CT reconstruction, the filter function of the Shepp and Logan was used. The parameters for reconstruction are as follows,

Table 4 Parameters for reconstruction of CT images

Data sampling number per one projection angle	420
Interval of sampling data (μm)	120
Number of pixcels of reconstructed image	400x400
Distance between pixcels of reconstructed image (μm)	125

In the case of NTV CT, the procedure and the time needed for 100 projections in TOSPIX-2 were as follows,
1) Record the modified pattern of the neutron beam through the sample imaged by the TV camera in the video-tape recorder (40 min)
2) Input the projection data into the TOSPIX-2 (35min),
3) Obtain a CT image with the computer (order of min. depending on the scales of reconstructed CT images).

The processing time for imaging and data inputting of the NTV CT was only about 1/40 of that of the former NR CT, though the resulting CT images were slightly inferior to those by NR CT. In this case, the CT processing was carried out by off-line. The necessary CPU time for TOSPIX-2 was still rather longer compared with that for a large scale computer. If the speed of the CPU of TOSPIX-2 could be much high, the on-line CT processing would be possible and much attractive.

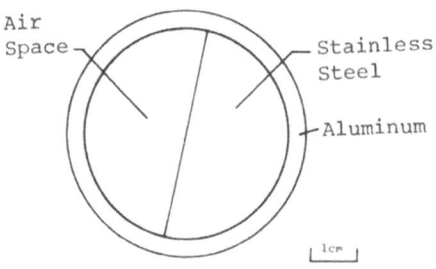

Fig.3 Sharp-edge test sample A

Fig.6 NCT of sample A by film
method

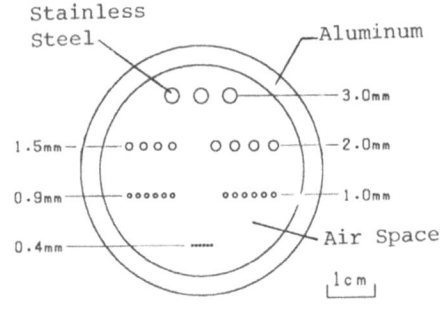

Fig.4 Resolution test sample B

Fig.7 NCT of sample B by film
method

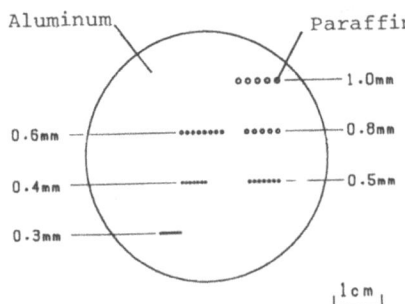

Fig.5 Resolution test sample C

Fig.8 NCT of sample C by film
method

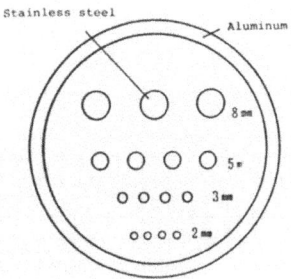

Fig.9 Resolution test sample D

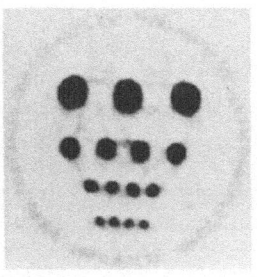

Fig.12 NCT of sample D by TV
method

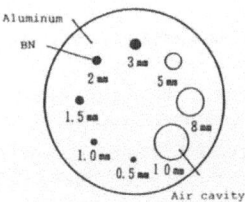

Fig.10 Contrast and resolution
test sample E

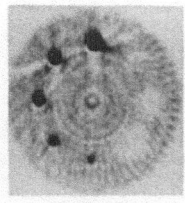

Fig.13 NCT of sample E by TV
method

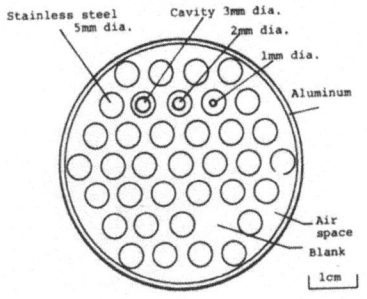

Fig.11 Regular array sample F

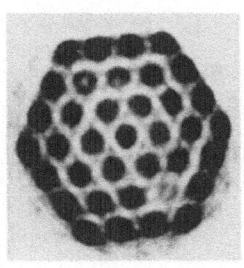

Fig.14 NCT of sample F by TV
method

RESULTS AND DISCUSSION

Various samples were tested. For NR CT, Figs.3 to 5 shows the drawings of them. The reconstructed CT images are shown in Figs.6 to 8. The spatial resolution power of the CT images are given in Table 5.

Table 5 Spatial resolution power of the CT images by NR CT

Object	Measurement item	Value (μm)
Air-SUS	FWHM of line spread function	130
Air-SUS	Resolution power	400
Al-Paraffin	Resolution power	400

For NTV CT, the samples are shown in Figs.9 to 11, and the reconstructed CT images are shown in Figs.12 to 14. In Fig.13, 0.5mm dia. BN rod in a Al block can be clearly observable.

From these test, the purposes of the present work, that is, the reduction of experimental time and the improvement of CT images, are considerably but still partly accomplished. We are very happy if this our work can contribute to the broad application of NCT.

COMPUTED NEUTRON TOMOGRAPHY FROM REAL-TIME RADIOGRAPHIC IMAGES

M. F. Sulcoski and J. S. Brenizer

University of Virginia

Charlottesville, Virginia   USA

ABSTRACT

A real-time neutron radiography facility was constructed including the capability of neutron tomography. A tomography algorithm, using the convolution method, was programmed on an Intellect 100 Image Processing System. The method produced results near the theoretical resolution limits for a given number of projections. A tomographic resolution of at least 1.3 mm was demonstrated using 200 projections. Computer running time for the convolution method was found to be about 30 seconds for each projection used.

INTRODUCTION

Three-dimensional (3D) neutron radiography is a new field enjoying an increase in research and development. A review of available neutron tomography literature indicates a growing interest in the industrial and medical applications of neutron tomography (1-7).

Much of the early work in both the USA and abroad was severely limited by slow and tedious methods of optical densitometer scanning of static neutron radiographs to obtain data. The use of BF_3 counters for data collection was an improvement but was again relatively slow and tedious. As Matsumoto el al. (6) have stated, the next step for neutron tomography is to become a powerful diagnostic method using state-of-the-art imaging and image processing methods.

This paper describes the use of real-time neutron radiography for performing computed neutron tomography. A brief description of the imaging system and image processing hardware is given along with details of the implementation of computed neutron tomography using real-time neutron radiography.

MATERIALS AND METHODS

The neutron radiography facility at the University of Virginia consists of a beam port, neutron imaging devices, positioning devices and a dedicated image processor. The major modifications to the existing one of the reactors existing beam ports to obtain an acceptable radiography beam were described in an earlier paper.(7) The neutron beam characteristics are given in reference 8.

Real-time neutron radiography is performed using a Precise Optics Neutron Image Intensifying System (neutron camera) which incorporates a Thompson CSF intensifier. A description of this intensifier is available in the literature.(9) The neutron camera provides a standard RS-170 video signal which is fed to a time date generator, video monitor, video recorder, and image processor.

The image processor used was a Science Applications, Inc. (SAI) Intellect 100. This is a DEC LSI 11/23 based system with one 512 x 512 x 8 bit (256 gray scale) framestore, multiple hard disks, and two floppy disk drives. The Intellect 100 is used as a development system for real-time radiography image processing software.

A convolution-backprojection tomography algorithm (10) was programed on the Intellect 100 image processor. The tomography software was divided into four separate FORTRAN programs, each performing a different function; data acquisition and Intellect 100 initialization (CONV1), projection data scaling and storage (CONV2), tomographic reconstruction and image storage (CONV3), and image scaling and display (CONV4).

The normal procedure is to run the convolution programs in sequence, however, several variations can be used. If multiple tomographs are to be taken and maximum utilization of reactor beam time is desired, then only CONV1 need be run to take data. After each set of projections is taken, the data can be stored for later processing. In this way, no beam time is wasted when processing each tomograph. At a later time, the desired data file is read into the projection data file and processing is performed by running CONV2, CONV3, and CONV4 in sequence. The reconstruction data stored in a file by CONV3 can also be copied into another file (named by the operator) and be saved for future processing. This file need only be used as input to CONV4 for display. The final image displayed by CONV4 is present in the Intellect 100 framestore and thus may be stored on disk as a digital image or on video tape as an analog image.

Neutron tomography experiments were performed using a parallel neutron beam geometry. The objects to be imaged were mounted on a stepping motor (0.9° per step) controlled by a Timex Sinclair 1000 microprocessor. The Sinclair 1000 was used to step the object thorough a predetermined set of angles while projection data was taken from real-time radiographs using the neutron imaging system and the Intellect 100.

RESULTS AND DISCUSSION

Neutron tomography was performed using three test objects. The first two objects are shown in Figures 1a and 1b. The first object (Test Object #1) consisted of a 12.7-mm x 9.5-mm plywood piece mounted in a 25.4-OD lead cylinder. It should be noted that the lead cylinder wall were not uniform in thickness. The second object (Test Object #2) was a 15.9-mm diameter ash wood dowel with five holes of various sizes drilled axially in one end. This object was designed to help determine the resolution capability of the neutron tomographs. The third object (Test Object #3) consisted of Test Object #2 inserted into the lead cylinder used in Test Object #1.

The first set of tests investigated the image quality versus the number of projections. A series of six tomographs were taken of Test Object #1 using 4,10, 20, 50, 100, and 200 projections. The 10, 50, 100 and 200 projection tomographs are shown in Figure 2 a, b, c, and d respectively. These results indicated that

image formation began at approximately 10 projections
and continued to improve as the number of projections
used increased to the present maximum of 200
projections. It should be noted that at 200
projections the glue joints in the plywood began to
become visible in the video tomograph.

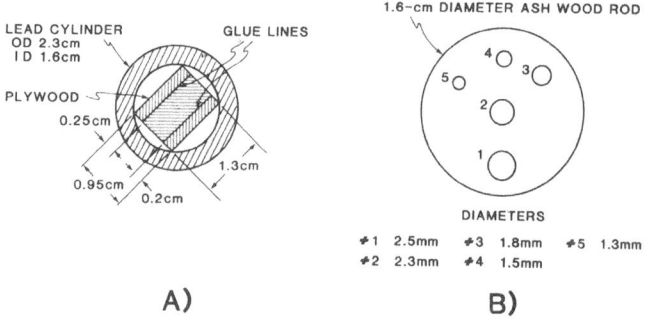

Figure 1. Top view of a) the plywood test object in
lead cylinder (Test Object #1) and b)
drilled wooden dowel rod (Test Object #2).

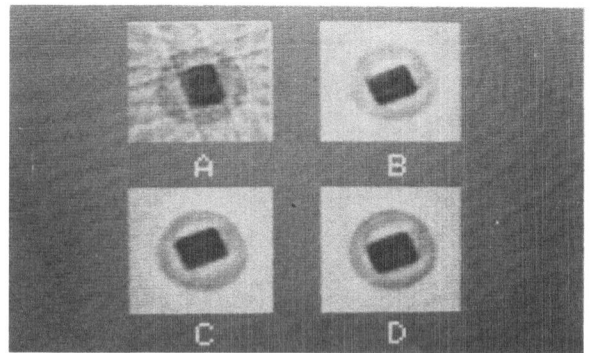

Figure 2. Tomograph of Test Object #1 using a) 10, b)
50, c) 100, and d) 200 projections.

Figure 2 d was also examined to determine if spacial distortion was present. The Intellect 100 software was used to measure (in pixels) the length of the diagonals of the plywood piece and the outside diameter of the lead cylinder. No distortion was found. The tomographs can thus be calibrated to yield the dimensions of the internal structure of an object.

The resolution capability of the neutron tomographs was investigated by calculating tomographs of Test Object #2 (shown in Figure 1b). A series of six tomographs were taken of this object. The 10, 50, 100, and 200 projection tomographs are shown in Figure 3 a, b, c, and d respectively. The use of only four projections yielded no useful information. The general shape of the object (low spatial frequencies) became visible at ten projections, however, no internal structure was visible. The 20 projection tomograph yielded a less noisy image but still lacking the high spatial frequencies necessary to resolve any of the voids with confidence, although an argument could be made that the largest void is visible. The largest void began to appear conclusively in the 50 projection tomograph. The 100 projection tomograph clearly shows all five voids with some distortion in the hole definition indicating a void resolution of at least 1.3 mm. The 200 projection tomograph had lower contrast definition than the 100 projection tomograph although the image is much less noisy. This result is believed

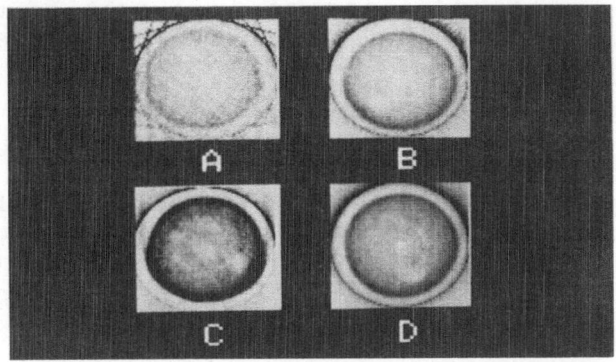

Figure 3. Tomograph of Test Object #2 using a) 10, b) 50, c) 100, and d) 200 projections.

to be due to non-uniform object rotation for this particular case.

Budinger and Gullberg (11) report that if a reconstructed image is to be uniformly resolved to a resolution, d, of a completely asymmetrical object of dimension, D, the number of discrete views, n, must be at least:

$$n \sim \pi D/d$$

Thus, for 200 projections of a 15.9-mm diameter object (Test Object #2) the tomograph should theoretically be capable of 0.25-mm resolution.

The tomographs of Test Object #2 generally conform to the resolution criterion as given by Budinger and Gullberg. The largest hole in the test object (2.5 mm) conforms very closely to the expected resolution for 20 projections. The 20 projection tomograph was interpreted as showing the 2.5-mm hole but not with great confidence. The 50 projection tomograph should show all of the holes if the theoretical resolution were achieved and, indeed, the four largest holes are resolved with only the 1.3-mm hole not present.

Test Object #3 consisted of the previous multihole dowel placed inside the same lead cylinder used in Test Object #1. A series of three tomographs using 20, 50 and 100 projections respectively were taken to investigate the effect of the lead cylinder on the resolution. These tomographs are shown in Figure 4 a, b, and c. The addition of a neutron scattering material (lead) around Test Object #2 had a serious effect on the resolution of the tomographs. The 20 and 50 projection tomographs show the 2.5-mm hole for 20 projections and the 2.5-mm and 2.3-mm holes for 50 projections. The 100 projection tomograph shows four of the five holes, the 2.5-mm, 2.3-mm, 1.8-mm, and 1.5-mm holes are visible with some distortion of the hole shapes expected from the relatively low number of projections. This resolution is again in good agreement with the theoretical values.

CONCLUSIONS

Neutron computed tomography was performed using a real-time neutron radiography system and a small computer system. The existing neutron radiography facility needed no modification thus allowing neutron tomography to coexist with normal facility function. A convolution-backprojection algorithm implemented on a small computer yielded 80 x 80 pixel reconstructions in about 30 seconds for each projection used. Although the present maximum number of projections utilized is limited to 200 increasing this number is only a matter of turntable improvement. The limited range (256 gray scale) of the digitized projection data which can result in low signal to noise ratio and ring-like artifacts did not prevent near theoretical tomographic resolution of approximately 1.3 mm (200 projections) from being achieved. The software was designed to allow the introduction of either previous projection data files or reconstructed images for image processing.

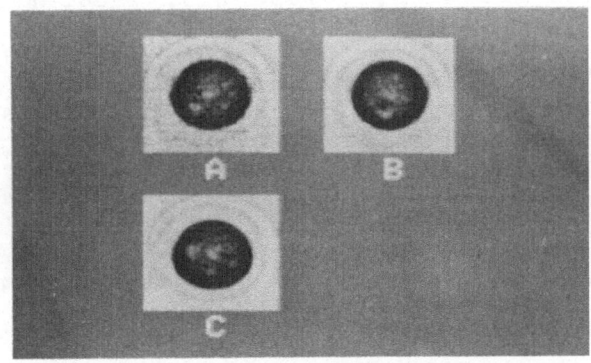

Figure 4.   Tomograph of Test Object #3 using a) 20, b) 50, and c) 100 projections.

REFERENCES

1.   Berger, H., Neutron Laminography for Inspection of Nuclear Fuel Subassemblies, Eighth World Conference on Non-Destructive Testing, CONF-760903-6, 1973, pages 214-223.

2.   Tow, D. M., "Post Irradiation Examination using Neutron Tomography," World Conference on Neutron Radiography, CONF 811215-6, Dec., 1982.

3.   Koeppe, R., R. Brugger, G. Schdapper, G. Larsen, and R. Jost, "A Demonstration of Filtered Neutron

Beam Computed Tomography," <u>Journal of Computer Assisted Tomography</u>, Volume 5, 1981, pages 79-88.

4. DeVolpi, A., and E. A. Rhodes, "Neutron and Gamma Ray Tomographic Imaging of LMFBR SAREF-Program Safety - Test Fuel Assemblies," <u>Materials Evaluation</u>, Volume 40, 19880, pages 1273-1279.

5. Matsumoto, G., and S. Krata, "The Neutron Computer Tomography," <u>Neutron Radiography Proceedings of the First World Conference</u>, 1981, pages 899-906.

6. Yamamoto, S., K. Yoneda, S.S. Hayashi, K. Kobayashi, I. Kimura, T. Suzuki, H. Nishihara, and S. Kanazawa, "Application of Iron-Filtered Neutrons to Radiography of a Copper Step within a Large Iron Block and to Computer Tomography of Metallic Cylinders," <u>Nuclear Instrument Methods in Physics Research</u>, Volume 225, 1984, pages 439-444.

7. Brenizer, J. S., and Sulcoski, M. F., "Real-Time Neutron Radiography at the University of Virginia," published in, <u>Use and Development of Low and Medium Flux Reactors</u>, a supplement to Vol. 44 (1984) of Atomkernenergie Kerntechnick, pp. 958-963.

8. Sulcoski, M. F., "Neutron Computed Tomography Using Real-Time Neutron Radiography," PhD. Thesis, Univesity of Virginia, January 1986.

9. Verat, M., H. Rougeot, and B. Driard, "Neutron Image Intensifier Tubes," <u>Neutron Radiography Proceedings of the First World Conference</u>, 1981, pages 601-607.

10. Shepp, L. A., and B. F. Logan, "The Fourier Reconstruction of a Head Section," <u>IEEE Transactions on Nuclear Science</u>, Volume NS21, June, 1974.

11. Budinger, T. F., and G. T. Gullberg, "Three-Dimensional Reconstruction in Nuclear Medicine Emission Imaging," <u>IEEE Transactions on Nuclear Science</u>, Volume NS21, June 1974, pages 2-20.

MACROSCOPIC DENSITY FLUCTUATIONS IN THE NIOBIUM-HYDROGEN

SYSTEM STUDIED BY NEUTRON RADIOGRAPHY AND COMPUTERIZED

TOMOGRAPHY

G. Steyrer[+x] and J. Peisl[x]
[+]Institut Laue-Langevin, 38042 Grenoble,
Cedex, France
[x]Sektion Physik der Universität München,
Geschwister Scholl Platz, FRG

ABSTRACT

By means of neutron radiography and computerized tomography the hydrogen density distribution was determined
during a phase transition of hydrogen in niobium. In a
coherent lattice macroscopic density modes are shown to
exist. The loading with hydrogen and the final distribution depend strongly on crystal imperfections. The formation of cracks under internal stress is followed in the
bulk of the sample.

INTRODUCTION

The interstitial metal-hydrogen alloys have been
extensively studied under a great variety of aspects.
These range from fundamental solid-state physics to technological applications (hydrogen storage, hydrogen embrittlement). The niobium-hydrogen system is an example to understand the properties of defects in metals. Niobium has a
simple body-centered cubic lattice and atomic hydrogen
interstitials reach rapidly thermal equilibrium because of
their high mobility.

We used neutron radiography combined with tomographic
reconstruction and show that this method is ideally suited
to study as well a phase transition of dissolved hydrogen
as more practical questions like the cracking of single
crystals under internal stress caused by the defects.

THE NIOBIUM-HYDROGEN SYSTEM

Hydrogen dissolves in niobium up to high concentrations of about one interstitial per metal atom. One striking property of these light defects is their high mobility, which is comparable to the mobility of particles in a liquid. Hydrogen occupies tetrahedral sites in the host lattice and displaces the metal atoms around it from their positions in the undisturbed crystal. This leads to a volume expansion of a hydrogen loaded crystal proportional to the concentration. The expanded region around a defect is felt by the other defects and gives rise to a weak attractive interaction between them. This 'elastic interaction' is thus transmitted by strain fields in the crystal. It is responsible for a phase transition of the defects (1). The metal lattice may be regarded as the component which provides the space where the hydrogen moves as a 'lattice-gas'.

The α-α' transition is very similar to a gas-liquid phase transition, where the two phases differ only in concentration (see Fig.1).

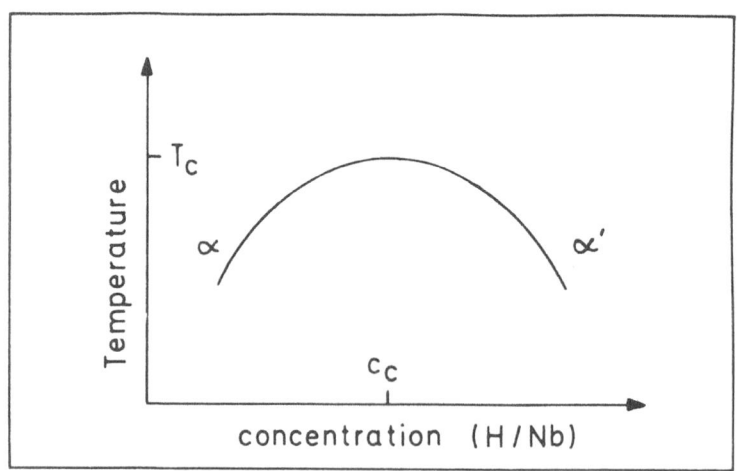

Fig. 1 Phase diagram of hydrogen in niobium
      (schematically)

A critical point is found at T = 171°C and c_c = 0.31 H/Nb (2) . The 'liquid' (α') phase may be regarded as droplets of higher concentration precipitated in the low-concentration 'gas' (α) phase. Stresses arise in a coherent lattice ('coherency stresses'). The difference between a coherent and incoherent lattice is shown in Fig. 2.

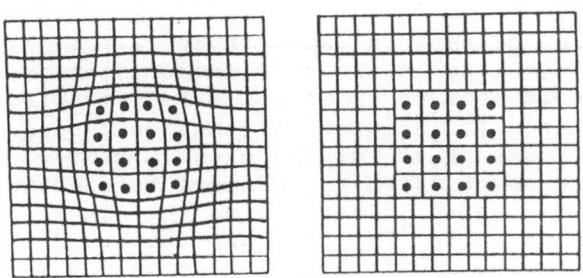

Fig. 2 Coherent and incoherent lattice

In order to get the phase diagram shown in Fig. 1 the coherency stresses have to be released by plastic deformations, leading to an incoherent lattice. In a single crystal this is only possible when the stresses attain the critical yield stress. In a coherent lattice, a completely different phase transition occurs.

THE COHERENT PHASE TRANSITION

Any rapid variation of hydrogen concentration produces high stresses and rise the free energy of the system. Therefore critical fluctuations and the α-α' phase transition is suppressed. A theoretical treatment of the region in the vicinity of the critical point was given by Wagner (3) and Horner (4) . They found that the free energy in a coherent crystal is lowered by a combination of 'macroscopic modes', which are static concentration fluctuations with wavelengths in sample dimensions. The long range of the elastic interaction leads to a dependence on boundary conditions (i.e. sample shape) of the macroscopic modes. They have been calculated for an isotropic sphere (3) and a rotational ellipsoid (5). The stability limit is given by 'spinodal temperatures' associated to the individual modes. The highest spinodal corresponds to a mode which creates a linear concentration gradient along the sample. Above this line the concentration is a constant.

The onset of the incoherent phase separation is determined by the lattice imperfections in a real crystal. As soon as plastic deformations are possible, the metastable coherent fluctuation is distroyed by the precipitation of the α' phase in a now incoherent lattice.

NEUTRON RADIOGRAPHY

The method of neutron radiography is ideally suited to determine the local hydrogen concentration in NbH_x. The niobium host lattice is practically transparent to neutrons compared to the high attenuation caused by the incoherent scattering of the hydrogen. Moreover, we measured the cross-section of hydrogen in niobium as a function of wavelength (6) and found that it varies in between the free value (20 barn) and the bound value (80 barn) in the range of thermal neutrons. Therefore, the contrast of the radiographs can be varied by the choice of an appropriate wavelength in dependence of sample size and concentration. We used neutrons of 2.4 $\overset{\circ}{A}$, the corresponding cross-sections are:

$$\sigma_{Nb} = 2.1 \text{ barn} \qquad \sigma_H = 65.5 \text{ barn}$$

The value for niobium is the cross-section for a single crystal which is not scattering coherently. It is essentially the absorption cross-section; the incoherent scattering of niobium is negligible (7).

A collimation of 0.2° of the neutron beam was provided by a neutron-guide; the selection of the wavelength was made by reflection on a perfect germanium crystal. We used monochromatic neutrons in order to give the hydrogen concentration on an absolute scale. Complications would arise with the white beam because of the spectral intensity distribution and the strong dependence on wavelength of the hydrogen cross-section.

SAMPLES AND EXPERIMENTAL SET-UP

The samples were niobium single crystals of cylindrical shape (110 direction along the z-axis). They were loaded in-situ on the instrument used for radiography (D13c at the ILL). The sample dimensions were about $\emptyset$ = 1 cm and h = 2 cm. The loading was performed in the 'gas' phase at $600^{\circ}C$. We could follow for the first time the process of hydrogen penetration into the bulk of the sample.

The area of the homogeneous beam was 3 x 5 cm^2. With the gadolinium converter technique an exposure time of 40 min was necessary for radiography with monochromatic neutrons.

The processed films were analyzed on a microdensitometer (Service de Microdensitometrie du C.N.R.S., Université Paris-Sud, Orsay). The spatial resolution was measured at a sharp edge and gave a value of 0.2 mm. This may be improved by the use of neutrons with shorter wavelength (better collimation in the neutron-guide).

A tomographic reconstruction method based on Fourier-backprojection (8) gave the hydrogen concentration distribution in arbitrary planes of the samples.

THE HYDROGEN LOADING

Following the loading process in several samples two completely different results showed up. In some samples the concentration increased homogeneously in the whole volume and no gradients appeared even in the early stages. In Fig. 4 the concentration along the z-axis is displayed at time intervals of about 1 hour in between each radiograph.

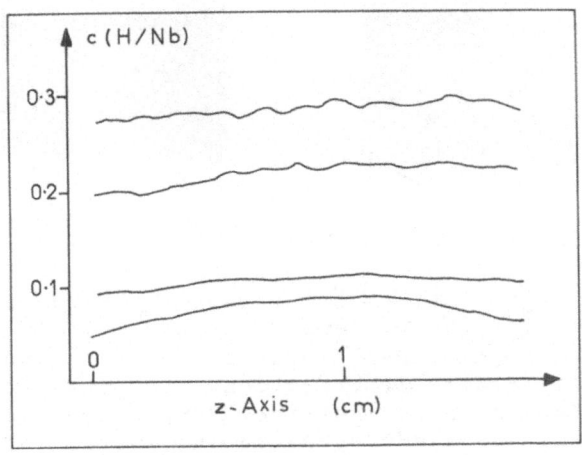

Fig. 4 Homogeneous loading

On the other hand, an inhomogeneous concentration distribution was observed in several other samples. Two examples of the final distribution are shown in Fig. 5a,b. There dark pixels correspond to high transmission. The transmission is proportional to the concentration, but a varying sample thickness in the direction perpendicular to the axis has to be taken into account. No homogeneous redistribution was possible, even at high temperatures (700°C).

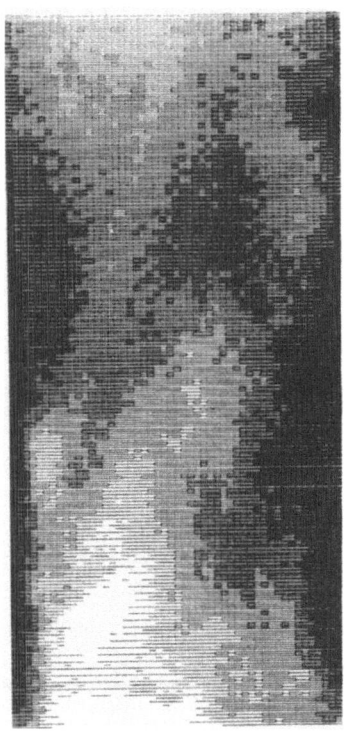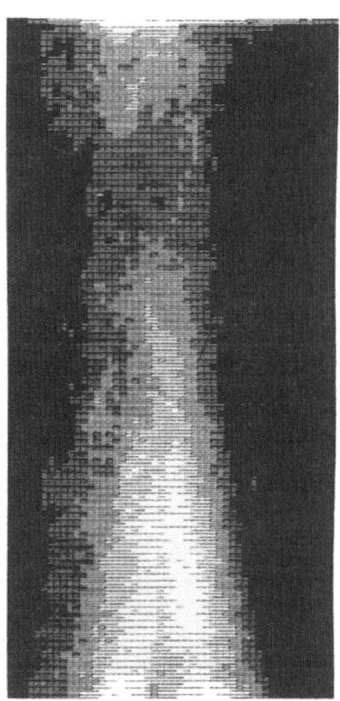

Fig. 5 Inhomogeneously loaded samples

We conclude that an homogeneous loading of a niobium single crystal with a volume of about 2 cm^3 is only possible if the hydrogen can dissolve on the whole surface. Severe lattice imperfections may prevent the macroscopic diffusion in the lattice. Sample treatment like UHV-degassing and electroerosive cutting in general has little influence on the final hydrogen distribution.

THE COHERENT PHASE TRANSITION

The sample was homogeneously loaded with c = 0.295 H/Nb. It was cooled down below the transition line in steps of 0.3°C. A serie of 4 projections were taken at fixed temperatures, after a rotation of the sample of 45° around its z-axis in between each radiograph. A slowly varying concentration distribution showed up below 167.9°C. The amplitude of the fluctuation increased with decreasing temperature to a maximal value of Δc = 0.18 H/Nb. In Fig. 5 the transmission patterns (arb.units) before and after the phase transition are compared. Levels of constant transmission are separated by regions of

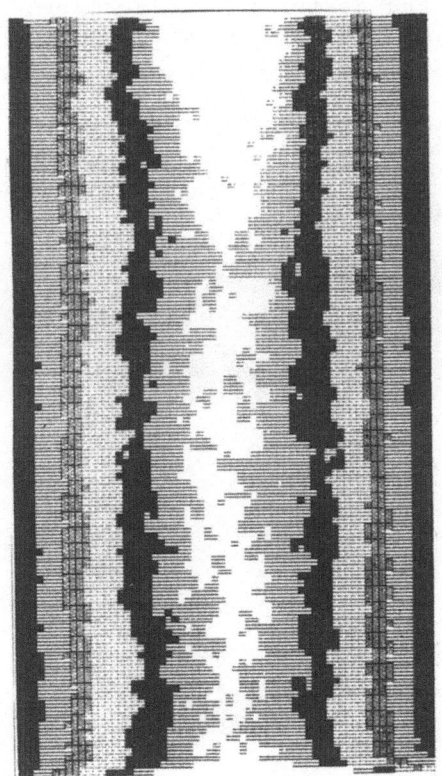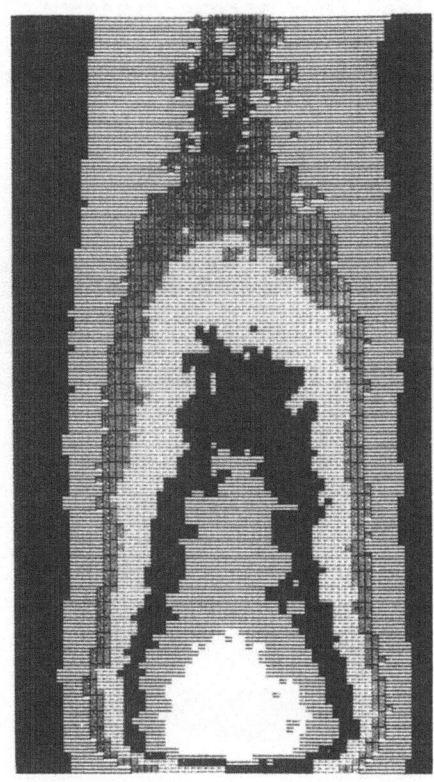

Fig. 5 Transmission patterns before and after
the phase transition in a coherent
lattice (arbitrary units)

different gray-values. In Fig. 5a the cylindrical shape
of the sample with a constant concentration is reflected.

The fluctuation was nearly symmetric around the z-
axis, therefore tomographic reconstruction was possible
with only 4 projections.

The concentration distribution in two planes along
the z-axis, perpendicular to each other, are shown in
Fig. 6a. Low concentration is represented by a dark shading. In the two cuts along the lines A and B perpendicular to the z-axis (Fig. 6b), the variation of the concentration in radial direction is reconstructed. At one
end of the sample (A)an influence of higher order modes
lead to a strong gradient from the center to the surface.
The nearly symmetric distribution around the axis is
evident.

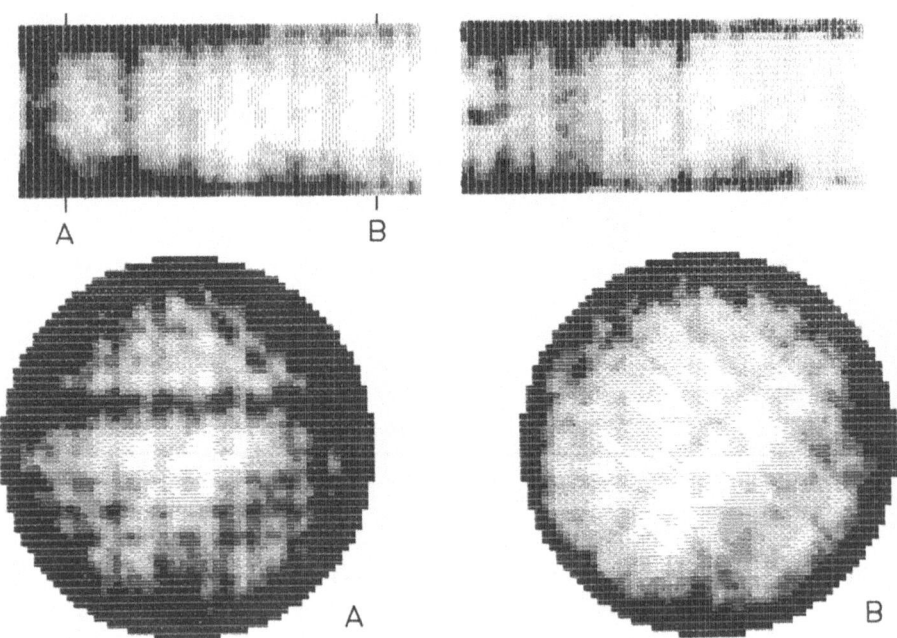

Fig. 6 Tomographic reconstruction of the con-
        centration distribution at T = 166.2°C
        top: planes along the axis
        bottom: planes perpendicular to the
                axis along A and B

A comparison of the reconstructed concentration distribution on a cut along the axis with the predictions of theory is given in Fig. 7. The calculations for the rotational ellipsoid included 4 macroscopic modes with the highest spinodal temperatures(9).

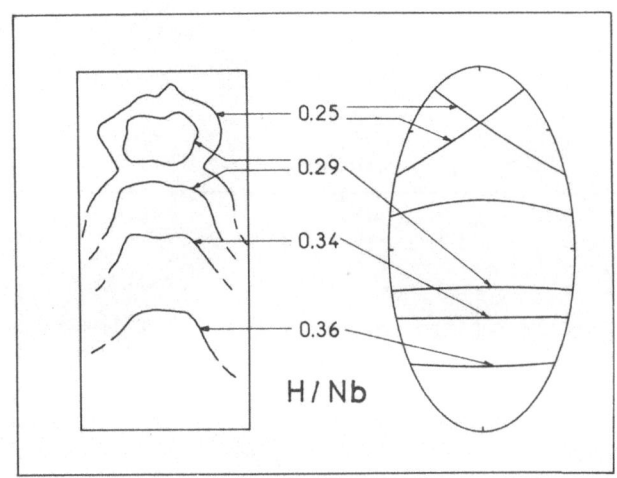

Fig. 7 Comparison of theory and tomographic
reconstruction at T = 166.2°C

The spatial form and the absolute values are in good agreement, even the asymmetric saddle-point at one end of the sample is reproduced in the experiment.

This is the first direct evidence of macroscopic modes in the inhomogeneous coherent phase, extending over the whole sample. Coherency was maintained over a temperature range of 3°C.

THE INCOHERENT α-α' PHASE SEPARATION

The high stresses in a sample with a coherent fluctuation are finally released by severe plastic deformations. We could observe the macroscopic cracking of single crystals under internal stress. Tomographic reconstructions from 18 projections of an incoherent sample were made. The propagation of the cracks, which showed up on the surface could be followed into the bulk of the sample. We found that cracking occurred along the {100} cleavage planes of bcc lattices and {110} planes giving rise to a network separating large crystallites. Fig 8 shows two cuts perpendicular to the z-axis with the intersections of crack planes.

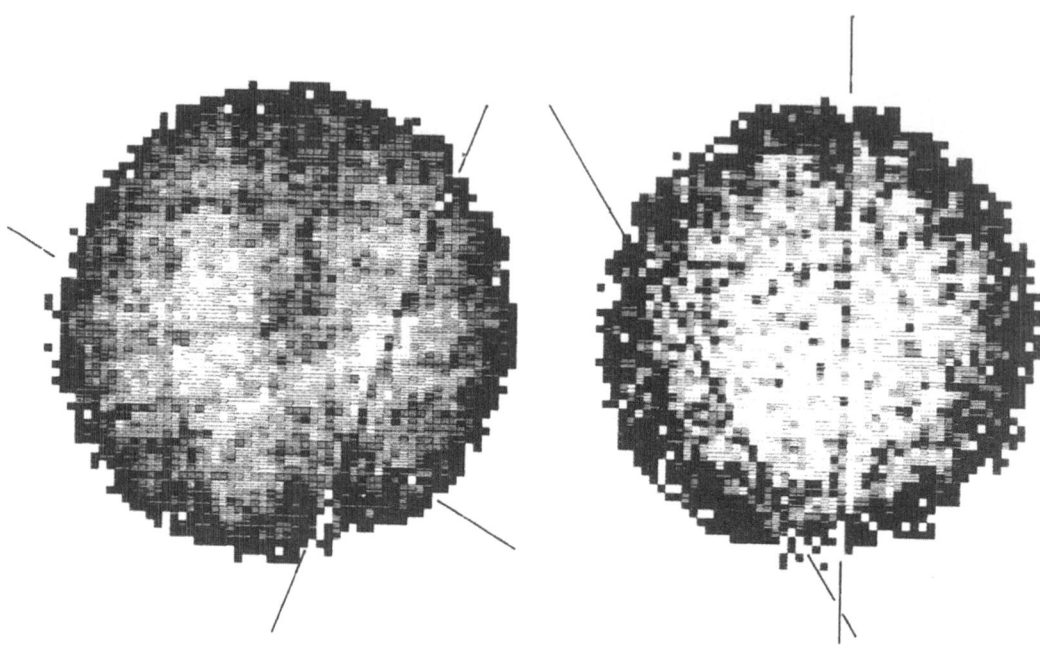

Fig. 8 Tomographic reconstruction of planes
perpendicular to the z-axis with
intersection of crack- planes

REFERENCES

1 G. Alefeld, Phys. Stat. Sold. 32, 67 (1969)

2 H. Zabel, J. Peisl, J. Phys. F: Metal Phys. 9, 1461
   (1979)

3 H. Wagner, H. Horner, Adv. Phys. 23, 587 (1974)

4 H. Horner, H. Wagner, J. Phys. C: Solid State Phys.
   7, 3305 (1974)

5 H. Maier-Bötzel, W. Platzer , H. Wagner, Phys. Rev.
   B27, 5173 (1983)

6 G. Steyrer, J. Peisl, to be published

7 S.F. Mughabghab, M. Divadeenam, N.E.Holden, "Neutron
   Cross Sections", Academic Press

8 L.S. Shepp, B.F. Logan, IEEE Trans.Nucl. Sci.21 (1974)

9 H. Maier-Bötzel, private communication

PROGRESS IN NEUTRON TOMOGRAPHY FOR NON-DESTRUVTIVE TESTING

P. Maier, G. Pfister, G. Stier, W. Kegreiss, G. Hehn
Institut für Kernenergetik/ Universität Stuttgart
Stuttgart, West Germany

ABSTRACT

A neutron computer tomograph scanner was developed and constructed. Our equipment is transportable and we can use different neutron sources: A 14 MeV (d on be) neutron source at essen and the
3.5 MeV (d on be) source of the dynamitron in Stuttgart. Further
studies will be done with cf-252 neutrons and fission neutrons
form a nuclear reactor. Since each neutron source contains gamma
radiation as well we are able to get simultaneously a neutron and
a gamma CT image. In non-destructive material testing, fast neutrons can be used especially to examine thick metallic samples or
to show small differences in the concentration of light nuclei.
Several image reconstruction algorithems are available for samples
with different size and material composition. A spatial resolution
of 0.5 mm and a density resolution of about 5 % could be achieved
in our first CT images with fast neutrons. Neutron CT images show
nearly no artifacts compared to x-ray scans. Therefore the neutron
CT technique can be used with advantage to examine strongly inho-
mogneous samples. Up to now we used the single beam technique in
all our measurements. The Fanbeam technique is under study to
reduce the measuring time.

1. INTRODUCTION

The tomographic technique gives 1:1 cross-section pictures of
the examined probe. Without extensive interpretations of the
measured signals (as with ultrasonic or radiographic methode)
cracks, inclusions and other inhomogenities can be seen directly
in the image.

771

After the great success of this method in the medical diagnostic it is now obvious to make it available for the non-destructiv material testing. For this purpose tomography scanners with x-
and gamma-rays havew been developed world-wide and since a few
years they are used in industrial applications /1-5/.

The great advantage of fast neutrons for CT measurements is
the fact that they interact with the nuclei and not with the electrons of the atoms. The differences in the absorption koefficients
of different materials are not as large as for electromagnetic
radiation (like x- or gamma-rays). Therefore neutron CT images
show nearly no artefacts and neutrons are qualified to examine
thick probes and materials with high density as well as for
composit materials.

We have developed a scanner to produce CT images with fast
neutrons. In the following we will describe the scanner and
present some results.

2. DESCRIPTION OF THE SCANNER

At the "Institut für Kernenergetik" (university of Stuttgart)
a neutron tomography scanner was developed. The scanner can be
transported an therefore be used at different neutron sources:

- Universitätsklinikum Essen, Prof. Rassow
  14 MeV cyclotron (deuteron on be-target) /6/

- Institut für Strahlenphysik, Stuttgart, Prof. Hoffman
  4 MeV dynamitron (deuteron on be-target) /7, 8/

The devices of a system used to measure neutron CT images are
listed in the following:

- A neutron source and components for shilding and for
  collimating a narrow pencil beam.

- Neutron detectors with high efficiency and good neutron/
  gamma dicrimination properties.

- Mechanical units for translation and rotation of the
  examined probe through the neutron beam.

- A computer system for the control of the scans and for
  the data aquisition and a first evaluation of the data.

- Peripheral equipment for data storage (floppy, tape-
  recorder) and for supervising the measurements (sprinter,
  graphic-terminal).

- A Computer with fast graphic support for the on-line-reconstruction.

An overview of the system and it's components is given in Fig. 1. With this equipment we can get CT images either with pure neu trons or with a mixes neutron/gamma-field. The CT images can be displayed on a graphic screen parallel to the measurements.

3. NEUTRON COUNTING

With regard to a high efficiency of the neutron counting system we use szintillating detectors (NE 213) with fast photo-multipliers (2.5 nsec risetime). The signals of these detectors can be analyzed by a pulse-shape discrimator to discriminate bet-ween neutrons and gammas /9/. Therefore we can get a neutron- and a gamma-image from the same measurement simultaneously.

The electronic devices used for our CT measurements at Essen and Stuttgart are all comercial and generally used in neutron physies. In addition to the detector counts the target current and the measuring time are recorded.

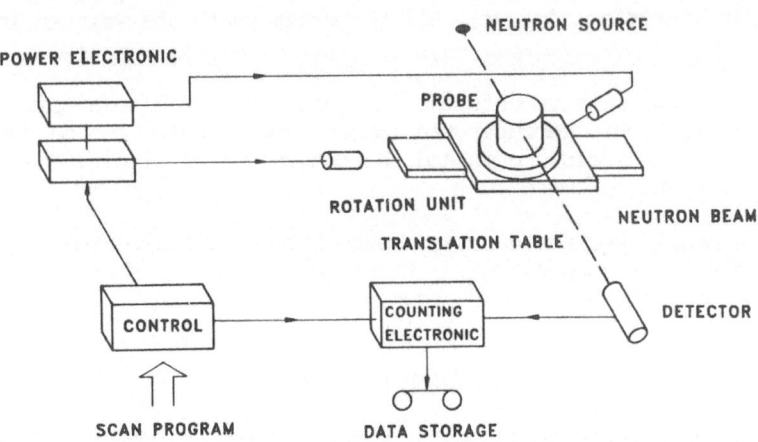

FIG. 1: SCHEMATIC ARRANGEMENT OF THE DEVICES USED IN THE IKE FAST NEUTRON SCANNER

4. CONTROL AND DATA-AQUISITION

The most important device of the system is a computer-system
for the control and the data-aquisition during the CT measurements. This computer, developed by IKE, is build in 19"-technique.
It includes a Z80 microprozessor, a supervising electronic and the
following interfaces:

- Two serial ports (RS 232) for a terminal and for data
  transfer to a host computer or to a tape-recorder.

- A BCD-parallel port to connect the neutron counters
  (Ortec-printing-loop).

- Parallel port for the control signals to the mechanical
  units. This port is equiped with relays and optocouplers to
  seperate the power-electronic from the microprozessor
  system.

- An IEEE-488 interface to connect a floppy-drive,
  a printer and other devices.

The software of the prozessor system includes a complet program development system. For this purpose a FORTH compiler was installed, including a screen-editor, an assembler and debug possibilities /10/. The kernal of the control-software is written in
assembler (interface driver). All the other parts are written in
FORTH.

During the measurements the data are transfered to a host
computer. With this computer the image-reconstruction can be done
parallel to the measurements and the images can be displayed on a
graphic screen.

The nearly perfect correspondence between the experimental
pictures and our computer simulations shows that it is possible to
use fast neutrons in CT measurements.

5. COMPUTER PROGRAMMS

Several computer programs were specialized ore developed to
calculate radiation fields and to simulate CT-scans. Different
image reconstruction algorithms are available and adjusted to the
requirements of technical probes. The sophisticated modular program system RSYST /11/ is used for graphical display of the CT-
images. A flow-chart showing the measuring and data processing in
neutron tomography is shown in Fig. 2.

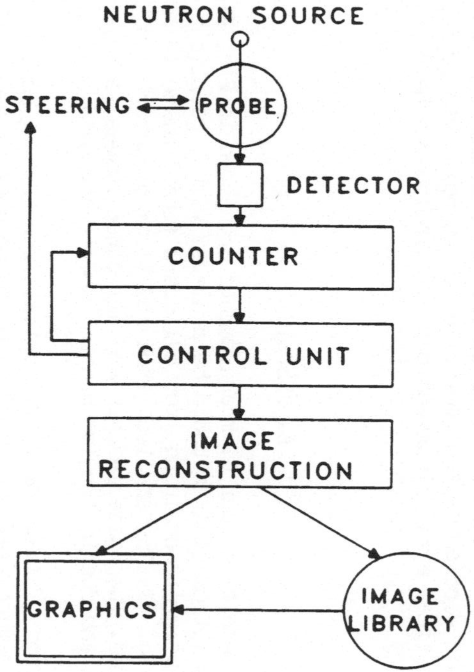

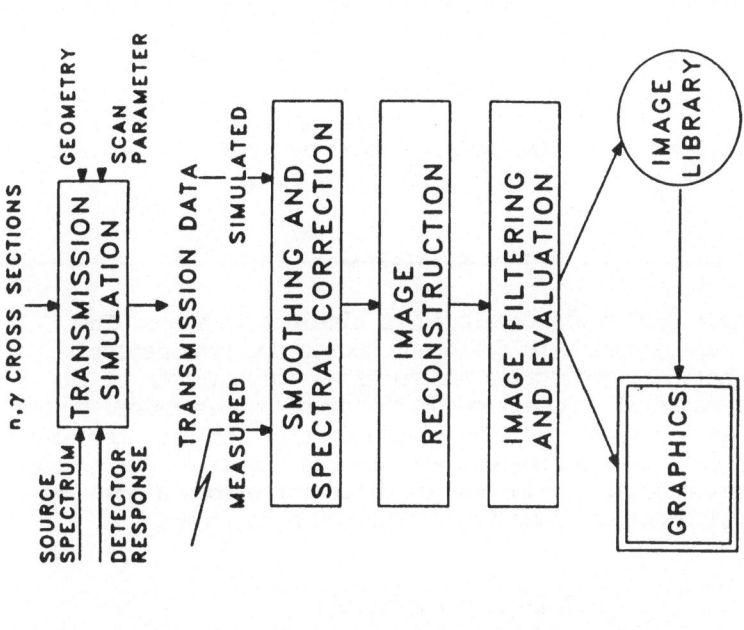

FIG. 3: SIMULATION AND PROCESSING OF
TRANSMISSION DATA TO PRODUCE THE
NEUTRON CT-IMAGE

FIG. 2: FLOW-CHART SHOWING THE
MEASURING AND DATA PROCESSING
IN NEUTRON TOMOGRAPHY

5.1 Transport calculations

One and two-dimensional transport codes (ANISN, DOT 4.2) /12/ are available to calculate the neutron and gamma fields in CT with fast neutrons. The calculations are done to optimize the neutron and gamma collimation and to find the best shielding of the detectors. Studies can be done to determine the spectral shift in the transmitted radiation and the portion of scattered radiation. The multigroup nuclear cross section libraries for neutrons and gammas used are based on ENDF/B-V point data /13/.

5.2 CT-Simulation

A program system was developed to simulate CT-scans. The simulation can be done with different source spectra, detector response functions, geometries and material compositions of the probes. The material composition of the probe is determined by the nuclear cross sections used. The simulations are helpful to determine the optimal scan parameters and the best spectral correction of transmission data. Simulations of different probes allows the further development of image reconstruction algorithms.

5.3 Image reconstruction

A flow chart for image reconstruction is shown in Fig. 3. The measured or simulated transmission data can be smoothed by different algorithms to reduce the statistical noise. A spectral correction must be done to compensate the spectral shift of the transmission data due to different transmitted lengths of the probe. The central part are different image reconstruction algorithms to calculate the attenuation coeffizient distribution in the considered cross section of the probe. The image noise can be filterd. Statistical evaluations and contour finding can be done. The CT-images are stored in an image library and displayed on a color TV-screen.

6. RESULTS

First results of our CT measurements are shown in Fig. 4-6. A spatial resolution of 1 mm could be achieved with a plastic probe (145 mm diameter) containing iron cylinders from 1 mm to 15 mm diameter (Fig. 4). The smallest iron rod (1 mm diameter) could be well detected. The noise in this and all other images is relatively high and due to the low neutron source strength available.

A neutron CT-image of a cylindrical iron testing module with arteficial defect /14/ - as used in ultrasonic testing - is shown in Fig. 5. The diameter of the probe was 70 mm, the diameter of the defect 10 mm. The density resolution of the defect is about 5 %, because the hight of the defect was 1 mm and our beam hight 20 mm. The position and shape of the defect could be well detected.

A cross section through a concrete probe is presented in Fig. 6. Two iron rods of different diameters, several pebbles and air cavities are seen within the concrete region. The cylindrical plastic container of the probe is shown by the outer green ring. All contours are cornered because we reconstructed this image only out of 23 scans.

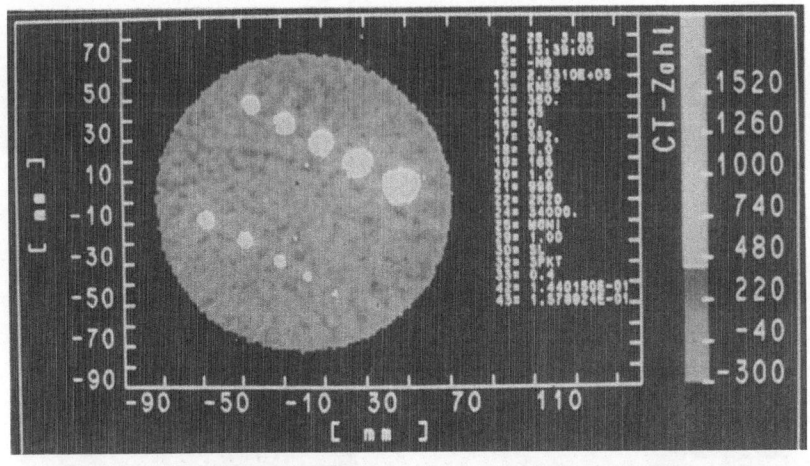

FIG. 4: NEUTRON CT–IMAGE OF A PLASTIC/IRON RESOLUTION PROBE

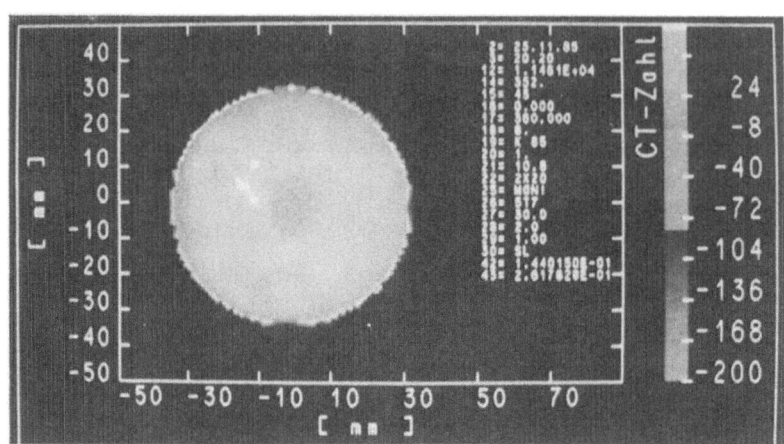

FIG. 5: NEUTRON CT-IMAGE OF THE IRON TESTING MODULE WITH ARTIFICIAL DEFECT

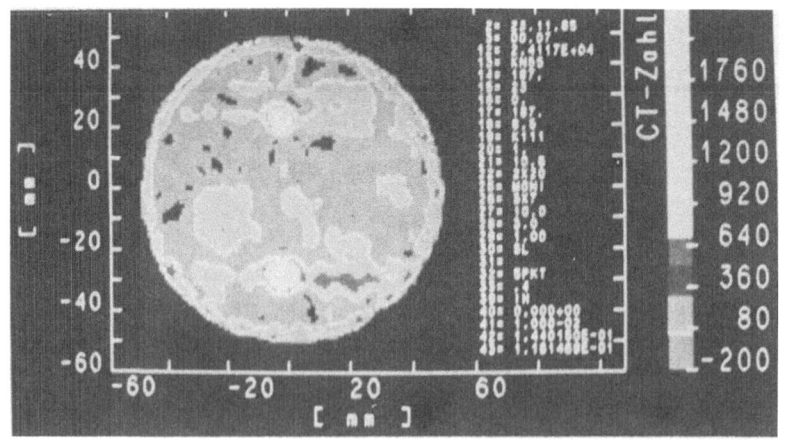

FIG.6: NEUTRON CT-IMAGE OF A CONCRETE PROBE

7. Conclusions

Our results show that it is possible to produce good CT images
with fast neutrons. Because of the large free-mean path of fast
neutrons they can be used with advantage in the examination of
thick metallic probes. This technique can complement other methods
of non-destructiv material testing. An important field of application is the testing of components in the aero-engine industrie
as well as the examination of nuclear reactor components. The
spacial resolution in the measurements is limited by our collimators to about 0.5 mm. The density resolution is a function of
the neutron source strength and of the measuring time. A resolution better than 1 % is possible.

For measurements witch requires a better spacial resolution
thermal neutrons and neutrons in an energie range less than about
100 keV are now under study.

LITERATUR

/1/ Reimers, P., J. Goebbels, H.-P. Weise, K. Wilding: Some
Aspects of Industrial Non-Destructive Evaluation by X- and
-Ray Computed Tomography. Nucl. Instr. and Meth. 221 (1984)

/2/ Habermehl, H., H.-W. Ridder: Mobile Tomographie mit Radio-
nukliden. In: Schopka, H.-J., (Herausg): Medizinische-Physik
'78. Heidelberg: Dr. Alfred Hüthig Verlag 1978, S. 245-250

/3/ Gilboy, W.B.: X- and -Ray Tomography in NDE Applications.
Nucl. Instr. and Meth. 221 (1984)

/4/ Onoe, M., J.W. Tsao, H. Yamada, H. Nakamura, J. Kogure,
H. Kawamura: Computed Tomography for Measuring the Annual
Rings of a Live Tree. Nucl. Instr. and Meth. 221 (1984)

/5/ Hanson, K.M., J.N. Brandbury, R.A. Koeppe, R.J. Macek,
D.R. Machen, R. Morgado, M.A. Paciotti, S.A. Sandford,
V.W. Steward: Proton Computed Tomography of Human Specimen.
Phys. Med. Biol. 27 (1982) 25

/6/ Rassow, J., G. Hüdepohl, E. Maier, P. Meissner: CIRCE,
Cyclotron Isocentric Neutron Therapy Facility Radiation
Physics. 3rd Symp. on Neutron Dosimetry in Biology and
Medicine, EUR 5848 (1978)

/7/ Hammer, J.W., H.M. Schupferling, E. Bergandt, T. Pflaum: Beam
Transport System for a 4 MV Dynamitron Accelerator. Nucl.
Instr. and Meth. 128 (1975) 409

/8/ Hammer, J.W., B. Fischer, H. Hollick, H.P. Trautwetter, K.U. Kettner, C. Rolfs, M. Wiescher: Beam Properties of the 4 MV Dynamitron Accelerator at Stuttgart. Nucl. Instr. and Meth. 161 (1979)

/9/ Bulski, G., N. Grum, J.W. Hammer, H. Postner, G. Schleussner, E. Speller: Neutron Analyzing Power and Elastic Differential Cross Section for Carbon-12 and some Medium and Heavy Elements (Si, S, Ca, Cu, La, Pb, Bi and U). Nuclear Data for Science and Technology, Proc. of the Int. Conf., Sept. 1982, EUR 8355, Antwerpen: D. Reidel Publishing Company, 1983, p. 783-791

/10/ Wilson, D.: Z80-Fig-Forth 1.1 d. FORTH-Interest Group. San Carlos: 1982

/11/ Schlecht, B.: Graphik in modularen Systemen. (IKE-Bericht in Vorbereitung)

/12/ Rhoades, W.A., et al: The DOT-IV Two Dimensional Discrete Ordinates Transport Code ORNL/TM-6529, 1979

/13/ Kinsey, R.: ENDF/B Summary Documentation ENDF-201, 3rd edition (ENDF/B-V), BNL-NCS-17541 (July 1979)

/14/ Mayer, H.G., G. Haufler: Diffusionsgeschweißte Testkörper mit künstlichen Fehlern für die Weiterentwicklung zerstörungs- freier Prüfverfahren in der Kerntechnik

PART XII

SPECIAL TECHNIQUES

LOAD TESTING OF NEUTRON ABSORBENT MATERIALS BY USING

NEUTRON RADIOGRAPHIC PICTURES

G. BAYON, IRDI/DERPE/SPS
A. LAPORTE, IPSN/DSMN/SITN
Commissariat à l'Energie Atomique (France)

ABSTRACTS

By using standards whose characteristics are known, quantitative measurement of the neutronographic negatives density, makes possible the estimation of a neutrophagic constituent proportion in each point of a given material. This presentation describes the methodology, the automatic equipment developed at Saclay and the method performance characteristics applied to mass-produced components.

I - INTRODUCTION

In imaging techniques using radiographic films, quantitative measurements are relatively rare. The process is generally restricted to qualitative testing. Neutron Radiography is no exception to this rule. Nevertheless, the characteristics of certain photographic emulsions help to consider loads measurements (mass per unit area) of absorbent materials that are homogeneous or present within other components (clad, matrix, etc.) which are "transparent" to the used radiation.

A study conducted in this field culminated in the setting of an automated densitometer for scanning large films. At the present time, the testing of the Boron 10 distribution in fuel element assembly plates is one industrial application of this process. This paper describes the work carried out for such a control.

II - FUNCTION AND EDGE PLATES MANUFACTURING

The main function of edge plates is to maintain the fuel plate spacing for mechanical purposes (rigidity) to guarantee the normal fuel cooling with the water flow. To minimise the movements of the shim rods during a reactor operating cycle, and to flatten the axial flux distribution, the edge plates contain a burnable poison (Boron). The neutron flux and thermal requirements have led to define several of specifications regarding the Boron distribution in the plates.

Taking into account the relatively low Boron load (*) these plates are made by a co-rolling technique, with the borated "core" inserted in an aluminium (AG3) "sandwich".

At this step, neutron radiographic testing is used to locate accurately the borated zone within the rough plate edges.

Dimensional measurements are used for two purposes :
. to discard plates not meeting the specifications,
. to help proceeding with finishing machining operations and to align the borated core in relation to the uranium containing zone of the fuel plates.

Saclay Test Facility (testing by travelling scan) serves to obtain the complete neutron radiographic image of the borated core on the same film (18 cm x 102 cm) (Fig. 1). Once this operation is completed, quantitative measurements are taken from the Boron distribution on the neutron radiographic picture, by means of an automated densitometer.

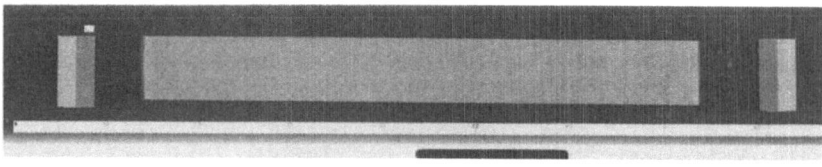

Fig. 1 - Neutron radiography of an edge plate (CEA/CEN/Saclay)

(*) Boron load $\simeq$ 1 mg . cm^{-2}

III - PHYSICAL PRINCIPLES AND APPLICATION

III.1 - Neutron beam absorption

The principle of neutron radiographic testing is based on the two-dimensional visualisation of the neutron beam transmitted by the plate. For the Boron 10 load contained in the Al/Boron mixture, it is assumed that the beam transmission obeys to a decreasing exponential law, exclusively related to the Boron load.

This can be expressed by a simplified form as follows :

$$\emptyset_{(T)} = A \, \emptyset_{(o)} \exp \, (-\mu \, T) \qquad\qquad /1/$$

where $\emptyset_{(T)}$ = neutron fluence transmitted by the plate for a Boron 10 content T

A = correction factor taking into account effects in aluminium (constant)

$\emptyset_{(o)}$ = incident neutron fluence

μ = average value of mass attenuation coefficient for the used spectrum

T = Boron 10 content (mass per unit area).

III.2 - Response of silver halide film

At the Test Facility, the direct technique is used by associating a back converter screen (solid Gadolinium sheet - thickness : 250 µm) and a high definition silver halide film (Kodak Industrex : single-coated type R) in a vacuum cassette which guarantees tightness to light. The silver halide film response is generally characterised by measuring the optical density obtained after chemical processing.

By definition, the optical density is : $D = lg \, (I_0/I)$

where I_0 : measurement of incident light beam intensity normal to the film
I : measurement of light beam intensity transmitted by the film.

In a more or less broad density range, certain radiographic emulsions display a linear response as a function of the fluence received (Fig. 2). By making sure to locate the measurement in this zone, and by assuming that the efficiency of the converter is identical over its entire effective surface area, equation /1/ can be written as follows :

$$\emptyset_{(T)} / A \cdot \emptyset_{(o)} = \exp(-\mu T) = [\, D_{(T)} - Db \,] / [\, D_{(o)} - Db \,] \qquad /2/$$

where $D_{(T)}$: optical density measured on the film for a sample with content T

$D_{(o)}$: optical density measured on the film for a sample with zero content

Db : basic gross fog (chemical fog + support)

By developing equation /2/, the following equation is obtained :

$$I_{(T)} = Ib \exp[\, -C \cdot \exp(-\mu T)\,] \qquad /3/$$

where $C : Ln[\, Ib/I_{(o)}\,]$

A curve obtained in the experimental study is presented in Figure 3.

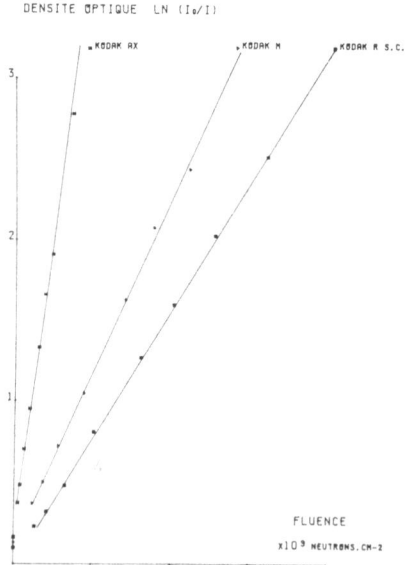

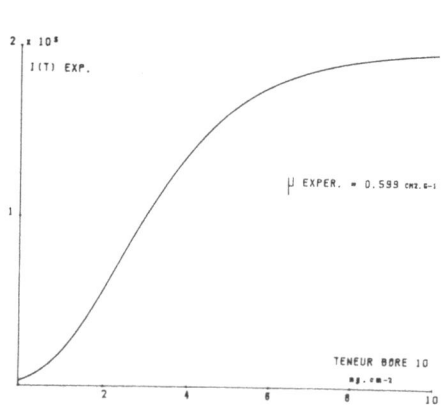

Fig. 2 - Different response radiographic films

Fig. 3 - Experimental curve

III.3 - Application to automatic control

The value of C and the measurements of $I_{(T)}$ are characteristic of several parameters which may change considerably :
. uniformity and reproductibility of the radiation dose received by the film,
. stability of chemical processing in time.

A definitive calibration carried out at a given moment may prove defective in the absence of perfect control of all parameters involved in the process of the photographic image formation on the film. This difficulty was circumvented by preparing standard samples whose Boron content was representative of the limits allowed by fabrication specifications. These small standard plates were placed on all the plates subjected to neutron radiographic testing, and their image is therefore present on the film. Calibration can therefore be carried out on each film, thus totally eliminating the consequences of certain parameters fluctuations on time, such as :
. reproductibility of fluence,
. stability of chemical processing,
. difference in the film batches speed.

As to the uniformity of the neutron fluence, the horizontal gradient of the beam is compensated by means of irradiation scanning at a constant speed in front of a neutron guide (25 x 100 mm^2). On the other hand, the vertical gradient is preserved (film width direction). Density measurements are read along 5 distinct lines (26 points/line) and calibrations are performed for each line, with the standards image covering the effective distance. In this way, the density gradient is no longer a hindrance.

An addition parameter was identified during this study. Since the films were developed in an automatic machine, random variations in optical density occurred along the film length (102 cm) in despite of travel at constant speed during irradiation. This defect was attributed to the process for regenerating the chemical baths in the automatic machine. At the present time, this problem has not yet been solved. This is the reason why, for each plate tested, a 2-series of standards were placed on either side of the borated core. This helps estimating the relative error due to the measurements, as this inaccuracy may be different from one film to another.

The use of standards (1.2 and 0.8 times the nominal content) helped to further simplify the formulation of equation /3/. In fact in this load range, the exponential can be compared to a line, leading to extremely simple digital processing.

IV - CHARACTERISTICS OF THE AUTOMATIC DENSITOMETER

The instrument developed at Saclay is of relatively simple design, but it displays certain characteristics adapted to this type of testing (Fig. 4 and 5). The film is placed on a bracket

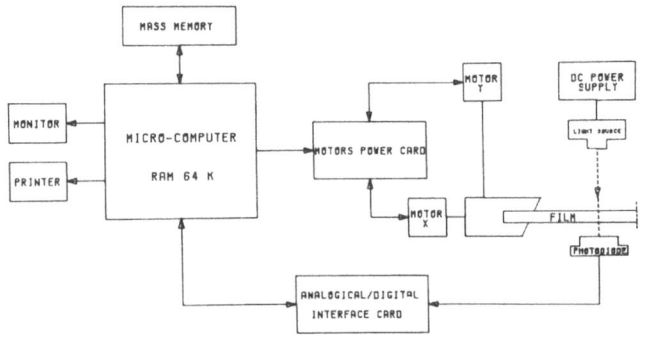

Fig. 4 - Instrument principle

Fig. 5 - Photograph of the densitometer (CEA-CEN/Saclay)

fixed to the frame of the mechanical unit. The mobil carriage (maximum travel X = 100 cm, Y = 15 cm) comprises a clamp supporting the detection cell and the source of light transmitted by an optical fibre. The analog signal (voltage 0 to 5 volts) is digitized by a programmable gain analog-to-digital converter.

The density range of the used films is 3 ($Ln [I_0/I]$). To limit the effects of particle size and small film defects, a cell with a large sensitive area (1 cm^2) has been choosen. The light spot can be optimized by using interchangeable collimators. Carriage travel is actuated by two step-by-step motors controlled by a micro-computer. The travels/acquisition sequence (15 minutes for 150 measurement points) is immediately followed by digital processing.

V - CONCLUSION

The quantitative estimation of the absorbent materials content can be achieved by neutron radiography. The accuracy of the measurements, currently in the neighbourhood of 10 %, can be substantially improved by a thorough analysis of the density fluctuations of the films processed in an automatic machine. This should help to approach the accuracies reached in neutron radiometry ($\sim$ 1 to 2 %).

The method offers the advantage of preserving a complete neutron image of the examined parts, thus meeting quality assurance requirements. This technique can also find a wide variety of applications, including measurements of enrichment and moisture content.

BIBLIOGRAPHY

"Use of a neutron guide in Industrial Neutron Radiography" A. LAPORTE - WCNR-1, San Diego, USA, 1981

"Actions des rayonnements sur les couches sensibles" J. SULTAN - Nucleus, n°1, volume 11, 1970

"Les causes de dispersions en radiographie industrielle" P.A. RUAULT - Revue Pratique de Contrôle Industriel, n°133 bis Septembre 1985

"Evolution du contrôle des éléments combustibles laminés" C. HAMEAU - COGEMA (Rapport Interne)

ENHANCEMENT OF IMAGE CONTRAST BY COMBINATION OF

NEUTRON AND X-RAY RADIOGRAPHY

Hisao KOBAYASHI, Kenji TOMURA

Institute for Atomic Energy, Rikkyo University,
2-5-1 Nagasaka, Yokosuka, Kanagawa, 240-01, JAPAN.

Hiromi WAKAO, Shinichiro SUZUKI, Tomomitsu HIGASHI

Department of Oral Radiology, Kanagawa Dental College,
82 Inaoka, Yokosuka, Kanagawa, 238,
JAPAN.

ABSTRACT

Images of X-ray radiography and thermal neutron radiography are theoretically analyzed for an object with local irregularities in atomic composition. Physical meanings of the new combined images are discussed. The image combination method is applied to resolve the measured images of enamel and dentin in a slice of tooth.

INTRODUCTION

The attenuation coefficients of X-rays and thermal neutrons show a remarkably different dependence on the atomic number of objects. Then, X-ray radiography (XR) and neutron radiography (NR) complement each other. For instance, in the mixture of substances with high atomic number and hydrogenous substances such as water and bonding agents, the comparison of photographs XR and NR has been proved effective to distinguish these substances (1-4). However, experimental researches in direct combination of images XR and NR have not been well developed and few theoretical considerations have been so far made.

This paper describes (1) theoretical consideration of direct comparison of images XR and NR, (2) a technical development of new image combination from XR and NR, and (3) experimental results for dental applications.

THEORY OF IMAGE RECONSTRUCTION

Generally, samples for XR and NR testing have complex configurations and local irregularities in atomic composition. The attenuation coefficient of X-rays, μ_X, depends monotonically and the neutron attenuation coefficient, μ_N, depends irregularly on atomic numbers. XR images presents more informations of thickness rather than atomic composition relative to NR images. On the contrary, NR images give more informations on specific nuclides rather than informations of thickness. There are some possibilities to extract more fruitful informations from combination of the two images than individual XR and NR images. We will first examine what physical quantities can be derived by the combination of XR and NR images and will develope a technique in which image contrast is enhanced in comparisen with each single image by processing and rearranging XR and NR images by a computer.

Although an object of radiography has no simple atomic composition usually, linear attenuation coefficients μ_X and μ_N for X-rays and neutrons are postulated to be additive functions of the coefficients of constituent atoms,

and

$$\mu_X = \rho \sum w_i \cdot \mu_i, \qquad (1a)$$

$$\mu_N = N\rho \sum w_i \cdot \sigma_i / A_i, \qquad (1b)$$

where ρ is density of the object (g/cm^3), w_i weight fraction of i'th atomic element, μ_i (cm^2/g) mass absorption coefficient for X-rays, N Avogadro's number, σ_i (cm^2) total cross sections for thermal neutrons, and A_i atomic mass.

When X-rays and neutrons incident uniformly over the field of exposure and the effects by scattering are neglected, total exposures, E_X and E_N, on an imaging screen are

$$E_{X,N} = I_{X_0,N_0} \cdot t_{X,N} \cdot \exp(1 - \mu_{X,N} \cdot d) \qquad (2a,b)$$

where I_{X_0} and I_{N_0} are incident fluxes, t_X and t_N are exposure time for X-rays and neutrons, respectively, and d sample thickness. To avoid redundant expressions in equations, E_X or E_N, I_{X_0} or I_{N_0}, t_X or t_N, etc. are hereinafter abbreviated as $E_{X,N}$, I_{X_0,N_0}, $t_{X,N}$, etc. as shown in eqs. (2).

Net optical film densities subtracted the base background

densities, $D_{x,N}$, are proportional to $E_{x,N}$ when total exposures are not very large and when sample size and thickness are both small enough to neglect scattering effects (5). If ratios of $E_{x,N}$ to $D_{x,N}$ are k_x and k_N, respectively, the optical film densities by X-rays and neutrons are

$$D_{x,N} = D_{xo,No} \cdot \exp(-\mu_{x,N} \cdot d) \qquad (3a,b)$$

where $D_{xo,No}$ correspond to the optical film density on the locations in the absence of object (100 % transmission) and are represented by the following equations;

$$D_{xo,No} = k_{x,N} \cdot I_{xo,No} \cdot t_{x,N} . \qquad (4a,b)$$

From equations (3),

$$\mu_{x,N} = (1/d) \cdot (\ln D_{xo,No} - \ln D_{x,N})$$
$$= (1/d) \cdot \ln(D_{xo,No} / D_{x,N}). \qquad (5a,b)$$

Thus, μ_x and μ_N, are obtained from XR and NR images themselves, if d is known over the whole object.

It is often necessary to have information of local composition in an object having complex configurations and compositions. We can draw a new information about this by the combination of XR and NR images. Linear attenuation coefficient ratio of X-rays to neutrons can be calculated for each corresponding points of the two images by an equation

$$R = (\mu_x/\mu_N) = \{\ln(D_{xo}/D_x)\} / \{\ln(D_{No}/D_N)\}, \qquad (6)$$

which is derived easily from eqs. (5). The ratio of X-rays to neutrons is independent of the sample thickness, d, in this method. Moreover, if we can measure the optical film densities at 100 % transmission, D_{xo} and D_{No}, we can obtain the ratio without knowing incident fluxes, exposure time, and constants $k_{x,N}$ characterized by film properties.

Next, we shall discuss a comparison of image contrast among single XR and NR images and the combination of them. From subtracting eq. (3b) from (3a),

$$S = (\mu_x - \mu_N) \cdot d = \ln\{(D_{xo} \cdot D_N) / (D_{No} \cdot D_x)\}, \qquad (7)$$

is obtained. Now, let $\Delta\mu_x$ and $\Delta\mu_N$ denote the difference of attenuation coefficient between two arbitrarily selected locations for X-ray and neutrons, respectively. When $(\Delta\mu_x) \cdot (\Delta\mu_N) < 0$, relation

$$\Delta(\mu_x - \mu_N) > \Delta\mu_x \text{ or } \Delta\mu_N \qquad (8)$$

is valid. Hence, the contrast is enhanced in the newly reconstructed image by eq. (7) in comparison with single XR or NR images, by eqs. (3). When the sample thickness, d is not so large, eq. (7) can be described as

$$S \simeq (D_{x_o} - D_x) - (D_{N_o} - D_N), \qquad (9)$$

and it is clear that the image treatment can be made more easily than that by eq. (7).

EXPERIMENTAL

An aluminum vertical type exposure facility was set on the graphite reflector of 100 kW TRIGA-II reactor. This facility has a diverging type collimator and an effective imaging area of 100 mm inner diameter. Detailed characteristics and configurations were described in another paper presented in this Proceedings (6). Table 1 shows the feature characteristics.

NR images were photographed on a single coated KODAK industrex R (SR-5) film coupled with 25 μm thick gadolinium screen in which Gd_2O_3 was evaporated on a 2 mm thick aluminum plate. The inherent image unsharpness in this technique was evaluated to be 48 μm (6).

For XR, we used a 150 kV X-ray generator equipped with a tungsten tube. The effective size of the X-ray source of the generator, D, was 0.8 mm x 0.8 mm and the distance of the source to the screen, L, was 100 cm, i.e. the L/D reached to 1250.

Table 1. Characteristics of 100 mm φ vertical type collimator of Rikkyo university TRIGA-II reactor.

REACTOR POWER ·············	100 kW
SOURCE ······ TOP OF THE REFLECTOR	
SIZE ·········· 100 mm φ x 5265 mm	
L / D ················ 110.9 ± 1.2	
n_{th} FLUX ······· 6.9×10^5 n/cm^2-s	
GAMMA DOSE RATE ·········· 4.1 R/h	
n-GAMMA RATIO···6.1×10^5 n/cm^2/mR	
Cd RATIO ····················· 4.3	

The XR images were recorded on a single coated KODAK XTL-2 film, through a 1.5 mm aluminum filter exposing 1 to 6 seconds on the operating condition of 50 kV and 100 mA. Mean effective X-ray energy of the operating condition was estimated to be 27 keV by a half value layer measurement. The inherent unsharpness on this case was estimated to be 57 μm by the same method by Kobayashi et al. (6)

A computerized microdensitometer (SAKURA PDM-5) was employed to convert photographed image into numerical data for a computer input. The microdensitometer can read two-dimensional picture image having a pixel of 0.1 mm x 0.1 mm typically to be equal to slit size.

The numerical data of images were treated by a 16 bit, 10 MHz microprocessor having 640 kB of users' memory space. This computer system was equipped with four 1 MB floppy disks and a 20 MB hard disk as external memories which are used to store and to treatment of the image data.

A linear relationship must be held between $D_{X,N}$ and $E_{X,N}$ for eqs. (5) to (8) to be valid. Figures 1 and 2 illustrate experimental results of the relation for X-rays and thermal neutrons, respectively. From these results, we can confirm the linearity for both KODAK XTL-2 and KODAK SR-5 in the range of $D_{X,N} \lesssim 3$.

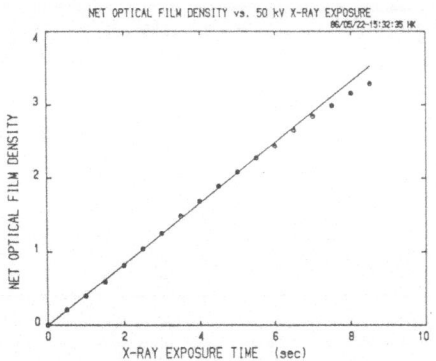

Fig.1 Optical film density as a function of 50 kV X-ray exposure time.

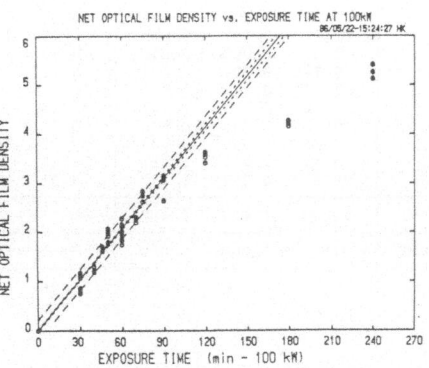

Fig.2 Optical film density as a function of neutron exposure time.

APPLICATIONS TO DENTAL SAMPLES

Enamel and dentine in tooth differ remarkably from each other in atomic compositions(7) shown in Tables 2(a) and (b). The weight fraction of hydrogen in dentin is 6 times as much as in enamel and that of calcium in dentin is 0.6 times in enamel. Therefore, teeth samples are thought to be suited to verify the theoretical constitutions mentioned in the preceding paragraph. Figures 3(a) and (b) show typical XR and NR results of a 3.5 mm thick slice of tooth, respectively. It is evident from Table 2 that enamel gives larger μ_X and absorbs more strongly X-rays. This was confirmable by experimental results as shown in Figs.3; the outer part of XR

Table 2(a). Cross section for thermal neutron (2200 m/s) and mass attenuation coefficient for 27 keV X-ray

ELE-MENT	AT. No.	AT. wt.	C. SEC. sc (barn)	C. SEC. abs (barn)	C. SEC. tot (barn)	ATT(27keV) (cm**2/g)
H	1	1.01	38.000	0.332	38.332	0.360
C	6	12.01	4.800	0.003	4.803	0.292
N	7	14.01	10.000	1.880	11.880	0.318
O	8	16.00	4.200	0.000	4.200	0.463
F	9	19.00	3.900	0.010	3.910	0.563
Na	11	22.99	4.000	0.536	4.536	0.954
Mg	12	24.31	3.600	0.063	3.663	1.179
P	15	30.97	5.000	0.200	5.200	2.263
S	16	32.06	1.100	0.520	1.620	2.821
Cl	17	35.45	16.000	33.600	49.600	3.243
K	19	39.12	1.500	2.070	3.570	4.582
Ca	20	40.08	3.200	0.440	3.640	5.465

Table 2(b). Linear attenuation coefficient of enamel, dentin, and cementum in a typical tooth structure for 2200 m/s thermal neutron and for 27 keV X-ray. Ratios of attenuation coefficient for X-ray .to that for neutron are also shown.

ELE-MENT	ENAMEL wt.%	(1/cm) n-ATT	X-ATT	DENTIN wt.%	(1/cm) n-ATT	X-ATT	CEMENTUM wt.%	(1/cm) n-ATT	X-ATT
H	1.72	1.150	0.018	9.90	5.125	0.081	11.97	5.785	0.091
C	1.50	0.011	0.013	8.06	0.044	0.053	7.54	0.038	0.046
N	0.20	0.003	0.002	1.14	0.013	0.008	1.47	0.016	0.010
O	42.61	0.197	0.575	42.45	0.152	0.444	48.31	0.161	0.471
F	0.01	0.000	0.000	0.02	0.000	0.000	0.00	0.000	0.000
Na	0.36	0.001	0.010	0.16	0.000	0.003	0.00	0.000	0.000
Mg	0.43	0.001	0.015	0.78	0.002	0.021	0.00	0.000	0.000
P	17.06	0.050	1.127	12.20	0.028	0.624	9.74	0.021	0.465
S	0.00	0.000	0.000	0.02	0.000	0.001	0.02	0.000	0.001
Cl	0.24	0.006	0.023	0.06	0.001	0.004	0.00	0.000	0.000
K	0.13	0.000	0.017	0.05	0.000	0.005	0.00	0.000	0.000
Ca	35.64	0.057	5.687	24.91	0.031	3.077	20.95	0.024	2.416
TOTAL	99.90	1.476	7.488	99.75	5.395	4.322	100.00	6.045	3.501
RATIO	————	1.000	5.072	————	1.000	0.801	————	1.000	0.579
DENSITY	2.92	————	————	2.26	————	————	2.11	————	————

(a)   (b)

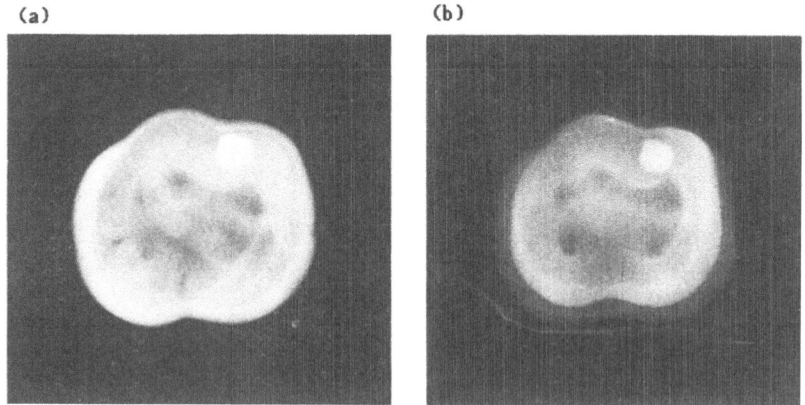

Fig.3 (a) 50 kV X-ray and (b) thermal neutron radiography images of a sliced tooth (3.5 mm thick).

image (enamel) was clear and that of NR was dark conversely. A clear circular image in the photographs was due to a silver alloy pin (65% Ag, 20% Sn, 8% Hg, 6% Cu and 1 % Zn) inserted into the tooth.

The two-dimensional XR and NR images of the sliced tooth shown in Figs. 3 were analyzed by the computerized microdensitometer as three-dimensional data with x, y-axis for the image position and z-axis for five different analysis of the physical quantities. Results are shown as stereograms in Figs. 4. Figures 4(a) and (b) show the optical film density, D_x and D_N, respectively. The values D_{xo}-D_x and D_{No}-D_N obtained by reversing the images in Figs. 3 are shown in Figs. 4(c) and (d), respectively. Figures 4(e) and (f) illustrate the stereograms of values $\mu_x d$ and $\mu_N d$ calculated by eqs. (5). Figures 4(g) and (h) show the stereograms reconstructed using R and S values in eqs. (6) and (7), respectively. A sharp projection in these figures is due to the silver alloy pin. From these figures, we also found that the image contrast was more enhanced in the order, Figs. 4(a)-(b), (c)-(d), (e)-(f), and (h). Some irregular thickness of sample reflected the unevenness in the part of dentin in Figs. 4 (e), (f) and (h). On the contrary, Fig. 4(g) supported that the R values were uniform over the dentin as expected from the above mentioned theoretical consideration.

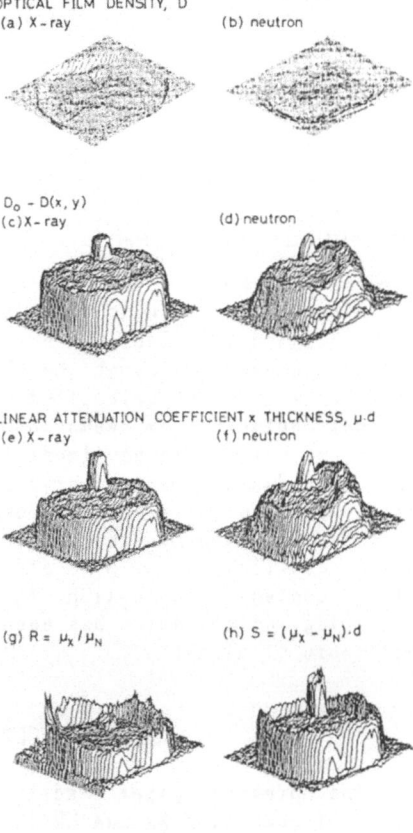

OPTICAL FILM DENSITY, D
(a) X-ray   (b) neutron

$D_o - D(x, y)$
(c) X-ray   (d) neutron

LINEAR ATTENUATION COEFFICIENT x THICKNESS, $\mu \cdot d$
(e) X-ray   (f) neutron

(g) $R = \mu_X / \mu_N$ (h) $S = (\mu_X - \mu_N) \cdot d$

Fig.4 (a) Stereographic images of optical film density, D, of a sliced tooth for 50 kV (27 keV) X-ray and
(b) for thermal neutron.
(c) Reversing optical film density, D_o-D, for X-ray and
(d) for thermal neutron.
(e) Linear attenuation coefficient multiplied by thickness, μd, for X-ray and
(f) for thermal neutron.
(g) R value; ratio of linear attenuation coefficient for X-ray to that for thermal neutron (μ_X/μ_N), and
(h) S value; difference between linear attenuation coefficient multiplied by thickness of a sliced tooth for X-ray and for thermal neutron.

Table 3. MEASURED LINEAR ATTENUATION COEFFICIENT FOR THERMAL NEUTRON AND 50 kV X-RAY[*] (μ_N and μ_X), R($=\mu_X/\mu_N$), AND S($=[\mu_X-\mu_N] \cdot$d) VALUES.

	NEUTRON		X-RAY		R		S
	obs.(cm⁻¹)	calc.(cm⁻¹)	obs.(cm⁻¹)	calc.(cm⁻¹)	obs.	calc.	obs.
ENAMEL	0.9±0.3(0.7)[**]	1.5(0.3)	3.5±1.0(1.7)	7.4(1.7)	5.6±2.2(3.3)	5.1(6.4)	0.7±0.1(2.4)
DENTIN	1.3±0.4(1.0)	5.4(1.0)	2.1±0.5(1.0)	4.3(1.0)	1.7±0.1(1.0)	0.8(1.0)	0.3±0.1(1.0)
PIN	1.9±0.1(1.5)	4.0(0.7)	13.8±1.2(6.6)	48.4(11.3)	2.8±0.6	12.1	1.0±0.2

[*] EFFECTIVE ENERGY = 27 keV.
[**] PARENTHESES: RELATIVE VALUES TO DENTIN.

A summary of results in this work was shown in Table 3. The values signed ± in Table 3 show the surface unevenness of enamel, dentine and pin. Although the observed μ-values disagreed about a factor of 2 with the calculated values from Table 2, the general tendency between them seems to be satisfactory. However, the μ-values were found to be almost the same as other six teeth samples. The disagreement of μ-values between calculation and observation may be attributed to use 2200 m/sec values as neutron cross-sections and monochromatic values at 27 keV as X-rays attenuation coefficients. Some errors are likely to be resulted from water content fluctuation in tooth and from the multiple scattering effect which has been neglected in this paper based on Whittemore et al. (5)

CONCLUSION

The present paper described a physically significant technique of combining XR and NR images. Some applications were made for sliced tooth samples, and the values, μ_X/μ_N and $(\mu_X-\mu_N)$d were estimated quantitatively. It was also proved that the technique was effective to enhance image contrast. The quantitative XR and NR technique is applicable to many other samples and has a possibility to be extended to more complex imaging technique such as a computed tomography method.

REFERENCES

1. R. L. Newacheck, "Neutron Radiography", J. P. Barton and P. v. d. Hardt eds. (D. Reidel Publ. Co., Dordrecht, 1983) p. 77.
2. P. A. Gillespie and T. Wall, ibid. p. 85.
3. J. J. Veenema, D. Mesman, and H. P. Leeflang, ibid. p. 127.
4. C. L. Leighty, ibid. p. 153.
5. W. L. Whittemore, G. Schlueter, and J. Shoptaugh, ibid. p. 885.
6. H. Kobayashi et al. (In this Proceedings, JP. 5, 1986).
7. Y. Suzuki et al. Jpn. Soc. Dental Rad. 16, 63(1976) (in Japanese).

POTENTIAL SAFEGUARDS APPLICATIONS OF FAST

NEUTRON HODOSCOPE RADIOGRAPHY[*]

A. DeVolpi

Argonne National Laboratory

Argonne, Illinois USA

The accurate determination of fissile content is essential to domestic and international safeguards of nuclear fuel. Experience gained with the fast-neutron hodoscope at the U.S. Department of Energy reactor TREAT indicates a sensitivity of 1 g of fuel out of a kg. The actual unanticipated absence of less than 1 g of mixed oxide was found in one instance; in fact, all prior forms of quality control, including mass measurements, x-ray radiography, and neutron radiography had failed to detect the deficit. Although the means utilized for uncovering this particular deficiency is a large fixed installation, the generic principles of using a neutron-induced source for digital radiography with an array of fast-neutron detectors operated in a scanning mode are applicable to many other situations. For example, a modest-size detector array could be installed at reactor or neutron sources to achieve equivalent sensitivity with longer operating times.

[*]Work supported by the U.S. Department of Energy, Technology Support Program, under Contract W-31-109-Eng-38.

WORK WAS DONE AT TREAT REACTOR

- U.S. DEPARTMENT OF ENERGY FACILITY
- ARGONNE NATIONAL LABORATORY, IDAHO
- PULSED REACTOR
- OPEN SLOT THROUGH THE BIOLOGICAL SHIELD AND CORE

HODOSCOPE AT TREAT (SEE FIGURE)

- FUEL-MOTION MONITORING SYSTEM
- CINERADIOGRAPHIC
- DIGITAL RADIOGRAPHS
- FAST-REACTOR TRANSIENT SAFETY EXPERIMENTS
- DESTRUCTIVE AND NON-DESTRUCTIVE EXPERIMENTS
- FISSILE AND NON-FISSILE MATERIAL MONITORING

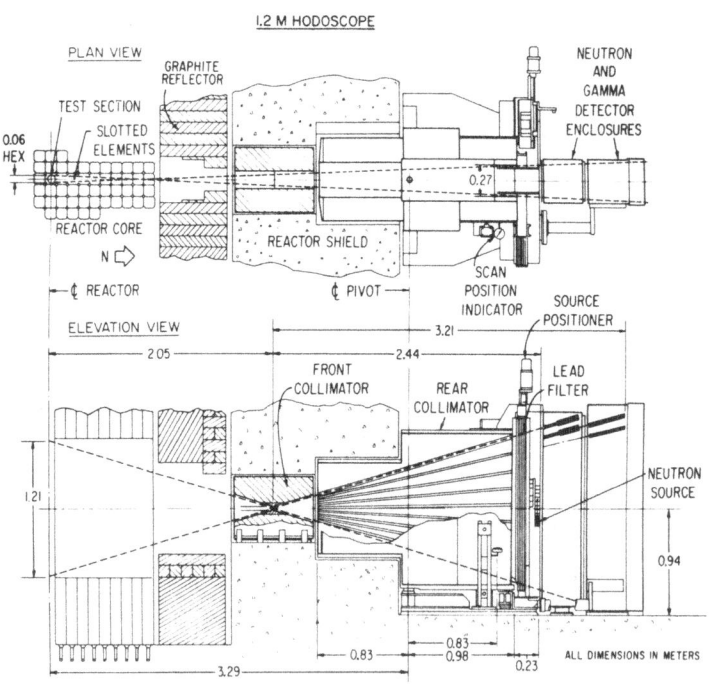

FIGURE: HODOSCOPE DIGITAL SCANNING SYSTEM

- MULTICHANNEL COLLIMATOR SYSTEM
- 360 COLLIMATOR SLOTS
- COLLIMATOR PLATES MADE OF STEEL
- DISTANCE FROM TEST FUEL TO DETECTORS IS OVER 4 M
- TWO ENCLOSURES FOR THREE DETECTOR ARRAYS

- PROPORTIONAL-COUNTER FAST NEUTRON
- HORNYAK-BUTTON FAST NEUTRON
- SODIUM-IODIDE GAMMA-RAY
- ALL THREE TYPES OF DETECTORS IN TANDEM
- CAPACITY FOR ABOUT 1000 DETECTORS
- ABOUT 700 DETECTOR CHANNELS INSTRUMENTED
- COLLIMATOR CAN BE PIVOTED FOR
  - HORIZONTAL SCANNING
  - VERTICAL SCANNING

HODOSCOPE RADIOGRAPHY

- BEFORE AND AFTER DESTRUCTIVE EXPERIMENTS
- DIGITAL DATA COLLECTION
- FAST NEUTRON AND GAMMA
- 700 COLLIMATED DETECTORS

OBJECTS BEING INSPECTED (TYPICAL)

- 1 TO 7 FUEL PINS
- ABOUT 160 G HEAVY METAL PER PIN
- U AND/OR PU
- VARYING FISSILE FRACTIONS
- WITHIN TEST CONTAINER
- THICK WALLS (SEE FIGURE)

ALIP

ZIRCONIA FILLER
(25.5 mm I.D, 30 mm O.D)

FUEL PELLETS
(4.95 mm O.D, 1.5 mm I.D)

FLUTED FLOW TUBE
(23 mm I.D, 25 mm O.D)

(WEST)

OUTER ADIABATIC WALL
(30.5 mm I.D, 32 mm O.D)

LOOP WALL
(33.5 mm I.D, 50.8 mm O.D)

HODOSCOPE COLUMN

HODOSCOPE
(NORTH)

FIGURE: TEST VEHICLE CROSS-SECTION

- 7-PIN CONTAINER
- FITS IN 10 X 10 CM OR 10 X 20 CM SLOT IN REACTOR
  CORE
- SOME MINOR STRUCTURE OMITTED FROM DRAWING
- PLACED AT CENTER OF TREAT REACTOR

SAFEGUARDS APPLICATIONS

HODOSCOPE RADIOGRAPHIC OPERATION

- HORIZONTAL AND VERTICAL SCANNING OF COLLIMATOR
- LOW, STEADY-STATE REACTOR POWER
- FISSION INDUCED BY REACTOR NEUTRONS
- REMOTE COMPUTER-CONTROLLED OPERATION

FISSION-RATE DEFICIENCY (SEE FIGURE)

- DEFICIENCY IN LOWER-RIGHT-HAND-CORNER
- TOTAL MASS OF FUEL: 1000 G HEAVY METAL
- 24% FISSILE/FERTILE FRACTION
- 1 6 FISSILE PU DEFICIENCY (0.9 ± 0.4 G)
- DATA HIGHLY SIGNIFICANT AND SELF-CONSISTENT
- ~ 1 HR SCANNING OPERATION

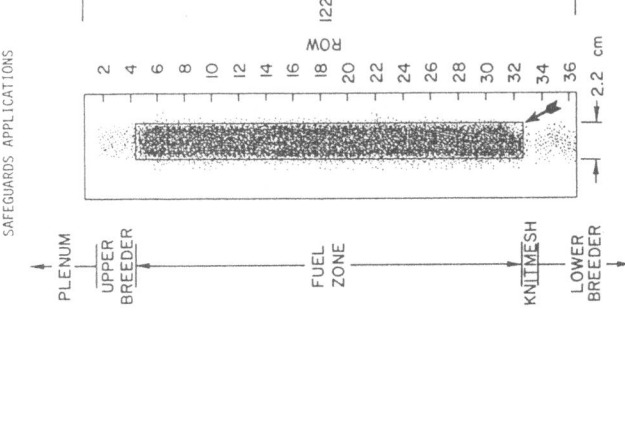

FIGURE: HODOGRAPH RECONSTRUCTION OF DIGITAL DATA

- PU-OXIDE IN FUEL ZONE
- DEPLETED U IN UPPER AND LOWER BREEDER ZONES
- NO FUEL IN PLENUM
- NO FUEL IN KNITMESH
- NOTE ABSENCE OF FUEL AT ARROW

SAFEGUARDS APPLICATIONS

DEFICIENCY SURPRISING

• NOT ANTICIPATED
• NOT DETECTED BY GRAVIMETRIC METHODS
• NOT NOTICED IN INDIVIDUAL PIN RADIOGRAPHS
  - X-RAY
  - THERMAL-NEUTRON
  - IN-RESONANCE NEUTRON
• NEVER OBSERVED IN PRIOR EXPERIMENTS

BUILT-IN CALIBRATIONS

• 2-CM UNFUELED REGION (KNITMESH) BELOW FUEL
• 1% FISSILE BREEDER PELLETS ABOVE AND BELOW FUEL
• KNOWN GEOMETRY

EXPLANATION FOR DEFICIENCY

• ONE PELLET MISSING FROM ONE PIN IN STACK (BUT WITHIN SPECS)
• SAME PIN PUSHED UP THE HEIGHT OF ABOUT TWO PELLETS

REASON FOR HODOSCOPE SENSITIVITY

• MEASURES FISSILE RATE
• PENETRATING RADIATION
• RELATIVELY INSENSITIVE TO OTHER MATERIALS
• MINIMAL BACKGROUND

LIMITATIONS OR INADEQUACIES OF ALTERNATIVE METHODS

• RADIATION FROM IRRADIATED FUEL
• ACCOMMODATION OF AS-BUILT ASSEMBLIES
• EFFECT OF THICK CONTAINER WALLS
• SENSITIVITY TO EVEN-ISOTOPES OF PU

ADAPTATION TO OTHER REACTORS

• SINGLE-CHANNEL COLLIMATOR
• VALIDATION OF OTHER METHODS
• APPLICATION TO SPECIAL SITUATIONS

SUMMARY

• SENSITIVITY: 1 G OF 1KG HEAVY METAL
• SURROUNDINGS: THICK CONTAINERS, AS-BUILT
• FUEL: PREIRRADIATED OR FRESH
• DETECTION RADIATION: FISSION NEUTRONS
• INSTRUMENT: HODOSCOPE DIGITAL SCANNER

REFERENCES

A. DEVOLPI AND E.A. RHODES, "REACTOR IN-SITU TEST-FUEL NEUTRON RADIOGRAPHY," TRANSACTIONS, FIRST WORLD CONFERENCE ON NEUTRON RADIOGRAPHY, SAN DIEGO, CALIF. (1981).

NEUTRON RADIOGRAPHIC TECHNIQUES FOR NUCLEAR

ARMS CONTROL APPLICATIONS[*]

A. DeVolpi

Argonne National Laboratory

Argonne, Illinois USA

The verification of nuclear arms control agreements is likely to entail provisions for detection, identification, and imaging of objects that are known to have or might contain fissile masses. Determining the number of nuclear warheads in a MIRV or finding out whether a cruise missile has a nuclear or conventional war head are examples. To do so requires techniques that make use of either inherent radiation associated with fissile materials used in weapons or their interrogation with external sources. Because of the possibility of evasive measures, some application of more intrusive active interrogation techniques might be necessary, in which case the minimum delivery of radiation doses would be preferred or mandated. Fast neutrons are most promising for interrogation because of their penetrability and because of the distinctive characteristic radiation that can be induced in the object being inspected. The information that is collected, which might include a radiograph, must be tailored to the situation, providing neither too much nor too little information. These and other requirements that are public information are examined, particularly as they relate to potential terrestrial and space applications.

[*]Work supported by the U.S. Department of Energy, Technology Support Programs, Contract W-31-109-Eng-38.

PROPOSED APPLICATION OF RADIOGRAPHIC EXPERIENCE

• EXPERIENCE GAINED IN FISSILE-MATERIAL MEASUREMENT
• DATA BASE ACCUMULATED WITH FAST-NEUTRON HODOSCOPE
• CONCEPT OF APPLICATION TO ARMS-CONTROL TREATY
  VERIFICATION

NUCLEAR ARMS-CONTROL AGREEMENTS ON STRATEGIC NUCLEAR WEAPONS

• EXISTING TREATIES AND AGREEMENTS
  - SALT-I: SETS LIMITS ON LAUNCHERS AND RE-ENTRY VEHICLES
  - SALT-II: NEW LIMITS ON LAUNCHERS AND RE-ENTRY VEHICLES
  - OUTER SPACE TREATY: FORBIDS NUCLEAR WEAPONS IN ORBIT
  - SEABED TREATY: FORBIDS NUCLEAR WEAPONS ON THE SEABED
• POSSIBLE FUTURE TREATIES OR AGREEMENTS
  - START: STRATEGIC ARMS REDUCTIONS
  - INF: REDUCTIONS IN INTERMEDIATE RANGE MISSILES
  - OTHER????

STRATEGIC NUCLEAR WEAPONS

• INTERCONTINENTAL BALLISTIC MISSILES
  - SINGLE WARHEADS
  - MULTIPLE WARHEADS (MIRV)
• INTERMEDIATE RANGE MISSILES
  - MOBILE MISSILES
  - CRUISE MISSILES

VERIFICATION OF STRATEGIC-WEAPON LIMITS

• NTM: NATIONAL TECHNICAL MEANS
  - SATELLITE RECONNAISSANCE
  - HUMAN INTELLIGENCE
• IN-COUNTRY MONITORING
  - PORTAL/PERIMETER MONITORS
  - OTHER DEVICES
• ON-SITE INSPECTION
  - VISUAL CONFIRMATION
  - NON-INTRUSIVE INSTRUMENTS

NON-INTRUSIVE INSTRUMENTS

• DETERMINE NUMBER OF NUCLEAR WARHEADS
• WITHOUT REMOVING MISSILE NOSE CONE
• WITHOUT RISK TO MISSILE OPERATION
• IN THE FIELD OR SILO
• HIGH RELIABILITY
• NOT SUBJECT TO SPOOFING
• NOT OVERLY INTRUSIVE

POSSIBLE TREATY PROVISIONS

• COUNT NUMBER OF NUCLEAR WARHEADS IN A MISSILE PAYLOAD
  - MIRV MULTIPLICITY CAN BE UP TO ABOUT 15
  - DUMMIES OR DECOYS CAN BE SUBSTITUTED
• DISTINGUISH BETWEEN NUCLEAR OR CONVENTIONAL WARHEAD
  - CRUISE MISSILES
  - OTHER WEAPON SYSTEMS

ARMS CONTROL APPLICATIONS

POSSIBLE NUCLEAR WARHEAD SIGNATURES

- FISSION TRIGGER
    - INHERENT RADIATION
    - EXTERNAL-SOURCE INTERROGATION
- FUSION COMPONENTS
    - TRITIUM
    - EXTERNAL-SOURCE INTERROGATION

COMPLICATIONS

- CHEATING
- USE IN THE FIELD
- RELIABILITY
- BILATERAL DISCLOSURE OF INSTRUMENTS AND DATA
- NOT SENSITIVE TO UNNECESSARY DETAILS

FAST-NEUTRON INTERROGATION

- PENETRABILITY
- SENSITIVITY TO FISSILE MATERIALS
- PRODUCES UNIQUE SIGNATURE REACTIONS
- MINIMUM COLLATERAL EFFECTS

RADIOGRAPHIC TECHNIQUES

- FAST-NEUTRON SOURCES
    - RADIOACTIVE SOURCES
    - NEUTRON-GENERATORS
- FAST-NEUTRON DETECTORS
    - HODOSCOPE ARRAYS
    - RESOLUTION CONTROL
    - DIGITAL DATA AND RECONSTRUCTION

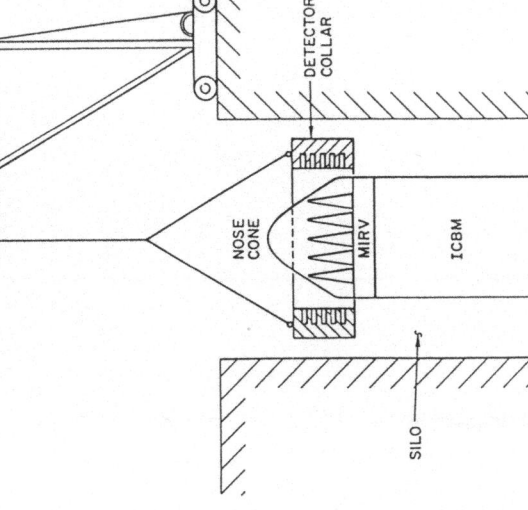

SKETCH: PROPOSED PHYSICAL ARRANGEMENT

- MIRV-ICBM MEASUREMENT SYSTEM
- IN MISSILE SILO
- TOMOGRAPHIC RECONSTRUCTION
- RESOLUTION LIMITED BY NUMBER OF DETECTORS
- DOES NOT REQUIRE REMOVAL OF NOSE CONE

REFERENCE

A. DE VOLPI, BULLETIN OF ATOMIC SCIENTISTS (APR. 1970).

CHEMICAL AND EXPLOSIVE IDENTIFICATION

USING ADVANCED NEUTRON RADIOGRAPHIC TECHNIQUES[*]

A. DeVolpi

Argonne National Laboratory

Argonne, IL 60439 USA

The detection of chemicals or explosives for weapons is of technological interest. Radiographic techniques that take advantage of unique characteristics of such substances might have a role in their detection and identification. Imaging of an object is the first step in evaluating its identity, but specific detection of its chemical composition is another feature of importance. Because we must generally consider extended objects of mixed composition, active interrogation must be done with penetrating radiation. Although both fast neutrons and gamma rays might be candidates, only the neutrons can yield useful reaction radiation that is adequately characteristic of an object. Therefore, a combination of fast-neutron radiography and nuclear interrogation methods appears to be needed. Neutrons for nuclear interrogation have usually been provided by accelerators or radioactive sources. Some ideas derived from first-hand experience will be discussed.

[*]Work supported by the U.S. Department of Energy, "Technology Support Programs" under Contract W-31-109-Eng-38.

DETECTING EXPLOSIVES

- INTERNATIONAL PROBLEMS IN ILLEGAL TRANSPORT AND USE
- DIFFICULTIES IN DETECTION BY EXISTING MEANS

APPLICATIONS OF EXISTING TECHNOLOGY BASE

- MATERIALS-DETECTION IN REACTOR-SAFETY PROGRAM
  - FISSILE MATERIALS
  - STEEL
  - HYDROGEN AND SODIUM
- FAST-NEUTRON HODOSCOPE EXPERIENCE AND DATA BASE

NEUTRON RADIOGRAPHY HAS POTENTIAL ROLE

- SEVERAL LEVELS OF INFORMATION
  - DETECTION
  - IMAGING
  - IDENTIFICATION
- NEUTRONS CAN PRODUCE INFORMATIVE REACTIONS IN OBJECTS

THE PROBLEM

- DETECTION OF SELECTED SUBSTANCES
- UNDER ADVERSE CONDITIONS
- WITH A RELIABLE SYSTEM
- WITHOUT DEPENDENCE ON HUMAN JUDGEMENT
- WITH MINIMAL COLLATERAL EFFECTS
- AT REASONABLE COST

PART OF THE SOLUTION

- INJECT PENETRATING NEUTRAL RADIATION
  - LARGE OBJECTS
  - VARYING MATERIALS
- DETECT CHARACTERISTIC REACTIONS IN OBJECT
  - TRANSMISSION
  - SCATTERING

  - REACTION
- MAXIMIZE DATA COLLECTION
- COMPUTERIZE DATA ANALYSIS
- USE HIGH ENERGY NEUTRONS
  - DEEP PENETRATION
  - MINIMAL ACTIVATION RESIDUE
  - UNIQUE SIGNATURE REACTIONS
  - USABLE CROSS SECTIONS
- ANALYZE MULTIPLE CORRELATIONS
  - RADIATION PARTICLES
  - RADIATION ENERGY
  - RADIATION MULTIPLICITY
  - RADIATION ANGLES

PASSIVE AND ACTIVE INTERROGATION

- A COMPREHENSIVE PROPOSAL
- BASED ON ARGONNE TECHNOLOGY
- A NEW METHOD HAS BEEN INVENTED: Cf^{252} CORRELATION HODOSCOPE
- DETECTION, IDENTIFICATION, AND IMAGING SIMULTANEOUSLY

REASONS FOR NEW METHOD

- LIMITATIONS OF OTHER TECHNIQUES
- HIGH-ENERGY RADIOGRAPHY
- CHEMICAL/NUCLEAR IDENTIFICATION
- WITHOUT AN ACCELERATOR
- FIELDABLE
- MANY DERIVATIVE APPLICATIONS

FEATURES OF THE CF-252 CORRELATION METHOD

- MAXIMUM INFORMATION
- PENETRATING RADIATION
- MINIMUM ACTIVATION
- MULTIPLE CORRELATION ANALYSIS
- CHARACTERIZATION BY CROSS-SECTION AND REACTION PRODUCTS
- SIMULTANEOUS TRANSMISSION AND REACTION
- COPES WITH HIGH AND VARIABLE BACKGROUND
- INEXPENSIVE, FIELDABLE SOURCE
- MODULAR, FIELDABLE SYSTEM
- TOMOGRAPHIC IMAGING

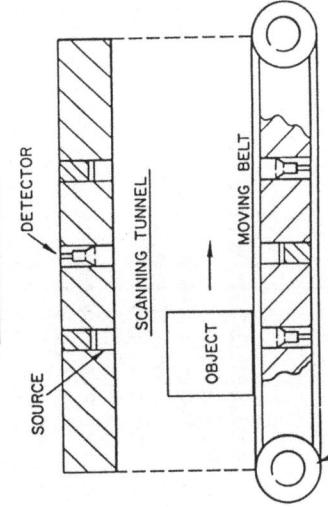

SCANNING STATIONS

SOURCE   DETECTOR

SCANNING TUNNEL

OBJECT   MOVING BELT

BELT DRIVE

FIGURE: SCANNING STATIONS

- MOVABLE BELT
- DETECTOR ARRAYS AT ONE OR MORE POSITIONS

THREE MAJOR FEATURES

- A MULTIPLICITY OF Cf^{252} FISSION COUNTERS
- A MULTIPLICITY OF GAMMA DETECTORS
- THE RECORDING AND ANALYSIS OF A MULTIPLICITY OF CORRELATIONS

γ, n
TRANSMISSION
DETECTOR

OBJECT

^{252}Cf
SOURCE

γ, n REACTION
DETECTOR

SCHEMATIC: MULTIPLE CORRELATION HODOSCOPE

- CF-252 SPONTANEOUS SOURCE
- IMAGING TRANSMISSION DETECTOR
  - GAMMA RAYS
  - FISSION NEUTRONS
- IMAGING REACTION DETECTOR
  - SCATTERED NEUTRONS
  - INELASTIC AND CAPTURE GAMMAS

A SWITCHABLE RADIOACTIVE NEUTRON SOURCE

- (ALPHA,N) SOURCE
- REDUCED OPERATOR EXPOSURE OR RISK FROM NEUTRONS
- SUBSTITUTE FOR ACCELERATOR NEUTRON SOURCE
- HIGH SWITCHING RATIO
- REMOTELY OPERATED
- INTENDED TO BE TRANSPORTABLE AND LICENSABLE
- LESS COMPLICATED, MORE RELIABLE, MORE FIELDABLE THAN ACCELERATOR

SUMMARY

- NEUTRON RADIOGRAPHY MIGHT BE USEFUL
- IN DETECTING CHEMICAL EXPLOSIVES
- IN CHECKED LUGGAGE AND AIR FREIGHT
- PROVIDED A FIELDABLE
- FAST-NEUTRON SYSTEM
- IS DEVELOPED

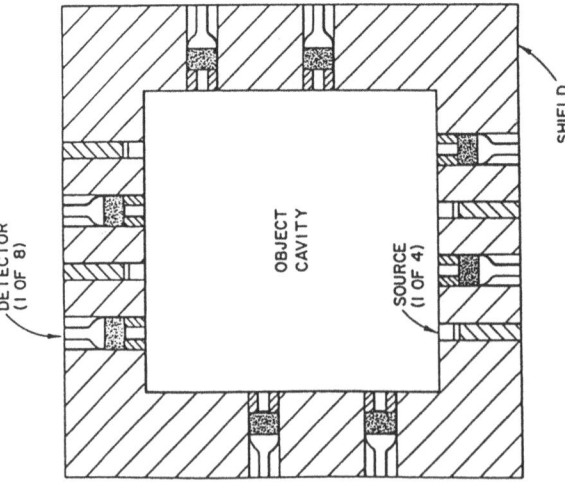

DETECTOR
(1 OF 8)

OBJECT
CAVITY

SOURCE
(1 OF 4)

SHIELD

FIGURE: END-VIEW OF SCANNING STATION

- OPPOSED SOURCES AND DETECTORS
- DETECTORS ON ALL FOUR SIDES
- TWO PAIRS OF TRANSMISSION DETECTORS (VERTICAL)
- TWO PAIRS OF REACTION DETECTORS (HORIZONTAL)

TEMPERATURE SENSITIVE CONTRAST AGENTS FOR NEUTRON RADIOGRAPHY

J. S. Brenizer[1], R. W. Jenkins, Jr.[2], D. D. McRae[2],
J. B. Paine, III[2] and M. F. Sulcoski[1]

[1]University of Virginia, Charlottesville, VA 22901
[2]Philip Morris USA, Richmond, VA 23261

ABSTRACT

Gadolinium salts have been used as contrast agents from the
early days of neutron radiography because of the extremely
high neutron absorption cross section of gadolinium.
Gadolinium salts, such as the chloride and the nitrate, are
generally very soluble in water but are also often
hygroscopic and corrosive. This work reports a series of
gadolinium chelates based on gadolinium acetylacetonate that
were adapted for use as contrast agents. The chelates are
soluble in a variety of volatile organic solvents such as
methylene chloride and acetone. In addition, some are
thermally stable in the absence of oxygen and water, and by
observing when they melt or vaporize, they can be used as
temperature indicators.

INTRODUCTION

The use of contrast agents in neutron radiography is well
known although often not covered in detail in the literature. For
example, Ref. 1 makes only brief mention of contrast agents. Of
the contrast agents used in neutron radiography, those containing
gadolinium are the most common since this element has the highest
neutron absorption cross section of all the elements. Several
papers at the First World Conference on Neutron Radiography
discussed the use of gadolinium oxide or salts to enhance the
detection of ceramic mould material left in gas cooled turbine
blades (2-4). Other applications of gadolinium contrast agents

include use in penetrants for leak and crack detection, and for doping of explosives and teeth (5-8). Other contrast agents discussed in the literature are dysprosium metal and helium-3 (9-11). Deuterium, which has a lower scattering cross section than hydrogen, has been used as a negative contrast agent in hydrogen transport studies and in biological systems (12,13). By comparison, development of contrast agents for X-ray radiography is an active field of research. The X-ray research is primarily driven by the need for specific and/or nontoxic contrast agents for medical imaging. The analogous medical application for neutron radiography does not exist for live human subjects (1). In addition, since neutron radiography is frequently used to image hydrogenous materials in metals as a complementary method to X-ray radiography, the need to develop new contrast agents for neutron imaging has not been as great.

This paper reports the development of a series of gadolinium chelates, specifically, gadolinium β-diketonates, for use as new contrast agents in neutron radiography. The β-diketonates of several lanthanide elements, including gadolinium, are commonly used as NMR shift reagents (14). One of the more useful aspects of the gadolinium chelates as contrast agents is that they are soluble in a variety of volatile organic solvents such as methylene chloride and acetone. Gadolinium salts, such as the chloride and the nitrate, are generally very soluble in water but are also often hygroscopic and corrosive. A second aspect of these chelates is that they can be used as temperature indicators in the absence of oxygen and water by observing their melting points. In the case of one chelate, a second temperature indication is provided when it vaporizes.

Although they will not be discussed further in this paper, the sodium-, potassium- and ammonium-gadolinium salts of EDTA (ethylenediaminetetracetic acid) were also made. They are very easy to synthesize and are useful as contrast agents in aqueous solutions where the EDTA sequesters the gadolinium ion preventing undesired reactions. Dysprosium, europium and samarium also have large neutron capture cross sections and have uses or potential uses as contrast agents. Since they are in the lanthanide series, their chemistry is practically identical to the chemistry of gadolinium. Thus, any chelate made with gadolinium can be made from these other lanthanides. Terbium chelates, while not useful as neutron contrast agents, fluoresce under ultraviolet light and can be used as visual tracers when mixed with the gadolinium chelates. The lanthanides, because of their high atomic numbers, also make reasonably good contrast agents for X-ray radiography. A recent patent discusses this application in more detail and indicates that they can be used as contrast agents in NMR imaging as well (15).

EXPERIMENTAL

The gadolinium β-diketonates were prepared essentially by literature methods (16-21). Generally, an aqueous or aqueous-alcoholic solution of gadolinium chloride or nitrate was treated with an aqueous or aqueous-alcoholic solution of the ammonium or sodium salt of the β-diketone. For large scale synthesis, it was convenient and economically desirable to dissolve Gd_2O_3 in reagent grade HCl adjusting the pH with NH_4OH if necessary. The gadolinium chelates of the following β-diketones were prepared: 2,4-pentanedione (acetylacetone, ACAC) (16); 2,2,6,6-tetramethyl-3,5-heptanedione (TMHD) (17-19); 1,1,1,5,5,5-hexafluoro-2,4-pentanedione (hexafluoroacetylacetone, HFAA) (19,20); and 1,3-diphenyl-1,3-propanedione (DPPD) (21). All of the β-diketones are available commercially. The syntheses are relatively easy except for the HFAA chelate which requires more care. The structures of the chelates are shown in Table I.

Experimental trials of the thermal behavior of the gadolinium β-diketonates in a neutron beam were performed by placing pellets of GdHFAA, GdTMHD and GdDPPD separately into the tops of 3-mm holes drilled vertically through a 12-mm aluminum rod. The rod was heated in a furnace made by wrapping a 15-mm quartz tube with a nichrome wire heating element. The furnace was placed in the neutron beam in front of the real-time neutron camera. As the temperature increased, each of the pellets melted and drained out of its hole. The temperature of the rod was measured with a thermocouple with a digital temperature display. Because the chelates are sensitive to water and oxygen at elevated temperatures, the furnace was continually flushed with nitrogen at a low flow rate.

The neutron radiography facility is part of the 2 MW research reactor at the University of Virginia Department of Nuclear Engineering and Engineering Physics (22,23). The neutron beam has a flux of 1×10^7 neutrons/cm^2-sec with a beam diameter of 15 cm. The facility has a real-time video neutron imaging camera produced by Precise Optics and an advanced video system (24). The video system includes a Micro Consultants Intellect 100 digital image processor based on a DEC LSI-11/23 computer and a 512 x 512 pixel, 8 bit frame store. The Intellect 100 has a complete range of image processing software including noise reduction, picture integration and differentiation, line densitometer, filtering masks for image enhancement, contrast enhancement and false color. Other features of the video system are a high resolution video tape recorder, a video timer and a special effects generator. The latter permits the insertion of a video image from a second video camera into the image from the neutron camera. In this work, the second camera was used to observe the digital temperature display so that both the temperature and the neutron image of the sample were synchronously recorded in the video picture.

TABLE I

Names and structures of the gadolinium β-diketonates.

Tris (acetylacetonate) gadolinium abbreviated as GdACAC	
Tris (2,2,6,6-tetramethyl-3,5-heptanedionato) gadolinium abbreviated as GdTMHD (available commercially)	
Tris (1,3-diphenyl-1,3-propanedionato) gadolinium abbreviated as GdDPPD	
Tris (1,1,1,5,5,5-hexafluoro-2,4-pentanedionato) gadolinium abbreviated as GdHFAA	

RESULTS AND DISCUSSION

Table II presents the melting and vaporization temperatures of the gadolinium β-diketonates that were synthesized. The melting points for GdTMHD, GdDPPD and GdHFAA were determined in a melting point apparatus. The results from the tube furnace agreed well with data from the melting point apparatus. As shown in Table II, only GdTMHD vaporizes. GdACAC does not melt or vaporize and while GdDPPD and GdHFAA melt, they seem to decompose at higher temperatures.

TABLE II

Melting and vaporization temperatures of the gadolinium β-diketonates.

Chelate	Melting Point	Vaporization Point
GdACAC	- - -	- - -
GdTMHD	190°C	220°C
GdDPPD	245°C	- - -
GdHFAA	125°C	- - -

Figures 1-4 show radiographs of the sequential melting of GdHFAA, GdTMHD and GdDPPD as they were heated in the tube furnace. The temperature is shown in the upper left-hand corner. In Fig. 1 all of the chelates are solid at 113°C. At 129°C in Fig. 2 the GdHFAA has melted and drained out of its hole. Figure 3 shows that the GdTMHD has melted and drained out at 188°C. Finally in Fig. 4, the GdDPPD has melted at a temperature of 249°C.

As was noted above, the gadolinium β-diketonates are soluble in a variety of volatile organic solvents. Table III lists some of the solvents tested. The chelates do not seem to be soluble in nonpolar solvents such as hexane. The solubility of these compounds in organic solvents is a very useful characteristic. The contrast agent solutions dry quickly and the use of water can be avoided.

TABLE III

Solvents for the gadolinium β-diketonates.*

Solvent	GdACAC	GdTMHD	GdDPPD	GdHFAA
Acetone	Yes	Yes	Yes	Yes
Methanol	Yes	Yes	Yes	Yes
Hexane	No	No	No	No
Chloroform	Yes	Yes	Yes	No
Methylene Chloride	Yes	Yes	Yes	No

*Solubilities may be limited.

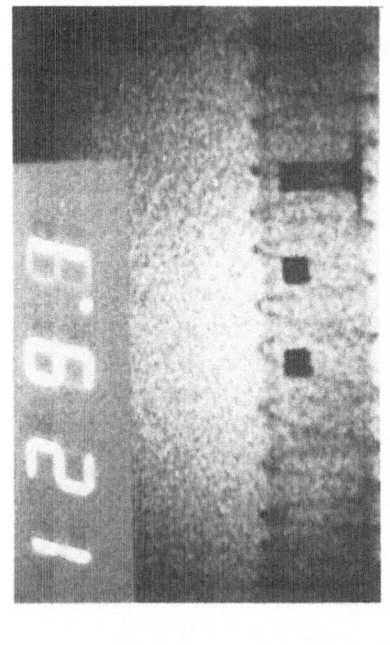

Figure 1. 113°C. All chelate samples are solid. From left to right: GdDPPD, GdTMHD and GdHFAA.

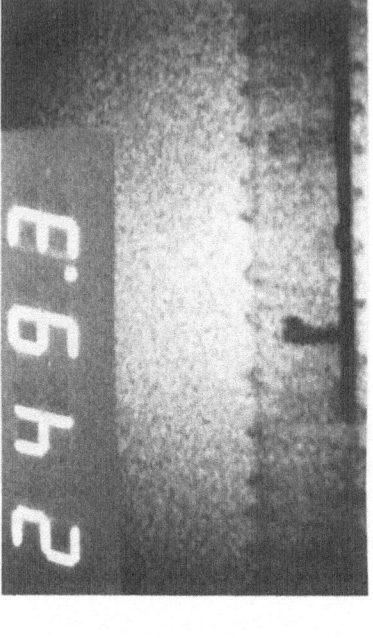

Figure 2. 129°C. The GdHFAA has melted.

Figure 3. 188°C. The GdTMHD has melted.

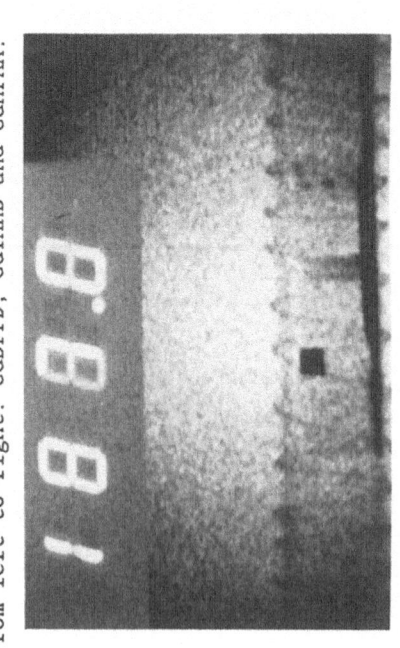

Figure 4. 249°C. The GdDPPD has melted.

CONCLUSIONS

The gadolinium β-diketonates were originally investigated for possible use in systems undergoing combustion and pyrolysis. Because of their sensitivity to water and oxygen at elevated temperatures, this use proved impractical. However, there are many occasions in nondestructive testing where these chelates could be useful. Possible applications would include contrast enhancement in objects sensitive to water or susceptible to corrosion. The chelates could be used to indicate if an object had been subjected to heat. Finally, these chelates could prove useful as contrast agents for not only neutron radiography, but also X-ray radiography and nuclear magnetic resonance imaging.

REFERENCES

1. P. von der Hardt and H. Röttger, Eds., Neutron Radiography Handbook, D. Reidel, Dordrecht, Holland, 1981.

2. R. L. Newacheck, in Neutron Radiography, p. 77, J. P. Barton and P. von der Hardt, Eds., D. Reidel, Dordrecht, Holland, 1983.

3. D. J. Taylor, in Neutron Radiography, p. 145, J. P. Barton and P. von der Hardt, Eds., D. Reidel, Dordrecht, Holland, 1983.

4. C. E. Leighty, in Neutron Radiography, p. 153, J. P. Barton and P. von der Hardt, Eds., D. Reidel, Dordrecht, Holland, 1983.

5. G. Bayon, L. Laporte and J. Le Gall, in Neutron Radiography, p. 67, J. P. Barton and P. von der Hardt, Eds., D. Reidel, Dordrecht, Holland, 1983.

6. P. A. Gillespie and T. Wall, in Neutron Radiography, p. 85, J. P. Barton and P. von der Hardt, Eds., D. Reidel, Dordrecht, Holland, 1983.

7. K. G. Golliher, in Neutron Radiography, p. 339, J. P. Barton and P. von der Hardt, Eds., D. Reidel, Dordrecht, Holland, 1983.

8. J. M. Vulcain, J. Tamisier, J. H. Espie, R. Masse and A. Laporte, in Neutron Radiography, p. 555, J. P. Barton and P. von der Hardt, Eds., D. Reidel, Dordrecht, Holland, 1983.

9. H. Hausen and R. Lölgen, in Neutron Radiography, p. 369, J. P. Barton and P. von der Hardt, Eds., D. Reidel, Dordrecht, Holland, 1983.

10. S. C. Johnson and L. W. Dahlke, Rev. Sci. Instr. 49 (2), 242 (1978).

11. G. B. Holland, in *Neutron Radiography*, p. 333, J. P. Barton and P. von der Hardt, Eds., D. Reidel, Dordrecht, Holland, 1983.

12. H. Rauch and A. Zeilinger, *Atomic Energy Rev.* 15 (2), 249 (1977).

13. A. R. Spowart, *J. Phys. E*, 5 (6), 497 (1972).

14. A. F. Cockerill, G. L. O. Davies, R. C. Harden and D. M. Rackham, *Chem. Rev.* 73 (6), 553 (1973).

15. A.-G. Schering, Belg. BE 898,709 (Cl. A61K), 16 May 1984, DE Appl. 3,302,410, 21 Jan 1983.

16. J. G. Stites, C. N. McCarty and L. L. Quill, *J. Amer. Chem. Soc.* 70, 3142 (1948).

17. G. S. Hammond, D. C. Nonhebel and C-H. S. Wu, *Inorg. Chem.* 2, 73 (1963).

18. K. G. Eisentraut and R. E. Sievers, *J. Amer. Chem. Soc.* 87, 5254 (1965).

19. E. W. Berg and J. J. Chaing Acosta, *Anal. Chim. Acta.* 40, 101 (1968).

20. F. Halverson, J. S. Brinen and J. R. Leto, *J. Chem. Phys.* 40, 2790 (1964).

21. A. Perrotto and R. G. Charles, *J. Inorg. Nucl. Chem.* 26, 373 (1964).

22. J. S. Brenizer and M. F. Sulcoski, in *Use and Development of Low and Medium Flux Research Reactors*, p.958, O. K. Harling, L. Clark, Jr. and P. von der Hardt, Eds., Karl Thiemig Graphische, W. Germany, 1984.

23. M. F. Sulcoski, *Neutron Computed Tomography Using Real Time Neutron Radiography*, Ph.D. Dissertation, University of Virginia, Charlottesville, Virginia, U.S.A., 1985.

24. J. S. Brenizer, B. Hosticka, R. W. Jenkins, Jr. and D. D. McRae, Second World Conference on Neutron Radiography, Paris, France, 1986.

COMPUTERIZED NEUTRON GAGING ADDS A NEW

DIMENSION TO NEUTRON RADIOGRAPHY

R. L. Newacheck, I. E. Lamb

Aerotest Operations, Inc., San Ramon, Ca. USA

M. C. Anderson, Explosive Technology, Inc., Fairfield, Ca. USA

ABSTRACT

Neutron gaging provides an excellent alternative to neutron radiography when it is necessary to determine material density or small differences in thickness. The neutron gage is virtually free of the "noise" inherent in the neutron radiographic process. The problem of variable neutron source intensity when using a nuclear reactor for gaging has been solved by a unique detection system. The system has been used successfully for production analysis of small assemblies on an automated sample changer, and for long explosive cords using a constant speed drive.

Sensitivities have been measured using the neutron gage and neutron radiography under carefully controlled tests. The data are compared to illustrate the results obtainable from the two processes. Comparative results, including destructive test data, are also discussed for an actual production inspection.

Neutron radiography has proven to be an excellent nondestructive inspection technique for specific combinations of materials where X-radiography fails or is inadequate. It is invaluable for detection of cracks, voids and discontinuities; however, its usefulness for measuring small differences in attenuation is limited by inherent variations in the exposure and developing process, particularly in regions of high attenuation.

Some of the reasons for film density variations not associated with "real" attenuation include neutron scattering, (both object and facility), distance the object is separated from the cassette, neutron exposure variation across the film, gamma exposure, conversion screen variations, film fog, film processing and object size. Some of these "noise" variables occur over the area of a single exposure, whereas others are more pronounced when comparing densities from exposure to exposure. Neutron gaging is a complementary inspection technique for measuring or comparing neutron attenuation. The neutron gage is mostly unaffected by the uncontrolled variables noted above.

For various reasons, neutron gaging systems have been restricted in their application. For one thing, neutrons are non-ionizing so detection is inherently more complex. Also, available neutron sources emit high energies which must be moderated or slowed for efficient detection. This has the effect of enlarging the "source" physically, thus reducing the available flux significantly.

Nuclear reactors can provide large slow neutron fluences at a sufficient distance to allow collimation, but produce a stability problem for precision gaging. The flux of the reactor is continuously shifting in distribution and intensity at any given location due to temperature fluctuations, Xenon (poison) distribution and other factors.

Using the Aerotest Research and Radiography Reactor (ARRR), a method was devised to virtually eliminate the effect of these beam strength variations and permit precision gaging. This method has been used (1) to measure neutron attenuation of materials to thicknesses far greater than can be attained by radiographic techniques, (2) for static gaging of discrete assemblies, and (3) for dynamic gaging of long explosive linear shaped charges. The device developed for this method is called the Dual Detector Neutron Gage.

Originally, the system was operated manually and all the data was collected and computed by hand. Because of the mass of data and the complexity of the calculations, the use of the gage was limited. In the current configuration, a microcomputer and a matching interface for the scalers are used and all need for manual operations other than setup and teardown is eliminated. Immediately after the data is taken, the raw data is printed along with a statistical analysis to flag any variations outside of preset limits. An expanded scale graph can also be printed for visual analysis of data and trends. The raw data are stored on the computer disk for further analysis if required.

The Dual Detector Neutron Gage is comprised of three basic components: (1) a collimated thermal neutron beam, (2) a beam monitor detection channel, and (3) a signal neutron detection channel.

A key feature of the dual detector is the beam monitoring detector which monitors the beam before it passes through the collimator and specimen. The instrumentation channels are connected such that the signal channel is active only as long as the monitoring channel is active. A fixed count is set on the monitoring channel for repetitive measurements. The count on the signal detector becomes directly proportional to the beam attenuation for a specific sample, regardless of beam fluctuations during the measurements. Thus, beam strength is eliminated as a variable for attenuation measurement.

The signal counter indicates the neutron beam attenuation through the specimen. The measurement is affected by the material mass within the neutron beam area and is equivalent to density only if the volume is constant. An explosive material discontinuity, such as a crack or high density area, may not result in a detector output variance if the average mass in the neutron beam is unchanged. For this reason the neutron attenuation gage is complementary to, and not a replacement for, neutron radiography. Discontinuities are better indicated in a neutron radiograph. On the other hand, the gage is far superior to neutron radiography for explosive density determination because of variables, previously mentioned, that are inherent in the neutron radiography process.

The remainder of this paper will describe programs and tests which illustrate the differences between the capabilities of neutron radiography and neutron gaging.

TEST I

Purpose: To compare attenuation data from the neutron gage with data from the neutron radiographic process.

Description: Copper absorbers were used for N-gage neutron beam attenuation measurements to a thickness of 15.24 cm (6 in). A single neutron radiograph was exposed using the same copper absorbers. For the neutron gage, the signal counts penetrating the absorber were divided by the signal counts for no absorber. For the neutron radiograph, the optical film density in the area shielded by the absorber was divided by the unshielded background (bkgd) density. The resulting fraction provides a value normalized for bkgd variation.

Conclusions: Comparative results are shown quite clearly on Figure 1. The slope or steepness of the curve indicates the sensitivity to material thickness change. The "transmission" curve for the neutron radiograph shows that all sensitivity to change is lost at about 23% of bkgd. In effect, this is the lowest neutron radiograph density attainable since only 80% of the film darkening is by thermal neutrons. On the other hand, the transmission curve for the neutron gage extends to well under 1% since more than 99.7% of the unattenuated signal counts are due to thermal neutrons. Thus, for a copper absorber, the neutron gage can detect small thickness changes in a 12cm absorber with the same sensitivity as a neutron radiograph through a 3.8cm absorber.

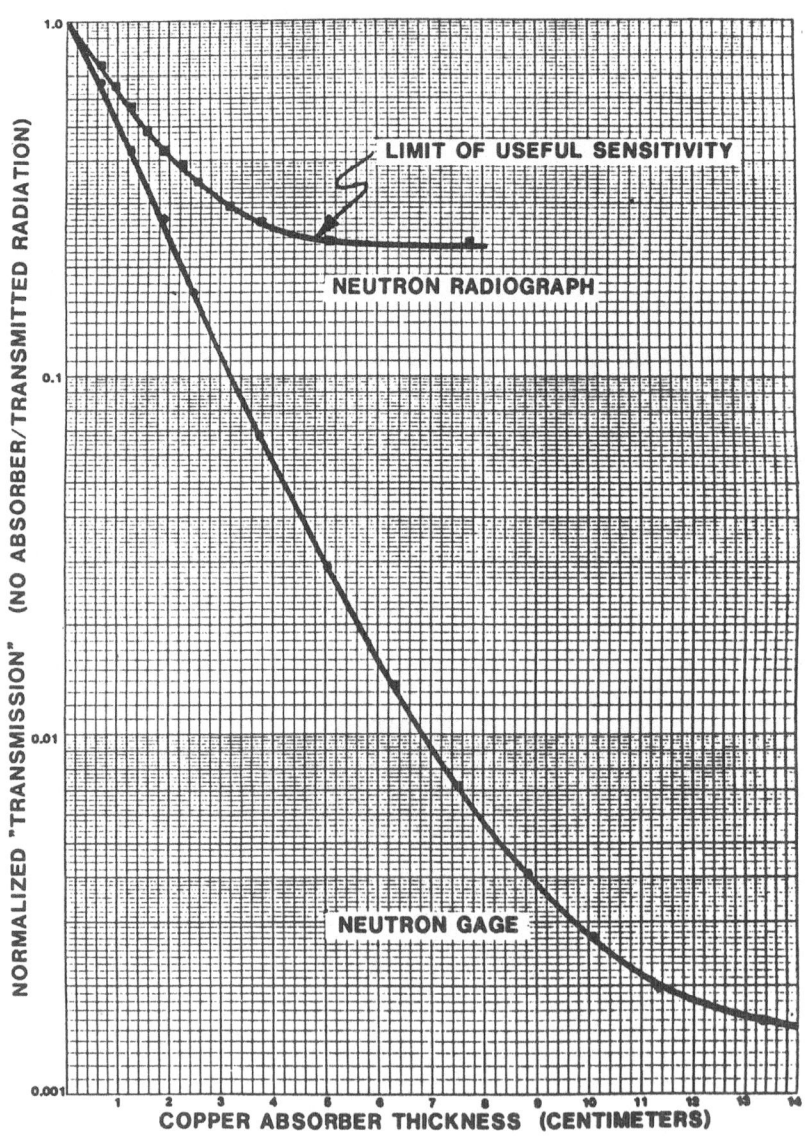

Figure 1 - Neutron "Transmission" Curves

TEST II

Purpose: To compare attenuation measurement stability and reproducibility of the neutron gage and neutron radiographs.

Description: Nine copper absorbers, 0.635 cm (0.25 in) thick were used for this test. The attenuation of the nine absorbers was determined with the neutron gage and with neutron radiographs. For the neutron radiograph reproducibility tests, three films were exposed using identical equipment, materials and techniques. The first film was exposed and processed early in the day with the second immediately thereafter. The third film was exposed and processed at the end of the day to see if the automatic processer characteristics remained constant after the day's production of about 50 films. Density measurements of the film shielded by the absorbers were divided by the adjacent background density to compensate for any background variation across the film.

The absorbers were measured with the neutron gage twice, at the beginning of the day and again at the end of the day. The comparison fractions were obtained in the same manner as in Test I.

Conclusions: The variations for the neutron radiographs were 2.4% (maximum - minimum divided by average) from sample to sample in different film locations and 1.2% for the same sample in the same location from film to film. The variations in the neutron gage readings were 1.09% and 0.94% respectively. It must be understood that the neutron radiograph exposures and processing were carefully controlled. One should not expect this type of accuracy over a long period. The neutron gage does retain this magnitude of accuracy over longer periods as indicated by Test III.

TEST III

Purpose: To compare the results of neutron radiography and the neutron gage for a set of prepared explosive device density standards.

Description: A number of standards were manufactured by filling aluminum cups with RDX explosive and pressing to specific pressures. Four of the standards were selected for analysis by neutron radiograph and the neutron gage. In addition, the standards were measured physically to determine explosive density, (gm/cc). The standards were counted twice, several days apart. Several neutron radiograph exposures were made. Two were with the samples placed closely together (2 cm separation) near the center of the radiograph where the background density variation is negligible. Three exposures were made with the standards separated approximately 16cm to preclude any interaction between parts. In each exposure, the order of the standards was changed to determine the effect of background variation or, in the case of the centered samples, interaction due to object scattering.

Conclusions: The data are plotted in Figure 2. As in the previous tests, the vertical axis is the signal divided by background. The results of the neutron radiograph appear very erratic when compared to the neutron gage results. Test II, described previously, confirmed that the normal variation for identical samples is at least 2.5%, and the total attenuation change for these samples is only 3%. Consequently, devices such as these standards, which

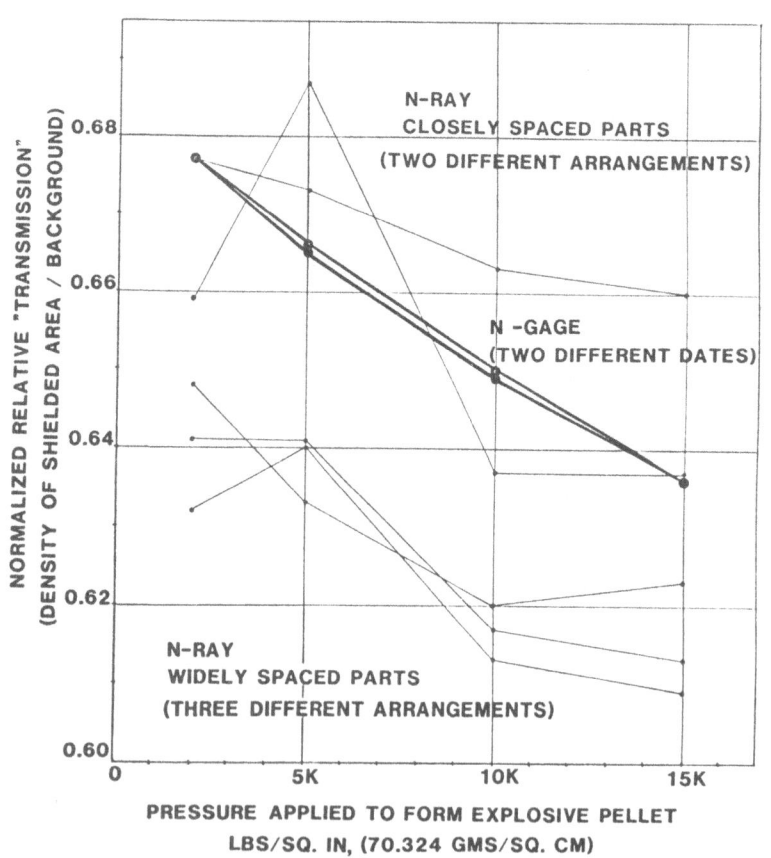

Figure 2 - Reproducibility Comparisons for N-ray and N-gage Systems

826

do not change in attenuation more than a few percent are not good candidates for neutron radiograph studies. On the other hand, the neutron gage is able to reliably determine the attenuation within plus or minus 0.75%. The neutron gage measurements correlated very well to the pressure used to press the explosive into the cups and to the actual density.

TEST IV

Purpose: Production analysis of 21 foot (6.37 meters) long, 30 grains per foot (66mg per cm) Linear Shaped Charge (LSC) to detect explosive loading variation.

Description: This "test" is one of a number of production analyses of LSC. Each 21 ft. (6.37 meter) length of LSC was neutron radiographed and neutron gaged. This particular length of LSC was shown by the neutron gage to have a short section with a low explosive load. The neutron radiograph was then retrieved and analyzed with a densitometer. Finally, to confirm the findings, a destructive analysis was performed on the section of line under question. In interpreting the data, one should realize that the minimum and maximum readings for both the N-Ray data and the gage data were normalized to the destructive analysis data and as such will appear to be of the same magnitude in change.

Conclusion: The results of the three analyses are shown in Figure 3. Visually, the section of line with the low explosive load was not obvious from the neutron radiograph until the neutron gage isolated the position. The lightly loaded section showed clearly in the neutron gage data as the Figure 3 plot shows. The neutron radiograph densities were taken with two aperture sizes, one that simulated, approximately, the size of the neutron gage aperture used (4mm) and the second was the smallest size available (1 mm) in an attempt to have all the light beam transmitted through the explosive image. The smaller aperture gave considerably more erratic results and these data are not plotted.

One advantage of the digital data acquisition is that the data have a known statistical validity i.e. they can be analyzed on the basis of random statistics. A pair of horizontal lines has been placed on the counting data plot showing the +/- 3 sigma value of the average count value of the line. For multiple counts of a single point, the probability of an individual count falling outside of the +/- 3 sigma value is only 0.0027. A series of points falling outside the lines thus represents a definite change from the norm.

It should be noted that the transmission path through the line for gaging and for the radiograph are 90 degrees to one another since the LSC configuration requires a coil laying flat for the minimum number of exposures and the gage was set up for vertical examination of the 'v' shaped cord.

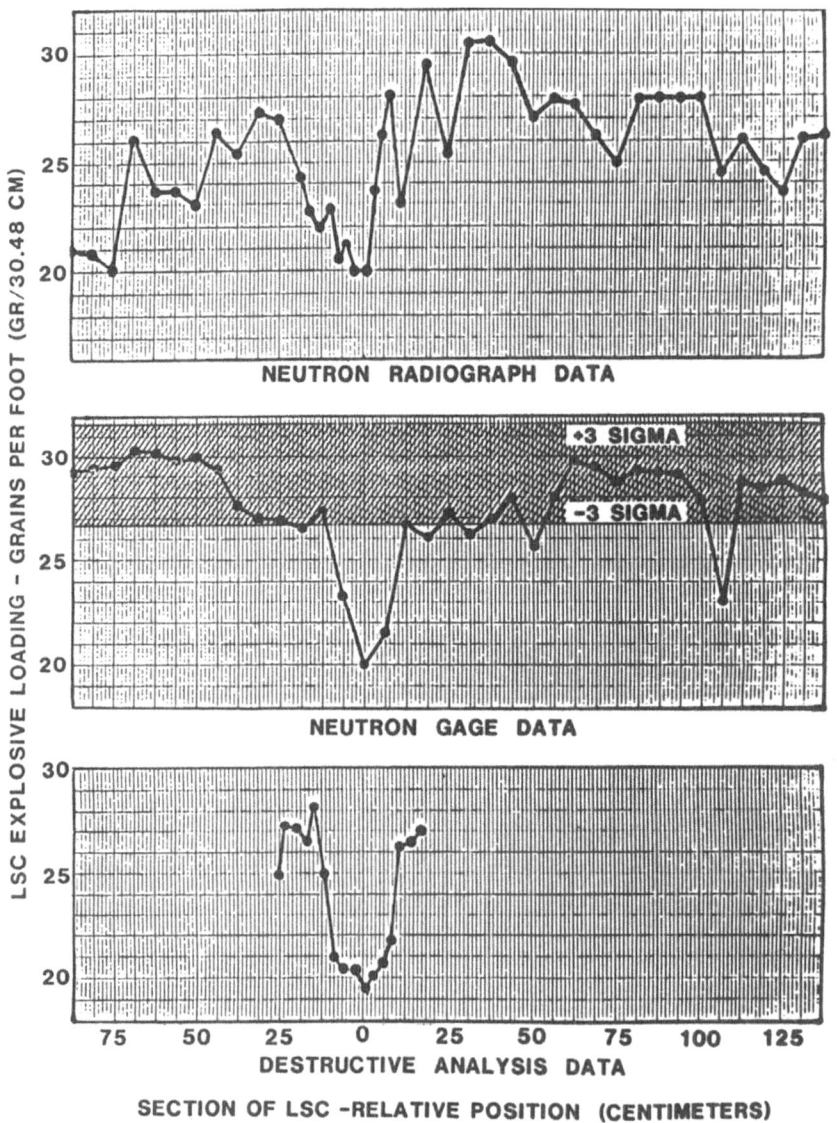

Figure 3 - LSC Detectability Comparisons

A MULTIWIRE PROPORTIONAL COUNTER FOR NEUTRON RADIOGRAPHY

M. Wrobel, D. Bünemann, L. Greim

GKSS-Research Center, Geesthacht FRG

ABSTRACT

Neutron radiography with a position sensitive multi-wire proportional counter enables testing at low intensity neutron beams.

The sensitive area of the described multiwire counter is 250 mm x 250 mm. Three planes of wires are provided, two cathodes and one anode. Readout of position is done by charge division. The neutrons are detected by the ^{3}He(n,p)T reaction in the gas filling. The detector efficiency for thermal neutrons is about 50 %.

Neutron radiographic images of some test objects are shown and discussed with respect to contrast and resolution.

INTRODUCTION

Multiwire proportional counters (MWPC) are used for position determination of charged particles in the gas filling of the chamber. For position sensitive detection of thermal neutrons the chamber contains either a boron or gadolinium foil, which is fixed inside the counter housing at the top and the bottom (1, 2, 3), or by a neutron absorbing gas filling like ^{3}He or ^{10}BF, (4, 5, 6). Working with foil converters the detection efficiency is about 5 %, while a neutron absorbing gas filling enables a detection efficiency of about 50 % for thermal neutrons.

For the position detection there exist different methods, e.g. the charge division method (7) and the rise-time method (8). The counter described here has a gas filling with ^{3}He and operates with the charge division method.

MWPC are suited for neutron radiography at low neutron intensity in cases where no high position resolution is required, e.g. for the detection of hydrogen containing fillings.

DESIGN OF THE MULTIWIRE CHAMBER

The MWPC has 3 wire planes in an aluminium housing. The x-coordinate is delivered by the central anode-plane, the y-coordinate by one of the cathode planes. These wires are orthogonal to the anode wires. The mechanical dimensions of this MWPC are given in table 1. The optimal choice of the electric resistance for the charge division method is described by Alberi and Radeka (7). The electronic parameters of this counter are listed in table 2.

As neutron sensitive gas we have chosen ^{3}He instead of ^{10}BF$_3$, because it is not aggressive and easier to handle. The poor stopping power of ^{3}He requires additional heavy atoms or molecules for shortening the ranges of proton and triton because the possible position resolution of the MWPC is limited mainly by the ranges of both particles in the counter gas. The gas mixture consists of ^{3}He, Xe and CO_2 (5, 6).

NEUTRON RADIOGRAPHY SET UP

The block diagram of the electronic circuit of the position sensitive MWPC working by charge division is shown in fig. 1, where only the wires of the anode plane are drawn.

When detecting a neutron in the counter gas the charge shares between the two preamplifiers at both ends of each wire plane according to the electric resistance between the position of the center of charge and both preamplifiers. After amplification both pulses are added in the sum amplifier, whose output is proportional to the energy of proton and triton deposited in the counter gas by ionization. The division of one single pulse by the summed pulse in the divider gives a signal which is proportional to the x- resp. y-coordinate of the detected charge. For background elimination the division is only performed if the sum impulse exceeds the discriminator level which is set at the divider.

830

Table 1. General Design Parameters

Sensitive area:	250 mm x 250 mm
Wire spacing:	5 mm
Wire diameter:	25 μm
Distance anode-cathode:	10 mm
Thickness of the chamber:	30 mm
Filling gas:	190 kPa ^{3}He, 95 kPa Xe, 15 kPa CO_2
Neutron detection reaction:	^{3}He(n,p)T
Reaction energy:	764 keV
Number of ion pairs:	$2.8 \cdot 10^4$
Proton energy:	573 keV
Proton ranging:	5.5 mm
Triton energy:	191 keV
Triton ranging:	3.5 mm
Distance between the position of ^{3}He(n,p)T reaction and the center of charges:	1.9 mm

Table 2. Electronic Design

Anode potential:	+ 2800	V
Cathode potential:	− 350	V
Gas amplification factor:	100	
Capacitance of a wire plane:	63	pF
Resistance of a wire plane:	10	kΩ
Energy discrimination for background elimination:	350 keV	
Position readout:	by charge division	
Pulse storage in a 64 x 64 multichannel-analyzer	(ND 620)	
Picture processing:	with a HP 9845 C computer	

In the case of coincidence both pulses from the x- resp. y-divider are stored two-dimensional in the multichannel-analyzer. For later data processing and interpretation the picture is transferred to a computer and stored on a type cartridge.

The measurements were done with an Am-Be-source of the source-strength $1.8 \cdot 10^6$ s^{-1}. It was installed in the bore-hole of a polyethylene-moderator (fig. 2). The thermal neutron flux at the counter position 1 m beneath the neutron exit area of the moderator is 0.79 cm^{-2}s^{-1}, the measured L/D-ratio of the neutron field is 19. Fig. 3 is a photograph of the set up.

TEST MEASUREMENTS

All measurements were done with Cd-difference method. The potentials of anode and cathode and the resulting gas multiplication factor are listed in table 2.

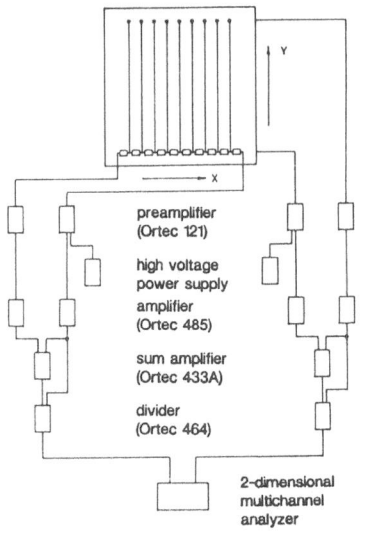

Figure 1. Electronic Circuit

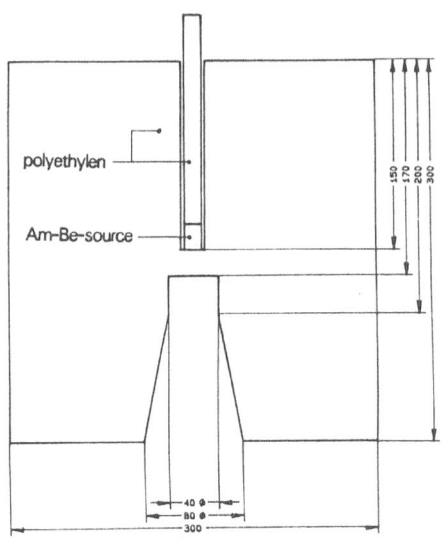

Figure 2. Neutron Source, Dimensions in mm

Figure 3. Neutron Radiography Set Up

The spectrum of the energy deposition (in keV) in the MWPC due to the ³He(n,p)T-reaction of the detected neutrons is shown in fig. 4. Pulses below 350 keV are suppressed by discriminator setting at the divider. With ionization events at the edges of the anode plane a pulse height enhancement is observed. Therefore, pulses from border regions with differing gas multiplication are suppressed by lower- and upper-level-discriminator setting at both ADC of the two-dimensional multichannel-analyzer.

The measurements of position resolution were done in one dimension and separately for the x- and y-direction. Stripes of a 3 mm thick boron containing foil are put on the top of the counter housing parallel to the wires of the measured plane. The result of such a measurement for the y-direction is shown in fig. 5. The measured position resolution is 7.9 mm with 1 mm being represented by 12.8 channels. An analogous measurement for the x-direction shows a position resolution of 5.7 mm. The better resolution of the anode-plane is due to the higher pulses coming from this plane.

The γ-sensitivity of the MWPC was measured by use of two different γ-sources (table 3).

EXAMPLES OF NEUTRON RADIOGRAPHS AND IMAGE PROCESSING

Finally some examples of neutron radiographic images are shown. The local difference of the detected neutron intensity is presented in form of a gray-shade-picture and as an isodensity-curve. The scaling at the sides of fig. 6 to fig. 11 is given in mm.

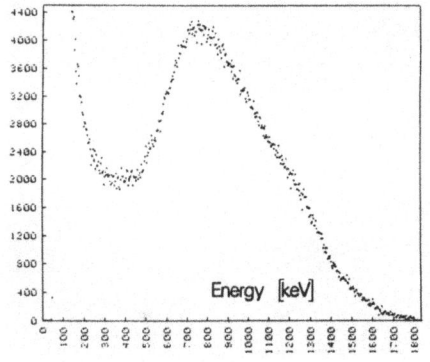

Figure 4. Pulse Height Spectrum

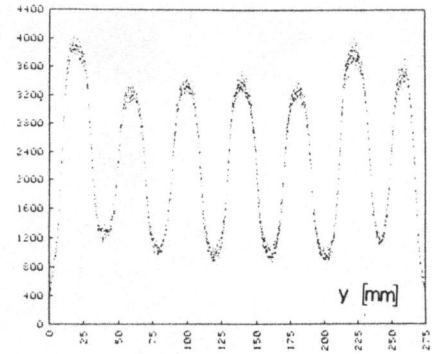

Figure 5. Position Resolution

Table 3. Sensitivity and Background

The following parameters are obtained with an energy discrimination
at 350 keV in the pulse height spectrum

Sensitivity for thermal neutrons: 50 %

Position resolution:
    x-coordinate (anode):   5.7 mm
    y-coordinate (cathode):   7.9 mm

γ-sensitivity:
    $1.8 \cdot 10^{-5}$ / incident γ (^{137}Cs, E_γ =   662 keV)
    $6.8 \cdot 10^{-5}$ / incident γ (^{60}Co, E_γ = 1173 keV and 1332 keV)

    Fig. 6 and fig. 7 show the neutron radiographic pictures of
discs in circlular, square, and equilateral triangle shapes whose
characteristic lengths are 50 mm. The objects in the left consist of
steel, the others are made from polyethylene. The thickness of all
objects amounts to 5 mm. The figures demonstrate the reproduction of
geometric forms and angles. Fig. 7 shows the rounded edges of the
squares and equilateral triangles due to the position resolution of
the MWPC.

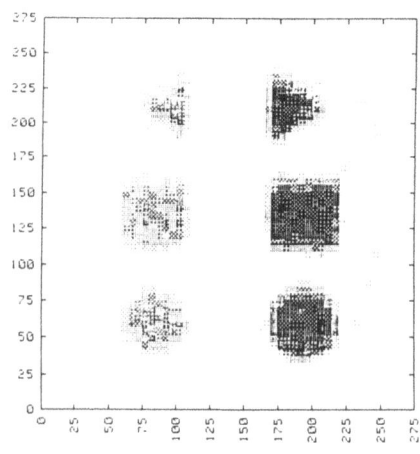

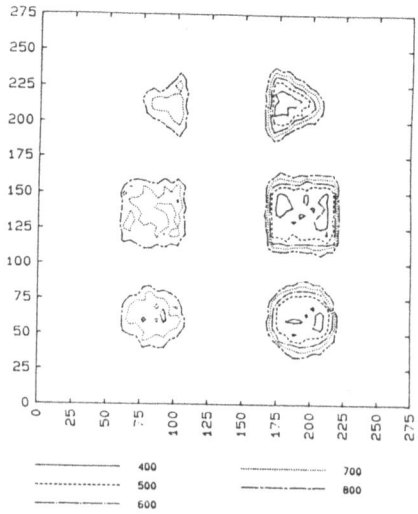

Figure 6.   Figure 7. Isodensity lines

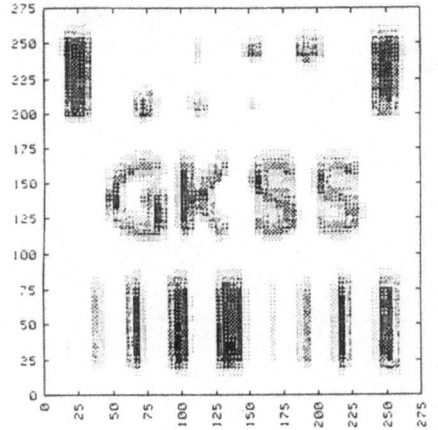

Figure 8.

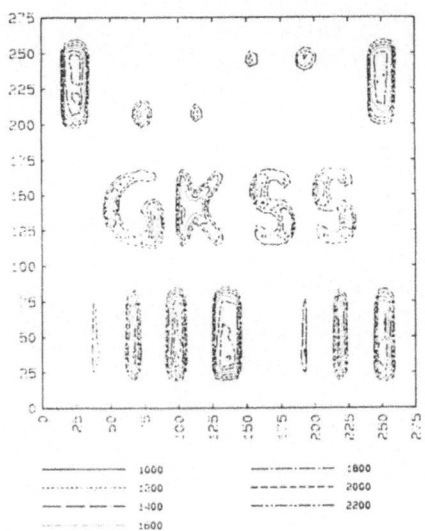

Figure 9. Isodensity picture

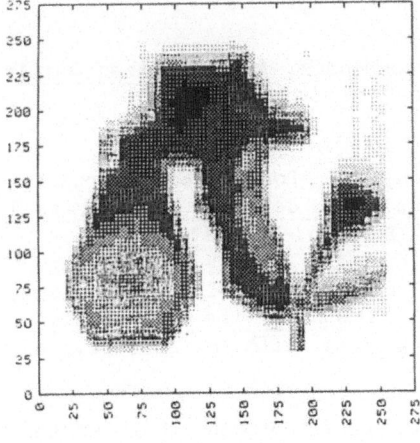

Figure 10. Neutron radiograph
of a spray pistol

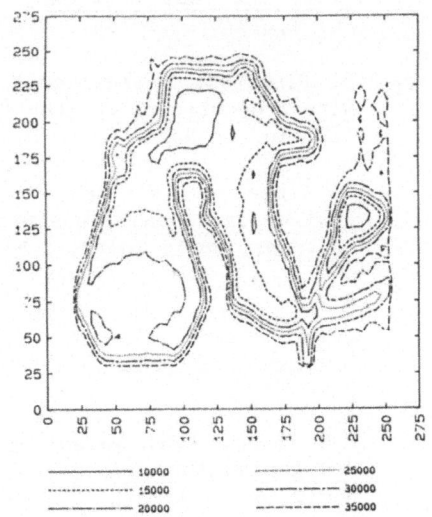

Figure 11. Isodensity picture of
the spray-pistol

Fig. 8 and fig. 9 present the neutron radiographic image of a 1 mm thick Cd-pattern with slits, holes and letters. The width of slits amounts to 2, 5, 10, 15 and 20 mm, the diameter of holes 5, 10, 15 and 20 mm. The figures show that small object details cannot be seen in the image, e.g. the 5 mm-holes, which are missing in both figures or the 2 mm-slit at x = 168 mm, which is weakly recognizable in fig. 8.

The neutron radiographic image of a spray-pistol shown in fig. 10 and fig. 11 demonstrates a possible technical application. The local blackening in fig. 10 corresponds to the amount of neutron absorption. In both figures the different material sizes of the machine, empty paint container, plug and cable are recognizable.

REFERENCES

1. K.H. Valentine, The development of a multiwire proportional chamber imaging system for neutron radiography. Department of Physics, Lawrence Berkeley Laboratory and Department of Nuclear Engineering, University of California, Berkeley, April 1974.

2. K.H. Valentine, S. Kaplan, V. Perez-Mendez, L. Kaufman, A multiwire proportional chamber for imaging thermal, epicadmium, and fast neutrons, IEEE Tr. on Nucl. Sc., Vol. NS-21, No. 1, 1974, 178-183.

3. B. Director, S. Kaplan, V. Perez-Mendez, A pressurized multiwire proportional chamber for neutron imaging, IEEE Tr. on Nucl. Sc., Vol. NS-25, No. 1, Febr. 1978, 558-561.

4. J. Alberi, J. Fischer, V. Radeka, L.C. Rogers, B. Schoenborn, A two-dimensional position-sensitive detector for thermal neutrons, Nucl. Instr. & Meth. 127 (1975) 507-523.

5. R.K. Abele, G.W. Allin, W.T. Clay, C.E. Fowler, M.K. Kopp, Large-area proportional counter camera for the U.S. national small-angle neutron scattering facility, IEEE Tr. on Nucl. Sc., Vol. NS-28, No. 1, Febr. 1981, 811-815.

6. M.K. Kopp, K.H. Valentine, L.G. Christophorou, J.G. Carter, New gas mixture improves performance of ^{3}He neutron counters, Nucl. Instr. & Meth. 201 (1982) 395-401.

7. J.L. Alberi, V. Radeka, Position sensing by charge division, IEEE Tr. on Nucl. Sc., Vol. NS-23, No. 1, Febr. 1976, 251-258.

8. E. Mathieson, K.D. Evans, W. Parkes, P.F. Christie, Signal location by shorted RC line, Nucl. Instr. & Meth. 121 (1974) 139-149.

PART XIII

STANDARDS

CAN NEUTRON BEAM COMPONENTS
AND RADIOGRAPHIC IMAGE QUALITY
BE DETERMINED BY THE USE OF
BEAM PURITY AND SENSITIVITY INDICATORS?

J. C. Domanus

Risø National Laboratory
DK-4000 Roskilde, Denmark

Abstract

   In the Euratom Neutron Radiography Working Group Test Program beam
purity and sensitivity indicators, as prescribed by the ASTM E 545-81 were
used together with the NRWG beam purity indicator-fuel and calibration fuel
pin. They were radiographed together at neutron radiography facilities of
the European Community. The direct, transfer and track-etch methods using
different film recording materials were used. Neutron beam components were
calculated from film density measurements under the beam purity indicators
and radiographic image quality was assessed by visual examination of the
sensitivity indicator. Results obtained under the NRWG Test Program are
summarized and compared.

1. INTRODUCTION

At present there are only two national standards dealing with the problem of radiographic image quality in neutron radiography: The first one is the ASTM E545 |1| and the second the AFNOR A09-220 |2|.

In the first issue of the ASTM E 545 standard from 1975 four types of sensitivity indicators were listed. In the second issue (from 1981) only one, different, sensitivity indicator is recommended for use.

The ASTM Standard from 1981 is now under revision. A draft E545-85 of this revision |3| was taken into account in preparing of this paper.

Both in the 1975 as well as the 1981 edition of E545 the use of the beam purity indicator was limited to radiography with metal conversion screens and silver halide·film. In the 1985 draft the whole E545 standard is limited to the direct thermal neutron radiography and the use of the BPI to metallic conversion screens and single emulsion silver halide films. In all the three editions of E545 it is stated the the "use of alternative detection systems may produce densitometric readings which are not directly comparable to the formulas" given in the standard.

No doubt the nitrocellulose film falls into the category of "alternative detection systems".

No such restrictions are given to the use of the sensitivity indicator, so according to the 1985 E545 draft it can be used within the direct thermal neutron radiographic testing, also with double emulsion silver halide films.

The ASTM standard |1| explicity states that "requirements expressed in this method are intended to determine the quality of the neutron radiographic images". The sensitivity indicator SI is described as "a device for quantitative determination of radiographic quality". The "sensitivity level is determined by visually analyzing the image of the sensitivity indicator".

This is done by "evaluating the image of the sensitivity indicator and assessing a numerical value" to it.
The whole procedure consists of determining the number of visible consecutive holes on the neutron radiographs of the SI and the number of visible gaps (Al shims between acrylic steps).

The number of visible holes (H) together with the number of visible gaps (G) (Al shims) is thereafter used to determine the neutron radiographic category.

Further information necessary for the determination of those categories is obtained from densitometric measurements under various parts of the Beam Purity Indicator-BPI.

There is still another information which ought to be retrieved from neutron radiographs of the SI by visual examination. It is the number of visible holes (Pb) in the shim under the lead steps D of the SI.

According to 7.14 of the ASTM E545-85: "Visually inspect the image of the lead steps in the sensitivity indicator. If the 0.25 mm holes are not visible, the exposure contribution from gamma radiation is very high. An evaluation should be made to determine if the radiographic quality is sufficient for the required inspection".

The ASTM standard describes the purpose of using of the beam purity indicator in the following way (§3.1): "The BPI is designed to yield information concerning neutron beam and image system parameters that contribute to film exposure and thereby affect overall image quality. In addition the beam purity indicator can be used to verify the day-to-day consistency of neutron radiographic quality". Furthermore formulas are given in the standard to calculate neutron beam constituents from density measurements of the BPI.

This ASTM method of determining neutron radiographic image quality by the use both of the beam purity indicator (BPI) and sensitivity indicator (SI) seems rather awkward and complicated.
In other fields of industrial radiography (X-and gamma-radiography) different image quality indicators are used.
The assessment of radiographic image quality by the use of those IQI's is, however, limited to visual examination of wires, holes or steps visible on radiographs. No film density measurements are prescribed and no calculations are performed therefrom. Although in X- and gamma-radiography the radiographic image quality depends of such factors as e.g. the spectral composition of the radiation beam, and of the scattered radiation reaching the radiograph, those factors are not determined by the use of the IQI's. The user of a radiograph is interested only of the final radiofraphic quality, whereas the information about the factors influencing this quality are of interest only to the producer of the radiograph.

2. NEUTRON BEAM COMPONENTS

As mentioned above, according to ASTM E545, the beam purity indicator (BPI) serves to determine the neutron beam components. In the 1981 version of the E545 the use of BPI was restricted to the use of silver halide film and metal conversion screens.

Now, according to the 1985 draft of the E545 it is further restricted to the direct method and single coated silver halide film.

The NRWG Test Program was started before this 1985 draft was issued. It was decided then to test the usefulness of the BPI. As the NRWG is interested only in neutron radiography of nuclear reactor fuel, which is mainly radiographed in post-irradiation examination, a modification of the ASTM BPI was designed by the NRWG. It is the beam purity indicator-fuel (BPI-F), not meant to determine the gamma-ray content of the neutron beam (as the irradiated fuel is itself a source of strong gamma-radiation).

Both the BPI and BPI-F (together with the SI and CFP-El) were neutron radiographed at all NR facilities participating in the NRWG Test Program. Both the direct as well as the transfer method were used with metal converter screens. Although the ASTM method is restricted only to silver halide film exposures were made also on nitrocellulose film. Those radiographs were thereafter copied on high contrast silver halide film and also viewed through polarizing filters.
In both instances the images of the nitrocellulose film looked similar

to those made on silver halide film. Therefore it was worth comparing the neutron beam component findings using all those different recording methods.

2.1. Thermal neutron content measured by the direct method

Throughout the NRWG Test Programn three brands of silver halide film were used both for the direct as well as the transfer method. They were the single coated Kodak Industrex SR film, double coated Agfa-Gevaert Structurix D4 and Kodak Industrex M film.

In all instances where the direct method (with 0.25 μm Gd) was used highest content of thermal neutrons was always calculated from density measurements of the SR film, and lowest for the D4 film. For the BPI this difference could be as great as 41% whereas for the BPI-F the greatest difference was 32%. The thermal neutron content values calculated from densities under the BPI-F were always greater than those from the BPI.

2.2. Thermal neutron content measured by the transfer method

The same phenomenon was observed when using the transfer method (100 μm Dy) and the same silver halide film. Here also the highest values of thermal neutron content were calculated for the SR film and lowest for D4 film.

In general the percent differences were lower than those for the direct method. As with the direct metod the absolute values calculated for the BPI-F were in general higher than those for the BPI.

2.3. Thermal neutron content measured by the track-etch method

As mentioned before neutron radiographs made on nitrocellulose film and copied on high contrast silver halide film or viewed through polarizing filters can be compared with those taken on silver halide film.

Whereas the viewing through polarizing filters gives quite erroneous results as regards the measurements of thermal neutron content, the copies of nitrocellulose film on high contrast silver halide film can be used for that purpose.

During the NRWG Test Program different etching conditions were investigated. At Risø the nitrocellulose films were etched at 20° C for 21 h and at 50° C for 45 min. In most instances the second etching mode gave slightly higher values of the thermal neutron content than the first mode.

With the track-etch method two film/converter combinations were used for the nitrocellulose film. In the first the CN85 B film coated on both sides with converter was used. The second consisted of the CN85 film sandwiched between two BN1 converters.

The CN85+BN1 film/converter combination gave higher values of the thermal neutron content.

As the nitrocellulose film can be compared with the silver halide film used with the transfer method such comparison was made between the SR/Dy silver halide film/converter combination of the transfer method and the

CN85+BN1 nitrocellulose film/converter combination, copied on high contrast silver halide film. They both give the highest readings of thermal neutron content for the exposure methods under comparison. In many instances the results reached for the SR/Dy transfer method and the CN85+BN1 track-etch method were very similar.

2.4. Thermal neutron content measured by the direct method on single coated silver halide film.

As mentioned before the draft ASTM E545-85 standard limits the use of the BPI to the direct method using single coated silver halide film and metal converter.

To check the consistency of measuring results by the above method 15 BPI neutron radiographs, taken on SR film with the 25 μm Gd converter were compared. Those radiographs were taken at the same DR1 reactor at Risø with the same fluence but at different period of times. The results of this comparison, reported in [4], are shown on fig. 1.

As can be seen the readings of the thermal neutron content are always higher for the BPI than for the BPI-F, and the difference of those varies, as does the mean background film density

The highest variation between the BPI readings is about 20%, for the BPI-F only 9% and for the mean background film density as large as 1.4. Although it is possible that at different operational periods of the DR1 the composition of the neutron beam was varied, it is improbable that the variation could reach 20%.

The variations in the background film densities were no doubt caused instead by the changing neutron beam fluence and not the thermal neutron content. One could, perhaps, look for a correlation between the thermal neutron content readings and those of the background film density. A. Laporte and G. Bayon have already found such a correlation. According to [5] the readings for the thermal neutron content decrease with increasing film density. This effect could perhaps also explain the variations of thermal neutron content as shown in fig. 1.

3. SENSITIVITY OF DETAIL VISIBILITY

The ASTM sensitivity indicator (SI) was used (exposed together with the BPI and BPI-F) to determine the sensivity level. According to the ASTM one shall visually determine the number of visible Al shims and holes in the acrylic shims of the SI. The visual examination of the SI performed during comparison mentioned above has shown that in all instances the 7 Al shims are visible on all neutron radiographs. This wholly supports the statement made in [5] that the SI "is not selective as changes in image quality....do not correspond to the readings of the sensitivity indicator".

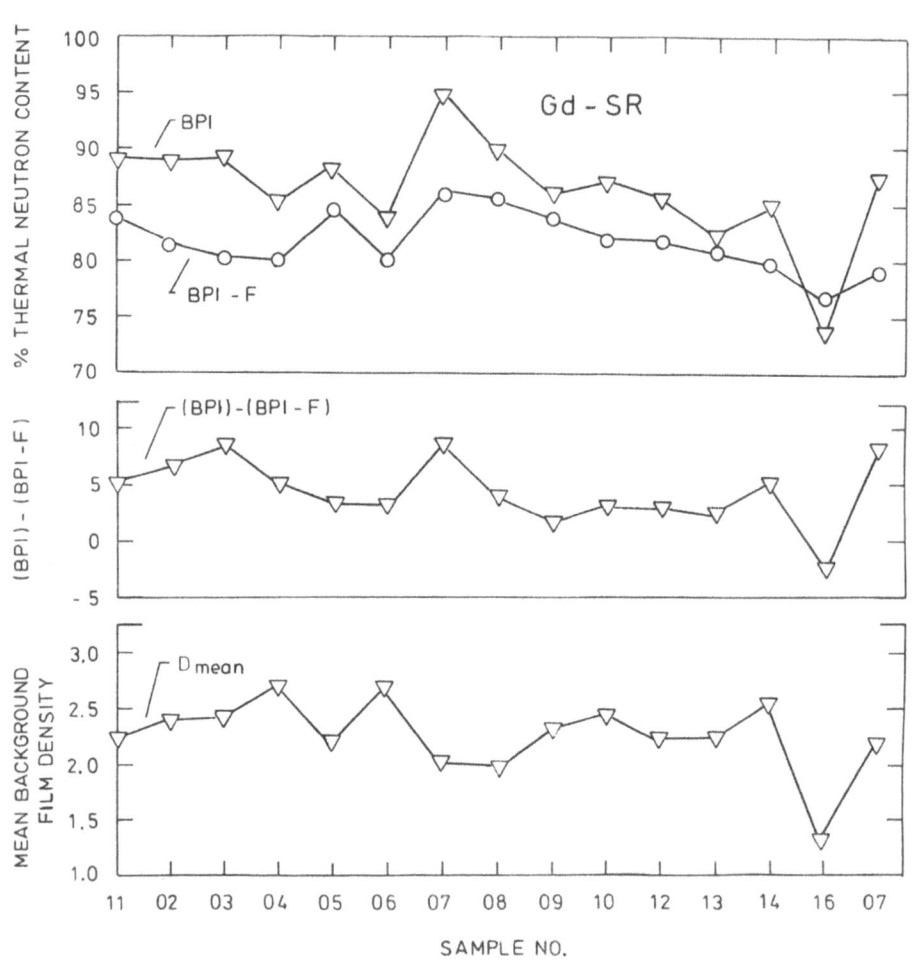

Fig. 1. Thermal neutron content and mean background film
densities

Regarding the second information about the sensitivity of detail visibility, which according to ASTM can be retrieved from the image of the SI, one can say that the visibility of the holes in the acrylic shims (under acrylic step-wedge) is more selective than that of the Al shims. Although the L/D ratio has changed considerably from one reactor to another (from 75 to 500) the relation between the hole visibility and the L/D was not so direct as one could expect. Obviously other factors (e.g. scattered and gamma radiation) were of importance.

4. RADIOGRAPHIC CATEGORIES

The ASTM designation of quality level includes thermal neutron content and sensitivity level, and according to it neutron radiographic categories are specified. When no designation of quality level is specified by the user the ASTM requires that neutron radiographs shall be at least category III, which means that the thermal neutron content shall be at least 55%, that at least 5 consecutive holes and 5 Al shims ought to be seen in the neutron radiograph of the SI.

However, in the 1985 draft of the ASTM this has been changed admitting also categories IV and V.

Looking from that point of view at the results obtained under the NRWG test program one can see e.g. that at Risø category III could never be reached as 5 consecutive holes could never be seen. The same can be said about the other 8 NR facilities contributing to the NRWG Test Program. There, only in very few instances, the 5th hole could be seen, although the L/D ratio was sometimes as high as 500.

5. GAMMA RAY CONTENT

There are further, rather vague, requirements in the ASTM standard regarding the image quality. It is, e.g. required to visually inspect the image of the lead steps in the SI. If the 0.25 mm holes are not visible the exposure contribution from gamma radiation is considered very high. All the four 0.25 mm holes under Pb steps were never seen on either of the NRWG neutron radiographs, although, e.g. in the Risø DR1 facility the gamma contribution in the neutron beam is negligible. It proves that the assessment of the gamma contribution from the measurements performed by the use of the BPI and visual examination of the SI are incompatible.

6. L/D RATIO

Another vague requirement of te ASTM standard relates to the judgement of the L/D ratio from the visual comparison of the sharpness of the cadmium rods of the BPI. Such visual examination is very subjective and hard to perform.Only by using a travelling microdensitometer the problem can be solved, but as this instrument is relatively expensive it is unavailable in most of the NR facilities. Therefore it is not realistic to recommend its use in a standard.

7. CALIBRATION FUEL PIN

As described above for the NRWG Test Program three indicators: BPI, BPI-F and SI were produced at Risø. But even the possesion of those three indicators seemed to be insufficient for the image quality control of nuclear radiographs of nuclear fuel. Therefore, the NRWG has followed the statement of the ASTM E545 that: "It is recognized that the only truly valid sensitivity indicator is a material or component, equivalent to the part being radiographed, with a known standard discontinuity (reference standard comparison part)". Such a "reference standard comparison part" was designed by the NRWG for nuclear fuel. It was produced at Risø as a calibration fuel pin (CFP-El) and distributed among all the participants of the NRWG. It includes such "standard discontinuities" as pellet-to-pellet and pellet-to-cladding gaps (calibrated gaps from 50 to 300 μm).

Neutron radiographs of the calibration fuel pin were used for the assessment of radiographic image quality by the visual method |6|.

The calibration fuel pin CFP-El in its present form cannot directly be recommended for use as an IQI for neutron radiography of nuclear reactor fuel. However, the results reached under the NRWG Test Program, regarding the accuracy of dimensional measurements from the CFP-El, together with the results quoted in |6| will form a basis for designing a proper image quality indicator for neutron radiographs of nuclear reactor fuel.

8. DETERMINING IMAGE QUALITY

The title of ASTM E545 implies that by the use of it one can determine radiographic image quality in thermal neutron radiography. For that purpose both the BPI and SI must be used.

As demonstrated above the results obtained with the BPI depend on the exposure and detecting method in use and give different results for the same NR facility.

The indications obtained by the use of the SI about radiographic sensitivity are not selective enough and give not consistent information in that respect.

Besides, the results obtained with the BPI and SI, regarding the gamma-ray content of the neutron beam are often contradictory and the judgement of the L/D ratio from the examination of the image of the BPI is practically impossible in routine neutron radiography.

It is not clear why an analysis of nuclear beam components must be included into the procedure of assessing the radiographic image quality of a neutron radiograph.

The final assessment prescribed by the ASTM E545, by determining neutron radiographic categories is also ambiguous. It requires e.g. that when no designation of thermal neutron content (from the BPI readings) and hole and gaps visibilities (from visual assessment of the SI) is specified by the user then the neutron radiographs shall be at least category III, IV, or V (category V being the worst of all).

The problem of the usefulness of the ASTM E545 method was also investigated by others.

In the more extensive study A. Laporte and G. Bayon |5| have examined three types of X-ray films, exposed with 25 μm GD foil by the direct technique and processed both manually and automatically. The conclusions

846

from those investigations performed using the BPI and SI were the following:

- It is impossible to determine the neutron beam components themselves by the use of BPI or BPI-F, because the results of film density measurements give indications about the effect of neutron radiation on the whole irradiation and detection system (collimator, film cassette, film, converter, processing) and do not permit calculation of the neutron beam components separately themselves. It is, e.g. quoted in |5| that the percentage of thermal neutron content can vary by as much as 83% when measured with X-ray films of different speed and it also depends on the film density and the processing mode.

- The IS is not selective enough to determine the image quality. As reported in |5| on all neutron radiographs, disregarding the different film brands used, their density and processing mode, all the slits of the SI were visible. Furthermore, the conclusions drawn about the gamma-ray content derived from the visual examination of holes in the lead shim of the SI contradict the calculations of gamma-ray content from the BPI. Similar conclusions were also formulated in |7|.

A. Laporte and G. Bayon |5| have also investigated the findings obtained with the ASTM BPI from the point of view of film speed and density, as well as the mode of processing and have come to the conclusion that "the text of the standard is wholly unsuited to the purpose for which it is intended both from a fundamental point of view and as far as details are concerned". This was also proved by other results quoted in |4|. In |5| the same conclusions are also given for determining scattered neutron, gamma and pair production content.

9. CONCLUSIONS

1) Both the BPI and BPI-F cannot serve the purpose of determining the neutron beam components of a NR facility. Therefore, they cannot be used for the purpose of comparing various facilities.

2) The assessment of radiographic image quality by the evaluation of the components of the SI is not selective enough if the visible gaps are counted (all are always seen). At the same time it gives an incorrect impression of image quality from the evaluation of visible holes (even with very good L/D more than 4 holes can very seldom be seen).

3) The assessment of the gamma-ray content from the densities under BPI and visual examination of te SI are imcompatible.

4) The ASTM standard (even as modified by the NRWG) is unsuited to image quality control in neutron radiography of nuclear fuel.

5) After obtaining detailed results of measurements performed now on the calibration fuel pin one will be able to draw conclusions about its usefulness for determining the image quality of neutron radiographs of nuclear fuel.

REFERENCES

|1| ASTM E545-81. Standard method for determining image quality in thermal neutron radiographic testing.

|2| AFNOR A09 - 220. Neutronographie industrielle. Détermination des charactéristiques des installations pour controle des pieces non radioactives. Norme experimentale. Septembre 1982.

|3| ASTM Draft E545-85. Standard method for determining image quality in direct thermal neutron radiographic testing.

|4| J. C. Domanus, P. Gade-Nielsen, J. Olsen - How good are the standards for the image quality in neutron radiography of nuclear fuel? Proceedings of the 7th International Conference on Non Destructive Evaluation in Nuclear Industry. Grenoble, 29.1 - 1.2.1985.

|5| A. Laporte, G. Bayon. determination of image quality in industrial neutron radiography. CEN-FAR. Service des Piles de Saclay. Section d'Exploatation TRITON. 6.203.2.

|6| J. C. Domanus. Assessment of radiographic image quality by visual examination of neutron radiographs of the calibration fuel pin. Risø-M-2578. April 1986. (To be presented at the 2nd World Conference on Neutron Radiography, Paris, 16.-20.06.86).

|7| H. P. Leeflang. Preliminary assessment of the NRWG test program performed with the Petten neutron radiography facility. ECN-83-074. Petten. May 1983.

ASSESSMENT OF RADIOGRAPHIC IMAGE QUALITY
BY VISUAL EXAMINATION OF NEUTRON RADIOGRAPHS
OF THE CALIBRATION FUEL PIN

J. C. Domanus

Risø National Laboratory
DK-4000 Roskilde, Denmark

Abstract

Up till now no reliable radiographic image quality standards exist for neutron radiography of nuclear reactor fuel. Under the Euratom Neutron Radiography Working Group (NRWG) Test Program neutron radiographs were produced at different neutron radiography facilities within the European Community of a calibration fuel pin.

The radiographs were made by the direct, transfer and track-etch methods using different film recording materials. These neutron radiographs of the calibration fuel pin were used for the assessement of radiographic image quality. This was done by visual examination of the radiographs and assessing their radiographic image quality on an arbitrary scale.

1. INTRODUCTION

Up till now only one standard was published in which the problem of determining image quality of neutron radiographs is discussed. It is the ASTM E545-81 |1|. In both the first issue of the E545 from 1975 as well as in the issue of 1981 the method was limited to neutron radiography performed with metallic converter screens and silver halide film. Therefore, it could be used for the assessment of radiographic image quality of irradiated nuclear fuel (radiography performed with the transfer method using, e.g. Dy converter and X-ray film).

This E545 standard is now under revision and in the 1985 draft |2| it is now limited to the direct method and single emulsion silver halide film. It is therefore not suitable for neutron radiography of irradiated nuclear fuel, where either the transfer method and silver halide film or nitrocellulose film are used.

The need for a separate method of assessment of the radiographic image quality in neutron radiography of nuclear fuel was recognised long ago. Already in 1976 a calibration fuel pin was designed and produced at Risø. It was used to check the accuracy of dimensional measurements from neutron radiographs |3|.

This calibration fuel pin was thereafter used at Risø for the visual comparison of image quality of nuclear fuel neutron radiographs |4|. It was also used for the assessemnt of the nitrocellulose film |5| by the visual method.

When the Euratom Neutron Radiography Working Group (NRWG) was constituted in 1979 a new design of the calibration fuel pin |6| was adapted and is now used in the NRWG Test Program |7|.

The control of radiographic image quality in neutron radiography of nuclear fuel was also discussed in |8, 9, 10|.

Under the NRWG Test Program the sensitivity level is visually evaluated from neutron radiographs of the sensitivity indicator SI, as prescribed by the ASTM E545.

The conclusions from this visual examination were already drawn in |8| and thereafter were formulated in |9| as follows: "The assessment of radiographic image quality by the evaluation of the components of the SI is not selective enough if the visible gaps are counted (all are always seen). At the same time it gives an incorrect impression of image quality from the evaluation of visible holes (even with good L/D more than 4 holes can be seen very rarely).
Similar conclusions were drawn by A. Laporte and G. Bayon from their investigation described in |11|.

Since the ASTM method of determining the sensitivity level from the visual examination is not selective enough another approach was adopted at Risø to determine the radiographic image quality.

2. VISUAL EXAMINATION OF NEUTRON RADIOGRAPHS

Under the NRWG Test Progam neutron radioqraphs of the calibration fuel pun CFP-El were made using different film/converter combinations (as described in |7| and |12| where the NRWG Test Program markings are shown). Those radiographs were thereafter visually examined.

The quality of the radiographic image of the CFP-El can be best assessed by judging the sharpness of the pellet-to-clad calibrated gaps by eye. Due to the geometry of the test object (calibration fuel pin – rund object) it is very difficult to get a sharp image of the small gaps on a radiograph. The sharpest radiographic image of the gaps of the calibration fuel pin could be obtained by X-rays. Therefore an X-ray radiograph (taken at 140 kV) was used as reference for assessing image quality on neutron radiographs.

An arbitary scale was used (from 1 to 5) where the sharpest picture produced by X-rays was given the value of 5. Three persons were visually examining the neutron radiographs of the CFP-El and were assessing the image quality of those radiographs by using the 1 to 5 arbitrary scale.

The results of this examination are summarized in fig. 1. Here the assessment of each of the three persons is marked by a black line. The dotted markings on fig. 1 mean, that the findings of all three persons were the same. In further evaluation of the above results only those dotted findings were taken into consideration.

The method used in this investigation proves to be useful as it shows differences in quality for neutron radiographs obtained by different methods using different recording material. At the same time in can differentiate between neutron radiographs taken by the same method on the same recording material but at different neutron facilities with different L/D ratios.

3. EVALUATION OF EXAMINATION RESULTS

The results of the visual examination, summarized in fig. 1 can be used for different comparisons of the radiographic image quality of neutron radiographs of the CFP-El taken by various exposure and recording techniques.

3.1. General evaluation

If one will look for the best results obtained among all techniques and recording materials one will see that for each NR facility best radiographic quality was always reached by the direct method using a single-coated, fine grain, silver halide film and 25 μm Gd converter.

This is quite obvious and could be expected in advance. In those facilities (Mol BR2, Petten HB8 and PO), where the direct method could not be used, best radiographic image quality was reached by using the fine grain, single-coated silver halide film for the transfer method using a 100 μm Dy converter.

As such general comparison of all exposure techniques and recording materials is meaningless, below other comparisons were made of methods and recorders which can be used optionally for specific applications (radioactive or non-radioactive objects).

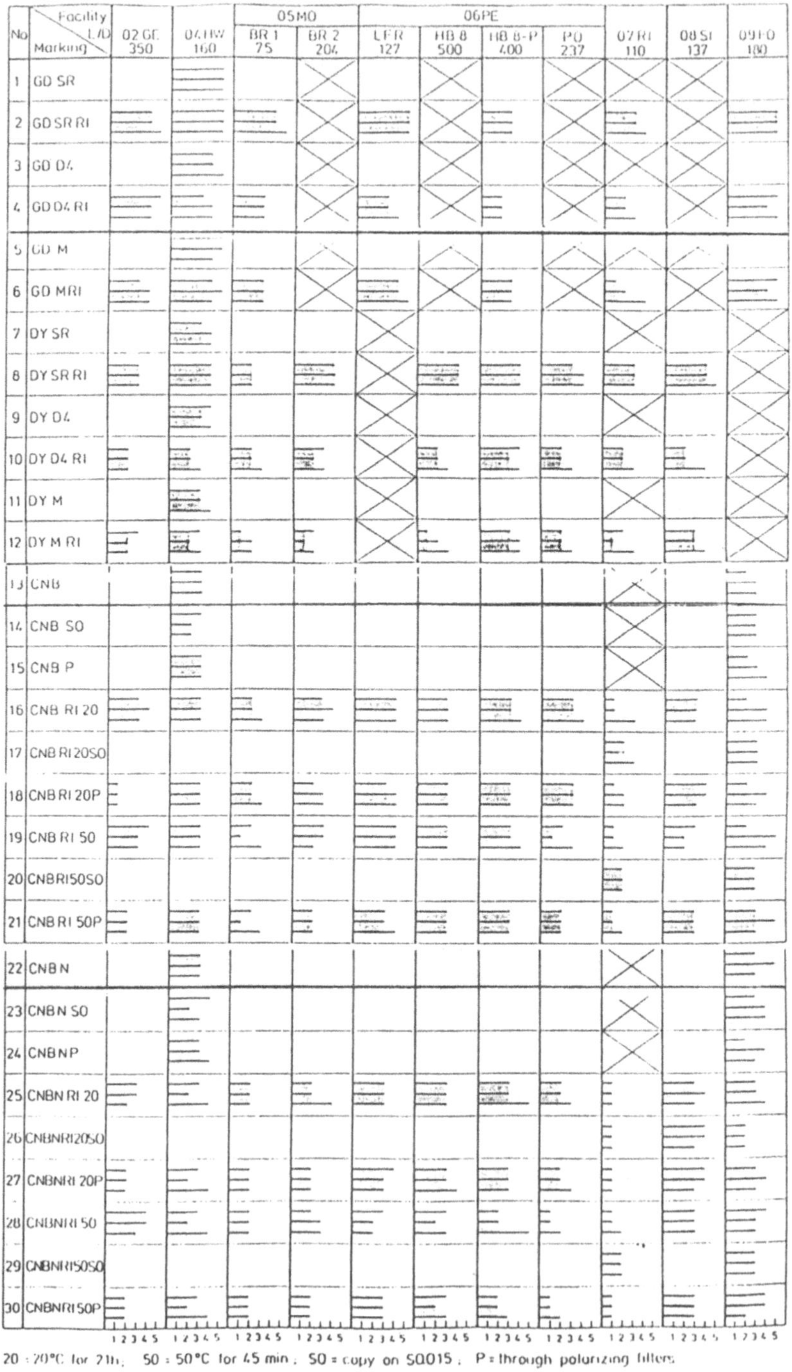

No	Facility I/D Marking	02 GE 350	04 HW 160	05MO		06PE				07 RI 110	08 SI 137	09 FO 180
				BR 1 75	BR 2 204	LFR 127	HB 8 500	HB 8-P 400	PO 237			
1	GD SR											
2	GD SR RI											
3	GD D4											
4	GD D4 RI											
5	GD M											
6	GD MRI											
7	DY SR											
8	DY SR RI											
9	DY D4											
10	DY D4 RI											
11	DY M											
12	DY M RI											
13	CNB											
14	CNB SO											
15	CNB P											
16	CNB RI 20											
17	CNB RI 20SO											
18	CNB RI 20P											
19	CNB RI 50											
20	CNBRI50SO											
21	CNB RI 50P											
22	CNB N											
23	CNBN SO											
24	CNBN P											
25	CNBN RI 20											
26	CNBNRI20SO											
27	CNBNRI 20P											
28	CNBNRI 50											
29	CNBNRI50SO											
30	CNBNRI50P											

20 = 20°C for 21h, 50 = 50°C for 45 min, SO = copy on SO015, P = through polarizing filters.

Fig. 1. Visual assessment of radiographic image quality of the CFP-El

3.2. Silver halide vs. nitrocellulose film

When examining radioactive objects (e.g. spent nuclear fuel) either the transfer method with silver halide film or the track-etch method with nitrocellulose film can be used. Therefore those two methods will be compared now.

While looking on the results, listed on fig. 1 under numbers 7 to 12 (for the transfer method with a 100 μm Dy converter) and numbers 13 to 30 (for the track etch method) one can do such a comparison.

Here the best radiographic image quality was reached for the transfer method with the Dy converter using a single coated, fine grain silver halide film.

The results reached in the comparison of the silver halide and nitrocellulose film were analyzed in detail in |13|.

Looking again on fig. 1 one can see, that for some NR facilities equally good results were reached with the direct method using either a single coated as well as a double-coated silver halide film.

3.3. Track-etch method with nitrocellulose film

It will be interesting not only to compare the transfer with the track-etch method but also to look closer into the different recording and evaluating methods possible with the nitrocellulose film.

Here two possibilities are open. One can use a double coated CN85 B film (numbers 13 to 21 in fig. 1) or the CN85 film sandwiched between two BN1 converters (numbers 22 to 30 in fig. 1).

Both combinations can be viewed either directly (numbers 13, 16, 19, for CN85B and 22, 25, 28 for CNB + BN1), viewed from copies on high contrast duplicating film (numbers 14, 17,20 for CN85B and 23, 26, 29 for CN85 + BN1), or viewed through polarizing filters (numbers 1, 18, 21 for CN85B and 24, 27, 30 for CN85 + BN1).

3.3.1. <u>Direct viewing.</u> It usually gives the same results of radiographic image quality whether judged from the double-coated CN85B of the CN85 film used with two BN1 converters.

3.3.2. <u>Viewing from copies.</u> On high contrast silver halide film it can sometimes improve the radiographic image quality. However, the main advantage of using such copies is their easier interpretation, as they look like neutron radiographs taken on silver halide film.

3.3.3. <u>Viewing through polarizing filters.</u> This technique is used for the same purpose as the former, i.e. to facilitate viewing. Also here the neutron radiographs on nitrocellulose film appear as those on silver halide film.

In most instances the radiographic image quality was equally good when viewed through polarizing filters as viewed directly without them.

3.3.4. **Etching conditions.** Neutron radiographs taken at different NR facilities were thereafter etched at Risø both at 20o C for 21 h (number 16 for CN85B and 25 for CN85 + BN1) and 50o C for 45 min (number 19 for CN85B and 28 for CN85 + BN1).

Here only in few instances the 20o C, 21 h etching gave slightly better image quality results than the 50o C, 45 min. procedure.

3.3.5. **Influence of the L/D ratio.** The last factor which could be studied from the results summarized in fig. 1 was the influence of the L/D ratio on the radiographic image quality.

The L/D ratios of NR facilities used in the NRWG Test Program varied from 75 to 500.

Here no direct connection between the L/D ratio and the radiographic image quality could be found. It is probably due to the fact that the L/D radio itself is not the only decisive factor in that respect. Other factors, as e.g. the gamma ray content in the neutron beam have also influence on the overall radiographic image quality.

4. CONCLUSIONS

After analyzing the partial results of the NRWG Test Program (summarized in fig. 1) one can come to a general conclusion that for such different NR facilities as listed in fig. 1 (with different L/D ratios, different gamma-ray content in the neutron beam and different exposure geometries) recommendations as to the use of a best exposure and recording technique can be given only whe analyzing the results reached for each particular facility.

However, many interesting general informations about the radiographic image quality can be found in the comparison performed in chapter 3 above.

It must be stressed that the analysis performed in this paper has taken into account only the subjective evaluation of radiographic image quality by the visual examination of neutron radiograph of a calibration fuel pin. In this analysis such factors as the relative speed of different exposure and recording methods were not taken into account. They can, however, be the decisive factors for some applications.

The problem of accuracy of dimensional measurments from neutron radiographs was also outside of the scope of this analysis. This problem was taken into account in |13| when comparing the properties of silver halide and nitrocellulose film.

REFERENCES

|1| ASTM E545-81. Standard method for determining image quality in thermal neutron radiographic testing.

|2| ASTM Draft E545-85. Standard method for determining image quality in direct thermal neutron radiographic testing.

|3| DOMANUS J. C., Accuracy of dimension measurements from neutron radiographs of nuclear fuel pins. Risø-M-1860.
26.03.1976 (also as paper 3L8 presented at the Eighth World Conference on Nondestructive Testing, Cannes, 6-11.09.1976.

|4| DOMANUS, J. C., Comparison of image quality of nuclear fuel neutron radiographs taken on silver halide and nitrocellulose film. Risø-M-2170, April 1979 (also as paper 2BDD-1 of the Ninth World Conference on Non-Destructive Testing, Melbourne, 18-23.11.1979).

|5| J. C. Domanus. How good is the nitrocellulose film for neutron radiography? 729-736 in the Proceedings of the 1st WCNR.

|6| J. C. Domanus. Calibration fuel pin CFP-E1. Risø Report B-499. Metallurgy Department, Risø National Laboratory. February 1981.

|7| J. C. Domanus. Euratom test program for image quality and accuracy of dimensions. 1025-1033 in Proceedings of the 1st WCNR.

|8| J. C. Domanus. Control of radiographic image quality in neutron radiography of nuclear fuel. Proceedings of the 6th ASM International Conference on NDE in the Nuclear Industry. Zürich, Switzerland. 28.11.-2.12.1983, 447-451.

|9| J. C. Domanus, P. Gade-Nielsen & J. Olsen. How good are the standards for the image quality control in neutron radiography of nuclear fuel? Proceedings of the 7th International Conference on NDE in the Nuclear Industry. Grenoble, France. 29.1-1.2.1985. 325-328.

|10| J. C. Domanus. Activities and achievements of the Euratom Radiography Working Group. Materials Evaluation, Vol. 44, No 1, January 1986, 114-119.

|11| A. Laporte & G. Bayon. Determination of image quality in industrial neutron radiography. CEN-FAR. Service des Piles de Saclay - Section d'exploatation TRITON. 6.203.2.

|12| J. C. Domanus. Euratom Neutron Radiography Working Group. Proceedings of the 2nd WCNR.

|13| J. C. Domanus. In "Neutron radiography on nitrocellulose film". EUR report to be published in 1986.

INTERCOMPARISON OF NEUTRON RADIOGRAPHY INDICATORS USING KUR

Kohsuke Katsurayama, Keiji Kanda, Kenji Yoneda,
Shigenori Fujine, Hiroshi Yamagata, Otohiko Aizawa[1],
Gen-ichi Matsumoto[2], Hisao Kobayashi[3], Takao Tsuruta[4],
Takeo Niwa[4], Nobuo Wada[5], Ei-ichi Hiraoka[6]

Research Reactor Institute, Kyoto University
Kumatori-cho, Sennan-gun, Osaka 590-04, Japan

1) Atomic Energy Research Laboratory, Musashi Institute of
   Technology
   Ouzenji, Asou-ku, Kawasaki, Kanagawa 215, Japan

2) Dept. of Nuclear Engineering, Nagoya University
   Furo-cho, Chikusa-ku, Nagoya 464, Japan

3) Institute for Atomic Energy, Rikkyo University
   Nagasaka, Yokosuka, Kanagawa 240-01, Japan

4) Atomic Energy Research Laboratory, Kinki University
   Kowakae, Higashi-Osaka, Osaka 577, Japan

5) Japan Atomic Energy Research Institute
   Oarai-machi, Ibaraki-ken 311-13, Japan

6) Radiation Center of Osaka Prefecture
   Shinge-cho, Sakai, Osaka 593, Japan

ABSTRACT

ASTM and EURATON standard indicators have been used
in order to determine image quality in thermal neutron
radiographic testing, however it is thought that they
have been some limitations for current application of
precise measurements.
Nineteen standard indicators from eight institutes
were tested at the Kyoto University Reactor (KUR).
From the results, some inadequacies of the current standard
indicators were confirmed, particularly it is thought
that image quality must be determined automatically by
a machine.

INTRODUCTION

Neutron radiography is relatively recent technology developed mainly for non-destructive testing. One of the applications of this technique is the pre- and post-irradiation testing of nuclear fuel elements.

In any method of testing, the standard indicators are needed to determine the quality and limitation of the method of testing. Neutron radiography is no exception.

Since Barton first presented a Visual Image Quality Indicator in 1971, many indicators have been presented. At present there are the standard indicators prepared by the American Society for Testing and Materials (ASTM) and the European ·Atomic Energy Community (EURATON) for determining image quality in thermal neutron radiographic testing in 1981. These standard indicators have been widely used, however it is thought that they have some limitations for current application of precise measurements. The present paper describes the inadequacy of the current standard indicators and the proposal for solving them.

The determination of the quality of a neutron radiography system is based upon evaluation of images obtained from the standard indicators. In the present study, three types of standard indicators were put to a test:

(1) Sensitivity Indicator (SI),
(2) Beam Purity Indicator (BPI),
(3) Beam Purity Indicator Fuel (BPI-F).

While sensitivity level was determined with ASTM-E545-81[1], only when determining exposure contributors by the BPI, the following calculations[2] were used:

Thermal neutron content
$$= [D_5-(\text{higher value of } D_1 \text{ and } D_2)]/D_5 \times 100,$$

Scattered neutron content $= (D_1-D_2)/D_5 \times 100,$

Gamma content $= [D_6-(\text{lower value of } D_3 \text{ and } D_4)]/D_5 \times 100,$

Pair production contribution $= (D_3-D_4)/D_5 \times 100,$

where D_1 : density under the lower boron nitride disk,
D_2 : density under the upper boron nitride disk,
D_3 : density under the lower lead disk,
D_4 : density under the upper lead disk,
D_5 : background film density in the center of the hole,
D_6 : film density through the teflon body of the BPI.

The quality level of a neutron radiography system is determined mainly by the values of NC-H-G (NC: thermal neutron content). In Category I (the highest grade), the values of NC-H-G are more than 65-6-6, respectively. The values of NC-H-G of the neutron radiography facility where the experiments were performed are 79-7-7. This facility belongs to Category I and has capability enough to intercompare the standard indicators.

EXPERIMENTAL

Table 1. shows the standard indicators used in the experiment.

The experiments were performed at the E-2 experimental hole of the Kyoto University Reactor (KUR) as shown in Fig. 1. The property of the neutron radiography facility and the experimental condition are shown in Table 2.

One kind of standard indicators from various institutions were tested in a radiation field simultaneously. After changing location of the standard indicators, they were tested again in order to keep out the influence of non-uniformity in the field of the neutron beam. One of these films is shown in Fig. 2.

The images obtained from the sensitivity indicators were visually inspected by 9 persons, and film densities under different parts of the BPI and the BPI-F were measured by a diffuse transmission densitometer.

RESULTS AND DISCUSSION

The sensitivity level with the SI at the E-2 experimental hole of the KUR are shown in Table 3. The results of experiment are summarized in Table 4.

From these results, the followings are observed.

(1) Only in two sensitivity indicators, the 0.25-mm holes of the lead steps were not visible. It may therefore be thought that those of the lead steps in these indicators are filled with one like a dust.

(2) The value of H differed both among the sensitivity indicators and by individual readers of the films.

(3) The value of G was always seven. The value of seven means that the smallest gap is visible, so that we could not determine even if smaller gap is visible in this irradiation field.

(4) The BPI indicated negative numbers of gamma content under two conditions, because non-uniformity of film density was more than the attenuation of gamma ray by 2-mm lead disk.

(5) The BPI-F indicated lower number of the scattered neutron content than the BPI, because the distance between the upper and the lower BN disks is shorter in the BPI-F than

in the BPI.

To solve these flaws, the follow improvements can be considered.

(1) Quality of the SI must be guaranteed.
(2) The value of H must be automatically determined by a machine.
(3) Smaller gaps are needed in the SI.
(4) The thicker lead disk is needed in the BPI.
(5) The distance between the upper and the lower BN disks should be longer in the BPI-F.

Figure 3 shows the neutron transmission through 2mm-Cd and 2mm-BN that are used in the BPI-F. The epithermal neutron content is calculated by using the difference of the transmission between 2mm-Cd and 2mm-BN. However, the transmission of 2mm-Cd is lower than that of 2mm-BN under 0.5 eV. Then the epithermal neutron content is evaluated lower than a real one. In order to keep out this effect, thickness of Cd and BN disks should be improved.

Now the authors are designing new indicators with the above view.

ACKNOWLEDGMENT

These experiments were performed under the Joint Use Program in Research Reactor Institute Kyoto University in 1985.
The authors express their thanks to A. Tsuruno of JAERI and H. Yamamoto of Komazawa University for valuable discussions and participating the experiment.

REFERENCES

1. ASTM E545-81. Standard method for determining image quality in thermal neutron radiographic testing (1981).

2. J. C. Domanus, "EURATON Test Program for Image Quality and Accuracy of Dimensions", Proc. 1st world Conf. Neutron Radiography, (California, 1981) 1025.

3. H. Berger et al., Atomic Energy Review, 15 (1977) 123.

4. K. Kanda, "Development of an Online Neutron Radiography System of High Resolution for Nuclear Materials", Proc. 1st world Conf. Neutron Radiography, (California, 1981) 219.

Table 1. The Standard Indicators used in the Experiment.

Research Institute [Symbol]	ASTM-E545-81 (RISφ) SI,BPI,BPI-F			(Research Chemicals) SI,BPI,BPI-F	
Kyoto University [KUR]	o	o	o		
Radiation Center of Osaka [ORC]				o	o
Kinki University [UTR]	o	o	o		
Nagoya University [NAG]	o	●			
Musashi Institute of Technology [MIT]	o	o			
Rikkyo University [RUR]				o	o
JAERI Tokai [JRR]	o	o			
JAERI Oarai [R I]	●	●	●	(made in Japan)	

Table 2. The Property of Neutron Radiography Facility at the KUR and the Experimental Condition.

Reactor type	M T R
Reactor power (kW)	5000
Peak Φ_{th} in core $(cm^{-2}s^{-1})$	6×10^{13}
Moderator	D_2O
Tube type	Tangential
Bi filter (cm)	1
Diameter of aperture (cm)	5
Range of L (cm)	500
L/D	100
Diameter of radiation field (cm)	16
Φ_{th}, at film $(cm^{-2}s^{-1})$	1.2×10^6
Cadmium ratio	400
Gamma dose rate (R/hr)	4.2
n/γ ratio $(cm^{-2}mR^{-1})$	10^6
Conversion screen	Gd-25μ
Film	KODAK/SR
Developing method	20°C 5min tank
Exposure time (min)	25

Table 3. The sensitivity level with the SI at the KUR E-2.

Name	A			B			C			D			E			F			G			H			I		
	*	H	G	*	H	G	*	H	G	*	H	G	*	H	G	*	H	G	*	H	G	*	H	G	*	H	G
KUR	-	-	7	-	9	-	x	9	7	o	10	7	o	9	7	o	9	7	o	9	7	o	9	7	o	8	7
	o	7	7	-	8	-	x	8	7	o	7	7	o	10	7	o	7	7	o	8	7	o	7	7	o	-	7
ORC	-	-	7	-	8	-	o	7	7	o	8	7	o	7	7	o	7	7	o	8	7	o	7	7	o	7	7
	-	-	-	-	-	-	-	-	-	-	-	-	-	-	-	-	-	-	-	-	-	-	-	-	-	-	-
UTR	-	-	7	-	9	-	o	9	7	o	10	7	o	10	7	o	9	7	o	7	7	o	9	7	o	7	7
	o	8	7	-	9	-	o	7	7	o	9	7	o	11	7	o	9	7	o	8	7	o	11	7	o	-	7
NAG	-	-	7	-	9	-	x	7	7	x	10	7	x	10	7	x	10	7	x	7	7	x	9	7	x	7	7
	x	7	7	-	9	-	x	6	7	x	10	7	x	10	7	x	6	7	x	6	7	x	9	7	x	-	7
MIT	-	-	7	-	9	-	x	7	7	x	9	7	x	9	7	x	9	7	x	8	7	x	8	7	o	8	7
	o	8	7	-	10	-	o	7	7	o	10	7	x	10	7	o	8	7	o	8	7	o	8	7	o	-	7
RUR	-	-	7	-	9	-	o	8	7	o	10	7	o	10	7	o	7	7	o	8	7	o	8	7	x	7	7
	o	8	7	-	9	-	o	8	7	o	8	6	o	8	7	o	8	7	o	8	7	o	8	7	o	-	7
JRR	-	-	7	-	8	-	x	7	7	x	7	7	x	7	7	x	7	7	x	7	7	x	7	7	x	7	7
	x	7	7	-	9	-	x	7	7	x	7	7	x	7	7	o	7	7	x	7	7	x	7	7	x	-	7
R I	-	-	7	-	10	-	o	7	7	o	8	7	o	9	7	o	9	7	o	8	7	o	10	7	o	8	7
	o	8	7	-	9	-	o	9	7	o	9	7	o	9	7	o	9	7	o	8	7	o	9	7	o	-	7

*) All holes of lead steps are visible (o) or not (x).

Table 4.   The Results of Experiment at the KUR E-2.

SI	*	H	G
KUR	o	8.4±1.0	7
ORC	o	7.4±0.5	7
UTR	o	8.9±1.3	7
NAG	x	8.3±1.6	7
MIT	o	8.5±1.0	7
RUR	o	8.3±0.9	7
JRR	x	7.2±0.5	7
R I	o	8.7±0.8	7
Average		8.2±1.1	

BPI	
Thermal neutron content	79.7±0.6
Scattered neutron content	2.14±1.02
Gamma content	−0.19±1.39
Pair production contribution	0.52±2.66

BPI-F	
Thermal neutron content	83.5±0.8
Scattered neutron content	1.2±0.4
Epithermal neutron content	0.0±1.1

* All holes of lead steps are visible (o) or not (x)

C : Core
Hy : Hydrolic Conveyor
Pn : Pneumatic tubes (x3)
V : Vertical exposure tube
DC : Heavy water tank
BI : Bismuth Shield
GR : Graphite layers
Sh : Concrete shutter
TC : Graphite thermal column
DR : Door
B₁,B₂,B₃,B₄ : Beam tubes
E₁,E₂,E₃,E₄ : Exposure tubes
Th : Through tubes
Tk : Reactor tank

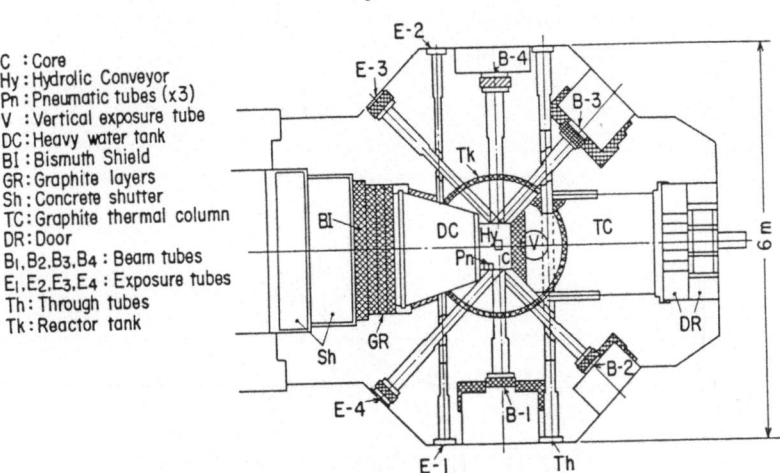

Fig. 1. Plan View of the Beam-tube layout of the KUR[4].

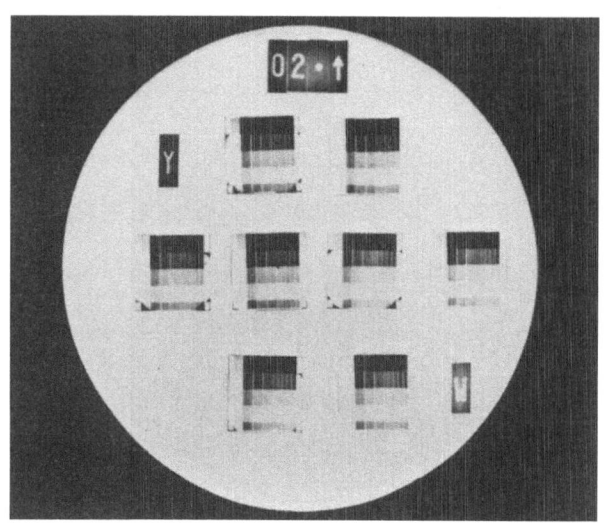

Fig. 2.   Film Obtained from the Sensitivity Indicators at the KUR
E-2.

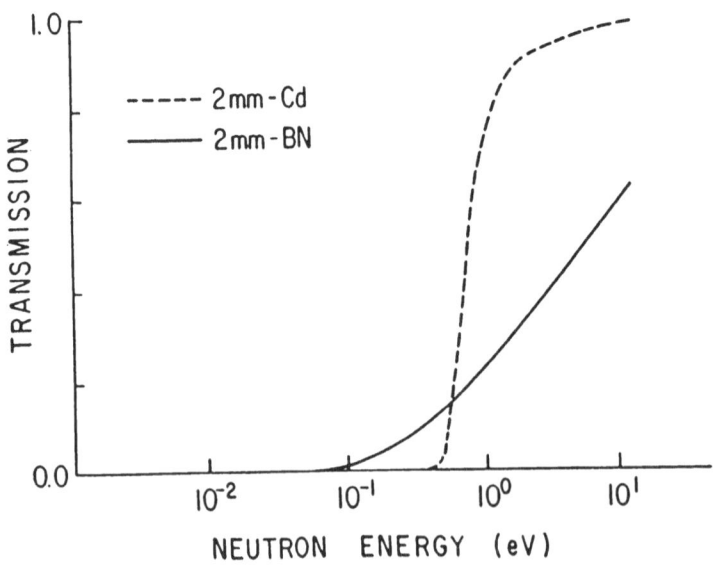

Fig. 3.   Transmissions of neutron by 2mm-Cd and 2mm-BN disks.

NORMES FRANCAISES DE NEUTRONOGRAPHIE INDUSTRIELLE

E. Julliard et G. Bayon

Association Française de Normalisation, Paris La Défense, France

CEA/IRDI/DERPE/SPS, Saclay, France

RESUME

L'Association Française de Normalisation (AFNOR) vient d'homologuer deux textes de normes dans le domaine de la radiographie aux neutrons.
Ces deux textes se limitent à l'examen de pièces et objets non radioactifs en utilisant la technique directe associant un écran convertisseur en gadolinium et un film argentique.

I - INTRODUCTION

En 1980, certains utilisateurs industriels des installations de radiographie aux neutrons ont émis le souhait de pouvoir se référer à des textes officiels de normes concernant cette technique. Le but de cette démarche consistait à pouvoir connaître les performances de toutes les installations françaises et étrangères susceptibles de répondre à leurs besoins. Les grandeurs et paramètres définis par ces textes devaient être compréhensifs par des personnes étrangères au domaine nucléaire.

L'Association Française de Normalisation a donc constitué, sous son égide, deux Groupes de Travail qui ont travaillé à la rédaction de ces normes.

II - GENERALITES

L'Association Française de Normalisation tire une partie de ses ressources financières de la vente des textes et normes qu'elle publie.

Il nous est donc impossible de reproduire ici l'intégralité des deux normes existantes. Les personnes intéressées par ces textes peuvent les obtenir en adressant leur demande à :

Service de Diffusion de l'AFNOR - Tour Europe
Cédex 7 - 92080 Paris La Défense (France) - Télex : 611 974 F.

L'objet de cette communication est de dégager certains aspects généraux qui en ont guidé la réalisation.

III - NORME CONCERNANT DES FILMS ARGENTIQUES UTILISABLES

NF A 09-213 : Neutronographie industrielle - Détermination de la sensibilité et du contraste moyen des systèmes récepteurs d'image à base de films argentiques, au moyen d'électrons d'énergie inférieure à 200 keV.

Cette norme a été construite sur le même modèle que les normes AFNOR utilisées pour la radiographie industrielle X et gamma. La difficulté de définir une source de neutrons thermiques normalisée nous a conduit à lui préférer une source électronique reproduisant au mieux le phénomène de réémission d'électrons de conversion interne du gadolinium soumis au rayonnement neutronique. Le carbone 14 a été choisi pour effectuer les irradiations de film et en déduire ainsi les caractéristiques essentielles des émulsions argentiques, à savoir :

. la sensibilité S,
. le constraste moyen G.

La norme définit les conditions expérimentales permettant, pour chaque type de film, d'estimer ces deux valeurs caractéristiques.

IV - NORME DE CARACTERISATION DES INSTALLATIONS DE RADIOGRAPHIE AUX NEUTRONS

NF A 09-220 : Neutronographie industrielle - Détermination des caractéristiques des installations pour le contrôle de pièces non radioactives.

Cette norme est utilisée pour chiffrer certains paramètres types d'une installation de radiographie aux neutrons : les grandeurs géométriques du faisceau :

. rapport moyen de collimation L/D,
. angle de divergence,

ainsi que les termes de vocabulaire spécifique utilisé pour cette technique.

Dans la suite du texte, on préconise une méthodologie permettant de définir des caractéristiques complémentaires telles que :

. homogénéité du faisceau,
. contribution du rayonnement gamma parasite, etc.

Les caractéristiques sont évaluées non pas à l'aide d'Indicateurs de Qualité d'Image mais grâce au matériel couramment utilisé sur l'installation et par mesurage de densités optiques sur des films obtenus par des irradiations dans le faisceau suivant diverses configurations.

L'AFNOR ne définit pas d'indicateur de qualité d'image normalisé, contrairement à ce qui existe en radiographie industrielle X ou gamma.

Par contre, une annexe non homologuée recommande l'utilisation d'indicateur de qualité d'image spécifique au type de pièce contrôlée. Ces IQI reproduisent au mieux le type de défaut recherché et la géométrie globale de la pièce.

Cette annexe présente également une mire en gadolinium à fentes très fines qui permet, en faisant varier la distance IQI-détecteur, de pratiquer des analyses micro-densitométriques sur des clichés et de juger ainsi les performances globales de chaque paramètre influant la qualité de l'image produite.

V - CONCLUSIONS

Les deux textes normatifs publiés en France pour la radiographie aux neutrons montrent une originalité dans la démarche suivie, qui se veut proche du souci des utilisateurs de cette technique.

Ils ne préconisent pas d'IQI "universels" mais encouragent une approche spécifique au type de pièces contrôlées et aux défauts recherchés.

En ce sens, l'AFNOR se démarque nettement de l'ASTM.

PART XIV

WORKSHOP REPORTS

Workshop on "Non-Reactor Neutron Sources for Radiography"

Co-Chairmen : S. Cluzeau and M.R. Hawkesworth

This workshop was attended by about 40 people, 25 % of the conference, reflecting the still growing interest in non-reactor sources for neutron radiography. Discussion was vigorous and wide ranging, so the following summary can merely be a personal recollection of some of the comments which received support.

Many kinds of accelerators are available for neutron radiography - sealed-tube, Van de Graaff, linac, RFQ, cyclotron - and all can now be used to drive transportable systems. So, what properties should be seen as most important in guiding the customer's choice : beam intensity, neutron spectrum, exposure time, reliability, safety, cost?

Of course, the customer's primary concern is that the radiographs produced are adequate for his purpose at a price he can afford. But, few accelerator-based systems actually exist at present, and none are transportable systems using the latest technology. So, unless the customer is prepared to trust the technical design data and radiographs taken with reactors at realistic L/D ratios, he will have to wait until accelerator manufacturers are prepared to spend their own funds to develop a system which can be seen to work. This could take some time.

The meeting supported the view that beam intensity should not be the only criterion in system selection. For example, in testing aircraft honeycomb structures for corrosion, the greater manoeuvrability of a sealed-tube system could well be more important than the greater throughput available with a cyclotron - even at the same cost. Another factor often overlooked is the customers' psychology; he has his customs and proven non-destructive testing methods. Why should he change?

The meeting speculated on the possible market for a proven system. At present no "commercial" non-reactor systems are operational in Europe, and less than ten based on accelerators are operating in the USA, Japan and Korea. Widely diverging opinions were expressed and no consensus emerged, but clearly the market will be small unless the new manoeuvrable systems or ones based on powerful, 10^{13} n s^{-1}, accelerators provide radiographs which revolutionize aircraft maintenance routines.

Perhaps two basic types of system will find a market; one based on highly mobile small accelerators, the other based on more powerful machines with fewer degrees of freedom.

Projects to develop a sealed-tube accelerator and a mobile cyclotron with outputs in the range $10^{13} - 10^{14}$ n s^{-1} were outlined. It was observed, however, that apart from proving the basic design, target reliability and safety will also be concerns which have to be adequately resolved before customers can be sought. It was also noted that the electrical power requirements will be high for the more powerful machines, ranging from 30 kW for a super-conducting magnet cyclotron to, perhaps, over 100 kW for a sealed-tube source. Moreover, the weight of machine and moderator that will have to be moved closely around valuable aircraft will be high, perhaps 2000 kg or more, unless the cyclotron beam can be reliably passed some distance along an articulatable beam tube. It was generally felt that superconducting cyclotrons held great promise nevertheless, because of the great reliability shown by fixed cyclotrons with only slightly inferior performance over many years.

The general view of the meeting was that isotopic sources held little promise, since ^{252}Cf was not available outside the USA at a competitive cost. In any case, its transportation was increasingly difficult because of new safety regulations. Imaging systems were also little discussed since "Electronic Imaging" was to be the topic for discussion the following day. The meeting closed after 1 hr 20 min.

872

Workshop on "Industrial Applications"

Co-Chairmen : J.P. Barton and J.P. Bouloumié

Summary (by J.P. Barton)

The session was attended by about twenty participants.
Application Areas : In the USA and the UK, the inspection of aircraft
turbine blades forms a major part of the workload. Mr. Newacheck
provided details. However, Mr. G. Bayon reported that essentially no
routine turbine blade neutron radiography inspection was performed in
France. The difference is notable because France does have a major
aircraft industry, and does perform much neutron radiography (typically
200 films per month at SACLAY, of which eighty per cent are
pyrotechnics.
Cassette Types : The aluminium cassettes routinely used at the Aerotest
Inc. facility differ from the SACLAY cassettes which use a magnesium
window and a thich epoxy base to minimize neutron activation. During
discussion of why the epoxy base does not cause backscatter, it was
noted that the SACLAY team use gadolinium foils of thickness 250 microns :
much thicker than the converters used at Aerotest, Inc.. This combined
with the pure, low energy neutron beam from the ORPHEE guide tube could
result in elimination of any scatter problem.
Image Intensifiers : Mr. Rougeot, the specialist in neutron image
intensifiers from Thomson-CSF, raised the question of whether it was
safe to use such high voltage equipment near pyrotechnics. The ensuing
discussion was not conclusive but it was pointed out by Dr. Orphan of
SAIC - USA, that pyrotechnics were often positioned for inspection near
x-ray generators including powerful linear accelerators.
It was the impression of this reviewer that each of the above topics
deserve follow-up consideration.

Résumé (par J.P. Bouloumié)

Au niveau de la participation, on peut noter la présence exclusive de
représentants américains. Ce fait a limité considérablement les échanges
techniques, ceci d'autant plus qu'en l'absence des interprètes J.P. Barton
a du se substituer à eux.

Parmi les problèmes soulevés dans cette intercomparaison de l'utilisation du contrôle neutronographique au niveau industriel, on peut relever :

- Un prix du cliché sensiblement équivalent.
- Un pourcentage très différent sur les différents produits contrôlés en France et aux U.S.A. :
    - France 90 % produits pyrotechniques
    - U.S.A. 45 % produits pyrotechniques
              30 % contrôles de corrosion.
- L'utilisation directe du contrôle neutronographique par les industriels sous une forme intégrée au sein même des sociétés (AEROTEST chez O.E.A.).
- Une utilisation accrue des études quantitatives de type microdensito-métriques.
- L'utilisation d'une source unique : le réacteur.
- Un plan de charge quasi continu des installations de contrôle aux U.S.A.

Workshop on "Electronic Imaging"

Chairman : J. Rogers

A shortened discussion session was held on electronic imaging. The large
number of papers at this conference which referred to real-time system
was noted.

Smith gave a brief description of the neutron imaging system used at
ORNL for difraction studies on crystals and on turbine blade materials.
Matusomoto discussed the studies of different converter screens
(including those which incorporate BN and LiF). Barton raised the
question of neutron flux levels which would not adversely affect the
camera of a Thomson tube-based system used in line directly in the beam.
Steinbock described a solid state system being investigated using a line
array of 1024 x 1 pixels (pixel size 25 mm) fibre optically coupled to a
converter screen and raised the possibility of the converter being
deposited directly on the surface of this array.

Kedem discussed loss of resolution and signal/noise losses in recording
of data which involved A/D and D/A conversion through several stages.
McRae indicated a high resolution video recorder (3/4" format) which has
been used together with camera time-based correction for reducing these
losses.

A discussion was held on the requirements for storage of large
quantities of video data. Optical unit – only disc systems of 500 mbyte.
Heidt described a large capacity storage requirements for a base catalog
for weld inspection. Images of 512 x 512 panels were stored on optical
disc. The storage time (long-term) of these discs was questioned and it
would seem that these discs should be copied at intervals of
approximately two to three years.

Image processing requirements for real-time radiography was discussed.
Dance described some of the image subtraction methods used in
radiography on the DIDO cold beams at Harwell.

Kedem raised the question of penetrameter positioning in real-time
imaging. It was clear that a new code of practice for IOI's needed to be
discussed.

Requirements for possible future improvements in spatial resolution for real-time imaging systems were also discussed. In general, it was felt that such improvements were not required.

The study of rapid transients were touched on. Barton referred to NR work with a high speed camera at Oregon using the pulsed TRIGA reactor. The requirements for fast transient studies are high flux plus efficient and fast converter screen.

Workshop on "Computed Tomography"

Co-chairmen : C.F. Barton and G. Ducros

The workshop was attended by sixteen participants. Reviews of neutron radiography computed tomography projects were presented by P. Rizo, CENG, France, J. Brenizer, University of Virginia, USA, G. Matsumoto, Nagoya University, Japan.

Much progress has been made in the raw data analysis for algorithm reconstruction by taking into account some second order effects such as neutron scattering and resonance. Progress has also been made in electronic imaging as a technique more flexible than conventional film radiography. There was some discussion on the need for the next generation of real-time neutron radiography C.T.. What are the applications which might require the cost and effort for developing real-time? Brenizer mentioned some real-time phenomena such as combustion. J. Walker from University of Birmingham presented some results on positron tomography as a promising alternative technique for three dimensional image analysis.

Workshop on "Standards"

Co-Chairmen : H. Berger and J.C. Domanus

A workshop on Standards for neutron radiography was held on Wednesday, June
18, 1986. There were discussions of priorities for needs for neutron
radiography standards (direct, transfer, real-time, non-reactor source
systems) and comments on the differences obtained by various laboratories
for sensitivity values using the ASTM E545 type test pieces. The European
results appear to be different from those obtained in the US and Japan. The
European results show unusually high effective thermal neutron content with
the Beam Purity Indicator (BPI) and an unusually low value of H (hole
visibility in the Sensitivity Indicator (SI). In the latter case, the
European results usually show only four (4) holes or less versus visibility
of seven (7) or more holes for US and Japanese tests.

In our attempts to resolve these differences, ASTM has already exchanged
BPI units with the European Neutron Radiography Working Group (NRWG).
Tests of the NRWG BPI in the USA showed that the unit is all right and
gives comparable results to US-produced BPI units. In further attempts
to resolve the differences, it was agreed that an SI assembly from
Europe would be sent to the US for test (Domanus to send to Newacheck).
It was also agreed that an SI assembly from the USA would be sent to
Europe to test (Newacheck to send to Domanus). Also, since film
differences were among the possibilities, it was agreed that we would
send some SR film to NRWG (Newacheck to send to Domanus).

There was considerable interest in the development of standards for
real-time radiography. In an attempt to get all interested people
throughout the world working together from the beginning on this
important problem, it was agreed that we would organize the flow of
information through three people from around the world, as follows :

Europe :	Japan :	USA :
J.C. Domanus	Prof. K. Katsurayama	Harold Berger
Risø National Laboratory	Kyoto University	Industrial Quality Inc.
DK-4000 Roskilde	Noda, Kumatori-cho	P.O. Box 2397
Denmark	Sennan-Gun Osaka	Gaithersburg,
		MD 20879-0397

CLOSURE OF THE CONFERENCE

CLOSING REMARKS

J.P. Barton - Co-chairman

During this conference, the opportunity has been taken to receive input
from all concerned about a possible future conference in the series
which started in San Diego in 1981.

The following summarizes the outcome of these discussions :

1. There will be a Third World Conference on Neutron Radiography.

2. The location will be Japan, if that can be arranged. Failing that, it
will be in California, USA. The logic of these choices is the belief
that most people prefer to travel to regions of concentrated activity in
the field.

3. The time interval will be three to four years.

4. A committee consisting of J.P. Barton (Chairman), G. Farny, G. Matsumoto
and H. Röttger (for complete addresses see list of participants) will be
responsible for ensuring that the Third World Conference on Neutron
Radiography is organized in a way consistent with the input from the
previous conferences and the recommendations of the individual neutron
radiographers.

CONFERENCE PROGRAMME SUMMARY

Posters to the various sessions were presented Tuesday, Wednesday and
Thursday from 08.30 to 19.30 hrs. All contributions to session II "Reactor
Facilities", session XI "Tomography" and session XII "Special Techniques"
were presented as posters. Session III B "Non-Reactor Sources - Isotopic
Sources" had only oral presentations.

Tuesday, June 17th, 1986

- Opening of the Conference
- Session I   - Neutron Radiography Programmes
- Session IV   - Industrial Applications
- Session III A   - Non Reactor Sources - Accelerators
- Workshop (IIIA+B) - Non Reactor Neutron Sources for Radiography
- Workshop (IV)   - Industrial Applications

Wednesday, June 18th, 1986

- Session VII   - General Applications
- Session V   - Nuclear Applications
- Session IX   - Track-Etch Imaging
- Session X   - Dimensional Measurements
- Session XIII   - Standards
- Workshop (VIII) - Electronic Imaging
- Workshop (XI)   - Computed Tomography
- Workshop (XIII) - Standards

Thursday, June 19th, 1986

- Session VIII - Electronic Imaging
- Session VI   - Corrosion Inspection
- Session III B - Non Reactor Sources - Isotopic Sources
- Closure of the Conference

Editors' note : The sessions do not correspond completely with the main
                chapters (parts) of the proceedings.

N O T E S, C O M M E N T S A N D C O N C L U S I O N

J. L. PERSON - H. RÖTTGER

During the past 18 months, it was really a pleasure to participate in the organization of this meeting and to follow the evolution of the project.

The 16th June was an unforgettable day for the team to welcome more than 160 delegates from 24 differents countries. Besides the presence of about fifty French attendees, we could notice an important Japanese delegation, a fairly good number of European and American representatives, and also the participation of new countries as China, Greece, Norway, Portugal, South Africa, Taïwan and Turkey.

Although the delegates of Hungaria, India, Iraq, Iran, Malaysia, Tchecoslovakia and Soviet Union were unable to attend, the number of participants increased to 30 % compared to WCNR1 held in 1981 : it was indeed for the great majority of them a "première" as only 26 persons (less than 16 %) attended the first meeting in San Diego. This underlines the importance of the development of this method around the world.

The items of the agenda were chosen during the Programme-Committee Meeting in December 1985, 6 months before the Conference : 112 papers were reviewed and selected at that time ; 95 communications have been finally presented (52 oral + 43 poster presentations).

One can see that there is a slight decrease compared to the 140 papers presented in San Diego, but among those, a large number related facilities or work carried out since the origin of the method ; these items are well known nowadays ; that is why the reactor facilities (part III) included only 7 papers (poster session).

The initial choice to spread the hundred communications over 3 days was decided for practical and financial reasons : it is possible that the amount of conference fee have led several potential attendees to give up ; we must underline that only the effective costs invoiced have been included in the registration fee : rooms and equipment booking, lunchs, documents and proceedings.

It can easily be understood by everyone that, taking into account the whole organization and the expenditures of simultaneous interpretation, the organisers could not provide individual claimers with a special financial support.

Nevertheless, at the beginning of the project and to decrease those expenditures, we thought of organizing the Conference at the "Centre d'Etudes Nucléaires de Saclay" : a brief investigation led us to reject such a possibility, as less than 100 persons would have been interested. On another hand, considerable difficulties would have increased : the transportation problems between Paris and Saclay, with large consequences on the final Programme, thus inevitably made it skimpy.

On the contrary, the location at "Palais des Congrès" in Paris allowed us to welcome all delegates (even late comers...) and speakers to present their papers in more comfortable conditions.

Programming parallel sessions could not be held : we prefered to set up workshops with specific items. During this kind of conference where most of the attendees are expert technicians, simultaneous oral presentations would have been less beneficial, because it would have prevented many attendees to participate.

Besides, a few attendees would have liked that more time be dedicated to poster sessions : adding one more day or shortening allocated time for oral presentations became a dilemma which would, even now, be impossible to resolve.

However, all the participants seemed rather satisfied and the very large scale of communications covering so many different fields as nuclear fuel control, aluminium corrosion, inspection in aircraft industries and examination of ancient art pieces of work and even more dentistry tools studies could be presented to the audience.

Participants' comments are quite unanimous : the promises coming from non-reactor (and mobile) sources and the new electronic imaging methods are at the root of NR development in the future.

There is no doubt that the number of users will increase in connection with better performances and implement ability of transportable systems associated with real time imaging. Moreover, as we could also foresee it, participants were interested by tomography and these last items are generally quoted in their replies.

Below, we let some of the attendees express themselves directly and give their personal comments and estimations :

"The inspection of aircraft is one of the most promising application of NR".

A. ITAYA

"I found of most interest the fact that there was a shift of attention away from reactor-based radiography sources towards an enhanced interest in smaller sources which might even be considered mobile. This allowed the attendees to focus on the possibilities for moving the practice of neutron radiography into the workplace : there, neutron radiography may have an opportunity to be applied in ways that we cannot conceive even today(...). The papers presented at the 2nd WCNR which dealt with this new aspect of NR were the most interesting communications for I believe the presage an exciting renaissance in NR".

J. ANTAL

"I my opinion, some of the highlights (...) were the advent and future possibilities of small accelerators which are transportable(...). If the aircraft industry accepts and adopts the neutron sensitive TV image intensifier, other industries will do likewise and the field will rapidly expand - a healthy situation".

H.G. SMITH

"A presentation regarding 'Sealed Neutron Tubes' is very interesting ; it can easily be moved and used more widely in NDT : the operation repair and maintenance of the facility are very convenient for users".

MA ZHEN ZE

"I hope to know when the smallest portable reactor for NR would come out ?".

ZHOU YONG MAO

"The most interesting sessions were those which covered about Mobile NR".

M. DUBOUCHET

"I was especially interested in the session on real-time NR and its applications to the study of fluid flow problems".

H.C. ADERHOLD

"The sessions on corrosion inspection, non-reactor sources, accelerators and electronic imaging were particulary interesting because of present needs for major projects".

J.P. BARTON

"I found the papers and posters on industrial applications corrosion inspection especially for aircraft and on electronic imaging most useful".

L.G.I. BENNETT

"En tant qu'utilisateur industriel, j'ai été plus particulière-ment attiré par les sessions (...) contrôle de corrosion (...), sources distinctes des réacteurs. Pour ce qui est de la méthode de contrôle non destructif elle-même, les principaux éléments retenus sont les suivants : (...) le développement des moyens portables (...), le développement des techniques dynamiques de contrôle (vidéo)".

J.P. BOULOUMIE

"I found the sessions on electronic imaging and tomography to be the most interesting. I would have liked to have seen the tomography session as an oral session rather than a poster session.
The addition of real-time NR imaging was quite valuable but in future conferences, provisions should be made to emphasis the use of video equipment to demonstrate the research results".

J.S. BRENIZER, Jr

"Regarding the different presentations, I think all of them were very interesting, especially lectures or posters about digital image processing and dynamic imaging methods (...)".

T. BUCHBERGER

"The most interesting were those concerned with mobile accelerator NR and electronic imaging systems (...). Likewise the emphasis at the Conference on electronic imaging reconfirms our early premise that this technology is vital to realizing the full potential of NR".

W.E. DANCE

"Improvements in delivered flux by various sources was of particular interest to me, especially the status of sealed-tube generators and the applications of cyclotrons (...) The growing range of applications is of importance, especially in the aerospace industries".

A. DE VOLPI

"What I learn professionally from such a Conference, is to meet the researchers involved and discuss the problems directly with them. Here, at the University of Chicago, for example, I am involved with digital imaging. Especially from the poster sessions, I learnt a lot about this technique with application to NR".

E. HEIBERG

"(...) Paper US.3 and YU.1 examined the detectabilities of aluminium corrosion quantitatively : these studies should be very useful for practical applications in the nondestructives inspections of aluminium products".

H. KOBAYASHI

"The main areas of interest from WCNR.2 were (...) corrosion inspection, electronic imaging (...). My interest was detection of corrosion in aircraft structures with the emphasis on mobile units and the papers provided me with a "state-of-the-art" in current technology".

A. RIDAL

"Electronic imaging (...), tomography, corrosion inspection (...), were some of the most interesting communications during the oral and/or poster-board presentations".

V. STULENS

"The most interesting communications were all papers presented at session III (corrosion inspection) because the detection of hidden corrosion is very important from an aeroplane security viewpoint".

N. WADA

"(...)the development and use of cyclotron neutron sources for
NR is new, encouraging and possibly a solution for one type of
NR application. It will be very desirable to develop lighter
weight cyclotron (...)".

W.L. WHITTEMORE

In fact, if the technical descriptions have been fully
presented (about 90 % of the referenced subject), and most of the time
with many details, the economical analysis was mentioned only in one
paper, which was noticed by some attendees :

"Economics of NR for commercial aviation was one of the most
interesting ones for me".

A. ATAYA

"Il manquait également, comme il a été très justement remarqué
dans les conclusions, des éléments économiques. Il est bien
clair que les questions des industriels seront toujours :
combien cela coûtera-t-il et combien cela rapportera-t-il ?
Cela peut paraître une préoccupation bien mercantile, mais les
entreprises ont le devoir de ne pas s'engager dans des voies
sans issue. L'exposé de M. BARTON, bien que très intéressant et
ayant le mérite d'évoquer le problème, ne répond pas aux questions précises que peuvent se poser les utilisateurs.
L'exposé qui m'a le plus intéressé était celui de John LINDSAY
car c'est celui d'un prestataire de service (...). Sans donner
de chiffres, il pose clairement le problème du rapport coût/
service pour le consommateur d'images".

P. MORO

"The future of NR, to apply to actual field as airplane,
depends on the appearance of a new economical source with
neutron intensity comparable to nuclear reactor, more compact
and in some cases transportable".

S. TASAWA - T. NAKANII

It is not so surprising if the economical points highly
important to industrial people, could not really be discussed, because
of their low participation : this was mainly noticeable on the European
part :

"Au niveau de la participation, on peut regretter l'absence de
représentants des industries aérospatiales européennes alors
que cette nouvelle méthode de contrôle non destructif concerne
de plus en plus ce secteur industriel".

J.P. BOULOUMIE

Perplexity ? Reserve ? Lack of knowledge ? The reason is probably due, even now, to the last supposition.

At least, we hope that those people who attended were interested, otherwise convinced, to promote on a large scale a method which remains too restricted only to researchers or students, because it did not came off the sites.

The realization of non reactor sources, really mobile, will be able to remedy such a situation, even if a great doubt remains on the importance of the investments and operating costs : this should be considerably clarified during the coming years.

Besides the above comments regarding mostly half of the replies, the following items have been pointed out :

a) Special applications (so various and so surprising) :
   "L'exposé de l'odontologue français était lui aussi très intéressant cas placé du côté de l'utilisateur de l'image et non de celui qui la met au point".

                                                        P. MORO

   "The sessions on general applications and papers FR10 (VULCAIN), SD1 (LARSSON), US19 (BOYNE) and WG7 (FISHER) were the most interesting from the point of view they open up whole new horizons".

                                                        J.P. BARTON

   "The most interesting communication (...) was the oral presentation by W.L. WHITTEMORE (US.19 : Delineations of pathologic intraosseous lesions by neutron photographic images)".

                                                        J.C. DOMANUS

   "An autoradiography of oil paintings was fascinating. The presenter obviously loved his work. The paper combined an elegant technique with really beautiful subject matter. Also, bringing the worlds of art and science is very pleasing.
   I think that the most interesting communication for countries not involved in power reactor programs were the ones on industrial, medical and/or applications as :
      . delineations of pathologic intraosseous lesions,
      . autoradiography paintings,
      . NR applied to ancient art.
   It is now important to put more emphasis on possibilities of applications in various fields which would help promoting NR and justifying funds for such installations around the world"

                                                        N.N. PAPADOPOULOS

"The medical and biological applications of NR are encouraging. The fine work by the French dentist group was highly interesting and indicative of useful applications as was the work on cancer of the human mandible.

Although the splendid work on Rembrandt's "Man in the Golden Helmet" was not NR per se, this was a beautiful piece of work. It was proper to include it".

W.L. WHITTEMORE

b) **Nitrocellulose films** (of which manufacturing and marketing has been confirmed by KODAK-PATHE) :

"One very specific benefit of the Conference was to alert users to the fact that KODAK PATHE will lease its sale of tracketch foils in 1987 (...) In my project, there is no present substitute for the non gamma ray sensitivity and more particulary for the inceased resolution compared to the usual transfer foils".

W.L. WHITTEMORE

"In my opinion, all communications were very helpful in giving me a survey about the status in NR activities. Especially, the contribution about the experimental study of yields of (n,alpha) converters was very interesting".

W. SCHULZ

"Paper on experimental study of yields of (n, alpha) converters shows many graphs and tables of the film characteristics for combinations of many different converters and nitrocellulose film. These data are practically useful".

H. KOBAYASHI

c) **Standard and dimensional measurements** (which are and will be necessarily to be examined to increase the credibility of the method to users who are more and more demanding) :

"Pour ce qui est de la méthode de contrôle non destructif elle-même, les principaux éléments retenus sont les suivants :
. une importante action de normalisation et de standardisation au niveau international,
. une utilisation qui s'étend de plus en plus du domaine qualitatif au domaine quantitatif (métrologie dimensionnelle, microdensitométrie)...

J.P. BOULOUMIE

"A few theoritical approches were made to describe the dimensional measurements (cf. A.A.HARMS). Their analysis essentially based on an idealized geometrical configuration of converter imaging film system (...). Extensive efforts on the analysis of the unsharpness will be valuable to extensive studies based on more practical converter imaging film systems".

K. KOBAYASHI

Although we could be satisfied with the whole participants' encouraging comments concerning the good running of that Conference, some of them propose their suggestions which seem very useful to mention here, mainly for the benefit of the next Conference.

"L'iconographie de plusieurs orateurs laissait grandement à désirer et dans bien des cas cela allait de pair avec un exposé peu dynamique, assez mal structuré et difficile à suivre. Il est clair que tous les sujets ne se prêtent pas à une présentation de diapositives ou de projections flatteuses mais plusieurs orateurs auraient été mieux écoutés s'ils avaient fait des efforts de ce côté ('God is in the details'). Il manquait un exposé résumant en quelques chiffres l'état de l'art : combien de pays, combien d'équipes, combien de clients, combien de clichés ou d'images ? etc."

P. MORO

"I should have liked to have seen the tomography session as an oral session rather than a poster session (...) In future Conference, provision should be made to emphasize the use of video equipment".

J.S. BRENIZER,Jr

"Unfortunatly, very bad presentations of many interesting papers have rendered their subjects quite incomprehensible".

J.C. DOMANUS

"It would be helpful, in order to increase the number of participants and give a chance also to colleagues which cannot afford to pay high registration fees, if the fee of such meetings could be reduced as the one of 500 FF at the International Meeting in Orleans (...)".

N.N. PAPADOPOULOS

"One general remark should be made : the Conference was too
expensive concerning conference fee and all other costs : that
gives an unpleasant reduction of interested people".

H. RAUCH

"A problem for me as member of a University was to find financial support and with respect to future conferences, I want to
ask if the conference couldn't be held in an University or
Research Centre where lecture-rooms are mostly gratuitous available and thus the conference fee possibly could be lower".

E. STEICHELE

"We thought that perhaps parallel sessions may have been used
to advantage in order to shorten the length of time required
for the formal presentations programme and therefore allow more
time discussions around the posters, which necessary had to be
confined to lunch and coffee breaks. We also thought that a lot
of valuable information imparted by the Japanese contingent may
have been lost because of the language problem. In view of the
large number of Japanese who attended the Conference, it was a
pity that there was not at least one interpreter available for
both their formal presentations and informal discussions around
the posters.
We think that in the future, the Conference could be held a
little more often, for example every 2 to 3 years, since it
provides an excellent international forum for the exchange of
information concerning not only the availability of NR
facilities, especially portable equipment but also the variety
of applications currently being explored. This latter topic
would be extremely valuable to industrial research groups, like
ourselves, who do not have their own NR facility but who are
interested in the potential ranges of applications for which
the technique could be used from which specific examples may be
relevant to their own studies".

P. SWIFT - P. ATTWOOD

Above, we have already replied to most of the aspects of the above
questions : many others, perhaps, have not been transmitted (30 % among
the participants have sent their comments). That is why we suggest to
the future organisers, to include a question-sheet in the attendees'
folders to unable them giving their remarks on the meeting site : it is
quite difficult to get their opinions when the meeting is over.

892

On another hand, it would be very judicious and useful to request together with the registration form, a couple of each attendee's photographs (passeport size), first one to be included in the final programme and 2nd one sticked on the conference badges.

WCNR.2 is over.

We wish a good success to WCNR.3, already so promising.

LIST OF CHAIRMEN

Opening and Session I	J.P. Barton	G. Farny
Session II	G. Bayon	J.-L. Person
Session III (A+B)	J.J. Antal	E. Hiraoka
Session IV	R. Newacheck	J. Walker
Session V	U. Bergenlid	J. Markgraf
Session VI	J.P. Barton	J.J. Rant
Session VII	H. Rauch	
Session VIII	S. Fujine	V. Orphan
Session IX	M. Fantini	K. Katsurayama
Session X	A.A. Harms	R. Ruault
Session XI	G. Ducros	G. Matsumoto
Session XII	L. Greim	K. Okamoto
Session XIII	P. Underhill	J. Markgraf
Closure	J.P. Barton	G. Farny
Workshop III(A+B)	S. Cluzeau	M. Hawkesworth
Workshop IV	J.P. Barton	J.P. Bouloumié
Workshop VIII	J. Rogers	
Workshop XI	J.P. Barton	G. Ducros
Workshop XIII	H. Berger	J.C. Domanus

M. Howard C. ADERHOLD
CORNELL UNIVERSITY
Ward Laboratory
U.S.A. - ITHACA, N.Y., 14853

M. Otohiko AIZAWA
Atomic Energy Research Lab.
Musashi Institute of Tech.
OZENJI 971, ASAO-KU
KAWASAKI
215 JAPAN

M. Henri AMAURY
CESTA Bordeaux
BP 2 LE BARP
33830 BELIN - BELIET
FRANCE

M. John ANTAL
Army Materials Tech. Lab.
SICMT-OMM
Arsenal St
WATERTOWN MA 02172-0001
U.S.A.

M. Philip ATTWOOD
SHELL RESEARCH Ltd
Thornton Research Centre
P.O. Box 1 CHESTER CH 1354
GREAT BRITAIN

M. Peter BARBONUS
INTERATOM GmbH
Werk 1 : Friedrich-Ebert-Straße
5060 BERGISCH GLADBACH 1 (BENSBERG)
FED. REP. GERMANY

M. John BARTON
N-RAY ENGINEERING CO.
5709 Waverly Avenue
LA JOLLA CA.92037
U. S. A.

M. Guy BAYON
SPS SEARP
C. E. N. Saclay
91191 GIF sur Yvette Cedex
FRANCE

M. Ahmet BAYÜLKEN
Org. Nat. pour L'En. Atomique
C. N. A. E. M.
P.K. 1 Havaalani
ISTANBUL
TURKEY

M. Les BENNETT
Department of Chemistry and
Chemical Engineering
ROYAL MILITARY COLLEGE OF CANADA
KINGSTON, ONTARIO
CANADA K7K 5L0

M. Ulf BERGENLID
Studsvik Energiteknik
61182 Nyköping
SWEDEN

M. Harold BERGER
Industrial Quality Inc.
P.O. Box 2397
GAITHERSBURG, MD 20879
U. S. A.

M. Esnö Johan BLEEKER
E. C. N. PETTEN
Westerduinweg 3
PETTEN
THE NETHERLANDS

M. Jean BORDO
Centre Commun de Recherche de la
Commission des Communautés européennes
Etablissement de Petten
Boite postale 2
1755 ZG PETTEN
THE NETHERLANDS

M. Jean-Pierre BOULOUMIE
C. N. E. S.
18, avenue Edouard Belin
31055 TOULOUSE CEDEX
FRANCE

M. Jean-Louis BOUTAINE
C.E.A./S.A.R.
C.E.N./SACLAY
91191 GIF/YVETTE
FRANCE

M. Jack BRENIZER Jr
UNIVERSITY OF VIRGINIA
Nuclear Reactor Facility
CHARLOTTESVILLE, Va. 22901
U. S. A.

M. Knut BRYHN-INGEBRIGTSEN
Institute for Energy Technology
P.O. Box 40
N. 2007 KJELLER
NORWAY

M. Thomas BUCHBERGER
ATOMINSTITUT WIEN
Schüttelstr. 115
A. 1020 WIEN
AUSTRIA

M. Ciro CANDELA
E. N. E. A.
Via V. Brancati, 48
00144 ROMA
ITALIA

M. Paul CAPIOO
SODERN
1, av. Descartes
94450 LIMEIL-BREVANNES CEDEX
FRANCE

M. Claudio CAPPABIANCA
Enea Vel. Tecn. Matecn
V. Anguillarese
KM 1+300 - 00100 ROMA
ITALY

M. Eugenio CAPURRO
Ansaldo DIV. NIRA
Via dei Pescatori, 35
16129 GENOVA
ITALY

M. Albert CHEVALIER
UNIV.CATHOLIQUE DE LOUVAIN
Centre de Recherches du Cyclotron
2, Chemin du Cyclotron
1348 LOUVAIN LA NEUVE
BELGIUM

M. Serge CLUZEAU
SODERN
1, avenue Descartes
94450 LIMEIL-BREVANNES CEDEX
FRANCE

M. Guy COLOMB
SRSC/C.E. VALDUC
21120 IS-sur-TILLE
FRANCE

M. Lucien COUTEL
KODAK PATHE
8 et 26, rue Villiot
75594 PARIS CEDEX 12
FRANCE

M. Sylvain CRESPIN
C.E.A./CEN Bruyères-le-Châtel
BP 12
91680-BRUYERES-LE-CHATEL
FRANCE

Mrs Verginia Reis CRISPIM
COPPE/UFRJ
C.P. 68.509
RIO DE JANEIRO
RJ - BRAZIL

M. William E. DANCE
LTV Aerospace and Defence Co.
P.O. Box 650003, MS EM-16
DALLAS, TEXAS 75265 - 0003
U. S. A.

M. GU DENING
China Nuclear Energy Industry Co
P.O. Box 2139
BEIJING
CHINA

M. Jacques DEPOITIER
INSTITUT NATIONAL DES
RADIOELEMENTS
FLEURUS
BELGIUM

M. Alex DEVOLPI
Building 208
Argonne National Laboratory
ARGONNE, IL, 60439
U.S.A.

M. Jozef C. DOMANUS
Risø National Laboratory
DK-4000 ROSKILDE
DENMARK

M. Michel DUBOUCHET
SODERN
1, avenue Descartes
94450 LIMEIL-BREVANNES CEDEX
FRANCE

M. Gérard DUCROS
C. E. A. - C. E. N. G.
85 X
38041   GRENOBLE CEDEX
FRANCE

M. Prof. Eckhart DÜHMKE
Göttingen University
Robert-Koch-Str.40
Department of Radiotherapy
D-3400 GöTTINGEN
FED. REP. OF GERMANY

M. Joachim EGELHOFER
TRANSNUKLEAR
D. 6450 HANAU 11
Postfach 110030
FED. REP. GERMANY

M. Marcel FANTINI
69 rue de Fontenay
94130 Nogent sur Marne
FRANCE

M. Gérard FARNY
SPS/SEARP
C. E. N. Saclay
91191 GIF sur Yvette Cedex
FRANCE

M. Otto FICKEL
Brown Boveri Reaktor GmbH
Dudenstr. 44
D-68 MANNHEIM 31
FED. REP. OF GERMANY

M. Carl O. FISCHER
HAHN-MEITNER-INSTITUT
P.O. Box 390128
1000 BERLIN 39
FED. REP. GERMANY

M. Georges FOREST
E. D. F.
3, rue de Messine
75008 PARIS
FRANCE

M. Ichiro FUJII
Nuclear Business Dev. Div.
Sumitomo Heavy Industries Ltd
MITOSHIRO-CHO, KANDA,
CHIYODA-KU, TOKYO 101
JAPAN

M. Shigenori FUJINE
Research Reactor Institute
Kyoto University
KUMATORI-CHO, SENNAN-GUN
OSAKA 590-04
JAPAN

M. Keiji FURUYA
The Kansai Atomic Conference
1-8-4- Utsuo Hommachi
NISHI-KU OSAKA
JAPAN

M. WU FUXING
China Nuclear Energy Industry Co
P.O. Box 2139
BEIJING
CHINA

M. Frédéric GIBIAT
C. E. A.
C.E.N. CADARACHE
13115 SAINT PAUL LEZ DURANCE
FRANCE

M. Bernard GLEIZES
C.E.N. GRENOBLE
85 X
38041 GRENOBLE CEDEX
FRANCE

M. Jean-Jacques GRAF
C. E. A.
33, rue de la Fédération
75752 PARIS CEDEX 15
FRANCE

M. Ludwig GREIM
G.K.S.S.-GEESTHACHT GMBH
Postfach 1160
2054 GEESTHACHT
FED. REP. OF GERMANY

Mrs. Muebeccel GREIM
G.K.S.S.-GEESTHACHT GMBH
Postfach 1160
2054 GEESTHACHT
FED.REP. OF GERMANY

M. Gérard GRENIER
C.E.A./CEN Bruyères-le-Châtel
BP 12
91680 - BRUYERES-LE-CHATEL
FRANCE

M. Archie HARMS
Mc Master University
HAMILTON, ONTARIO
CANADA L 85 4MI

M. Dennis HARRIS
U. K. A. E. A.
AERE HARWELL
OXON OXII ORA
GREAT BRITAIN

M. Michael HAWKESWORTH
University of Birmingham
Dept of Physics
P.O. Box 363
BIRMINGHAM B 15 2TT
GREAT BRITAIN

M. Masaki HAYASHI
Fukui Prefectural Office
3-17-1. OTE
FUKUI CITY
JAPAN 12

M. Alain Hauducoeur
CEA/C.E. Vaujours
B.P. 7
77181 COURTRY
FRANCE

M. Edvard HEIBERG
UNIVERSITY OF CHICAGO
Physics Department ; GHJ-116
5747 S. Ellis Ave.
CHICAGO, IL 60637
U.S.A.

M. Heinrich HEIDT
BUNDES.FÜR MATERIAL PRÜFUNG
Unter den Eichen 87
D 1000 BERLIN 45
FED. REP. OF GERMANY

M. Eiichi HIRAOKA
Radiation Center of Osaka
Prefecture
704 Shinzaike-cho
SAKAI-SHI, OSAKA
JAPAN

M. Jean-Benoit HOURST
C. E. A. Centre VALDUC
21120 IS-sur-TILLE
FRANCE

M. Shuichi IGAKASHI
Fukui Prefectural Inst. of Public Health
4-39 Harame-cho
FUKUI CITY
JAPAN

M. Yasushi IKEDA
Fac. of Eng., Nagoya Univ.
Furô-chô, Chikusa-ku,
NAGOYA
464 JAPAN

M. Akira ITAYA
NIPPON KOKAN K.K. (NKK)
1.1.2. MARUNOUCHI, CHIYODA-KU
TOKYO 100
JAPAN

M. Jean-Claude JAUREGUY
E. T. C. A.
16 bis, av. Prieur de la Côte d'Or
94114 ARCUEIL CEDEX
FRANCE

M. Alain JEGAT
C. E. A.
CEN Grenoble Pi EDTI
avenue des Martyrs
FRANCE - 38041 GRENOBLE CEDEX

M.John D. JONES
University of Michigan
Phoenix Laboratory
2301 Bonisteel Bvd
ANN ARBOR MI 48109
U. S. A.

M. Hiromi KAJISHIMA
Mitsubishi Corporation
5-30 Shinrogane-cho
TSURUGA-SHI, FUKUI
JAPAN

M. Osamu KASHIMURA
Safety Division Power Reactor & Nuclear
Fuel Development corporation
1-9-13 AKASAKA, MINATO-KU
TOKYO
JAPAN

M. Kosuke KATSURAYAMA
Kyoto University
Noda Kumatori-cho
SENNAN - GUN OSAKA
JAPAN

Dr. Dan KEDEM
Kedem Technologies
P.O. Box 1219
Rehovot 76267
ISRAEL

M. Ekkart KNORR
Industrieanlagen-
Betriebsgesellschaft mbH
D. 8012 Ottobrunn b. München
Einsteinstraße Geb. 21
FED. REP. OF GERMANY

M. Hisao KOBAYASHI
Inst. Atomic Energy
RIKKYO UNIVERSITY
2-5-1 NAGASAKA
YOKOSUKA, KANAGAWA, 240-01
JAPAN

M. KUSMINARTO
Faculty of Science and Mathematics
Gadjar Mada University
Sekip Unit III
YOGYAKARTA
INDONESIA

M. Pierre LANNEAU
INTERCONTROLE
4 à 10 place Vauban
94583 RUNGIS Cedex
FRANCE

M. André LAPORTE
C.E.A. IPSN/DSMN/SITN
B.P. 6
92265 FONTENAY AUX ROSES
FRANCE

M. Börje LARSSON
Uppsala University
The Gustaf Werner Institute
Box 251/751 21
UPPSALA
SWEDEN

M. Alain LECART
KODAK-PATHE
8 et 26, rue Villiot
75594 PARIS CEDEX 12
FRANCE

Melle Isabelle LEFESVRE
SODERN
1, av. Descartes
94450 LIMEIL-BREVANNES CEDEX
FRANCE

M. Henri LEHN
Direction
TECHNICATOME
91191 GIF-sur-YVETTE
FRANCE

M. André LESEUR
C. E. A.
C.E.N. SACLAY-LECI
91191 GIF-sur-YVETTE
FRANCE

M. Samuel H. LEVINE
Pennstate University
231 Sackett Blvd
Nuclear Eng. Dept.
PENNSTATE UNIV. UNIVERSITY PARK
PA 16802 U.S.A.

Dr John LINDSAY
Phoenix Mem. Laboratory
2301 Bonisteel Blvd
ANN ARBOR Michigan 48109
U. S. A.

M. Akihiko MAEDA
The Japan Atomic Power Company
1, Myojin-cho
TSURUGA-SHI, FUKUI
JAPAN

M. Peter MAIER
Institut fur Kerntechnik und
Energiewandlung eV
Pfaffenwaldring 31
7000 STUTTGART 80
FED. REP. OF GERMANY

M. Daniel MALYS
C. E. A./ C. E. de Vaujours
Etablissement T
B.P. n° 7
77181 COURTRY
FRANCE

M. Joachim MARKGRAF
JOINT RESEARCH CENTRE OF CEC
P.O. Box 2
NL 1755 ZG PETTEN N.H.
THE NETHERLANDS

Dr Hans-Ulrich MAST
Industrieanlagen-Betriebsgesellschaft mbH
D. 8012 Ottobrunn b. München
Einsteinstraße Geb. 21
FED. REP. GERMANY

M. Gen-ichi MATSUMOTO
Nagoya University
Furo-cho, Chikusa-ku,
NAGOYA 464
JAPAN

M. Tetsuo MATSUMOTO
Atomic Energy Research Lab.
Musachi Institute of Tech.
OZENJI 971, ASAO-KU
KAWASAKI
215 JAPAN

M. MA ZHEN-ZE
Southwest Inst. of Nuclear
Physics & Chemistry
P.O. Box 515
Chengdu
CHINA - Szechuan

M. Douglas Mc RAE
PHILIP MORRIS
P.O. Box 26583
RICHMOND VA 23261
U. S. A.

M. Gabriel MEYRAND
C.G.R./M.E.V.
B.P. 34
78530 BUC
FRANCE

M. Hartmut MIEMCZYK
INTERATOM
Postfach
D. 5060 Bergisch Gladbach 1
FED. REP. GERMANY

M. Philippe MORISSEAU
INTERCONTROLE
4 à 10, place Vauban
SILIC 433,
94583 RUNGIS CEDEX
FRANCE

M. Paul MORO
NARDEUX
Avenue d'Islande
Z.A. de Courtaboeuf
91940 LES ULIS
FRANCE

902

M. Raymond L. MOSS
Commission of the EUR. COMM.
JOINT RESEARCH CENTRE OF CEC
P.O. Box 2
NL 1755 ZG PETTEN N.H.
THE NETHERLANDS

M. Dietrich MUNDT
INTERATOM GmbH
Werk 1 : Friedrich-Ebert-Straße
5060 BERGISCH GRADBACH 1 (BENSBERG)
FED. REP. GERMANY

M. Kazuo NAKAMURA
Department non-destructive
Inspection Co, Ltd
2-32 KITA-KYUHOJI MINAMI-KU
OSAKA
JAPAN

M. Takeshi NAKAMURA
The Kansai Electric Power
Niu Mihama-cho
FUKUI PREFECTURE
JAPAN

M. Takehiko NAKANII
Tokyo Div. Sumitomo Heavy
Industries Ltd.
799-13 JAPAN

M. Paul NATHANSON
ATOMIC ENERGY CORPORATION
COMMISSION
Private Bag X256
PRETORIA 0001
SOUTH AFRICA

M. Daniel NERWER
PALIDENT
POB 1044
Ramat Asharon 47100
ISRAEL

M. Richard NEWACHECK
Aerotest Operations, Inc.
3455 Fostoria Way
SAN RAMON, CA 94583
U. S. A.

Dr Yoram NIR-EL
SOREQ Nuclear Res. Centre
YAVNE
ISRAEL 70600

Mr. Koichi OKAMOTO
International Atomic Energy
Agency (I. A. E. A.)
Wagramerstr. 5, P.O. Box 100
A-1400 VIENNA
AUSTRIA

M. Atsuo ONO
College of Liberal Arts
KOBE University
NADAKU, KOBE
657 JAPAN

M. Victor ORPHAN
SAIC
10401 Roselle Street
SAN DIEGO, CA 92121
U.S.A.

M. Chen-Chun OU-YANG
A. S. P. E. C. T.
9, avenue Matignon
75008 PARIS
FRANCE

M. Neophytos PAPADOPOULOS
Greek Atomic Energy Comm.
Nuclear Research Center
Demokritos
Aghia Pavaskevi
ATTIKI 15310
GREECE

M. Michel PELLETIER
C. E. A.
C.E.N. CADARACHE
13115 SAINT PAUL LEZ DURANCE
FRANCE

M. François PERRET
C. E. A.
C.E.N. SACLAY-L.E.C.I.
91191 GIF-sur-YVETTE CEDEX
FRANCE

M. Jean-Louis PERSON
SPS/SEARP
C. E. N. Saclay
91191 GIF sur Yvette Cedex
FRANCE

M. Claude PIRART
INSTITUT NATIONAL DES
RADIOELEMENTS
FLEURUS
BELGIUM

M. Joseph RANT
Reactor Department
Institute Joseph Stefan
Jamova 39 - P.O. Box 199
YUGOSLAVIA - 61001 LJUBLJANA

M. Helmut RAUCH
Atominstitut
Schuttelstrasse 115
A - 1020 WIEN
AUSTRIA

Dr Holger REITER
Fraunhofer Institut für
Zerstörungsfreie Prufrerfahren
Universität, Gebaüde 37
0-6600 SAARBRÜCKEN
FED. REP. OF GERMANY

M. Alan RIDAL
A. A. E. C.
PMB, SUTHERLAND
NSW 2232
AUSTRALIA

M. Philippe RIZO
C. E. N. GRENOBLE
85 X
FRANCE - 38041 GRENOBLE CEDEX

M. John ROGERS
SILTON
P.O.B. 3
BRISTOL
GREAT BRITAIN

M. Heinz RÖTTGER
Commission Europ. Communities
JRC Petten
P.O. Box 2
1755 CE Petten, THE NETHERLANDS

M. Henri ROUGEOT
THOMSON-C.S.F.
38, rue Vauthier
92100 BOULOGNE-BILLANCOURT
FRANCE

M. Alain ROUYER
Service Radiochimie
C.E.A./CEN Bruyères-le-Chatel
BP 12
91680- BRUYERES-LE-CHATEL
FRANCE

M. Pierre RUAULT
COFREND
32, Bd. de la Chatelle
75880 PARIS Cedex 18
FRANCE

M. Mario RUSCEV
GIERS SCHLUMBERGER
12, place des Etats-Unis
92124 MONTROUGE CEDEX
FRANCE

Dr Kurt SAUERWEIN
ISOTOPEN-TECHNIK
Bergische Straße 16
Postfach 1354
D. 5657 HAAN/RHEINL 1
FED. REP. GERMANY

M. Karl Albrecht SCHMIDT
Im Zeitvogel 5
7500 KARLSRUHE 41 - DURLACH
FED. REP. OF GERMANY

M. Wolfang SCHULZ
PREUSSEN-ELEKTRA AKTIENG.
Tresckow str. 5
3000 HANNOVER 91
FED. REP. OF GERMANY

M. Wilhelm SCHUMACHER
G.K.S.S.-GEESTHACHT GMBH
Postfach 1160
2054 GEESTHACHT
FED. REP. OF GERMANY

M. William SEYMOUR
U. K. A. E. A.
AERE HARWELL
D.DCOT, OXON OXII ORA
GREAT BRITAIN

M. Boris SHAPIRO
E. C. N.
P.O. Box 1
1755 ZG PETTEN
THE NETHERLANDS

M. Heitor SILVA
INSTITUTO DE CIENCIAS
E ENGENHARIA NUCLEARES
Estrada Nacional n° 10
2685 SACAVEM
PORTUGAL

M. Harold SMITH
Oak Ridge National Lab.
OAK RIDGE, TN 37830
U.S.A.

M. Michel SORS
CEA/IRDI/DTech/SECS/SELECI
CEN SACLAY Bât. 605
91191 GIF-sur-YVETTE
FRANCE

M. Harry SPENCE
Armed Forces Radiobiology
Institute
NMC/NCR
BETHESDA,   MD 20814-5145
U. S. A.

M. Jürgen STADE
BUNDES .FüR MATERIAL PRüFUNG
Unter den Eichen 87
1000 BERLIN 45
FED. REP. OF GERMANY

M. Erich STEICHELE
TECHN. UNIVERSITÄT MüNCHEN
FACHBEREICH PHYSIK E 21
8046 GARCHING
FED. REP OF GERMANY

M. Lothar STEINBOCK
C. E. N. CADARACHE
DERS/SES
B.P. N° 1
13115 ST PAUL-LEZ-DURANCE
FRANCE

M. Kenneth STEPHENSON
SCHLUMBERGER
EMR Photoelectric
P.O. Box 44, PRINCETON NJ
U. S. A. 08542

M. Gunnar STEYRER
Institut Max Von Laue - Paul Langevin
156 X
38042 GRENOBLE CEDEX
FRANCE

M. Victor STULENS
SCK/CEN-MOL
BOERETANG 200
2400 MOL
BELGIUM

M. J. SULTAN
KODAK-PATHE
71102 CHALON-sur-SAONE CEDEX
FRANCE

M. Peter SWIFT
SHELL RESEARCH Ltd
Thornton Research Centre
P.O. Box 1, CHESTER CH 1354
GREAT BRITAIN

M. Jean-Louis SZABO
C.E.A./C.E.N.
SACLAY
ORIS/SAR BP 21
91191 GIF/YVETTE
FRANCE

Mrs. Marie-Hélène TAILLEUR
CESTA Bordeaux
BP 2 LE BARP
33830 BELIN - BELIET
FRANCE

M. Ryoichi TANIGUCHI
Radiation Center of Osaka Prefecture
SHINKE-CHO, SAKAI-SHI
593-JAPAN

Mme Betty TATTEGRAIN
CEA/CEN CADARACHE
B.P. n° 1
13115 St PAUL-LEZ-DURANCE
FRANCE

M. Shuichi TAZAWA
Industrial Machinery Division,
SUMITOMO HEAVY INDUSTRIES CO, Ltd
SOBIRAKI-CHO, NIIHAMA-SHI,
792 JAPAN

M. Paul UNDERHILL
AEROTEST OPERATIONS INC.
3455 Fostoria Way
SAN RAMON - CALIFORNIA 94583
U. S. A.

M. Karel VAN OTTERDYK
E. C. N.
P.O. Box 1
1755 ZG PETTEN
THE NETHERLANDS

M. Maurice VERAT
THOMSON-C.S.F.
38, rue Vauthier
92100 BOULOGNE-BILLANCOURT
FRANCE

M. René VERRECCHIA
C. E. A.
C.E.N .SACLAY-L.E.C.I
91191 GIF-sur-YVETTE CEDEX
FRANCE

M. Jean-Marie VULCAIN
Université de Rennes
2, place Pasteur
35000 RENNES
FRANCE

M. Nobuo WADA
Japan Atomic Energy
Research Institute
Oarai-machi, Higashiibaraki-gun
IBARAKI-KEN 311-13
JAPAN

M. John WALKER
University of Birmingham
Department of Physics
PO Box 363
BIRMINGHAM B.15 2TT
GREAT BRITAIN

M. William WHITTEMORE
G.A. TECHNOLOGIES, INC
SAN DIEGO CALIFORNIA 92138
P.O. Box 81608
U. S. A.

M. Martin WILSON
OXFORD INSTRUMENTS
Osney Mead
OXFORD OX2 0DX
GREAT BRITAIN

M. Matthias WROBEL
G.K.S.S.-GEESTHACHT GMBH
Postfach 1160
2054 GEESTHACHT
FED.REP. OF GERMANY

M. Shinichi YAMAKI
The Japan Steel Works, Ltd
4 - CHATSU-MACHI
MURORAN, HOKKAIDO
JAPAN

M. Kenji YONEDA
Research Reactor Institute
Kyoto University
KUMATORI-CHO, SENNAN-GUN
OSAKA (590.04)
JAPAN

M. Zhou YONGHAO
INSTITUTE OF ATOMIC ENERGY
P.O. Box 275(56)
BEIJING,
CHINA

M. Jean-Pierre YTHIER
SODERN
1, av. Descartes
94450 LIMEIL-BREVANNES CEDEX
FRANCE

INDEX BY COUNTRY AND CENTRE/COMPANY/ORGANIZATION

UNITED STATES OF AMERICA

YUGOSLAVIA

COMMISSION OF THE EUROPEAN COMMUNITIES

INDEX BY AUTHOR

INDEX BY SUBJECT

N.B. The page number refers to the first page of the relevant contribution

922